Lecture Notes in Mathematics

Edited by A. Dold and B. Eckmann

771

Approximation Methods for Navier-Stokes Problems

Proceedings of the Symposium Held by the International Union of Theoretical and Applied Mechanics (IUTAM) at the University of Paderborn, Germany, September 9 – 15, 1979

Edited by R. Rautmann

Springer-Verlag
Berlin Heidelberg New York 1980

Editor

Reimund Rautmann
Gesamthochschule Paderborn
Fachbereich Mathematik-Informatik
Warburger Straße 100, Gebäude D
D-4790 Paderborn

AMS Subject Classifications 34C35, 35BXX, 35Q10, 65MXX, 65NXX, 73K25, 76D05, 76EXX, 76FXX, 82A70

ISBN 3-540-09734-1 Springer-Verlag Berlin Heidelberg New York
ISBN 0-387-09734-1 Springer-Verlag New York Heidelberg Berlin

Library of Congress Cataloging in Publication Data
Symposium on Approximation Methods for Navier-Stokes
Problems, University of Paderborn, 1979.
Approximation methods for Navier-Stokes problems.
(Lecture notes in mathematics; 771)
Bibliography: p.
Includes index.
1. Navier-Stokes equations--Congresses. 2. Fluid dynamics--Congresses.
I. Rautmann, R., 1930- II. International Union of Theoretical and Applied Mechanics.
III. Title. IV. Series: Lecture notes in mathematics (Berlin); 771.
QA3.L28 no. 771 [QA911] 510s [532'.05'0151535] 79-28682
ISBN 0-387-09734-1

Printing and binding: Beltz Offsetdruck, Hemsbach/Bergstr.
2141/3140-543210

Math
Sep.

Extract from a speech after the Conference dinner

I am sure you will all have noticed the remarkable emblem of Paderborn which is reproduced on our Conference programmes: this depicts three hares, each having two ears, but each ear being shared between two hares. Each hare is moreover in pursuit of the tail of the hare in front, an activity that appears to be both futile and painful on account of the centrifugal forces sustained by the ears! And yet we at this meeting are not unlike these three hares - we represent three different approaches to problems of fluid mechanics - existential, numerical and asymptotic - and we have a natural tendency to drift apart into areas which yield most easily to our respective techniques. At this meeting, through listening to each other, we have been drawn back by our ears to some of the hard-core problems concealed in the Navier-Stokes equations, which lie at the centre of the circle of pursuit. The local organizing committee deserves our thanks, not only for the warmth of their hospitality here in Paderborn, but also for the success in drawing these three groups together to promote the stimulating cross-fertilisation of ideas from which we have all so greatly benefited. The triangular pursuit of the hares of Paderborn is perhaps more fruitful and rewarding than might at first be supposed!

H.K. Moffatt.

SCIENTIFIC COMMITTEE

ACKNOWLEDGEMENT

The organizers are indebted to the following organizations for their effective help in the preparation of the Symposium:

International Union of Theoretical and Applied Mechanics (IUTAM)

University of Paderborn

Gesellschaft für Angewandte Mathematik und Mechanik (GAMM)

Deutsche Forschungsgemeinschaft (DFG)

Firma Nixdorf Computer AG

PREFACE

Recently Navier-Stokes problems have found growing interest:

- Hilbertspace methods (or more general function-space approaches) in connection with embedding theorems open new ways to existence and regularity theory and lead to new theorems on (non-)uniqueness, asymptotical decay and constructive approximation of the solutions.
- Semigroup methods result in existence and regularity theorems in the framework of different function-spaces.
- Group theoretic methods provide us with a systematic approach to bifurcation.
- Bifurcation methods lead to a new theory of hydrodynamic stability.
- Dynamical systems offer geometric models for the unfolding in time of Navier-Stokes solutions and for the transition to turbulence.
- Refined potential theoretic approaches enable us to asymptotically describe flows behind obstacles and flows in regions having non-compact boundaries.
- Refined finite element methods (including error-estimates), difference methods (satisfying suitable stability conditions), spectral methods and fast Stokes solvers result in numerical solutions of flow problems in complicated two-dimensional and even in three-dimensional geometries.
- Refined modeling ideas of flows, e.g. at high Reynolds numbers, lead again to new formulations of promising mathematical questions.

The exchange of ideas about these new aspects and approximations was the aim of the Symposium on "Approximation Methods for Navier-Stokes Problems", which the International Union of Theoretical and Applied Mechanics (IUTAM) held in the Department of Mathematics, University of Paderborn, September 9-15, 1979.
More than 70 German or foreign mathematicians, fluid dynamicists and numerical analysts took part. Thirty-five invited lectures, additional short communications and a round-table discussion on present-day research trends gave a vivid insight into the state of the art and led to stimulating interdisciplinary contacts.
The discussions demonstrated lively interaction among the different working areas in the common interest field of the Navier-Stokes equations. Therefore the following papers submitted by the invited lecturers purposely had not been grouped into special areas. The many-faceted cross-connections which, according to the unanimous opinion of the participants, became evident during the symposium could thus be best maintained.-

I would like to thank the members of the Scientific Committee very cordially for the good cooperation in preparing and conducting the meeting. Special thanks should be given to the colleagues and co-workers in the Department of Mathematics at the University of Paderborn. Without their help this Symposium would not have been possible. Thanks should also be given to all participants who contributed to the success of the Symposium! Also I would like to thank the editors of the Lecture Notes in Mathematics and the Springer-Verlag for their friendly assistance during the planning stages and the speedious completion of this volume.

R. Rautmann

VORWORT

Navier-Stokes-Probleme haben in den letzten Jahren wachsendes Interesse gefunden:

- Hilbertraum-Methoden (oder Ansätze in allgemeineren Funktionenräumen) zusammen mit Einbettungssätzen eröffnen neue Zugänge zur Existenz- und Regularitätstheorie und führen zu neuen Sätzen über (Nicht-) Eindeutigkeit, asymptotisches Verhalten und konstruktive Approximierbarkeit von Lösungen.
- Halbgruppenmethoden ergeben neue Existenz- und Regularitätssätze in unterschiedlichen Funktionenräumen.
- Gruppentheoretische Methoden ermöglichen die systematische Behandlung von Verzweigungsproblemen.
- Bifurkationsmethoden führen zu einer neuen Theorie der hydrodynamischen Stabilität.
- Dynamische Systeme bieten geometrische Modelle der zeitlichen Entwicklung Navier-Stokes'scher Lösungen und des Übergangs in Turbulenz.
- Verfeinerte potentialtheoretische Methoden ermöglichen die asymptotische Beschreibung von Strömungen im Nachlauf von Körpern und in Gebieten mit nichtkompakten Rändern.
- Verfeinerte finite Elemente-Verfahren (mit Fehlerabschätzungen), Differenzenverfahren (mit entsprechenden Stabilitätsbedingungen), Spektralmethoden und schnelle Stokes-Solver ermöglichen die numerische Lösung komplizierter zweidimensionaler Strömungsaufgaben und grundsätzlich auch schon dreidimensionaler Probleme.
- Verfeinerte Modellvorstellungen von Strömungen, wie z.B. bei hohen Reynoldszahlen, führen ihrerseits zu neuen und vielversprechenden mathematischen Fragestellungen.

Dem Gedankenaustausch über diese vielfältigen neuen Aspekte und Approximationen diente das Symposium über "Approximationsmethoden für Navier-Stokes-Probleme", das von der Internationalen Union für Theoretische und Angewandte Mechanik (IUTAM) vom 9. bis 15. September 1979 im Fachbereich Mathematik der Gesamthochschule Paderborn veranstaltet wurde. Am Symposium nahmen mehr als 70 deutsche und ausländische Mathematiker sowie Fachleute der Strömungslehre und ihrer numerischen Methoden teil. 35 eingeladene Vorträge, zusätzliche Kurzvorträge und ein Rundgespräch über aktuelle Forschungsrichtungen vermittelten einen lebendigen Einblick in den derzeitigen Wissensstand und führten zu anregenden interdisziplinären Kontakten.

Die fachlichen Gespräche zeigten die rege Wechselwirkung zwischen den verschiedenen Arbeitsrichtungen im gemeinsamen Gebiet der Navier-Stokesschen Gleichungen. Bewußt sind daher die folgenden Beiträge der eingeladenen Referenten nicht nach Spezialgebieten gruppiert worden. Die vielfältigen Querverbindungen, die sich nach der wohl einhelligen Meinung der Tagungsteilnehmer ergeben haben, dürften so am besten zum Ausdruck kommen.

Den Mitgliedern des Wissenschaftlichen Komitees möchte ich sehr herzlich für die gute Zusammenarbeit bei der Vorbereitung und Durchführung der Tagung danken. Ein besonderer Dank gilt den Kollegen, Mitarbeiterinnen und Mitarbeitern im Paderborner Fachbereich Mathematik, ohne deren Hilfe die Tagung nicht hätte durchgeführt werden können. Gedankt sei auch allen Tagungsteilnehmern, die zum Gelingen des Symposiums beigetragen haben! Besonders möchte ich auch den Herausgebern der Lecture Notes in Mathematics und dem Springer-Verlag für das freundliche Entgegenkommen bei der Vorbereitung und für die schnelle Fertigstellung dieses Bandes danken.

R. Rautmann

CONTENTS

Amick, C.J.:

Steady solutions of the Navier-Stokes equations representing plane flow in channels of various types 1

Babenko, K. I.:

*On properties of steady viscous incompressible fluid flows 12

Basdevant, C.:

Parameterization of subgrid-scale motion in numerical simulation of 2-dimensional Navier-Stokes equation at high Reynolds number.......... 43

Bemelmans, J.:

$C^{o+\alpha}$-semigroups for flows past obstacles and for flows with capillary surfaces 59

Bristeau, M. O., Glowinski, R., Mantel, B., Periaux, J., Perrier, P., Pironneau, O.:

A finite element approximation of Navier-Stokes equations for incompressible viscous fluids. Iterative methods of solution.......... 78

Cannon J. R. and DiBenedetto, E.

*The initial value problem for the Boussinesq equations with data in L^p..129

Dervieux, A. and Thomasset, F.

A finite element method for the simulation of a Rayleigh-Taylor instability 145

Deville, M. and Orszag, S. A.:

Spectral calculation of the stability of the circular Couette Flow 159

Fasel, H. F.:

Numerical solution of the complete Navier-Stokes equations for the simulation of unsteady flows 177

Foias, C.:

A survey on the functional dynamical system generated by the Navier-Stokes equations... 196

Gresho, P. M., Lee, R. L., Chan, S. T., Sani, R.L.:

Solution of the time-dependent incompressible Navier-Stokes and Boussinesq equations using the Galerkin finite element method 203

Heywood, J. G.:

Auxiliary flux and pressure conditions for Navier-Stokes problems 223

Heywood, J.G.:

Classical solutions of the Navier-Stokes equations 235

Joseph, D. D.:
Direct and repeated bifurcation into turbulence 249
(With photographs of flow phenomena on pp. 265 - 271)

Kaniel, S.:
Approximation of the hydrodynamic equations by a transport process 272

Knightly, G. H.:
Some decay properties of solutions of the Navier-Stokes equations ... 287

Kovenya, V. M. and Yanenko, N. N.:
The implicit difference schemes for numerical solving the Navier-Stokes equations 299

Krause, E. and Bartels, F.:
Finite-difference solutions of the Navier-Stokes equations for axially symmetric flows in spherical gaps 313

Kravchenko, V.I., Shevelev, Yu. D., Shchennikov, V. V.:
Numerical investigation of unsteady viscous incompressible flow about bodies for varying conditions of their motion 323

Masuda, K.
On the regularity of solutions of the nonstationary Navier-Stokes equations 360

Moffatt, H. K.:
The asymptotic behaviour of solutions of the Navier-Stokes equations near sharp corners.. 371

Orszag, S. A. and Gottlieb, D.:
High resolution spectral calculations of inviscid compressible flows... 381

Prouse, G.:
Analysis of Navier-Stokes type equations associated to mathematical models in fluid dynamics 399

Rannacher, R.:
On the finite element approximation of the nonstationary Navier-Stokes problem 408

Rautmann, R.:
On the convergence rate of nonstationary Navier-Stokes approximations.. 425

Roux, B., Bontoux, P., Daube, O., Phuoc Loc, T.:
Optimisation of Hermitian methods for Navier-Stokes equations in the vorticity and stream-function formulation 450

Rubin, S.G. and Khosla, P.K.:
Navier-Stokes calculations with a coupled strongly implicit method. Part II: Spline deferred-corrector solutions 469

Ruelle, D.:
Strange attractors and characteristic exponents of turbulent flows 489

Sattinger, D. H.:
Selection mechanisms in symmetry breaking phenomena 494

Stewartson, K.:

High Reynolds-number flows 505

Taylor, T. D. and Murdock, J. W.:

Application of spectral methods to the solution of Navier-Stokes equations 519

Wahl, W. von:

Regularity questions for the Navier-Stokes equations 538

Wesseling, P. and Sonneveld, P.:

Numerical experiments with a multiple grid and a preconditioned Lanczos type method 543

Zandbergen, P. J.:

New solutions of the Karman problem for rotating flows 563

Papers submitted to the editor, but not presented during the symposium.

PARTICIPANTS

Agarwal, R. P. Dr.	Institute of Mathematical Sciences, Madras 6ooo2o INDIA
Amick, D. J. Dr.	University of Cambridge, Department of Pure Mathematics and Mathematical Statistics, 16 Mill Lane, Cambridge CB2 1SB, U. K.
Arker, H. Dr.	Gesamthochschule Paderborn, Fachbereich Mathematik-Informatik, Warburger Str. 100, 4790 Paderborn, W-Germany
Basdevant, C. Dr.	Laboratoire de Météorologie Dynamique, 24 Rue Lhomond, 75231 Paris Cedex o5, FRANCE
Bauer, P. Dipl.-Ing.	VOEST-ALPINE AG, Abteilung FAT, Postfach 2, 4o1o Linz, AUSTRIA
Bemelmans, J. Dr.	Mathematisches Institut der Universität, Wegelerstraße 10, 5300 Bonn, W-GERMANY
Bhatnagar, R. K. Prof. Dr.	Institut of Mathematics, State University of Campinas, Caiza Postal 117o, 131oo Campinas(S.P.), BRASIL
Bontoux, P. Dr.	Université D'Aix-Marseille, Institut de Mécanique des Fluides, 1, Rue Honnorat, 13oo3 Marseille, FRANCE
Brancher, J. P. Prof. Dr.	Institut National Polytechnique de Lorraine, Laboratoire D'Energetique et de Mecanique Theoretique et Appliquée, Rue de la Citadelle, B.P. 85o, 54o11 Nancy Cedex, FRANCE
Bulgarelli, U. Dr.	Consiglio Nazionale delle Ricerche, Istituto per le Applicazioni del Calcolo "Mauro Picone", Viale del Policlinico 137, Roma, ITALY
Deville, M. Dr.	Unité de Mécanique Appliquée, Université Catholique de Louvain, Faculté des Sciences Appliquées, Bâtiment Simon Stévin, Place du Levant, 2 B-1348 Louvain-la-Neuve, BELGIUM
DiBenedetto, E. Prof. Dr.	University of Texas at Austin, Department of Mathematics, Austin, Texas 78712, USA
Dijkstra, D. Dr.	Department of Mathematics, Technische Hogeschool Twente, Postbus 217, Enschede, NETHERLANDS
Fasel, H. Dr.	Institut A für Mechanik, Universität Stuttgart, Pfaffenwaldring 9, 7ooo Stuttgart 8o, W-GERMANY
Foias, C. Prof. Dr.	Université de Paris-Sud, Centre D'Orsay, Mathématique, Bâtiment 425, 914o5 Orsay, FRANCE

Fromm, J. Dr.	Deutsche Forschungs- und Versuchsanstalt für Luft- und Raumfahrt, Linder Höhe, 5000 Köln 90, W-GERMANY
Fujita, H. Prof. Dr.	Department of Mathematics, University of Tokyo, Hongo, Tokyo, Japan 113, JAPAN
Gamst, A. Dipl.-Math.	Institut für Schiffbau der Universität, Lämmersieth 9o, 2ooo Hamburg 6o, W-GERMANY
Gersten, K. Prof. Dr.	Institut für Thermo- und Fluiddynamik der Universität, Universitätsstraße 50, Gebäude B, 4630 Bochum-Querenburg, W-GERMANY
Girault, V. Prof. Dr.	Université Paris VI, Analyse Numérique, Tour 55 5 E, 9,Quai Saint-Bernard, Paris 5e, FRANCE
Glowinski, R. Prof. Dr.	Institut de Recherche, D'Informatique et D'Automatique IRIA, Domaine de Voluceau-Rocquencourt 7815o Le Chesnay, FRANCE
Gresho, P. M. Dr.	Lawrence Livermore Laboratory, University of California, P.O. Box 8o8, Livermore, California 9455o, USA
Hebeker, F.-K. Dipl.-Math.	Gesamthochschule Paderborn, Fachbereich Mathematik-Informatik, Warburger Straße 1oo, 479o Paderborn, W-GERMANY
Heywood, J. G. Prof. Dr.	Department of Mathematics, University of British Columbia, 2o75 Wesbrook Mall, Vancouver B.C., CANADA
Jirman, M. Dipl.-Math.	Fachbereich Mathematik der Universität, Schloßgartenstraße 7, 61oo Darmstadt, W-GERMANY
Joseph, D. D. Prof. Dr.	Department of Aerospace Engineering and Mechanics, University of Minnesota, 11o Union Street, S. E. Minneapolis, Minnesota 55455, USA
Kambe, T. Prof. Dr.	Faculty of Engineering 36, Department of Applied Science, Kyushu University, Hakozaki, Fukuoka 812, JAPAN
Kaniel, S. Prof. Dr.	Department of Mathematics, The Hebrew-University of Jerusalem, Institut of Mathematics, Jerusalem, ISRAEL
Knightly, G. H. Prof. Dr.	University of Massachusetts, Department of Mathematics and Statistics, GRC Tower, Amherst 01003, USA
Kordulla, W. Dr.	Institut für Theoretische Strömungsmechanik, Deutsche Forschungs- und Versuchsanstalt für Luft- und Raumfahrt, Bunsenstraße 10, 3400 Göttingen, W-GERMANY
Kovenya, V.M. Dr.	Institute of Pure and Applied Mechanics, Academy of Sciences, Novosibirsk 63oo9o, USSR

Krause, E. Prof. Dr. — Aerodynamisches Institut der Rheinisch-Westfälischen Technischen Hochschule Aachen, Templergraben 55, 51oo Aachen, W-GERMANY

Kreth, H. Dr. — Institut für Angewandte Mathematik, Universität, Bundesstraße 55, 2ooo Hamburg, W-GERMANY

Lange, H. Prof. Dr. — Mathematisches Institut der Universität, Weyertal 86-9o, 5ooo Köln 41, W-GERMANY

Lütcke, H. Dr. — Mathematisches Institut der Universität Düsseldorf, Universitätsstraße 1, 4ooo Düsseldorf 1, W-GERMANY

Martensen, E. Prof. Dr. — Mathematisches Institut II der Universität, Englerstraße 2, 75oo Karlsruhe 1, W-GERMANY

Masuda, K. Prof. Dr. — University of Tokyo, Department of Pure and Applied Sciences, 3-8-1, Komaba, Meguro-ku, Tokyo, 153 Japan, JAPAN

Meyer-Spasche, R. Dr. — Max-Planck-Institut für Plasmaphysik, 8o46 Garching bei München, W-GERMANY

Moffatt, H. K. Prof. Dr. — School of Mathematics, University of Bristol, University Walk, Bristol, U. K.

Orszag, S. A. Prof. Dr. — Massachusetts Institute of Technology, Department of Mathematics, M.I.T. 2-347, Cambridge, Mass. o2139, USA

Ôtani, M. Prof. Dr. — Department of Mathematics, Tokai University, 1117, Kitakaname, Hiratsuka, Kanagawa, Japan, 259-12, JAPAN

Periaux, J. Dr. — Avions Marcel Dassault - Breguet Aviation 42, Allée de Saint-Cucufa, B.P. 32 9242o Vaucresson, FRANCE

Peyret, R. Prof. Dr. — Department de Mathématiques, Université de Nice, Avenue Valrose, o6o34 Nice Cedex, FRANCE

Potsch, K. Dr. — Institut für Gasdynamik und Thermodynamik der Technischen Universität Wien, Karlsplatz 13, 1040 Wien, AUSTRIA

Prouse, G. Prof. Dr. — Istituto di Matematica del Politecnico, Piazza Leonardo da Vinci, 32, 2o133 Milano, ITALY

Rannacher, R. Dr. — Universität Bonn, Institut für Angewandte Mathematik, Beringstr. 4-6, 5300 Bonn 1, W-GERMANY

Rautmann, R. Prof. Dr. — Gesamthochschule Paderborn, Fachbereich Mathematik-Informatik, Warburger Straße 1oo, 479o Paderborn, W-GERMANY

Roux, B. Prof. Dr. — Université D'Aix-Marseille, Institut de Mécanique des Fluides, 1, Rue Honnorat, 13oo3 Marseille, FRANCE

Rubin, S. G. Prof. Dr.	Department of Aerospace Engineering and Applied Mechanics, Rhodes Ha-1, University of Cincinnati, Cincinnati, Ohio 45221, USA
Ruelle, D. Prof. Dr.	Institut des hautes êtudes scientifique, 35 Route de Chartres, 9144o Bures-Sur-Yvette, FRANCE
Rues, D. Prof. Dr.	Institut für Theoretische Strömungsmechanik, Deutsche Forschungs- und Versuchsanstalt für Luft- und Raumfahrt, Bunsenstraße 10, 3400 Göttingen, W-GERMANY
Sattinger, D. H. Prof. Dr.	University of Minnesota, School of Mathematics, 127 Vincent Hall, 2o6 Church Street S.E. Minneapolis, Minnesota 55455, USA
Schilling, R. Dr.-Ing.	Institut für Strömungslehre und Strömungsmaschinen, Universität Karlsruhe (TH), Kaiserstraße 12, 75oo Karlsruhe 1, W-GERMANY
Schröck-Pauli, C. Dr.	Institut für Festkörperforschung der Kernforschungsanlage Jülich, 517 Jülich, W-GERMANY
Shevelev, Yu. D. Prof. Dr.	Institute for Problems in Mechanics, Prospect Vernadskogo 1o1, 117526 Moscow, USSR
Socolescu, D. Dr.	Institut für Angewandte Mathematik der Universität Karlsruhe, Englerstraße 2, 75oo Karlsruhe, W-GERMANY
Socolescu, R. Dr.	Institut für Angewandte Mathematik der Universität Karlsruhe, Englerstraße 2, 75oo Karlsruhe, W-GERMANY
Sonneveld, P. Dr.	Department of Mathematics, Julianalaan 132, Delft University of Technology, Delft, NETHERLANDS
Sohr, H. Prof. Dr.	Gesamthochschule Paderborn, Fachbereich Mathematik-Informatik, Warburger Straße 1oo, 479o Paderborn, W-GERMANY
Sprekels, J. Dr.	Institut für Angewandte Mathematik der Universität Hamburg, Bundesstraße 55, 2ooo Hamburg 13, W-GERMANY
Stephan, E. Dr.	Fachbereich Mathematik, Technische Hochschule, Schloßgartenstraße 7, 61oo Darmstadt, W-GERMANY
Stewartson, K. Prof. Dr.	University College London, Department of Mathematics, Gower Street, London WCIE 6 BT, U. K.
Strampp, W. Dr.	Gesamthochschule Paderborn, Fachbereich Mathematik-Informatik, Warburger Straße 1oo, 479o Paderborn, W-GERMANY
Swaminathan, K. Dr.	Department of Mathematics, Indian Institute of Technology, Madras 6ooo36,INDIA

Takeshita, A. Prof. Dr.	Chikusa-Ku, Nagoya, Nagoya-University, Nagoya, Japan 464, JAPAN
Taylor, T. D. Prof. Dr.	Aerospace Corporation, P.O. Box 92951, Los Angeles, California 9ooo9, USA
Thomasset, F. Dr.	Institut de Recherche, D'Informatique et D'Automatique IRIA, Domaine de Voluceau-Rocquencourt, 7815o Le Chesnay, FRANCE
Valli, A.	Dipartimento di Matematica e Fisica, Libera Università di Trento, 38o5o Povo(Trento),ITALY
Van de Vooren, A.I. Prof. Dr.	Rijksuniversiteit te Groningen, Mathematisch Instituut, Postbus 8oo, Hoogbou WSN, Universiteitscomplex Paddepoel, NETHERLANDS
Vasanta Ram, V. Dr.-Ing.	Institut für Thermo- und Fluiddynamik der Universität, Universitätsstraße 50 Gebäude B, 463o Bochum-Querenburg, W-GERMANY
Veldman, A.E.P. Dr.	National Aerospace Laboratory NLR, P.O. Box 9o5o2, 1oo6 BM Amsterdam, NETHERLANDS
Verri, M. Dr.	Istituto di Matematica del Politecnico, Piazza Leonardo da Vinci, 32, 2o133 Milano, ITALY
Wachendorff, R. Dipl.-Math.	INTERATOM GmbH, Friedrich-Ebert-Straße, 5o7 Bergisch Gladbach 1 (Bensberg), W-GERMANY
Wahl, W. von, Prof. Dr.	Lehrstuhl für Angewandte Mathematik der Universität, Postfach 3oo8, 858o Bayreuth, W-GERMANY
Weiland, C. Dr.	Institut für Theoretische Strömungsmechanik, Deutsche Forschungs- und Versuchsanstalt für Luft- und Raumfahrt Göttingen, Bunsenstraße 1o, 3400 Göttingen, W-GERMANY
Wesseling, Prof. Dr.	Department of Mathematics, Julianalaan 132, Delft University of Technology, Delft, NETHERLANDS
Zandbergen, P. J. Prof. Dr.	Technische Hogeschool Twente, Postbus 217, Enschede, NETHERLANDS

STEADY SOLUTIONS OF THE NAVIER-STOKES EQUATIONS REPRESENTING PLANE FLOW IN CHANNELS OF VARIOUS TYPES

C. J. Amick*
St. John's College, Cambridge, England.

1. Introduction

Recent work of Heywood [1], Ladyzhenskaya and Solonnikov [2], and others [3], [4], [5], [6] has drawn attention (a) to questions of uniqueness of Navier-Stokes solutions for certain unbounded domains Ω in $\mathbb{R}^n$ that can be regarded as models of channels, tubes, or conduits of some kind, and (b) to the importance of prescribing not merely the fluid velocity u on the boundary $\partial\Omega$, but also some quantity like the flux **M** (that is, the total volumetric flow rate, defined for n=2 by (1.3b) below). However, the existence theory for such domains (which have non-compact boundaries) seems somewhat sparse relative to that for bounded and exterior domains, and it is this existence problem to which we address ourselves. (By an exterior domain we mean a connected open set that is the complement of a compact set.) The differences between two- and three-dimensional problems with unbounded domains are evident from the results known for the exterior problem; for n=3, the existence of classical solutions which approach a prescribed constant vector at infinity are known for any positive value of the kinematic viscosity ν, and the work of Babenko [7] essentially completes the picture by precisely describing the asymptotic form of the velocity at infinity. For n=2, considerably less is known; although the existence of a weak solution was shown by Leray, it is only in recent years that substantial progress has been made [8], [9] and the behavior of the weak solution at infinity still remains unknown. (We hasten to mention the results in [10] which completely solve the two-dimensional exterior problem for sufficiently large viscosity.) The difference between the cases n=2 and n=3 for problems in unbounded domains is due to the space of functions in which the solutions are sought; the case n=3 allows one to 'control' weak solutions at infinity far more easily than the case n=2. Hence, the problem of steady Navier-Stokes flow in channels $\Omega \subset \mathbb{R}^3$ is almost certainly more tractable than the case n=2, and so the present work deals with the latter.

By a channel, we mean an unbounded domain $\Omega \subset \mathbb{R}^2$ that (a) is simply connected and (b) has a boundary $\partial\Omega$, of class C^∞, consisting of two unbounded components Γ_+ and Γ_- (the channel walls) such that $\mathrm{dist}(\Gamma_+,\Gamma_-)>0$. We seek a solution (u,p) of the steady Navier-Stokes equations

$$-\nu D^2 u + (u\cdot D)u = -Dp + f \qquad \text{in } \Omega, \qquad (1.1)$$

$$\mathrm{div}\, u = D\cdot u = 0 \qquad \text{in } \Omega, \qquad (1.2)$$

such that

*Research supported by a post-doctoral Fellowship of the United States National Science Foundation. Current address: Department of Mathematics, University of Chicago.

$$u = 0 \text{ on } \partial\Omega, \qquad \int_{\gamma} u\cdot n = M>0, \qquad (1.3a,b)$$

where $D = (\partial/\partial x, \partial/\partial y)$ is the gradient operator, D^2 is the Laplacian, ν denotes the kinematic viscosity, M the flux, γ is a smooth simple arc in $\bar{\Omega}$, from Γ_- to Γ_+, and f is a smooth function which goes to zero sufficiently fast at infinity. All of the results in this paper hold for suitable f, but, for simplicity, we take $f = 0$. The Reynolds number R is defined by $R = M/\nu$. If the channel width tends to infinity far upstream and downstream, then we demand that

$$|u(z)| \to 0 \quad \text{as} \quad |z| \to \infty \text{ in } \Omega. \qquad (1.4a)$$

When the channel width tends to a finite limit, we demand that

the appropriate Poiseuille velocities be approached. (1.4b)

Let S denote the strip $\mathbb{R} \times (-1,1)$ and let F denote a one-to-one conformal map of $\bar{S}$ onto $\bar{\Omega}$ such that Γ_- and Γ_+ are the images of $\mathbb{R} \times \{-1\}$ and $\mathbb{R} \times \{1\}$, respectively:

$$z = F(\zeta), \quad z = x + iy \in \Omega \text{ and } \zeta = \xi + i\eta \in S,$$

with

$$\frac{dz}{d\zeta} = F'(\zeta) = h\, e^{i\theta} \qquad (h = |F'|\,). \qquad (1.5)$$

We write $\nabla = (\partial/\partial\xi, \partial/\partial\eta)$ for the gradient operator in S.

The results in this paper will be stated for various types of channels, and to this end, we need the following

DEFINITION 1.1. *Let* $h : \bar{S} \to (0,\infty)$ *be the arc-length function introduced in* (1.5). *We shall say that a channel* Ω *is*

(a) *of type* I *if* $1/h$ *and* $|\nabla(1/h)|$ *belong to* $L_2(S)$;

(b) *of type* I' *if it is of type* I *and for each* $\xi_0 \in \mathbb{R}$,

$$0<c\leq \frac{h(\xi,\eta)}{h(\xi_0,-1)} \leq d \qquad \forall(\xi,\eta) \in (\xi_0-1,\xi_0+1) \times (-1,1),$$

where the constants c *and* d *are independent of* ξ_0. *Furthermore,* $\kappa(\zeta)/h(\zeta)$, $\lambda(\zeta)/h(\zeta)$ $\to 0$ *as* $|\zeta| \to \infty$, *where* $\kappa = h_\xi/h$ *and* $\lambda = h_\eta/h$;

(c) *of type* II *if* $|\nabla \log h(\xi,\eta)| \to 0$ *pointwise as* $|\xi| \to \infty$, *uniformly with respect to* η, *and* $|\nabla(1/h)| \in L_2(S)$;

(d) *of type* III *if it is of type* II *and also* $|\nabla \log h| \in L_2(S)$, $|\nabla h| \in L_r(S)$ *for some* $r>2$, *and* $h(\xi,\eta) \leq \text{const.}\,|\xi|^{\frac{1}{2}}$ *for* $|\xi| \geq 1$.

Our main results will be stated for domains of type I' and III. As an example of this definition, let

$$\Omega_f = \{ (x,y)\in \mathbb{R}^2 : x\in \mathbb{R}, -f(x)<y<f(x) \}, \tag{1.6}$$

where

$$f(x) = \begin{cases} \text{const. } (-x)^{k_1} & \text{for } x\leq -L_1 - 1, \\ \text{const. } (x)^{k_2} & \text{for } x\geq L_2 + 1, \end{cases}$$

for certain k_j, $L_j \geq 0$. One can show that Ω_f is of type I' if $k_j>1/3$ and of type III if $k_j\leq 1/3$ $(j = 1,2)$; between them, they include many reasonable kinds of channels and justify our intention to state the main results (particularly in section 3) for these two types.

The existence theorems in section 2 are different for types I' and III as are the methods of proving (1.4) in section 3. A non-rigorous argument which shows the difference between these two types can be seen by examining the pressure p for channels Ω_f, as in (1.6), for $k_j<1$ $(j =1,2)$. For these, the distinctions between $h\, d\xi$ and dx, between $h(\xi,\eta)$ and $f(x)$, and between η and $y/f(x)$ are not important at large distances, and a physicist would not hesitate to assert that the longitudinal velocity component u_1 has the asymptotic form $g(y/f(x))/f(x)$ as $|x|\to \infty$, by conservation of mass (div u = 0) and because $f(x)$ is the only 'natural length' at large distances. Now the Navier-Stokes equations state that at a channel wall the tangential pressure gradient, essentially $-\partial p/\partial x >0$ in the present case, balances the normal gradient of viscous shear stress, essentially $-\partial^2 u_1/\partial y^2$ here; hence, $-\partial p/\partial x$ is asymptotically proportional to $1/f(x)^3$. For a channel Ω_f, this is integrable on $\mathbb{R}$ if and only if $k_j>1/3$, and then Ω_f *is* of type I'. We shall show rigorously in section 3 that the pressure has finite limits at infinity for type I' channels and is unbounded at infinity for those of type III.

By necessity, the calculations and estimates needed in this paper are exceedingly long and technical, and so only the briefest of proofs will be given. The results represent joint work with L. E. Fraenkel, and the detailed results and proofs may be found in [11].

2. Existence of weak solutions

Let $J(\Omega) = C_0^{\infty,\text{sol}}(\Omega\to \mathbb{R}^2)$ denote the set of infinitely differentiable vector fields $v = (v_1,v_2)$ that are solenoidal (div v = 0) and have compact support in Ω. We write

$$Dv:Dw = \sum_{i,j=1,2}^{2} (D_i v_j)(D_i w_j) \; , \; |Dv|^2 = Dv:Dv \; ,$$

and let $H(\Omega)$ denote the completion of $J(\Omega)$ in the norm $\|\cdot\|$ corresponding to the inner product

$$\langle v,w\rangle = \int_\Omega Dv:Dw \quad .$$

Elements of $H(\Omega)$ are weakly solenoidal, vanish on $\partial\Omega$ in the sense of a trace, and carry zero flux. The following lemma will be useful in making certain estimates.

LEMMA 2.1. Let Ω be a channel. Then, for every $v \in H(\Omega)$,

$$\int_\Omega \frac{|v|^2}{h^2} = \int_S |v|^2 \leq \frac{4}{\pi^2}\|v\|^2 \quad , \tag{2.1}$$

$$\int_\Omega \frac{|v|^2}{h^2(\eta+1)^2} = \int_S \frac{|v|^2}{(\eta+1)^2} \leq 4\|v\|^2 \quad , \tag{2.2}$$

and (2.2) holds with η replaced by $-\eta$.

Proof. It suffices to prove the result for $v \in J(\Omega)$. The equalities in (2.1) and (2.2) follow since the Jacobian of our map is merely h^2, and the inequalities then follow from standard results for functions of compact support in the strip and the relation

$$\int_S |\nabla v|^2 = \int_\Omega |Dv|^2 \equiv \|v\|^2 \quad .$$

q.e.d.

For suitable functions u, v, and w, we define

$$\{u,v,w\} = \int_\Omega u\cdot(v\cdot D)w = \sum_{i,j=1,2}^{2} \int_\Omega u_i v_j D_j w_i \, .$$

DEFINITION 2.2. For any channel Ω, a vector field will be called a flux carrier if it belongs to $C^\infty(\bar{\Omega}\to \mathbb{R}^2)$ and satisfies (1.2) to (1.4). A velocity field $u = g + v$ is a weak solution of (1.1) to (1.4) if g is a flux carrier, $v \in H(\Omega)$ and

$$\nu\int_\Omega Dw{:}Du + \{w,u,u\} = 0, \qquad \forall w \in J(\Omega) \, , \tag{2.3}$$

or, equivalently

$$\nu\langle w,v\rangle + \{w,g+v,v\} + \{w,v,g\} = -\nu\int_\Omega Dw{:}Dg \; - \{w,g,g\} \tag{2.4}$$

for all $w \in J(\Omega)$.

The following lemma will be used in Theorem 2.5 to prove the existence of a weak solution for type I channels for any Reynolds number R>0.

LEMMA 2.3. For channels of type I and any Reynolds number $M/\nu \in (0,\infty)$, there exists a flux carrier g such that, if $v \in H(\Omega)$ and

$$\nu \|v\|^2 + \{v,v,g\} = -\nu \int_\Omega Dv{:}Dg - \{v,g,g\} \quad , \tag{2.5}$$

then

$$\|v\| \leq \text{const.}, \tag{2.6}$$

where the constant depends only on the data Ω, ν, and M.

Proof. For any $\varepsilon>0$, let $\mu(\cdot,\varepsilon) \in C^\infty([0,\infty) \to [0,1])$ be the usual mollifier used in Navier-Stokes problems for extending boundary-value functions; $\mu(t,\varepsilon) = 1$ at, and sufficiently near, $t = 0$; $\mu(t,\varepsilon) = 0$ for $t\geq\varepsilon$; and for $t>0$

$$\mu(t,\varepsilon)\leq\varepsilon/t, \quad 0\leq -\mu'(t,\varepsilon)\leq\varepsilon/t \quad , \tag{2.7}$$

where μ' denotes the t derivative. We define $g = (G_y, -G_x)$, where

$$G(\zeta,\varepsilon) = -\tfrac{1}{2}M\,\{\mu(\eta+1,\varepsilon) - \mu(1-\eta,\varepsilon)\} \quad ,$$

and claim that g is a flux carrier.

A calculation ensures that

$$|g| = -\tfrac{1}{2}M\,\{\mu'(\eta+1,\varepsilon) + \mu'(1-\eta,\varepsilon)\}/h(\zeta) \quad . \tag{2.8}$$

An integration by parts ensures that

$$- \{v,v,g\} = \{g,v,v\} = \int_\Omega g\cdot(v\cdot D)v$$

$$\leqslant \tfrac{1}{2}M\varepsilon \int_\Omega \frac{1}{h}\left(\frac{|v|}{\eta+1} + \frac{|v|}{1-\eta}\right)|Dv| \leqslant 2M\varepsilon\|v\|^2 \tag{2.9}$$

by (2.2), (2.7), and (2.8). We choose $\varepsilon = \varepsilon_0$ in (0,1) such that $2M\varepsilon_0 \leqslant \frac{1}{2}\nu$, and let $g = g(\cdot,\varepsilon_0)$ henceforth; then the left side of (2.9) is not less than $\frac{1}{2}\nu\|v\|^2$.

A calculation gives

$$\int_S |\nabla g|^2 = \int_S \{(\lambda G_\eta)^2 + 2(\kappa G_\eta)^2 + (G_{\eta\eta} - \lambda G_\eta)^2\}/h^2 \quad ,$$

where $\kappa = h_\xi/h$ and $\lambda = h_\eta/h$. Since $\varepsilon = \varepsilon_0$ is fixed, the functions G_η and $G_{\eta\eta}$ are uniformly bounded pointwise, whence

$$\int_S |\nabla g|^2 \leqslant \text{const.} \int_S \{1/h^2 + |\nabla(1/h)|^2\} \leqslant \text{const.},$$

since $1/h \in W_2^1(S)$ because Ω is of type I. Thus $|\nabla g| \in L_2(S)$ and so $|Dg| \in L_2(\Omega)$. Accordingly,

$$-\nu\int_\Omega Dv{:}Dg \leqslant \nu\|v\|\ |Dg|_{0,2,\Omega} = c_1\|v\| \ , \text{ say,}$$

$$-\{v,g,g\} \leqslant \text{const.}\int_\Omega \frac{|v|}{h}|Dg| \leqslant \text{const.}\|v\| = c_2\|v\| \ , \text{ say,}$$

where we have used the Schwarz inequality and (2.1). It follows that $\|v\| \leqslant 2(c_1+c_2)/\nu$. q.e.d.

Note that the crucial step in the proof of Lemma 2.3 is to first choose ε sufficiently small so that the quadratic term $-\{v,v,g\}$ is less than $\frac{1}{2}\nu\|v\|^2$, and then to use the fact that $(1/h) \in L_2(S)$ to ensure that the right-hand side of (2.5) defines a bounded linear functional for $v \in H(\Omega)$.

For channels of type III, this approach is impossible since $h(\xi,\eta) \leqslant \text{const.}|\xi|^{\frac{1}{2}}$ for $|\xi| \geqslant 1$, whence $1/h \notin L_2(S)$. For such channels and for channels of type II, a different form of flux carrier g must be used: the desired function g has the form $g = (G_y, -G_x)$, where $G(\xi,\eta)$ is equal to $Q(\eta) = \frac{3}{4}M(\eta - 1/3\eta^3)$ for all sufficiently large $|\xi|$. Hence, for such distant points, g has the magnitude $\frac{3}{4}M(1 - \eta^2)/h(\zeta)$ and direction $D\xi(z)$ in Ω.

This special form of g allows one to show that the right side of (2.5) defines a bounded linear functional for $v \in H(\Omega)$. However, the absence of a small term ε as in the proof of Lemma 2.3 prevents us from controlling the term $\{v,v,g\}$ as before. The proof of the following lemma is analogous to arguments in [3].

LEMMA 2.4. Let $W = W(S)$ denote the completion of $C_0^\infty(S \to \mathbb{R})$ in the norm

$$\|\phi\|_W^2 = \int_S (\nabla^2\phi)^2 = \int_S \{\phi_{\eta\eta}{}^2 + 2\phi_{\eta\xi}{}^2 + \phi_{\xi\xi}{}^2\} \quad ,$$

so that W is equivalent to the Hilbert-Sobolev space $\overset{\circ}{W}{}_2^2(S)$, and define

$$\nu_0 = \sup_{\phi \in W \setminus 0} \int_S Q_\eta(\phi_\eta\phi_{\eta\xi} - \phi_\xi\phi_{\eta\eta}) / \|\phi\|_W^2 \quad ,$$

where $Q(\eta) = \frac{3}{4}M(\eta - 1/3\eta^3)$ is the stream function of Poiseuille flow in S. For channels of type II and $\nu > \nu_0$, there exists a flux carrier g, different at large distances from that in Lemma 2.3, such that (2.5) implies (2.6) for any $v \in H(\Omega)$.

In [4], careful analysis of the functional whose supremum appears above, followed by numerical computation, shows that $\nu_0 = M/116.5$. If only odd functions of η are admitted, as is legitimate when one seeks symmetrical velocity fields in a symmetric channel, then $\nu_0 = M/194.6$.

THEOREM 2.5. The problem (1.1) to (1.4) has a weak solution u for each Reynolds number $R = M/\nu \in (0,\infty)$ if Ω is of type I, and for each $R < M/\nu_0$ if Ω is of type II (ν_0 being as in Lemma 2.4).

Proof. Let $\{\Omega_m\}$, m = 1, 2, ..., be an expanding sequence of simply-connected bounded subdomains of Ω such that $\Omega_m \to \Omega$ as $m \to \infty$ and $\partial\Omega_m$ is of class C^∞. Consider the problem of finding a solution (u^m,p_m) of (1.1) and (1.2) with $u^m = g$ on $\partial\Omega_m$. It is known that this problem has a weak solution $u^m = g + v^m$, defined as in Definition 2.2, but with Ω_m replacing Ω; in particular, $v^m \in H(\Omega_m)$. Since (for fixed m) Ω_m is bounded, $H(\Omega_m)$ is embedded in $L_q(\Omega_m \to \mathbb{R}^2)$ for all $q \geq 1$, and this allows the analogue of (2.4) to be extended to all test functions $w \in H(\Omega_m)$. Putting $w = v^m$ and noting that $\{v^m, g+v^m, v^m\} = 0$, there results

$$\nu \|v^m\|^2 + \{v^m,v^m,g\} = -\nu\int_\Omega Dv^m{:}Dg - \{v^m,g,g\} \quad .$$

Since $v^m \in H(\Omega_m) \subset H(\Omega)$, it follows that each v^m satisfies (2.5).

By Lemmas 2.3 and 2.4, $\|v^m\|$ is bounded independently of m. By relabelling the sequence if necessary, we may assume that $v^m \to v$ weakly in $H(\Omega)$. For any given $w \in J(\Omega)$, we have supp $w \subset \Omega_k$ for some k, so that v^m satisfies (2.4) for that w if $m \geq k$. By letting

$m \to \infty$ and using the compact embedding of $W_2^1(\Omega_k)$ in $L_q(\Omega_k)$, a standard argument ensures that v satisfies (2.4). q.e.d.

We remind the reader that there is no question of our solution being trivial since $u = g + v$ carries flux $M>0$ due to g being a flux carrier and v being an element of $H(\Omega)$. Standard theory allows us to state that there exists a pressure p (unique modulo a constant) [12] such that (u,p) satisfies (1.1) and (1.2) as distributions. Regularity theory [11], [12] proves that $u \in C^\infty(U \to \mathbb{R}^2)$ and $p \in C^\infty(U \to \mathbb{R})$ for all bounded domains $U \subset \Omega$. Hence, (u,p) satisfies (1.1) to (1.3) pointwise, and to complete our work, we next show that (1.4) is satisfied. Before tackling this problem in section 3, we give the following a priori estimate for solutions in type III channels.

THEOREM 2.6. <u>Let</u> Ω <u>be of type</u> III, <u>let</u> $\nu > \nu_0$, <u>and let</u> $u = g + v$ <u>be a weak solution of</u> (1.1) <u>to</u> (1.4); <u>here</u> g <u>is constructed as in the remarks before Lemma</u> 2.4. <u>Then</u>

$$\int_\Omega \tilde{h}^2 |Dv|^2 = \int_S h^2 |\nabla v|^2 < \infty, \tag{2.10}$$

<u>where</u> $\tilde{h}(z) = h \circ F^{-1}(z)$.

The proof of Theorem 2.6 is tortuous and the details may be found in [11]. Since $\tilde{h}$ and h are unbounded at infinity in most cases, this theorem gives a considerably stronger result than knowing previously that $v \in H(\Omega)$ and $|\nabla v| \in L_2(\Omega)$.

3. <u>Behavior at infinity in channels of type</u> I' <u>and</u> III

We shall show later that Theorem 2.6 allows the use of standard integral estimates [13] to prove that v goes to zero uniformly at infinity for type III channels; since $u = g + v$ and g satisfies (1.4), it will then follow that u satisfies (1.4). For type I' channels, the flux carrier g satisfies (1.4), but since (2.10) need not hold for $v = u - g$, we can not use the forementioned integral estimates. A different approach is needed, and this is done with a one-sided maximum principle for the total head pressure $\Phi = p + \frac{1}{2}|u|^2$; the usefullness of this quantity was shown in the work by Gilbarg and Weinberger [8], [9] on the two-dimensional exterior problem.

Until further notice, we shall deal with channels of type I'. If we write $u = (u_1, u_2)$ and (the vorticity) $\omega = u_{2x} - u_{1y}$, then a calculation using (1.1) and (1.2) gives

$$\nu D^2\Phi - u_1\Phi_x - u_2\Phi_y = \nu\omega^2 \geq 0 \quad \text{in } \Omega,$$

and so if U is a suitable bounded open set in Ω, then

$$\max_{z \in \bar{U}} \Phi(z) \leq \max_{z \in \partial\bar{U}} \Phi(z),$$

and by changing variable to the strip S, it follows that the same one-sided maximum principle holds there.

To use this principle, we need the following

LEMMA 3.1. *Let* $\Omega \subset R^2$ *be a channel of type* I'. *There exist constants* B_1 *and* B_2 *such that the pressure* p *satisfies*

$$|p(\zeta) - B_1|,\ |p(\zeta) - B_2| \to 0 \quad \text{as } |\zeta| \to \infty \text{ in } S$$

upstream and downstream, respectively.

Proof. The detailed proof of this result in [11] consists of nine long lemmas, and so we can merely state the main steps here. It suffices to work downstream, $\xi \to \infty$, and since the pressure may be altered by a constant, we shall take $B_2 = 0$. The chief steps are as follows:

(i) there exists a number $a \in (0,1)$ such that

$$\int_{-a}^{a} p(\xi,\eta)\, d\eta \to 0 \quad \text{as} \quad \xi \to \infty ;$$

(ii)
$$\int_{-1}^{1} |p(\xi,\eta)|\, d\eta \to 0 \quad \text{as} \quad \xi \to \infty;$$

(iii) $p(\xi,\pm 1) \to 0$ as $\xi \to \infty$;

(iv) $p(\xi,\eta) \to 0$ uniformly as $\xi \to \infty$.

To indicate the type of arguments needed, we now sketch the proof of (i). Recall that $u = g + v$, where $|\nabla v| \in L_2(\Omega)$ since $v \in H(\Omega)$. The proof of Lemma 2.3 shows that $|\nabla g| \in L_2(\Omega)$, whence $|\nabla u| \in L_2(\Omega)$. Since $\omega = u_{2x} - u_{1y}$, we have $\omega \in L_2(\Omega)$, and so a change of variables gives

$$\int_S h^2 \omega^2 = \int_\Omega \omega^2 < \infty.$$

Hence,

$$\int_0^1 d\eta \int_{-\infty}^{\infty} d\xi\ \{h(\xi,\eta)^2 \omega(\xi,\eta)^2 + h(\xi,-\eta)^2 \omega(\xi,-\eta)^2\} = |h\omega|^2_{0,2,S} < \infty,$$

and the mean-value theorem gives the existence of a number $a \in (0,1)$ such that

$$\int_{-\infty}^{\infty} d\xi\ \{h(\xi,a)^2 \omega(\xi,a)^2 + h(\xi,-a)^2 \omega(\xi,-a)^2\} = |h\omega|^2_{0,2,S} . \tag{3.1}$$

Equations (1.1) and (1.2) imply

$$\nu\omega_\eta - (u_1 u_{2\eta} - u_2 u_{1\eta}) = -p_\xi \quad \text{in } S, \tag{3.2}$$

and so

$$\frac{d}{d\xi} \int_{-a}^{a} p(\xi,\eta)\, d\eta = -\nu\{\omega(\xi,a) - \omega(\xi,-a)\} + \int_{-a}^{a} \{u_1 u_{2\eta} - u_2 u_{1\eta}\} d\eta. \tag{3.3}$$

Since $1/h \in W_2^1(S)$ by the definition of type I domains, the standard theory of trace

yields
$$\int_{-\infty}^{\infty} \frac{d\xi}{h(\xi,\eta)^2} \leq \text{const.}\ |1/h|^2_{1,2,S} \quad \forall\eta \in (-1,1), \tag{3.4}$$

and the constant is independent of η. If we combine this inequality with (3.1) and use the Schwarz inequality, there results

$$\int_{-\infty}^{\infty} \{ |\omega(\xi,a)| + |\omega(\xi,-a)| \}\, d\xi < \infty .$$

It follows that the first term on the right of (3.3) is an element of $L_1(-\infty,\infty)$. Next, by the Schwarz inequality and the estimate

$$\int_{-1}^{1} f^2\, d\eta \leq \frac{4}{\pi^2}\int_{-1}^{1} (f_\eta)^2\, d\eta \quad \forall f \in \overset{\circ}{W}{}^1_2(-1,1) ,$$

there results

$$\int_{-a}^{a} \{u_1 u_{2\eta} - u_2 u_{1\eta}\} d\eta \leq \int_{-1}^{1} \{ |u_1||u_{2\eta}| + |u_2||u_{1\eta}| \} d\eta \leq \frac{2}{\pi}\int_{-1}^{1} |\nabla u|^2\, d\eta ,$$

and this last term is in $L_1(-\infty,\infty)$ since $|\nabla u| \in L_2(S)$. It follows that

$$\frac{d}{d\xi}\int_{-a}^{a} p(\cdot,\eta)\, d\eta \in L_1(-\infty,\infty) ,$$

and so $\int_{-a}^{a} p(\xi,\eta)\, d\eta$ has a limit as $\xi \to \infty$. Altering p by a constant gives (i). q.e.d.

We are now in a position to prove (1.4) for type I' channels. We first claim that for each $v \in H(\Omega)$,

$$\int_{-1}^{1} |v(\xi,\eta)|^2\, d\eta \to 0 \quad \text{as} \ |\xi| \to \infty . \tag{3.5}$$

To see this let $v \in J(\Omega)$, so that

$$\int_{-1}^{1} |v(\xi_0,\eta)|^2\, d\eta = -2\int_{-1}^{1} d\eta \int_{\xi_0}^{\infty} v\cdot v_\xi\, d\xi \leq \int_{S_0} (|v|^2 + |v_\xi|^2) \leq \frac{4}{\pi^2}\int_{S_0} |\nabla v|^2 ,$$

where $S_0 = \{\zeta \in S : \xi > \xi_0\}$. Equation (3.5) then follows by continuity. Since $1/h \in L_2(S)$, equation (2.8) gives $g \in L_2(S)$. Furthermore, $|\nabla u| \in L_2(S)$, and so for each positive integer m, there exists $\xi_1(m) \in (m,m+1)$ such that

$$\int_{-1}^{1} d\eta\, \{ |g(\xi_1,\eta)|^2 + |u_\eta(\xi_1,\eta)|^2 \} = \int_{m}^{m+1} d\xi \int_{-1}^{1} \{ |g|^2 + |u_\eta|^2 \} d\eta \to 0 \tag{3.6}$$

as $m \to \infty$. Combining (3.5) and (3.6) gives

$$\int_{-1}^{1} d\eta\, \{ |u(\xi_1,\eta)|^2 + |u_\eta(\xi_1,\eta)|^2 \} \to 0 \text{ as } m \to \infty .$$

The embedding of $W^1_2(-1,1)$ into $C[-1,1]$ is bounded, and so

$$\max_{\eta\in[-1,1]} |u(\xi_1(m),\eta)| \to 0 \text{ as } m\to\infty.$$

By Lemma 3.1, we may assume that $p(\xi,\eta)\to 0$ uniformly as $\xi\to\infty$. Let $\varepsilon>0$ and choose a positive integer $N = N(\varepsilon)$ such that $m>N$ implies that

$$|u(\xi_1(m),\eta)|^2 \leq \varepsilon \text{ for all } \eta\in[-1,1], \tag{3.7}$$

$$|p(\xi,\eta)|\leq\varepsilon \quad \text{for all } (\xi,\eta)\in(N,\infty)\times[-1,1]. \tag{3.8}$$

If $A_m = (\xi_1(m),\xi_1(m+1))\times(-1,1)$, then $\Phi\leq 3/2\varepsilon$ on ∂A_m by (3.7), (3.8), and the fact that u vanishes on ∂S. The one-sided maximum principle for Φ ensures that $\Phi = p + \frac{1}{2}|u|^2 < 3/2\varepsilon$ in A_m, whence (3.8) yields $|u(\zeta)|^2\leq 5\varepsilon$ in A_m. Since m and ε may be taken arbitrarily large and small, respectively, we have proved that (1.4a) holds for type I' channels. Combining this result with Theorem 2.5 gives the existence of a classical solution of (1.1) to (1.4) in type I' channels for every Reynolds number $R>0$.

The behavior at infinity for solutions in type III channels is fairly easy to determine because of Theorem 2.6. Recall that $u = g + v$, where $g = (G_y,-G_x)$ was constructed a priori, and $v\in H(\Omega)$. One introduces a scalar-valued stream function $\Psi = G + \psi$, where $u = (\Psi_y,-\Psi_x)$ and $v = (\psi_y,-\psi_x)$. The function ψ vanishes on ∂S along with its gradient and Theorem 2.6 ensures that $\psi\in\overset{\circ}{W}{}^2_2(S)$. The stream function equation arising from (1.1)

$$\{\nu D^2 - (\Psi_y\partial/\partial x - \Psi_x\partial/\partial y)\}\, D^2\Psi = 0 \text{ in } \Omega \tag{3.9}$$

may be rewritten as an equation $(3.9)^*$ in S. Since G is known, equation $(3.9)^*$ may be considered as one for ψ in S, where $\psi\in\overset{\circ}{W}{}^2_2(S)$. The estimates of Agmon [13] with the methods in [4] may be applied to $(3.9)^*$, and it then follows easily that $|\nabla\psi(\zeta)|$ goes to zero as $|\zeta|\to\infty$ in S. By translating this result back to Ω, it follows that $|v(z)|\to 0$ as $|z|\to\infty$ in Ω. Since $u = g + v$ and g satisfies (1.4), it follows that the same is true for u. Combining this result with Theorem 2.5 ensures that for suitable Reynolds numbers, there exists a classical solution of (1.1) to (1.4) in channels of type III. With regard to the pressure p, we state the following result:

$$-p(\xi_0,\eta_0) \Big/ \int_0^{\xi_0} d\xi \int_{-1}^{1} 1/h^2\, d\eta \to \tfrac{3}{4}M\nu$$

uniformly as $|\xi_0|$ goes to infinity. Since channels of type III satisfy $h(\xi,\eta)\leq\text{const.}\,|\xi|^{\frac{1}{2}}$ for $|\xi|\geq 1$, it follows that the pressure is unbounded at infinity.

References

1 Heywood, J. G., On uniqueness questions in the theory of viscous flow. Acta math., 136 (1976), 61-102.

2 Ladyzhenskaya, O. A. and Solonnikov, V. A., On the solvability of boundary and initial-boundary value problems for the Navier-Stokes equation in

domains with non-compact boundary. Vestnik Leningrad. Univ., 13 (1977), 39-47.

3 Amick, C. J., Steady solutions of the Navier-Stokes equations in unbounded channels and pipes. Ann. Sc. norm. super. Pisa, (4) 4 (1977), 473-513.

4 Amick, C. J., Properties of steady Navier-Stokes solutions for certain unbounded channels and pipes. Nonlinear Anal., Theory, Methods Appl., 2 (1978), 689-720.

5 Fraenkel, L. E., On a theory of laminar flow in channels of a certain class. Proc. Cambridge philos. Soc., 73 (1973), 361-390.

6 Fraenkel, L. E. and Eagles, P. M., On a theory of laminar flow in channels of a certain class. II. Math. Proc. Cambridge philos. Soc., 77 (1975), 199-224.

7 Babenko, K. I., On stationary solutions of the problem of flow past a body by a viscous incompressible fluid. Math. USSR, Sbornik, 91 (1973), 3-26.

8 Gilbarg, D. and Weinberger, H. F., Asymptotic properties of Leray's solution of the stationary two-dimensional Navier-Stokes equations. Russ. math. Surveys, 29 (1974), 109-123.

9 Gilbarg, D. and Weinberger, H. F., Asymptotic properties of steady plane solutions of the Navier-Stokes equations with bounded Dirichlet integral. Ann. Sc. norm. super. Pisa, (4) 5 (1978), 381-404.

10 Finn, R. and Smith, D. R., On the stationary solutions of the Navier-Stokes equations in two dimensions. Arch. rat. Mech. Analysis, 25 (1967), 26-39.

11 Amick, C. J. and Fraenkel, L. E., Steady solutions of the Navier-Stokes equations representing plane flow in channels of various types. Acta math., to appear.

12 Temam, R., Navier-Stokes equations: theory and numerical analysis. North-Holland, 1977.

13 Agmon, S., The L_p approach to the Dirichlet problem. Ann. Sc. norm. super. Pisa, (3) 13 (1959), 405-448.

ON PROPERTIES OF STEADY VISCOUS INCOMPRESSIBLE FLUID FLOWS

K.I. Babenko
Keldysh Institute of Applied Mathematics
Academy Nauk of the USSR
Moscow, USSR

Introduction

Let us consider a steady flow of viscous incompressible fluid past a body or a set of bodies. The number of bodies is unessential for the problems considered below. Therefore, we restrict ourselves to the case when there is only one body $T \subset R^3$ bounded by the surface S satisfying the Ljapunov's condition. Let u, p be the dimensionless velocity vector and pressure respectively (density may not be taken into account since, by virtue of its being constant, we can assume $\rho \equiv 1$). Let the origin be inside T and let the axes be directed so that $u_\infty = (1,0,0)$. As a length unit we take the diameter of T . The steady fluid flow is defined by the solution of the boundary value problem

$$(u \cdot \nabla) u + \operatorname{grad} p = R^{-1} \Delta u, \quad \operatorname{div} u = 0,$$

$$u\big|_S = u_0, \quad \lim_{|x| \to \infty} u(x) = u_\infty. \tag{0.I}$$

As to the function $u_0(x)$, we suppose that

$$\int_S u_0 \cdot n \, d\sigma = 0, \tag{0.2}$$

where n is a normal to S , and $d\sigma$ is the Lebesgue measure on S .

We remind that the existence of the solution of problem (0.I) was obtained first by J. Leray [I] , [2] , who assumed the boundary condition at infinity to be fulfilled only for $u_\infty = 0$, but in a generalized sense when $u_\infty \neq 0$. Other versions of the proof were given by R. Finn [2] , O. Ladyzenskaja [3] , and H. Fujita. The fulfilment of the condition at infinity for an arbitrary solution was established by R. Finn [3] .

The behaviour of the solution for large $|x|$ is of considerable interest. Indeed, according to the boundary-layer concept, the flow is potential outside the close vicinity of the body and the wake extending to infinity. Inside the wake the flow is essentially vortical and the order of convergence to zero of the difference $u - u_\infty$ is different than outside the wake. It would obviously be highly desirable to determine what classes of solutions of the problem have the same structure and to substantiate the boundary layer concept. The first step in this direction should be to elucidate the structure of the solution of the flow problem, and first of all to elucidate the asymptotics of $u(x) - u_\infty$ for $|x| \to \infty$.

As it is shown below, the problem on the asymptotics of the above difference is a key one in a great number of problems. From these one should point out the following: the establishment of the uniqueness solution theorem for problem (O.I) at small Reynolds numbers; the construction of a perturbation theory for $R \to 0$, and the substantiation of the known asymptotical results; the investigation of the spectrum of linearized problem (O.I), and the construction of a bifurcation theory of stationary or periodic solutions.

In this paper we shall touch upon some problems that are as yet to be solved, but a most important and basic problem which we shall consider is as follows: In the case of plain flows we have not got yet the solution of the problem on the nature of the decay of the difference $u(x) - u_\infty$ with $|x| \to \infty$ for any natural classes of solutions, e.g. the class of solutions with the finite Dirichlet integral. This implies the fact that the above mentioned problems still remain unsolved. Certainly, we could restrict ourselves to considering the PR class suggested by R. Finn [5], but in this case we should have a theorem of the existence of solutions of this class at any Reynolds number.

When the problem on the onset of turbulent flows is under consideration, the investigation of some specific flows, i.e. the investigation of the spectrum of the linearized problem, the establishment of the fact of loss of stability, the study of the problem on bifurcation and the onset of secondary stationary and periodic flows, is of particular importance. A plane flow round a circular cylinder is a remarkable illustration of such flows.

It would be extremely exciting to clear up the question whether the Karman's vortex street is an example of a secondary periodic flow. It is evident that this question can be solved only with the aid of computer. The computations, however, required to solve this problem need proper theoretical foundation which we cannot provide yet, since the corresponding problems of the theory are still open.

I. Some auxiliary propositions

The Oseen's system

$$\Delta u - R\frac{\partial u}{\partial x_1} - \operatorname{grad} p - \lambda u = f, \quad \operatorname{div} u = 0 \tag{I.I}$$

is of great importance in our further discussion. If $f(x)=\delta_i\,\delta(x)$, $\delta_i=(\delta_{i1},\delta_{i2},\delta_{i3})'$, and δ_{ij} is the Kronecker symbol, then the fundamental solution of system (I.I) has the form

$$H_i(x)=(H_{i1},H_{i2},H_{i3})', \quad p_i(x),$$

with

$$H_{ij}(x)=\left(\delta_{ij}\Delta-\frac{\partial^2}{\partial x_i \partial x_j}\right)\phi \ , j=1,2,3, \quad p_i(x)=\frac{\partial}{\partial x_i}\frac{1}{4\pi|x|}, \tag{I.2}$$

where

$$\phi=\frac{1}{16\pi^2}\int \exp\left[\alpha(x_1-\xi_1)-\beta|x-\xi|\right]\frac{d\xi}{|\xi||x-\xi|}, \qquad \alpha=\frac{1}{2}R, \ \beta=\sqrt{\alpha^2+\lambda}. \tag{I.3}$$

The prime in the last formulae means transposition. The integration in (I.3) is implemented over the whole of the space R^3. The verification of the validity of formula (I.2) is carried out in a simple way. Integral (I.3) converges if $Re\beta \geqslant \alpha$ or

$$Re\lambda \geqslant -R^{-2}(Im\lambda)^2. \tag{I.4}$$

The boundary value problems for equation (I.I)can be solved by means of the potential theory [6]. Let $\mathcal{D}=R^3 \setminus T$, n be the normal to S outward with respect to $\mathcal{D}$, and let $x \in \mathcal{D}\setminus S$, $y \in S$. Let us put

$$H_{ij}(x,y)=H_{ij}(x-y), \ p_i(x,y)=p_i(x-y),$$

$$K_{ij}(x,y)=\sum_{k=1}^{3}\left(\frac{\partial}{\partial y_k}H_{ij}(x,y)+\frac{\partial}{\partial y_j}H_{ik}(x,y)+\delta_{kj}p_i(x,y)\right)n_k(y)+$$
$$+\frac{R}{2}n_1(y)H_{ij}(x,y)$$

and denote by $K(x,y)$ a matrix of elements $K_{ij}(x,y)$. By $L(x,y)$ let us denote a vector with components

$$L_i(x,y)=\sum_{k=1}^{3}\left(\frac{\partial}{\partial y_k}p_i(x,y)+\frac{\partial}{\partial y_i}p_k(x,y)-R\delta_{ik}p_1(x,y)\right)n_e(y)+$$
$$+\frac{R}{2}n_1(y)p_i(x,y).$$

Let $\varphi: S \to R^3$, $\varphi=(\varphi_1,\varphi_2,\varphi_3)'$, $\varphi \in C[S]$. The integrals

$$v(x)=\int_S K(x,y)\varphi(y)d\sigma_y\ ,\ p(x)=\int_S L(x,y)\cdot\varphi(y)d\sigma_y\ ,$$

where $d\sigma_y$ is the Lebesgue measure on S, will be called double layer potentials. To avoid certain difficulties which can arise in attempts to solve the exterior Dirichlet problem for system (I.I) for $\lambda=0$, $R=0$, we shall modify the expression for the double layer potentials. Let $H(x,y)$ be a matrix of elements $H_{ij}(x,y)$, and $P(x,y)$ be a vector with components $p_i(x,y)$. Let us put

$$\mathcal{K}(x,y)=K(x,y)-H(x,y),\quad \mathcal{L}(x,y)=L(x,y)-P(x,y)$$

and applying the integrals

$$v(x)=\int_S \mathcal{K}(x,y)\varphi(y)d\sigma_y\ ,\ p(x)=\int_S \mathcal{L}(x,y)\cdot\varphi(y)d\sigma_y \tag{I.5}$$

define the modified double layer potential. For homogeneous equation (I.I) in $\mathcal{D}$ let us consider the boundary value problem

$$u\big|_S = u_0(x),\ \lim_{|x|\to\infty} u(x) = 0 \tag{I.6}$$

The solution of this problem can be sought in the form of (I.5). Applying the common technique we reduce the problem to an integral equation whose unique solvability can be established by the same reasoning as in the classical case when $R=0$, $\lambda=0$. As a result, we find out that the condition of the unique solvability reduces to the inequality

$$Re\,\lambda \geqslant C R^{-2}(Im\,\lambda)^2, \tag{I.7}$$

where C is a constant depending only on S the boundary of the body T. Thus, of the two inequalities (I.4) and (I.7) the sharpest one is considered valid.

While considering boundary value problem (I.I), (I.6), when $u_0(x) = -H_{ij}(x,y)$, where y is a parameter, we construct the Green's matrix $G(x,y) = (G_{ij}(x,y))$ and the corresponding "pressure" vector $R(x,y) = (R_1(x,y), R_2(x,y), R_3(x,y))'$. The results obtained by Odqvist [6] imply the following proposition

<u>Proposition I</u>. Let $S \in C^{2+\delta}$, $\delta > 0$. Then for $x, x^1, y \in \mathcal{D}$ the inequalities

$$|G(x,y)| \leq C|x-y|^{-1}, \left|\frac{\partial}{\partial x_k} G(x,y)\right|, |R(x,y)| \leq C|x-y|^{-2}, \quad k = 1,2,3;$$

$$\left|\frac{\partial^2}{\partial x_k \partial x_\ell} G(x,y)\right|, \left|\frac{\partial}{\partial x_k} R(x,y)\right| \leq C|x-y|^{-3}, \quad k,\ell = 1,2,3;$$

$$\left|\frac{\partial}{\partial x_\ell} G(x^1,y) - \frac{\partial}{\partial x_\ell} G(x,y)\right| \leq C|x^1-x| \ln^2|x^1-x| \left[|x^1-x|^{-3} + |x-y|^{-3}\right],$$

hold, if $|x-x^1|$ is small. These inequalities hold uniformly in R, λ, if $(R,\lambda) \in [0,R_0] \times \mathcal{C}_\rho$, where R_0, ρ are any quantities $< \infty$, and $\mathcal{C}_\rho$ is the intersection of the disk $\{\lambda : |\lambda| \leq \rho\}$ with the domain defined by inequality (I.7).

The properties of $G(x,y)$ and $R(x,y)$ as functions of R and λ are given by the following propositions.

<u>Proposition 2</u>. Let $\lambda = 0$, $x \neq y$. The functions $G(x,y)$ and $R(x,y)$ are analytical functions of R in any interval $[0,R_0]$, $R_0 < \infty$.

<u>Proposition 3</u>. Let $x \neq y$. The functions $G(x,y)$ and $R(x,y)$ are analytical functions of (R,λ) in $(0,R_0) \times \hat{\mathcal{C}}_\rho$, where $\hat{\mathcal{C}}_\rho = \mathcal{C}_\rho \setminus \partial \mathcal{C}_\rho$.

The rest of the properties of the Green's matrix will be given below in case of need.

2. On the asymptotics at infinity of stationary viscous incompressible fluid flows

I. The natural class of solutions of problem (0.I) is a class of solutions with the finite Dirichlet integral

$$D[u] = \left(\int_{\mathcal{D}} \sum_{k,l} \left(\frac{\partial u_k}{\partial x_l} \right)^2 dx \right)^{1/2} < \infty . \tag{2.I}$$

It is exactly in this class that the existence theorem was established, and conditions (2.I) are reasonable from physical standpoint as well. The principal fact concerning the asymptotics of the solutions of the class under consideration, which is that for any solution of problem (0.I) there holds the relation

$$u(x) - u_\infty = O(|x|^{-1}), \tag{2.2}$$

was proved by the author in I972 [7]. Earlier R. Finn [8] showed that the fulfilment of the weaker condition

$$u(x) - u_\infty = O\left(|x|^{-\frac{1}{2}-\varepsilon}\right), \quad \varepsilon > 0 \tag{2.3}$$

was sufficient to obtain the asymptotics of the difference

$$u(x) - u_\infty = H(x,0)F + O\left(|x|^{-\frac{3}{2}+\delta}\right), \quad \delta > 0, \tag{2.4}$$

where F is the vector force exerted by the flow on the body.

Putting in (I.3) $\lambda = 0$, we obtain

$$\phi(x) = -\frac{1}{4\pi R} \int_0^{\frac{1}{2}Rs} (1 - e^{-\alpha}) \frac{d\alpha}{\alpha}, \quad s = |x| - x_1 .$$

Taking into account (I.2), it is immediately seen from (2.4) that there is far from the body a paraboloidal wake region in the direction u_∞, and that the decay of the difference $u(x) - u_\infty$ inside it is the slowest. Indeed, from the estimate

$$|H_{ij}(x)| < C\,|x|^{-1}[s+1]^{-1}$$

it follows that $u(x) - u_\infty$ decreases likewise $|x|^{-2}$ outside any cone with the vertex lying at the origin and having the axis directed along the x_1 axis and any arbitrarily small opening. Inside the wake the decay is only $\sim |x|^{-1}$.

K.I. Babenko and M.M. Vasil'ev [9] have corrected formula (2.4) and shown that if the vector force F is collinear with respect to the velocity u_∞, then

$$u(x) - u_\infty = H(x,0)F + \sum_j \frac{\partial}{\partial x_j} H(x,0)\,a_j + O\left[|x|^{-2+\varepsilon}(s+1)^{-\frac{1}{2}}\right], \quad \varepsilon > 0, \tag{2.5}$$

where a_j are some constants. If the vector F is directed in an arbitrary way, then, as it was shown by M.M. Vasil'ev [10], in the terms of the order of $-\frac{3}{2}$ there appears the logarithmic factor

$$u(x) - u_\infty = H(x,0)F + \left[\sum_j \frac{\partial}{\partial x_j} H(x,0)\,b_j\right] \ln|x| + O\left(|x|^{-\frac{3}{2}}\right), \tag{2.6}$$

where b_j are some constant vectors. These asymptotic formulae are obtained by highly complex and non-trivial calculations. As it is shown in [8], from the above formulae it follows that the vortex exponentially decreases outside the wake, and more exactly that

$$\frac{\partial u}{\partial x_j} = \frac{\partial}{\partial x_j} H(x,0)F + O\left[|x|^{-2}(s+1)^{-1}\ln^4|x|\right], \quad j = 1, 2, 3, \tag{2.7}$$

$$\omega=\frac{R}{8\pi}(\nabla s\times F)|x|^{-1}exp(-\frac{R}{2}s)+O\{|x|^{-2}exp[-(\frac{R}{2}-\varepsilon)s]\}, \qquad (2.8)$$
$$\omega = rot\, u,$$

where ε is any small quantity. Relation (2.8), independantly of the authors of paper [8], was also proved by D. Clark.

Let us consider the case of plane flows. If we assume together with R. Finn, that in the plane case there holds the relation

$$u(x)-u_\infty = O(|x|^{-\frac{1}{4}-\varepsilon}), \quad \varepsilon>0,$$

then one can obtain several terms of the asymptotic expansion for this difference. D. Smith [I2] and R. Finn and D. Smith [I3] obtained the main term of this difference; and K.I. Babenko [I4] obtained the expansion up to the terms $O(|x|^{-3/2})$. Denoting by u and v the Cartesian components of the velocity vector we have in the main terms

$$v+iu= i\,a_{1/2}\left(1+\frac{Ry\ln r}{2r}Re\, b_1\right)r^{-1/2}exp\left[\frac{R}{2}(x-r)\right]+$$
$$+\frac{b_1}{z}+O\left\{r^{-1}exp\left[\frac{R}{2}(x-r)\right]+r^{-\frac{3}{2}+\varepsilon}\right\},$$
$$\varepsilon>0, \qquad (2.9)$$

where $r=\sqrt{x^2+y^2}$, $z=x+iy$, and $a_{1/2}$ differs from the drag only by the factor. The formula for the vortex, analogous to formula (2,5), was established in [II] and [I4].

2. Formula (2.4) immediately leads to a highly important conclusion. Suppose we have a stationary solution in R^3 with finite Dirichlet integral and satisfying the condition $\lim_{|x|\to\infty} u(x)= u_\infty, u_\infty\neq 0$, at infinity. Then $u(x)\equiv u_\infty$. Indeed, since $F=0$, then $u(x)=u_\infty+O(|x|^{-3/2+\varepsilon})$, where $\varepsilon>0$ is any arbitrarily small quantity. Since $p(x)=O(|x|^{-1})$, the fact established by R. Finn [8], then multiplying equation (0.I)

by u and integrating with respect to a sphere of large radius, we easily find that $D[u] = 0$. Q.E.D. These arguments fail when $u_\infty = 0$, since (2.2) is established under assumption that $u_\infty \neq 0$.

Therefore, the question whether the solution of equation (0.I) in R^3 with finite Dirichlet integral and $u(x) \to 0$ for $|x| \to \infty$ differs from zero is still open.

In general, what are those minimal conditions one must impose on the solution in the whole of the space R^3 so that the uniqueness of the solution should take place whenever it is known that $u(x) \to 0$ for $|x| \to \infty$.

It is a striking fact that in the plane case the last question admits of an elementary solution, since for the vortex there exists the maximum principle. There arises the natural question whether there is any image of the maximum principle for the vortex vector $\omega = (\omega_1, \omega_2, \omega_3)'$ in the three-dimensional case. For example, whether the mapping $\omega : \mathcal{D} \to R^3$ is open. This problem was put forth by D. Gilbarg in his conversation with the author.

The analogy of relation (2.2) for plane flows has the form

$$|u(x) - u_\infty| \leq C |x|^{-1/2} . \tag{2.10}$$

It is not known until now whether this inequality is true. The question seems to be very difficult (and highly important!) G. Vineberger and D. Gilbarg [15] have shown that for the solutions of the two-dimensional flow problem, obtained by the Leray procedure, inequality (2.I) involves the boundedness of the velocity and pressure and that there takes place the convergence of velocities on the average.

3. On the asymptotics at small Reynolds numbers

I. It was long ago realized that the perturbation theory in flow problems was a theory of singular perturbations, even at small Reynolds numbers. The singularity of the theory is due to the non-compactness of the domain and manifests itself first of all in the fact that in the theory there arise two characteristic scales, namely the diameter of the body and the viscosity scale $1/R$. The first investigators of the perturbation theory did not quite properly understand the situation which resulted in some miscalculations and even errors quite natural for those times, though. Nevertheless, quite a number of remarkable results was obtained, among which one should note first of all the Stokes approximation and the Stokes formula for the force exerted by a viscous fluid flow on a slowly moving sphere, and also the Oseen's approximation and the Oseen's asymptotics of the flow in the vicinity of a point at infinity.

With the development of the method of the matching of asymptotic expansions a considerable progress was achieved in the perturbation theory at small Reynolds numbers, and in particular, it was shown how to calculate the successive terms of the asymptotic expressions in the formula for the force exerted on the body. Since it is not possible to list the extensive bibliography here we mention only the works by Proudman and Pearson [16] and also by Brenner and Cox [17] . Unfortunately, in all the works devoted to the above problem constructions are purely formal and the results established by one or another method lack any reasoning.

The construction of the perturbation theory safely allowing to get new results was given in papers [18] , [19] , [20] .

From now on we shall assume that for the solution under consideration the Dirichlet integral is finite and denote the class of such solutions by D. Putting $u = u_\infty + v$ we obtain non-homogeneous system (I.I) for $\lambda = 0$ with the right-hand side

$$f = R \sum_{k=1}^{3} v_k \frac{\partial v}{\partial x_k} \tag{3.I}$$

and the boundary value problem for this system

$$v|_S = v_0 = u_0 - u_\infty, \quad \lim_{|x| \to \infty} v(x) = 0. \tag{3.2}$$

Let us put $R = \varepsilon$ in (3.I) and let ε be a small parameter, the parameter R in the left-hand side of (I.I) is left unchanged. Then the solution of boundary value problem (I.I), (3.2) can be represented in the form of the following series

$$v(x,R) = \sum_{\ell=0}^{\infty} \varepsilon^\ell v^\ell(x,R),$$
$$p(x,R) = \sum_{\ell=0}^{\infty} \varepsilon^\ell p^\ell(x,R). \tag{3.3}$$

Then assuming that $\varepsilon = R$, we obtain the solution of the flow problem, provided that series (3.3) converge. If we substitute expansions (3.3) into equation (I.I) and compare the coefficients at the same powers of ε, we obtain

$$v^\ell(x,R) = \int_{\mathcal{D}} G(x,y) \sum_{k=1,j=0}^{3,\ell-1} v_k^j \frac{\partial v^{\ell-1-j}}{\partial y_k} dy, \tag{3.4}$$

$$p^\ell(x,R) = \int_{\mathcal{D}} \sum_{i=1,k=1,j=0}^{3,3,\ell-1} R_i(x,y) v_k^j \frac{\partial v_i^{\ell-1-j}}{\partial y_k} dy. \tag{3.5}$$

Assume that $\psi_\alpha(x) = |x|^{-1}(s+1)^{-\alpha}$, $0 < \alpha \leq 1$. By using (3.4) we obtain the estimate

$$|v^{\ell}(x,R)| \leq AB^{\ell-1}\ell^{-3/2}\lambda\psi_{\alpha}(\lambda x),\ x\in\mathcal{D},\ \ell\geq 1, \tag{3.6}$$

where A and B are some constants. The way of the estimation of the derivatives $v^{\ell}(x,R)$ and $p^{\ell}(x,R)$ is somewhat less wieldly. As a result we have the following theorem.

Theorem I. If the conditions of proposition I are satisfied and $u_0 \in C[S]$, then there exists a Reynolds number R_* such that the solution of flow problem (0.I) is given by series (3.3) converging for $R < R_*$.

This theorem was proved by R. Finn [5], who obtained less perfect estimates for v^{ℓ} and its derivatives, however. The solution given by series (3.3) satisfies inequality (2.2).

Solution (3.3) constructed allows us to prove the uniqueness theorem in the class D. As a matter of fact, if for $R < R_*$ there exists the solution $\tilde{v}$, $\tilde{p}$ from the class D differing from solution (3.3), then for the difference $w = v - \tilde{v}$, by virtue of asymptotic formulae (2.5), (2.6) and taking into account the fact that $\tilde{p} = O(|x|^{-2})$, $p = O(|x|^{-2})$, it is not difficult to derive the inequality

$$|\nabla w|_2^2 \leq R\,|w|_6^2\,|\nabla v|_{3/2}, \tag{3.7}$$

where we use the notation

$$|f|_r = \left(\int_{\mathcal{D}} |f(x)|^r dx\right)^{1/r}.$$

On the other hand, from the estimates obtained by the author [I8] it follows that

$$|\nabla v|_{3/2} \leq A(R)\,|\ln R|^{3/2},$$

where $A(R) \leq C(1 - BR|\ln R|)^{-1}$. By the Sobolev inequality

$|w|_6 \leq k|\nabla w|_2$, and so inequality (3.7) is non-contradictory for $|\nabla w|_2 \neq 0$ whenever

$$1 \leqslant C k^2 R |\ln R|^{3/2} (1 - BR|\ln R|)^{-1}.$$

Theorem 2. There exists a Reynolds number R_u, depending on S and u_0, such that the solution of problem (0.I) with finite Dirichlet integral is unique. If S and u_0 belong to some compact sets then $\inf R_u > 0$, where the lower bound is taken with respect to these sets.

2. The uniqueness theorem permits us, when investigating the asymptotics for $R \to \infty$, to confine ourselves to the solution of (3.3). Thus

$$u(x) = u_\infty + \sum_{\ell=0}^{\infty} R^\ell v^\ell(x, R),$$

$$p(x) = \sum_{\ell=0}^{\infty} R^\ell p^\ell(x, R). \tag{3.8}$$

From series (3.8) one can now obtain the asymptotic expansion in the Stokes domain $\mathcal{D}_S = \{x : |x| \ll R^{-1}\} \setminus T$, in the Oseen domain $\mathcal{D}_0 = \{x : |x| \gg R^{-1}\}$, and in the intermediate domain as well. For this, by virtue of estimate (3.6), it is sufficient to retain a finite number of terms in series (3.8) and then to refactor them. In the Stokes domain the expansion has the form

$$u(x) = u_\infty + \sum_{\nu=0}^{\infty} \varepsilon_\nu(R) \, \mathcal{U}_\nu(x),$$

where $\varepsilon_\nu(R) = R^\nu P_\nu(\ln R)$, P_ν being a polynomial of a degree n_ν. If we put $\xi = Rx$ in the Oseen domain then the expansion will have the form

$$u(x) = u_\infty + \sum_{\nu=0}^{\infty} \delta_\nu(R) \, \mathcal{V}_\nu(\xi),$$

where $\{\delta_\nu(R)\}$ is an asymptotic sequence similar to the previous one. By this very method there was obtained in [2I] the formula

$$u(x) = u_\infty + v^\circ(x,R) +$$
$$+ R\Big[\int_{\mathcal{D}} H^s(x-y)\sum v_k^s \frac{\partial v^s}{\partial y_k}\, dy - w_0(x)\Big] +$$
$$+ \frac{R^2}{4}\ln\frac{2}{R}\big[B - w_1(x)\big] + O(R^2), \tag{3.9}$$

where H^s is the fundamental matrix for the Stokes equation (equation (I.I) for $R=0$, $\lambda=0$), v^s is the solution of the Stokes system with boundary conditions (3.2), and w_0 and w_1 are the solutions of the boundary value problem for the Stokes equation with some special boundary conditions. B is a constant vector, which can be efficiently calculated. Note that, by virtue of the results of § I, in the Stokes domain $v^\circ(x,R)$ is the analytical function of R at $R=0$.

3. The proof of the validity of the formula for the force exerted on the body is as follows. From the standard formula for the force it is not difficult to get the following formula

$$F_i = F_i^0 - \int_{\mathcal{D}} \sum_{k,j} w_j^i v_k \frac{\partial v_j}{\partial x_k}\, dx, \tag{3.I0}$$

where F_i is the projection of the force on the x_i axis and F_i^0 is the projection of the force evaluated in the Oseen approximation; w^i is the solution of a boundary value problem for the system adjoint to (I.I), when $\lambda=0$. The vector w^1 has the explicit mechanical meaning, namely it equals to the velocity vector of a flow past a body, evaluated in the Oseen approximation, when the velocity vector has the opposite direction at infinity. The force exerted on the body in this case will be denoted by $(F_i^0)^*$. By using (3.8) one can transform equation (3.10). For this one should note that if F^s is the force, evaluated in the Stokes approximation, then $F_i^s = f_i/R$, where f_i does not depend on R. Let us denote by v^s, w^s the disterbance velocity vectors, evaluated in the Stokes approximation, respectively for the direct and the opposite flows past the body. Let us put

$$A = \int_{\mathcal{D}} \sum_{k,\ell} w_k^s v_\ell^s \frac{\partial v_k^s}{\partial x_\ell} dx.$$

Rather complex calculations yield the desired formula

$$F_1 = F_1^0 - A - \frac{1}{2} R \ln R \sum_{i,j,k} C_{ijk} f_i^* f_j f_k + \\ + BR + O(R^2 \ln^2 R), \tag{3.11}$$

where C_{ijk} are absolute constants and B is a constant, defined by a formula which we do not give here. Note, that if the body is symmetric with respect to the plane $x_1 = const$, then $A = 0$. If the body is a body of revolution with respect to the axis x_1 then

$$F_1 = F_1^0 - A - \frac{R \ln R}{160 \pi^2} f_1^* f_1^2 + BR + \\ + O(R^2 \ln^2 R). \tag{3.12}$$

If the body under consideration is a sphere then $f_1 = -f_1^* = 6\pi$ and so

$$F_1 = F_1^0 \left(1 + \frac{9}{40} R^2 \ln R\right) + BR + O(R^2 \ln^2 R). \tag{3.13}$$

Thus the famous Stokes formula and its generalizations are completely proved.

The similar theory for plane flows has not been developed yet. If we apply the above calculations to the expansion constructed in [I3] and similar to equation (3.8) then we can prove the validity of the Lamb formula and obtain two more terms in the asymptotic expansion for the force. Unfortunately, the asymptotic expansion will be carried out in this case in powers of $1/\ln R$ and such a series expansion is hardly of any interest. This program has not been carried out yet; paper [23] gives the proof of the existence of the limit

$$\lim_{R\to 0} F(R)\ln\frac{1}{R}$$

and the way of its calculation as well.

4. The construction of the perturbation theory at non-zero Reynolds numbers

I. Assume that we know the solution of flow problem (0.I) from the class D at the Reynolds number R_0. Consider the procedure of getting a solution of the flow problem at a Reynolds number close to R_0 by means of the perturbation theory. Putting $R = R_0 + \varepsilon$ we shall seek the solution of the problem

$$\begin{gathered} R^{-1}\Delta v - \frac{\partial v}{\partial x_1} - (v\cdot\nabla)v - \operatorname{grad} p = 0, \\ \operatorname{div} v = 0, \\ v|_S = 0, \quad \lim_{|x|\to\infty} v(x) = 0, \end{gathered} \tag{4.I}$$

in the form

$$v(x)=\sum_{n=0}^{\infty}\varepsilon^n v^n(x)\ ,\quad p(x)=\sum_{n=0}^{\infty}\varepsilon^n p^n(x). \tag{4.2}$$

Acting in a formal way, we get for v^n, p^n the boundary value problems

$$\Delta v^n - R_0\frac{\partial v^n}{\partial x_1} - R_0\mathcal{B}v^n - \operatorname{grad} p^n = \mathcal{F}_n,$$
$$\operatorname{div} v^n = 0,$$
$$v^n\big|_S = 0\ ,\quad \lim_{|x|\to\infty} v^n = 0\ ,\quad n=0,1,\dots, \tag{4.3}$$

where

$$\mathcal{B}v = (v^0\cdot\nabla)v + (v\cdot\nabla)v^0,$$

and

$$\mathcal{F}_n = \frac{\partial v^{n-1}}{\partial x_1} + R_0\sum_{\ell=1}^{n-1}(v^\ell\cdot\nabla)v^{n-\ell} + \sum_{\ell=0}(v^\ell\cdot\nabla)v^{n-1-\ell}$$

Therefore

$$v^n - Kv^n = G_n\ ,\quad n=0,1,\dots, \tag{4.4}$$

where

$$Kv = \int_{\mathcal{D}} G(x,y)\,\mathcal{B}v(y)\,dy\ ,\quad G_n = \int_{\mathcal{D}} G(x,y)F_n(y)\,dy. \tag{4.5}$$

Let $\overset{\circ}{W}{}_2^1(\mathcal{D})$ be the Sobolev class, which is the closure of functions smooth in $\mathcal{D}$ and vanishing on S. This class is a Hilbert space with the scalar product

$$(u,v) = \int_{\mathcal{D}} \sum_{k,j=1}^{3} \frac{\partial u_k}{\partial x_j}\frac{\partial v_k}{\partial x_j}\,dx\ .$$

Let us denote the norm in this space by $\|\cdot\|$. We shall show that $K:\overset{\circ}{W}{}_2^1(\mathcal{D}) \to \overset{\circ}{W}{}_2^1(\mathcal{D})$ and that K is a compact operator. Since the domain $\mathcal{D}$ is unbounded, the compactness of the operator K is guaranteed by the structure of the operator $\mathcal{B}$ namely, by the corresponding asymptotics of the values v^0 and ∇v^0. This asymptotics is of great importance here too! If the operator $I-K$, where I is a unit operator, is reversible and $G_n \in \overset{\circ}{W}{}_2^1(\mathcal{D})$, then one can prove the convergence of the series. If $I-K$ is irreversible, then it is obvious that the bifurcation of the solution is possible when $R=R_0$. Assuming that the invariant subspace of the operator K is one-dimensional, we can construct a secondary flow branching off from the given stationary one; to do this one should apply the procedure of expanding the solution in series in a small parameter $|R-R_0|^{1/2}$.

2. Now we give some simple auxiliary statements concerning the integral operator

$$\mathcal{G} f = \int_{\mathcal{D}} G(x,y) f(y)\, dy.$$

Consider a somewhat more general case and assume that $\lambda \neq 0$ and that inequality (I.7) holds. Let us introduce the following notations: if $f: \mathcal{D}_0 \to R^3$ then

$$|f|_{2,\mathcal{D}_0} = \Big(\int_{\mathcal{D}_0} |f(x)|^2 dx \Big)^{1/2},$$

where $|f(x)|$ denotes some norm of the vector $f(x)$. If $\mathcal{D}_0 = \mathcal{D}$ or $\mathcal{D}_0 = R^3$ we simply write $|f|_2$. Similarly

$$|\nabla f|_{2,\mathcal{D}_0} = \Big(\int_{\mathcal{D}_0} \Big[\sum_{i,k} \Big(\frac{\partial f_i}{\partial x_k} \Big)^2 \Big]^{2/2} dx \Big)^{1/2},$$

and $|\nabla f|_2 = |\nabla f|_{2,\mathcal{D}_0}$ if $\mathcal{D}_0 = \mathcal{D}$ or $\mathcal{D}_0 = R^3$.

Proposition 4. There holds the inequality

$$\Big| \frac{\partial}{\partial x_j} (\mathcal{G} f)(x) \Big|_\sigma \le A_2(\lambda, R) |f|_2,$$

where $\tau \leq 4$, $\sigma^{-1} = \tau^{-1} - 1/4$; if $1 < \tau < 2$ then

$$\left|\frac{\partial}{\partial x_1}(\mathcal{G}f)(x)\right|_\tau \leq B_\tau(\lambda, R)\,|f|_\tau .$$

The constants A_τ , B_τ depend on λ , R , S .

Proposition 5. Let $f \in L^\tau(\mathcal{D}) \cap L^{\tau_1}(\mathcal{D})$, $2 < \tau < 3$, $\tau_1 = 6\tau/(6+\tau)$. Then

$$\left|\frac{\partial}{\partial x_j}\mathcal{G}(f)(x)\right|_\sigma \leq A_{\tau,\tau_1}(\lambda, R)\left[|f|_\tau + |f|_{\tau_1}\right],$$

where $\sigma^{-1} = \tau^{-1} - 1/3$. If $\tau > 3$, $\tau_1 < 2$ then

$$\left|\frac{\partial}{\partial x_j}\mathcal{G}(f)(x)\right|_\infty \leq A_{\tau,\tau_1}(\lambda, R)\left[|f|_\tau + |f|_{\tau_1}\right].$$

Consider the question concerning the smoothness of the function $\mathcal{G}(f)(x)$. Let ω be an arbitrary vector and let $\Delta_\omega \varphi(x) = \varphi(x+\omega) - \varphi(x)$. Let $\mathcal{D}_*$ denote a subdomaim $\mathcal{D}$, obtained by removing a small vicinity of the manifold S . We shall take $|\omega|$ below small enough so that $x + \omega \in \mathcal{D}$, whenever $x \in \mathcal{D}_*$. To prove the following proposition one should make use of Proposition I.

Proposition 6. Let $0 < \alpha < 1$. Assume that $f \in L^\tau(\mathcal{D}) \cap L^{\tau_1}(\mathcal{D})$, $3/(3-\alpha) < \tau < 3/(1-\alpha)$, $\tau_1 = 3\tau/(3+\alpha\tau) < 2$. Then, if

$$g(x) = \frac{\partial}{\partial x_j}\mathcal{G}(f)(x),$$

$$|\Delta_\omega g|_{\sigma,\mathcal{D}_*} < C_{\tau,\alpha}\left[|f|_\tau + |f|_{\tau_1}\right]|\omega|^\alpha \left(\ln\frac{1}{|\omega|}\right)^{2\alpha},$$

where $\sigma = 3\tau/(3 - \tau(1-\alpha))$. If $\tau > 3/(1-\alpha)$, $\tau_1 < 2$, then

$$|\Delta_\omega g|_{\infty,\mathcal{D}_*} < C_{\tau,\tau_1}\left[|f|_\tau + |f|_{\tau_1}\right]|\omega|^\alpha \left(\ln\frac{1}{|\omega|}\right)^{2\alpha}.$$

The following proposition is essentially based on asymptotic formulae (26), (27).

Proposition 7. When $r > 12/11$ there holds the inequality

$$|\mathcal{B}v|_r \leqslant C|\nabla v|_2\left[|v^\circ|_\beta + |\nabla v^\circ|_\gamma\right],$$

where $\beta = 2r/(2-r)$, $\gamma = 6r/(6-r)$.

Proposition 8. The operator K is compact.

Proof. Let $\mathcal{D} = \mathcal{D}_0 \cup \mathcal{D}_1 \cup \mathcal{D}_2$, where $\mathcal{D}_0$ is the intersection of $\mathcal{D}$ with the δ-vicinity of S, $\mathcal{D}_2 = \mathcal{D} \cap \{x : |x| \geqslant N\}$ and $\mathcal{D}_1 = \mathcal{D} \setminus (\mathcal{D}_0 \cup \mathcal{D}_2)$. Put

$$u_j(x) = \int_{\mathcal{D}_j} G(x,y)\,\mathcal{B}v\,dy$$

Proposition 7 all the same holds even if it is applied not to the domain $\mathcal{D}$ but to any of the domains $\mathcal{D}_j$. Therefore

$$\|u_j\| \leqslant C\|v\|\left[|v^\circ|_{4,\mathcal{D}_j} + |\nabla v^\circ|_{\frac{12}{7},\mathcal{D}_j}\right].$$

Since $v^\circ \in L^6(\mathcal{D})$, $\nabla v^\circ \subset L^2(\mathcal{D})$, and for $|x| \to \infty$ formulae (2.6), (2.7) are valid, then, when δ and N^{-1} are small

$$\|u_0\| + \|u_3\| \leqslant \varepsilon\|v\|. \tag{4,6}$$

Let δ_x be a distance from the point x to S, measured along the normal to S and passing through the point x. Assume that x belongs to some small vicinity of S. Let us construct a function $\varphi \in C^2$ satisfying the following conditions: 1) $\varphi(x) = 0$, if $\delta_x \leqslant \frac{1}{2}\delta_0$; 2) $\varphi(x) \equiv 1$, if $\delta_x \geqslant \delta_0$; 3) $\operatorname{grad}\varphi = O(\delta_0^{-1})$; and 4) $0 \leqslant \varphi \leqslant 1$. Since $u_1(x)|_S = 0$, then by the well known inequality we have

$$\int_{\delta_x \leqslant \delta_0} |\operatorname{grad}\varphi|^2 |u_2(x)|^2 dx \leqslant C \int_{\delta_x \leqslant \delta_0} \left|\frac{u_2(x)}{\delta_x}\right|^2 dx \leqslant C_1 \int_{\delta_x \leqslant \delta_0} |\nabla u_2(x)|^2 dx,$$

where C_1 depends only on the maximum of the curvature of the surface of S. By virtue of Propositions 4 and 7 we have $|\nabla u_2|_3 \leq C_3 \|v\|$, and so, by Hölder's inequality,

$$\int_{\delta_x \leq \delta_0} |\operatorname{grad} \varphi|^2 |u_2(x)|^2 dx \leq C \|v\| \delta_0^{1/3}. \tag{4.7}$$

Extend $\varphi(x)$ so that it should equal to I in the domain $\mathcal{D}_1$, and demand that the following conditions be satisfied: I) $\varphi(x) \equiv \equiv 0$, when $|x| \geq 2N_1$; 2) $\varphi(x) \equiv 1$, if $|x| \leq N_1$ and $\delta_x \geq \delta_0$; and 3) $0 \leq \varphi \leq 1$. Note, that if $u = Kv$, then $|u|_\sigma \leq C_4 |\nabla u|_s$, where $\sigma = 3s/(3-s)$, and so, by Propositions 4, 7, $|u|_\sigma < C_\sigma \|v\|$ whenever $4/3 < s < 3$. Thus for $p > 12/5$

$$|u|_p < C_p \|v\|. \tag{4.8}$$

Hence

$$\int_{N_1 \leq |x| \leq 2N_1} |\operatorname{grad} \varphi|^2 |u_2(x)|^2 dx \leq C N_1^{-1} |u_2|_3 \leq C_1 N_1^{-1} \|v\|^2. \tag{4.9}$$

Let us put $w = \varphi u_2$ and estimate the distance between u_2 and w. Using inequalities (4.7), (4.9), we obtain

$$\|u_2 - w\| \leq C \|v\| \left(N_1^{-\frac{1}{2}} + \delta_0^{1/6}\right) + C\left(\int_{|x| \geq N_1} |\nabla u_2|^2 dx\right)^{\frac{1}{2}} \tag{4.10}$$

Since $\mathcal{D}_1$ is a bounded set, then for large $|x|$ the behaviour of u_2 is defined by the asymptotics of the Green function. Therefore

$$|\nabla u_2| \leq C_N |\nabla H(x)| \|v\| \leq C_N \|v\| |x|^{-\frac{3}{2}} (s+1)^{-1},$$

where the constant C_N depends only on N. Thus

$$\left(\int_{|x| \geq N_1} |\nabla u_2|^2 dx\right)^{1/2} \leq C_2(N) \|v\| N_1^{-1/2}$$

and taking into account inequality (4.10), we obtain $\|u_2 - w\| < \varepsilon\|v\|$ if N_1 is large enough. It remains to be noted that, by virtue of Proposition 6,

$$\|\Delta_\omega \nabla w\|_2 \leq C_3\left(|\omega| \ln^2|\omega|\right)^\alpha \|v\| .$$

Thus, if v are elements of some bounded set $\mathcal{V}$, then the image of $\mathcal{V}$ is compact when $v \mapsto w$. Hence there exists a finite ε-net for this image and so for the set $\{u : u = Kv,\ v \in \mathcal{V}\}$ too there exists a finite 3ε-net, which proves the compactness of K.

3. By making use of the established auxiliary propositions, we can now prove the convergence of series (4.2).

Theorem 3. If the operator $I - K$ has an inverse then series (4.2) converge in $L^\infty(\mathcal{D})$ and define the classical solution of problem (4.1).

The proof of this theorem is not simple at all so the reader is referred to paper [18]. In accordance with what was said above, we can get the solution of problem (0.1) by analytically continuing it with respect to Reynolds number, starting with $R = 0$. The process will break, if for some R_0 homogeneous equation (4.4) has a non-trivial solution or if $\|v(x, R)\| \to \infty$ for $R \to R_0$. The latter possibility does not come true if condition (0.2) is satisfied. The above propositions permit to solve rather simply the question about the bifurcation of the solution in the case when equation $(I - K)v = 0$ has non-trivial solutions in $\overset{\circ}{W}{}_2^1$, of course under assumptions imposed by the theory. There is also an opposite proposition. Assume that for $R = R_0$ problem (0.1) has the only solution u°, p°, but in the vicinity of the value of R_0 the uniqueness is violated, i.e. $\forall\, \varepsilon > 0$, there exist such values of R in the interval $[R - \varepsilon, R + \varepsilon]$, for which problem (0.1) has no less than two solutions.

Theorem 4. If the previuos hypothesis is valid then the equation $v - Kv = 0$ has non-trivial solutions in $\overset{\circ}{W}{}_2^1$, where K is an operator defined by the solution u°, p°, of problem (0.I). The proof of this theorem, which is far from being simple, is given in paper [I8].

5. On the spectral problem for the linearized Navier-Stokes equation

I. When investigating the stability of a stationary flow v^*, p^*, past a body, one should analyze the spectral problem for linearized equations (0.I)

$$\Delta v - R\frac{\partial v}{\partial x_1} - R\mathcal{B}v - \operatorname{grad} p - \lambda v = 0,$$
$$\operatorname{div} v = 0 \tag{5.I}$$
$$v|_S = 0, \quad \lim_{|x|\to\infty} v(x) = 0,$$

in the space $\overset{\circ}{W}{}_2^1(\mathcal{D})$. In view of the fact that the domain $\mathcal{D}$ is unbounded and differential expression (5.I) is not self-adjoint the structure of the spectrum is far from being clear. How does there arise a discrete spectrum in this problem? Does the continuous spectrum influence in any way the stability of the flow? etc. These are questions requiring a careful study. Below we shall give answers to some of these questions. Instead of problem (5.I) we shall consider the problem

$$v - K(\lambda)v = 0, \tag{5.2}$$

where the dependence on the spectral parameter λ is explicit. In the domain $\Lambda_0 = \{\lambda : \lambda = \sigma + i\tau, \sigma \geqslant -C(R)\tau^2\}$, by Proposition 3, the operator $K(\lambda)$ is an analytical function of λ. Hence, by virtue of the compactness of $K(\lambda)$, equation (5.2) can have non-trivial solutions only for discrete values of $\lambda \in \Lambda_0$. Such values of λ will be called eigenvalues. But one must make sure first that equation (5.2) has only trivial solutions for at least one value $\lambda_0 \in \Lambda_0$. Assume that (5.2) has a non-trivial solution in $\overset{\circ}{W}{}_2^1(\mathcal{D})$; then it will be a classical solution of equation (5.I) and its decrease at infinity is defined by the fact that v and p belong to the respective classes $\mathcal{L}^2(\mathcal{D})$. Using Propositions 4, 5, 7, it is not difficult to show that (5.I) involve the energy relation

$$\int_{\mathcal{D}} [|\nabla v|^2 + \lambda |v|^2]\, dx + R \int_{\mathcal{D}} [\frac{\partial v}{\partial x_1} \bar{v} + \mathcal{B} v \cdot \bar{v}]\, dx = 0.$$

Separating the real and imaginary parts, we obtain

$$\| v \|^2 + \sigma |v|_2^2 + R \,\mathrm{Re} \int_{\mathcal{D}} [(v \cdot \nabla) v^\circ] \cdot \bar{v}\, dx = 0$$

$$\tau |v|_2^2 + R \,\mathrm{Im} \int_{\mathcal{D}} [\frac{\partial v}{\partial x_1} \bar{v} + \mathcal{B} v \cdot \bar{v}]\, dx = 0,$$

where $\sigma = \mathrm{Re}\,\lambda$, $\tau = \mathrm{Im}\,\lambda$. It follows from these relations that

$$\sigma + \left(\frac{|\tau|}{AR} - 1\right)^2 < AR,$$

where A is some constant whose value is defined by the stationary solution. Thus we have

Theorem 5. In the domain $\Lambda = \{\lambda : \lambda = \sigma + i\tau,\ \sigma \geqslant -\alpha\tau^2, \alpha > 0\}$ equation (5.2) can have non-trivial solutions only for a finite number of values of λ . Each of these values of is an eigenvalue of differential operator (5.I). The corresponding eigenfunctions are classical solutions belonging not only to $\overset{\circ}{W}{}_2^1(\mathcal{D})$ but to $\mathcal{L}^2(\mathcal{D})$ as well.

From what has been said above it immediately follows that the eigenvalues are discrete and the only possible point of accumulation is $\lambda = 0$. A more careful consideration when the compactness of $K(0)$ and the behaviour of $K(\lambda)$ in the vicinity of $\lambda = 0$ are taken into account allows to show that there is a finite number of the eigenvalues in Λ.

Theorem 6. If $R < R_c$ then all the eigenvalues lie in the half-plane $\{\lambda : Re\lambda < 0\}$.

2. The operator $K(\lambda, R)$ analytically depends on R. Therefore the eigenvalues unceasingly depend on R, and for $R = R_c$, if $R_c < \infty$, there is a certain number of the eigenvalues present on the imaginary axis $Re\lambda = 0$. Assume that $\lambda = 0$, for $R = R_c$, is not an eigenvalue and that the eigenvalue $\lambda = i\omega$ is simple (see [22]), and the invariant subspace of the operator $K(i\omega)$ is one-dimensional. Under these conditions one can construct a periodic solution branching off from the stationary one. The construction of such a solution can be carried out by ex panding in series in the powers of the parameter $\varepsilon = \sqrt{|R - R_c|}$. So we must seek a periodic, with respect to time, solution, having the period $2\pi | C$, of the following boundary value problem

$$
\begin{gathered}
R\frac{\partial v}{\partial t} + R\frac{\partial v}{\partial x} + R\mathcal{B}v + (v\cdot\nabla)v + \operatorname{grad} p = \Delta v, \\
\operatorname{div} v = 0, \\
v|_S = 0, \quad \lim_{|x|\to\infty} v(x) = 0.
\end{gathered}
\tag{5.3}
$$

The operator $\mathcal{B}$ is defined, as above, by

$$\mathcal{B}v = (v^*\cdot\nabla)v + (v\cdot\nabla)v^*,$$

where, by Theorem 3, v^* is an analytical function of R in the vicinity of the point R_c. Putting $t = c\tau$, $R = R_c + \theta\varepsilon^2$, $\theta = \pm 1$, we shall seek the solution of problem (5.3) in the following form

$$v=\sum_{n=0}^{\infty}\varepsilon^{n+1}v^{n}, \quad p=\sum_{n=0}^{\infty}\varepsilon^{n+1}p^{n},$$

$$Rc = R_c c_0 + \sum_{k=1}^{\infty} c_k \varepsilon^k. \tag{5.4}$$

In addition, it should be noted that

$$v^{*}=\sum_{n=0}^{\infty}v_n^{*}\,\theta^{n}\varepsilon^{2n},$$

where $v^{n}=v^{n}(\tau,x)$, $p^{n}=p^{n}(\tau,x)$ are 2π-periodic functions of τ, and

$$v^{n}\big|_S=0, \quad \lim_{|x|\to\infty} v^{n}(\tau,x)=0.$$

Carrying out substitution in (5.3) and equalizing coefficients at the same powers of ε, we immediately find

$$v^{n}=\mathrm{Re}\sum_{k=0}^{n+1}v_{nk}e^{ik\tau}, \quad p^{n}=\mathrm{Re}\sum_{k=0}^{n+1}p_{nk}e^{ik\tau}.$$

If φ_0 is an eigenfunction corresponding to the eigenvalue of $\lambda = i\omega$, then one should assume $v^{0}=\gamma_0\mathrm{Re}(e^{i\tau}\varphi_0)$, $c_0=\omega$, where γ_0 is some positive constant. The functions v_{nk}, p_{nk} are defined as the solutions of the boundary value problems

$$\mathcal{L}_k v_{nk} + \mathrm{grad}\, p_{nk} = F_{nk},$$

$$\mathrm{div}\, v_{nk}=0, \tag{5.5}$$

$$v_{nk}\big|_S=0, \quad \lim_{|x|\to\infty} v_{nk}(x)=0;$$

$$\mathcal{L}_k = ik\omega R_c + R_c\frac{\partial}{\partial x_1} + R_c\left[v_0^{*}\cdot\nabla + (\cdot\nabla)v_0^{*}\right]-\Delta,$$

where F_{nk} is defined by v_{mj}, $m\le n-1$. Thus we get a triangular system. If $ik\omega$ is an eigenvalue only for $k=1$, then homogeneous problem (5.5) is degenerate when $k=1$, and to solve it one must require that

$$(F_{n1}, \zeta) = 0, \tag{5.6}$$

where ζ is an eigenfunction of the problem adjoint in $L^2(\mathcal{D})$. The latter condition is satisfied for odd n if we take $C_n = 0$. If condition (5.6) can be satisfied for $n = 2$ by the proper choice of C_2, γ_0, and θ, then it is easy to show that it can be satisfied for $n > 2$ as well. The convergence of the series is proved here by the same technique as in Theorem 3 (see [18]).

3. In conclusion we should like to note that the extension of the above results to the case R^2 is of great importance. True, one may not wait for the theory to be developed but attack the problem on the stability of a flow past a circular cylinder numerically to clear up the question whether the periodic Karman flow is a bifurcation of the initial stationary one. The numerical procedure might be based on the above series. This method would be safe for Reynolds numbers not very much differing from R_c.

BIBLIOGRAPHY

1. J. Leray, Étude de diverse équations, intégrales non linéaires et de quelques problèmes que pose l'hydrodynamique, J. Math. Pures Appl. 12 (1933), 1-82.

2. J. Leray, Les problèmes non linéaires, Enseignement Math. 35 (1936), 139-151.

3. R. Finn, On the steady-state solutions of the Navier-Stokes equations, III, Acta Math. 105 (1961), 197-244.

4. O. Ladyzenskaja, Investigation of the Navier-Stokes equation for a stationary flow of an incompressible fluid, Uspehi Math. Nauk 14 (1959), no. 3(87), 75-97.

5. R. Finn, On the exterior stationary problem for the Navier-Stokes equations, and associated perturbation problems, Arch. Rational Mech. Anal. 19 (1965), 363-406.

6. F.K.G. Odqvist, Uber Randwertaufgaben der Hydrodynamik zäher Flussigkeiten, Math. Zeit. 32 (1930), 329-375.

7. K.I. Babenko, On stationary solutions of the problem of flow past a body of a viscous incompressible fluid, Mat. Sbornik Tom 91(133) (1973), 3-26.

8. R. Finn, Estimates at infinity for stationary solutions of the Navier-Stokes equations, Bull. Math. Soc. Sci. Math. Phys. R.P. Roumaine 3(51) (1959), 387-418.

9. K.I. Babenko, M.M. Vasil'ev, Asymptotic behavior of the solution of the problem of the flow of a viscous fluid around a finite body, Preprint No. 84, Inst. Appl. Math. Moscow, 1971.

10. M.M. Vasil'ev, On the asymptotic behavior of the velocity, and forces, exerted on a body, in a stationary viscous fluid flow, Preprint Inst. Appl. Math. No. 50, (1973), Moscow.

II. D.C. Clark, The vorticity at infinity for solutions of the stationary Navier-Stokes equations in exterior domains, Indiana Univ. Math. J. 20 (I970/7I), 633-654.

I2. D.R. Smith, Estimates at infinity for stationary solutions of the Navier-Stokes equations in two dimensions, Arch. Rational Mech. Anal. 20 (I965), 34I-372.

I3. R. Finn, D.R. Smith, On the linearized hydrodynamical equations in two dimensions, Arch. Rational Mech. Anal. 25 (I967), I-25.

I4. K.I. Babenko, On the asymptotic behavior of the vorticity at large distance from a body in the plane flow of a viscous fluid, Prikl. Math. Mech. 34 (I970), 9II-925.

I5. D. Gilbarg, H.F. Weinberger, Asymptotic properties of Leray's solution of the stationary two-dimensional Navier-Stokes equations, Uspehi Math. Nauk Tom XXIX, 2 (I974) I09-I229.

I6. I. Proudman, J.R.A. Pearson, Expansions at small Reynolds numbers for the flow past a sphere and a circular cylinder, J. Fluid Mech. 2, 237 (I957).

I7. H. Brenner, Cox R.G., The resistance to a particle of arbitrary shape in translational motion at small Reynolds number, J. Fluid Mech. I7, 56I-595, (I963).

I8. K.I. Babenko, The perturbation theory of stationary flows of a viscous incompressible fluid at small Reynolds numbers, Preprint Inst. Appl. Math. Moscow, No. 79, (I975).

I9. K.I. Babenko, The perturbation theory of stationary flows of a viscous incompressible fluid at small Reynolds numbers, Dokl. Akad. Nauk SSSR No. 227 No. 3 (I976).

20. K.I. Babenko, On stationary solutions of the problem of a viscous incompressible fluid flow past a body, Proc.

All-Union Conf. Partial Differential Equations, Moscow Univ. Publ. House (1978).

21. M.M. Vasil'ev, On a viscous incompressible fluid flow in close vicinity of a body at small Reynolds numbers, Preprint Inst. Appl. Math. No. 116, Moscow (1975).

22. M.V. Keldysh, On the completeness of eigenfunctions of some classes of not self-adjoint linear operators, Uspehi Math, Nauk Tom XXYI, 4, (1971).

23. I-Dee Chang, R. Finn, On the solutions of a class of equations occurring in continuum mechanics, with application to the Stokes paradox, Arch. Rational Mech. Anal. 7 (1961), 388-401.

PARAMETERIZATION OF SUBGRID-SCALE MOTION IN NUMERICAL SIMULATION OF 2-DIMENSIONAL NAVIER-STOKES EQUATION AT HIGH REYNOLDS NUMBER

C. BASDEVANT
Laboratoire de Météorologie Dynamique
24, rue Lhomond
75231 Paris Cedex 05 / France

INTRODUCTION

One of the main problems occuring in numerical simulations of high Reynolds number turbulence in dimension two is the problem of inter - actions between subgrid scale motion and the motion at larger scales through the nonlinear term of the equation of motion. In fact, for high Reynolds number flows, the theory of 2-D turbulence predicts the existence of a k^{-3} inertial range in the one-dimensional energy spectrum where enstrophy cascades from large scales to smaller scales until it reaches the dissipation scales. The extent of this inertial range is also estimated by the theory : if k_D is a wavenumber in the dissipation range and k_I is typical of the energy injection range there are related to the Reynolds number R by :

$$(k_D/k_I)^2 \sim R$$

This relation shows that current simulations, for which the ratio of the largest to the smallest wavelength computed is hardly one thousand, are unable to describe the full spectrum of motion including the dissipation range for Reynolds numbers of several millions. Consequently in these codes the cutoff wavenumber which separates subgridscales from resolved scales fall in the enstrophy cascading range, where the real flow dynamics transfer enstrophy toward the subgrid scales at a constant rate. Moreover, in most numerical models energy and enstrophy conservations are built in to allow correct transfers among explicit scales. For instance in spectral galerkin approximations for periodic flows triad interactions between explicit scales are treated with the exact coefficient of interaction and thus formal detailed conservations hold. Consequently in numerical simulations energy and enstrophy are trapped into the explicit scales by the nonlinear dynamics, and a small viscosity (or high Reynolds number) will have very little effect on the behaviour of the spectrum : the enstrophy transfer toward smaller scales, together with the blocking effect of the cutoff, will result in spite of such a small dissipation, in an accumulation of enstrophy in large wavenumbers, leading to the well known equipartition of the absolute equilibrium spectrum. This misbehaviour asks for a solution of the parameterization problem ; what is needed is a model of the nonlinear mechanisms involving both explicit and subgrid scales, which by transfering enstrophy through the cutoff, prevent any unrealistic accumulation of enstrophy. Most of the numerous attempts to solve this problem have been based on the eddy viscosity concept, one drawback of which is that it cannot describe the detailed modal structure of the nonlinear transfers is the inertial range. In fact the parameterization problem arises because of the uncertainty about the excitations of the subgrid scales and is therefore closely related to the predictability problem.

In two dimensional flows, it takes an infinite time for a perturbation initially confined to infinitely small scales to contaminate the whole spectrum of the motion. In that sense the theoretical predictability time of two-dimensional turbulent flows is infinite. There is however, a resolution-dependent predictability time associated to any numerical model : the time it takes for the initial uncertainty in the subgridscales to destroy any deterministic prediction of the resolved scales. This implies that the parameterization problem can be solved only in the statistical sense ; that is by the formulation of a statistical forcing on the modal excitations that would preserve at least the statistical properties of the flow. In fact, we have to distinguish here between short-, medium- and long-range predictions, relative to the predictability time limit of the model. For the short time prediction one can expect that with very mathematically accurate methods and large computers we will be able to obtain accurate deterministic results. For medium and long range calculations the parameterization problem becomes important : we know that our information about subgrid scales is purely statistical at $t = 0$ and will remain so no matter what, just because the excitations there are not memorized. On the other hand, our deterministic knowledge about explicit scales, which we may suppose perfect at $t = 0$, will gradually disappear and give way to pure statistical information when the predictability time is over. Therefore a parameterization which is designed for long-range simulations, beyond the predictability time scale, like for instance in view of climatic studies, can be based on a pure statistical model of the nonlinear interactions ; on the other hand, for medium-range experiments, whatever deterministic information we have in explicit scales should be utilized in parameterizing the transfers.

We restrict our study to the long range problem, this choice means an important simplification of the task. It is then natural to study the problem first on a stochastic model of two-dimensional turbulence, where for simplicity we also assume homogeneity and isotropy.

Now, what is the interest of long-range simulation if one cannot expect any deterministic prediction after the predictability time scale? The answer is : if we actually produce a long term numerical solution reproducing the observed statistics of real flows, we shall have made an important step in our comprehension of the turbulence problem : such a model will be an important verification tool for the theory. Furthermore, we may consider more sophisticated models where the 2-dimensional motion interacts with various physical mechanisms like topography effect, heat transfers, water vapor trans - port, etc., as actually happens in models of the general circulation of the atmosphere. Such models are far too heretogeneous to give way easily to statistical treatment of interactions between the various processes involved. One has therefore to simulate the long-term statistics (or climate) of the system, through a long-range integra - tion of a "deterministic" model easier to handle. Such a use requires of course a correct treatment of the statistical dynamics, of the kind underlined here.

I PARAMETERIZATION OF NONLOCAL TRANSFERS

Using the quasinormal eddy-damped Markovian approximation the equation of evolution for the one dimensional energy spectrum reads :

$$\left(\frac{\partial}{\partial t} + 2\nu k^2 \right) E(k,t) = \iint_{\Delta(k)} S(k,p,q)\, dpdp \qquad (1)$$

where $E(k,t)$ is defined in such a way that

$$\int_0^\infty E(k,t)\, dk = \langle u^2 \rangle$$

the symbol $\Delta(k)$ refers to a domain in the p - q plane such that k, p, q can be the sides of a triangle, and $S(k, p, q)$ is defined by:

$$S(k,p,q) = \theta_{k,p,q}(t)\; b(k,p,q)\left[kE(p,t)E(q,t)-pE(q,t)E(k,t)\right]$$

where $b(k,p,q)$ is a geometrical coefficient and $\theta_{k,p,q}$ is the relaxation time for triple correlations.

The main idea to derive an operator solving the parameterization problem is to take advantage of the fact that in the enstrophy cascading range nonlocal interactions dominate, that is the flux of enstrophy through a fixed wavenumber located in the inertial range is due for its main part to interactions within triads which are such that the ratio of the smallest to largest interacting wavenumbers is very small. This fact enables us to say : if the cutoff lies in the inertial range, where nonlocal interactions dominate, we shall recover most of the subgrid scale effects if we are able to represent the part of subgrid scale effects due to nonlocal interactions. Now we have to study nonlocal interactions and this can be done by means of an expansion in a small parameter $\underline{a}$ measuring nonlocalness — we shall precisely call nonlocal all triad interactions such that the ratio of two interacting wave-number is less than $\underline{a}$. Then the transfer term

$$T(k) = \iint_{\Delta(k)} S(k,p,q)\, dpdq$$

can be split into two parts,

$$T(k) = T_L(k) + T_{NL}(k) \qquad (2)$$

where T_L includes all interactions that are local with respect to $\underline{a}$, and T_{NL} all nonlocal interactions. Our purpose is to evaluate T_{NL}, this is done by first expanding the corresponding nonlocal enstrophy flux

$$Z_{NL}(k) = \int_k^{\infty} k'^2 \; T_{NL}(k') \, dk'$$

in powers of $\underline{a}$. In fact due to detailed conservation properties this flux can be split into two parts :

$$Z_{NL}(k) = Z_{NL}^{+}(k) - Z_{NL}^{-}(k)$$

Z_{NL}^{+} is the flux of enstrophy due to nonlocal triads (k',p', q') such that :

$$q' \ll p' \lesssim k \lesssim k'$$

Z_{NL}^{-} is the flux of enstrophy due to nonlocal triads (k',p',q') such that :

$$k' < k \ll p' \sim q'$$

It turns out that in the enstrophy cascading range Z_{NL}^{-} is negligible compared to Z_{NL}^{+} (in fact Z_{NL}^{-} dominates at very large scales where the energy spectrum falls to zero for idealized turbulence on an infinite domain). As we are interested in transfers within the enstrophy cascading range the only need to obtain the leading term of Z_{NL}^{+} in its expansion in powers of $\underline{a}$. This leads to the following results :

$$T_{NL}(k) = k^{-2} \frac{\partial}{\partial k} \left(k^3 \frac{\partial \phi}{\partial k}\right) = \frac{\partial}{\partial k} \left[k^{-1} \frac{\partial}{\partial k} (k^2 \phi) \right] \qquad (3)$$

where $\phi(k) = \frac{1}{8} \, kE(k) \int_0^{ak} \theta_{kkq} \; q^2 E(q) \, dq$

this form is energy-conserving as well as enstrophy-conserving, provided the energy spectrum decreases fast enough at infinity. The corresponding energy and enstrophy fluxes read

$$\Pi_{NL}(k) = -k\frac{\partial\phi}{\partial k} - 2\,\phi(k)$$

$$Z_{NL}(k) = -k^3\frac{\partial\phi}{\partial k} \qquad (4)$$

An important property of this form is that in the stationary case, when one assumes a constant enstrophy flux $Z_{NL}(k) = \beta$, the total energy flux vanishes exactly ; furthermore it implies a stationary spectrum of the form

$$E(k) = C\,\beta^{2/3}k^{-3}\left[\mathrm{Log}\left(\frac{k}{k_1}\right)\right]^{-1/3}$$

It should be noticed that this new term T_{NL} is a second-order diffusion term in spectral space, and therefore needs boundary conditions at both ends of the spectrum. From (4) the natural boundary conditions are

$$\begin{cases} \phi = 0 \quad \text{at} \quad k = 0 \\ \partial\phi/\partial k = 0 \text{ at } k = \infty \end{cases}$$

the first condition imposes a vanishing energy flux (hence a vanishing enstrophy flux) at $k = 0$, while the second imposes a vanishing enstrophy flux (hence a vanishing energy flux) at $k = \infty$. Now, in numerical calculations, where the spectral domain is bounded by cutoff wavenumbers k_{min} and k_{max}, we can still choose $\phi = 0$ as a proper boundary condition at $k = k_{min}$ because T_{NL} will in any case be negligible in the vicinity of $k = 0$; but we cannot take $\partial\phi/\partial k = 0$ as the appropriate boundary condition at $k = k_{max}$, except if k_{max} belongs to the dissipation range where the enstrophy flux actually vanishes. If the enstrophy injection rate is known, say β, a possible method consists in imposing a subgrid scale enstrophy flux equal to β. The boundary condition would then read

$$-k^3\frac{\partial\phi}{\partial k}\bigg|_{k = k_{max}} = \beta \qquad (5)$$

This method, however, is not general enough : apart from the fact that β is not usually known in practice, it relies on the assumption that the flux is stationary. A better approach is based on the remark that what is really known to a good approximation in all cases appears to be the energy flux rather than the enstrophy flux. It is well known and verified again on formula (4) that the energy flux vanishes in the enstrophy cascading range once the stationary regime has been obtained ; the boundary condition $\Pi_{NL}(k_{max}) = 0$, or, from (4)

$$k\frac{\partial\phi}{\partial k} + 2\phi\Big|_{k = k_{max}} = 0 \tag{6}$$

is thus exact in the stationary case, while it has over (5) the advantage that it does not require any knowledge of β.

Another advantage of (6) over (5) is that (6) is still a satisfactory approximation in the nonstationary case, since the energy flux through k_{max} must be negligible compared to the energy fluxes occuring at $k \ll k_{max}$.

Using (4) again, we rewrite (6) in terms of the enstrophy flux at k_{max}, i.e.,

$$Z_{NL}(k_{max}) = 2\, k_{max}^2 \phi(k_{max}) \tag{7}$$

an important consequence of which is that there is no enstrophy flux at the cutoff, which means no loss of enstrophy for the whole system, as long as there is no excitation at k_{max}.

The use of T_{NL} given by (3) with equation (2) in numerical simulations asks for a discretization which clearly separates, in wavenumber space, local interactions, treated explicitly, from nonlocal interactions which are entirely parameterized by (3) : this is the case when using logarithmic discretization in wavenumber space to integrate the master equation (1). We have performed numerical integrations of the spectral equation (1) which demonstrate that the parameterization operator (3) with boundary condition (6) allow accurate simulation of high Reynolds number turbulence even with low-resolution models, when the cutoff wavenumber falls within the enstrophy cascading range ; some results are given in Figure 1.

II NONLOCAL TRANSFERS ACROSS THE CUTOFF

Unfortunately, a logarithmic discretization in wavenumber space is not well suited to direct simulation of the Navier-Stokes equations. In direct simulations there is not a clearcut distinction between what is local and what is nonlocal, in fact most of the nonlocal transfers, except those across the cutoff wavenumber, are explicitly taken into account in the truncated model. It follows that, for the master equation, the transfer term must be split in the following way :

$$T(k) = T_{DS}(k) + T'_{NL}(k)$$

where $T_{DS}(k)$ stands for all transfers computed explicitly in the direct simulation and $T_{NL}(k)$ is the parameterization of nonlocal transfers across the cutoff. In turn the parameterization $T'_{NL}(k)$ must be formulated as a difference :

$$T'_{NL}(k) = T_{NL}(k) - T^{*}_{NL}(k)$$

between the term $T_{NL}(k)$ described in (3) and (6) representing all nonlocal transfers, and a new term $T^{*}_{NL}(k)$ representing the nonlocal transfers that are already computed in the explicit truncated model. The derivation of T^{*}_{NL}, which depends on both $\underline{a}$ and k_{max} follows exactly the derivation of T_{NL} except that all interacting wavenumbers are constrained to lie below k_{max} and that there is no need of boundary condition for T^{*}_{NL}, the dynamics of a truncated model being of course self-contained.

The parameterized energy transfer due to subgrid scales finally reads

$$T'_{NL}(k) = k^{-2} \frac{\partial}{\partial k} (k^3 \frac{\partial}{\partial k} \phi') = \frac{\partial}{\partial k} (k^{-1} \frac{\partial}{\partial k} \left[k^2 \phi'\right])$$

$$\phi'(k) = kE(k) \; (w-w^{*}) + k_{max}E(k_{max}) \; w^{*}$$

$$w(k) = \frac{1}{8} \int_0^{ak} \theta_{kkq} \, q^2 \, E(q) \, dq \tag{7}$$

$$w^{*}(k) = \frac{1}{8} \int_0^{ak} \theta_{kkq} \, q^2 \, E(q) \, H \left(\frac{k_{max}-k}{q}\right) dq$$

where

$$H(t) = \begin{cases} 1 & \text{if } t \geq 1 \\ \frac{2}{3\pi}\left[3 \text{ Arcsin } (t) + (5-2t^2)\, t\sqrt{1-t^2}\right] & \text{if } 0 \leq t \leq 1 \end{cases}$$

together with the vanishing of the energy flux at $k = k_{max}$.

The corresponding enstrophy flux

$$Z'_{NL}(k) = -k^3 \frac{\partial \phi'}{\partial k}$$

presents a strong cusp behavior which means that the subgrid-scales extract enstrophy mainly from scales close to k_{max}.

The energy flux can be split into two parts

$$\Pi'_{NL} = \Pi'^{+}_{NL} - \Pi'^{-}_{NL}$$

where $\Pi'^{+}_{NL} = k^{-2}\, Z'_{NL}$ represents the extraction of energy which goes along with the extraction of enstrophy in the vicinity of k_{max}. On the other hand Π'^{-}_{NL} is a correct estimate of that part of the negative viscosity effect which comes from the subgrid scales.

III PARAMETERIZATION IN DIRECT SIMULATIONS

The model of nonlinear transfers due to subgrid-scale motion is now ready to be applied in arbitrary discretization in wavenumber space, we have coupled it with a direct simulation program.

The direct simulation code was a spectral galerkin code of the two-dimensional Navier-Stokes equation in a square box, periodic in both directions. We choose a spectral galerkin rather than pseudospectral or finite difference code to insure exact triad interactions among explicit modes. The spectral simulation involved all wave vectors within a circle $k \leq k_{max}$ in order to remain as close to isotropy as possible.

The method is implemented as follows.

First, starting from a streamfunction distribution $\Psi(\vec{k},t)$ the truncated Navier-Stokes equation is integrated over one time step as usual, for instance by the leapfrog technique, this provides temporary updated values of the streamfunction we denote $\Psi_1(\vec{k},t+\delta t)$, they define a temporary energy distribution $E_1(k,t+\delta t)$. The forcing on the energies resulting from the parameterized transfers is then obtained from (7) using a semi implicit backward technique :

$$E(t+\delta t) = E_1(t+\delta t) + T'_{NL}\ (E(t+\delta t)\ ;\ w(t),\ w^*(t)\ )$$

the method needs to be implicit because the onedimensional distribu - tion in wavenumber, resulting from the regular two-dimensional grid of wavevectors, is highly irregular with an average density increasing as k, this would lead for a forward explicit scheme to a timestep $\delta t \sim k^4$ decreasing rapidly with resolution. On the contrary, the frequencies w and w^* need not be updated at every time step, because their time variation is slow.†

The last remaining step consists in inferring a forcing term for the streamfunction Ψ. It is in that final step that the limitations of the present model become obvious. Since the method is based on a stochastic formulation of two-dimensional isotropic turbulence, it gives no information whatsoever on the phase distribution of the

† There is a little cost in using an implicit time extrapolation scheme because we are dealing with a onedimensional distribution of energies for which the inversion of the diffusion operator is trivial.

forcing term. We take the simplest form

$$\psi(\vec{k},t+\delta t) = \left[\frac{E(k,t+\delta t)}{E_1(k,t+\delta t)}\right]^{1/2} \psi_1(\vec{k},t+\delta t)$$

In that case the parameterization reduces to a (positive or negative) eddy viscosity, depending on $k = |\vec{k}|$ only, which just affects the amplitudes of the modes.

The general features of our numerical tests are as follows. A source is located at wavenumber $k = \sqrt{10}$, and acts as an isotropic forcing term. The initial conditions correspond to a spectrum strongly peaked at $k = \sqrt{10}$. Three different resolutions are compared, where the cutoff circle is inscribed in a 16x16, 20x20 and 32x32 point square respectively. Figure 2 displays a comparison of runs corresponding to the three cases in terms of the energy spectrum, averaged over a time interval corresponding to 2000 time steps for the higher resolution. The parameter a in this case has been chosen equal to 1. The spectra are seen to be in fairly good accordance with one another ; there is no visible dissipation zone in any of them, the k^{-3} shape extending up to the cutoff wavenumber in all three cases. In figure 3 we show a time average of parameterized enstrophy flux Z'_{NL} together with a time average of the parameterized energy flux Π'^{-}_{NL} corresponding to the negative eddy viscosity. We notice the strong cusp behaviour of Z'_{NL} near k_{max} and its systematically negative values for $k \ll k_{max}$, in accordance with the negative viscosity effect described by Π'^{-}_{NL}.

In these experiments a is a true parameter which determines the degree of nonlocality of the transfers we choose to parameterize : it must be chosen in such a way that these account for the quasi-totality of the flux of enstrophy across k_{max}. The choice a = 1 is obviouslycorrect in that respect. On the other hand, the accuracy of our expansions becomes dubious when a is not small. A compromise has to be found and some optimum value determined, small enough to guarantee the accuracy of the expansions, yet large enough to account for as many of the transfers as necessary.

CONCLUSION

Starting from a simplified model of nonlocal interactions in two-dimensional isotropic homogeneous turbulence we have determined an operator parameterizing the statistical effects of subgrid-scale eddies on the energy distribution at explicit scales. The average enstrophy transfer near and across the cutoff proves to be accurately simulated. It is modeled as a purely nonlocal effect governed by the large scales and operating mainly in the close vicinity of k_{max}. In the limit of large wavenumbers, the main effect of this parameterization would be to transfer enstrophy locally from k^-_{max} to k^+_{max}. The method designed for long-range direct simulation experiments is able to actually produce k^{-3} spectra up to the cutoff wavenumber without artificial dissipation range. The negative viscosity effect which has been included balances exactly the energy flux through the cutoff and thus provides exact energy conservation. On the other hand being based on a stochastic model, the method only yields average tendencies for the modal energies ; one cannot avoid therefore, in direct simulations, some arbitrariness in the process of de-averaging. The most general way to proceed would be to generate the subgrid scale forcing as a random process : however, the statistical properties of such a random forcing, which may take into account the predictability of the flow, cannot be entirely determined by the present approach.

REFERENCE

C. BASDEVANT, M. LESIEUR and R. SADOURNY, 1978 : Subgrid-scale modeling of enstrophy transfer in two-dimensional turbulence. J.Atmos.Sci., 35, 1028-1042.

FIGURE CAPTION

Figure 1

Numerical integration of equation (1) using a logarithmic spacing for wavenumbers. Nonlocal interactions are parameterized by (3) and (6), in that case $\underline{a} = 0.26$. For the local term we used the conservative scheme given by Leith. The figure displays energy spectrum of a forced turbulence with Reynolds number $Re=8x10^{26}$ at t=400. The arrows point toward the wavenumber where the forcing is applied. Three spectra corresponding to three different resolutions have been superimposed in solid line and are undistinguishable from one another :

(a) $k_{max} = 2.8 \times 10^{13}$ $(=Re^{1/2})$

(b) $k_{max} = 2.6 \times 10^{5}$

(c) $k_{max} = 100$.

Note that, except in curve (a) where the cutoff is in the dissipation range, the enstrophy cascading range is well developped up to the cutoff wavenumber in low resolution cases. Curve (d) in dotted line corresponds to an experiment where the parameterization of the nonlocal term was missing, it emphasized the fact that local interactions are unable to spread out enstrophy from the injection range at the correct rate. (LogxLog scale).

Figure 2

Time averaged energy spectra (LogxLog scale) of a forced turbulence from a direct simulation with parameterization. The vertical arrow points toward the wavenumber where the forcing is applied. The three spectra (a,b,c) correspond respectively to the resolutions 16x16, 20x20, 32x32. They are superimposed at the bottom. The k^{-3} slope is indicated.

Figure 3

Time averaged fluxes in the experiment corresponding to curve (c) in Figure 2, as functions of log K. Curve (a) shows the parameterized enstrophy flux due to subgrid-scale forcing (Z'_{NL} in the text) and curve (b) shows the part of the parameterized energy flux which corresponds to turbulent negative viscosity, counted positively when oriented towards the subgrid scales ($-\Pi'^{-}_{NL}$ in the text.)

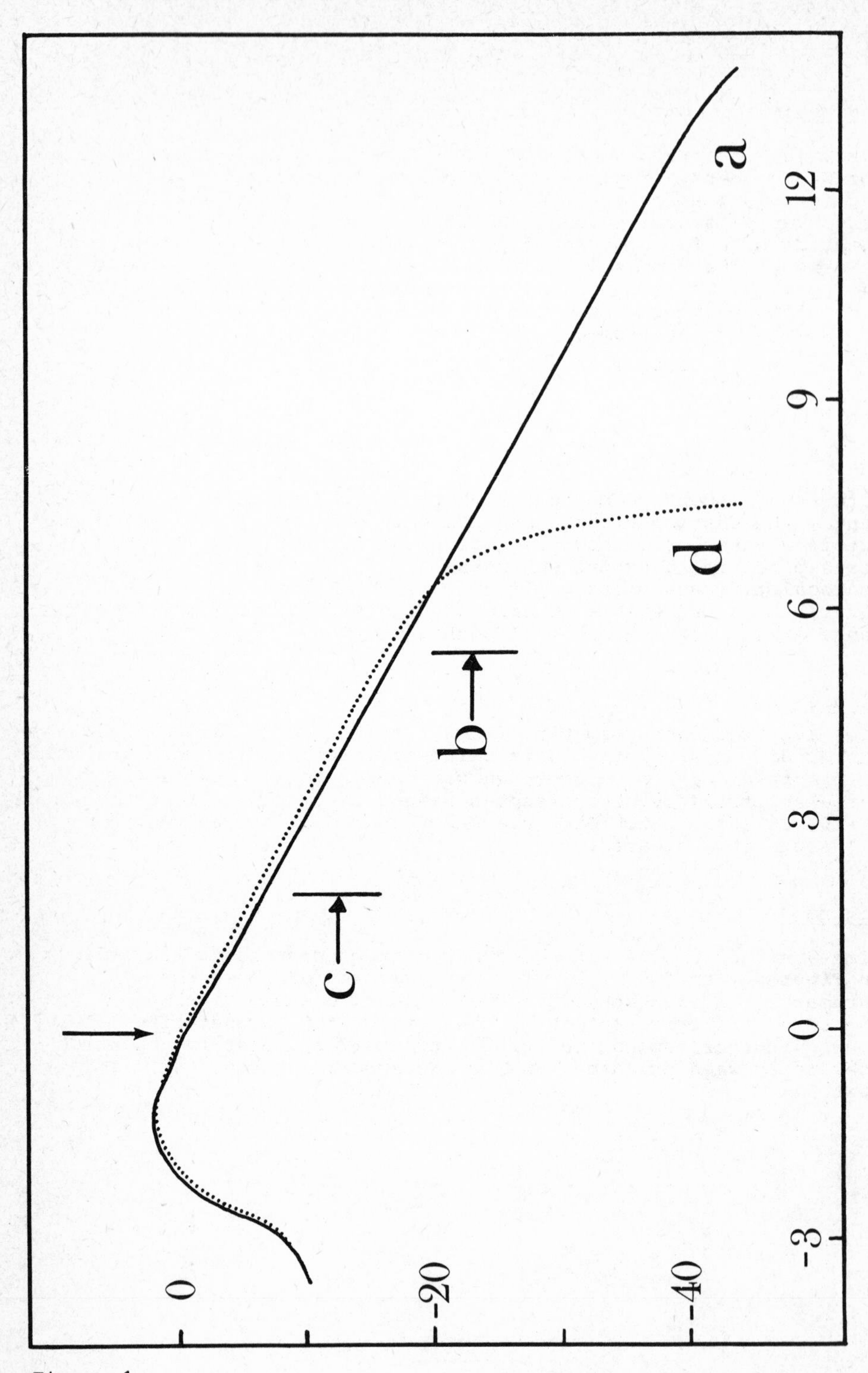

Figure 1

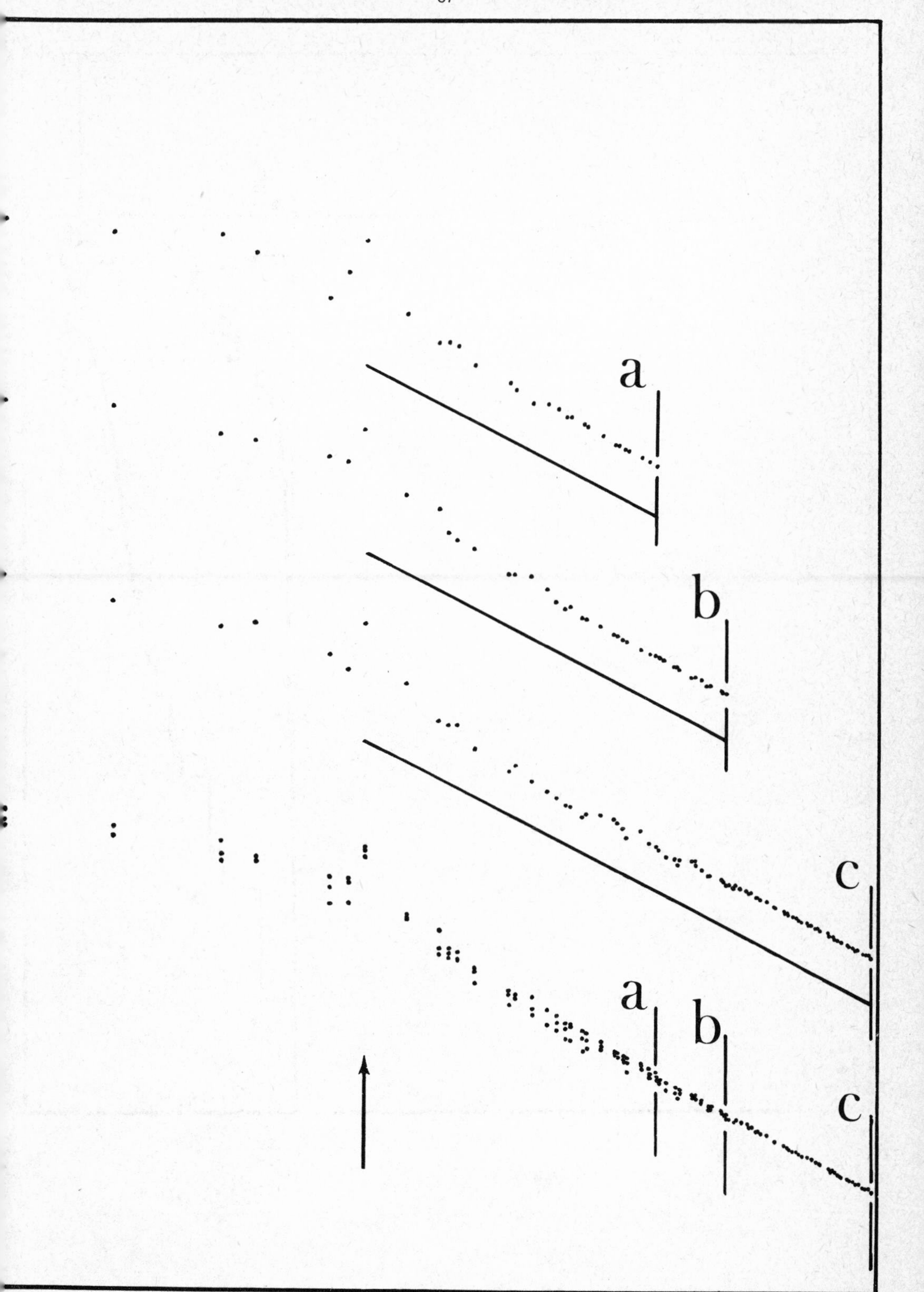

Figure 2

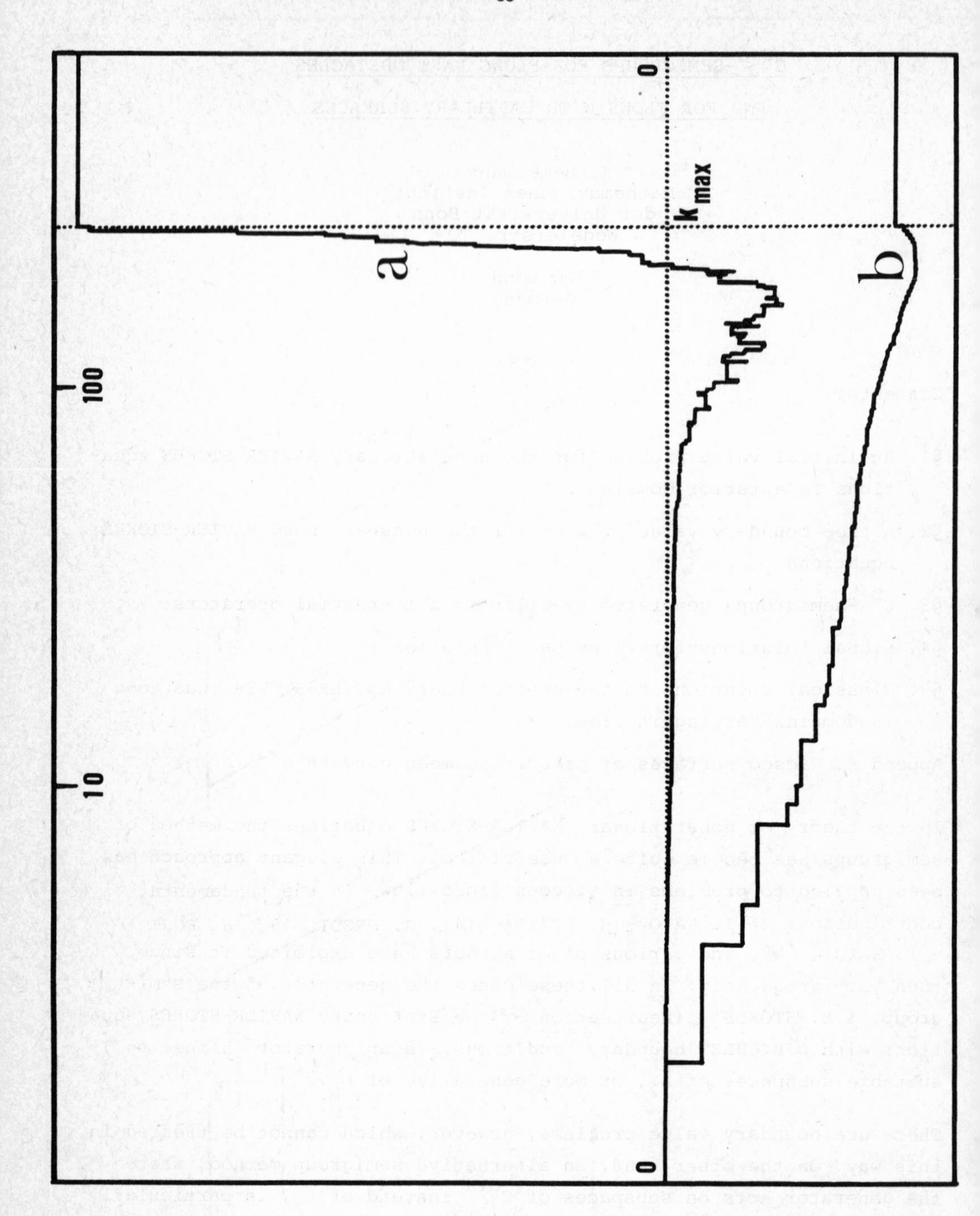

Figure 3

$C^{0+\alpha}$-SEMIGROUPS FOR FLOWS PAST OBSTACLES AND FOR FLOWS WITH CAPILLARY SURFACES

J. Bemelmans
Mathematisches Institut
der Universität Bonn
Wegelerstr. 10

5300 Bonn 1
Germany

Contents:

§1. An initial value problem for the nonstationary NAVIER-STOKES equations in exterior domains

§2. A free boundary value problem for the nonstationary NAVIER-STOKES equations

§3. $C^{0+\alpha}$-semigroups generated by elliptic differential operators

§4. Global solutions for flows past obstacles

§5. Classical solutions to the nonstationary NAVIER-STOKES equations in domains varying in time

Appendix. Closed surfaces of prescribed mean curvature.

In the theory of nonstationary NAVIER-STOKES equations the method of semigroups has become quite a useful tool. This elegant approach has been applied to problems in viscous fluid flow in the fundamental contributions of T. KATO - H. FUJITA [14], G. PRODI [19], H. FUJITA - T. KATO [6], and various other authors have exploited it since then very frequently. In all these cases the generator of the semigroup, i.e. STOKES' linearisation of the stationary NAVIER-STOKES equations with DIRICHLET boundary conditions, is an operator defined on suitable subspaces of L_2, or more generally, of L_p.

There are boundary value problems, however, which cannot be treated in this way. On the other hand, an alternative semigroup method, where the generator acts on subspaces of $C^{0+\alpha}$ instead of L_p, is particularly appropriate to them. The first problem of this kind, outlined in §1 below, is the flow past a body immersed in an infinite reservoir of fluid and moving with prescribed velocity. Then in §2 we formulate a class of free boundary value problems for the NAVIER-STOKES equations, which can also be treated by the method of semigroups in HÖLDER spaces. These $C^{0+\alpha}$-semigroups, in the general case generated by elliptic operators of order 2m, have been introduced by W. von WAHL [23], we recall their basic properties in §3. In §4 we report on our paper [2], which contains a global existence theorem for time dependent flows past obstacles. The free boundary value problems of §2 lead to existence and regularity problems for the NAVIER-STOKES equations in domains that vary with time. In §5 the solution to these questions is indicated. In addition one needs existence theorems for the free surface which is assumed to be governed by surface tension. Naturally, these questions centered around surfaces of prescribed mean curvature require totally different methods. We therefore restrict ourselves in this paper to a short appendix and refer for further information to our work that will be published elsewhere.

The research for this paper was supported by SONDERFORSCHUNGBEREICH 72 at the University of Bonn.

§1. An initial value problem for the nonstationary NAVIER-STOKES equations in exterior domains

We consider a body that is immersed in an infinite reservoir of fluid and that moves with prescribed velocity $\underline{U}(t)$. Denoting the velocity and the pressure of the fluid by $\underline{v}$ and p resp., the kinematic viscosity by ν, and the external force density by $\underline{f}$, the fluid motion is described by the following initial value problem

$$(1)\quad \begin{aligned} \underline{v}_t - \nu\Delta\underline{v} + \nabla p + (\underline{v}\cdot\nabla)\underline{v} &= \underline{f} + \underline{U}_t \\ \nabla\cdot\underline{v} &= 0 \end{aligned} \qquad \text{in } \mathfrak{E}\times(0,T)$$

$$(2)\quad \underline{v}(x,t) = 0 \quad \forall x\in\Sigma\ \ \forall t\in[0,T]$$

$$(3)\quad \underline{v}_t(x,t) \to \underline{U}_t(t)\ ,\ \text{as } |x|\to\infty$$

$$(4)\quad \underline{v}(x,0) = \underline{v}_0(x) \quad \forall x\in\mathfrak{E}$$

Here the coordinate system is attached to the body that occupies the simply connected, compact region $\Omega\subset\mathbb{R}^3$. By Σ we denote the smooth boundary of Ω, and $\mathfrak{E} = \mathbb{R}^3 \setminus (\Omega\cup\Sigma)$ is the region that is filled with fluid. The velocity field at some initial instant of time is $\underline{v}_0$. Before one can attack this problem, one has to specify condition (3). Of special interest is the following case

$$(5)\quad \begin{aligned} &\underline{U}(0) = 0,\ \underline{U}(t) = \underline{U} = \text{const.} \quad \forall t\geq T^* \\ &\underline{U}(t) \text{ smooth and} \\ &|\underline{v}(x,t) - \underline{U}| \leq C|x|^{-1} \quad \forall |x| \text{ large enough} \\ &\text{and for all } t\geq T^*. \end{aligned}$$

It describes the flow past a body that is at rest at time t = 0, is accelerated during some finite interval $(0,T^*)$ and is kept at constant velocity $\underline{U}$ afterwards. At infinity we have imposed the condition that occurred first in the associated stationary problem

$$(6)\quad \begin{aligned} -\nu\Delta\underline{v} + \nabla p + (\underline{v}\cdot\nabla)\underline{v} &= \underline{f} \\ \nabla\cdot\underline{v} &= 0 \end{aligned} \qquad \text{in } \mathfrak{E}$$

$$(7)\quad \underline{v}\big|_\Sigma = 0$$

$$(8)\quad |\underline{v}(x) - \underline{U}| \leq C|x|^{-1}\ ,\ |x| \text{ large enough.}$$

This boundary value problem has been investigated by R. FINN. He introduced solutions with the decay property (8) and termed them "physically reasonable" (PR-solutions), because they agree very well with real flows: cf. the existence of the paraboloidal wake region behind the body, energy and force relations etc.*). In [6] FINN proved that every PR-solution, such that the force extented on the body does not vanish, has infinite kinetic energy:

$$(9) \qquad \int_{\mathfrak{E}} |\underline{v}(x) - \underline{U}|^2 dx = +\infty \ .$$

This energy property was the starting-point for our investigation of the initial value problem (1)-(5) in [23]. For it is obvious from (9) that the required existence theorem cannot be based on energy estimates. Therefore, we worked with the OSEEN linearisation

$$(10) \qquad -\Delta\underline{v} + \nabla p + (\underline{U}\cdot\nabla)\underline{v} = 0$$
$$\nabla \cdot \underline{v} = 0$$

as generator of the semi-group and chose a suitable subspace of $C^{0+\alpha}(\overline{\mathfrak{E}})$ as underlying function space, hence avoiding any integral norm at all, especially the L_2-norm.

*) For a detailed discussion see the survey articles [7],[8] by R. FINN as well as his original publications.

§2. A free boundary value problem for the nonstationary NAVIER-STOKES equations

The dynamical stability of spherical drops furnishes a second example for an existence proof where the method of $C^{0+\alpha}$-semigroups is used. We consider a drop of viscous, incompressible fluid under the influence of constant atmospheric pressure p_0 and of surface tension*). If the fluid is at rest the equilibrium configuration will be a sphere. For let $\mathfrak{x}: B \to \mathbb{R}^3$ be a parametric representation of the closed surface Σ that separates the fluid from the atmosphere, with $B \subset \mathbb{R}^2$ as parameter domain, then the mean curvature H of Σ equals the normal component of the stress vector

$$(11) \quad H = n^i T_{ij} n^j \equiv n^i \left\{ (p-p_0)\delta_{ij} + \nu\left(\frac{\partial v^i}{\partial x^j} + \frac{\partial v^j}{\partial x^i}\right)\right\} n^j \quad .$$

As we assume $\underline{v} \equiv 0$, we arrive at

$$(12) \quad H = p = \text{const.} \; ,$$

for $\underline{n} = (n^1, n^2, n^3)$ is a unit vector, and the pressure is determined up to a constant only. Now it is known that surfaces of constant mean curvature are critical points+) to the functional

$$(13) \quad E(\mathfrak{x}) \equiv \int_B |\mathfrak{x}_\alpha \wedge \mathfrak{x}_\beta| d\alpha d\beta + \tfrac{4}{3} H \int_B \mathfrak{x}\cdot(\mathfrak{x}_\alpha \wedge \mathfrak{x}_\beta) d\alpha\, d\beta \; ,$$

and in the case of closed surfaces the spheres are even strict minima to $E(\mathfrak{x})$. Finally we have to adjust the constant in (12), such that $\mathfrak{x}$ encludes precisely the volume of the initially given mass of fluid.

Now there are various ways to investigate the stability of the resulting sphere. One can study the second variation of the integral $E(\mathfrak{x})$, for instance. Results of this type belong to <u>hydrostatics</u>, because dynamical properties of the fluid do not enter into the investigation at all. In contrast to these problems we are interested in the <u>dynamical stability of spherical drops</u>: we start with a perturbation Σ_0 of

---

*) In this example we neglect gravity forces: both an exterior gravity field as well as the attraction of the fluid particles.

+) Strictly speaking this variational approach which is due to E. HEINZ [11] has been used in the study of PLATEAU's problem for surfaces of constant mean curvature.

the sphere and with a velocity field $\underline{v}_0$, given in Ω_0, the domain inside of Σ_0. We assume that $\underline{v}$ and p fulfil the NAVIER-STOKES system and that the shape of the domain occupied by the fluid is determined by (11). The sphere is called dynamically stable, if the solution $\underline{v}(x,t)$ of the equations of motion tends to zero as $t \to \infty$ and if the boundary $\Sigma(t)$ approaches the sphere with $t \to \infty$.

Hence we get a free boundary value problem for the nonstationary NAVIER-STOKES equations. The first part (with which we deal exclusively here) of its solution consists of the determination of a family of closed surfaces, when their mean curvature is prescribed, and the solution of the NAVIER-STOKES equations in a given domain that varies with time. An iteration procedure then leads to the solution for the free boundary value problem. We now state the two problems to be considered in a form that is related to our method of solving them.

We seek the unknown boundary as critical point of the functional*)

$$(14) \qquad E^*(\mathfrak{x}) = \int_B |\mathfrak{x}_\alpha \wedge \mathfrak{x}_\beta| \, d\alpha d\beta + \int_B Q(\mathfrak{x}) \cdot \mathfrak{x}_\alpha \wedge \mathfrak{x}_\beta \, d\alpha d\beta$$

in the class $\mathcal{L}$ of closed surfaces which are $C^{2+\alpha}$-perturbations of the unit sphere. $Q(\mathfrak{x})$ is defined by

$$(15) \qquad Q(\mathfrak{x}) = \tfrac{4}{3}\Big(\int_0^{x^1} H(s,x^2,x^3)\,ds,\ \int_0^{x^2} H(x^1,s,x^3)\,ds,\ \int_0^{x^3} H(x^1,x^2,s)\,ds\Big),$$

where H denotes the prescribed mean curvature.

Let $Q \equiv \bigcup_{0\le t\le T} \Omega(t)$ be a given domain in $\mathbb{R}^3 \times [0,T]$. We assume that Q can be mapped onto a cylindrical domain $\Omega \times [0,T]$ by a $C^{3+\alpha}$-diffeomorphism $\xi = \phi(x,t)$, such that $\phi(\Omega(t)) = \Omega \times \{t\}$. We transform the dependent variables such that divergence-free vector fields in Q are divergence-free in $\Omega \times [0,T]$, too:

$$(16) \qquad \tilde{v}^i = \frac{\partial \xi^i}{\partial x^j} \tilde{v}^j \quad , \ \tilde{p} = p \ .$$

The NAVIER-STOKES equations then have the following form

*) The functional $E^*(\mathfrak{x})$ has been introduced by S. HILDEBRANDT [12] in his contributions to PLATEAU's problem for surfaces of prescribed variable mean curvature.

$$\frac{\partial \tilde{v}^i}{\partial t} - \nu \sum_{j,k} \frac{\partial}{\partial \xi^j} \left(g^{jk} \frac{\partial \tilde{v}^i}{\partial \xi^k}\right) + L^i(\xi, \underline{\tilde{v}}, \nabla \underline{\tilde{v}}) \tag{17}$$
$$+ N^i(\xi, \underline{v}) + g^{ij} \frac{\partial \tilde{p}}{\partial \xi^j} = F^i \quad .$$

By definition $g^{ij} = \sum_k \frac{\partial \xi^i}{\partial x^k} \frac{\partial \xi^j}{\partial x^k}$, L^i is a linear operator in $\underline{v}$ and its derivatives with coefficients depending on $\underline{\phi}$, $N^i(\xi, \underline{v})$ is the transformed nonlinearity

$$N^i = \sum_j \tilde{v}^j \frac{\partial \tilde{v}^i}{\partial \xi^j} + \sum_{j,k} \Gamma^i_{jk} \tilde{v}^j \tilde{v}^k \quad , \tag{18}$$

where the Γ^i_{jk} are the CHRISTOFFEL symbols for the "metric" g_{ij}. Reducing the problem to homogeneous data yields that the normal component of the velocity and the tangential components of the stress vector vanish on the lateral boundary of Q; finally, an initial distribution for the velocity field is given, and these conditions induce by means of (16) conditions on the boundary of $\Omega \times [0,T]$.

In the iteration scheme to solve the free boundary value problem in each step, the solution of (14) enters into the data of (17) and vice versa. Therefore, the solution of the variational problem must meet the regularity properties that are imposed on the data of the initial value problem, i.e. $\underline{\phi} \in C^{3+\alpha}$. To get a smooth solution to (14), H must be regular, which means in view of (11) that $\underline{v}$, p are required to be classical solutions. In order to establish them we use again $C^{0+\alpha}$-semigroups.

§3. $C^{0+\alpha}$-semigroups generated by elliptic differential operators

Let L be an elliptic operator with smooth coefficients which are defined on some domain G, such that the spectrum for the DIRICHLET problem

(19) $\quad Lu - \lambda u = f \qquad \text{in } G$

(20) $\quad u = 0 \qquad \text{on } \partial G$

is contained in a sector S of the left half-plane. Here L is an operator in $L_p(G)$, $p > 1$. For the construction of fractional powers of L the resolvent estimate of S. AGMON is essential:

(21) $$\|(L - \lambda)^{-1}\|_{L_p} \leq \frac{C}{|\lambda|} \qquad \forall\lambda \notin S.$$

If L acts on $C^{0+\alpha}(\bar{G})$ however, the following estimate holds

(22) $$\|(L - \lambda)^{-1}\|_{C^{0+\alpha}} \leq \frac{C}{|\lambda|^{1-\alpha/2}} \qquad \forall\lambda \notin S.$$

This inequality was proved by W. von WAHL [23]*) and he showed in addition that it is sharp with respect to the decay in $|\lambda|$. Based on (22) von WAHL introduced $C^{0+\alpha}$-semigroups

(23) $$e^{-tL} = \frac{1}{2\pi i} \int_\Gamma e^{\lambda\tau} (L - \lambda)^{-1} \, d\lambda,$$

and proved the basic estimates for them. As e^{-tL} operates on HÖLDER spaces it may be called $C^{0+\alpha}$-semigroup. Further estimates concern the fractional powers of L, defined by

(24) $$L^{-\gamma}u = \frac{1}{\Gamma(\gamma)} \int_0^\infty e^{-sL} s^{\gamma-1} \, ds ,$$

where $\gamma > \frac{\alpha}{2}$ (otherwise the integral would not converge), and

$$L^{\gamma}u = (L^{-\gamma})^{-1} u$$

if γ is positive. Fractional powers yield interesting interpolation theorems:

---

*) There L is allowed to be of order 2m with time dependent coefficients.

(25) $$\|u\|_{C^{1+\beta}} \leq C \, \|L^{\gamma} u\|_{C^{0+\alpha}} ,$$

where

$$1 - \frac{\alpha}{2} > \gamma > 1 - \frac{2-2\beta-\alpha^2}{4+2\alpha} \quad \text{and} \quad 0 < \beta < 1-\alpha(1+\alpha).$$

Inequality (25), as well as

(26) $$\|e^{-tL} u\|_{C^{0+\alpha}} \leq C \, \frac{e^{-at}}{t^{\alpha/2}} \|u\|_{C^{0+\alpha}} ,$$

C, a suitable constants , show, if compared with L_p - analogues, a singular behavior: one has for instance the above restrictions on the powers γ and on the HÖLDER spaces on which the L^{γ} are defined; then for $t \downarrow 0$ e^{-tL} does not approach the identity, as follows from (26).

These concepts were used in several contributions by W. von WAHL regarding regularity and existence of solutions of parabolic problems.

Another idea that we use in the next chapter is due to H. KIELHÖFER [15], who investigated semilinear initial value problems. A $C^{0+\alpha}$-semigroup is not strongly continuous, cf. (26). So when he considered in the BANACH space $\mathfrak{X} = C^{0+\alpha}(\bar{G})$ the evolution equation

$$D_t u(t) = -Lu(t) + F(u(t)), \; u(0) = u_0 ,$$

he replaced D_t, the differentiation with respect to the norm $\| \;\; \|_{\mathfrak{X}}$, by a weaker version, where a continuous semi-norm is involved. One then needs

(27) $$\lim_{t \downarrow 0} \|e^{-tL} u - u\|_{C^0} = 0.$$

This property is not known for elliptic operators in general, but has to be shown in each single case.

§4. Global solutions for flows past obstacles

In our approach to the initial value problem

$$(28)\qquad \begin{aligned} \underline{v}_t - \nu\Delta\underline{v} + \nabla p + (\underline{v}\cdot\nabla)\underline{v} &= \underline{f} \\ \nabla\cdot\underline{v} &= 0 \end{aligned} \qquad \text{in } \mathfrak{E} \times [0,T]$$

$$(29)\qquad \underline{v}(x,t) = 0 \quad \forall x \in \Sigma \quad \forall t \in [0,T]$$

$$(30)\qquad |\underline{v}(x,t) - \underline{U}| \leq C|x|^{-1}, \text{ as } |x| \to \infty$$

$$(31)\qquad \underline{v}(x,0) = \underline{v}_0(x) \quad \forall x \in \mathfrak{E}\ ,$$

the first step will be the estimate (22) for the resolvent of OSEEN's operator, i.e.

$$(32)\qquad \begin{aligned} -\nu\Delta\underline{v} + \nabla p + (\underline{U}\cdot\nabla)\underline{v} + \lambda\underline{v} &= \underline{f} \\ \nabla\cdot\underline{v} &= 0 \end{aligned} \qquad \text{in } \mathfrak{E}$$

$$(33)\qquad \underline{v}\big|_\Sigma = 0 \quad , \ |\underline{v}(x)| < C|x|^{-1}, \text{ as } |x| \to \infty.$$

To use the OSEEN linearisation was suggested by the method of FINN to which we referred earlier already. For the boundary value problem we can prove

<u>Theorem 1</u>: Let $\lambda = \alpha + i\beta$ be a point outside the domain P that is bounded by the parabola $|\alpha| = c(\underline{U},\nu)\beta^2$, $-\infty < \alpha \leq 0$. Then (32), (33) is uniquely solvable, and the solution decays at infinity like $C|x|^{-1}$ if $\lambda = 0$ and like $C|x|^{-3}$ if $\lambda \neq 0$.

For the resolvent we have the estimate

$$(34)\qquad \begin{aligned} &|\lambda|\,\|\underline{v}\|_0 + |\lambda|^{1-\alpha/2}\|\underline{v}\|_{0+\alpha} + |\lambda|^{1/2}\|\underline{v}\|_1 + |\lambda|^{1/2-\alpha/2}\|\underline{v}\|_{1+\alpha} \\ &+ \|\underline{v}\|_2 + |\lambda|^{-\alpha/2}\|\underline{v}\|_{2+\alpha} \leq C\|\underline{f}\|_{0+\alpha},\ \lambda \notin P. \end{aligned}$$

For brevity we write $\|\cdot\|_{k+\alpha}$ instead of $\|\cdot\|_{C^{k+\alpha}(\overline{\mathfrak{E}})}$.

<u>Proof</u>: We start with the construction of a fundamental solution (E_{ij}, P_j) to the OSEEN system (32). This can be done in the same way as for $\lambda = 0$, namely

$$(35)\qquad E_{ij}(x,y;\lambda) = \left(\delta_{ij}\Delta - \frac{\partial^2}{\partial y^i \partial y^j}\right)\phi(x,y;\lambda),$$

where ϕ is fundamental solution to a scalar equation of fourth order. The resulting expression for E_{ij} is more complicated, however, than the one for $\lambda = 0$:

$$(36) \quad \phi(x,y;\lambda) = \frac{-e^{-\lambda(y^1-x^1)/U}}{4\pi U} \int_{y^1-x^1}^{\infty} \frac{e^{-\lambda t/U}}{\sqrt{t^2+s^2}} \left\{1 - e^{\sqrt{\lambda^+}\sqrt{t^2+s^2}-\frac{Ut}{2\nu}}\right\}dt,$$

where $\underline{U} = (U,0,0)$, $s^2 = (y^2-x^2)^2 + (y^3-x^3)^2$, $\lambda^+ = \frac{\lambda}{\nu} + \frac{U^2}{4\nu^2}$. E_{ij} (as well as P_j) behaves for small values of $|x-y|$ like the fundamental solution for the STOKES system. Therefore one can define hydrodynamic potentials, and, as in the classical potential theory, gets jump relations, that finally lead to FREDHOLM equations. For λ outside the parabola $|\alpha| = c(\underline{U},\nu)\beta^2$ (the constant $c(\underline{U},\nu)$ can be calculated from (36)) these equations admit a unique solution. This potential theoretic method is well known and was applied to linearisations of the NAVIER-STOKES equations already by F.K.G. ODQVIST [17] and H. FAXEN [5]. It can be carried over to the problem (32), (33) if one only has control over the behavior of (35) for $|x-y|$ small; for details see [2] pp. 150-158.

In this way we represent the solution by means of potentials and hence we get informations on the behavior of the solutions for large values of $|x-y|$ from the corresponding properties of E_{ij} and P_j.

For the last proposition, the inequality for the resolvent, we need more refined estimates on the fundamental solution. (34) can be regarded as an improved SCHAUDER estimate, because in addition to the usual $C^{2+\alpha}$-bounds it contains also how the constants depend on the parameter λ. For this purpose we prove

$$(36)' \quad \phi(x,y;\lambda) = \frac{1}{8\pi\nu} |x-y| \, e^{-\sqrt{\lambda}|x-y|} + \psi(x,y;\lambda) \qquad ,$$

where ψ does not affect the constants in (34). Then we get the desired inequality by examining the corresponding proof in AGMON-DOUGLIS-NIRENBERG [1].

Instead of repeating all the details we make the following observation: when we compare the estimates for the $C^{k+\alpha}$- and the $C^{k+1+\alpha}$-norms ($k = 0,1$; $\alpha = 0$ or $\alpha \in (0,1)$), the exponent of $|\lambda|$ diminishes by 1/2; the same is true for H_p^k-norms: $|\lambda| \, \|\underline{v}\|_{L_p} + |\lambda|^{1/2} \|\underline{v}\|_{H_p^1} + \|\underline{v}\|_{H_p^2} \leq C \|\underline{f}\|_{L_p}$.

This follows from a property of E_{ij}: differentiating E_{ij} once yields a factor $|\lambda|^{1/2}$, as can be seen from (36)'.The same expansion (36)' also explaines the different powers of $|\lambda|$ for C^k- and $C^{k+\alpha}$-norms for $\alpha > 0$. Because the HÖLDER seminorm consists of a fractional difference quotient of order α, we get the factor $(|\lambda|^{1/2})^\alpha$, when we replace C^k- by $C^{k+\alpha}$-norms. Finally we remark that the fundamental difference between the resolvent estimates in L_p and $C^{0+\alpha}$, which we mentioned in §3, (21)-(22) already, turns out to be an immediate consequence of (36)', too: the L_p- and $C^{0+\alpha}$-norms of thekernel E_{ij} differ by the factor $|\lambda|^{\alpha/2}$, because for the function $e^{-\sqrt{\lambda}|x|}$ these norms differ by $|\lambda|^{\alpha/2}$. q.e.d.

Theorem 1 now allows to define the semigroup e^{-tA}, generated by the OSEEN operator A, as well as fractional powers of A; they have the same properties as the semigroups of strongly elliptic operators, as we have outlined in §3.

<u>Theorem 2</u>: The initial value problem (28)-(31) is for $\underline{v}_0 \in D(A^\delta)$, δ suitable chosen, uniquely solvable. The solution is an element of $C^0((0,T),C^{2+\alpha}(\overline{\mathfrak{E}})) \cap C^1((0,T),C^{0+\alpha}(\overline{\mathfrak{E}}))$.

If $\nu/|\underline{U}|$ is large enough, or equivalently if $\| A^\delta \underline{v}_0 \|_{0+\alpha}$ is small enough, the solution exists for T arbitrarily large.

<u>Proof</u>: As usual we transform the problem to homogeneous boundary data and call the new unknown function $\underline{u}(x,t)$. According to the properties of the linearized problem we proved in Theorem 1 we investigate the non-linear integral equation

$$(36)'' \quad u(t) = e^{-tA}u_0 + \int_0^t e^{-sA}(f + (u\cdot\nabla)u)(s)ds$$

in the BANACH space $\mathfrak{X} = C^{0+\alpha}(\overline{\mathfrak{E}}) \cap \{\underline{u}\colon |\underline{u}(x)| \leq C|x|^{-1}, \text{ as } |x| \to \infty\}$. For the existence of a local solution we apply KIELHÖFER's theorem and therefore have to show, cf. (27), that

$$(37) \quad \lim_{t\downarrow 0} \|e^{-tA}u(t) - u(t)\|_{C^0} = 0$$

holds. In order to prove this property we exploit the fact that the resolvent A^{-1} of the OSEEN operator is given as an integral operator with the GREEN function $G = G_{ij}(x,g;\lambda)$ of the boundary value problem (32), (33) as kernel. Hence we get

$$(38) \quad e^{-tA}f = \frac{1}{2\pi i}\int_\Gamma e^{\lambda t}A_\lambda^{-1} fd\lambda = \frac{1}{2\pi i}\int_\Gamma e^{\lambda t} \int_{\mathfrak{E}} G(x,y;\lambda)f\, dy\, d\lambda,$$

and from this we can deduce (37), using a result of V.A. SOLONNIKOV [21].

With the representation (38) we can also show that e^{-tA} maps $\mathfrak{X}$ into itself. We know from theorem 1 the asymptotic behavior of $G_{ij}(x-y;\lambda)$ for $|x-y| \to \infty$, and we can apply an estimate of FINN. This yields

$$|e^{-tA}u| \leq c|x|^{-1}$$

for all u such that $|\underline{u}| \leq C|x|^{-1}$.

The final step for the a priori estimate consists in an inequality for the nonlinearity $\| (\underline{u}\cdot\nabla)\underline{u} \|_{0+\alpha}$ in terms of $\| A^{\delta}\underline{u} \|_{0+\alpha}$. We use the interpolation theorem (25) and calculus inequalities. The fact that (25) is valid only for certain values of α and δ leads to restrictions in the theorem above. q.e.d.

Our starting-point in §1 was the problem of accelerating the body and of the attainability of PR-solutions. The existence of a solution during the acceleration process can be proved by the methods of the next chapter, if we describe the problem by equations in a noncylindrical domain.

§5. Classical solutions to the nonstationary NAVIER-STOKES equations in domains varying in time

NAVIER-STOKES equations in non-cylindrical domains were investigated by various authors: J.O. SATHER [20], H. FUJITA-N. SAUER [10], O.A. LADYZHENSKAYA [16], A. INOUE-M. WAKIMOTO [13], D.N. BOCK [4] and M. ÔTANI-Y. YAMADA [18]. These contributions differ both in methods and results, but they do not treat the regularity of the solutions*).

We prove the existence of classical solutions for the initial value problem (17), also for DIRICHLET data, cf. [3], in a way similar to the one described in §4. The problem (17) is easier in so far as Ω is a bounded domain, but on the other hand the linear operator in (17) that generates a semigroup has variable coefficients. Hence the proof of the resolvent estimates needs additional considerations. We start with the construction of a fundamental solution for the system

$$
\begin{aligned}
-a_{ij} \frac{\partial^2}{\partial x^i \partial x^j} v^k + b_{ik} \frac{\partial}{\partial x^i} p + \lambda v^k &= 0 \\
\nabla \cdot \underline{v} &= 0
\end{aligned} \tag{39}
$$

where a_{ij}, b_{ik} are positive definite matrices with constant coefficients. Using OSEEN's device we seek the solution in the form

$$E_{kl} = (\delta_{kl} \Delta - \frac{\partial^2}{\partial x^l \partial x^k}) \phi$$

with some scalar function ϕ. After normalizing one obtains $b_{ik} = \delta_{ik}$, and P_l is of the form

$$P_l = \frac{\partial}{\partial x^l} (a_{ij} \frac{\partial^2}{\partial x^i \partial x^j} - \lambda) \phi .$$

ϕ can be determined by the equation

$$\Delta (a_{ij} \frac{\partial^2}{\partial x^i \partial x^j} - \lambda) \phi = \delta$$

As the fundamental solution of $(a_{ij} \frac{\partial^2}{\partial x^i \partial x^j} - \lambda) \psi = \delta$ is of the form

$$\psi = \frac{-1}{\sqrt{\lambda} \sqrt{\det a^{ij}}} \psi \left(\sqrt{\lambda} \sqrt{a^{ij} (x^i - y^i)(x^j - y^j)} \right) ,$$

*) The fact that the authors restrict themselves to DIRICHLET data is not really important; the case that $\underline{v} \cdot \underline{n}$ and $\underline{\tau}^i \cdot \underline{\underline{T}} \cdot \underline{n}$ ($\underline{\tau}^i$ span the tangent plane to $\Sigma(t)$, $\underline{\underline{T}}$ is the stress tensor) are prescribed can be handled along the same lines.

ψ being a solution of BESSEL's equation, the characteristic exponent of λ appears if we make an expansion at $x = y$.

Next we extend this result to the case that the coefficients in (39) are no longer constants and that lower order terms are added. As the diffeomorphism $\underline{\phi}$ that enters into the coefficients, cf. (16), (17), is of class $C^{3+\alpha}$, the a_{ij}, b_{ij} and the coefficients in L are at least HÖLDER-continuous. Hence we can apply classical methods and derive the existence and the singular behavior of the fundamental solution to the linearisation of (17) by an integral equation argument. The proof of the resolvent estimates follows closely the one given in [2].

Appendix: Closed surfaces of prescribed mean curvature

As no general methods for the construction of closed surfaces of prescribed mean curvature are developed it is tempting especially with respect to the stability problem of §2, to seek the solution as a perturbation of an explicitly known surface which is in our case the unit sphere. Let $\mathfrak{x}^0$ be the unit sphere, $\mathfrak{n}$ its normal; we then denote the perturbed surface by

$$(40) \qquad \mathfrak{x} = \mathfrak{x}^0 + f\,\mathfrak{n} .$$

Calculation of the mean curvature $H = 1 + h$ of $\mathfrak{x}$ leads to

$$(41) \qquad \Delta^* f + 2f = N(f,h) \quad ,$$

where Δ^* is the LAPLACE-BELTRAMI operator on $\mathfrak{x}^0$. Although the nonlinearity $N(f,h)$ is small in the sense that the method of successive approximations can be applied formally, the fact that 2 is a triple eigenvalue and N consists of numerous terms makes the bifurcation equations very complicated.

Therefore, we base our approach on the fact that $\mathfrak{x}^0$ is a strict relative minimum for (13), provided we identify spheres that differ by a translation only. The main argument is the following perturbation result of F. TOMI [22]:

Theorem 3: (F. TOMI)

Let $\mathfrak{X}, \mathcal{Y}$ be BANACH spaces, $\mathfrak{X}$ continuously embedded into $\mathcal{Y}$, $\phi_0 \in C^1(\mathfrak{X}_R, \mathbb{R})$, $\Phi_0 \in C^1(\mathfrak{X}_R, \mathcal{Y})$ and β_0 a positive, continuous bilinear form on $\mathcal{Y} \times \mathcal{Y}$ such that

$$(42) \qquad D\phi_0(x) = \beta_0(\cdot, \Phi_0(x)) \quad \forall x \in \mathfrak{X}_R \equiv \mathfrak{X} \cap \{|x| < R\} \qquad .$$

Let 0 be strict relative minimum of ϕ_0, and $D\Phi_0(0)$ be a FREDHOLM operator of index zero.

Conclusion: For every $\varepsilon > 0$ there exists a $\delta(\varepsilon) > 0$ with the following property: If $\phi \in C^1(\mathfrak{X}_R, \mathbb{R})$ is a functional, $\Phi \in C^1(\mathfrak{X}_R, \mathcal{Y})$ is a mapping, and β is a continuous positive bilinear form on $\mathcal{Y} \times \mathcal{Y}$ such that

$$(43) \qquad D\phi(x) = \beta(\cdot, \Phi(x)) \quad \forall x \in \mathfrak{X}_R$$

and moreover

(44) $\max\{\|\phi-\phi_0\|_{C^0}, \|\Phi-\Phi_0\|_{C^1}, \|\beta-\beta_0\|\} \leq \delta(\varepsilon)$,

then ϕ possesses at least one critical point in $\mathfrak{X}_\varepsilon$.

If we denote by $E^*(f;H)$ the functional $E^*(\mathfrak{x})$ in (14) with $\mathfrak{x} = \mathfrak{x}^0 + f\mathfrak{n}$, we can prove the following propositions:

(i) $D_f E^*(f,H) = 0$ iff $\mathfrak{x} \equiv \mathfrak{x}^0 + f\mathfrak{n}$ is a surface of mean curvature H.

This means that the surfaces with mean curvature H are critical points to $E^*(f;H)$.

(ii) There exists a mapping ψ_H, such that

$$D_f E^*(f;H)g = \int_{\partial B_1} \psi_H(f)g\, d\sigma .$$

(iii) $D_f\psi_H(f)\big|_{f=0}\, g = \Delta^* g + 2g$

We now perturbe $H \equiv 1$ such that $H = 1$ on $\mathfrak{x}^0$ and such that $\mathfrak{x}^0$ is a strict minimum to $E^*(f,H)$. This can be proved under some conditions on H by suitable comparison arguments. Then with this H we take $E^*(f,H)$ as functional ϕ_0 in TOMI's theorem, ψ_H from (ii) as Φ_0 and the L_2-scalar product on $\mathfrak{x}^0$ as the form β_0. The perturbed quantities are the ones where H is replaced by the given mean curvature which has to fulfil an additional condition.

For details of this construction as well as for the proof of uniqueness and (higher) regularity we refer to a forthcoming publication.

References

[1] AGMON, S. - DOUGLIS, A. - NIRENBERG, L.
Estimates Near the Boundary for Solutions of Elliptic Partial Differential Equations Satisfying General Boundary Conditions, II. Comm. Pure Appl. Math. 17 (1964), 35-92.

[2] BEMELMANS, J.
Eine Aussenraumaufgabe für die instationären Navier-Stokes-Gleichungen. Math. Z. 162 (1978), 145- 173.

[3] BEMELMANS, J.
Klassische Lösungen der instationären Navier-Stokes-Gleichungen in Gebieten mit beweglichen Rändern. To appear.

[4] BOCK, D.N.
On the Navier-Stokes Equations in Noncylindrical Domains. J. Diff. Equ. 25 (1977), 151-162.

[5] FAXÉN, H.
Fredholmsche Integralgleichungen zu der Hydrodynamik zäher Flüssigkeiten, I. Arkiv för Mat., Astr. och Fys. 21A (1928/29), 1-40.

[6] FINN, R.
An Energy Theorem for Viscous Fluid Motions. Arch. Rat. Mech. Anal. 6 (1960), 371-381.

[7] FINN, R.
Stationary Solutions of the Navier-Stokes Equations. Proc. Symp. Appl. Math. 17 (1965), 121-153.

[8] FINN, R.
Mathematical Questions Relating to Viscous Fluid Flow in an Exterior Domain. Rocky Mountain J. Math. 3 (1973), 107-140.

[9] FUJITA, H. - KATO, T.
On the Navier-Stokes initial value problem I. Arch. Rat. Mech. Qual. 16 (1964), 269-315.

[10] FUJITA, H. - SAUER, N.
On existence of weak solutions of the Navier-Stokes equations in regions with moving boundaries. J. Fac. Sci. Univ. Tokyo Sect. I 28 (1970), 403-420.

[11] HEINZ, E.
Über die Existenz einer Fläche konstanter mittlerer Krümmung bei vorgegebener Berandung. Math. Ann. 127 (1954), 258-287.

[12] HILDEBRANDT, S.
Randwertprobleme für Flächen mit vorgeschriebener mittlerer Krümmung und Anwendungen auf die Kapillaritätstheorie, I: Fest vorgegebener Rand. Math. Z. 112 (1969), 205-213.

[13] INOUE, A. - WAKIMOTO, M.
On existence of solutions of the Navier-Stokes equations in a time dependent domain. J. Fac. Sci. Univ. Tokyo Sect. IA, 24 (1977), 303-320.

[14] KATO, T. - FUJITA, H.
On the nonstationary Navier-Stokes system. Rend. Sem. Mat. Univ. Padova 32 (1962), 243-260.

[15] KIELHÖFER, H.
Halbgruppen und semilineare Anfangs-Randwertprobleme. manuscripta math. 12 (1974), 121-154.

[16] LADYZHENSKAYA, O.A.
Initial-boundary problems for Navier-Stokes equations in domains with time-varying boundaries. Sem. in Math. V.A. Steklov Math. Inst. 11 (1970), 35-46.

[17] ODQVIST, F.K.G.
Über die Randwertaufgaben der Hydrodynamik zäher Flüssigkeiten. Math. Z. 32 (1930), 329-375.

[18] ÔTANI, M. - YAMADA, Y.
On the Navier-Stokes equations in non-cylindrical domains: an approach by the subdifferential operator theory. J. Fac. Sci. Univ. Tokyo Sect. IA, 25 (1978), 194-205.

[19] PRODI, G.
Teoremi die tipo locale per il sistema di Navier-Stokes e stabilità delle soluzioni stazionarie. Rend. Sem. Mat. Univ. Padova 32 (1962), 374-397.

[20] SATHER, J.O.
The Initial Boundary Value Problem for the Navier-Stokes Equations in Regions with Moving Boundaries. Ph. D. Thesis, Univ. of Minnesota, 1963.

[21] SOLONNIKOV, V.A.
On bounadry value problems for linear parabolic systems of differential equations of general form. Proc. Steklov Inst. Math. 83 (1965).

[22] TOMI, F.
A Perturbation Theorem for Surfaces of Constant Mean Curvature. Math. Z. 141 (1975), 253-264.

[23] WAHL, W. von
Gebrochene Potenzen eines elliptischen Operators und parabolische Differentialgleichungen in Räumen hölderstetiger Funktionen. Nachr. Akad. Wiss. Göttingen Math.-Phys. Kl. II 11 (1972), 231-258.

A FINITE ELEMENT APPROXIMATION OF NAVIER-STOKES EQUATIONS FOR INCOMPRESSIBLE VISCOUS FLUIDS. ITERATIVE METHODS OF SOLUTION.

M.O. Bristeau
IRIA-LABORIA
B.P. 105, 78150 Le Chesnay
France

R. Glowinski
Université Paris VI, L.A. 189
4, Place Jussieu, 75230 Paris

B. Mantel, J. Periaux, P. Perrier
AMD/BA, 78 Quai Carnot, B.P. 300
92214 St-Cloud, France

O. Pironneau
Université Paris-Nord
Place du 8 Mai 1945, 93200 St-Denis
France

We present in this paper a method for the numerical solution of the steady and unsteady Navier-Stokes equations for incompressible viscous fluids. This method is based on the following techniques :

. A mixed finite element approximation acting on a pressure-velocity formulation of the problem,
. A time discretization by finite differences for the unsteady problem,
. An iterative method using - via a convenient nonlinear least square formulation - a conjugate gradient algorithm with scaling ; the scaling makes a fundamental use of an efficient Stokes solver associated to the above mixed finite element approximation.

The results of numerical experiments are presented and analyzed. We conclude this paper by an appendix introducing a new upwind finite element approximation ; we discuss in this appendix the solution by this new method of $-\varepsilon\Delta u+\underset{\sim}{\beta}\cdot\underset{\sim}{\nabla}u = f$ on Ω, $u=0$ on $\partial\Omega$ (Ω : bounded domain of $\mathbb{R}^2$), but we plan to apply it to the solution of Navier-Stokes problems.

0. INTRODUCTION

The *numerical solution* of the *Navier-Stokes equations* for *incompressible viscous fluids* has motivated so many authors that giving a complete bibliography has become an impossible task. Restricting therefore our attention to only very recent contributions making use of *finite element approximations* we shall mention among many others [1] - [11] (see also the references therein).

We would like to discuss in the present paper a method for the effective solution of the above Navier-Stokes equations in the steady and unsteady cases. The basic ingredients of the method to be described are the following :

- A *mixed finite element approximation* - based on a new variational principle - of a *pressure-velocity* formulation of the original problem.
- *Time discretizations* of the unsteady problem by *finite differences* ; several schemes will be presented.
- *Iterative methods*, using - via convenient *nonlinear least square* formulations - *conjugate gradient algorithms* with *scaling* ; the scaling operation is based on an efficient *Stokes solver*, derived from the very remarkable algebraic properties of the above mixed finite element approximation.

To illustrate the possibilities of our methods we present the results of various numerical experiments concerning non trivial two-dimensional flows. To conclude this paper we introduce in an Appendix a new upwind finite element scheme ; the model problem under consideration is much simpler than Navier-Stokes problems since it is

$$\begin{cases} -\varepsilon\Delta u + \underset{\sim}{\beta}\cdot\underset{\sim}{\nabla} u = f \text{ } in \text{ } \Omega , \\ u = 0 \text{ } on \text{ } \partial\Omega, \end{cases}$$

where Ω is a bounded domain of $\mathbb{R}^2$, $\underset{\sim}{\beta}$ a constant vector of $\mathbb{R}^2$ and ε a "possibly small" positive parameter. We have the feeling that the method described in this appendix has a good potential for solving some Navier-Stokes problems (involving possibly *compressible fluids*).

The content of our paper is as follows :

1. *Formulation of the steady und unsteady Navier-Stokes equations for incompressible viscous fluids.*
2. *A mixed finite element method for the Stokes and Navier-Stokes problems.*
3. *Time discretization of the unsteady Navier-Stokes problem.*
4. *Least square formulation and iterative solution by conjugate gradient with scaling.*
5. *A Stokes' solver.*
6. *Numerical experiments.*
7. *Further comments. Conclusion.*

Appendix : A finite element method with upwinding for second order problems with "large" first order terms.

References.

1. FORMULATION OF THE STEADY AND UNSTEADY NAVIER-STOKES EQUATIONS FOR INCOMPRESSIBLE VISCOUS FLUIDS.

Let us consider a *newtonian, viscous* and *incompressible* fluid. If Ω and Γ denote the region of the flow and its boundary, respectively, then this flow is governed by the Navier-Stokes equations

$$(1.1) \qquad \frac{\partial \underset{\sim}{u}}{\partial t} - \nu\Delta\underset{\sim}{u} + (\underset{\sim}{u}\cdot\underset{\sim}{\nabla})\underset{\sim}{u} + \underset{\sim}{\nabla} p = \underset{\sim}{f} \text{ } in \text{ } \Omega,$$

(1.2) $\nabla \cdot u = 0$ *in* Ω (*incompressibility condition*),

which in the *steady* case reduce to

(1.3) $-\nu \Delta u + (u \cdot \nabla) u + \nabla p = f$ *in* Ω,

(1.4) $\nabla \cdot u = 0$ *in* Ω.

In (1.1)-(1.4)

u is the *flow velocity*,
p is the pressure,
ν is the *viscosity* of the fluid ($\nu = 1/Re$, Re : Reynold's number),
f is a density of *external* forces.

Boundary conditions have to be added ; for example in the case of the flow around the airfoil $\underline{B}$ of Fig. 1.1, since the fluid is viscous we have the following *adherence conditions*

(1.5) $u = 0$ *on* $\partial \underline{B} = \Gamma_{\underline{B}}$.

Typical conditions at infinity are

(1.6) $u = u_\infty$

where u_∞ is a *constant* vector (at least in space).
Finally, for the time dependent problem (1.1),(1.2) an *initial condition* such as

(1.7) $u(x,0) = u_o(x)$ *a.e. on* Ω,

where u_o is *given*, is usually prescribed.
Other boundary and/or initial conditions may be prescribed (periodicity in space and/or time, non homogeneous boundary conditions, etc...).

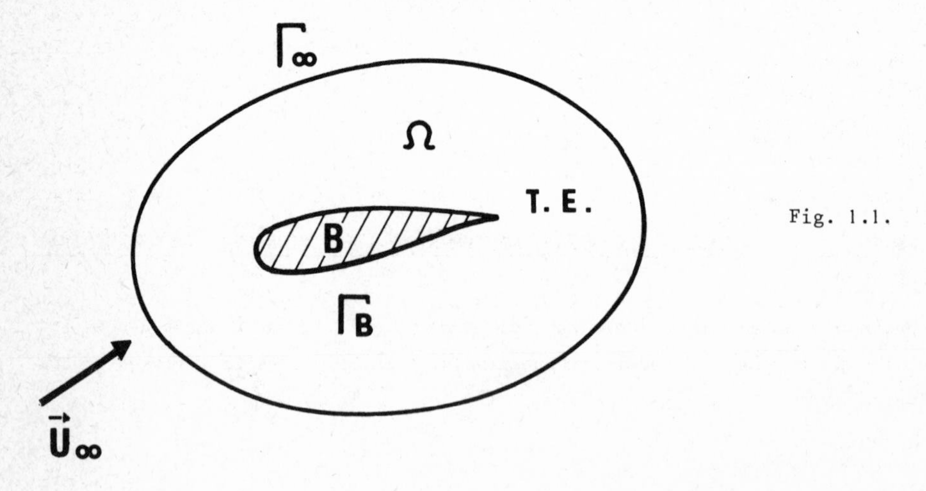

Fig. 1.1.

In two dimensions it may be convenient to formulate the Navier-Stokes equations using a *stream function-vorticity* formulation (see, e.g., [2, Sec. 4], [7],[8]).
To conclude this section let us mention that a *mathematical analysis* of the Navier-Stokes equations for incompressible viscous fluids can be found in, e.g., [12]-[15].

2. - A MIXED FINITE ELEMENT METHOD FOR THE STOKES AND NAVIER-STOKES PROBLEMS.

2.1. Synopsis.

We discuss in this section a mixed finite element approximation of the Navier-Stokes problems which have been considered in Sec. 1. For simplicity we shall begin our discussion with the approximation of the *steady Stokes problem* for incompressible viscous flows, i.e.

$$(2.1)\qquad \begin{cases} -\nu\Delta \underset{\sim}{u} + \underset{\sim}{\nabla} p = \underset{\sim}{f} \textit{ in } \Omega, \\ \underset{\sim}{\nabla}\cdot\underset{\sim}{u} = 0 \textit{ in } \Omega \ ; \end{cases}$$

as boundary conditions we choose (with $\int_\Gamma \underset{\sim}{g}\cdot\underset{\sim}{n}\, d\Gamma = 0$, $\underset{\sim}{n}$ being the unit vector of the outward normal at Γ)

$$(2.2)\qquad \underset{\sim}{u} = \underset{\sim}{g} \textit{ on } \Gamma.$$

Also for simplicity we suppose in the following that Ω is a *bounded polygonal domain* of $\mathbb{R}^2$, but the following methods are easily extended to domains with curved boundary in $\mathbb{R}^2$ and $\mathbb{R}^3$.

2.2. A mixed variational formulation of the Stokes problem (2.1),(2.2).

We follow the discussion in [11].

2.2.1. Some functional spaces. Standard formulation of the Stokes problem.

The following (*Sobolev*) spaces play an important role in the sequel (we refer to [18]-[21] for a general discussion on Sobolev spaces) :

$$H^1(\Omega) = \{\phi \in L^2(\Omega),\ \frac{\partial\phi}{\partial x_i} \in L^2(\Omega)\quad \forall\ i=1,\dots N\}\ ,$$

$$H^1_o(\Omega) = \overline{\mathcal{D}(\Omega)}^{H^1(\Omega)} = \{\phi \in H^1(\Omega),\ \phi = 0 \textit{ on } \Gamma\}\ ,$$

$$H^2(\Omega) = \{\phi \in H^1(\Omega),\ \frac{\partial^2\phi}{\partial x_i \partial x_j} \in L^2(\Omega),\quad \forall\ i,j\ ,\ 1 \le i,j \le N\}\ ,$$

with

$$\mathcal{D}(\Omega) = \{\phi \in C^\infty(\overline{\Omega})\ ,\ \phi \textit{ has a compact support in } \Omega\}\ .$$

From these spaces we define also

$$(2.3)\qquad V_g = \{\underset{\sim}{v} | \underset{\sim}{v} \in (H^1(\Omega))^N,\ \underset{\sim}{\nabla}\cdot\underset{\sim}{v} = 0 \textit{ in } \Omega,\ \underset{\sim}{v} = \underset{\sim}{g} \textit{ on } \Gamma\}\ .$$

Suppose that $f \in (H^{-1}(\Omega))^N$, where $H^{-1}(\Omega) = (H^1_o(\Omega))'$ = *dual space* of $H^1_o(\Omega)$, and that $\underset{\sim}{g} \in (H^{1/2}(\Gamma))^N$, $\int_\Gamma \underset{\sim}{g}\cdot\underset{\sim}{n}\, d\Gamma = 0$; it follows then from [12]-[15] that (2.1), (2.2) has a *unique solution* $\{\underset{\sim}{u},p\} \in V_g \times (L^2(\Omega)/\mathbb{R})$. If $\underset{\sim}{f} \in (L^2(\Omega))^N$, and $\underset{\sim}{g} \in (H^{3/2}(\Gamma))^N$, then

$\{\underset{\sim}{u},p\} \in (V_g \cap (H^2(\Omega))^N \times (H^1(\Omega)/\mathbb{R})$ if Γ is sufficiently smooth. The above $\underset{\sim}{u}$ is also the unique solution of the following variational problem

$$(2.4)\quad \begin{cases} \textit{Find } \underset{\sim}{u} \in V_g \textit{ such that} \\ \nu \int_\Omega \nabla \underset{\sim}{u} \cdot \nabla \underset{\sim}{v}\, dx = \langle \underset{\sim}{f}, \underset{\sim}{v} \rangle \quad \forall \underset{\sim}{v} \in V_o \end{cases}$$

where $\langle \cdot , \cdot \rangle$ denotes the duality between $(H^{-1}(\Omega))^N$ and $(H^1_o(\Omega))^N$.

2.2.2. A new variational formulation of the Stokes problem (2.1), (2.2).

We suppose for simplicity that $\underset{\sim}{f} \in (L^2(\Omega))^N$; we define then $W_g \subset (H^1(\Omega))^N \times H^1_o(\Omega)$ by

$$(2.5)\quad \begin{cases} W_g = \{\{\underset{\sim}{v},\phi\} \in (H^1(\Omega))^N \times H^1_o(\Omega),\ \underset{\sim}{v}|_\Gamma = \underset{\sim}{g},\ \int_\Omega \underset{\sim}{\nabla}\phi \cdot \underset{\sim}{\nabla} w\, dx = \int_\Omega \underset{\sim}{\nabla} \cdot \underset{\sim}{v}\, w dx \\ \forall w \in H^1(\Omega)\} . \end{cases}$$

We consider now the following variational problem

$$(\mathrm{P})\quad \begin{cases} \textit{Find } \{\underset{\sim}{u},\psi\} \in W_g \textit{ such that} \\ \nu \int_\Omega \nabla \underset{\sim}{u} \cdot \nabla \underset{\sim}{v}\, dx = \int_\Omega \underset{\sim}{f} \cdot (\underset{\sim}{v} + \underset{\sim}{\nabla}\phi) dx \quad \forall \{\underset{\sim}{v},\phi\} \in W_o. \end{cases}$$

In [11] one has proved the following

Theorem 2.1 : Problem (P) *has a unique solution* $\{\underset{\sim}{u},\psi\}$ *such that*

$$(2.6)\quad \psi = 0,$$

(2.7) $\quad \underset{\sim}{u}$ *is the solution of the Stokes' problem* (2.1),(2.2) (and (2.4)).

Remark 2.1 : The potential ϕ introduced above is not at all mysterious. Indeed the formulation (P) can be interpreted as follows : if $\underset{\sim}{v} \in (H^1(\Omega))^N$ and if Γ is sufficiently smooth, there exist $\phi \in H^2(\Omega) \cap H^1_o(\Omega)$ and $\underset{\sim}{\omega} \in (H^1(\Omega))^N$ with $\underset{\sim}{\nabla} \cdot \underset{\sim}{\omega} = 0$, such that

$$(2.8)\quad \underset{\sim}{v} = -\underset{\sim}{\nabla}\phi + \underset{\sim}{\omega},$$

and the decomposition (2.8) is unique.

In the formulation (P) instead of directly imposing $\underset{\sim}{\nabla} \cdot \underset{\sim}{v} = 0$ we try to impose $\phi = 0$; these procedures are equivalent in the continuous case but not in the discrete case as it will be seen below.

2.3. A mixed finite element approximation of the steady Stokes' problem (2.1),(2.2).

As mentioned before Ω is a bounded polygonal domain of $\mathbb{R}^2$. We follow in this section [1] and [11].

2.3.1. Triangulation of Ω. Fundamental discrete spaces

Let $\{\mathcal{T}_h\}_h$ be a family of *regular triangulations* of Ω such that $\overline{\Omega} = \bigcup_{K \in \mathcal{T}_h} K$. We set $h(K)$ = length of the greatest side of K, $h = \max_{K \in \mathcal{T}_h} h(K)$ and we suppose that

$$(2.9)\qquad \frac{h}{\min_{K\in\mathcal{T}_h} h(K)} \le \beta \quad \forall \mathcal{T}_h .$$

We define then the following finite element spaces

$$(2.10)\qquad H_h^1 = \{\phi_h \in C^o(\bar{\Omega}),\ \phi_h|_K \in P_1 \quad \forall K \in \mathcal{T}_h\} ,$$

$$(2.11)\qquad H_{oh}^1 = \{\phi_h \in H_h^1,\ \phi_h|_\Gamma = 0\} = H_h^1 \cap H_o^1(\Omega),$$

$$(2.12)\qquad V_h = \{\underset{\sim}{v}_h \in (C^o(\bar{\Omega}))^2,\ \underset{\sim}{v}_h|_K \in P_2 \times P_2 \ \forall K \in \mathcal{T}_h\},$$

$$(2.13)\qquad V_{gh} = \{\underset{\sim}{v}_h \in V_h,\ \underset{\sim}{v}_h|_\Gamma = \underset{\sim}{g}_h\}$$

(where in (2.13), $\underset{\sim}{g}_h$ is a convenient approximation of g ; if $g = \underset{\sim}{0}$ one takes $\underset{\sim}{g}_h = \underset{\sim}{0}$),

$$(2.14)\qquad W_{gh} = \{\{\underset{\sim}{v}_h,\phi_h\} \in V_{gh}\times H_{oh}^1 ,\ \int_{\Omega} \underset{\sim}{\nabla}\phi_h \cdot \underset{\sim}{\nabla} w_h \, dx = \int_{\Omega} \nabla\cdot \underset{\sim}{v}_h w_h \, dx \quad \forall w_h \in H_h^1\}.$$

We shall also use the variants of V_{gh}, W_{gh} obtained from

$$(2.15)\qquad V_h = \{\underset{\sim}{v}_h \in (C^o(\bar{\Omega}))^2,\ \underset{\sim}{v}_h|_K \in P_1 \times P_1 \quad \forall K \in \tilde{\mathcal{T}}_h\}$$

where $\tilde{\mathcal{T}}_h$ is the triangulation obtained from $\mathcal{T}_h$ by subdivision of each triangle $K \in \mathcal{T}_h$ into 4 subtriangles (by joining the mid-sides ; see Fig. 2.1). In the above definitions P_k denotes the space of the polynomials in two variables of degree $\le k$.

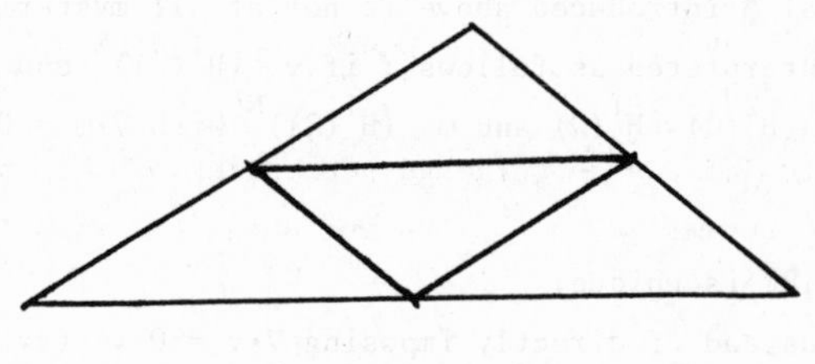

Figure 2.1

2.3.2. Definition of the approximate problems. Characterization of the approximate solutions.

We approximate (P) of Sec. 2.2.2 (therefore the Stokes' problem (2.1),(2.2)) by

$$(P_h)\qquad \begin{cases} \textit{Find } \{\underset{\sim}{u}_h,\psi_h\} \in W_{gh} \textit{ such that} \\ \nu\int_\Omega \underset{\sim}{\nabla}\underset{\sim}{u}_h \cdot \underset{\sim}{\nabla}\underset{\sim}{v}_h \, dx = \int_{\Omega} \underset{\sim}{f}\cdot(\underset{\sim}{v}_h + \underset{\sim}{\nabla}\phi_h)dx \quad \forall \{\underset{\sim}{v}_h,\phi_h\} \in W_{oh} . \end{cases}$$

One may find in [11] the proof of the following

Theorem 2.2 : Problem (P_h) *has a unique solution* $\{\underset{\sim}{u}_h,\psi_h\}$ *which is characterized by the existence of a discrete pressure* $p_h \in H_h^1$ *such that*

(2.16) $$\int_\Omega \nabla p_h \cdot \nabla w_h \, dx = \int_\Omega \underset{\sim}{f} \cdot \nabla w_h \, dx \quad \forall w_h \in H_{oh}^1 ,$$

(2.17) $$\nu \int_\Omega \nabla \underset{\sim}{u}_h \cdot \nabla \underset{\sim}{v}_h \, dx = \int_\Omega (-\nabla p_h + \underset{\sim}{f}) \cdot \underset{\sim}{v}_h \, dx \quad \forall \underset{\sim}{v}_h \in V_{oh} ,$$

(2.18) $$\{\underset{\sim}{u}_h,\psi_h\} \in W_{gh}.$$

2.3.3. Uniqueness of the discrete pressure. Convergence of the approximate solutions.

We introduce the notation

$$|q|_{1,\Omega} = \left(\int_\Omega |\nabla q|^2 dx\right)^{1/2} , \quad |\underset{\sim}{v}|_{o,\Omega} = \|\underset{\sim}{v}\|_{L^2(\Omega)\times L^2(\Omega)} .$$

The following lemma (proved in [11]) plays a role, fundamental to the study of the convergence.

Lemma 2.1 : Suppose that the angles of $\mathcal{T}_h$ *are bounded below by* $\theta_o > 0$, *independent of* h. *Suppose also that* $\forall K \in \mathcal{T}_h$, K *has at most two vertices belonging to* Γ. *Then if* H_h^1 *and* V_{oh} *are defined by either* (2.11) *and* (2.12), (2.13), *or* (2.11) *and* (2.15), (2.13) *(with* $\underset{\sim}{g}_h = \underset{\sim}{0}$ *in* (2.13)), *there exists* C_o *independent of* h *such that*

(2.19) $$\forall q_h \in H_h^1, \; |q_h|_{1,\Omega} \le C_o \max_{\underset{\sim}{v}_h \in V_{oh}-\{\underset{\sim}{0}\}} \frac{\int_\Omega \underset{\sim}{v}_h \cdot \nabla q_h \, dx}{|\underset{\sim}{v}_h|_{o,\Omega}} .$$

A first consequence of Lemma 2.1 is given by the following

Theorem 2.3 : Suppose that (2.19) *holds. Then the discrete pressure* p_h *occuring in* (2.16)-(2.18) *is unique in* $H_h^1/\mathbb{R}$ *(i.e. up to an arbitraru constant) ;*

for a proof see [11].

A fairly complete discussion of the convergence properties of the approximate solutions is done in [11] ; we summarize in the two following theorems the results which have been obtained there :

Theorem 2.4 : Suppose that Ω *is a bounded, polygonal, convex domain of* $\mathbb{R}^2$ *and that* $(\mathcal{T}_h)_h$ *obeys* (2.9) *and the statement of Lemma 2.1 ; then the following error estimates hold (if* $\underset{\sim}{g}$ *has been conveniently approximated by* $\underset{\sim}{g}_h$*), where* $\{\underset{\sim}{u}_h,\psi_h\}$ *is the solution of* (P_h) *and* p_h *the corresponding discrete pressure :*

(i) If H_h^1 *and* V_h *are defined by* (2.10), (2.12) *and if* $\underset{\sim}{u} \in (H^3(\Omega))^2$, $p \in H^2(\Omega)/\mathbb{R}$ *we have*

(2.20) $$\|\underset{\sim}{u}_h - \underset{\sim}{u}\|_{(H^1(\Omega))^2} \le Ch^2 \left(\|\underset{\sim}{u}\|_{(H^3(\Omega))^2} + \|p\|_{H^2(\Omega)/\mathbb{R}} \right) ,$$

(2.21) $$\|p_h - p\|_{H^1(\Omega)/\mathbb{R}} \le Ch \left(\|\underset{\sim}{u}\|_{(H^3(\Omega))^2} + \|p\|_{H^2(\Omega)/\mathbb{R}} \right),$$

(2.22) $$\|\psi_h\|_{H^1(\Omega)} \le Ch^2(\|\underset{\sim}{u}\|_{(H^2(\Omega))^2} + \|p\|_{H^2(\Omega)/\mathbb{R}}).$$

(ii) If H_h^1 *and* V_h *are defined by either* (2.10),(2.12) *or* (2.10),(2.15) *and if* $\underset{\sim}{u} \in (H^2(\Omega))^2$, $p \in H^1(\Omega)/\mathbb{R}$ *we have*

(2.23) $$\|\underset{\sim}{u}_h - \underset{\sim}{u}\|_{(H^1(\Omega))^2} \le Ch(\|\underset{\sim}{u}\|_{(H^2(\Omega))^2} + \|p\|_{H^1(\Omega)/\mathbb{R}}),$$

(2.24) $$\|\psi_h\|_{H^1(\Omega)} \le Ch(\|\underset{\sim}{u}\|_{(H^2(\Omega))^2} + \|p\|_{H^1(\Omega)/\mathbb{R}}).$$

In (2.20)-(2.24), C *denotes various quantities independent of* $\underset{\sim}{u}$,p *and* h.

<u>*Theorem 2.5*</u> *: If the hypotheses on* Ω *and* $(\mathcal{T}_h)_h$ *are those of Theorem 2.4, then the following error estimates hold*

(i) If H_h^1 *and* V_h *are defined by* (2.10),(2.12) *and if* $\underset{\sim}{u} \in (H^3(\Omega))^2$, $p \in H^2(\Omega)/\mathbb{R}$ *, we have*

(2.25) $$\|p_h - p\|_{L^2(\Omega)/\mathbb{R}} \le Ch^2(\|\underset{\sim}{u}\|_{(H^3(\Omega))^2} + \|p\|_{H^2(\Omega)/\mathbb{R}}),$$

(2.26) $$\|(\underset{\sim}{u}_h + \underset{\sim}{\nabla}\psi_h) - \underset{\sim}{u}\|_{(L^2(\Omega))^2} \le Ch^3(\|\underset{\sim}{u}\|_{(H^3(\Omega))^2} + \|p\|_{H^2(\Omega)/\mathbb{R}}).$$

(ii) If H_h^1 *and* V_h *are defined by either* (2.10),(2.12) *or* (2.10),(2.15) *and if* $\underset{\sim}{u} \in (H^2(\Omega))^2$, $p \in H^1(\Omega)/\mathbb{R}$, *we have*

(2.27) $$\|p_h - p\|_{L^2(\Omega)/\mathbb{R}} \le Ch(\|\underset{\sim}{u}\|_{(H^2(\Omega))^2} + \|p\|_{H^1(\Omega)/\mathbb{R}}),$$

(2.28) $$\|(\underset{\sim}{u}_h + \underset{\sim}{\nabla}\psi_h) - \underset{\sim}{u}\|_{(L^2(\Omega))^2} \le Ch^2(\|\underset{\sim}{u}\|_{(H^2(\Omega))^2} + \|p\|_{H^1(\Omega)/\mathbb{R}}).$$

In (2.25)-(2.28), C *denotes various quantities independent of* $\underset{\sim}{u}$,p *and* h.

For more information concerning the convergence, see [11] where the extension to finite element approximations of (2.1),(2.2) by quadrangular elements is also considered (in Sec. 7).

2.4. <u>A mixed finite element approximation of the steady Navier-Stokes equations</u>.
We consider now the *steady Navier-Stokes equations*

(2.29) $$-\nu\Delta\underset{\sim}{u} + (\underset{\sim}{u}\cdot\underset{\sim}{\nabla})\underset{\sim}{u} + \underset{\sim}{\nabla}p = \underset{\sim}{f} \text{ in } \Omega,$$

(2.30) $$\underset{\sim}{\nabla}\cdot\underset{\sim}{u} = 0 \text{ in } \Omega,$$

(2.31) $$\underset{\sim}{u} = \underset{\sim}{g} \text{ on } \Gamma.$$

Using the notation of Sec. 2.3 we approximate (2.29)-(2.31) by

$$(2.32)\quad \begin{cases} \textit{Find } \{\underset{\sim}{u}_h,\psi_h\} \in W_{gh} \textit{ such that } \forall\, \{\underset{\sim}{v}_h,\phi_h\} \in W_{oh} \textit{ we have} \\ \nu\displaystyle\int_\Omega \nabla\underset{\sim}{u}_h\cdot\nabla\underset{\sim}{v}_h\,dx + \int_\Omega (\underset{\sim}{u}_h\cdot\nabla)\underset{\sim}{u}_h\cdot(\underset{\sim}{v}_h+\nabla\phi_h)\,dx = \int_\Omega \underset{\sim}{f}\cdot(\underset{\sim}{v}_h+\nabla\phi_h)\,dx. \end{cases}$$

It is proved in [9] that under suitable hypotheses on $\underset{\sim}{f}$ and $\underset{\sim}{g}$ then (2.32) has a unique solution. It is then a trivial task to prove that to each solution of (2.32) we can associate a *discrete pressure* $p_h \in H_h^1$ such that

$$(2.33)\quad \int_\Omega \nabla p_h\cdot\nabla w_h\,dx + \int_\Omega (\underset{\sim}{u}_h\cdot\nabla)\underset{\sim}{u}_h\cdot\nabla w_h\,dx = \int_\Omega \underset{\sim}{f}\cdot\nabla w_h\,dx \quad \forall w_h \in H_h^1\,,$$

$$(2.34)\quad \nu\int_\Omega \nabla\underset{\sim}{u}_h\cdot\nabla\underset{\sim}{v}_h\,dx + \int_\Omega (\underset{\sim}{u}_h\cdot\nabla)\underset{\sim}{u}_h\cdot\underset{\sim}{v}_h\,dx + \int_\Omega \nabla p_h\cdot\underset{\sim}{v}_h\,dx = \int_\Omega \underset{\sim}{f}\cdot\underset{\sim}{v}_h\,dx \quad \forall \underset{\sim}{v}_h \in V_{oh}.$$

It is also proved in [9] that if some reasonable assumptions on the smoothness of $\underset{\sim}{u}$ and p are satisfied then the estimates (2.20)-(2.25) and (2.27) still hold ; it is very likely that (2.26),(2.28) also hold but proving these estimates is still an open problem.

2.5. A semi-discrete approximation of the unsteady Navier-Stokes equations.

Using suitable finite element basis for the spaces defined in Sec. 2.3.1 it is possible to reduce the *unsteady Navier-Stokes equations* to a system of ordinary differential equations. Suppose that the time dependent problem under consideration is

$$(2.35)\quad \frac{\partial \underset{\sim}{u}}{\partial t} - \nu\Delta\underset{\sim}{u} + (\underset{\sim}{u}\cdot\nabla)\underset{\sim}{u} + \nabla p = \underset{\sim}{f} \textit{ in } \Omega,$$

$$(2.36)\quad \nabla\cdot\underset{\sim}{u} = 0 \textit{ in } \Omega,$$

$$(2.37)\quad \underset{\sim}{u} = \underset{\sim}{g}\ (=\underset{\sim}{g}(x,t)) \textit{ on } \Gamma,$$

$$(2.38)\quad \underset{\sim}{u}(x,o) = \underset{\sim}{u}_o(x) \textit{ in } \Omega;$$

we approximate then (2.35)-(2.38) by

$$(2.39)\quad \begin{cases} \textit{Find } \{\underset{\sim}{u}_h(t),\psi_h(t)\} \in W_{gh}(t) \textit{ a.e. in } t\textit{, such that} \\ \displaystyle\int_\Omega \frac{\partial \underset{\sim}{u}_h}{\partial t}\cdot\underset{\sim}{v}_h\,dx + \nu\int_\Omega \nabla\underset{\sim}{u}_h\cdot\nabla\underset{\sim}{v}_h\,dx + \int_\Omega (\underset{\sim}{u}_h\cdot\nabla)\underset{\sim}{u}_h\cdot(\underset{\sim}{v}_h+\nabla\phi_h)\,dx = \int_\Omega \underset{\sim}{f}\cdot(\underset{\sim}{v}_h+\nabla\phi_h)\,dx \\ \forall\, \{\underset{\sim}{v}_h,\phi_h\} \in W_{oh}, \\ \underset{\sim}{u}_h(o) = \underset{\sim}{u}_{oh} \in V_h \textit{ given } \quad (\underset{\sim}{u}_{oh} : \textit{approximation of } \underset{\sim}{u}_o)\ ; \end{cases}$$

in (2.39), the space $W_{gh}(t)$ is defined as follows :

$$(2.40) \qquad W_{gh}(t) = \{\{\underset{\sim}{v}_h,\phi_h\} \in V_{gh}(t) \times H^1_{oh},\ \int_{\Omega^\sim} \nabla\phi_h \cdot \nabla w_h \, dx = \int_{\Omega^\sim} \nabla\cdot \underset{\sim}{v}_h w_h \, dx \quad \forall w_h \in H^1_h\}$$

$$(2.41) \qquad V_{gh}(t) = \{\underset{\sim}{v}_h \in V_h,\ \underset{\sim}{v}_h|_\Gamma = \underset{\sim}{g}_h(t)\},$$

$\underset{\sim}{g}_h(t)$ being a suitable approximation of $\underset{\sim}{g}(t)$.

The approximate problem (2.39) is not directly suitable for computation and for this purpose a convenient time discretization is required in order to obtain a fully discrete approximate problem ; several such time discretization schemes are given in the following Sec. 3.

3. - TIME DISCRETIZATION OF THE UNSTEADY NAVIER-STOKES PROBLEM.

We follow the presentation in [1] (see also [9],[16]) ; in the sequel $k = \Delta t$ is the *time discretization step*. We just consider *fully implicit schemes* since the stability condition of the semi-implicit schemes that we have tested on realistic problems seems to be a severe limitation for the existing computers (a description of such semi-implicit schemes is given in [1],[9],[16]).

3.1. An ordinary implicit scheme.

This scheme is defined as follows ($\underset{\sim}{u}^n_h$ approximates $\underset{\sim}{u}_h(nk)$)

$$(3.1) \qquad \underset{\sim}{u}^o_h = \underset{\sim}{u}_{oh} \quad (\text{see } (2.39)),$$

then for $n \geq 0$, *we obtain* $\underset{\sim}{u}^{n+1}_h$ *from* $\underset{\sim}{u}^n_h$ *by solving*

$$(3.2) \qquad \begin{cases} \displaystyle\int_\Omega \frac{\underset{\sim}{u}^{n+1}_h - \underset{\sim}{u}^n_h}{k} \cdot \underset{\sim}{v}_h \, dx + \nu \int_\Omega \nabla \underset{\sim}{u}^{n+1}_h \cdot \nabla \underset{\sim}{v}_h \, dx + \int_\Omega (\underset{\sim}{u}^{n+1}_h \cdot \nabla) \underset{\sim}{u}^{n+1}_h \cdot (\underset{\sim}{v}_h + \nabla\phi_h) \, dx \\ = \displaystyle\int_\Omega \underset{\sim}{f}^{n+1} \cdot (\underset{\sim}{v}_h + \nabla\phi_h) \, dx \quad \forall \{\underset{\sim}{v}_h,\phi_h\} \in W_{oh}, \quad \{\underset{\sim}{u}^{n+1}_h, \psi^{n+1}_h\} \in W^{n+1}_{gh} \end{cases}$$

where (see (2.40),(2.41)) $W^m_{gh} = W_{gh}(mk)$.

To obtain $\underset{\sim}{u}^{n+1}_h$ from $\underset{\sim}{u}^n_h$ in (3.2) we have to solve a finite dimensional nonlinear problem very close to the steady discrete Navier-Stokes problem (2.32). The above scheme has a time truncation error in $O(\Delta t)$ and appears to be *unconditionally stable*.

3.2. A Crank-Nicholson implicit scheme.

This scheme is as follows

$$(3.3) \qquad \underset{\sim}{u}^o_h = \underset{\sim}{u}_{oh},$$

then for $n \geq 1$ *we obtain* $\underset{\sim}{u}_h^{n+1}$ *from* $\underset{\sim}{u}_h^n$ *by solving*

$$(3.4)\quad \begin{cases} \displaystyle\int_\Omega \frac{\underset{\sim}{u}_h^{n+1}-\underset{\sim}{u}_h^n}{k}\cdot \underset{\sim}{v}_h dx + \nu\int_\Omega \nabla\underset{\sim}{u}_h^{n+1/2}\cdot\nabla\underset{\sim}{v}_h\, dx + \int_\Omega (\underset{\sim}{u}_h^{n+1/2}\cdot\underset{\sim}{\nabla})\underset{\sim}{u}^{n+1/2}\cdot(\underset{\sim}{v}_h+\underset{\sim}{\nabla}\phi_h)dx = \\ = \displaystyle\int_\Omega \underset{\sim}{f}^{n+\frac{1}{2}}\cdot(\underset{\sim}{v}_h+\underset{\sim}{\nabla}\phi_h)dx \quad \forall\ \{\underset{\sim}{v}_h,\phi_h\}\in W_{oh},\ \{\underset{\sim}{u}_h^{n+1/2},\psi_h^{n+1/2}\}\in W_{gh}^{n+1/2}\ \ (=W_{gh}((n+\tfrac{1}{2})k)), \end{cases}$$

where in (3.4)

$$(3.5)\qquad \underset{\sim}{u}_h^{n+1/2} = \frac{1}{2}(\underset{\sim}{u}_h^{n+1}+\underset{\sim}{u}_h^n).$$

Since $\underset{\sim}{u}_h^{n+1} = \underset{\sim}{u}_h^n + 2(\underset{\sim}{u}_h^{n+1/2}-\underset{\sim}{u}_h^n)$ we can eliminate $\underset{\sim}{u}_h^{n+1}$ in (3.4) and therefore reduces also this problem to a variant of the discrete Navier-Stokes problem (2.32). The above scheme has a time truncation error in $O(|\Delta t|^2)$ and appears to be *unconditionally stable*.

3.3. A two-step implicit scheme.

The scheme is defined by

$$(3.6)\qquad \underset{\sim}{u}_h^o,\underset{\sim}{u}_h^1\ \textit{given},$$

then for $n \geq 1$, *we obtain* $\underset{\sim}{u}_h^{n+1}$ *from* $\underset{\sim}{u}_h^n,\underset{\sim}{u}_h^{n-1}$ *by solving*

$$(3.7)\quad \begin{cases} \displaystyle\int_\Omega \frac{3\underset{\sim}{u}_h^{n+1}-4\underset{\sim}{u}_h^n+\underset{\sim}{u}_h^{n-1}}{2k}\cdot \underset{\sim}{v}_h dx + \nu\int_\Omega \nabla\underset{\sim}{u}_h^{n+1}\cdot\nabla\underset{\sim}{v}_h\, dx + \int_\Omega (\underset{\sim}{u}_h^{n+1}\cdot\underset{\sim}{\nabla})\underset{\sim}{u}_h^{n+1}\cdot(\underset{\sim}{v}_h+\underset{\sim}{\nabla}\phi_h)dx = \\ = \displaystyle\int_\Omega \underset{\sim}{f}^{n+1}\cdot(\underset{\sim}{v}_h+\underset{\sim}{\nabla}\phi_h)dx \quad \forall\ \{\underset{\sim}{v}_h,\phi_h\}\in W_{oh},\ \{\underset{\sim}{u}_h^{n+1},\psi_h^{n+1}\}\in W_{gh}^{n+1}. \end{cases}$$

To obtain u_h^1 from u_h^o we may use either one of the two schemes discussed in Sec. 3.1, 3.2 or one of the semi-implicit schemes described in [1] ; scheme (3.6),(3.7) appears to be *unconditionally stable* and his truncation error is in $O(|\Delta t|^2)$.

4. - LEAST SQUARE FORMULATION AND ITERATIVE SOLUTION BY CONJUGATE GRADIENT WITH SCALING.

4.1. Generalities.

The various discrete Navier-Stokes problems of Secs.2,3 lead to the solution of nonlinear problems in finite dimension which may be viewed as obtained by space discretization of the following family of nonlinear problems (with $\alpha \geq 0$)

$$(4.1)\quad \begin{cases} \alpha\underset{\sim}{u}-\nu\Delta\underset{\sim}{u}+(\underset{\sim}{u}\cdot\underset{\sim}{\nabla})\underset{\sim}{u}+\underset{\sim}{\nabla}p = \underset{\sim}{f}\ \textit{in}\ \Omega, \\ \underset{\sim}{\nabla}\cdot\underset{\sim}{u} = 0\ \textit{in}\ \Omega, \\ \underset{\sim}{u}|_\Gamma = \underset{\sim}{g}\ \textit{with}\ \displaystyle\int_\Gamma \underset{\sim}{g}\cdot\underset{\sim}{n}\, d\Gamma = 0. \end{cases}$$

whose *variational formulation* is

$$\alpha\int_\Omega \underset{\sim}{u}\cdot\underset{\sim}{v}\, dx + \nu\int_\Omega \nabla\underset{\sim}{u}\cdot\nabla\underset{\sim}{v}\, dx + \int_\Omega \underset{\sim}{v}\cdot(\underset{\sim}{u}\cdot\nabla)\underset{\sim}{u}\, dx = \int_\Omega \underset{\sim}{f}\cdot\underset{\sim}{v}\, dx \quad \forall \underset{\sim}{v}\in V_o,\ \underset{\sim}{u}\in V_g \tag{4.2}$$

(see (2.3) for the definition of V_g).

Following [1],[2], [16] we shall describe in Sec. 4.2 a *least square formulation* of (4.1),(4.2) and in Sec. 4.3 a *conjugate gradient algorithm* for solving the least square problem. The discrete variants of those methods described in Sec. 4.2, 4.3 (making use of the finite element approximation of Sec. 2) are too much complicated to be described in this short paper ; we therefore refer to [1],[2],[16] for a complete description of these methods.

4.2. A least square formulation of (4.1),(4.2).

A convenient *least square formulation of* (4.1),(4.2) is

$$\underset{\underset{\sim}{v}\in V_g}{\text{Min}}\ J(\underset{\sim}{v}) \tag{4.3}$$

with

$$J(\underset{\sim}{v}) = \frac{\alpha}{2}\int_\Omega |\underset{\sim}{v}-\underset{\sim}{\xi}|^2 dx + \frac{\nu}{2}\int_\Omega |\nabla(\underset{\sim}{v}-\underset{\sim}{\xi})|^2 dx \tag{4.4}$$

where $\underset{\sim}{\xi}$ is a function of $\underset{\sim}{v}$ through the *state equation*

$$\begin{cases} \alpha\underset{\sim}{\xi}-\nu\Delta\underset{\sim}{\xi}+\nabla\pi = \underset{\sim}{f}-(\underset{\sim}{v}\cdot\nabla)\underset{\sim}{v} \ \textit{in}\ \Omega, \\ \nabla\cdot\underset{\sim}{\xi} = 0 \ \textit{in}\ \Omega, \\ \underset{\sim}{\xi} = \underset{\sim}{g} \ \textit{on}\ \Gamma. \end{cases} \tag{4.5}$$

Remark 4.1 : Solving the state equation (4.5) is a Stokes problem.

4.3. Conjugate gradient solution of (4.3)-(4.5).

4.3.1. Description of the algorithm.

A *conjugate gradient algorithm* for solving the least square problem (4.3)-(4.5) is

$$\underset{\sim}{u}^o \in V_g \ \textit{given}\ , \tag{4.6}$$

then compute $\underset{\sim}{z}^o$ *as the solution of the variational equation*

$$\begin{cases} \alpha\int_\Omega \underset{\sim}{z}^o\cdot\underset{\sim}{\eta}\, dx + \nu\int_\Omega \nabla\underset{\sim}{z}^o\cdot\nabla\underset{\sim}{\eta}\, dx = \langle J'(\underset{\sim}{u}^o),\underset{\sim}{\eta}\rangle \\ \forall\ \underset{\sim}{\eta}\in V_o,\ \underset{\sim}{z}^o\in V_o \end{cases} \tag{4.7}$$

and set

(4.8) $\qquad w^o = z^o.$

For $m \geq 0$, *assuming* u^m, z^m, w^m *known, compute* $u^{m+1}, z^{m+1}, w^{m+1}$ *by*

(4.9) $$\lambda^m = \arg \min_{\lambda \geq 0} J(u^m - \lambda w^m),$$

(4.10) $$u^{m+1} = u^m - \lambda^m w^m,$$

define then z^{m+1} *as the solution of*

(4.11) $$\begin{cases} \alpha \int_\Omega z^{m+1} \cdot \eta \, dx + \nu \int_\Omega \nabla z^{m+1} \cdot \nabla \eta \, dx = \langle J'(u^{m+1}), \eta \rangle \quad \forall \eta \in V_o, \\ z^{m+1} \in V_o , \end{cases}$$

then

(4.12) $$\gamma^{m+1} = \frac{\alpha \int_\Omega z^{m+1} \cdot (z^{m+1} - z^m) dx + \nu \int_\Omega \nabla z^{m+1} \cdot \nabla (z^{m+1} - z^m) \, dx}{\alpha \int_\Omega |z^m|^2 dx + \nu \int_\Omega |\nabla z^m|^2 \, dx},$$

and finally

(4.13) $$w^{m+1} = z^{m+1} + \gamma^{m+1} w^m ,$$

$m = m+1$, *go to* (4.9).

Remark 4.2 : The above algorithm uses the *Polak-Ribière* strategy to compute γ^{m+1} (see POLAK [23]).

Remark 4.3 : To obtain z^{m+1} via (4.11) we also have to solve a Stokes problem (written here in variational form).

4.3.2. Calculation of J' and z^{m+1}.

A most important step in order to use algorithm (4.6)-(4.13) is the calculation of $J'(u^{m+1})$; owing to the importance of this step we shall detail this calculation ; we have

(4.14) $$\delta J = \langle J'(v), \delta v \rangle = \alpha \int_\Omega (v - \xi) \cdot \delta (v - \xi) dx + \nu \int_\Omega \nabla (v - \xi) \cdot \nabla \delta (v - \xi) dx,$$

where $\delta\xi$ is from (4.5) the solution of

(4.15) $$\begin{cases} \alpha \int_\Omega \delta\xi \cdot \eta \, dx + \nu \int_\Omega \nabla \delta\xi \cdot \nabla \eta \, dx = - \int_\Omega \eta \cdot (\delta v \cdot \nabla) v \, dx - \int_\Omega \eta \cdot (v \cdot \nabla) \delta v \, dx \\ \forall \eta \in V_o, \ \delta\xi \in V_o . \end{cases}$$

Since $(\underset{\sim}{v}-\underset{\sim}{\xi})$ V_o in (4.14), it follows from (4.14), (4.15) that

$$(4.16)\quad \begin{cases} \langle J'(\underset{\sim}{v}),\underset{\sim}{\eta}\rangle = \alpha\int_\Omega (\underset{\sim}{v}-\underset{\sim}{\xi})\cdot\underset{\sim}{\eta}\, dx + \nu\int_\Omega \underset{\sim}{\nabla}(\underset{\sim}{v}-\underset{\sim}{\xi})\cdot\underset{\sim\sim}{\nabla\eta}\, dx + \int_\Omega (\underset{\sim}{v}-\underset{\sim}{\xi})\cdot(\underset{\sim}{v}\cdot\underset{\sim}{\nabla})\underset{\sim}{\eta}\, dx\, + \\ + \int_\Omega (\underset{\sim}{v}-\underset{\sim}{\xi})\cdot(\underset{\sim}{\eta}\cdot\underset{\sim}{\nabla})\underset{\sim}{v}\, dx \quad \forall\, \underset{\sim}{\eta}\in V_o\ . \end{cases}$$

It follows in turn from (4.16) that to compute $\underset{\sim}{z}^{m+1}$ from $\underset{\sim}{u}^{m+1}$ we solve (4.5) with $\underset{\sim}{v} = \underset{\sim}{u}^{m+1}$ which gives $\underset{\sim}{\xi}^{m+1}$; then from (4.16) we have $\forall \underset{\sim}{\eta}\in V_o$

$$(4.17)\quad \begin{cases} \langle J'(\underset{\sim}{u}^{m+1}),\underset{\sim}{\eta}\rangle = \alpha\int_\Omega (\underset{\sim}{u}^{m+1}-\underset{\sim}{\xi}^{m+1})\cdot\underset{\sim}{\eta}\, dx + \nu\int_\Omega \underset{\sim}{\nabla}(\underset{\sim}{u}^{m+1}-\underset{\sim}{\xi}^{m+1})\cdot\underset{\sim\sim}{\nabla\eta}\, dx\, + \\ + \int_\Omega (\underset{\sim}{u}^{m+1}-\underset{\sim}{\xi}^{m+1})\cdot(\underset{\sim}{u}^{m+1}\cdot\underset{\sim}{\nabla})\underset{\sim}{\eta}\, dx + \int_\Omega (\underset{\sim}{u}^{m+1}-\underset{\sim}{\xi}^{m+1})\cdot(\underset{\sim}{\eta}\cdot\underset{\sim}{\nabla})\underset{\sim}{u}^{m+1}\, dx\ . \end{cases}$$

Finally we obtain $\underset{\sim}{z}^{m+1}$ from (4.11), (4.17).

In conclusion at *each iteration* we have to solve *several Stokes problems*, namely :

- The Stokes problem (4.5) with $\underset{\sim}{v} = \underset{\sim}{u}^{m+1}$ to obtain ξ^{m+1} from $\underset{\sim}{u}^{m+1}$.
- The Stokes problem (4.11) to obtain $\underset{\sim}{z}^{m+1}$ from $\underset{\sim}{u}^{m+1}, \underset{\sim}{\xi}^{m+1}$ via (4.17).
- Moreover the solution of the one-dimensional problem (4.9) may require several evaluations of J i.e. several solutions of the state equation (4.5) (which is a Stokes problem).

Thus an efficient *"Stokes solver"* will be an important tool to solve the Navier-Stokes equations by algorithm (4.6)-(4.13) through the least square formulation (4.3)-(4.5). The description of such a solver is given in Sec. 5.

5. - A STOKES SOLVER.

5.1. The continuous case.

5.1.1. Synopsis.

It follows from Sec. 4 that an efficient solver for the Stokes' problem

$$(5.1)\quad \begin{cases} \alpha\underset{\sim}{u} - \nu\Delta\underset{\sim}{u}+\underset{\sim}{\nabla} p = \underset{\sim}{f} \text{ in } \Omega, \\ \underset{\sim}{\nabla}\cdot\underset{\sim}{u} = 0 \text{ in } \Omega, \\ \underset{\sim}{u} = \underset{\sim}{g} \text{ on } \Gamma, \end{cases}$$

will be very helpful for the numerical solution of the steady and unsteady Navier-Stokes equations. In (5.1) $\alpha=0$ for the steady case, $\alpha>0$ for the unsteady case. We shall describe a method for solving (5.1). This method reduces the solution of the Stokes' problem to the solution of a *finite number of Dirichlet's problems* and to the solution of a *boundary integral equation*. The finite element implementation of this method will be discussed in Sec. 5.2 (see also [1],[9],[16],[17]). We advise the readers mostly interested by applications (or afraid by some Functional Analysis) to skip the following of Sec. 5.1 and to go directly to Sec. 5.2.

5.1.2. Principles of the method.

The following holds in $\mathbb{R}^N$, $N \geq 2$; let us define

$$\mathcal{H}^{1/2}(\Gamma) = \{\mu \in H^{1/2}(\Gamma), \int_\Gamma \mu \, d\Gamma = 0\}.$$

For the definition and properties of the Sobolev spaces $H^s(\Gamma)$, $s \in \mathbb{R}$, see [18]-[21]. The decomposition properties of the Stokes' problem follows directly from

Theorem 5.1 : Let $\lambda \in H^{-1/2}(\Gamma)$ *; let* $A : H^{-1/2}(\Gamma) \to H^{1/2}(\Gamma)$ *be defined implicitly by the following chain of (Poisson) problems*

$$(5.2) \qquad \begin{cases} \Delta p_\lambda = 0 \text{ in } \Omega, \\ p_\lambda = \lambda \text{ on } \Gamma, \end{cases}$$

$$(5.3) \qquad \begin{cases} \alpha \underset{\sim}{u}_\lambda - \nu \Delta \underset{\sim}{u}_\lambda = -\underset{\sim}{\nabla} p_\lambda \text{ in } \Omega, \\ \underset{\sim}{u}_\lambda = 0 \text{ on } \Gamma, \end{cases}$$

$$(5.4) \qquad \begin{cases} -\Delta \psi_\lambda = \underset{\sim}{\nabla} \cdot \underset{\sim}{u}_\lambda \text{ in } \Omega, \\ \psi_\lambda = 0 \text{ on } \Gamma, \end{cases}$$

$$(5.5) \qquad A\lambda = -\frac{\partial \psi_\lambda}{\partial n}\Big|_\Gamma .$$

Then A *is an isomorphism from* $H^{-1/2}(\Gamma)/\mathbb{R}$ *onto* $\mathcal{H}^{1/2}(\Gamma)$. *Moreover the bilinear form* $a(\cdot,\cdot)$ *defined by*

$$(5.6) \qquad a(\lambda,\mu) = \langle A\lambda, \mu \rangle$$

(where $\langle \cdot,\cdot \rangle$ *denotes the duality pairing between* $H^{1/2}(\Gamma)$ *and* $H^{-1/2}(\Gamma)$*) is continuous, symmetric and* $H^{-1/2}(\Gamma)/\mathbb{R}$*-elliptic.*

A proof of this theorem will be given in [24] (a variant of it concerning the *biharmonic problem* is available in [25]).

Application of Theorem 5.1 to the solution of the Stokes' problem (5.1).

Let define $p_o, \underset{\sim}{u}_o, \psi_o$ solutions of, respectively,

$$(5.7)\qquad \begin{cases} \Delta p_o = \nabla\cdot f \; in\ \Omega, \\ p_o = 0 \; on\ \Gamma, \end{cases}$$

$$(5.8)\qquad \begin{cases} \alpha u_o - \nu\Delta u_o = f - \nabla p_o \; in\ \Omega, \\ u_o = g \; on\ \Gamma, \end{cases}$$

$$(5.9)\qquad \begin{cases} -\Delta\psi_o = \nabla\cdot u_o \; in\ \Omega, \\ \psi_o = 0 \; on\ \Gamma. \end{cases}$$

It is then fairly easy to prove the following

Theorem 5.2 : Let $\{u,p\}$ *be a solution of the Stokes' problem (5.1). The trace* $\lambda = p|_\Gamma$ *is the unique solution of the linear variational equation*

$$(E)\qquad \begin{cases} \lambda \in H^{-1/2}(\Gamma)/\mathbb{R}\ , \\ \langle A\lambda,\mu\rangle = \langle \frac{\partial\psi_o}{\partial n},\mu\rangle \quad \forall\mu \in H^{-1/2}(\Gamma)/\mathbb{R}\ . \ \blacksquare \end{cases}$$

It follows from Theorem 5.2 that the solution of the Stokes' problem has been reduced to 2N+3 *Dirichlet problems* (N+2 to obtain ψ_o, N+1 to obtain $\{u,p\}$ once λ is known) and to the solution of (E) which is a kind of *boundary integral equation*. The main difficulty in this approach is the fact that the (pseudo-differential) operator A is not explicitely known ; in fact the mixed finite element approximation of Sec. 2 has been precisely introduced (in [26],[27]) to overcome this difficulty, extending to the Stokes' problem an idea discussed in [25] for the biharmonic problem.

5.2. The discrete case.

5.2.1. Introduction. Formulation of the basic problem.

We clearly have from the above sections that efficient solvers for the discrete Stokes' problem

$$(5.10)\qquad \begin{cases} \textit{Find}\ \{u_h,\psi_h\} \in W_{gh}\ \textit{such that}\ \forall\{v_h,\phi_h\}\in W_{oh}\ \textit{we have} \\ \alpha\int_\Omega u_h\cdot v_h\,dx + \nu\int_\Omega \nabla u_h\cdot\nabla v_h\,dx = \int_\Omega f_{oh}\cdot v_h\,dx + \int_\Omega f_{1h}\cdot(v_h+\nabla\phi_h)\,dx \end{cases}$$

may be very helpful for solving the discrete *steady and unsteady Navier-Stokes problems* by the methods of the above sections. Actually the content of this Sec. 5.2 is the discrete analogue of Sec. 5.1 and follows closely [27],[17],[1],[16].

5.2.2. The space M_h.

Using the notation of Sec. 2, let us introduce M_h as a *complementary subspace* of H^1_{oh} in H^1_h i.e.

$$H^1_h = H^1_{oh} \oplus M_h \; ;$$

we set $N_h = \dim(M_h)$.

Various M_h may be used but in the following we shall only consider the most convenient which is defined by

$$(5.11) \quad \begin{cases} H^1_h = H^1_{oh} \oplus M_h, \; \mu_h \in M_h \Longrightarrow \mu_h|_K = 0 \quad \forall\, K \in \mathcal{T}_h \\ \textit{such that } \Gamma \cap \partial K = \emptyset \; . \end{cases}$$

The above space M_h is *uniquely* defined by (5.11) and we observe that if $\mu_h \in M_h$ then μ_h vanishes *outside* a neighborhood of Γ whose measure $\to 0$ with h. Therefore M_h can "almost" be considered as a *boundary space* and in fact it will play the role played by $H^{-1/2}(\Gamma)$ in Sec. 5.1. We observe also that $\mu_h \in M_h$ is completely determined by its *trace* on Γ, in fact by the values it takes at the *boundary nodes*.

5.2.3. Equivalence of (5.10) with a variational problem in M_h.

Let $\lambda_h \in M_h$; from λ_h we define $p_h, \underset{\sim}{u}_h, \psi_h$ as the solution of the chain of discrete variational problems

$$(5.12) \quad \begin{cases} \textit{Find } p_h \in H^1_h \textit{ such that } p_h - \lambda_h \in H^1_{oh} \textit{ and} \\ \int_\Omega \underset{\sim}{\nabla} p_h \cdot \underset{\sim}{\nabla} q_h \, dx = 0 \quad \forall q_h \in H^1_{oh} \; , \end{cases}$$

$$(5.13) \quad \begin{cases} \textit{Find } \underset{\sim}{u}_h \in V_{oh} \textit{ such that} \\ \alpha \int_\Omega \underset{\sim}{u}_h \cdot \underset{\sim}{v}_h \, dx + \nu \int_\Omega \underset{\sim}{\nabla} \underset{\sim}{u}_h \cdot \underset{\sim}{\nabla} \underset{\sim}{v}_h \, dx = - \int_\Omega \underset{\sim}{\nabla} p_h \cdot \underset{\sim}{v}_h \, dx \quad \forall \underset{\sim}{v}_h \in V_{oh}, \end{cases}$$

$$(5.14) \quad \begin{cases} \textit{Find } \psi_h \in H^1_{oh} \textit{ such that} \\ \int_\Omega \underset{\sim}{\nabla} \psi_h \cdot \underset{\sim}{\nabla} \phi_h \, dx = \int_\Omega \underset{\sim}{\nabla} \cdot \underset{\sim}{u}_h \, \phi_h \, dx \quad \forall \phi_h \in H^1_{oh} \; . \end{cases}$$

From (5.12)-(5.14) we define then $a_h : M_h \times M_h \to \mathbb{R}$ by

$$(5.15) \quad a_h(\lambda_h, \mu_h) = - \int_\Omega (\underset{\sim}{\nabla} \psi_h + \underset{\sim}{u}_h) \cdot \underset{\sim}{\nabla} \mu_h \, dx.$$

The discrete analogue of Theorem 5.1 (proved in [24]) is :

Theorem 5.3 : Suppose that Lemma 2.1 holds (see Sec. 2.3.3) then $a_h(\cdot,\cdot)$ *is bilinear symmetric and positive definite over* $(M_h/\mathbb{R}_h)\times(M_h/\mathbb{R}_h)$ *where*

$$\mathbb{R}_h = \{\mu_h \in M_h,\ \mu_h = \text{const. on } \partial\Omega\} \quad . \blacksquare$$

Back to Sec. 5.2.1., it is easily proved that (5.10) *has a unique solution* $\{\underset{\sim}{u}_h,\psi_h\}$ *characterized by the existence of* $p_h \in H_h^1$ *such that*

$$(5.16)\qquad \begin{cases} \displaystyle\int_\Omega \underset{\sim}{\nabla} p_h \cdot \underset{\sim}{\nabla}\phi_h \, dx = \int_\Omega \underset{\sim}{f}_{1h} \cdot \underset{\sim}{\nabla}\phi_h dx \quad \forall\ \phi_h \in H_{oh}^1, \\ p_h \in H_h^1 , \end{cases}$$

$$(5.17)\qquad \begin{cases} \displaystyle\alpha\int_\Omega \underset{\sim}{u}_h \cdot \underset{\sim}{v}_h \, dx + \nu\int_\Omega \underset{\sim}{\nabla} \underset{\sim}{u}_h \cdot \underset{\sim}{\nabla} \underset{\sim}{v}_h dx + \int_\Omega \underset{\sim}{\nabla} p_h \cdot \underset{\sim}{v}_h \, dx = \int_\Omega (\underset{\sim}{f}_{oh} + \underset{\sim}{f}_{1h})\cdot \underset{\sim}{v}_h \, dx \\ \forall \underset{\sim}{v}_h \in V_{oh},\ \underset{\sim}{u}_h \in V_{gh}, \end{cases}$$

$$(5.18)\qquad \{\underset{\sim}{u}_h,\psi_h\} \in W_{gh}.$$

The key result concerning Sec. 5.2 is the following Theorem 5.4 which relates the data $\underset{\sim}{g}_h (= \underset{\sim}{u}_h|_\Gamma)$, $\underset{\sim}{f}_{oh}$, $\underset{\sim}{f}_{1h}$ to the *trace of the discrete pressure* p_h :

Theorem 5.4 : Let p_h *be the above discrete pressure and* λ_h *the component of* p_h *in* M_h. *If the statement of Theorem 5.3 holds, then* λ_h *is the unique solution in* $M_h/\mathbb{R}_h$ *of the following linear variational problem (discrete analogue of* (E) *in* Sec. 5.1.2)

$$(E_h)\qquad \begin{cases} \lambda_h \in M_h/\mathbb{R}_h , \\ \displaystyle a_h(\lambda_h,\mu_h) = \int_\Omega (\underset{\sim}{\nabla}\psi_{oh} + \underset{\sim}{u}_{oh})\cdot \underset{\sim}{\nabla}\mu_h \, dx \quad \forall\ \mu_h \in M_h/\mathbb{R}_h , \end{cases}$$

where p_{oh}, $\underset{\sim}{u}_{oh}$, ψ_{oh} *are respectively solutions of*

$$(5.19)\qquad \begin{cases} \displaystyle\int_\Omega \underset{\sim}{\nabla} p_{oh} \cdot \underset{\sim}{\nabla} q_h dx = \int_\Omega \underset{\sim}{f}_{1h} \cdot \underset{\sim}{\nabla} q_h dx \quad \forall q_h \in H_{oh}^1 , \\ p_{oh} \in H_{oh}^1 , \end{cases}$$

$$(5.20)\qquad \begin{cases} \displaystyle\alpha\int_\Omega \underset{\sim}{u}_{oh} \cdot \underset{\sim}{v}_h dx + \nu\int_\Omega \underset{\sim}{\nabla} \underset{\sim}{u}_{oh} \cdot \underset{\sim}{\nabla} \underset{\sim}{v}_h \, dx = \int_\Omega (\underset{\sim}{f}_{oh} + \underset{\sim}{f}_{1h} - \underset{\sim}{\nabla} p_{oh})\cdot \underset{\sim}{v}_h \, dx \quad \forall\ \underset{\sim}{v}_h \in V_{oh}, \\ \underset{\sim}{u}_{oh} \in V_{gh}, \end{cases}$$

$$(5.21)\qquad \begin{cases} \displaystyle\int_\Omega \underset{\sim}{\nabla}\psi_{oh} \cdot \underset{\sim}{\nabla}\phi_h dx = \int_\Omega \underset{\sim}{\nabla}\cdot \underset{\sim}{u}_{oh} \underset{\sim}{\nabla}\phi_h dx \quad \forall\phi_h \in H_{oh}^1 , \\ \psi_{oh} \in H_{oh}^1 \ . \blacksquare \end{cases}$$

5.2.4. Solution of (E_h) by a direct method.

We follow [1], [16], [17], [27] (and in fact [25]).

5.2.4.1. Construction of a linear system equivalent to (E_h).

Generalities : The space M_h is defined by (5.11) ; let $B_h = \{w_i\}_{i=1}^{N_h}$ be a basis of M_h ; then $\forall\ \mu_h \in M_h$

$$\mu_h = \sum_{i=1}^{N_h} \mu_i w_i \,, \tag{5.22}$$

and from now we shall write

$$\underset{\sim}{r}_h \mu_h = \{\mu_1, \dots \mu_{N_h}\} \in \mathbb{R}^{N_h} \,. \tag{5.23}$$

In practice B_h is defined by

$$B_h = \{w_i\}_{i=1}^{N_h} \tag{5.24}$$

and

$$\begin{cases} \forall i=1,\dots N_h, \\ w_i(P_i) = 1, \\ w_i(Q) = 0 \quad \forall\ Q \text{ vertex of } \mathcal{C}_h,\ Q \neq P_i \end{cases} \tag{5.25}$$

where we assumed implicitly (but in practice it is not necessary) *that the boundary nodes are numbered first.*

With this choice for B_h, $\mu_i = \mu_h(P_i)$ in (5.22),(5.23).

Then (E_h) is equivalent to the *linear system*

$$\begin{cases} \displaystyle\sum_{i=1}^{N_h} a_h(w_j, w_i)\lambda_j = \int_\Omega (\underset{\sim}{\nabla}\psi_{oh} + \underset{\sim}{u}_{oh}) \cdot \underset{\sim}{\nabla} w_i \, dx, \\ 1 \leq i \leq N_h. \end{cases} \tag{5.26}$$

Let $a_{ij} = a_h(w_j, w_i)$, $\underset{\sim}{A}_h = (a_{ij})_{1 \leq i,j \leq N_h}$, $b_i = \int_\Omega (\underset{\sim}{\nabla}\psi_{oh} + \underset{\sim}{u}_{oh}) \cdot \nabla w_i \, dx$, $\underset{\sim}{b}_h = \{b_i\}_{i=1}^{N_h}$.

The matrix $\underset{\sim}{A}_h$ is *full* and *symmetric, positive semi-definite*. If (2.19) holds, then *zero* is a *single eigenvalue* of $\underset{\sim}{A}_h$;furthermore if B_h is defined by (2.54),(2.55) then

$$\operatorname{Ker}(\underset{\sim}{A}_h) = \{\underset{\sim}{y} \in \mathbb{R}^{N_h},\ y_1 = y_2 = \dots y_{N_h}\} \,. \tag{5.27}$$

Construction of $\underset{\sim}{b}_h$: To compute the right-hand side of (5.26) it is necessary to solve the 4 (5 if $\Omega \subset \mathbb{R}^3$) approximate Dirichlet problems (5.19)-(5.21). On account of the choice (5.24),(5.25) for B_h, the integrals in the right hand side of (5.26), involve functions whose supports are in the neighborhood of Γ only.

Construction of $\underset{\sim}{A}_h$: the matrix $\underset{\sim}{A}_h$ is constructed column by column according to the relation $a_{ij} = a_h(w_j,w_i)$. To compute the jth column of $\underset{\sim}{A}_h$ we solve (5.12)-(5.14) with $\lambda_h = w_j$ and compute a_{ij} from (5.15). Thus 4 discrete Dirichlet problems must be solved for each column (5 in $\mathbb{R}^3$). The matrix $\underset{\sim}{A}_h$ being *symmetric* one may restrict i to be greater or equal to j. Incidently the above observation about the choice of M_h and B_h still holds, thus the integrals occuring in the computation of a_{ij} are not very costly.

5.2.4.2. Solution of (E_h) by the Cholesky's method.

Suppose that (2.19) and (5.23),(5.24) hold. Then the sub-matrix $\tilde{\underset{\sim}{A}}_h = (a_{ij})_{1 \leq i,j \leq N_h - 1}$ is *symmetric* and *positive definite*. We may therefore proceed as follows :

Take $\lambda_{N_h} = 0$ *and solve*

$$(5.28) \qquad \tilde{\underset{\sim}{A}}_h \tilde{\underset{\sim}{r}}_h \lambda_h = \tilde{\underset{\sim}{b}}_h$$

(*where* $\tilde{\underset{\sim}{r}}_h \lambda_h = \{\lambda_1,\ldots\lambda_{N_h-1}\}$, $\tilde{\underset{\sim}{b}}_h = \{b_1\ldots b_{N_h-1}\}$) *by the Cholesky method via a factorization*

$$(5.29) \qquad \tilde{\underset{\sim}{A}}_h = \tilde{\underset{\sim}{L}}_h \tilde{\underset{\sim}{L}}_h^t \quad (\text{or } \tilde{\underset{\sim}{A}}_h = \tilde{\underset{\sim}{L}}_h \tilde{\underset{\sim}{D}}_h \tilde{\underset{\sim}{L}}^t)$$

with $\tilde{\underset{\sim}{L}}_h$ *lower triangular non singular (and* $\tilde{\underset{\sim}{D}}_h$ *diagonal).*

Let us review now the subproblems arising from the computation of $\{\underset{\sim}{u}_h,p_h\}$ via $(E_h)_h$ if the Cholesky method is used :

- The 4 approximate Dirichlet problems (5.19)-(5.21) to compute $p_{oh},\underset{\sim}{u}_{oh},\psi_{oh}$ and $\tilde{b}_h$ (5 if $\Omega \subset \mathbb{R}^3$),
- $4(N_h-1)$ approximate Dirichlet problems to construct $\tilde{A}_h$ ($5(N_h-1)$ if $\Omega \subset \mathbb{R}^3$),
- 2 triangular systems to compute λ_h : $\tilde{\underset{\sim}{L}}_h \tilde{\underset{\sim}{y}}_h = \tilde{\underset{\sim}{b}}_h$, $\tilde{\underset{\sim}{L}}_h^t \tilde{\underset{\sim}{r}}_h \lambda_h = \tilde{\underset{\sim}{y}}_h$,
- 3 approximate Dirichlet problems to obtain p_h and $\underset{\sim}{u}_h$ from λ_h (4 if $\Omega \subset \mathbb{R}^3$).

In practice the matrices of the approximate Dirichlet problems should be factorized once and for all. There are two symmetric positive definite matrices, one to approximate $-\Delta$ using *affine* elements, one to approximate $\alpha I - \nu\Delta$ using *quadratic* elements (or *affine* on $\tilde{\mathcal{C}}_h$ if (2.15) is used) ; an alternative is to use very efficient iterative solvers to solve these Dirichlet problems, like for example, those described in [28].

Remark 5.1 : One may find in [17],[24],[27] a conjugate gradient method for solving (E_h). This method *does not require the knowledge of* $\underset{\sim}{A}_h$ but 4 approximate Dirichlet problems have to be solved at each iteration (5 if $\Omega \subset \mathbf{R}^3$).

6. - NUMERICAL EXPERIMENTS.

We shall present in this section the results of some numerical experiments obtained using the methods of the above sections. Further numerical results obtained using the same methods may be found in [1], [9], [16].

We shall describe in Sec. 6.1 the results related to flows in a *channel* whose section presents a sudden enlargement due to a *step* ; in Sec. 6.2 the numerical results will be related to a flow, at Re = 250, *around* and *inside* an idealized *nozzle* at high incidence.

6.1. Flows in a channel with a step.

Following [16] we consider the solution of the Navier-Stokes equations for the flows of incompressible viscous fluids in the *channel with a step* of Fig. 6.1.

In order to compare our results with those of A.G. HUTTON [29] we have considered flows at Re = 100 and 191 ; the computational domain and the boundary conditions are also those of [29], i.e. $\underset{\sim}{u} = 0$ on the channel walls and *Poiseuille flow* upstream and downstream. We have used the *space discretization* associated to (2.10), (2.15) in Sec. 2, i.e. $\underset{\sim}{u}$ (resp. p) *piecewise linear* on $\tilde{\mathcal{T}}_h$ (resp. $\mathcal{T}_h$) ; both triangulations $\mathcal{T}_h$ and $\tilde{\mathcal{T}}_h$ are shown on Fig. 6.1 on which we have also indicated the number of *nodes, finite elements,* and *non zero elements* in the *Cholesky factors* of the discrete analogue of $-\Delta$ (resp. $\frac{I}{k}-\nu\Delta$) associated to $\mathcal{T}_h$ (resp. $\tilde{\mathcal{T}}_h$) ; as we can see from these numbers we are really dealing with big matrices ; the second of these matrices would have been bigger if one have used on $\mathcal{T}_h$ a *piecewise quadratic* approximation for the velocity. Figure 6.1 shows also the refinement of both triangulations close to the step corner and also behind the step.

The *steady state solutions* have been obtained via the *time integration* of the *fully discrete* Navier-Stokes equations using the Crank-Nicholson scheme of Sec. 3.2 with $k = \Delta t = .4$ (in fact convergence would have been faster with the simple backward implicit scheme of Sec. 3.1). We have indicated on Fig. 6.2 the stream lines of the computed solutions showing very clearly a *recirculation zone* whose size increases with Re ; if H is the height of our step, we observe that the length of the recirculation zone is approximately 6H at Re = 100 and 8H at Re = 191, in good agreement with the results obtained by other authors and particularly HUTTON in [29].

We have finally shown on Figure 6.3 the pressure distributions corresponding to the two above Reynold numbers.

6.2. Unsteady flow around and inside an idealized nozzle at high incidence.

We follow again [16] in this subsection ; the problem under consideration is the un-

TRIANGULATION $\mathcal{T}_h$ and $\tilde{\mathcal{T}}_h$

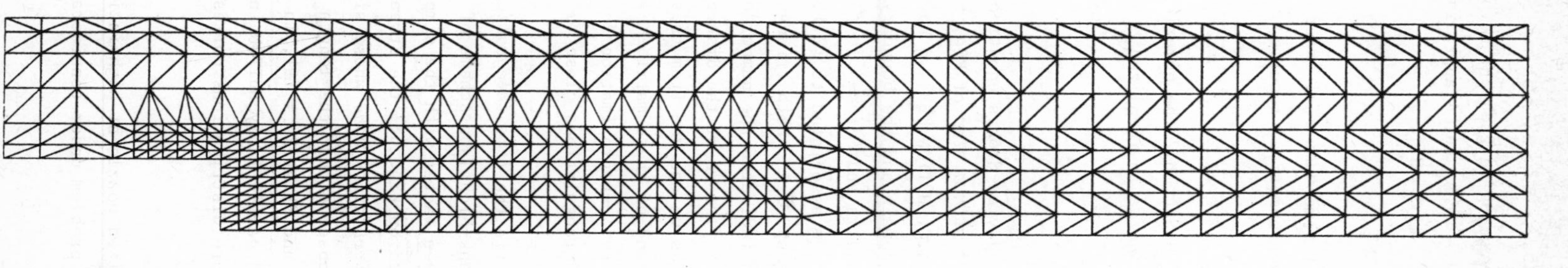

	$\mathcal{T}_h$	$\tilde{\mathcal{T}}_h$
NODES	619	2346
ELEMENTS	1109	4436
COEF. CHOL	21654	154971

Figure 6.1

Stream Lines

P1/P1 ISO P2 LIGNES DE COURANT
ESSAI DE LA NOUVELLE VERSION
INCIDENCE 0.00
MACH INFINI 0.00 REYNOLDS 100.0
CYCLE ITER 70
PAS DE TEMPS 0.40

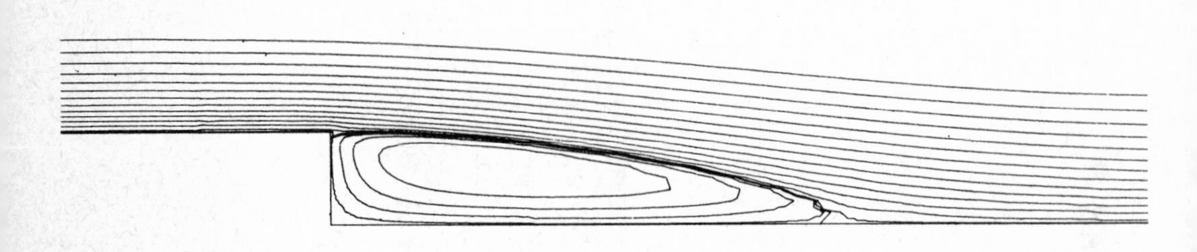

P1/P1 ISO P2 LIGNES DE COURANT
CALCUL MARCHE 2-D : CONDITIONS DE DIRICHLET SUR LES LIMITES
INCIDENCE 0.00
MACH INFINI 0.00 REYNOLDS 191.0
CYCLE ITER 70
PAS DE TEMPS 0.40

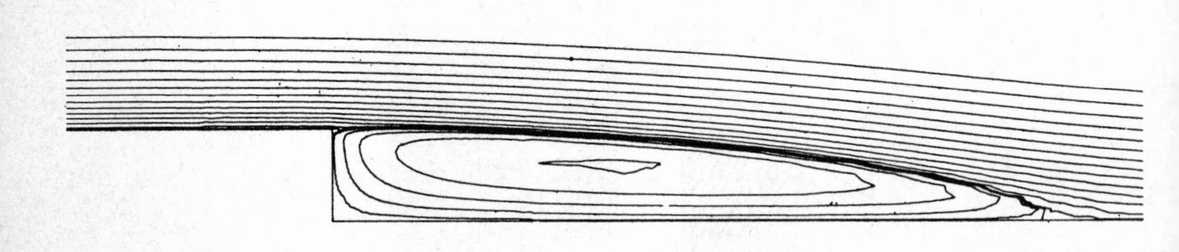

Figure 6.2

Iso-pressure Lines

P1/P1 ISO P2 LIGNES ISO-PRESSIONS
ESSAI DE LA NOUVELLE VERSION
INCIDENCE 0.00
MACH INFINI 0.00 REYNOLDS 100.0
CYCLE ITER 70
PAS DE TEMPS 0.40

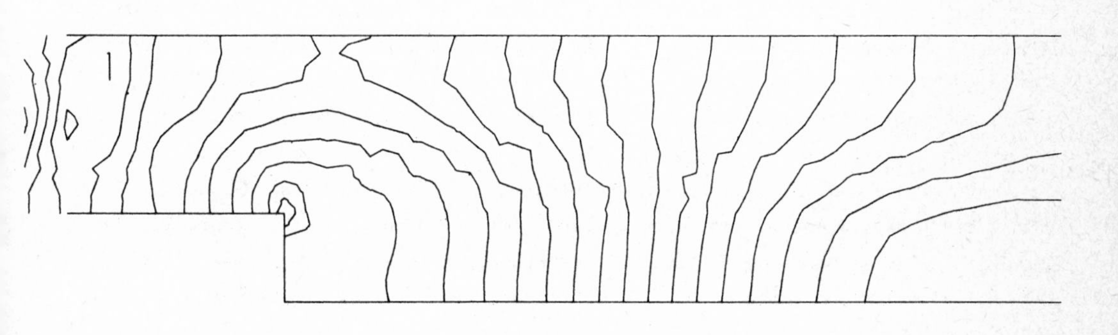

P1/P1 ISO P2 LIGNES ISO-PRESSIONS
CALCUL MARCHE 2-D ; CONDITIONS DE DIRICHLET SUR LES LIMITES
INCIDENCE 0.00
MACH INFINI 0.00 REYNOLDS 191.0
CYCLE ITER 70
PAS DE TEMPS 0.40

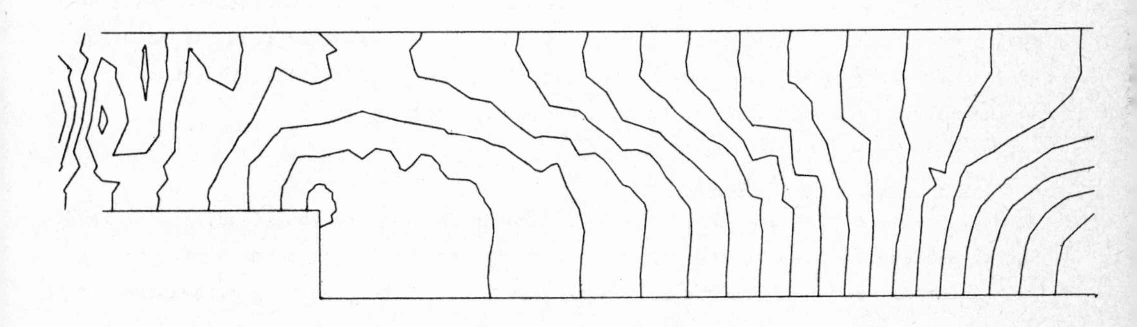

Figure 6.3

steady flow of an incompressible viscous fluid at Re = 250, around and inside and idealized nozzle at high incidence (Re has been computed using as characteristic length the distance between the two walls of the nozzle). Figures 6.4 and 6.5 show an enlarged view of the nozzle close to the air intake. The computational domain is clearly bounded and appropriate boundary conditions has been prescribed in the far field. The velocity vector at ∞ (in the actual problem) is *horizontal* and the angle of attack of the nozzle is 40° (which is quite large at that Reynold's number) ; another important feature of our problem is that a given flux has been prescribed (via a velocity distribution) on a cross section of the nozzle, in order to simulate a *suction* phenomenon due to an engine. We have used for the space and time discretizations the same methods than in Sec. 6.1 ; we have used a triangulation $\mathcal{T}_h$ (resp. $\tilde{\mathcal{T}}_h$) consisting of 1458 (resp. 5832) *finite elements*, 795 (resp. 3049) *nodes*. Details of $\mathcal{T}_h$ and $\tilde{\mathcal{T}}_h$ are shown on Figures 6.4 , 6.5 ; let us mention that the number of *non zero elements* of the *Cholesky factors*, corresponding to $\mathcal{T}_h$ and $-\Delta$ (resp. $\tilde{\mathcal{T}}_h$ and $(\frac{I}{k} - \nu\Delta)$) is 101370 (resp. 314685). More computational details about the present problem are given in [16]. As mentioned before, a Crank-Nicholson scheme has been used, taking as initial value the (discrete) solution of the corresponding Stokes' problem (creeping flow).

We have represented on Fig. 6.6-6.9 the distributions at various time steps of the velocity, stream function, pressure and vorticity, respectively. It is interesting to see the formation at the leading edges, and the propagation of eddies inside and outside the nozzle.

As it can be guessed, the computer time required for solving such a large time dependent nonlinear problem (about 7000 unknowns) on such a complicated geometry has to be important ; it actually requires several hours of IBM 370-168 to run 100 time steps. The results of computations at Re = 750, for the same geometry, will be shown elsewhere.

Let us mention that the convergence, at each time step, of our conjugate gradient algorithm is usually achieved in 3 to 5 iterations. Let us also mention to conclude that a very important saving in time and storage, can be expected from the solution of our very large linear systems by sophisticated iterative methods like those based on incomplete factorization and (again) conjugate gradient.

7. - FURTHER COMMENTS. CONCLUSION.

We have presented in this paper numerical methods for solving the steady and unsteady Navier-Stokes equations in the *pressure-velocity* formulation ; the machinery behind these methods may seems a bit complicated, but in fact the resulting methodology is outstandingly robust for the following reasons :

- The *finite element* formulation allows us to handle in a *stable* way (stable because a weak formulation is used) Navier-Stokes equations on complicated geometries ; we have used moreover finite element methods leading to convergence rates of

Figure 6.4

Figure 6.5

UNSTEADY FLOW AROUND AN INLET
Velocity Distribution

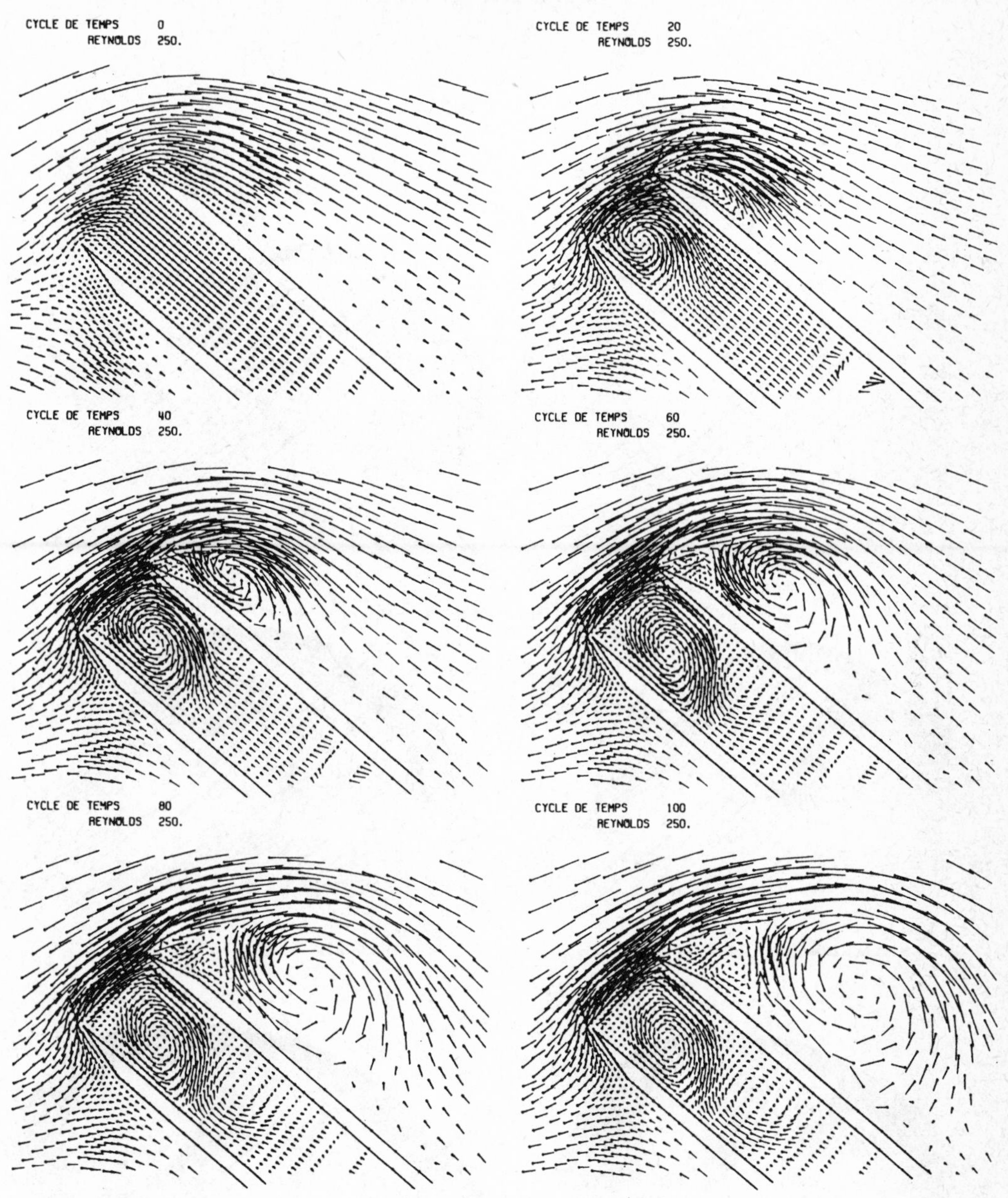

Figure 6.6

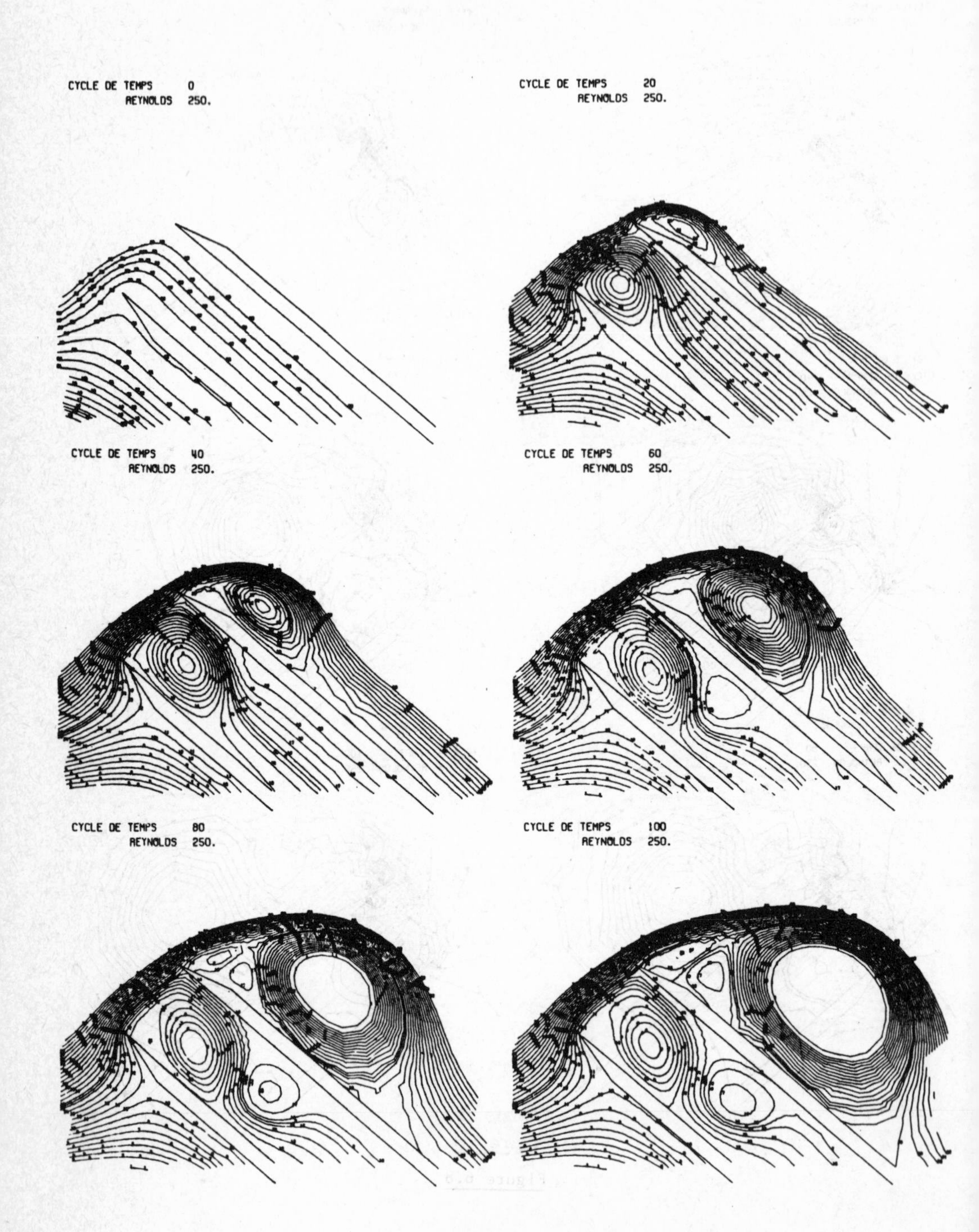
UNSTEADY FLOW AROUND AN INLET
Stream Lines
CYCLE DE TEMPS 0
REYNOLDS 250.
CYCLE DE TEMPS 20
REYNOLDS 250.
CYCLE DE TEMPS 40
REYNOLDS 250.
CYCLE DE TEMPS 60
REYNOLDS 250.
CYCLE DE TEMPS 80
REYNOLDS 250.
CYCLE DE TEMPS 100
REYNOLDS 250.

Figure 6.7

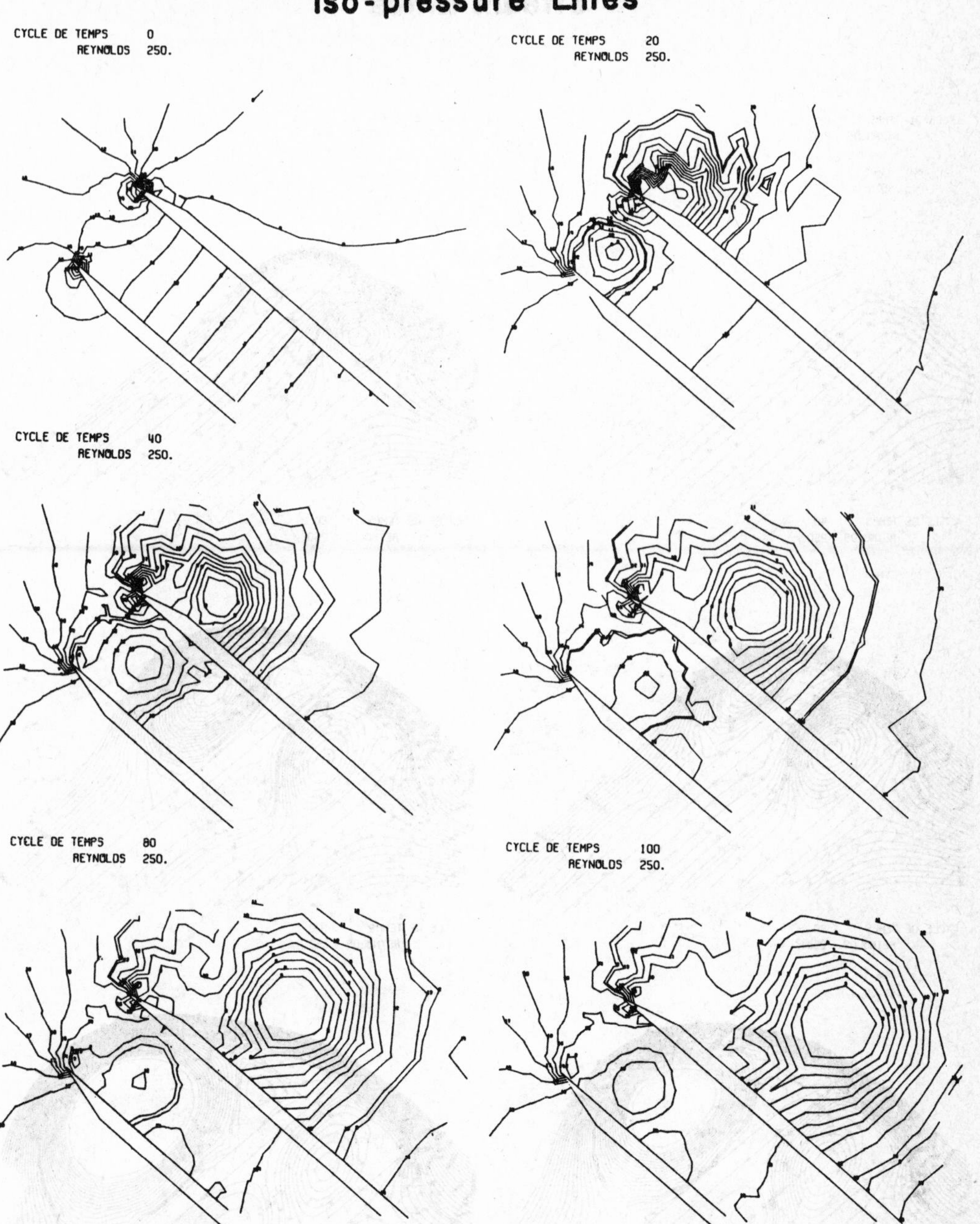
UNSTEADY FLOW AROUND AN INLET
Iso-pressure Lines
CYCLE DE TEMPS 0
REYNOLDS 250.
CYCLE DE TEMPS 20
REYNOLDS 250.
CYCLE DE TEMPS 40
REYNOLDS 250.
CYCLE DE TEMPS 80
REYNOLDS 250.
CYCLE DE TEMPS 100
REYNOLDS 250.

Figure 6.8

UNSTEADY FLOW AROUND AN INLET

Iso-vorticity Lines

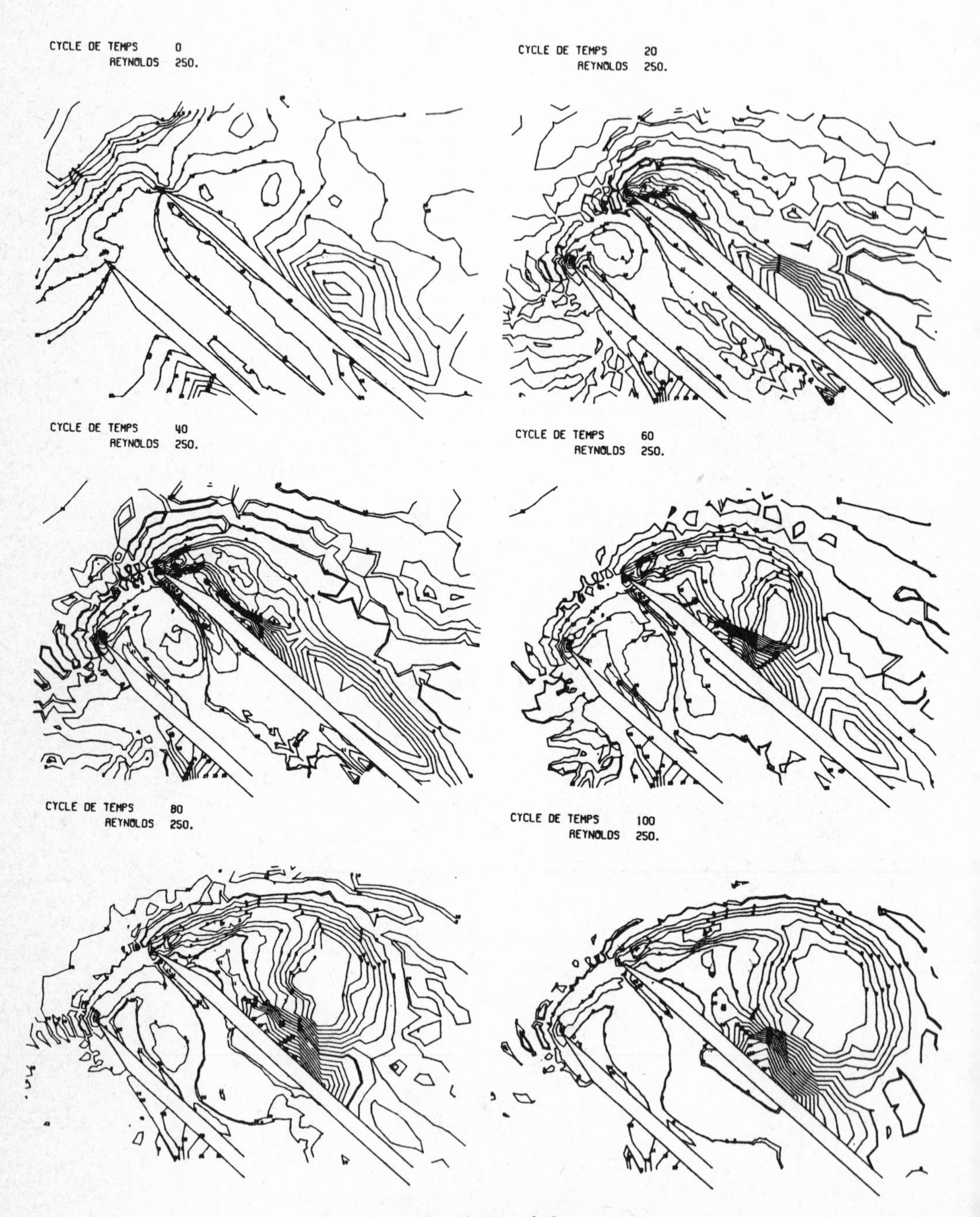

Figure 6.9

optimal order for both pressure and velocity.

- Fully implicit schemes are used for the time discretization since semi-implicit and explicit schemes resulted in very time consuming methods.
- The resulting nonlinear finite dimensional problems are solved, via a *nonlinear least square formulation*, by a *conjugate gradient method with scaling* (i.e. with *preconditioning*), the scaling operator being a *discrete Stokes operator*. Once again stability is reinforced since least square formulations act (locally at least) as *convexifier* of the original problem. Moreover conjugate gradient algorithms may have a *superlinear convergence* (practically *quadratic* in many cases).
- The only linear systems to be solved are related to the Stokes solver, and using the very particular finite element discretizations of Sec. 2 we have been able to factorize the solution of these discrete Stokes' problems in linear subproblems of smaller size associated to *symmetric, positive definite matrices*, which is also a guaranty of numerical stability (these matrices are obviously factored once for all if direct methods of solution are chosen).

From the above properties the resulting algorithms are very reliable and well-suited for computing separated flows with recirculation, in complicated geometries. There is no theoretical difficulties to apply the above techniques to the solution of 3-*dimensional* problems ; in this latter case, or in very complicated 2-dimensional problems, we definitely need for solving the very large linear systems involved in our methodology, very sophisticated iterative techniques resulting in a very important saving of storage and computational time. We are presently developping and testing such methods and consequently hope that in a near future we shall be able to present good numerical results for 3-dimensional flows in complicated geometries, modelled by Navier-Stokes equations.

In the present methodology *upwinding* has been avoided since we feared the corresponding *artificial viscosity* may "kill" some of the small scale phenomenon possibly associated to high Reynold's number flows ; one may find however in the following Appendix the description, with some numerical tests, of a new upwind finite element method of high accuracy which seems to have a promising future in Computational Fluid Dynamics.

Appendix. - A FINITE ELEMENT METHOD WITH UPWINDING FOR SECOND ORDER PROBLEMS WITH "LARGE" FIRST ORDER TERMS.

A.1. INTRODUCTION. - Upwinding finite element schemes has been a subject of very active researchs these last years ; we shall mention in that direction [30], [31], [32], [33] , (and the bibliography therein). We would like to describe in this appendix a method which can be viewed as an extension of the method introduced by Tabata in [30].

A.2. THE CONTINUOUS PROBLEM.

Let Ω be the domain indicated below (on Figure A.1), and $\Gamma = \partial\Omega$.

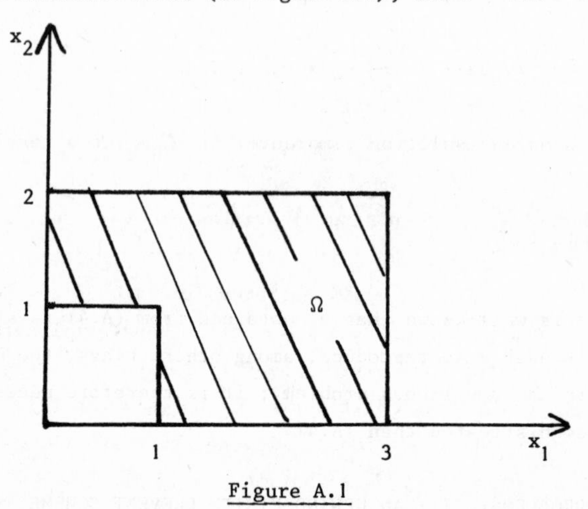

Figure A.1

We consider the problem (with $\varepsilon > 0$)

$$\text{(A.1)} \qquad \begin{cases} -\varepsilon\Delta u + \underset{\sim}{\beta}\cdot\underset{\sim}{\nabla} u = f \textit{ in } \Omega, \\ u = 0 \textit{ on } \Gamma, \end{cases}$$

where $\underset{\sim}{\beta} = \{\cos\theta, \sin\theta\}$.

We are mostly interested by solving (A.1) for small values of ε ; in the following we shall suppose that f is a *smooth* function and we shall use the notation

$$\text{(A.2)} \qquad \frac{\partial v}{\partial \beta} = \underset{\sim}{\beta}\cdot\underset{\sim}{\nabla} v .$$

Problem (A.1) has as *variational formulation*

$$\text{(A.3)} \qquad \begin{cases} \textit{Find } u \in H_0^1(\Omega) \textit{ such that} \\ \varepsilon\int_\Omega \underset{\sim}{\nabla} u\cdot\underset{\sim}{\nabla} v\, dx + \int_\Omega \frac{\partial u}{\partial\beta} v\, dx = \int_\Omega fv\, dx \quad \forall\, v \in H_0^1(\Omega), \end{cases}$$

from which we can easily prove, using Lax-Milgram theorem, the existence of a unique solution since $\int_\Omega \frac{\partial v}{\partial \beta} v \, dx = 0 \quad \forall v \in H^1_o(\Omega)$.

A.2. APPROXIMATE PROBLEMS. (I) A CENTERED APPROXIMATION.

Let $\mathcal{T}_h$ be a triangulation of Ω ; to approximate (A.1),(A.3) we use the space

$$H^1_{oh} = \{v_h | v_h \in C^o(\bar{\Omega}), \ v_{h|K} \in P_1 \quad \forall K \in \mathcal{T}_h, \ v_{h|\Gamma} = 0\} .$$

The obvious approximation of (A.3), using H^1_{oh}, is

$$\text{(A.4)} \qquad \begin{cases} \textit{Find } u_h \in H^1_{oh} \textit{ such that } \forall v_h \in H^1_{oh}, \\ \varepsilon \int_\Omega \nabla u_h \cdot \nabla v_h dx + \int_\Omega \frac{\partial u_h}{\partial \beta} v_h dx = \int_\Omega f \, v_h dx. \end{cases}$$

Problem (A.4) has a unique solution ; moreover if $(\mathcal{T}_h)_h$ is a regular family of triangulations we have

$$\lim_{h \to 0} \| u_h - u \|_{H^1_o(\Omega)} = 0.$$

If ε is "small" it is well-known that u_h obtained from (A.4) is afflicted with spurious oscillations and is unable to reproduce, among other things, the *boundary layer phenomenon* existing in the continuous problem ; it is therefore necessary to use an approximation more sophisticated than (A.4).

A.3. APPROXIMATE PROBLEMS. (II) AN UPWIND FINITE ELEMENT SCHEME.

We have in view an approximation of (A.1),(A.3) such that

$$\text{(A.5)} \qquad \begin{cases} \varepsilon \int_\Omega \nabla u_h \cdot \nabla v_h dx + (\frac{\partial_h u_h}{\partial \beta}, v_h)_h = \int_\Omega f v_h dx \quad \forall v_h \in H^1_{oh} , \\ u_h \in H^1_{oh} ; \end{cases}$$

in (A.5) the scalar product $(\cdot,\cdot)_h$ is defined by

$$(u_h, v_h)_h = \frac{1}{3} \sum_{K \in \mathcal{T}_h} \text{meas.}(K) \sum_{i=1}^{3} u_h(M_{iK}) v_h(M_{iK}) ,$$

where M_{iK}, i=1,2,3 are the vertices of K.

We look for an approximation $\frac{\partial_h u_h}{\partial \beta}$ of $\frac{\partial u}{\partial \beta}$, *backward with respect to* β, *second order accurate* (unlike the approximation in [30] which is only *first order* accurate) and defined at the nodes of $\mathcal{T}_h$. To define such an approximation we shall use an *interpolation* method defined as follows :

Let M_i be a node of $\mathcal{T}_h$ such that

(A.6) $\qquad M_i \notin \Gamma$.

Let S_i be the half line starting from M_i and directed by $-\underset{\sim}{\beta}$ (see Fig. A.2).

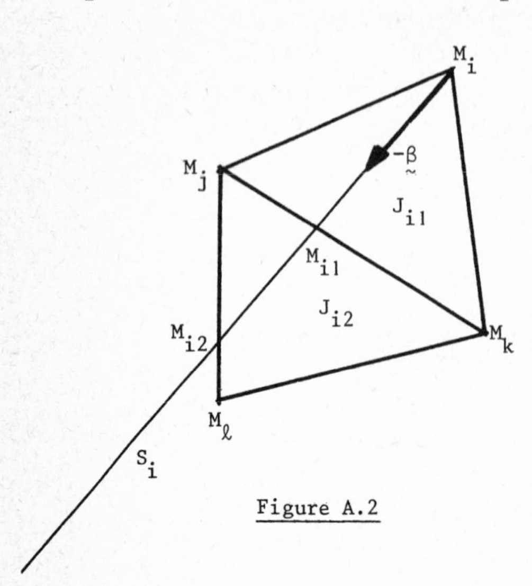

Figure A.2

To express $\frac{\partial_h u_h}{\partial \beta}(M_i)$ we look for two points of S_i belonging also to the edges of triangles close to M_i. Let us describe now the choice of these two points :

Step 1 : Among the triangles with M_i as a vertex we consider the triangle, denoted by J_{i1} which is crossed by S_i. Let M_j, M_k be the two other vertices of J_{i1} (taken in the trigonometric orientation) ; since (4.6) holds J_{i1} always exists.

Step 2 : Denote by M_{i1} the point at the intersection of S_i and of the side M_jM_k. If M_jM_k is supported by the boundary Γ *go to Remark A.1* ; if the contrary holds let J_{i2} be the triangle *adjacent* to J_{i1} along M_jM_k (see Fig. A.2) and let M_ℓ be the third vertex of J_{i2} ; denote finally by M_{i2} the point where S_i crosses M_jM_ℓ or $M_\ell M_k$.

Step 3 : We have constructed 3 points on S_i, which are M_i, M_{i1}, M_{i2}. But M_{i1}, M_{i2} may be very close one to each other and even coincide with either M_j or M_k ; if this unfortunate situation holds our second order approximation *degenerates* in a first order approximation. Thus we possibly need to define a different M_{i2} if S_i is "close" to one of the edges of J_{i1}. We shall proceed more precisely as follows :
We compute the two angles $(\overrightarrow{M_iM_j}, -\underset{\sim}{\beta})$ and $(\overrightarrow{M_iM_j}, \overrightarrow{M_iM_k})$ and consider then the following possibilities :

a) If $\frac{1}{4}(\overrightarrow{M_iM_j}, \overrightarrow{M_iM_k}) \leq (\overrightarrow{M_iM_j}, -\underset{\sim}{\beta}) \leq \frac{3}{4}(\overrightarrow{M_iM_j}, \overrightarrow{M_iM_k})$,

M_{i2} is defined as in Step 2.

b) If $(\overrightarrow{M_iM_j}, -\underset{\sim}{\beta}) < \frac{1}{4}(\overrightarrow{M_iM_j}, \overrightarrow{M_iM_k})$ denote $M_v = M_j$,

if $(\overrightarrow{M_iM_j}, -\underset{\sim}{\beta}) > \frac{3}{4}(\overrightarrow{M_iM_j}, \overrightarrow{M_iM_k})$ denote $M_v = M_k$.

If M_{i2} of Step 2 does not belong to M_vM_ℓ then M_{i1}, M_{i2} are not going to coincide and we can use the M_{i2} defined in Step 2 (see Fig. A.3).

If M_{i2} belongs to $M_v M_\ell$, $M_{i1}M_{i2}$ may be "very small" and we have to modify the definition of M_{i2} ; *go to Step 4* for such a modification.

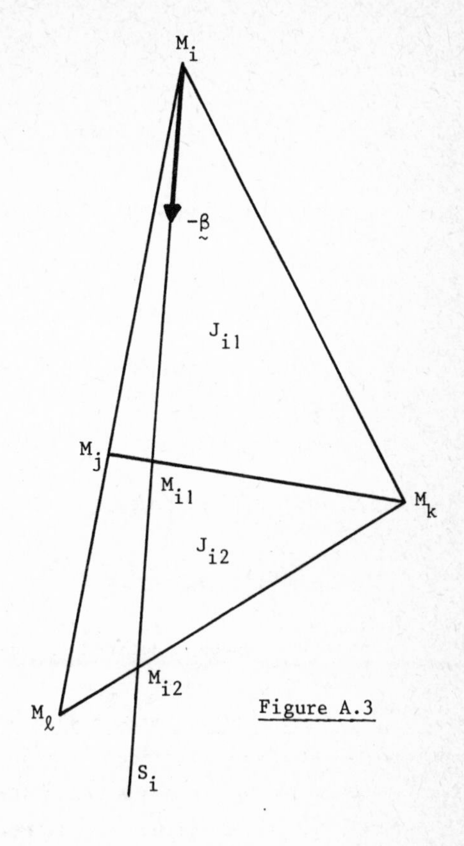

Figure A.3

Step 4 : We look for the triangle J_{i3} adjacent to J_{i2} along $M_v M_\ell$; let M_m be the third vertex of J_{i3} (see Fig. A.4.). If $M_v M_\ell$ is supported by Γ, go to Remark A.1.

If S_i crosses edge $M_\ell M_m$, we denote by M_{i2} the crossing point and one uses this point (with M_i and M_{i1}) to approximate $\frac{\partial u}{\partial \beta}$.
If S_i crosses $M_v M_m$ we follow with the process of this Step 4.

We finally obtain M_{i2} as the intersection of S_i and of the polygonal line, boundary of the polygonal domain consisting of the union of those triangles of $\mathcal{C}_h$ with M_v as a common vertex.

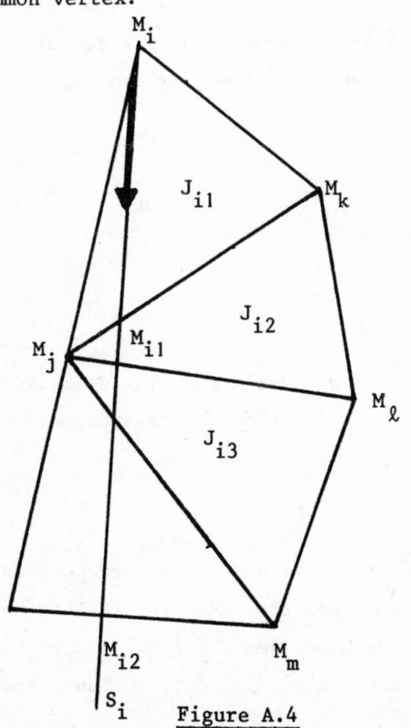

Figure A.4

Thus we have constructed 2 points M_{i1}, M_{i2} "close" to the vertex M_i. Since u_h is continuous on $\overline{\Omega}$ its values at M_i, M_{i1}, M_{i2} are known ; we denote

$$\text{(A.7)} \begin{cases} u_{io} = u_h(M_i), \; u_{i1} = u_h(M_{i1}), \\ u_{i2} = u_h(M_{i2}). \end{cases}$$

Let $h_{i1} = |M_i M_{i1}|$, $h_{i2} = |M_i M_{i2}|$; we define then $\frac{\partial_h u_h}{\partial \beta}(M_i)$ as the value at M_i of the derivative of a *second order polynomial*, defined on the half line S_i, coinciding with u_h at M_i, M_{i1}, M_{i2}. We obtain then

(A.8) $$\frac{\partial_h u_h}{\partial \beta}(M_i) = \frac{h_{i1}+h_{i2}}{h_{i1}h_{i2}} u_{io} - \frac{h_{i2}}{h_{i1}(h_{i2}-h_{i1})} u_{i1} + \frac{h_{i1}}{h_{i2}(h_{i2}-h_{i1})} u_{i2} .$$

Remark A.1 : If M_i is close to Γ and if we have not been able to define M_{i2} we can proceed as follows :

- either define a *centered approximation* of $\frac{\partial u}{\partial \beta}(M_i)$ - $O(h^2)$ accurate - by

(A.9) $$\frac{\partial_h u_h}{\partial \beta}(M_i) = \frac{1}{A_i} \sum_{K \in \mathcal{T}_i} A(K) \frac{\partial u_h}{\partial \beta}\Big|_K ,$$

where $\mathcal{T}_i$ is the subset of $\mathcal{T}_h$ consisting of those triangles with M_i as a common vertex ; A_i = meas. $(\bigcup_{K \in \mathcal{T}_i} K)$; $\frac{\partial u_h}{\partial \beta}\Big|_K = \underset{\sim}{\beta} . \nabla \underset{\sim}{u}_h\Big|_K$ (we recall that $\nabla \underset{\sim}{u}_h$ is *piecewise constant* on $\bar{\Omega}$).

- or use an *upwind first order approximation* (à la TABATA [30]), defined by

(A.10) $$\frac{\partial_h u_h}{\partial \beta}(M_i) = \frac{\partial u_h}{\partial \beta}\Big|_{J_{i1}} ,$$

where triangle J_{i1} has been defined in Step 1.

A.5. On the solution of the linear system obtained by upwinding.

Using the *upwind scheme* described in Sec. A.4, we obtain u_h via the solution of a *linear system* whose matrix has a bandwidth which is approximately *twice* the bandwidth of the matrix associated with the *centered* approximate problem (A.4) of Sec. A.3. The matrix of the above system is definitely *non symmetric*;in that preliminary stage of our numerical experiments we have chosen to solve this linear system, say

(A.11) $$\underset{\sim}{A}_h \underset{\sim}{U}_h = \underset{\sim}{b}_h$$

by the standard method consisting in solving the *normal equation*

(A.12) $$\underset{\sim}{A}_h^t \underset{\sim}{A}_h \underset{\sim}{U}_h = \underset{\sim}{A}_h^t \underset{\sim}{b}_h ,$$

by a *Cholesky method*, since the matrix in (A.12) is symmetric positive definite. We have in view the use of more sophisticated methods, like the *Lanczos-type methods* described in, e.g., WIDLUND [34].

A.6. NUMERICAL EXPERIMENTS.

The method of Sec. A.3 using a *centered approximation*, is producing very poor results as soon as $\varepsilon < 10^{-1}$; we therefore only present the results concerning the *upwind method* of Sec. A.4. ; for the nodes close to Γ we have used the first order approximation defined by (A.10). The domain Ω being the one of Fig. A.1, we have used for our tests the regular triangulations of Figures A.5, A.6 and also the fairly irregular triangulations of Figures A.7, A.8. We have then approximatively solved (A.1), using (A.5), with $f=1$ on Ω. The above tests are very severe since the solution contains very strong *boundary layer* and *free layer* (generated by the *corner) phenomenon.*

Using the notation $\underset{\sim}{\beta} = \{\cos\theta, \sin\theta\}$ we show on the following figures the equipotential lines of the approximate solutions corresponding to various values of ε and θ. The agreement between these results is good except if $\theta=0$, for which the results related to the regular triangulations $\mathcal{C}_h^1$, $\mathcal{C}_h^2$ are good (even very good) unlike those related to $\mathcal{C}_h^3$, $\mathcal{C}_h^4$. In fact the case $\theta=0$ is the most severe since it is the only case for which the *limit,* as $\varepsilon\to 0$, of the solution of (A.1) is a *discontinuous function* ; indeed it is easily shown that this limit is

$$\text{(A.13)} \quad \begin{cases} u(x_1,x_2) = x_1 \ \ \textit{if} \ 1 < x_2 < 2, \\ u(x_1,x_2) = x_1 - 1 \ \ \textit{if} \ 0 < x_2 < 1, \end{cases}$$

and therefore is *discontinuous along* the line $x_2 = 1$, with a jump of amplitude 1. This discontinuity is very well reproduced by $\mathcal{C}_h^1$ and $\mathcal{C}_h^2$, for $\varepsilon = 10^{-3}$ and 10^{-5}, but not by $\mathcal{C}_h^3$, $\mathcal{C}_h^4$, since *irregular meshes* introduce severe *phase distorsions* as shown by ENGQUIST and KREISS in, e.g. [35]. For the two other values of θ the limit solution as $\varepsilon\to 0$ is continuous (but not C^1) and therefore the numerical process is much less sensitive to phase distorsion.

A.7. CONCLUDING COMMENTS.

The upwind finite element scheme that we have introduced in Sec. A.4 of this appendix is fairly accurate and extremely robust ; it has also the possibility of handling fairly irregular geometries. The above scheme leads of course to a non trivial coding.

We are presently extending the ideas of Sec. A.4 to the solution of the *Navier Stokes equations for incompressible viscous flows* and also to the solution of *transonic flows for compressible inviscid fluids,* the ultimate goal being the numerical solution of *Navier-Stokes equations for compressible viscous fluids.*

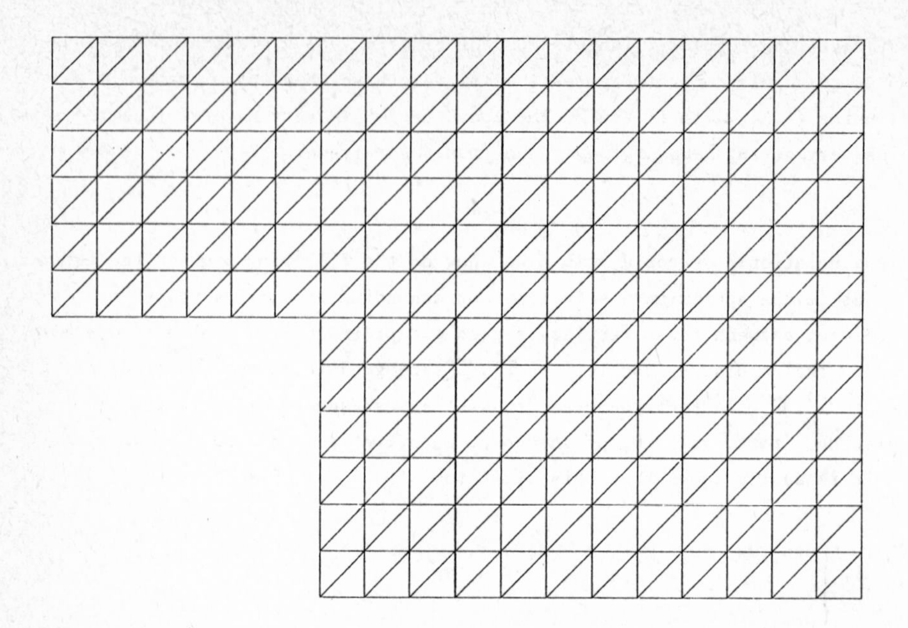

Figure A.5

Triangulation $\mathcal{C}_h^1$ - 360 triangles - 211 nodes.

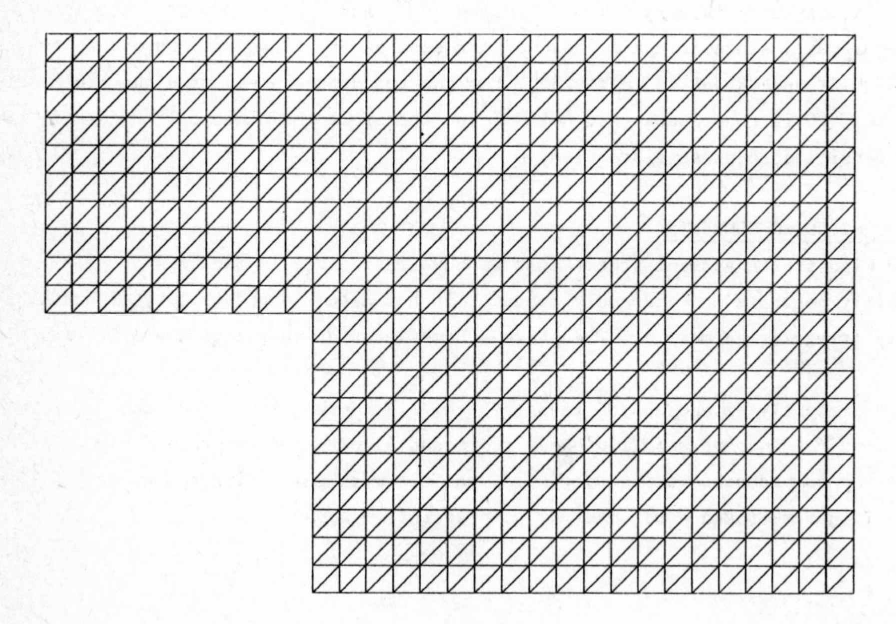

Figure A.6

Triangulation $\mathcal{C}_h^2$ - 1000 triangles - 551 nodes.

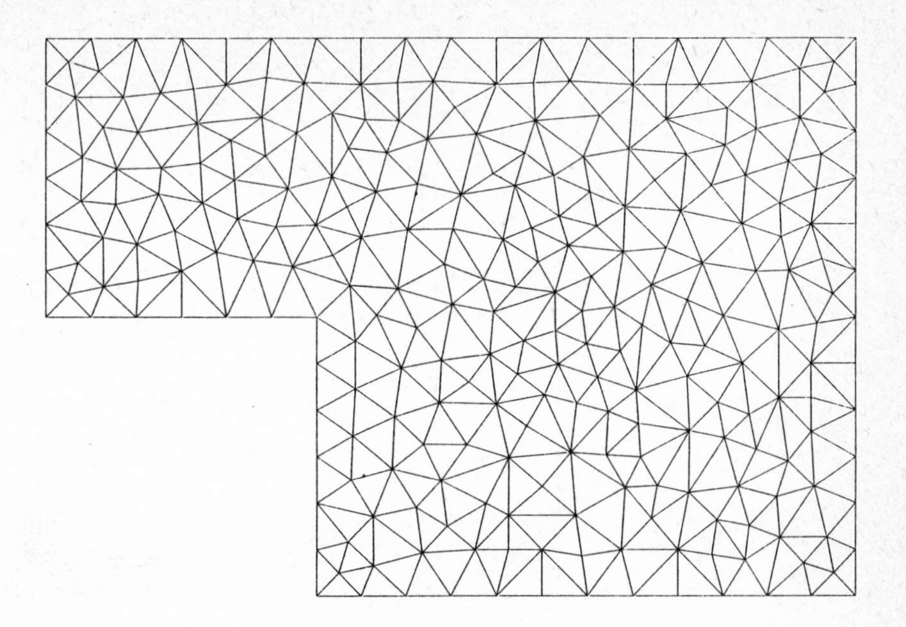

Figure A.7

Triangulation $\mathcal{C}_h^3$ - 472 triangles - 267 nodes.

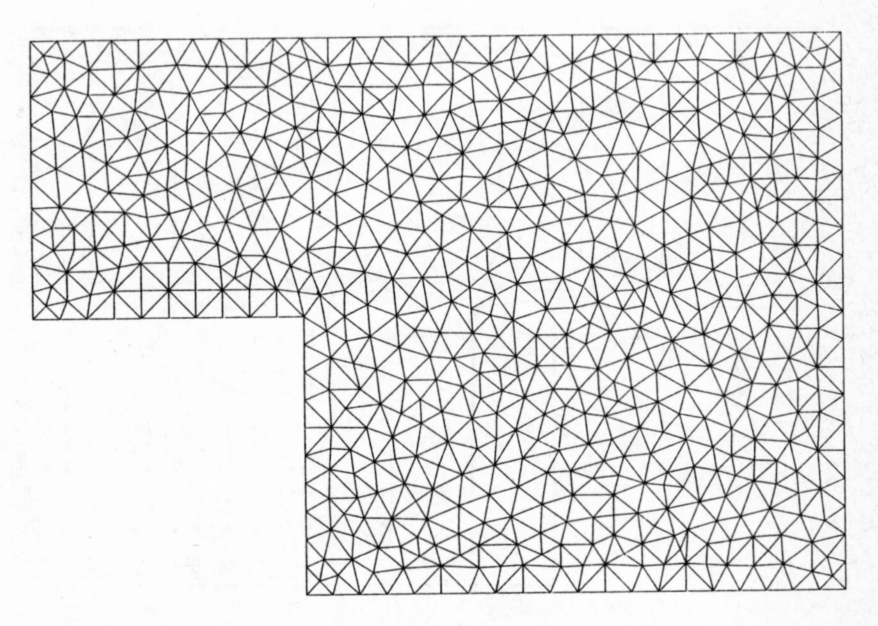

Figure A.8

Triangulation $\mathcal{C}_h^4$ - 1246 triangles - 674 nodes.

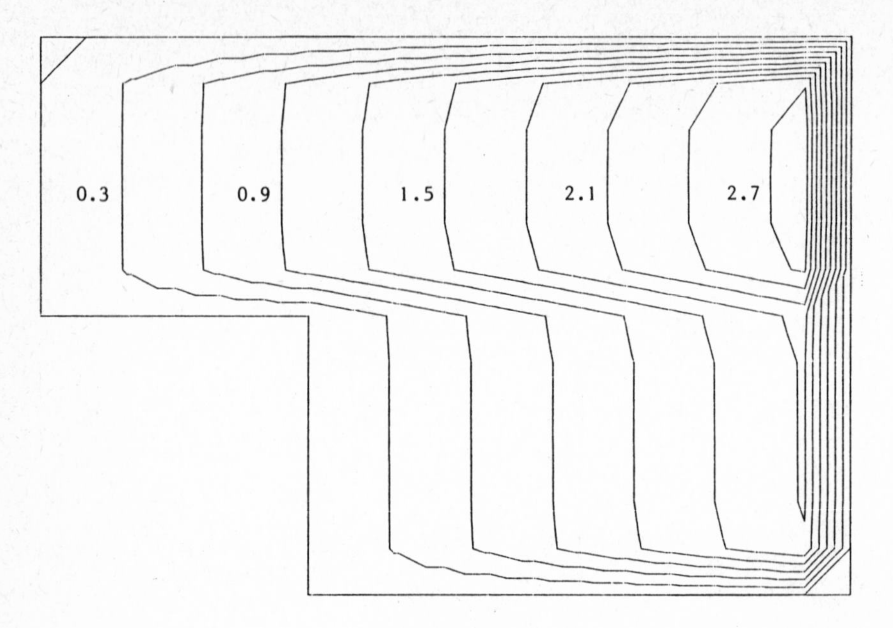

Figure A.9

Triangulation $\mathcal{C}_h^1$, $\varepsilon = 10^{-3}$, $\theta = 0$.

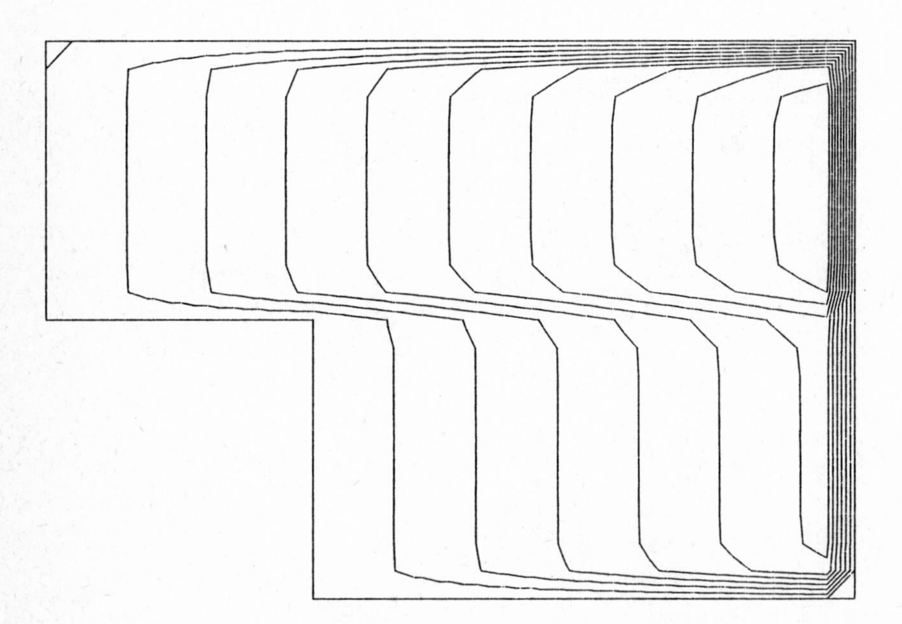

Figure A.10

Triangulation $\mathcal{C}_h^2$, $\varepsilon = 10^{-3}$, $\theta = 0$.

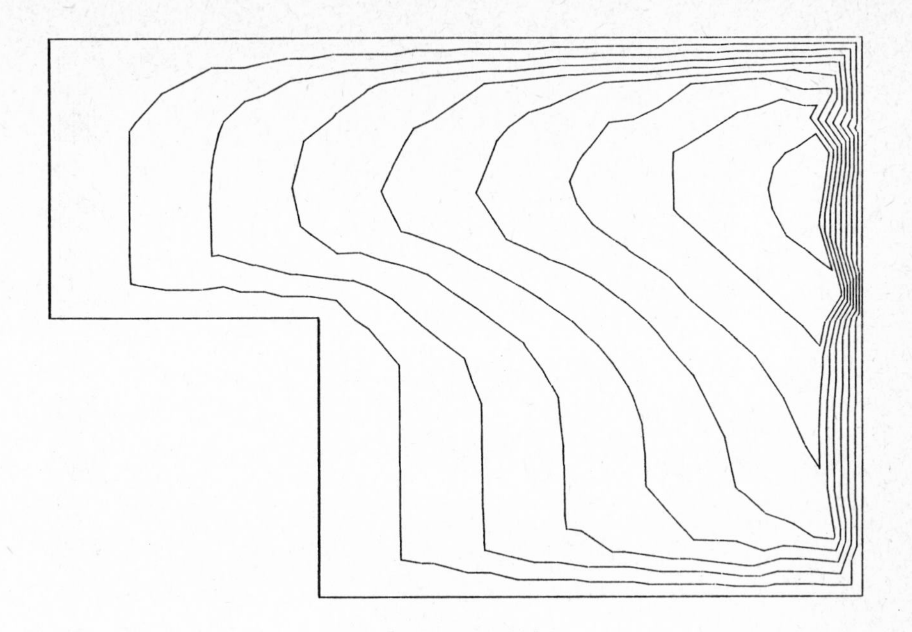

Figure A.11

Triangulation $\mathcal{T}_h^3$, $\varepsilon = 10^{-3}$, $\theta = 0$.

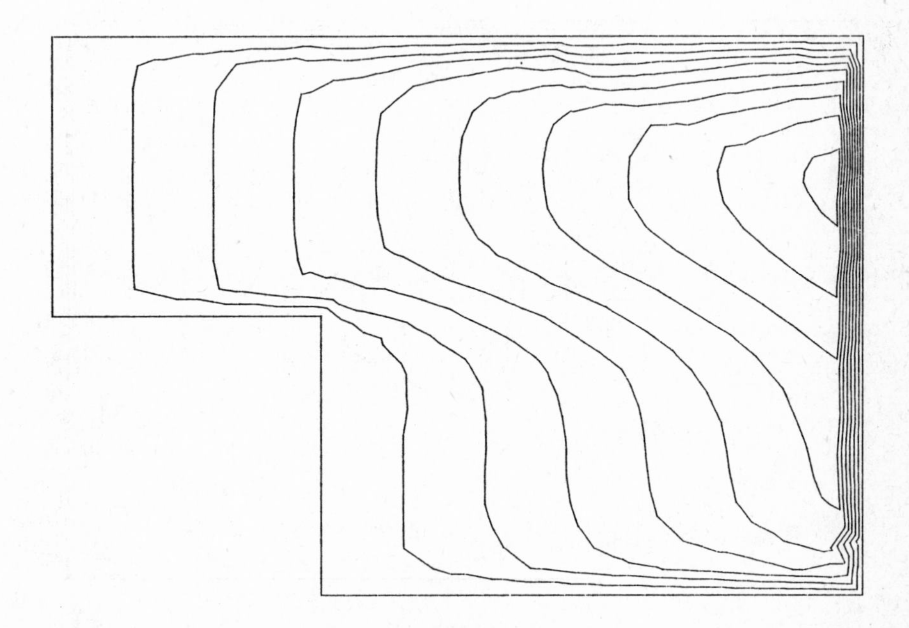

Figure A.12

Triangulation $\mathcal{T}_h^4$, $\varepsilon = 10^{-3}$, $\theta = 0$.

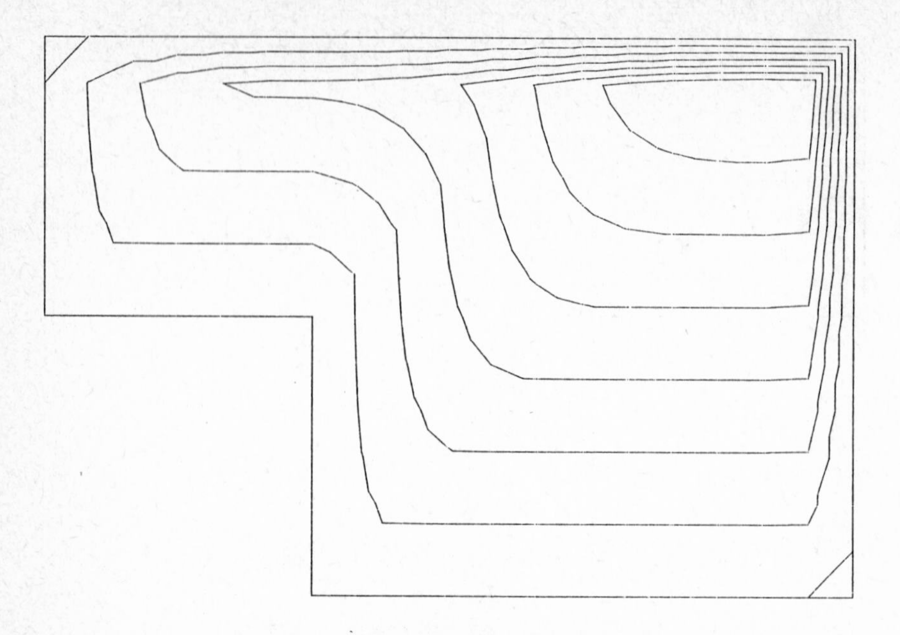

Figure A.13

Triangulation $\mathcal{C}_h^1$, $\varepsilon = 10^{-3}$, $\theta = \frac{\pi}{3}$.

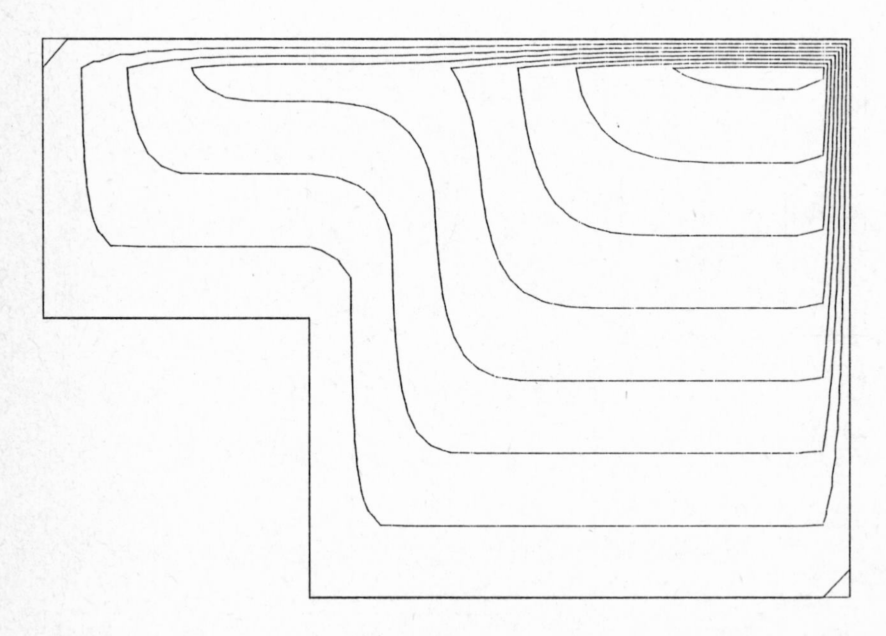

Figure A.14

Triangulation $\mathcal{C}_h^2$, $\varepsilon = 10^{-3}$, $\theta = \frac{\pi}{3}$.

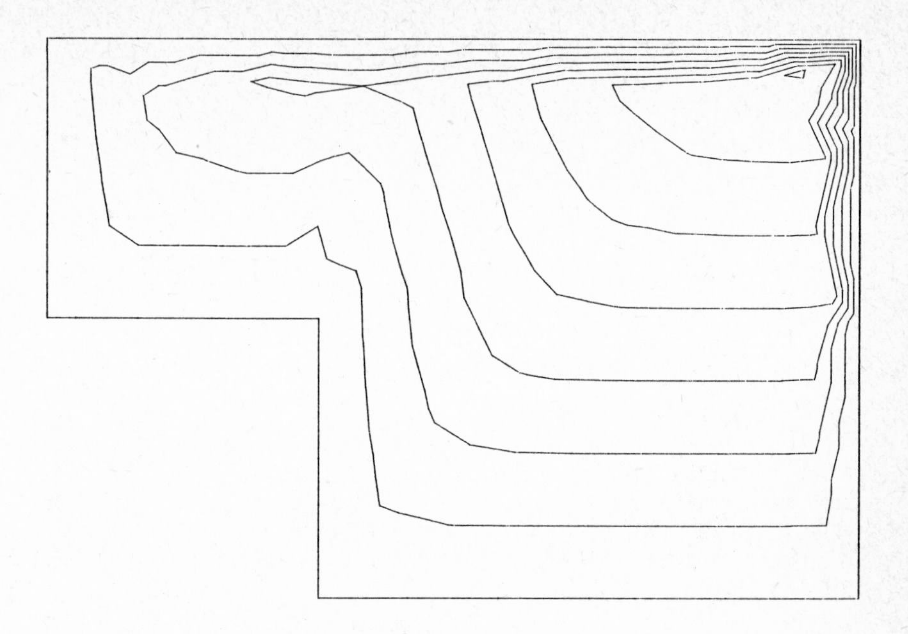

Figure A.15

Triangulation $\mathcal{C}_h^3$, $\varepsilon = 10^{-3}$, $\theta = \frac{\pi}{3}$.

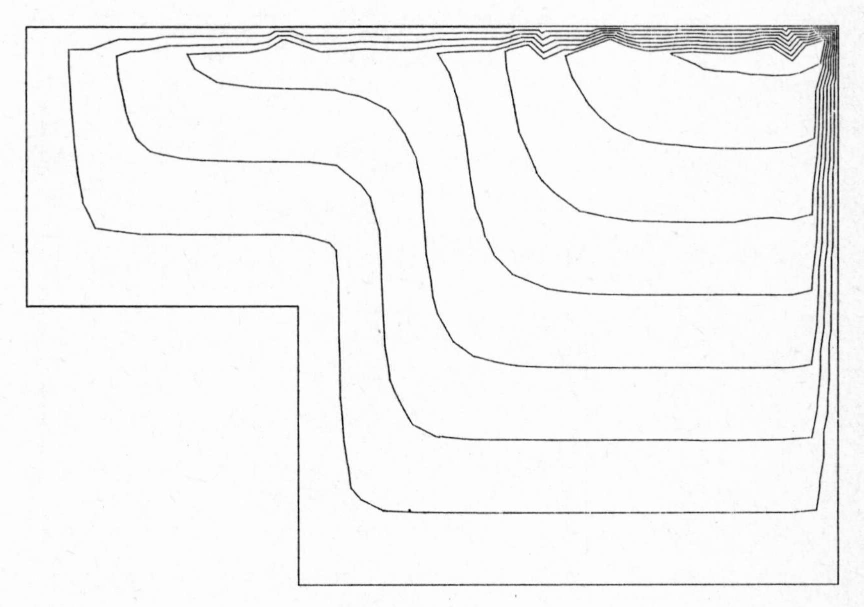

Figure A.16

Triangulation $\mathcal{C}_h^4$, $\varepsilon = 10^{-3}$, $\theta = \frac{\pi}{3}$.

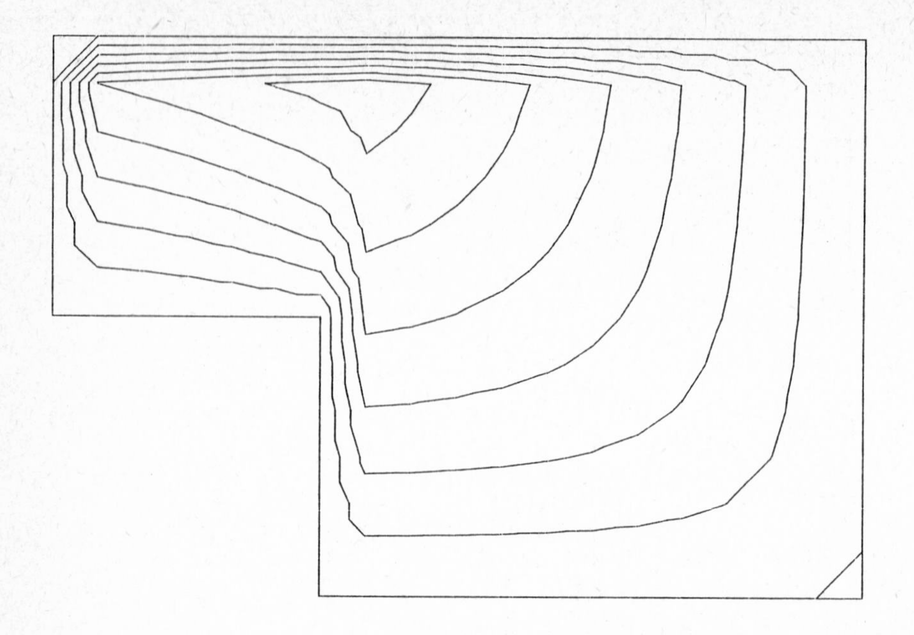

Figure A.17

Triangulation $\mathcal{T}_h^1$, $\varepsilon = 10^{-3}$, $\theta = \frac{3\pi}{4}$.

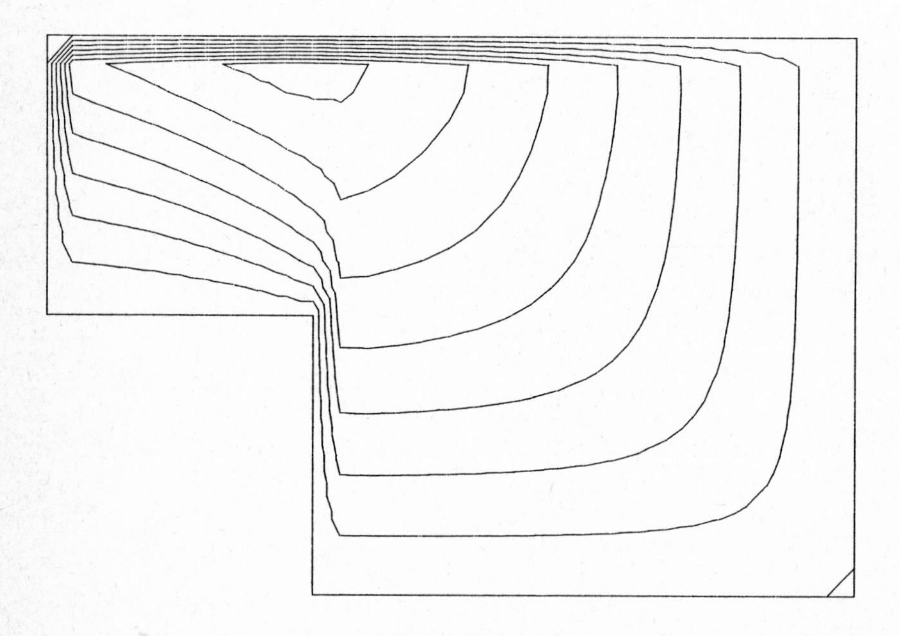

Figure A.18

Triangulation $\mathcal{T}_h^2$, $\varepsilon = 10^{-3}$, $\theta = \frac{3\pi}{4}$.

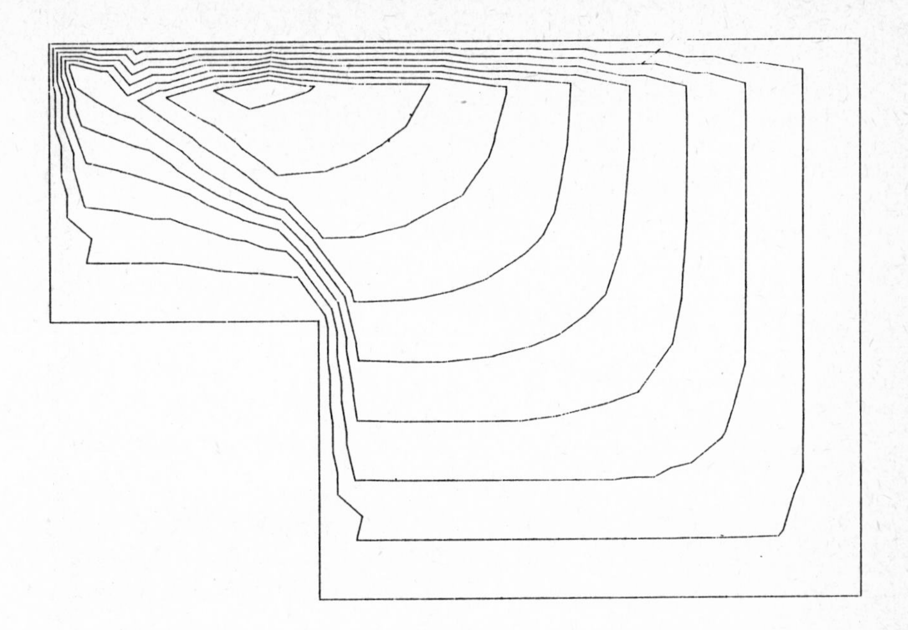

Figure A.19

Triangulation $\mathcal{C}_h^3$, $\varepsilon = 10^{-3}$, $\theta = \frac{3\pi}{4}$

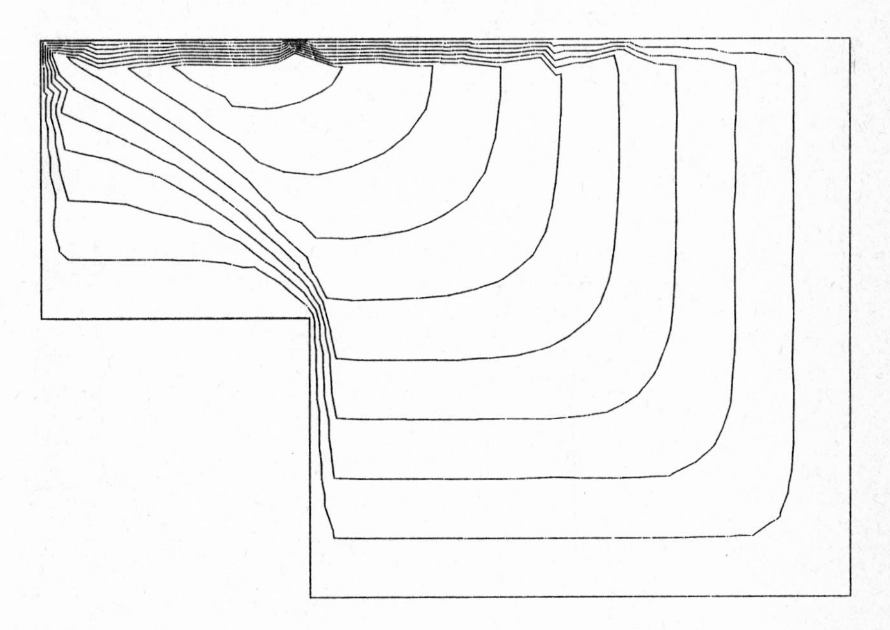

Figure A.20

Triangulation $\mathcal{C}_h^4$, $\varepsilon = 10^{-3}$, $\theta = \frac{3\pi}{4}$

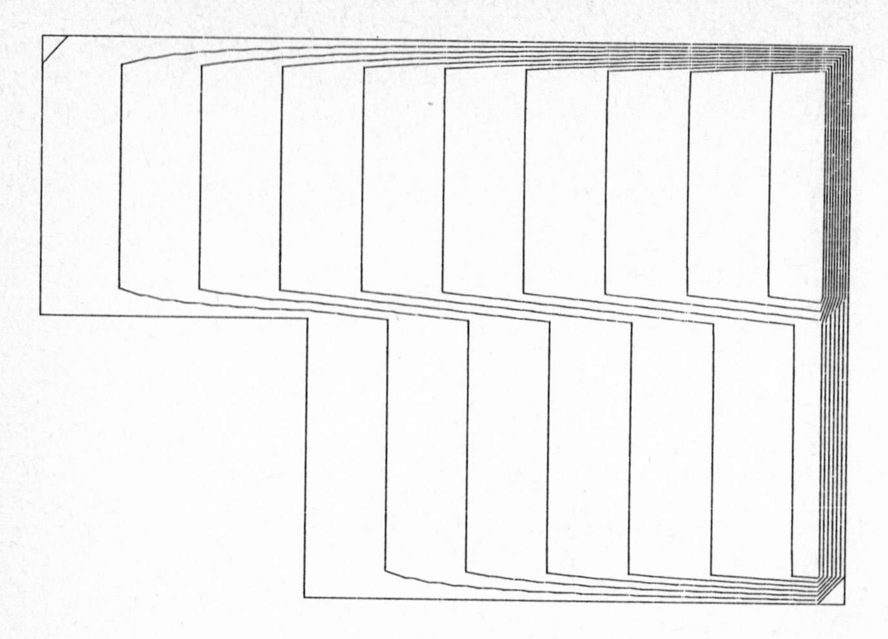

Figure A.21

Triangulation $\mathcal{C}_h^2$, $\varepsilon = 10^{-5}$, $\theta = 0$.

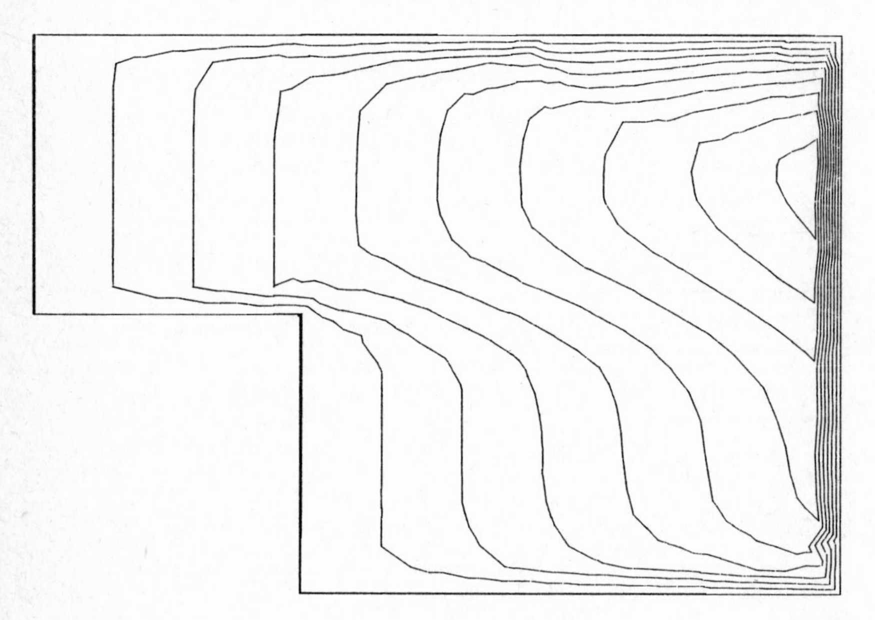

Figure A.22

Triangulation $\mathcal{C}_h^4$, $\varepsilon = 10^{-5}$, $\theta = 0$.

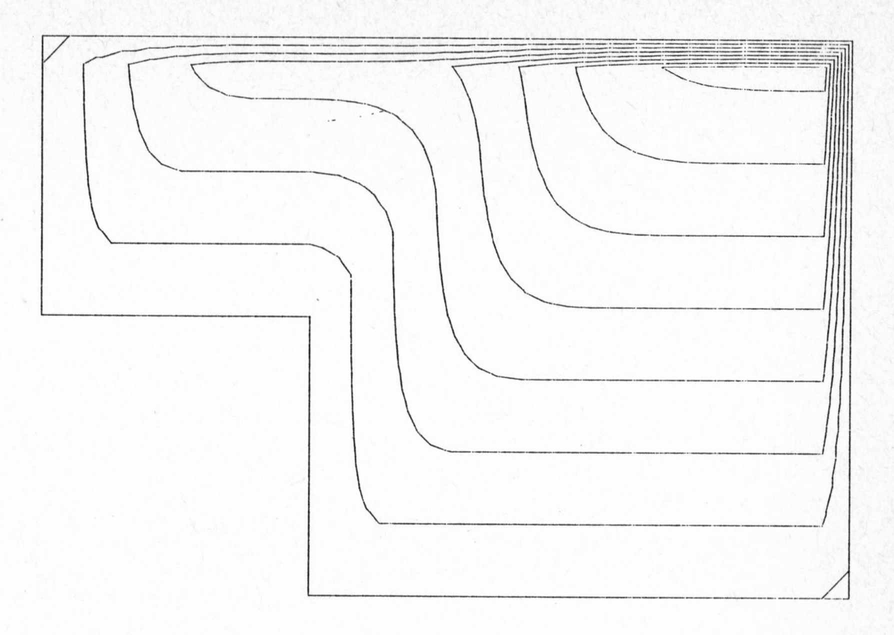

Figure A.23

Triangulation $\mathcal{C}_h^2$, $\varepsilon = 10^{-5}$, $\theta = \frac{\pi}{3}$.

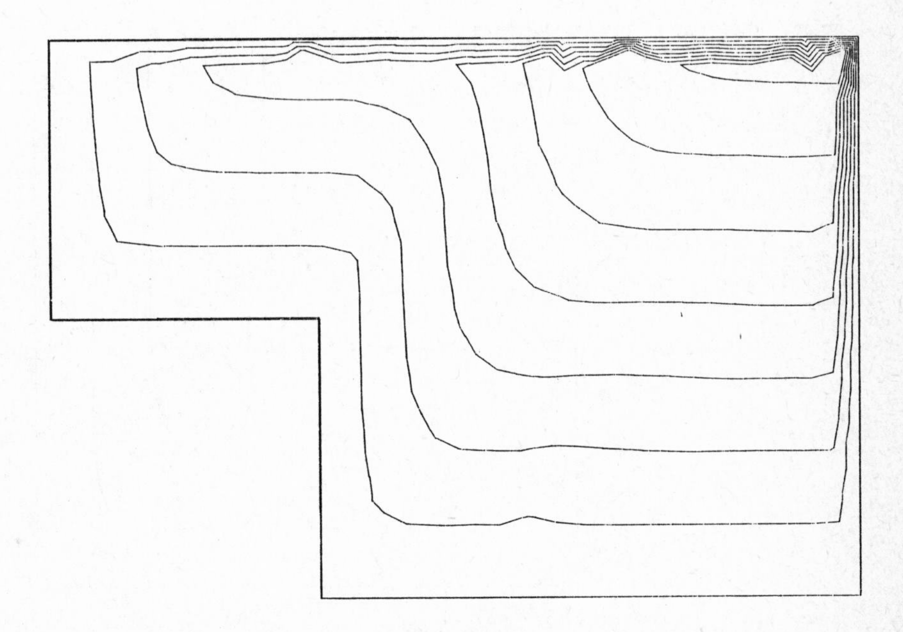

Figure A.24

Triangulation $\mathcal{C}_h^4$, $\varepsilon = 10^{-5}$, $\theta = \frac{\pi}{3}$.

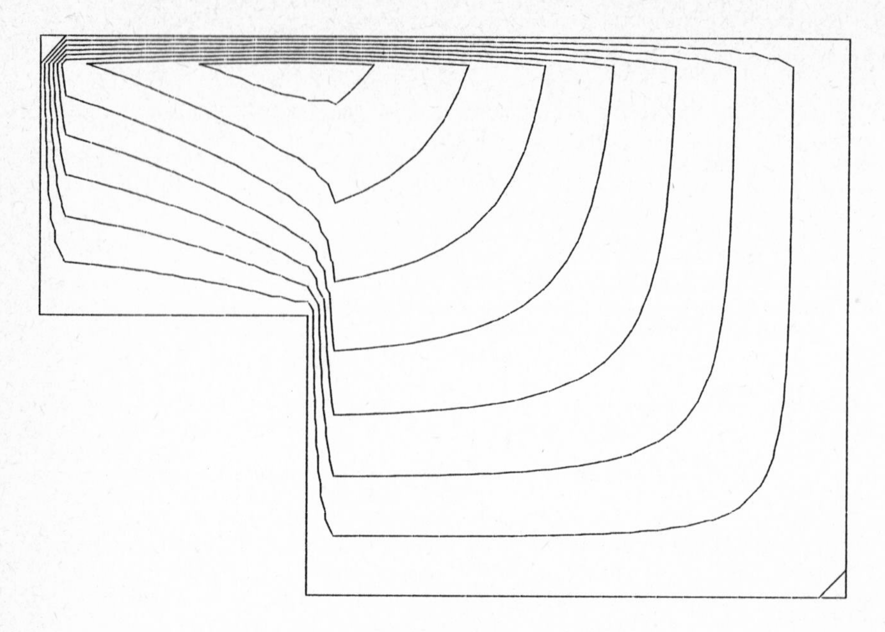

Figure A.25

Triangulation $\mathcal{T}_h^2$, $\varepsilon = 10^{-5}$, $\theta = \frac{3\pi}{4}$.

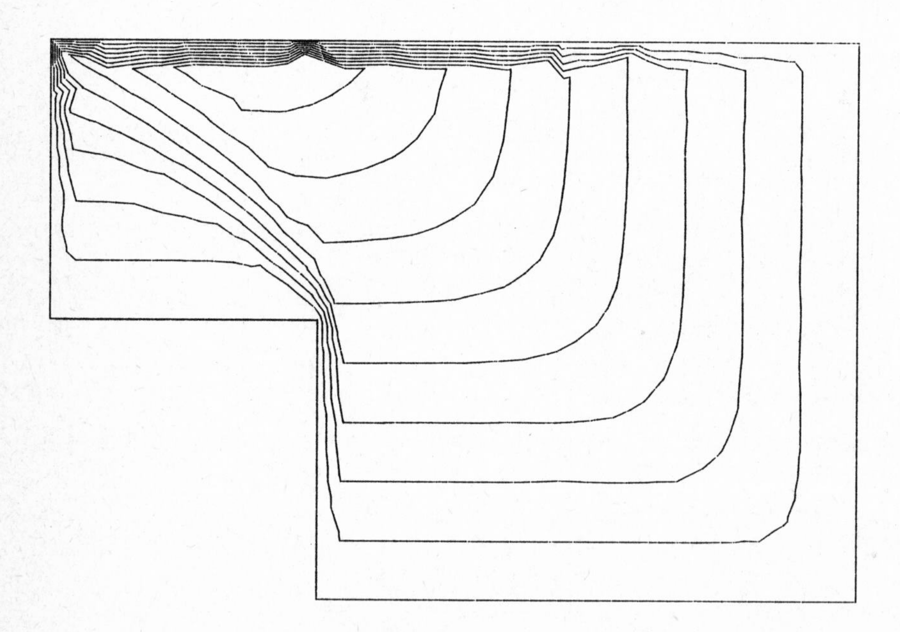

Figure A.26

Triangulation $\mathcal{T}_h^4$, $\varepsilon = 10^{-5}$, $\theta = \frac{3\pi}{4}$.

REFERENCES

[1] M.O. BRISTEAU, R. GLOWINSKI, J. PERIAUX, P. PERRIER, O. PIRONNEAU, G. POIRIER, Application of Optimal Control and Finite Element Methods to the calculation of Transonic Flows and Incompressible Viscous Flows, *Rapport de Recherche N° 294, Laboria,* Avril 1978 (also to appear in *Numerical Methods in Applied Fluid Dynamics,* B. Hunt ed., Acad. Press, London).

[2] M.O. BRISTEAU, R. GLOWINSKI, J. PERIAUX, P. PERRIER, O. PIRONNEAU, On the numerical solution of nonlinear problems in fluid dynamics by least squares and finite element methods (I) Least square formulations and conjugate gradient solution of the continuous problems, *Comp. Meth. Applied Mech. Eng., 17/18* (1979), p. 619-657.

[3] D.K. GARTLING, E.B. BECKER, Finite Element Analysis of Viscous Incompressible Fluid Flow, 1, *Comp. Meth. Applied Mech. Eng., 8* (1976), p. 51-60.

[4] D.K. GARTLING, E.B. BECKER, Finite Element Analysis of Viscous Incompressible Fluid Flow, 2, *Comp. Meth. Applied Mech. Eng., 8* (1976), p. 127-138.

[5] T.J.R. HUGHES, W.K. LIU, A. BROOKS, Finite Element Analysis of Incompressible Viscous Flows by the Penalty Function Formulation, *J. of Comp. Physics, 30,*(1979), p. 1.-60.

[6] M. BERCOVIER, M. ENGELMAN, A Finite Element for the Numerical Solution of Viscous Incompressible Flows, *J. of Comp. Physics, 30,* (1979), p. 181-201.

[7] M. FORTIN, F. THOMASSET, Mixed Finite-Element Methods for Incompressible Flow problems, *J. of Comp. Physics, 31,* (1979), p. 113-145.

[8] V. GIRAULT, P.A. RAVIART, *Finite Element Approximation of Navier-Stokes equations,* Lecture Notes in Math., Springer-Verlag (to appear).

[9] P. LE TALLEC, *Simulation Numérique d'Ecoulements Visqueux Incompressibles par des Méthodes d'Eléments Finis Mixtes.* Thèse de 3ème cycle, Université Pierre et Marie Curie, Paris, 1978.

[10] C. JOHNSON, A mixed finite element method for the Navier-Stokes equations, *R.A.I.R.O. Anal. Num., 12,* (1978), p. 335-348.

[11] R. GLOWINSKI, O. PIRONNEAU, On a mixed finite element approximation of the Stokes problem (I). Convergence of the approximate solutions, *Num. Math.* (to appear).

[12] J.L. LIONS, *Quelques méthodes de résolution des problèmes aux limites non linéaires,*Dunod, Paris, 1969.

[13] O.A. LADYSHENSKAYA, *The mathematical theory of viscous incompressible flows,* Gordon and Breach, 1969.

[14] R. TEMAM, *Theory and Numerical Analysis of the Navier-Stokes equations,* North-Holland, Amsterdam, 1977.

[15] L. TARTAR, *Topics in nonlinear analysis,* Public. Math. d'Orsay, 78.13, Université Paris-Sud, Dept. of Maths., 1978.

[16] J. PERIAUX, *Résolution de quelques problèmes non linéaires en Aerodynamique par des méthodes d'éléments finis et de moindres carrés fonctionnels,* Thèse de 3ème cycle, Université Pierre et Marie Curie, Paris, 1979.

[17] R. GLOWINSKI, O. PIRONNEAU, On Numerical methods for the Stokes problem. Chapter

13 of *Energy Methods in Finite Element Analysis*, R. Glowinski, E.Y. Rodin, O.C. Zienkiewicz Ed., J. Wiley and Sons, Chichester, 1979, p. 243-264.

[18] R.A. ADAMS, *Sobolev Spaces*, Acad. Press, New-York, 1975.

[19] J.L. LIONS; E. MAGENES, *Non-homogeneous boundary value problems and applications*, 1, Springer, New-York, 1972.

[20] E. NECAS, *Les Méthodes Directes en théorie des équations elliptiques*, Masson, Paris, 1967.

[21] J.T. ODEN, J.N. REDDY, *An introduction to the mathematical theory of finite elements*, J. Wiley and Sons, New-York, 1976.

[22] P. LE TALLEC, Convergence of a mixed finite element approximation of the Navier-Stokes equations (to appear).

[23] E. POLAK, *Computational Methods in Optimization*, Acad. Press, New-York, 1971.

[24] R. GLOWINSKI, J. PERIAUX, O. PIRONNEAU, On a mixed finite element approximation of the Stokes' problem (II). Solution of the approximate problems (to appear).

[25] R. GLOWINSKI, O. PIRONNEAU, Numerical methods for the biharmonic equation and for the two-dimensional Stokes problem, *SIAM Review*, *21*, 2, (1979), p. 167-212.

[26] R. GLOWINSKI, O. PIRONNEAU, Approximation par éléments finis mixtes du problème de Stokes en formulation vitesse-pression. Convergence des solutions approchées. C.R. Acad. Sc. Paris, T. 286 A, (1978), p. 181-183.

[27] R. GLOWINSKI, O. PIRONNEAU, Approximation par éléments finis mixtes du problème de Stokes en formulation vitesse-pression. Résolution des problèmes approchés. C.R. Acad. Sc. Paris, T. 286 A, (1978), p. 225-228.

[28] R. GLOWINSKI, J. PERIAUX, O. PIRONNEAU, An efficient preconditioning scheme for iterative numerical solutions of partial différential equations, *Applied Math. Modelling* (to appear).

[29] A.G. HUTTON, A general finite element method for vorticity and stream function applied to a laminar separated flow, *Central Electricity Generating Board* Report, Research Dept., Berkeley Nuclear Laboratories, August 1975.

[30] M. TABATA, A finite element approximation corresponding to the upwind finite differencing, *Memoirs of Numerical Mathematics* (Univ. of Kyoto and Tokyo), 4, (1977), p. 47-63.

[31] J.C. HEINRICH, P.S. HUYAKORN, O.C. ZIENKIEWICZ, A.R. MITCHELL, An "upwind" finite element scheme for two dimensional convective transport equation, *Int. J. Num. Meth. Eng.*, *11*, (1977), p. 131-143.

[32] I. CHRISTIE, A.R. MITCHELL, Upwinding of high order Galerkin methods in conduction-convection problems, *Int. J. Num. Meth. Eng.*, Vol. 12, N° 11, 1978, p. 1764-1771.

[33] C.V. RAMAKRISHNAN, An upwind finite element scheme for the unsteady convective diffusive transport equation, *Applied Math. Modelling*, *3*, (1979), p. 280-284.

[34] O. WIDLUND, A Lanczos method for a class of non symmetric systems of linear equations, *SIAM J. Num. Anal.*, *15* (1978), 4, p. 801-812.

[35] B. ENGQUIST, H.O. KREISS, Difference and Finite Element Methods for Hyperbolic Differential Equations, *Comp. Meth. Appl. Mech. Eng.*, *17/18*, (1979), p. 581-596.

THE INITIAL VALUE PROBLEM FOR THE BOUSSINESQ EQUATIONS WITH DATA IN L^p

J.R. Cannon and Emmanuele DiBenedetto
The University of Texas
Austin, Texas 78712

Existence and uniqueness of solutions in $L^{p,q}$ for the initial value problem for the Boussinesq equations that describe the flow of a viscous incompressible fluid subject to convective heat transfer is demonstrated via extensive use of the imbedding theorem for singular integral operators.

1. Introduction

In [3], Fabes, Jones and Riviere considered the initial value problem for the Navier-Stokes equations with initial data in L^p. Here, we shall consider the initial value problem for the Boussinesq equations in the infinite cylinder $S_T = R^n x(0,T]$. In other words, given $u_0(x) = (u_{10}(x),\ldots,u_{n0}(x))$ satisfying $\nabla \cdot u_0(x) \equiv \operatorname{div} u_0(x)$ $\equiv \sum_{j=1}^{n} (\frac{\partial}{\partial x_j}) u_{j0}(x) = 0$, $x \in \mathbb{R}^n$ and given $\psi(x)$, we seek a solution (u,θ,P), where $u = (u_1(x,t),\ldots,u_n(x,t))$, and $\theta = \theta(x,t)$, $P = P(x,t)$, such that

$$(1) \qquad \frac{\partial\theta}{\partial t} - \Delta\theta + u\cdot\nabla\theta = q(x,t), \qquad x \in \mathbb{R}^n, \qquad t \in (0,T],$$

$$(2) \qquad \theta(x,0) = \psi(x), \qquad x \in \mathbb{R}^n,$$

$$(3) \qquad \frac{\partial u}{\partial t} - \Delta u + u\cdot\nabla u + \nabla P = f(\theta(x,t)), \quad x \in \mathbb{R}^n, \qquad t \in (0,T],$$

$$(4) \qquad \nabla\cdot u = 0 \qquad x \in \mathbb{R}^n, \qquad t \in (0,T],$$

$$(5) \qquad u(x,0) = u_0(x), \qquad x \in \mathbb{R}^n$$

where q and f are known data functions, Δ denotes the Laplacian with respect to x, ∇ the gradient with respect to x, and $u\cdot\nabla$ denotes the differential operator

$$(6) \qquad u\cdot\nabla \equiv \sum_{j=1}^{n} u_j \frac{\partial}{\partial x_j} .$$

Since P is quickly eliminated below, we shall consider here solutions $U = (u,\theta)$ which are in the class

(7) $$L^{p,q}_{n+1}(S_T) = \{U : S_T \to R^{n+1} : \|U\|_{p,q} = \sum_{j=1}^{n+1} \{\int_0^T (\int_{R^n} |U_j|^p dx)^{\frac{q}{p}} dt\}^{\frac{1}{q}} < \infty,$$

$$p,q \geq 2; \quad \frac{n}{p} + \frac{2}{q} \leq 1, \quad n < p < \infty \quad ,$$

where $U_j = u_j$, $j = 1,\ldots,n$, and $U_{n+1} = \theta$. The conditions $\nabla \cdot u = \nabla \cdot u_0 = 0$ are to be regarded here as in the sense of distributions.

In Section 2 we begin our study of (1)-(5) by considering its weak formulation and showing that it is equivalent to solving the singular integral equation

(8) $$U(x,t) + B(U,U)(x,t) = U_0(x,t) + F(U)(x,t),$$

where B is a bilinear form defined below, U_0 involves the initial data u_0 and ψ, and F involves q and $f(\theta)$.

We gather some preliminary estimates into Section 3. Namely, we show that B is a continuous map of $L^{p,q}_{n+1}(S_T) \times L^{p,q}_{n+1}(S_T)$ into $L^{p,q}_{n+1}(S_T)$. Also, a mild generalization of q and f is allowed. This consists of assuming $q = q(x,t,U)$ and $f = f(x,t,U)$. It is shown that if $F = (f,q)$ is a continuous map from $L^{p,q}_{n+1}(S_T)$ into $L^{p_1,q_1}_{n+1}(S_T)$, where $1 < p_1 \leq p$, $1 < q_1 < q$ and $\frac{1}{q_1} + \frac{n}{2p_1} \leq \frac{1}{q} + \frac{n}{2p} + 1$, then $F(U)$ is a continuous map from $L^{p,q}_{n+1}(S_T)$ into $L^{p,q}_{n+1}(S_T)$. Similarly, it is shown that if $U_0 = (u_0,\psi) \in L^r_{n+1}(\mathbb{R}^n)$, where $0 < \frac{n}{r} < \frac{2}{q} + \frac{n}{p} \leq 1$, then U_0 is a continuous map of $L^r_{n+1}(\mathbb{R}^n)$ into $L^{p,q}_{n+1}(S_T)$. Hence, the essence of Section 3 is to show that $-B(U,U) + U_0 + F(U)$ is a continuous map of $L^{p,q}_{n+1}(S_T)$ into $L^{p,q}_{n+1}(S_T)$.

The estimates in Section 3 are used in Section 4 to show that $-B(U,U) + U_0 + F(U)$ is a contraction. Hence, existence and uniqueness follow immediately. Global existence and regularity of the solution are discussed in Sections 5 and 6.

This paper uses extensively the results of the Fabes, Jones and Riviere paper [3]. Hence, in what follows, we shall refer to it as the F-J-R paper.

2. A Non-linear Integral Equation

Consider equations (1) and (2). Let $S(\mathbb{R}^n)$ denote the space of rapidly decreasing functions [6, p. 168].

Definition. For $u \in L^{p,q}_n(S_T)$, which satisfies $\text{div } u = 0$ weakly, we say that

$\theta = \theta(x,t) \in L_1^{p,q}(S_T)$ is a weak solution of (1) and (2) if θ satisfies

(9)
$$\int_0^T \int_{\mathbb{R}^n} \theta(x,t)\{\phi_t + \Delta\phi + u\cdot\nabla\phi\}dxdt$$

$$= -\int_{\mathbb{R}^n} \psi(x)\phi(x,0)dx - \int_0^T\int_{\mathbb{R}^n} q(x,t)\phi(x,t)dxdt$$

for all $\phi \in \mathcal{S}(\mathbb{R}^{n+1})$ such that $\phi = 0$ for $t \geq T$.

Set

(10)
$$B_0(u,\theta)(x,t) = \int_0^t \int_{\mathbb{R}^n} u(y,s)\cdot\nabla\Gamma(x-y,t-s)\theta(y,s)dyds$$

and

(11)
$$g(x,t) = \int_{\mathbb{R}^n} \Gamma(x-y,t)\psi(y)dy + \int_0^t\int_{\mathbb{R}^n} \Gamma(x-y,t-s)q(y,s)dyds,$$

where

(12)
$$\Gamma(x,t) = (4\pi t)^{-\frac{n}{2}} \exp\{-\frac{|x|^2}{4t}\}, \quad t > 0, \quad |x|^2 = \sum_{i=1}^n x_i^2$$

is the fundamental solution of the heat operator $L = -\frac{\partial}{\partial t} + \Delta$. Consider the integral equation

(13)
$$\theta(x,t) + B_0(u,\theta)(x,t) = g(x,t).$$

The proof of the following proposition follows from an argument of [3], modulo minor modifications.

<u>Proposition 1.</u> Let $\psi \in L_1^r(\mathbb{R}^n)$ and $q \in L_1^{p_1,q_1}(S_T)$ for $1 \leq r < \infty$, $p_1, q_1 \geq 2$, and $p_1 < \infty$. Then, $\theta \in L_1^{p,q}(S_T)$ is a weak solution of (1) and (2) if and only if it is a solution of (13) for $u \in L_n^{p,q}(S_T)$ which satisfy div $u = 0$ in the weak sense.

Next, we consider equations (3), (4) and (5) and recall results from the F-J-R paper. Let $\mathcal{D}_n^T$ denote vectors $\phi = (\phi_1(x,t),\ldots,\phi_n(x,t))$ such that each $\phi_j \in \mathcal{S}(\mathbb{R}^{n+1})$ and $\phi_j \equiv 0$, $t \geq T$. Let D_n^T denote the linear subspace of $\mathcal{D}_n^T$ such that div $\phi = 0$.

<u>Definition.</u> A function $u = u(x,t) \in L_n^{p,q}(S_T)$ is a weak solution of (3), (4) and (5) if

$$(14) \qquad \int_0^T \int_{\mathbb{R}^n} \langle u, L^*\phi + u\cdot\nabla\phi\rangle\, dxdt$$

$$= -\int_{\mathbb{R}^n} \langle u_0(x), \phi(x,0)\rangle\, dx - \int_0^T \int_{\mathbb{R}^n} \langle f(\theta(x,t)), \phi(x,t)\rangle\, dxdt$$

is satisfied for every $\phi \in D_n^T$, where $\langle\cdot,\cdot\rangle$ denotes here the inner product in $\mathbb{R}^n$, and $\operatorname{div} u = 0$ in the sense of distributions. In the F-J-R paper the following results are shown. There exists an $n \times n$ symmetric matrix $E(x,t) = (E_{ij}(x,t))$, $i,j = 1,\ldots,n$, such that every row is a divergence-free fundamental solution of the heat operator L. The elements of this matrix have the form

$$(15) \qquad E_{ij}(x,t) = \delta_{ij}\Gamma(x,t) - R_i R_j \Gamma(x,t),$$

where δ_{ij} is the Kronecker symbol, Γ is as in (12), and R_i is the ith Riesz transform defined by

$$(16) \qquad R_i(f)(x) = \lim_{\varepsilon\to 0} c_i \int_{|x-y|>\varepsilon} \frac{x_i - y_i}{|x-y|^{n+1}} f(y)dy ,$$

i.e., R_i is a singular integral operator on $L^p(\mathbb{R}^n)$ with limit being taken in $L^p(\mathbb{R}^n)$. See [7]. Recall that R_i maps $L^p(\mathbb{R}^n) \to L^p(\mathbb{R}^n)$ for $1 < p < \infty$ and is of weak type for $p = 1$. On $L_n^{p,q}(S_T) \times L_n^{p,q}(S_T)$ define the bilinear form

$$(17) \qquad B(u,v)(x,t) = \int_0^t \int_{\mathbb{R}^n} \langle u(y,s), \nabla_x E(x-y, t-s)\rangle\, v(y,s)dyds,$$

where $\langle u(y,s), \nabla_x E(x-y,t-s)\rangle$ is the $n \times n$ matrix whose entries are

$$(18) \qquad c_{ik} = \langle u(y,s), D_{x_k} E_i(x-y,t-s)\rangle$$

and where here $E_i(\cdot,\cdot)$ denotes the ith row of $E(\cdot,\cdot)$. Set

$$(19) \qquad k(x,t) = \int_{\mathbb{R}^n} u_0(y)\Gamma(x-y,t)dy + \int_0^t \int_{\mathbb{R}^n} E(x-y,t-s)f(\theta(y,s))dyds.$$

<u>Proposition</u> 2. (F-J-R [3]) Let $u_0 \in L_n^r(\mathbb{R}^n)$ and $\theta \in L_1^{p,q}(S_T)$. Then $u \in L_n^{p,q}(S_T)$ is a weak solution of (3), (4) and (5) if and only if it is a solution of

$$(20) \qquad u(x,t) + B(u,u)(x,t) = k(x,t) .$$

At this point it is clear that the definition of weak solutions given above can be combined to yield one for (1)-(5) and that Proposition 1 and Proposition 2 can be employed to show that $(u,\theta) \in L^{p,q}_{n+1}(S_T)$ is a weak solution of (1)-(5) if and only if it is a solution of the system

$$(21)\qquad \begin{cases} u(x,t) + B(u,u)(x,t) = k(x,t), \\ \theta(x,t) + B_0(u,\theta)(x,t) = g(x,t). \end{cases}$$

Set

$$(22)\qquad U(x,t) = (u(x,t),\theta(x,t)),$$

$$(23)\qquad G(x,t) = \begin{vmatrix} E(x,t) & 0 \\ 0 & \Gamma(x,t) \end{vmatrix},$$

and

$$(24)\qquad \langle U(y,s),\nabla_x G(x-y,t-s)\rangle = \begin{vmatrix} \langle u(y,s),\ \nabla_x E(x-y,t-s)\rangle & 0 \\ \theta(y,s)\nabla_x \Gamma(x-y,t-s) & 0 \end{vmatrix}$$

Thus, the vector $\langle U(y,s),\ \nabla_x G(s-y,t-s)\rangle U(y,s)$ has its first n-components given by $\langle u(y,s),\ \nabla_x E(x-y,t-s)\rangle u(y,s)$ and the $(n+1)^{st}$ component by $\{u(y,s)\cdot\nabla_x\Gamma(x-y,t-s)\}\theta(y,s)$. Defining

$$(25)\qquad \mathcal{U}_0(x,t) = \int_{\mathbb{R}^n} (u_0(y),\psi(y))\Gamma(x-y)dy$$

and

$$(26)\qquad \mathcal{F}(U)(x,t) = \int_0^t \int_{\mathbb{R}^n} G(x-y,t-s)(f(\theta(y,s)),q(y,s))dyds$$

we see that the system (21) is equivalent to

$$(27)\qquad U(x,t) + \mathcal{B}(U,U)(x,t) = \mathcal{U}_0(x,t) + \mathcal{F}(U)(x,t),$$

where

$$(28)\qquad \mathcal{B}(U,V)(x,t) = \int_0^t \int_{\mathbb{R}^n} \langle U(y,s),\nabla_x G(x-y,t-s)\rangle V(y,s)dyds$$

is a bilinear form on $L^{p,q}_{n+1}(S_T) \times L^{p,q}_{n+1}(S_T)$. In the F-J-R paper it is shown that

$$(29)\qquad \frac{\partial}{\partial x_k} E_{ij}(x,t) = \delta_{ij}\frac{\partial}{\partial x_k}\Gamma(x,t) + \frac{1}{(4\pi)^{\frac{n}{2}}}\int_0^{\frac{1}{t}} \frac{\partial^3}{\partial x_k \partial x_i \partial x_j} e^{-\frac{s|x|^2}{4}} s^{\frac{n}{2}-\frac{1}{2}} ds.$$

A tedious but elementary argument shows that

$$\frac{\partial}{\partial x_k} G(x,t) \in L^1(S_T) \tag{30}$$

and that

$$\left|\frac{\partial}{\partial x_k} G(x,t)\right| \le \frac{C}{(|x| + \sqrt{t})^{n+1}}, \tag{31}$$

where here and below C will denote a constant depending at most upon n,p,q,p_1,q_1, and r, but not upon f, ψ, or u_0. From (28) and (30) it follows that $B(U,V)$ is a continuous map of $L^{p,q}_{n+1}(S_T) \times L^{p,q}_{n+1}(S_T)$ into $L^{p/2,q/2}_{n+1}(S_T)$. The purpose of the next section is to show that $B(U,V)$ maps $L^{p,q}_{n+1}(S_T) \times L^{p,q}_{n+1}(S_T)$ continuously into $L^{p,q}_{n+1}(S_T)$.

3. Preliminary Estimates

In this section extensive use of the following imbedding theorem will be made. See [2,7] for arguments.

Theorem 1. (Imbedding) Let $g \in L^{s_1}(\mathbb{R}^m)$. The transformation

$$Tg(x) = \int_{\mathbb{R}^m} \frac{g(y)}{|x-y|^{m-\alpha}}\, dy, \quad x \in \mathbb{R}^m, \quad 0 < \alpha < m, \tag{32}$$

maps $L^{s_1}(\mathbb{R}^m)$ continuously into $L^{s_2}(\mathbb{R}^m)$ provided that

$$0 < \frac{1}{s_2} = \frac{1}{s_1} - \frac{\alpha}{m}. \tag{33}$$

In other words, there is a constant C depending only upon m, s_1, and α, but not g such that

$$\| Tg\|_{L^{s_2}(\mathbb{R}^m)} \le C\| g\|_{L^{s_1}(\mathbb{R}^m)}. \tag{34}$$

We now proceed to show the continuity of $B(\cdot,\cdot)$.

Theorem 2. Let $U(x,t)$ and $V(x,t)$ belong to $L^{p,q}_{n+1}(S_T)$. Then, the form $B(\cdot,\cdot)$ is a continuous map of $L^{p,q}_{n+1}(S_T) \times L^{p,q}_{n+1}(S_T)$ into $L^{p,q}_{n+1}(S_T)$ provided that $\frac{n}{p} + \frac{2}{q} \le 1$ and $n < p \le \infty$. In particular, we have that if $\frac{n}{p} + \frac{2}{q} = 1$ and $n < p < \infty$, then

$$\|B(U,V)\|_{L^{p,q}_{n+1}(S_T)} \le C(n,p,q)\, \| U\|_{L^{p,q}_{n+1}(S_T)}\, \| V\|_{L^{p,q}_{n+1}(S_T)} \tag{35}$$

and that if $\frac{n}{p} + \frac{2}{q} < 1$ and $n < p \le \infty$, then

(36) $$\| B(U,V) \|_{L^{p,q}_{n+1}(S_T)} \le C(n,p,q) T^{1/2(1 - 2/q-n/p)} \| U \|_{L^{p,q}_{n+1}(S_T)} \| V \|_{L^{p,q}_{n+1}(S_T)}$$

<u>Proof</u>. The proof is similar to that in the F-J-R paper. We omit the few modifications needed.

Now, we consider the function $F(U)(x,t)$ and derive conditions under which it belongs to $L^{p,q}_{n+1}(S_T)$ and constitutes a continuous map from $L^{p,q}_{n+1}(S_T)$ into $L^{p,q}_{n+1}(S_T)$. The original formulation of the problem specified $f = f(\theta)$ and $q = q(x,t)$, but it does not complicate the analysis to assume that $f = f(x,t,U)$ and $q = q(x,t,U)$. Setting $F(x,t,U) = (f,q)$ in $F(U)(x,t)$, we have

(37) $$F(U)(x,t) = \int_0^t \int_{\mathbb{R}^n} G(x-y,t-s) F(y,s,U(y,s)) dyds.$$

For $F(U)$, we prove the following results.

<u>Theorem</u> 3. If $F(x,t,U(x,t))$ is a continuous map from $L^{p,q}_{n+1}(S_T)$ into $L^{p_1,q_1}_{n+1}(S_T)$; i.e. there exists a constant K independent of U such that

(38) $$\| F(\cdot,\cdot,U(\cdot,\cdot)) \|_{L^{p_1,q_1}_{n+1}(S_T)} \le K \| U \|_{L^{p,q}_{n+1}(S_T)},$$

and if

(39) $$1 < p_1 \le p, \quad 1 < q_1 < q \quad \text{and} \quad \frac{1}{q_1} + \frac{n}{2p_1} \le \frac{1}{q} + \frac{n}{2p} + 1 ,$$

then $F(U) \in L^{p,q}_{n+1}(S_T)$ and there exists a constant C independent of U such that

(40) $$\| F(U) \|_{L^{p,q}_{n+1}(S_T)} \le CT^{\gamma} \| F(\cdot,\cdot,U(\cdot,\cdot)) \|_{L^{p_1,q_1}_{n+1}(S_T)},$$

where

(41) $$\gamma = 1 + \frac{n}{2p} + \frac{1}{q} - \frac{n}{2p_1} - \frac{1}{q_1} .$$

<u>Proof</u>. Recalling (15) and that for $p > 1$ the Riesz transform maps $L^p(\mathbb{R}^n)$ into $L^p(\mathbb{R}^n)$, it suffices to demonstrate the result for

(42) $$H(x,t) = \int_0^t \int_{\mathbb{R}^n} \Gamma(x-y,t-s) F(y,s,U(y,s)) dyds.$$

If $p = p_1$, then

$$(43)\qquad \| H(\cdot,t)\|_{L^p_{n+1}(\mathbb{R}^n)} \le C\int_0^t \| F(\cdot,s,U(\cdot,s))\|_{L^p_{n+1}(\mathbb{R}^n)} ds$$

which implies that for all $q_1 \ge 1$, $H \in L^{p,\infty}_{n+1}(S_T) \subset L^{p,q}_{n+1}(S_T)$ and that

$$(44)\qquad \| H\|_{L^{p,q}_{n+1}(S_T)} \le CT^{1/q-1/q_1+1} \| F(\cdot,\cdot,U(.,.))\|_{L^{p_1,q_1}_{n+1}(S_T)} .$$

When $1 < p_1 < p$, we select $\sigma \in (0,1)$ so that

$$(45)\qquad 0 < \frac{1}{p} = \frac{1}{p_1} - (1-\sigma).$$

Since Γ satisfies the estimate

$$(46)\qquad \Gamma(x,t) \le \frac{C}{[|x|+t^{1/2}]^n} \le \frac{C_\sigma}{|x|^{\sigma n}\, t^{(1-\sigma)n/2}} .$$

For $\sigma \in (0,1)$, we see that

$$(47)\qquad |H(x,t)| \le C\int_0^t \frac{1}{|t-s|^{(1-\sigma)n/2}} \int_{\mathbb{R}_n} \frac{|F(y,s,U(y,s))|}{|x-y|^{n-n(1-\sigma)}} dyds .$$

An application of the Theorem 1 and the continuous version of the Minkowski inequality yields

$$(48)\qquad \| H(\cdot,t)\|_{L^p_{n+1}(\mathbb{R}^n)} \le C\int_0^t |t-s|^{-(1-\alpha)} \| F(.,s,U(.,s))\|_{L^{p_1}_{n+1}(\mathbb{R}^n)} ds ,$$

where

$$\alpha = 1 - \frac{n}{2}(1-\sigma) = 1 + \frac{n}{2p} - \frac{n}{2p_1} \ge \frac{1}{q_1} - \frac{1}{q} > 0$$

if $q_1 < q$. Another application of Theorem 1 yields

$$(49)\qquad \| H\|_{L^{p,q}_{n+1}(S_T)} \le C\| F(\cdot,\cdot,U(\cdot,\cdot))\|_{L^{p_1,q_1}_{n+1}(S_T)} .$$

Hence, (40) follows from (44) and (49).

We conclude our preliminary estimates with the following result concerning the initial condition $U_0(x)$.

<u>Theorem 4</u>. If $U_0(x) = (u_0(x),\psi(x)) \in L^r_{n+1}(\mathbb{R}^n)$, where

$$(50) \qquad 0 < \frac{n}{r} < \frac{2}{q} + \frac{n}{p} \leq 1,$$

then

$$(51) \qquad U_0(x,t) = \int_{\mathbb{R}^n} \Gamma(x-y,t)U_0(s)ds \in L^{p,q}_{n+1}(S_T)$$

and there exists a constant C independent of $U_0(x)$ such that

$$(52) \qquad \| U_0 \|_{L^{p,q}_{n+1}(S_T)} \leq CT^{\alpha} \| U_0 \|_{L^r_{n+1}(\mathbb{R}^n)},$$

where

$$(53) \qquad \alpha = \frac{1}{q} + \frac{n}{2p} - \frac{n}{2r} .$$

Proof. From (46), we see that for $\sigma \in (0,1)$,

$$(54) \qquad |U_0(x,t)| \leq Ct^{-\frac{n\sigma}{2}} \int_{\mathbb{R}^n} \frac{U_0(y)}{|x-y|^{n-n\sigma}} dy.$$

Applying Theorem 1 with $\alpha = n\sigma$ and σ satisfying

$$(55) \qquad 0 < \frac{1}{p} = \frac{1}{r} - \sigma ,$$

we obtain

$$(56) \qquad \| U_0(\cdot,t) \|_{L^p_{n+1}(\mathbb{R}^n)} \leq Ct^{-\frac{n\sigma}{2}} \|U\|_{L^r_{n+1}(\mathbb{R}^n)} .$$

From (50) and (55) it follows that $\frac{n\sigma q}{2} < 1$. Hence, we obtain (52) from a simple quadrature.

The following elementary estimate is also needed for the discussion of existence in the next section.

Lemma 1. Let $\{a_m\}$ be a sequence of positive numbers such that

$$a_m \leq \lambda_1 a^2_{m-1} + \lambda_2 a_{m-1} + a_0, \quad m \geq 1.$$

Then

$$a_m \le \frac{2\,a_0}{1-\lambda_2}$$

provided that

$$(1-\lambda_2)^2 > 4\,\lambda_1\,a_0.$$

and that

$$\lambda_2 < 1\ .$$

Proof. Assume that $a_m \le a_0\lambda^{-1}$ holds for all m. Then, from the recursion relation, we must have

$$\lambda_1(a_0\lambda^{-1})^2 + \lambda_2 a_0\lambda^{-1} + a_0 \le a_0\lambda^{-1}\ ,$$

or

$$\lambda_1 a_0 + (\lambda_2-1)\lambda + \lambda^2 \le 0\ .$$

So, λ must belong to $[\lambda_-,\lambda_+]$ where

$$\lambda_{\pm} = \frac{(1-\lambda_2)\pm\sqrt{(1-\lambda_2)^2 - 4\lambda_1 a_0}}{2}\ .$$

Select λ_+ and note that

$$a_0 \le \frac{2a_0}{(1-\lambda_2) + \sqrt{(1-\lambda_2)^2-4\lambda a_0}} \le \frac{2a_0}{(1-\lambda_2)}\ .$$

4. Existence

Let C in this section denote the largest constant appearing in the theorems in Section 3. Note that C depends only upon n,p,q,p_1,q_1 and r. We begin by demonstrating the following result.

Theorem 5. Suppose that $\frac{n}{p}+\frac{2}{q} \le 1$, $n < p < \infty$, that $U_0(x) \in L^r_{n+1}(\mathbb{R}^n)$, $0 < \frac{n}{r} < \frac{2}{q} + \frac{n}{p} \le 1$, that $F(x,t,U(x,t))$ is a continuous map from $L^{p,q}_{n+1}(S_T)$ into $L^{p_1,q_1}_{n+1}(S_T)$, $1 < p_1 \le p$, $1 < q_1 < q$, $\frac{1}{q_1} + \frac{n}{2p_1} \le \frac{1}{q} + \frac{n}{2p} + 1$, and that the continuity of F be Lipschitz; i.e., there exists a constant K such that

$$\| F(\cdot,\cdot,U_1(\cdot,\cdot)) - F(\cdot,\cdot,U_2(\cdot,\cdot))\|_{L^{p_1,q_1}_{n+1}(S_T)} \le K\| U_1-U_2\|_{L^{p,q}_{n+1}(S_T)} \tag{57}$$

for U_1 and $U_2 \in L^{p,q}_{n+1}(S_T)$. Then the integral equation

$$U + \mathcal{B}(U,U) = \mathcal{U}_0 + \mathcal{F}(U) \tag{58}$$

has a unique solution $U \in L^{p,q}_{n+1}(S_T)$ provided that

$$1 - KCT^{\gamma} > 4C^2T^{1/2(1-n/p-2/q}\left\{T^{\alpha}\| U_0\|_{L^r_{n+1}(\mathbb{R}^n)}\right\} \tag{59}$$

and that

$$C\{2MT^{1/2(1-n/p-2/q)} + KT^{\gamma}\} < 1, \tag{60}$$

where

$$M = \frac{2C\left[T^{\alpha}\| U_0\|_{L^r_{n+1}(\mathbb{R}^n)}\right]}{1-KCT^{\gamma}}, \tag{61}$$

γ is defined by (41) and α is defined by (53).

<u>Remark</u>. If α,γ and $1 - \frac{n}{p} - \frac{2}{q}$ are positive, then it is clear that (59) and (60) can be verified by selecting T sufficiently small. If $\gamma = 0$, but α and $4 - \frac{n}{p} - \frac{2}{q}$ are positive, then (59) and (60) are valid if T and K are sufficiently small. Clearly, Theorem 5 is <u>local in nature</u>. Further restrictions upon the data are forced if α and/or $1 - \frac{n}{p} - \frac{2}{q}$ vanish.

<u>Proof</u>. Set $V_0(x,t) = U_0(x)$ and define

$$V_m(x,t) = -\mathcal{B}(V_{m-1},V_{m-1})(x,t) + \mathcal{U}_0(x,t) + \mathcal{F}(V_{m-1})(x,t). \tag{62}$$

From Theorems 2,3 and 4, it follows that

$$\| V_m\|_{L^{p,q}_{n+1}(S_T)} \le CT^{1/2(1-2/q-n/p)}\| V_{m-1}\|^2_{L^{p,q}_{n+1}(S_T)} + CKT^{\gamma}\| V_{m-1}\|_{L^{p,q}_{n+1}(S_T)} + CT^{\alpha}\| U_0\|_{L^r_{n+1}(\mathbb{R}^n)}. \tag{63}$$

Setting

$$\lambda_1 = CT^{1/2(1-2/q-n/p)} \tag{64}$$

$$\lambda_2 = CKT^{\gamma} \quad , \tag{65}$$

and

$$a_0 = CT^{\alpha} \| U_0 \|_{L^r_{n+1}(\mathbb{R}^n)} \quad , \tag{66}$$

we see that if $\lambda_2 < 1$, Lemma 1 implies

$$\| V_m \|_{L^{p,q}_{n+1}(S_T)} \leq M \quad , \tag{67}$$

where M is given by (61). Consider now

$$V_{m+1} - V_m = B(V_{m-1}, V_{m-1}) - B(V_m, V_m) + F(V_m) - F(V_{m-1}). \tag{68}$$

Using the bilinearity of B, we obtain

$$V_{m+1} - V_m = B(V_{m-1}, V_{m-1} - V_m) + B(V_{m-1} - V_m, V_m) + F(V_m) - F(V_{m-1}) \quad . \tag{69}$$

Applying Theorems 3 and 4 and (67) we see that

$$\| V_{m+1} - V_m \|_{L^{p,q}_{n+1}(S_T)} \leq \{2MCT^{1/2(1-2/q-n/p)} + KCT^{\gamma}\} \| V_m - V_{m-1} \|_{L^{p,q}_{n+1}(S_T)} \quad . \tag{70}$$

From (60) we see that the mapping $-B(U,U) + U_0 + F(U)$ is a contraction of $L^{p,q}_{n+1}(S_T)$ into itself. Hence, we conclude that there exists a unique solution of the integral equation in $L^{p,q}_{n+1}(S_T)$.

5. Global Existence

We show now that for sufficiently small initial data existence for all T can be obtained in some cases. We begin the discussion by demonstrating the following result.

Theorem 6. Suppose that $\frac{n}{p} + \frac{2}{q} \leq 1$ and that $U_0(x) \in L^{r_1}_{n+1}(\mathbb{R}^n) \cap L^{r_2}_{n+1}(\mathbb{R}^n)$ with $\frac{n}{p} + \frac{2}{q} - \frac{n}{r_1} < 0 < \frac{n}{p} + \frac{2}{q} - \frac{n}{r_2}$. Then $U_0(x,t)$ belongs to $L^{p,q}_{n+1}(\mathbb{R}^n \times (0,\infty))$ and there exists a constant $C = C(p,q,n,r_1,r_2)$ such that

$$(71) \qquad \| U_0(x,t) \|_{L^{p,q}_{n+1}(\mathbb{R}^n \times (0,\infty))} \le C \| U_0 \|_{L^{r_1}_{n+1}(\mathbb{R}^n) \cap L^{r_2}_{n+1}(\mathbb{R}^n)}$$

where

$$(72) \qquad \| U_0 \|_{L^{r_1}_{n+1}(\mathbb{R}^n) \cap L^{r_2}_{n+1}(\mathbb{R}^n)} = \| U_0 \|_{L^{r_1}_{n+1}(\mathbb{R}^n)} + \| U_0 \|_{L^{r_2}_{n+1}(\mathbb{R}^n)} .$$

<u>Proof</u>. Since $\frac{n}{p} + \frac{2}{q} > \frac{n}{r_2}$, we obtain from Theorem 4 that

$$(73) \qquad \| U_0(x,t) \|_{L^{p,q}_{n+1}(S_1)} \le C \| U_0 \|_{L^{r_2}_{n+1}(\mathbb{R}^n)} .$$

Moreover, since $\frac{n}{p} + \frac{2}{q} < \frac{n}{r_1}$, the same argument used in the proof of Theorem 4 yields $\frac{n}{2}\sigma q > 1$ from which it follows that $t^{-(n\sigma/2)} \epsilon L^q_1(1,\infty))$. Therefore,

$$(74) \qquad \| U_0 \|_{L^{p,2}_{n+1}(\mathbb{R}^n \times [0,\infty))} \le \| U_0 \|_{L^{p,q}_{n+1}(S_1)} + \| U_0 \|_{L^{p,q}_{n+1}(\mathbb{R}^n \times [1,\infty))}$$

$$\le C \| U_0 \|_{L^{r_1}_{n+1}(\mathbb{R}^n) \cap L^{r_2}_{n+1}(\mathbb{R}^n)} .$$

Combining Theorems 5 and 6, we obtain the following result.

<u>Theorem 7</u>. Suppose that $\frac{n}{p} + \frac{2}{q} = 1$, $\frac{n}{2p} + \frac{1}{q} + 1 = \frac{n}{2p_1} + \frac{1}{q_1}$, and K and $\| U_0 \|_{L^{r_1}_{n+1}(\mathbb{R}^n) \cap L^{r_2}_{n+1}(\mathbb{R}^n)}$ are sufficiently small. Then, for each $T > 0$, there exists a unique function $U \epsilon L^{p,q}_{n+1}(S_T)$ which solves

$$U + B(U,U) = U_0 + F(U)$$

in S_T.

6. <u>Some Regularity Results</u>

We show here that when the initial data possesses a derivative with respect to x_k, so does the solution.

<u>Theorem 8</u>. Suppose that U is a solution of $U + B(U,U) = U_0 + F(U)$ in S_T, that $2 < q < \infty$, $n < p < \infty$, $\frac{n}{p} + \frac{2}{q} = 1$, $\frac{n}{2p} + \frac{1}{q} + 1 = \frac{n}{2p_1} + \frac{1}{q_1}$, $0 < \frac{n}{p_1} - \frac{2n}{p} < 1$, and

that $\frac{\partial}{\partial x_k} U_0 \in L^{p/2,q/2}_{n+1}(S_T)$, $k = 1,\ldots,n$. Then $\frac{\partial}{\partial x_k} U \in L^{p/2,q/2}_{n+1}(S_T)$.

Proof. Formally differentiating the integral equation, we obtain

$$\frac{\partial}{\partial x_k} U = -\frac{\partial}{\partial x_k} B(U,U) + \frac{\partial}{\partial x_k} U_0 + \frac{\partial}{\partial x_k} F(U).$$

From the L^p -theory of singular integrals of elliptic and parabolic type [1,4,5], it follows that $\frac{\partial}{\partial x_k} B(U,U) \in L^{p/2,q/2}_{n+1}(S_T)$ for $1 < \frac{p}{2}, \frac{q}{2} < \infty$. By assumption we have $\frac{\partial}{\partial x_k} U_0 \in L^{p/2,q/2}_{n+1}(S_T)$. Hence, it remains only to show that $\frac{\partial}{\partial x_k} F(U) \in L^{p/2,q/2}_{n+1}(S_T)$. From the definition of $F(U)$ and the estimate (31) on $\left|\frac{\partial G}{\partial x_k}\right|$, it follows that for each $\sigma \in (0,1)$,

$$(75) \qquad \left|\frac{\partial}{\partial x_k} F(U)(x,t)\right| \le C \int_0^t \int_{\mathbb{R}^n} \frac{F(y,s,U(y,s))\,dyds}{|x-y|^{(n+1)(1-\sigma)}|t-s|^{\sigma(n+1)/2}}$$

$$= C \int_0^t \int_{\mathbb{R}^n} \frac{F(y,s,U(y,s))\,dyds}{|x-y|^{n-[(n+1)\sigma-1]}|t-s|^{\sigma(n+1)/2}}$$

Select $\sigma \in (0,1)$ such that

$$(76) \qquad \frac{2}{p} = \frac{1}{p_1} - \frac{(n+1)\sigma-1}{n}.$$

Applying Theorem 1 with $\alpha = (n+1)\sigma-1$ yields

$$(77) \qquad \left\|\frac{\partial}{\partial x_k} F(U)(\cdot,t)\right\|_{L^{p/2}_{n+1}(\mathbb{R}^n)} \le C \int_0^t \frac{\|F(\cdot,s,U(\cdot,s))\|_{L^{p_1}_{n+1}(\mathbb{R}^n)}}{|t-s|^{\sigma(n+1)/2}}\,ds.$$

Solving for σ in (76), we can write

$$(78) \qquad \frac{n+1}{2}\sigma = 1 - \left(\frac{n}{p} - \frac{n}{2p_1} + \frac{1}{2}\right).$$

Hence,

$$(79) \qquad \frac{1}{q_1} - \left(\frac{n}{p} - \frac{n}{2p_1} + \frac{1}{2}\right) = \frac{2}{q}.$$

Another application of Theorem 1 yields

$$\left\| \frac{\partial}{\partial x_k} F(U) \right\|_{L_{n+1}^{p/2,q/2}(S_T)} \leq C \| F(\cdot,\cdot,U(\cdot,\cdot)) \|_{L_{n+1}^{p_1,q_1}(S_T)} .$$

Remark. Additional regularity results can be obtained if we require more smoothness of U_0 and F.

Set

$$D_x^\alpha U = \frac{\partial^{\alpha_1+\dots+\alpha_n} U}{\partial x_1^{\alpha_1},\dots,\partial x_n^{\alpha_n}} , \tag{81}$$

where $\alpha = (\alpha_1,\dots,\alpha_n)$, $\alpha_i \in \mathbb{N}$, is a multi-index with $|\alpha| = \alpha_1+\dots+\alpha_n$. The following result follows from Theorem (3.4) of the F-J-R paper.

Theorem 9. Suppose that $\frac{n}{p} + \frac{2}{q} \leq 1$ and $\frac{n}{2p} + \frac{1}{q} + 1 \geq \frac{n}{2p_1} + \frac{1}{q_1}$, that

$$D_x^\alpha D_t^j U_0 \in L_{n+1}^{p/(|\alpha|+2j+1),q/(|\alpha|+2j+1)}(S_T)$$

and

$$D_x^\alpha D_t^j F \in L_{n+1}^{p/(|\alpha|+2j+1),q/(|\alpha|+2j+1)}(S_T)$$

for $|\alpha| + 2j \leq k$, where k is a positive integer such that $k + 1 < p,q < \infty$.

Then

$$D_x^\alpha D_t^j U \in L_{n+1}^{p/(|\alpha|+2j+1),q/(|\alpha|+2jt+1)}(S_T) .$$

REFERENCES

[1] Calderon, A.P. and A. Zygumd. "On the existence of certàin singular integrals," Acta Math., 88(1952), 85-139

[2] Hedberg, L.I., "On certain convolution inequalities," Proceedings of the American Mathematical Society, Vol. 36, #2, Dec. 1972.

[3] Fabes, E.B., Jones, F.B. and N.M. Riviere, "The initial value problem for the Navier-Stokes equations with data in L^p," Archives for Rat. Mechanics and Analysis, 45(1972), 222-240

[4] Lewis, J.E., "Mixed estimates for singular integrals," Proceedings of Symposia in Pure Mathematics A.M.S., Vol. X.

[5] Riviere, N.M., "Singular integrals and multiplier operators," Arkiv for Matematik, #2, Dec. 1971, 243-308.

[6] Rudin, W., Functional Analysis, McGraw-Hill (1973).

[7] Stein, E.M., Singular integrals and differentiability properties of functions, Princeton Univ. Press, 1970.

A FINITE ELEMENT METHOD FOR THE SIMULATION OF A RAYLEIGH-TAYLOR INSTABILITY

A.Dervieux and F. Thomasset

IRIA-LABORIA

F-78150 Le Chesnay

1. INTRODUCTION

Consider the two phase fluid system shown on fig.1, the upper fluid being heavier (the volumic mass ρ^1 of fluid 1 is larger than ρ^2)

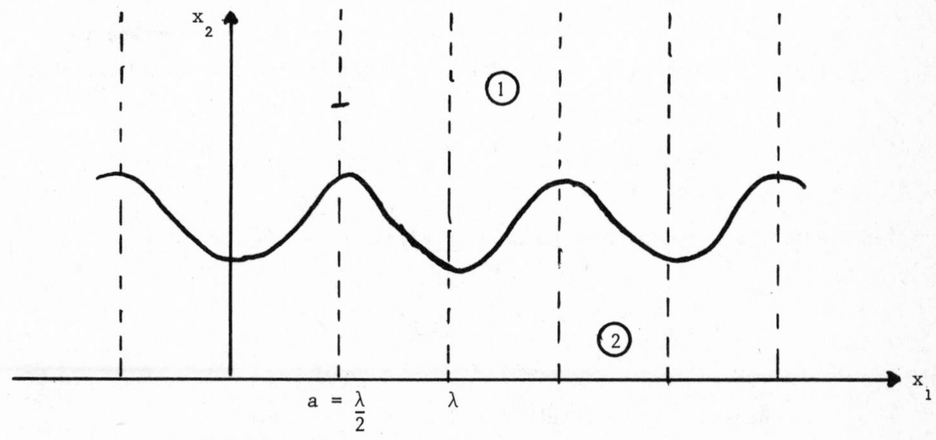

The problem is to predict the evolution of the interface, given some initial data (interface and velocity field).

The following assumptions are made :

. The fluids are non-miscible ;

. both fluids are incompressible, newtonian and viscous (dynamic viscosities : μ^1, μ^2) ;

. the problem is 2-dimensional in (x_1, x_2) space ;

. the problem is periodic in the x_1 direction, with period λ ;

. within a period interval $[o,\lambda]$ the solution is symmetric with respect to the line $x_1 = \lambda/2 \equiv a$.

. The motions beyonds "some" distance of the interface, are neglectible : therefore we can restrict our study to a rectangular box :

$$o \leq x_1 \leq a \ , \quad o \leq x_2 \leq L.$$

. The forces acting on a fluid element are :
inertia, gravity, viscous, and pressure forces. Surface tension effects are neglected in this paper.

Existing codes for two-(or multi-) phase flows usually come into one of three classes :

- Lagrangian methods which involve moving meshes ans some costly rezoning step when the mesh becomes distorded ;

- Eulerian methods where the boundaries or interfaces, when they are fitted, are approximated by piecewise linear continuous curve whose vertices are moved along with the fluid velocity (see for example BURSTEIN and TURKEL [1]

- Combinations of these two methods.

The reader is referred for more informations for example to the survey of TURKEL [13] .

The method proposed in this paper is of Eulerian type, in which the interface is not artificially fitted : we compute a "pseudo-density" function $\varphi(x_1, x_2 ; t)$ such that

$$\varphi > 0 \text{ in medium 1}$$
$$\varphi = 0 \text{ on the interface}$$
$$\varphi < 0 \text{ in medium 2 ;}$$

then the representation of the interface is naturally derived as the curve

$$\left\{ (x_1, x_2) \text{ s.t. } \varphi(x_1, x_2 ; t) = 0 \right\}.$$

In § 2, we derive a transport equation for function φ : therefore our methods are of "hybrid" type, i.e. density-method together with interface-fitting.

The weak formulation of the problem is obtained in § 3 .

In § 4, we explain which finite element mèthods have been used : conforming finite elements of degree 1 for φ, non conforming finite elements of degree 1 for $\vec{u}$.

2. THE PSEUDO-DENSITY AND TRANSPORT EQUATION.

Let χ be the characteristic function on medium 1 :

$$\chi(x_1, x_2 ; t) = \begin{cases} 1 \text{ in medium } \textcircled{1} \\ 0 \text{ in medium } \textcircled{2} \end{cases}$$

Owing to the mass conservation law (incompressibility condition) the displacement of the interface must be governed by :

$$\vec{u}.\vec{n} = \vec{v}.\vec{n} \tag{1}$$

with : $\vec{u}$ = fluid velocity on the interface ;
$\vec{v}$ = speed of interface displacement ;
$\vec{n}$ = unit normal vector to the interface.

From (1) it is well known that χ should satisfy in the distributions sense :

(2) $$\frac{\partial \chi}{\partial t} + \left(\vec{u}.\vec{\nabla}\right) \chi = 0$$

Equation (2) should be solved (coupled with the Navier Stokes equations), with initial data :

(3) $$\chi|_{t=0} = \chi_o (x_1 , x_2).$$

Now the numerical approximation of equation (2) with discontinuous initial data is far from being an easy matter. Therefore, let us consider a "smooth" function $\varphi_o(x_1 , x_2)$ satisfying :

(4) $$\chi_o = H \circ \varphi_o$$

where H is the Heaviside operator defined by :

(5) $$(H\circ \varphi)(x_1, x_2) = \begin{cases} +1 & \text{if } \varphi(x_1 , x_2) \geqslant 0 \\ 0 & \text{if } \varphi(x_1 , x_2) < 0 \ ; \end{cases}$$

Now for a given $\vec{u}$, let φ be the solution of the transport equation :

(6) $$\begin{cases} \dfrac{\partial \varphi}{\partial t} + \left(\vec{u}.\vec{\nabla}\right) \varphi = 0 \\ \varphi|_{t=0} = \varphi_0 \end{cases}$$

Then if we take χ as : $\chi = H\circ\varphi$ we can prove (for instance by an argument on the path lines) that χ satisfies the original equation (2).

Thus we will approximate equation (6) with smooth initial data. The interface will now be found as the line : $\varphi = 0$.

We set :

(7) $$\rho(x_1 , x_2 , t) = (\rho^1-\rho^2).\ \chi(x_1 , x_2 \ ; t) + \rho^2 \quad \text{that is to say :}$$
$$\rho(x_1 , x_2 ; t) = \begin{cases} \rho^1 & \text{if } \varphi(x_1 , x_2 \ ; t) > 0 \\ \rho^2 & \text{if } \varphi(x_1 , x_2 , t) < 0 \end{cases}$$

In the same way :

(8) $$\mu(x_1, x_2; t) = (\mu^1- \mu^2).\ \chi (x_1, x_2, t) + \mu^2$$

Finally we plug (7) and (8) into Navier Stokes equations as explained below.

3 THE MOMENTUM EQUATIONS.

We need the following notations : the index α will be related to the medium under consideration :

$\Omega_\alpha(t)$ = region filled by the fluid α , $\alpha = 1,2$

$\Omega = \Omega_1 \cup \Omega_2 =]0,a[\times]0,L[$

$\vec{u}^\alpha = (u_1^\alpha, u_2^\alpha)$ = velocity field in medium α

p^α = pressure in medium α

$\vec{u} = (\vec{u}^1 - \vec{u}^2)\chi + \vec{u}^2 \qquad p = (p^1 - p^2)\chi + p^2$

$\mathcal{L}(t)$ = the interface line between the two fluids

$\vec{n} = (u_1, u_2)$ = the unit normal vector to $\mathcal{L}$, oriented from Ω_1 to Ω_2 .

$\frac{d}{dt} = \frac{\partial}{\partial t} + u_j^\alpha \frac{\partial}{\partial x_j}$ (with the convention on repeated indices)

g = gravity acceleration $\vec{g} = \begin{pmatrix} 0 \\ -g \end{pmatrix}$

δ_{ij} = Kronecker symbol

Now, in each medium, we have the Navier Stokes equations :

$$\rho^\alpha \frac{du_i^\alpha}{dt} = \frac{\partial \sigma_{ij}^\alpha}{\partial x_j} - \rho^\alpha g \, \delta_{i2} \tag{9}$$

$$\sigma_{ij}^\alpha \text{ (= stress tensor) } = - p^\alpha \delta_{ij} + \mu^\alpha \left(\frac{\partial u_i^\alpha}{\partial x_j} + \frac{\partial u_j^\alpha}{\partial x_i} \right) \tag{10}$$

$$\operatorname{div} \vec{u}^\alpha = 0 \tag{11}$$

The boundary conditions on $\vec{u}$ are :

$$u_1 = 0 \text{ on } x_1 = 0 \text{ and on } x_1 = a \tag{12}$$

$$\frac{\partial u_2}{\partial x_1} = 0 \text{ on } x_1 = 0 \text{ and on } x_1 = a \tag{12.a}$$

$$u_1 = u_2 = 0 \text{ on } x_2 = 0 \text{ and on } x_2 = L \tag{13}$$

<u>Remark 1 :</u> We might impose a free-slip condition, instead of no-slip, on the top

and bottom boundaries .

On the interface, since we neglect the surface tension effects we get the conditions for the stresses :

(14) $$\sigma^{(1)}_{ij}\ n_j = \sigma^{(2)}_{ij}\ n_j$$

Of course we need an initial condition on $\vec{u}$: we usually took $\vec{u} = 0$ at $t = 0$.

Now we proceed to the Galerkin formulation of the problem ; let $\vec{v} = (v_1, v_2)$ be a smooth vector valued function such that :

(15) $$\text{div } \vec{v} = 0 \text{ in } \Omega$$

(16) $\vec{v}$ satisfies the boundary conditions (12)-(13) we multiply (9) throughout out by $\vec{v}$ and integrate over Ω_α

(17) $$\int_{\Omega_\alpha} \rho^\alpha \left(\frac{du^\alpha_i}{dt} + \delta_{i_2} g \right) v_i \, dx = \int_{\Omega_\alpha} \frac{\partial \sigma^\alpha_{ij}}{\partial x_j} v_i \, dx$$

We transform the right hand side with Green's fromula, using (16)

$$\int_{\Omega_1} \frac{\partial \sigma^1_{ij}}{\partial x_j} v_i \, dx = + \int_{\mathcal{L}} \sigma^1_{ij}\ n_j\ v_i \, ds - \int_{\Omega_1} \sigma^1_{ij} \frac{\partial v_i}{\partial x_j} dx$$

$$\int_{\Omega_2} \frac{\partial \sigma^2_{ij}}{\partial x_j} v_i \, dx = - \int_{\mathcal{L}} \sigma^2_{ij}\ n_j\ v_i \, ds - \int_{\Omega_1} \sigma^2_{ij} \frac{\partial v_i}{\partial x_j} dx$$

From (15) we eliminate the pressure

$$\sigma^\alpha_{ij} \frac{\partial v_i}{\partial x_j} = \frac{\mu^\alpha}{2} \left(\frac{\partial u^\alpha_i}{\partial x_j} + \frac{\partial u^\alpha_j}{\partial x_i} \right) \left(\frac{\partial v_i}{\partial x_j} + \frac{\partial v_j}{\partial x_i} \right)$$

Finally we add (17,α = 1) to (17,α = 2), using (14) the line integrals over $\mathcal{L}$ vanishs ; rearranging the terms yields :

(18) $$\int_\Omega \left[\rho \left(\frac{\partial u_i}{\partial t} + u_j \frac{\partial u_i}{\partial x_j} \right) v_i + \frac{\mu}{2} \left(\frac{\partial u_i}{\partial x_j} + \frac{\partial u_j}{\partial x_i} \right) \left(\frac{\partial v_i}{\partial x_j} + \frac{\partial v_j}{\partial x_i} \right) \right] dx = - \int_\Omega \delta_{i_2}\ g\ v_i \, dx$$

with $u_i \begin{cases} = u_i^1 & \text{in } \Omega_1 \\ = u_i^2 & \text{in } \Omega_2 \end{cases}$

and the notations (7), (8) , φ being the solution of the weak form of (6) :

$$(19) \qquad \int_\Omega \left[\frac{\partial \varphi}{\partial t} \, w + \frac{1}{2} \left(u_i \frac{\partial \varphi}{\partial x_i} \, w - u_i \varphi \frac{\partial w}{\partial u_i} \right) \right] dx = 0$$

(18) and (19) are required to be true for all w in $H^1(\Omega)$, and all $\vec{v}$ in $[H^1(\Omega)]^2$ satisfying (15)-(16) : For theorical results related to this problem we refer to KAJIKOV [6]

4. FINITE ELEMENT DISCRETIZATION.

In the method proposed in this paper the main points are the use of Finite Elements and a smooth pseudo-density function φ . We used triangular finite elements of degree one :

- conforming (that is the most standard ones [(1)])elements for φ ;
- non conforming elements for $\vec{u}$ [CROUZEIX and RAVIART [3]]

we do not claim that this is the only possible choice, we just found some advantages in it :

i) the use of conforming finite elements to compute the pseudo density φ is suitable to get a continuous interface : the interface is in fact piecewise linear, with at most one segment per triangle. Thus the method is attractive because it gives at neat answer for the position of the interface, and allows the possibility to take surface tension effects into account.

ii) As to the use of non conforming elements for the velocity, this is convenient to deal with the incompressibility condition ; we give some details on this point now :

First of all we notice that, if we defined on a triangular mesh a continuous, piecewise linear velocity field, the degrees of freedom would be the values at the vertices, and the conditions

$$\vec{u}_h = 0 \text{ at the boundary}$$

$$\operatorname{div} \vec{u}_h = 0 \text{ on each triangle}$$

---

(1) with the values of φ at the vertices as degrees of freedom.

would imply [(1)] $\vec{u}_h \equiv 0$ (cf. FORTIN [4]). Therefore we relax the continuity condition by bringing in more degrees of freedom so that the incompressibility constraint is likely to be satisfied. To be more specific, we require for the discrete velocity $\vec{u}_h$ to lie in the vector space V_h defined as follows ; given a triangulation of Ω, we demand that :

- $\vec{u}_h$ should be polymonial of degree at most one on each triangle ;
- $\vec{u}_h$ should be continuous at the mid side points : thus the degrees of freedom of $\vec{u}_h$ are its values at the mid side points , and $\vec{u}_h$ is not expected to be continuous along a side except at the mid point. CROUZEIX and RAVIART [3] introduced this finite element and proved the following order of convergence :

  $||u_h - u||_{1,\Omega} = O(h)$
  (see also TEMAM [9]).

Remark 2 : We also have to impose the boundary conditions (12)-(13) at the boundary nodes.

Now one has to deal with the discrete incompressibility condition :

$$\text{div } \vec{u}_h = 0 \quad \text{on each triangle} \tag{20}$$

This is achieved by the use of a divergence-free basis, generating the whole space V_h as described in THOMASSET [11]; the idea of this construction is due to CROUZEIX [2]: since $\vec{u}_h$ is of degree at most 1, requiring the vanishing of div $\vec{u}_h$ on triangle T is equivalent to :

$$\int_{\partial T} \vec{u}_h . \vec{n} \; ds = 0 \tag{21}$$

(∂T = boundary of T, $\vec{n}$ = unit normal vector to ∂T)

In view of (21) the divergence-free basis functions fall into one of the two classes (remind that they are defined by their values at the mid side nodes) :

. a function $\vec{w}_m$ of the first class is associated to a mid side node m :

(1) Except for special meshes.

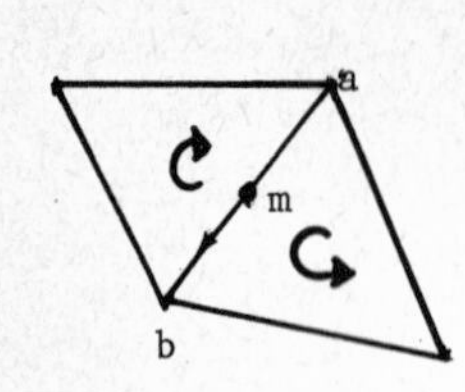

$$\begin{cases} \vec{w}_m(m) = \dfrac{\vec{ab}}{|ab|} \\ \vec{w}_m = 0 \text{ at all nodes different of } m \end{cases}$$

. a function $\vec{w}_v$ of the second class is associated to a vertex v :

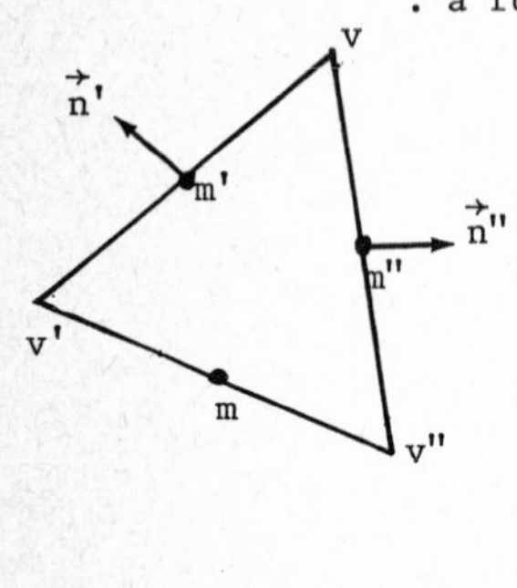

$w_v = 0$ at all mid side nodes which do not belong to a common triangle with vertex v

$w_v(m) = 0$

$w_v(m') = \dfrac{-\vec{n}'}{|vv'|}$

$w_v(m'') = \dfrac{+\vec{n}''}{|vv''|}$

. The case of an internal obstacle would require a special treatment for this, for the proof of completeness of the basis and for the pratical implementation we refer to THOMASSET [11,12] .

We sum up the situation :

. the number of degrees of freedom of u_h is : number of vertices + number of mid side nodes (notice that the construction of the basis is purely 2-dimensional ; in fact it is equivalent to a Ψ-method).

. The rest of the procedure is standard : we write that (18) should be true with $\vec{v}$ replaced by $\vec{w}_m$ or $\vec{w}_v$ and $\vec{u}$ by the expansion :

$$\vec{u}_h(x,t) = \sum_v U_v(t)\,\vec{w}_v(x) + \sum_m U_m(t)\,\vec{w}_m(x)$$

This yields a finite set of non linear equations in the U_v, U_m.

As to the φ-equation (19) we use the standard Galerkin procedure with finite elements of degree 1.

Remark : thus convective terms, in the momentum equations as well as in the φ-equation are discretized by a centered-in-space formula ; we expect that the results could be improved by the introduction of artificial viscosities as described in HUGUES et al [5], RAVIART [8].

For the time discretization we used Crank-Nicholson scheme with linearized convective terms ; we solved successively one equation in φ and one in $\vec{u}$ at each times step ; both equations were solved numerically using a LU-decomposition solver from the MODULEF library (cf. [14]).

As such, the scheme is centered in space and, as could be expected, did produce oscillations in φ. Substantial improvements can be obtained from the following remark : what we are interested in is the location of the interface and we can change the values of provided this does not move the interface. Therefore we adopted the following strategy : after the φ-equation has been solved and a new position of the interface has been determined, we compute at each discretization node (x_1, x_2), the distance to the interface : this defines the new modulus of $\varphi(x_1, x_2)$ [1].

This smoothing process does introduce some unwanted diffusion but we can choose to apply it only every N-Time steps (although we took N=1 in the following results).

5. NUMERICAL RESULTS.

We show there after some results in the box defined by :

$$a = \pi = 3.14159\ldots, \quad L = 3a.$$

The numerical data are :

$$\mu^1 = \mu^2 = 1$$

$$\rho^1 = \rho^2 = \frac{1}{10}$$

The initial data are :

$$\vec{u} = 0 \text{ at time } t = 0$$

and the initial interface is defined by the equation :

$$x_2 + \frac{1}{2}\cos x_1 - 4.6 = 0$$

We used a regular mesh ($\delta x = \delta y = \Pi/10$, that is 600 triangles) and a time increment $\delta t = 1s$; the system was integrated from time $t = 0$ to time $t = 70s$. The computing time for this run is about 70 minutes on CII-IRIS 80 ; the results at times 10, 40, 70 s are given on the figures 1, 2, 3.

We note that our computational box $(0,a) \times (0, 3a)$ is small, so that the assumption of rest at the top ans bottom boundaries is obviously violated from time $t \sim 30$, and speed of the interface cannot attain an asymptotic value.

5. CONCLUSION.

This method has the usual advantages of finite element methods : on particular it could deal with more complicated geometries and the incompressibility condition is automatically satisfied ; also various boundary conditions are easily handled.

---

[1] That is $\varphi^*(x_1, x_2) = \text{dist}(x_1, x_2; \mathcal{L}) \times \text{sign of } \varphi(x_1, x_2)$

Some numerical diffusion is introduced which can be controlled ; moreover it is an easy matter to take into account surface tension effects.

ACKNOLEDGEMENT :

This work was supported by a grant of the Commissariat à l'Energie Atomique (France).

REFERENCES.

[1] S.Z. BURSTEIN and E. TURKEL
"Eulerian Computations in domains with moving boundaries", Proc. Fifth Inter. Conf. Num. Meth. Fluid Dynamics, Springer Verlag, Lecture Notes in Physics, vol 59, pp 114-122, (1976).

[2] M. CROUZEIX
"Journées Eléments finis", Rennes, 1976.

[3] M. CROUZEIX and P.A. RAVIART
"Conforming and non conforming finite element methods for solving the stationary Stokes equations", RAIRO (R-3), pp. 33-75, Dec. 1973

[4] M. FORTIN
"Calcul numérique des écoulements des fluides de Bingham et des fluides Newtoniens incompressibles par la méthode des éléments finis", thèse de Doctorat d'Etat,Université Paris VI, 1972.

[5] T.J.R. HUGUES, W.K. LIU, A. BROOKS
"Finite Element Analysis of Incompressible Flows by the Penalty Function Formulation" J. Comp. Ph., 30, pp.1-60 (1979)

[6] A.V. KAJIKOV
"Resolution of boundary value problems for non homogeneous viscous fluids", Doklady Akad. Nank., 216, pp. 1008-1010 (1974)

[7] J.L. LIONS
"On some problems connected with Navier-Stokes equations", Colloque Madison 1977, and seminaire at Collège de France (Novembre 1977).

[8] P.A. RAVIART
"Approximation numérique des phénomènes de diffusion convection", Ecole

d'été d'Analyse Numérique (EDF/CEA/IRIA) (1979)

[9] R. TEMAM
"Navier Stokes equations", North Holland (1977)

[10] R. TEMAM and F. THOMASSET
"Numerical solution of Navier Stokes equations by a finite element method" Proc. Second Inter. Symposium on Finite Element Meth. in flow Problems, Santa Margherita Ligure (1976). Ed. Springer

[11] F. THOMASSET
"Numerical solution of the Navier Stokes Equations by Finite Element Methods", AGARD-VKI Lectures, N°86, Von Karman Institute (Rhode St Genèse, Belgium) 1976

[12] F. THOMASSET
"Modules NSNΦCΦ et NS NCST (Club MODULEF) Equations de Navier Stokes bidimensionnelles" to appear.

[13] E. TURKEL
"Order of accuracy and boundary conditions for large scale time dependent partial differential equations", Von Karman Institute Lectures series 1979-6 (Computational fluid dynamics, March 19-23, 1979) Rhode St Genèse, Belgium, 1979.

[14] D. BEGIS and A. PERRONNET, the Club MODULEF, IRIA-LABORIA, to appear at Dunod, Paris (1979).

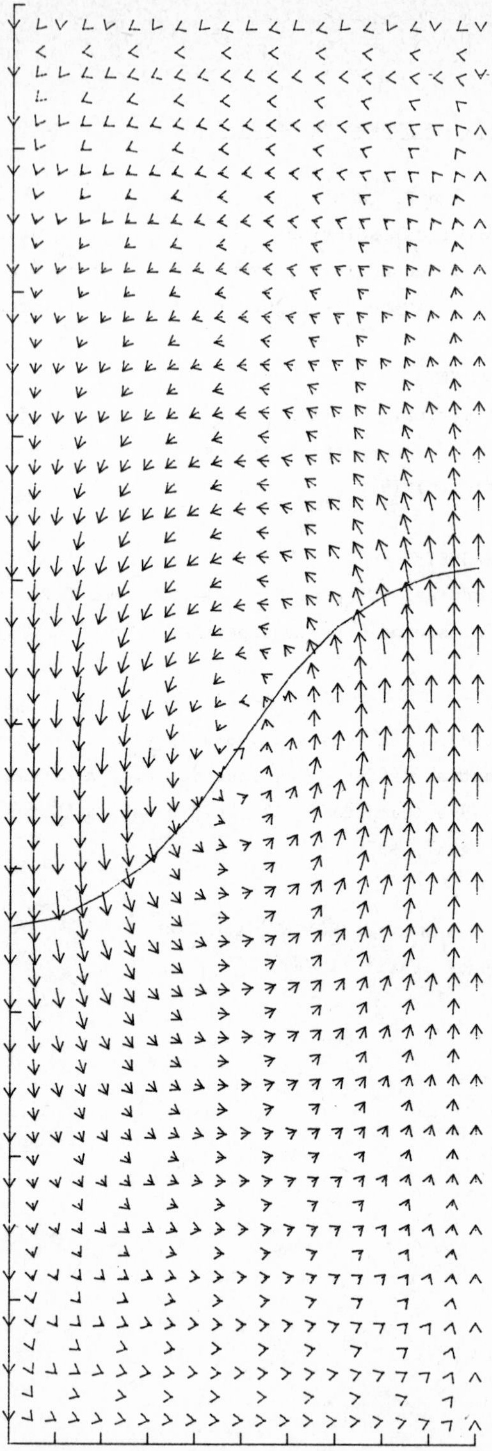

Figure 1.

TEMPS T = 0.100E+2 . VITESSE MAX. = 0.127E+0

Figure 2.

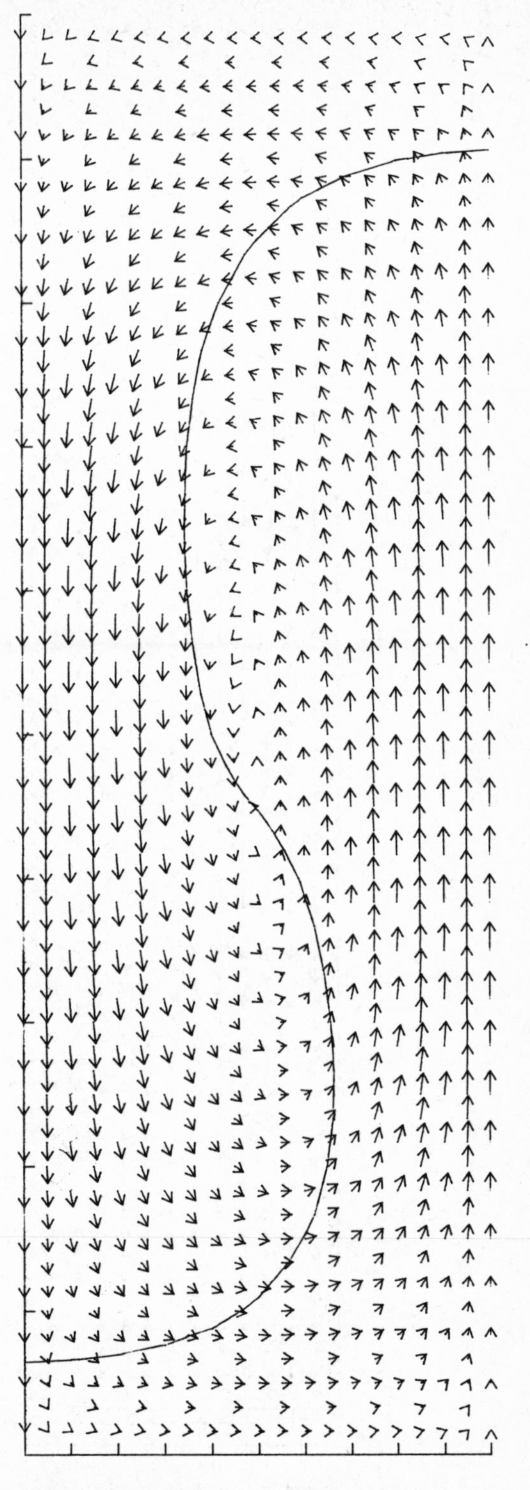

TEMPS T = 0.400E+2 . VITESSE MAX. = 0.243E+0

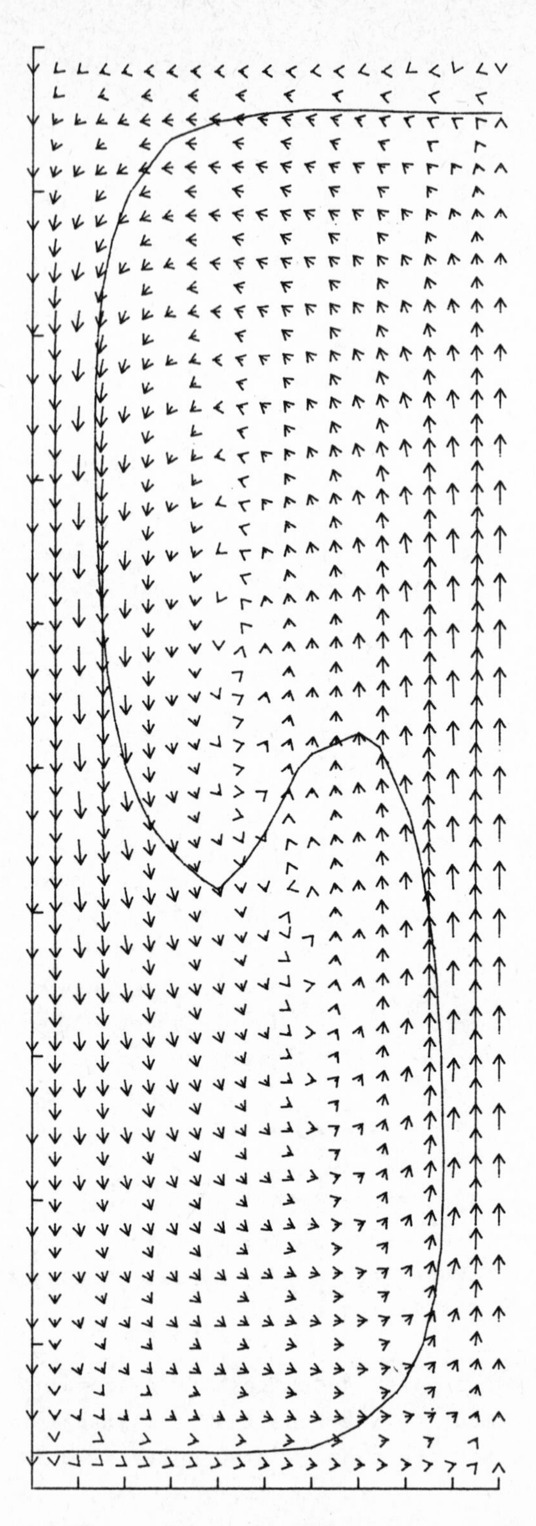

Figure 3.

SPECTRAL CALCULATION OF THE STABILITY OF THE CIRCULAR COUETTE FLOW

Michel Deville and Steven A. Orszag
Unité de Mécanique Appliquée
Université Catholique de Louvain
Louvain-la-Neuve - Belgium
Department of Mathematics
M.I.T.
Cambridge, Ma 02139 - USA

§ 1. Introduction.

The transition between laminar flow and turbulence in circular Couette flow between concentric rotating cylinders constitutes one of the richest phenomenon in fluid dynamics.

When the outer cylinder is at rest, one can define the dimensionless Taylor number T as $(\Omega_i R_i \ d/\nu)^2 \ (d/R_m)$, where $\Omega_i R_i$ is the azimuthal velocity of the inner cylinder, d the gap width, ν the kinematic viscosity and R_m the mean radius. Instability (Taylor 1923) occurs for T larger than a critical value T_{cr}. For $T > T_{cr}$, a secundary motion is set up having the form of toroidal vortices superposed on the modified circumferential flow. These vortices are periodically spaced in the axial direction. Two adjacent cells have opposite sense of rotation.

At a second critical Taylor number $T'_{cr} > T_{cr}$, the toroidal vortices themselves become unstable to wavy azimuthal disturbances, the axial wavenumber remaining fixed. As the Taylor number increases still more, the axial and azimuthal wavenumbers continue to change (Coles 1965). Other important experimental results have been reported by Gollub and Freilich (1976), Gollub and Swinney (1975). A review of these experimental data is given by Fenstermacher, Swinney and Gollub (1979).

Theoretical studies have been made by Davey (1962) for the Taylor transition at T_{cr} while the wave instability at T'_{cr} was studied by Davey, Di Prima and Stuart (1968) and Eagles (1971). Finally, numerical investigations have been reported by Meyer (1967 and 1969), Rogers and Beard (1969) and more recently by Majumdar and Spalding (1977) and by Fasel and Booze (1977).

The present paper describes the elaboration of a 3-D numerical algorithm for studying the Couette flow. With this code, we plan to examine the higher transitions of Couette flow in order to obtain a better picture of the mechanisms which generate turbulence in fluids. Our code is based on a representation of the velocity field by Fourier series in the circumferential and axial directions and Chebyshev polynomials in radius. The Navier-Stokes equations are integrated by means of a splitting method.

As the spatial accuracy of the spectral description is excellent, an explicit second order Adams-Bashforth scheme coupled with an implicit stabilization process is used to obtain second order accuracy for the time integration.

In section 2, we describe the general formulation of the problem. In section 3, we describe the numerical approximations. The numerical stability restrictions are discussed as well as the pressure calculation. In section 4, we present the sequential steps of the running code. The last section shows numerical results for two test cases. The first one is a small gap problem. Davey's (1962) theory shows some discrepancy with the numerical simulation. The computed torque value is very close to that found experimentally by Donnelly and Simon (1960). The second problem computes the temporal development of Taylor vortices in Coles' (1965) geometry for a Taylor number corresponding to the wavy boundary.

These preliminary results are particularly encouraging. All the critical values we have used in this study come from Di Prima and Eagles (1977).

§ 2. General formulation.

The three-dimensional flow considered here is bounded by inner and outer cylinders of radius R_i and R_o, respectively and by two parallel planes separated by a distance H normal to the symmetry axis (Fig.1). In the cylindrical coordinate system (r,θ,z), the flow is assumed to be periodic in the azimuthal and vertical directions.

The dynamical equations are the incompressible Navier-Stokes equations written in rotation form :

$$\frac{\partial \vec{v}}{\partial t} = \vec{v} \times \vec{\omega} - \nabla \Pi + \nu \nabla^2 \vec{v} \quad , \tag{1}$$

$$\operatorname{div} \vec{v} = 0 \quad . \tag{2}$$

Here $\vec{v}$ is the velocity field, $\vec{\omega}$ = curl $\vec{v}$ the vorticity, $\Pi = \frac{p}{\rho} + \frac{1}{2} v^2$

the total head and ν is the kinematic viscosity. Eq. (2) imposes the continuity constraint on the velocity field.

With lower indices indicating the corresponding components with respect to the coordinate axes, the Navier-Stokes and continuity equations (1) and (2) can be written as,

$$\frac{\partial v_r}{\partial t} = v_\theta \omega_z - v_z \omega_\theta - \frac{\partial \Pi}{\partial r} + \nu(\Delta v_r - \frac{v_r}{r^2} - \frac{2}{r^2}\frac{\partial v_\theta}{\partial \theta}) \quad , \tag{3}$$

$$\frac{\partial v_\theta}{\partial t} = v_z \omega_r - v_r \omega_z - \frac{1}{r}\frac{\partial \Pi}{\partial \theta} + \nu(\Delta v_\theta - \frac{v_\theta}{r^2} + \frac{2}{r^2}\frac{\partial v_r}{\partial \theta}) , \tag{4}$$

$$\frac{\partial v_z}{\partial t} = v_r \omega_\theta - v_\theta \omega_r - \frac{\partial \Pi}{\partial z} + \nu \Delta v_z \quad , \tag{5}$$

$$\frac{1}{r}\frac{\partial (r v_r)}{\partial r} + \frac{1}{r}\frac{\partial v_\theta}{\partial \theta} + \frac{\partial v_z}{\partial z} = 0 \quad . \tag{6}$$

In Eqs. (3-5), the vorticity components are defined in terms of velocity components by the following relationships,

$$\omega_r = \frac{1}{r}\frac{\partial v_z}{\partial \theta} - \frac{\partial v_\theta}{\partial z} \quad , \tag{7}$$

$$\omega_\theta = \frac{\partial v_r}{\partial z} - \frac{\partial v_z}{\partial r} \quad , \tag{8}$$

$$\omega_z = \frac{1}{r}[\frac{\partial (r v_\theta)}{\partial r} - \frac{\partial v_r}{\partial \theta}] \quad . \tag{9}$$

The Laplacian operator in cylindrical coordinates is,

$$\Delta = \frac{\partial^2}{\partial r^2} + \frac{1}{r}\frac{\partial}{\partial r} + \frac{1}{r^2}\frac{\partial^2}{\partial \theta^2} + \frac{\partial^2}{\partial z^2} \quad . \tag{10}$$

The pressure field is obtained from a Poisson equation derived by taking the divergence of (1),

$$\Delta \Pi = \operatorname{div}(\vec{v} \times \vec{\omega}) \quad . \tag{11}$$

The Couette flow problem is fully described by Eqs. (1), (2) and (11). It is solved using periodic boundary conditions in the azimuthal and axial directions and no-slip boundary conditions on the cylinders,

$$v_r = v_z = 0 \qquad r = R_i, R_o \tag{12.a}$$

$$v_\theta = \Omega_i R_i \quad \text{for } r = R_i \;\; ; \;\; v_\theta = \Omega_o R_o \quad \text{for } r = R_o \quad . \qquad (12.b)$$

§ 3. Spectral approximation and computational implementation.

a. Series expansion.

The azimuthal and axial periodicity makes the use of Fourier series very natural. As far as the radial geometry is concerned, the choice of Bessel functions may seem best. However, as shown by Gottlieb and Orszag (1977), the Fourier-Bessel expansion exhibits a Gibbs phenomenom at the boundary analogous to that for Fourier sine series, therefore reducing the rate of convergence. In contrast, Chebyshev series converge fast to smooth solutions regardless of their boundary conditions. That is the reason why the best spectral approximation for cylindrical geometry is achieved with Chebyshev polynomials in radius and Fourier series in the angular and axial directions. Thus, we assume the velocity field expanded in the series,

$$\vec{v}(r,\theta,z,t) = \sum_{n=0}^{N} \sum_{k_\theta,k_z} \vec{u}(k_\theta,k_z,n) e^{i(k_\theta\theta+k_z z)} T_n(r) \ , \qquad (13)$$

where the inner sum extends over wave vectors whose components are integral multiples of 2π in the angular direction and multiples of $2\pi/H$ in the axial direction. A similar expression holds for the pressure.

The truncation region of wavevector space sums in (13) are

$$|k_\theta| < K_\theta \quad \text{and} \quad |k_z| < K_z \ ,$$

$$\left|\frac{k_\theta}{K_\theta} \pm \frac{k_z}{K_z}\right| < \frac{4}{3} \ .$$

The spectral cutoffs K_θ and K_z are not equal.

Eq. (13) is not strictly correct, because the range of definition for Chebyshev polynomials is $|x| \leqslant 1$. However, we shall adopt the notation $T_n(r)$ with the meaning that the gap width $R_i \leqslant r \leqslant R_o$ is mapped onto the basic range of definition by the linear relationship,

$$r \equiv \alpha x + \beta = \frac{1}{2}(R_o - R_i)x + \frac{1}{2}(R_o + R_i) \ . \qquad (14)$$

b. Cylindrical geometry.

The Laplacian operator (11) involves geometrical factors r and r^2 in the denominator of two terms. Division by r or r^2 of variables expressed in terms of Chebyshev modes may be done efficiently by solving tridiagonal systems. Let $f(r) = \sum_{n=0}^{N} f_n T_n(r)$ a function to be divided by r and let us denote by $g(r) = \sum_{n=0}^{N} g_n T_n(r)$ the result of the division. One easily finds,

$$\frac{1}{2} c_{n-1} \alpha g_{n-1} + \beta g_n + \frac{1}{2} \alpha g_{n+1} = f_n \ , \ n=0 \ , \ \ldots, \ N, \qquad (15)$$

where $c_o=2$, and $c_n=1$ for $n>0$.

However, for ease of computing, it is better to multiply through the Laplacian by r^2 and to approximate the resulting equation. For example, Eq. (11) yields,

$$r^2 \Delta \Pi = r^2 \, \mathrm{div}(\vec{v} \times \vec{\omega}) \ .$$

The Chebyshev-tau approximation of the left hand side of the previous equation leads to the following system,

$$-p_{n-2} c_{n-2} \frac{k_z^2 \alpha^2}{4} - c_{n-1} p_{n-1} \alpha \beta k_z^2 + p_n \left[\frac{n^2}{c_n} - k_\theta^2 - k_z^2 \beta^2 - \frac{k_z^2 \alpha^2}{4}(c_n + c_{n-1})\right]$$
$$+ p_{n+1}\left[\frac{2\beta}{\alpha} (1+n) \left(2n + \frac{1}{c_n}\right) - \alpha \beta k_z^2 \right] - \frac{k_z^2 \alpha^2}{4} p_{n+2} +$$
$$\sum_{\substack{k=n+2 \\ k+n \text{ even}}}^{N} \frac{1}{c_n}\left[1 + \left(\frac{\beta}{\alpha}\right)^2\right] k(k^2-n^2) p_k + \frac{2\beta}{\alpha c_n} \sum_{\substack{k=n+3 \\ k+n \text{ odd}}}^{N} k(k^2-n^2) p_k \ , \quad 0 \leq n \leq N-2 \qquad (16)$$

The variables p_k are the Fourier-Chebyshev coefficients of the pressure head. The system (16) is completed by the boundary conditions applied to the radial pressure gradient at the inner and outer cylinders.

For $n=0$ and $k_\theta = k_z = 0$, the system (16) is singular. Nevertheless, it is easy to show that there is no need to compute the pressure for this particular choice of modes, because the radial velocity component vanishes by the continuity equation.

The same technique of multiplication by r^2 is also used for the implicit Laplacian operators in the viscous terms. In this case, no-slip wall boundary conditions are associated with the tau approximation.

c. Convective terms.

We use the Chebyshev collocation (or pseudospectral) method to evaluate the non-linear terms (Orszag 1972).

The collocation idea is to evaluate spectrally the derivatives occuring in the vorticity definitions (7-10), while the cross product $\vec{v} \times \vec{\omega}$ is locally computed in physical space. The fast Fourier transform (FFT) algorithm (Cooley-Tukey, 1965) facilitates the transformation from physical to wave space and vice-versa.

d. Fractional steps.

Time stepping is done by a fractional step or splitting procedure. Suppose we know the velocity components at time level $t = n\Delta t$; Δt is the time step and n an integer. One then computes a new velocity field $\vec{v}_1$,

$$\vec{v}_1 = \vec{v}^n + \frac{3\Delta t}{2}[(\vec{v} \times \vec{\omega}) + \nu\nabla^2\vec{v}]^n - \frac{\Delta t}{2}[(\vec{v} \times \vec{\omega}) + \nu\nabla^2\vec{v}]^{n-1} \quad (17)$$

using the second-order Adams-Bashforth scheme applied to the non-linear and viscous terms. The superscript indicates the time level of the variables.

Next, the pressure field is evalutated by solving

$$r^2\Delta\Pi = r^2 \operatorname{div} \vec{v}_1 \quad . \quad (18)$$

This gives the second fractional step :

$$\vec{v}_2 = \vec{v}_1 - \Delta t\nabla\Pi \quad . \quad (19)$$

In principle, $\vec{v}_2$ is the velocity field at time level $n+1$. However, the numerical stability restrictions on this scheme are quite severe. We ease these restrictions by second-order fractional step implicit method.

e. Numerical stability.

It was shown by Gottlieb & Orszag (1977) that application of an explicit scheme to the viscous diffusive terms in Chebyshev spectral methods are restricted by a stability condition of the form $\Delta t = O(1/N^4)$. This severe limitation is eliminated by a second-order (in Δt) fractional-step (splitting) method in which the viscous terms are treated implicitly (Orszag and Deville 1979).

Gottlieb and Orszag (1977) have shown that the stability condition for the Adams-Bashforth scheme used in (17) for the nonlinear terms is $\Delta t = O\ (1/N^2)$. In the Couette flow problem, the time differencing restriction comes essentially from the stability condition imposed on the azimuthal velocity component. A semi-implicit technique applied to these terms allows marching in time with bigger time steps.

In this semi-implicit scheme the first stage of the fractional step procedure is modified in the following way :

$$\vec{v}_1 + \Delta t\ \frac{V(r)}{r}\ \frac{\partial \vec{v}_1}{\partial \theta} = \vec{v}^n + \frac{3}{2}\ \Delta t\ [(\vec{v} \times \vec{\omega}) + \nu \nabla^2 \vec{v}]^n - \frac{\Delta t}{2}[(\vec{v} \times \vec{\omega}) + \nu \nabla^2 \vec{v}]^{n-1} + \Delta t\ \frac{V(r)}{r}\ \frac{\partial \vec{v}^n}{\partial \theta} \quad . \qquad (20)$$

The left hand side of (20) is treated implicitly in time, while the right hand side is treated explicitly. In Eq. (20), $V(r)$ denotes the analytical Couette solution.

f. Pressure boundary conditions.

Poisson's equation for the pressure field is classically solved with boundary conditions derived from the discretized version of the Navier-Stokes equations, where the velocity boundary conditions are taken into account (see for example, Harlow and Welch, 1965, for a finite difference approach). The same procedure applied to Chebyshev approximation leads to unconditional instability.

The right way to set the pressure boundary conditions is obtained from Eq. (19). As the velocity field must match the wall velocity, the pressure boundary conditions read,

$$\nabla \Pi = \frac{1}{\Delta t}(\vec{v}_1 - \vec{v}_{wall}) \quad .$$

In the calculations reported here, only the normal pressure boundary conditions have been used. The previous equation is now,

$$\frac{1}{\alpha} \sum_{n=0}^{N} (\pm 1)^n n^2 p_n = \frac{1}{\Delta t} \sum_{n=0}^{N} (\pm 1)^n v_{1rn} , \tag{21}$$

where v_{1rn} is the nth Chebyshev mode of the radial component of the $\vec{v}_1$ field.

§ 4. Computational Algorithm.

The section describes the sequential steps which are performed for a complete computational cycle.

a. Preprocessing.

As the matrices associated with Eq. (16) and the viscous splitting procedure have constant coefficients, they are inverted once for all in a preprocessing phase of the computation by a Gauss Jordan elimination with complete pivoting. The order of magnitude of the accuracy for N=32 is about 10^{-7}.

b. Computational cycle.

Suppose the initial data or the velocity components at the previous time level are given in wave space. Using a 3-D FFT, the velocity components are obtained in physical space. Taking the spectral derivatives of the velocity components, the vorticity components are immediately obtained in physical space by a repeated application of the 3-D FFT. The cross product of the convective terms is locally computed in the physical space and the inverse transform produces the spectral nonlinear terms.

The spectral computation of the viscous terms is easily done with the help of classical expansion formulae (Gottlieb and Orszag, 1977).

The inclusion of the terms needed by the semi-implicit technique requires the application of a one-dimensional FFT in the r direction to Eq. (20). This leaves us with modes in wave space for the θ and z directions, while the r dependence is expressed in physical space. The left hand side for example may be written as,

$$[1 + \Delta t \frac{V(r)}{r} ik_\theta]\vec{v}_1 \quad ,$$

which easily yields the $\vec{v}_1$ components. The $\vec{v}_1$ modes in wave space are obtained by a 1-D inverse FFT.

A straightforward computation provides the right member of (18), which, for each mode, is multiplied by the inverse matrix corresponding to the pressure field. The pressure boundary conditions are applied using a Chebyshev tau approximation. The pressure gradient is added to the previous velocity field (Eq. (19)). Finally, a simple explicit calculation produces the right hand side of the stabilizing process. Multiplication by the appropriate inverse matrices gives the final velocity at the new time level.

§ 5. Instability of Taylor cells.

Instability of Taylor cells has already been investigated experimentally, numerically, theoretically (see §1). The accumulation of these results constitutes a valuable background for the assessment of our computational code.

a. Initial data.

A perturbation velocity field is superimposed to the Couette solution and is gotten by solving the linear eigenvalue problem associated with the Taylor instability (Chandrasekhar, 1961). The set of equations is

$$\nu(DD_* - \lambda^2)^2 u - 2\lambda^2 \frac{V}{r} v = \sigma(DD_* - \lambda^2)u \quad , \qquad (22)$$

$$\nu(DD_* - \lambda^2)v - (D_*V)u = \sigma v \quad , \qquad (23)$$

$$u = v = Du = 0 \qquad \text{for } r = R_i, R_o \quad . \qquad (24)$$

In the eqns. (22-24), the symbol D represents d/dr and $D_* = d/dr + 1/r$. As the stability analysis is couched in terms of normal modes for the disturbance, solutions are obtained in the following forms,

$$v_r = e^{\sigma t} u(r) \cos \lambda z \quad ,$$
$$v_\theta = e^{\sigma t} v(r) \cos \lambda z \quad ,$$
$$v_z = e^{\sigma t} w(r) \sin \lambda z \quad ,$$

where λ is the wave number of the disturbance in the axial direction and σ is the growth rate.

The Chebyshev modes of u and v arise from the solution of a generalized eigenvalue problem by a matrix QR eigenvalue analysis or by an inverse iteration-Rayleigh quotient method, followed by an inverse iteration method for the computation of the eigenfuctions. Once the u and v modes are known, the w modes come from the relationship,

$$D_* u = -\lambda w \quad .$$

These Chebyshev modes are superposed onto the basic Couette flow in such a way that the maximum perturbed azimuthal component is 5.10^{-3} times the inner cylinder rotation velocity.

b. Time accuracy of results.

As spectral methods are spatially of "infinite order", the accuracy is mainly affected by time differencing errors. The comparison of the evolution of small-amplitude disturbances with their predicted behaviour according to the linear stability theory sheds light on the overall accuracy of the code.

Several tests made on small and wide gap geometries have shown that a time accuracy of the order of 1% is achieved.

For example, for a small gap problem (Ri/Ro = 0.95) at a Taylor number $T = 1.05T_{cr}$ and for the axial critical wavenumber, the predicted dimensionless linear growth rate $\sigma d^2/\nu$ is 0.6487. With 8 modes in the z direction and 17 Chebyshev polynomials inside the gap, the computed growth rate ($\Delta t = 10^{-3}$) is 0.6487 after 200 time steps and 0.6565 after 350 time steps. On a wide gap geometry (Ri/Ro = 0.5), at the critical wavenumber and for $T = 2T_{cr}$, the predicted linear growth rate is 11.365, while the computed growth rate using the previous spectral definition and $\Delta t = 0.02$ is 11.2125.

c. Comparison of numerical results with theory and experimental data.

This subsection reports numerical results obtained for two different problems, where the outer cylinder is kept fixed. The spectral definition is the same as in §5.b . The first case deals with a small gap geometry (R_i/R_o = 0.95), at a Taylor number 5% above the critical value. The computational results can be compared to the analytical calculation carried out by Davey (1962) using an expansion procedure developed by Stuart (1960) and Watson (1960). Summarily, it is supposed that the azimuthal velocity component v_θ can be expressed in terms of a shape function A(t) by the following relationship

$$v_\theta(r,z,t) = V(r) + A(t)\, v_{11}(r) \cos\lambda z +$$
$$A^2(t)[f_1(r) + v_{22}(r) \cos 2\lambda z] + A^3(t)[v_{31} \cos\lambda z + v_{33}(r) \cos 3\lambda z]$$
$$+ \;\ldots\ldots$$

The function $f_1(r)$ arises from nonlinear effects and represents the distortion of the mean motion by the Reynolds stress. The amplitude function A(t) is the solution of the nonlinear equation,

$$\frac{dA}{dt} = \sigma A + a_1 A^3 + a_2 A^5 + \ldots \qquad (25)$$

Davey's (1962) results are second order accurate in A because Eq. (25) is truncated at cubic terms. Furthermore, this analytical solution is developed within the framework of the small gap approximation ($d/R_i \to 0$), which induces an error of the order of d/R_i. Figures 2-6 exhibit the comparison of the analytical and numerical results. The continuous line depicts the theoretical solution by Davey (1962), with the corrected steady-state amplitude by Eagles (1971), who included quintic terms in the shape function equation and integrated the "full" equations instead of the "small-gap" equations. The circles show the numerical values. The physical problem has the following parameters : R_i = 5cm, R_o = 5.2632cm, H = 0.5288cm, $\nu = 0.05 cm^2/sec$ and $\Omega_i = 7.20332\ sec^{-1}$.

The theory underestimates slightly the radial velocity component and overestimates by a small amount the azimuthal velocity component .

The maximum discrepancy is of the order of 6.6% between the theory and the computation. There are two reasons for the disagreement :

(i) the numerical integration treats the full geometry;

(ii) the full nonlinear interaction of the convective part of the Navier-Stokes equations is completely incorporated in the spectral simulation, without any requirement of being close to the marginal state.

Figures 2 and 7 show that the jet moving radially outwards transfers higher momentum than the jet moving radially inwards. This phenomenon leads to the periodic axial distribution of the Taylor cells. From figure 6, one can see that the superposition of the perturbed velocity component onto the basic Couette flow at the center of the vorter cell tends to slow down the fluid adjacent to the inner cylinder and to accelerate the fluid close to the outer one. This trend would indicate that at higher Taylor numbers, the core of the Taylor vortices rotate nearly as rigid bodies around the longitudinal axis of the cylinders. Finally, as the axial velocity component distribution across the cell center (Fig.8) is almost linear near the mid gap, the vorticity component ω_θ (Eq.(8)) is constant. The Taylor vortex cores rotate with constant angular velocity around the cell centers.

The steady-state energy of the perturbed velocity components is $5.25\ 10^{-3}$ times the Couette energy, while the computed torque is 6.6% higher than the Couette torque. This last figure is in close agreement with the experiments of Donnelly and Simon, 1960.

The second problem solves the Taylor vortices development in Coles'(1965) geometry. The physical problem is described by the next factors : R_i = 5cm, R_0 = 5.72cm, H = 1.435cm, ν = 0.213cm^2/sec and Ω_i = 8.46084sec^{-1}. The Taylor number is that of the wave boundary but the flow is now restricted to be two-dimensional. Figures 9-10 show the time development of the radial velocity component. The curve t = 2.5 is very close to the steady-state.

Conclusions.

A three-dimensional Chebyshev-Fourier representation of the velocity field is used in cylindrical coordinates to simulate the spectral transition in the circular Couette flow between concentric cylinders.

The time integration of the Navier-Stokes equations is performed by a splitting procedure. The linear parts of the governing equations are treated by a Chebyshev-Tau approximation, while the nonlinear terms involve a collocation process. An Adams-Bashforth scheme is applied to both convective and viscous terms, with a viscous stabilization step to preserve numerical stability. A semi-implicit technique allows for acceptable time steps as far as the convective stability is concerned.

Two test cases were run. The small gap problem is compared with analytical results and experimental data, while the second problem deals with the time development of Taylor vortices in Coles' geometry.The numerical results agree very well with the experiments.

This research was initiated while one of us (M.D.) was visiting M.I.T. with a NATO postdoctoral fellowship.

We would like to acknowledge partial support by the Office of Naval Research under Contract N° N00014-77-C-0138 and the National Science Foundation under Grant ATM-78-17092.

Some of the calculations reported here were performed at the Computing Facility of the National Center for Atmospheric Research, Boulder, Colorado, which is sponsored by the National Science Foundation.

References

Chandrasekhar, S. 1961 Hydrodynamic and Hydromagnetic Stability, Oxford University Press, Oxford.

Coles, D. 1965 J. Fluid Mech., 21, 385.

Cooley, J.W. and Tukey, J.W. 1965 Math. Comp., 19, 297.

Davey, A. 1962 J. Fluid Mech., 14, 336.

Davey, A., Di Prima, R.C. and Stuart, J.T. 1968 J. Fluid Mech., 31, 17.

Di Prima, R.C. and Eagles, P.M. 1977 Phys. Fluids, 20, 171.

Donnelly, R.J. and Simon, N.J. 1960 J. Fluid Mech., 7, 401.

Eagles, P.M. 1971 J. Fluid Mech., 49, 529.

Fasel, H. and Booz, O. 1977 Second GAMM-Conference on Numerical Methods in Fluid Mechanics.

Fenstermacher, P.R., Swinney, H.L. and Gollub, J.P. 1979 J. Fluid Mech., 94, 103.

Gollub, J.P. and Freilich, M.H. 1976 Phys. Fluids, 19, 618.

Gollub, J.P. and Swinney, H.L. 1975 Phys. Rev. Letters, 35, 927.

Gottlieb, D. and Orszag, S.A. 1977 Numerical Analysis of Spectral Methods : Theory and Applications, NSF-CBMS Monograph n° 26, Soc. Ind. Appl. Math.

Harlow, F.H. and Welch, J.E. 1965 Phys. Fluids, 8, 2182.

Majumdar, A.K. and Spalding, D.B. 1977 J. Fluid Mech., 81, 295.

Meyer, K.A. 1967 Phys. Fluids, 10, 1874.

Meyer, K.A. 1969 Phys. Fluids, Supplement II, 12, II-165.

Orszag, S.A. 1972 Studies in Appl. Math., 51, 253.

Orszag, S.A. and Deville, M. 1979 to be published.

Rogers, E.H. and Beard, D.W. 1969 J. Comp. Phys., 4, 1.

Stuart, J.T. 1960 J. Fluid Mech., 9, 353.

Taylor, G.I. 1923 Phil. Trans. A 223, 289.

Watson, J. 1960 J. Fluid Mech., 9, 371.

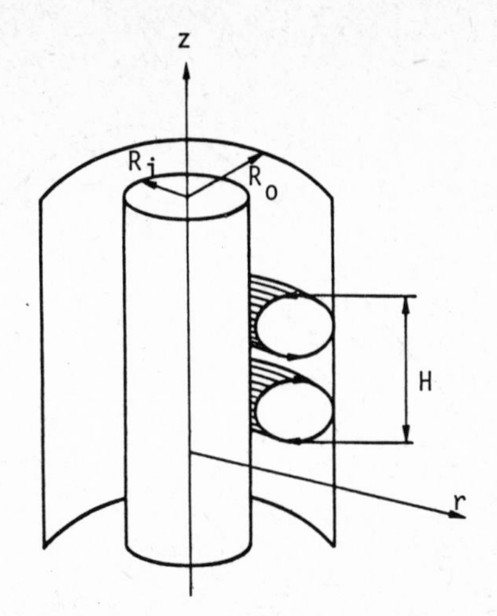

Figure 1. Geometrical Configuration.

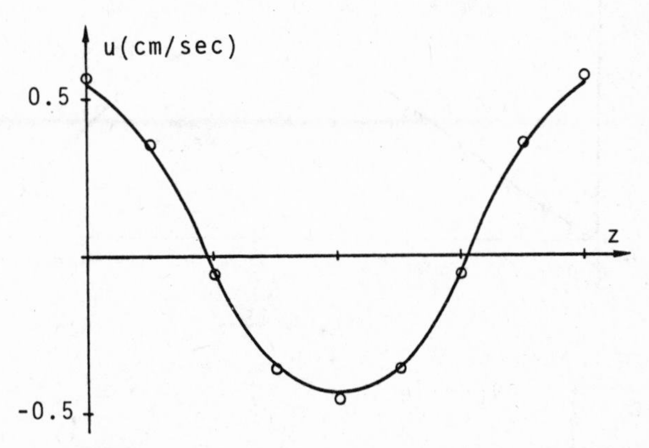

Figure 2. Axial distribution of the radial velocity component at midgap. — : Davey (1962); o : present results.

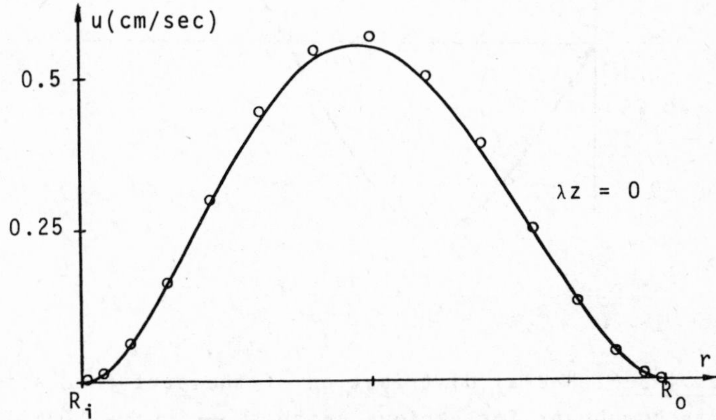

Figure 3. Radial velocity component profile at the bottom of the integration domain. — : Davey (1962); o : present results.

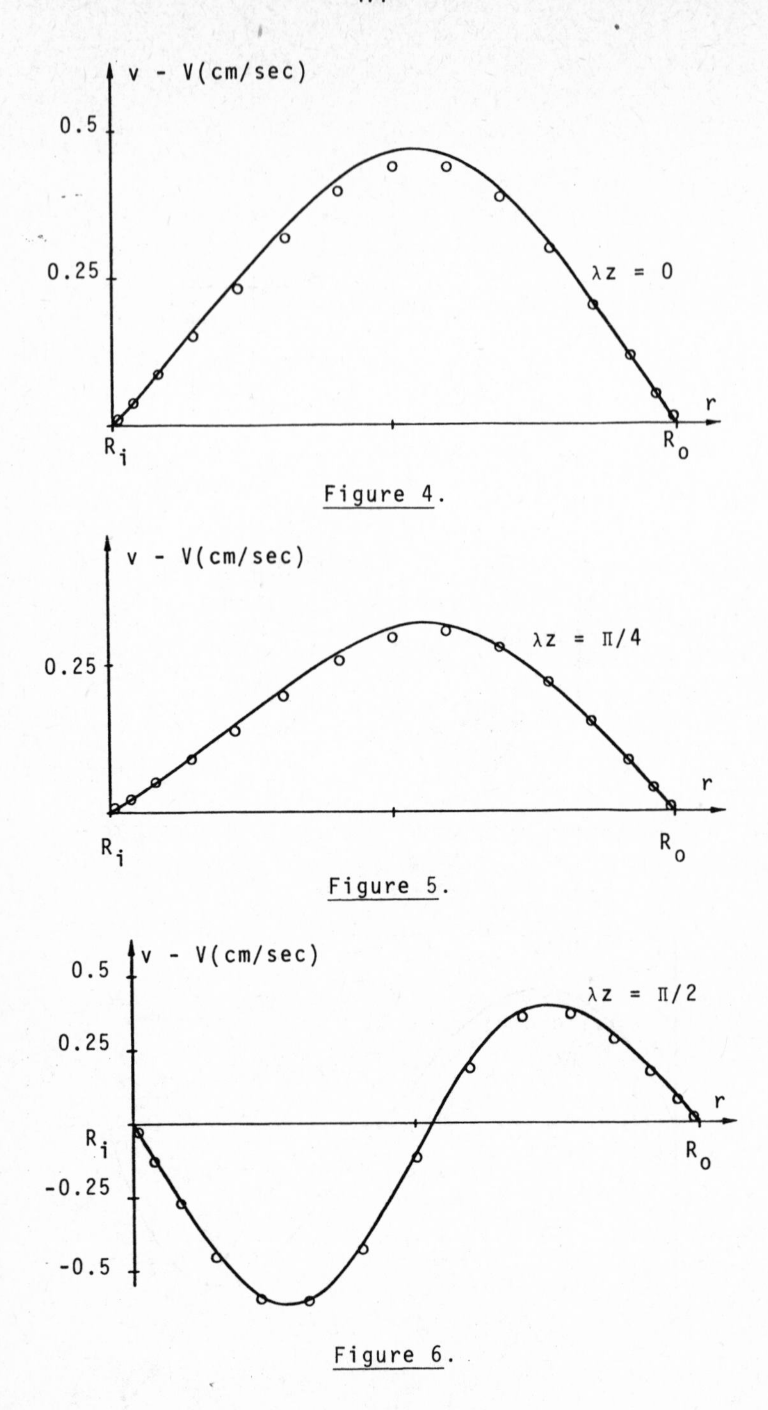

Figures 4-5-6. Radial distribution of the perturbed azimuthal velocity component for various heights. — : Davey(1962); o : present results.

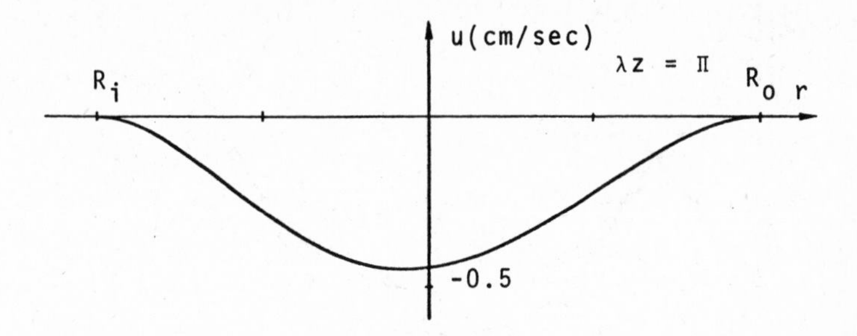

Figure 7. Computed radial velocity component at the interface between two adjacent cells.

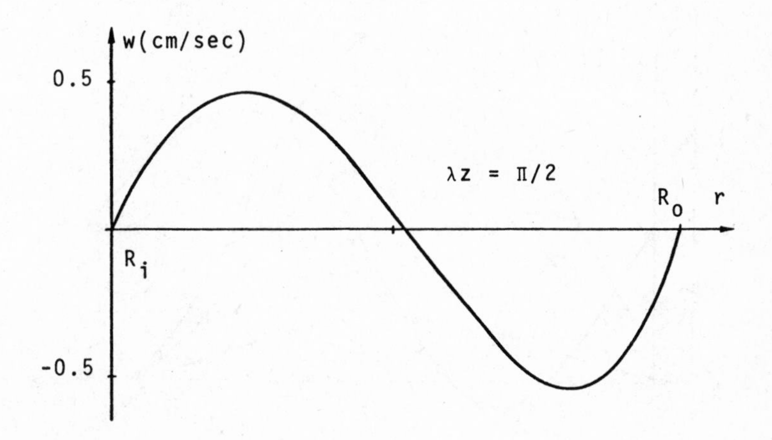

Figure 8. Radial distribution of the axial velocity component through the cell center.

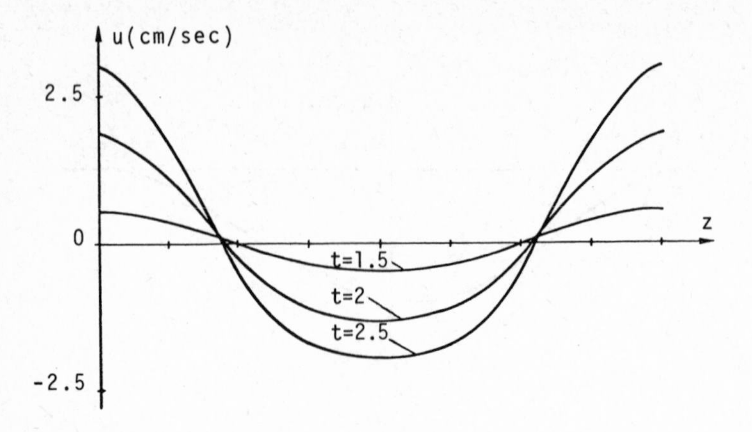

Figure 9. Time development of the radial velocity component at midgap in Coles' (1965) geometry.

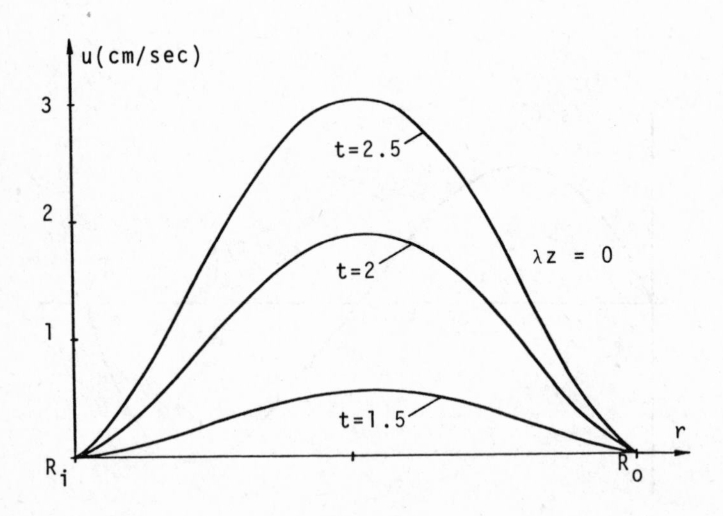

Figure 10. Time development of the radial velocity component across the gap at the bottom of the domain.

NUMERICAL SOLUTION OF THE COMPLETE NAVIER-STOKES EQUATIONS FOR THE SIMULATION OF UNSTEADY FLOWS

Hermann F. Fasel

Institut A für Mechanik
Universität Stuttgart
Stuttgart, Germany

I. INTRODUCTION

Today the development of a numerical method for the solution of the complete Navier-Stokes equations is still a considerable undertaking even with certain restrictions (incompressibility, low Reynolds numbers, etc.) or even if applications are to be limited to flows within simple geometries. Additionally, difficulties may increase considerably if for example the numerical method is to be applicable for higher Reynolds numbers, for flows in irregular boundaries or for realistic simulations of strongly unsteady flow phenomena. In these cases complications arise even for the two-dimensional and incompressible case and may become even more pronounced when applications to fully three-dimensional flows are considered.

The main difficulties in simulating flows at higher Reynolds numbers or flows of strongly unsteady behavior are due to the existence of extremely thin layers (especially near walls) of large gradients which have to be adequately resolved. Using finite-difference methods, which will be considered in the present paper, the space intervals must be small enough in the regions where the thin layers occur to assure sufficient accuracy. Thus large numbers of grid points are necessary, thereby resulting in large systems of equations to be solved.

Moreover, and in many cases more serious, the extremely small space intervals may for stability reasons require an even more disproportionate reduction of the time step, especially when explicit or weakly implicit methods are employed. Often for stability reasons a time-interval size may be required that is orders of magnitude smaller than that necessary for adequate resolution of the temporal gradients even if highly unsteady flow phenomena are to be simulated.

Due to their superior stability characteristics implicit methods are therefore generally preferable to explicit methods for the simulation of unsteady flow phenomena at higher Reynolds numbers. Of course an unconditionally stable scheme would be desirable, because then in selecting the time interval, only an adequate resolution of the temporal gradients would have to be considered.

In the present paper main aspects of a 'fully' implicit finite-difference method that has proven to be extremely stable are discussed. With judgement based on numerous numerical experiments it appears that the method may be unconditionally stable.

However, a stability analysis for the complete nonlinear equations will be required for final verification (a stability analysis for the linearized equations leads to unconditional stability).

The applicability of the numerical method to simulations of complicated unsteady flows at higher Reynolds numbers was demonstrated on several calculation examples. Some results presented in this paper are from calculations of unsteady flow phenomena that occur in the process of transition from laminar to turbulent flow. As these phenomena are of highly unsteady nature and occur only at relatively large Reynolds numbers (larger than the critical Reynolds number, which is for a boundary-layer flow 520, based on displacement thickness; and for plane Poiseuille flow about 5800, based on one half channel width) such calculations present a critical test of any numerical method. Thus, if a method is suitable for realistic simulations of such phenomena, it may be conjectured that the method is then also applicable for the calculation of other unsteady phenomena of equal or lesser degrees of complexity. This conjecture is supported by numerous applications of the present numerical method to other unsteady flow problems.

II. INTEGRATION DOMAIN

Using finite-difference methods the integration domain in the physical plane should be of finite size. Therefore flow fields that extend to infinity in the physical plane have to be made finite by introducing so-called artificial boundaries.

On the other hand, the use of transformations to map an infinite region in the physical plane into a finite region in the calculation plane would allow for simpler boundary conditions. From this viewpoint the latter approach would be preferable. However, as suggested by Thompson [1] this approach may give rise to serious difficulties, mainly associated with excessive artificial viscosity caused by the rapid grid expansion necessary to cover an infinite physical domain.

It is essential that proper boundary conditions for the artificial boundaries can be specified and implemented into the numerical scheme when, as in the present approach, domains are made finite. That is, boundary conditions have to be used allowing solutions that would be identical with exact solutions (which of course are not available for practical applications) in the originally larger or even unbounded domain. The limited storage capacity of computers available today allows for the selection of only a relatively small section (where the Navier-Stokes equations are solved) of the entire physical flow field. This applies to many flow problems of practical interest, especially for strongly unsteady flows at larger Reynolds numbers. The connection of this section with the remainder of the flow field, where governing equations much simpler than the Navier-Stokes equations frequently can be used, has to be accomplished with the boundary conditions. This context will become clearer with the examples discussed subsequently.

Here, for simplicity, only domains bounded by rectangular coordinate lines are considered, thus allowing the use of simple rectangular grid systems in the physical plane. However, the basic numerical method discussed in this paper can also be applied for flows in arbitrary boundary shapes, using the numerical transformation techniques as developed by Thompson [1]. For the examples given here, the rectangular domain may be a section of a boundary-layer flow on a flat plate as shown in Fig. 1 (a typical example for an external flow), a section of a channel flow (plane Poiseuille flow) between two parallel plates as shown in Fig. 2 (example for partially internal flow) or a section of a flow on a plate with a backward-facing step as in Fig. 3 (an example for a flow problem with the presence of a singularity).

The previous remarks already indicate that the selection of the integration domain is inherently tied to the entire complex of choosing and implementing proper boundary conditions. As the discussion in subsequent chapters will show, the question of what boundary conditions can actually be used is also directly dependent on the form of the Navier-Stokes equations on which the numerical method is based. Thus, indirectly, when the integration domain is selected, one has to keep in mind what formulation of the governing equations one intends to use.

III. GOVERNING EQUATIONS

In the present paper two different formulations of the Navier-Stokes equations to be used as the basis for finite-difference methods are discussed. For two-dimensional, incompressible flows, which are only considered here, they are as follows:

i) The customary vorticity-stream function (ω,ψ) formulation with the vorticity equation

$$(1) \qquad \frac{\partial \omega}{\partial t} + \frac{\partial \psi}{\partial y}\frac{\partial \omega}{\partial x} - \frac{\partial \psi}{\partial x}\frac{\partial \omega}{\partial y} = \frac{1}{Re}\Delta\omega \quad ,$$

and the Poisson equation for the stream function

$$(2) \qquad \Delta\psi = \omega \quad ,$$

where Δ is the Laplace operator, and the vorticity ω is defined as

$$(3) \qquad \omega = \frac{\partial u}{\partial y} - \frac{\partial v}{\partial x} \quad ,$$

and ψ defined as

$$(4) \qquad \frac{\partial \psi}{\partial y} = u \quad , \qquad \frac{\partial \psi}{\partial x} = -v \quad .$$

ii) A vorticity-velocity (ω,u,v) formulation with the vorticity equation

$$(5) \qquad \frac{\partial \omega}{\partial t} + u\frac{\partial \omega}{\partial x} + v\frac{\partial \omega}{\partial y} = \frac{1}{Re}\Delta\omega \quad ,$$

and two Poisson equations for the velocity components u and v (which are in x- and y-direction, respectively)

$$(6) \qquad \begin{cases} \Delta u = \dfrac{\partial \omega}{\partial y} \quad , \\ \Delta v = -\dfrac{\partial \omega}{\partial x} \quad . \end{cases}$$

Eqs. (6) are obtained from (3) by differentiation with respect to y and x, respectively, and subsequent use of the continuity equation

$$(7) \qquad \frac{\partial u}{\partial x} + \frac{\partial v}{\partial y} = 0 \quad .$$

All variables in Eqs. (1) to (7) are dimensionless; the dimensionless variables relate to their dimensional counterparts, denoted by bars, as follows:

$$x=\frac{\bar{x}}{L}\ ,\ y=\frac{\bar{y}}{L}\ ,\ u=\frac{\bar{u}}{U_0}\ ,\ v=\frac{\bar{v}}{U_0}\ ,\ t=\frac{\bar{t}U_0}{L}\ ,\ \omega=\frac{\bar{\omega}L}{U_0}\ ,\ Re=\frac{U_0L}{\nu}\ ,\ \psi=\frac{\bar{\psi}}{U_0L}\ ,$$

where L is a characteristic length, U_0 a reference velocity and Re a Reynolds number (ν kinematic viscosity).

Using the ω,ψ-formulation the continuity condition is automatically satisfied for the continuum equations. However this is not necessarily true for the discretized equations (with finite $\Delta x,\Delta y$) but can be readily assured if certain difference approximations are used for the space derivatives (for example with central differences). On the other hand for the ω,u,v-system, which obviously is of higher order than the ω,ψ-system, continuity is not automatically satisfied, not even for the continuum equations. However from Eqs.(6) it follows that $\Delta D=0$ is satisfied for this system, where D is the divergence $D=\partial u/\partial x+\partial v/\partial y$. Thus it follows (maximum principle) that $|D|$ is maximal on the boundary. Therefore continuity ($D\equiv 0$) is guaranteed for the continuum equations inside the boundary if $D\equiv 0$ is satisfied on the boundary itself. For the discretized equations it may be conjectured that (with central difference approximations) $|D|$ is always smaller inside the boundary than on the boundary itself. In implementing boundary conditions when the ω,u,v-formulation is used, it is therefore vital to assure that the continuity condition is satisfied on the boundaries to as high a degree of accuracy as possible.

These considerations suggest that the ω,u,v-system is more difficult to deal with than the ω,ψ-system, particularly in connection with the implementation of the boundary conditions. However, as discussed in Chapter IV, due to the higher order of the ω,u,v-system, less restrictive (or 'softer') boundary conditions may be used at the artificial boundaries than when the ω,ψ-formulation is used. This may be a decisive advantage for certain applications.

Both systems, the ω,ψ- and ω,u,v-system, may also be used in a conservative form where certain properties are conserved for finite interval sizes [2]. For example the vorticity equation (5) in conservative formulation takes the form

$$(8) \qquad \frac{\partial \omega}{\partial t} + \frac{\partial (u\omega)}{\partial x} + \frac{\partial (v\omega)}{\partial y} = \frac{1}{Re}\,\Delta\omega \quad ,$$

where the vorticity balance is guaranteed for finite $\Delta x,\Delta y$ when central differences are used for the spatial derivatives.

For the calculation of compressible flows the advantage of the conservative formulation over a nonconservative one appears to be a widely acknowledged fact and it is claimed that better accuracy can be obtained. However, for calculations of incompressible flows the relative merits of the conservative form is still a controversial subject. For the implicit method described in this paper, extensive comparison calculations have shown that practically identical results are obtained with either form, conservative or nonconservative. As the conservative form requires additional numerical operations and in the present numerical method caused a somewhat slower convergence of the iterative solution of the difference equations the nonconservative form was preferred.

For the calculation of unsteady flows it may sometimes be advantageous to decompose the total flow into a steady, time-independent basic flow and a time-dependent unsteady part such that

$$\begin{aligned} (9) \qquad & \omega(x,y,t)=\Omega(x,y)+\omega'(x,y,t) \ , \quad \psi(x,y,t)=\Psi(x,y)+\psi'(x,y,t) \ , \\ & u(x,y,t)=U(x,y)+u'(x,y,t) \ , \quad v(x,y,t)=V(x,y)+v'(x,y,t) \ , \end{aligned}$$

where the prime indicates the variables of the time-dependent, unsteady part and the capital letters denote the flow variables of the basic, time-invariant part. Substituting (9) into the original ω,ψ- or ω,u,v-formulation for the total flow, i.e., Eqs. (1) and (2) or Eqs. (5) and (6), these equations can be rewritten with the flow variables of the unsteady flow as dependent variables. Thus the vorticity equation (5), for the ω,u,v-formulation for example, leads to

$$(10) \qquad \frac{\partial\omega'}{\partial t} + (U + u')\frac{\partial\omega'}{\partial x} + u'\frac{\partial\Omega}{\partial x} + (V + v')\frac{\partial\omega'}{\partial y} + v'\frac{\partial\Omega}{\partial y} = \frac{1}{Re}\Delta\omega'$$

assuming that the basic flow also satisfies the Navier-Stokes equations.

Such a decomposition of the total flow into a steady and nonsteady part is often convenient when unsteady flow problems are considered where the unsteady deviations from a mean flow are of main interest. Typical examples are the wide field of oscillating flow phenomena such as unsteady wake flows behind blunt obstacles (Karman vortex street) or periodic oscillations that occur in the laminar-turbulent transition process. In such cases this decomposition has the advantage that the numerical solution yields directly those quantities which are of main interest and may therefore be closely observed during the course of calculations. Further, in this form the effects of the nonlinearities in the convective terms can easily be studied because here a possible linearization (if departure of the unsteady oscillating flow from the basic flow is small) can be conveniently switched on or off. This allows comparison of the calculations with linearized theories.

However the use of the governing equations with this decomposition also has disadvantages. The vorticity transport equation (10) contains in this case several additional terms (involving the variables of the unsteady part with those of the steady flow) which are not present in Eq. (5) for the total flow variables. Thus additional numerical operations are required and accordingly, more computation time. Further, because of the additional terms involving the basic flow, the grid values of the basic flow variables have to be kept in fast-access storage in order to avoid excessive computer times. If, on the other hand, the equations for the total flow variables are used the basic flow variables do not become involved in the solution algorithm. If the basic flow quantities are nevertheless required for the purpose of better analysis and better representation of the results, they then can be stored in mass storage of lower speed accessibility.

In this context it should be emphasized that the availability of sufficient fast-access storage is, even with the latest computer models, often a critical limitation for numerical simulations of unsteady flows. This is even more true for higher Reynolds numbers where large numbers of grid points are required for adequate resolution of the arising large gradients. In such cases the use of the governing equations in a form with the unsteady flow quantities as dependent variables may be prohibitive.

IV. BOUNDARY CONDITIONS

The discussions of Chapter II have indicated already that the specification of proper boundary conditions and their implementation into the numerical scheme represents one of the most important tasks in the development of a finite-difference method that is applicable for realistic simulations of unsteady flows. The choice of proper boundary conditions is irrevocably dependent on the selection of the integration domain and in addition strongly dependent on the form of the governing equations. Finally, one has to keep in mind that the conditions selected must be capable of being implemented into the prospective difference method.

For practical reasons in respect to required computation times the boundary conditions for the artificial boundaries (see Chapter II) have to allow for physically meaningful results when the integration domains are relatively small. The number of grid points (and thus computer storage requirements) and the amount of numerical operations necessary for a numerical solution are directly dependent on the physical size of the integration domain.

In this chapter boundary conditions are discussed which are applicable for the calculation of strongly unsteady flow phenomena. In these examples the unsteady flow results either from forced perturbations inserted at a fixed downstream location of the flow field (for the boundary layer and plane Poiseuille flow) or is caused by an abrupt geometric variation (for a flow along a plate with a backward-facing step). The domains of integration considered in these examples are those sketched schematically in Figs. 1 to 3 where the governing equations in the ω,ψ- or ω,u,v-formulation are used in the total variable form.

Forced perturbations may be introduced for example at the inflow boundary A-D of Figs. 1 to 3. This can be done by superimposing onto the profiles of the basic (steady) flow (for which Blasius profiles can be used for the cases considered in Figs. 1 and 3 and Poiseuille profiles for the case of Fig. 2) so-called perturbation functions which are solely dependent of y and t. Thus for the ω,ψ-formulations the conditions at the inflow boundary can be written as

$$(11)\quad \begin{cases} \omega(0,y,t) = \omega_B(y) + P_\omega(y,t) \ , \\ \psi(0,y,t) = \psi_B(y) + P_\psi(y,t) \ , \end{cases}$$

and for the ω,u,v-formulation as

$$(12)\quad \begin{cases} \omega(0,y,t) = \omega_B(y) + P_\omega(y,t) \ , \\ u(0,y,t) = u_B(y) + P_u(y,t) \ , \\ v(0,y,t) = v_B(y) + P_v(y,t) \ , \end{cases}$$

where subscript B denotes the variables of the basic flow and P_ω, P_ψ, .., etc. are the perturbation functions.

With the conditions (11) or (12) it is possible to simulate the unsteady reaction of flows to upstream disturbances that in physical experiments may be produced by some sort of a disturbance generator. An example would be the famous laboratory experiments by Schubauer and Skramstad [3] who used a vibrating ribbon to generate the so-called Tollmien-Schlichting waves that arise in the process of laminar-turbulent transition. Conditions (11) and (12) are of simple Dirichlet-type and therefore implementation into the numerical method can be readily accomplished.

At solid walls (which are considered to be impermeable and no-slip) such as boundaries A-B in Figs. 1 to 3 and C-D in Fig. 2, the velocity components vanish and ψ is constant

$$(13)\qquad u=0 \ , \quad v=0 \ , \quad \psi=\text{constant} \ , \quad \partial\psi/\partial y=0 \ .$$

For the ω,ψ-formulation vorticity at the walls is calculated from the relationship

$$(14)\qquad \omega = \frac{\partial^2\psi}{\partial y^2}$$

which follows from Eq. (2). For the ω,u,v-formulation the wall vorticity may be calculated using either

$$(15)\qquad -\frac{\partial\omega}{\partial x} = \frac{\partial^2 v}{\partial y^2}$$

which is derived from Eq. (6b) or

$$(16)\qquad \omega = \frac{\partial u}{\partial y}$$

which follows from Eq. (3).

For the calculation of the wall vorticity at the vertical part E-F of the step, relationships are used corresponding to those of Eqs. (14) to (16) for the horizontal parts. For the calculation of vorticity at the convex corner E (see Fig. 3) several methods have been suggested [2]. With the present numerical method satisfactory results are obtained when the corner is simply treated as if it would belong to wall A-E and therefore the corner vorticity is calculated from one of the Eqs. (14) to (15). At the concave corner F vorticity can simply be set to zero.

At the outflow boundary, conditions are desirable which are as 'soft' or as little 'restrictive' as possible, i.e., conditions that minimize the upstream influence of this artificial boundary. With the numerical method discussed here it is possible to use conditions that involve second derivatives with respect to x such as for the ω,ψ-formulation

$$(17)\quad \begin{cases} \dfrac{\partial^2\omega}{\partial x^2} = -\alpha^2(\omega-\Omega) \quad , \\[2ex] \dfrac{\partial^2\psi}{\partial x^2} = -\alpha^2(\psi-\Psi) \quad , \end{cases}$$

and for the ω,u,v-formulation

$$(18)\quad \begin{cases} \dfrac{\partial^2\omega}{\partial x^2} = -\alpha^2(\omega-\Omega) \quad , \\[2ex] \dfrac{\partial^2 u}{\partial x^2} = -\alpha^2(u-U) \quad , \\[2ex] \dfrac{\partial^2 v}{\partial x^2} = -\alpha^2(v-V) \quad . \end{cases}$$

Here, α is the local wave number of the unsteady flow at the outflow boundary. It may be conveniently introduced when the unsteady flow phenomena consist predominantly of spatially propagating fluctuations about a mean (basic) flow (denoted by capital letters in Eqs. 17 and 18). The mean flow may also vary as time evolves. However, the time-scale of the mean flow should be much larger than that of the fluctuating part. Typical examples are the unsteady vortex shedding behind blunt obstacles or flows perturbed by periodic disturbances. If reasonable approximate values for α (to be used in Eqs. 17 or 18) are not available, they may then be obtained by an iterative procedure which is performed during the actual Navier-Stokes calculations as discussed in [4].

For arbitrary unsteady phenomena, i.e., when distinct fluctuations about a mean flow are not present, it is best to use $\alpha=0$ in the conditions (17) or (18). These conditions with $\alpha=0$ are still superior to those that have to be frequently used in many other numerical methods, where only vanishing first derivatives with respect to x can be used. Boundary conditions with lower derivatives are more restrictive, i.e., the outflow boundary becomes less soft or less transparent, causing stronger upstream effects than when comparable boundary conditions involving higher derivatives are used.

There is no formal difference between conditions (17) for the ω,ψ-formulation and (18) for the ω,u,v-formulation because in both cases second derivatives are involved. However, a subtle difference does exist which in many practical applications may be of great importance. Due to the definition of ψ the use of condition (17b) is actually equivalent to prescribing a condition involving a first derivative for v

$$(19)\qquad \frac{\partial v}{\partial x} = +\alpha^2(\psi-\Psi) \quad .$$

Naturally this condition is more restrictive than condition (18c) which specifies a second derivative for v. The effects due to this difference between conditions (17) and (18) were clearly observed in comparison calculations for a boundary layer disturbed periodically at the inflow boundary A-D (see Fig. 1). In this case the use of conditions (17) with $\alpha=0$ causes reflections of the propagating waves at the outflow boundary, even when disturbance amplitudes are very small (with maximum of u'-fluctuation of 0.05% of free-stream velocity), while no such difficulties occur when conditions (18) are used instead. The reflections can be avoided if proper approximate values for α are used in condition (17). However, for larger disturbance amplitudes (larger than 1%), reflections occur again even with proper values for α. Using conditions (18) reflections may also appear for larger amplitudes if α is set to zero. However, with approximate α's the disturbance waves again pass through the downstream boundary without any noticeable reflections or other irregularities.

Numerical experiments have shown that conditions (17) and (18) are suitable for simulations of Tollmien-Schlichting waves. With these conditions physically reasonable results can be obtained already when the length X of the integration domain contains only three to four wave lengths (see [4]). However, with conditions (17) only small amplitudes are admissible (if proper α's are used) while with conditions (18) both small (even with α=0) and large amplitudes (with proper α's) can be treated.

At the (artificial) free-stream boundaries, such as boundary C-D in Fig. 1 of a boundary-layer flow, special considerations are also necessary in order that the integration domain be kept as small as possible. It is convenient in most cases to put this boundary at least far enough from the wall to assure that

(20) $\omega = 0$

and/or

(21) $\partial\omega/\partial y = 0$

at this boundary. This is reasonable as vorticity, which is produced on solid walls, decays rapidly with increasing distance from the wall and is practically zero at a few boundary-layer thicknesses from the wall. On the other hand, the magnitude of u,v, or ψ generally may still be substantial at such distances from the wall.

Often it is convenient to prescribe the free-stream velocity (corresponding to a given pressure distribution)

(22) $u = U_{fs}(x,t)$

which means that for the ω,ψ-formulation, the condition

(23) $$\frac{\partial\psi}{\partial y} = U_{fs}(x,t)$$

may be used. In this case for the ω,u,v-formulation the relation

(24) $$\frac{\partial v}{\partial y} = -\frac{\partial U_{fs}(x,t)}{\partial x} \quad ,$$

which follows from the continuity equation, can be used together with condition (22). Conditions (22) and (24) have been used, for example, in [5] to study the reaction of a boundary layer to disturbances of the free stream.

In many cases of practical interest, the free-stream velocity cannot be prescribed a priori. With the present numerical method in connection with the ω,u,v-formulation it is sufficient to only specify a normal gradient, for example

(25) $\partial u/\partial y = 0 \quad .$

and then Eq. (3) yields the condition for v

(26) $\partial v/\partial x = -\omega$.

For unsteady boundary-layer flows where the unsteadiness results from a periodic disturbance input within the boundary layer at some upstream location, it is advantageous to use conditions that prescribe asymptotic decay of the disturbances with increasing distance from the wall, such as for ψ of the ω,ψ-formulation

(27) $$\frac{\partial(\psi-\Psi)}{\partial y} = -\alpha(\psi-\Psi) \quad ,$$

and for the velocity components of the ω,u,v-formulation

(28) $$\begin{cases} \dfrac{\partial(u-U)}{\partial y} = -\alpha(u-U) \quad , \\ \dfrac{\partial(v-V)}{\partial y} = -\alpha(v-V) \quad . \end{cases}$$

Here, α is again a local wave number of the resulting disturbance waves. For the calculation of propagating disturbance waves, these conditions allow for a relatively small domain in y-direction. For example, in simulations of Tollmien-Schlichting waves, already about three δ (δ boundary-layer thickness) are sufficient, while with Dirichlet conditions (u', v' and ψ' equal to zero) a much larger domain would be required. In this flow the disturbances u', v', ψ' decay relatively slowly in y-direction and may at 3δ from the wall still amount to about 50% of the maximum amplitude.

V. NUMERICAL METHOD

The finite-difference method discussed here is 'fully' implicit. Although unconditional stability for the complete nonlinear equations was not proven, it can be assumed that for stability the time-step restriction is less severe than for equivalent explicit or weakly implicit schemes. This aspect is particularly important for the simulation of strongly unsteady flows at high Reynolds numbers where it is desirable that the time step be adjusted with respect to the physical needs of the unsteady flow variation rather than to satisfy a stability criterion. In such flows the spatial gradients may become very large thus calling for only extremely small time steps for stability reasons when explicit or weakly implicit methods are considered. In this case the overall computation time for the simulation of such flows may become considerably larger with explicit methods than with strongly implicit methods.

The 'full' implicity of the present method implies that all difference approximations and function values, for both the discretization of the governing equations and boundary conditions, are taken at the most recent time-level ℓ (for notation see Fig. 4). With the 3-point backward difference for the time-derivative

(29) $$\left.\frac{\partial\omega}{\partial t}\right|_{n,m}^{\ell} = \frac{1}{2\Delta t}\,(3\omega_{n,m}^{\ell} - 4\omega_{n,m}^{\ell-1} + \omega_{n,m}^{\ell-2}) + O(\Delta t^2) \quad ,$$

which is of second order accuracy, the method requires three time-levels where vorticity values have to be stored. However, for ψ in the ω,ψ-formulation or for u and v in the ω,u,v-formulation storage only has to be provided at the most recent time level.

Therefore this method requires storage for an equal (for the ω,ψ-formulation) or lesser (for the ω,u,v-formulation) number of dependent variables than an equivalent two-level method of the Crank-Nicholson type. For the latter method storage would be required for ψ or u and v, respectively, at two time levels.

For the space derivatives central differences of second order are employed exclusively. To maintain overall second-order accuracy extra care is necessary in implementing the boundary conditions into the numerical scheme. An exception would be the boundary conditions of Dirichlet-type used at the inflow boundary A-D (Figs. 1 to 3). All other boundaries require special considerations to obtain proper difference approximations.

For example, for the calculation of vorticity at the walls from Eq. (14) when the ω,ψ-formulation is used, or from Eq. (15), when the ω,u,v-formulation is used, difference approximations for the second derivatives with respect to y of ψ and v, respectively, are required. These difference approximations are derived by polynomial interpolation using node points as indicated in Fig. 5. The imaginary point (n,-1) can be eliminated for either formulation by exploiting $\partial\psi/\partial y=0$ (because u=0 at the wall) or $\partial v/\partial y=0$ (due to the continuity condition). Thus, for example, for the second derivative of ψ one obtains

$$(30) \qquad \left.\frac{\partial^2\psi}{\partial y^2}\right|_{n,0} = \frac{1}{2\Delta y^2}\,(-\,7\psi_{n,0} + 8\psi_{n,1} - \psi_{n,2}) + O(\Delta y^2)$$

and correspondingly an identical approximation for the second derivative of v. However, as suggested by Briley [6], for consistency reasons the approximation for $\partial\psi/\partial y$ (required for the vorticity equation) at the grid point adjacent to the wall also has to be derived from the same polynomial, yielding

$$(31) \qquad \left.\frac{\partial\psi}{\partial y}\right|_{n,1} = \frac{1}{4\Delta y}\,(-\,5\psi_{n,0} + 4\psi_{n,1} + \psi_{n,2}) + O(\Delta y^2) \quad .$$

Employing more grid points (and thus higher-order polynomials) difference approximations of higher than second-order accuracy can be derived. Higher accuracy for the vorticity calculation may be desirable for many applications, because the global accuracy of the solution in the interior depends strongly on the degree of accuracy achieved in the calculation of the wall vorticity. Using the Navier-Stokes equations in vorticity transport form, the flow problem can be conceived as being driven by the production of vorticity at the walls, which is then conducted and dissipated into the interior of the flow field. Accurate prediction of the wall vorticity is therefore premandatory. Using for example an additional point (n,3) yields the third-order accurate approximation

$$(32) \qquad \left.\frac{\partial^2\psi}{\partial y^2}\right|_{n,0} = \frac{1}{18\Delta y^2}(-85\psi_{n,0}+108\psi_{n,1}-27\psi_{n,2}+4\psi_{n,3}) + O(\Delta y^3) \quad .$$

Now, for consistency, the approximations for the two grid points adjacent to the wall have to be also derived from the same polynomial. Thus, for $\partial\psi/\partial y$, which is again needed for the convective terms in the vorticity equation of the ω,ψ-formulation, one obtains

$$(33)\quad \begin{cases} \left.\dfrac{\partial\psi}{\partial y}\right|_{n,1} = \dfrac{1}{18\Delta y}(-17\psi_{n,0}+9\psi_{n,1}+9\psi_{n,2}-\psi_{n,3}) + O(\Delta y^3) \quad , \\ \left.\dfrac{\partial\psi}{\partial y}\right|_{n,2} = \dfrac{1}{18\Delta y}(14\psi_{n,0}-36\psi_{n,1}+18\psi_{n,2}+4\psi_{n,3}) + O(\Delta y^3) \quad . \end{cases}$$

For the second derivatives of ψ or v needed in either formulation for the corresponding Poisson equations one obtains accordingly

$$(34)\quad \begin{cases} \left.\dfrac{\partial^2\psi}{\partial y^2}\right|_{n,1} = \dfrac{1}{18\Delta y^2}(29\psi_{n,0}-54\psi_{n,1}+27\psi_{n,2}-2\psi_{n,3}) + O(\Delta y^3) \ , \\ \left.\dfrac{\partial^2\psi}{\partial y^2}\right|_{n,2} = \dfrac{1}{18\Delta y^2}(11\psi_{n,0}-27\psi_{n,2}+16\psi_{n,3}) + O(\Delta y^3) \quad , \end{cases}$$

again with identical expressions for v.

However, the use of these third-order accurate formulas increases the band width of the coefficient matrices of the difference equations when an iterative solution procedure is used. This will be described subsequently. However, the solution algorithm can be easily modified to account for the additional nodes that are required for the treatment of the two grid points adjacent to the wall. For approximating $\partial\omega/\partial x$, when Eq. (15) is used for the calculation of wall vorticity, a three-point backward difference of second-order accuracy analogous to approximation (29) is employed.

The procedure of deriving difference approximations from polynomial interpolation by including an imaginary point beyond the boundary, may also be applied for the outflow boundary B-C (Figs. 1 to 3) or the free-stream boundary C-D (Figs. 1 and 3). Then, for the elimination of the imaginary points, difference relations resulting from the governing equations themselves and/or difference relations resulting from the continuity equation and from the vorticity definition may be employed. The resulting difference formulas are too lengthy to be presented here. With this procedure the implementation of the boundary conditions can be achieved with an accuracy that is equivalent to that in the interior of the domain.

The implicit scheme leads to systems of equations: two systems for the ω,ψ- and three systems for the ω,u,v-formulation. Because of the full implicity, the equation system resulting from the vorticity equation is coupled with the Poisson system for ψ (for the ω,ψ-formulation) or with the two Poisson systems for u and v (for the ω,u,v-formulation). Additional coupling exists between the vorticity system and the ψ-system due to the calculation of the wall vorticity (for the ω,ψ-formulation) from Eq. (14) or (for the ω,u,v-formulation) of the vorticity system with either the Poisson system for v or for u, depending on which relationship, Eq. (15) or (16), is used for the calculation of the wall vorticity.

Of course, the equation systems can be decoupled with the introduction of an iteration loop. In this case $\partial\psi/\partial x$, $\partial\psi/\partial y$ or u,v in the convective terms of the vorticity equation as well as $\partial^2\psi/\partial y^2$, $\partial^2 v/\partial y^2$ or $\partial u/\partial y$ needed for the wall vorticity calculation would have to be taken at the preceeding iteration level i-1 (with i denoting the current iteration level). The resulting equation systems for iteration level i can then be solved independently. For the solution of the individual equation systems any standard solution algorithm could be used, such as point- or line-SOR, ADI or direct methods (for the Poisson systems).

Unfortunately, this iteration loop introduced for the decoupling has a rather slow convergence behavior. For increasing Reynolds numbers the already poor convergence deteriorates even further and the iteration may eventually even fail to converge at all. Due to the poor convergence of this iteration the computation times for practical calculations would be excessive, even if for the solution of the individual systems efficient solution algorithms, such as direct methods [2] or multi-grid methods [7] were employed.

However, in the present method the convergence behavior of this interation loop could be improved enormously, after the ω-system was coupled with the respective Poisson system due to the calculation of the wall vorticity. For the equation systems coupled through the wall vorticity calculation, an effective solution algorithm based on line-iteration was developed. This line-iteration is organized such that the grid values of ω and of the dependent variable of the coupled Poisson system are determined directly on grid lines parallel to the axis normal to the wall (y-axis in Figs. 1 to 3) while proceeding iteratively in the direction parallel to the wall (x-axis). The coefficient matrices of the equation systems for the individual grid lines are then of tridiagonal form. For the solution of the tridiagonal systems, which of course are still coupled through the wall vorticity calculation, a direct elimination procedure is available [4] (a modified Thomas algorithm [8]).

The overall efficiency of the entire solution algorithm was improved considerably when attempts were successful to combine the iteration loop necessary for the decoupling (due to the nonlinear terms) with the iteration loop necessary for the line iteration. With this single iteration loop the difference equations for the ω,ψ-formulation can be written for example as follows (for notation see Fig. 5; integers n,m,ℓ denote the grid points in x-, y-, and t-directions, respectively):

$$
\begin{aligned}
(35)\qquad & \frac{\Delta y^2}{2\Delta t}\left(t_1\omega^{\ell,i}_{n,m} - t_2\omega^{\ell-1}_{n,m} + t_3\omega^{\ell-2}_{n,m}\right) \\
& + \frac{\Delta y}{4\Delta x}\left(\psi^{\ell,i-1}_{n,m+1} - \psi^{\ell,i-1}_{n,m-1}\right)\left(\omega^{\ell,i-1}_{n+1,m} + \underline{2\omega^{\ell,i}_{n,m} - 2\omega^{\ell,i-1}_{n,m}} - \omega^{\ell,i}_{n-1,m}\right) \\
& - \frac{\Delta y}{4\Delta x}\left(\psi^{\ell,i-1}_{n+1,m} - \psi^{\ell,i-1}_{n-1,m}\right)\left(\omega^{\ell,i}_{n,m+1} - \omega^{\ell,i}_{n,m-1}\right) \\
& - \frac{1}{Re}\frac{\Delta y^2}{\Delta x^2}\left(\omega^{\ell,i-1}_{n+1,m} - 2\omega^{\ell,i}_{n,m} + \omega^{\ell,i}_{n-1,m}\right) \\
& - \frac{1}{Re}\left(\omega^{\ell,i}_{n,m+1} - 2\omega^{\ell,i}_{n,m} + \omega^{\ell,i}_{n,m-1}\right) = 0 \quad ,
\end{aligned}
$$

$$(36) \qquad \frac{\Delta y^2}{\Delta x^2} (\psi^{\ell,i-1}_{n+1,m} - 2\psi^{\ell,i}_{n,m} + \psi^{\ell,i}_{n-1,m})$$

$$+ \psi^{\ell,i}_{n,m+1} - 2\psi^{\ell,i}_{n,m} + \psi^{\ell,i}_{n,m-1} - \omega^{\ell,i}_{n,m} = 0 \quad ,$$

where $1 \leq n \leq N-1$, $1 \leq m \leq M-1$, $\ell = 1,2,\ldots$,
and $t_1 = t_2 = 2$, $t_3 = 0$ for $\ell = 1$
$t_1 = 3$, $t_2 = 4$, $t_3 = 1$ for $\ell > 1$.

The underlined terms $2\omega^{\ell,i}_{n,m} - 2\omega^{\ell,i-1}_{n,m}$ in Eq. (35) have nothing to do with the discretization as such; the two terms cancel out when the iteration has converged. The terms are only added to increase the diagonal dominance of the tridiagonal systems. This is especially helpful for larger Reynolds numbers where, due to the large gradients, the intervals Δy have to be very small.

The solution algorithm just described can also be applied when the governing equations are transformed to allow for a variable mesh in the physical plane [1]. Also, for additional convergence acceleration an over-relaxation (for the line-iteration) can be implemented. Since the solution algorithm is basically identical for either formulation (ω,ψ- or ω,u,v-) both formulations can be used interchangeably in the same integration domain. This may be advantageous for some applications, an example of which is given in the next chapter. Moreover, the basic iterative solution procedure allows the use of 5-point central differences with fourth-order accuracy. In this case the equation systems for the line-iteration are pentadiagonal instead of tridiagonal as in the case of second-order accuracy. For the solution of the coupled pentadiagonal systems also a direct elimination procedure was developed corresponding to that one used for the coupled tridiagonal systems.

The efficiency of the entire algorithm for solving the difference equations may be best judged from a typical computation time. For a boundary-layer flow periodically disturbed at the inflow boundary a computation (with the second-order accurate method) using a 49 x 41 grid required only approximately five minutes CPU-time of a CDC 6600 for 260 time-steps (covering more than six disturbance periods). This is relatively little, considering that full implicity is retained in the numerical method.

VI. NUMERICAL RESULTS

In this chapter some typical results are presented which were obtained with the numerical method discussed in the previous chapters. The results are of calculations (with the second-order accurate method) for a boundary-layer flow and a plane Poiseuille flow (as depicted schematically in Figs. 1 and 2) which were disturbed periodically at the inflow boundary A-D. Additionally, results are shown of a calculation for a boundary-layer flow with a backward-facing step (step height approximately one displacement thickness).

For the boundary-layer and Poiseuille flow the Navier-Stokes equations were used in ω,u,v-formulation. At the outflow boundaries B-C in both cases the conditions (18) were employed, while at the free-stream boundary C-D for the boundary layer Eqs. (28) were used. In both cases eigenfunctions of linear stability theory were used for the generation of the periodic disturbance input at the left boundary. The parameters (Reynolds number and disturbance frequency) were those of amplified disturbances.

In Fig. 6 the instantaneous disturbance variables (obtained by subtracting the steady, non-fluctuating part from the total flow) are plotted in perspective representation. The Reynolds number for this case was 635 (based on displacement thickness and the free-stream velocity). The maximum disturbance amplitude (of the u'-fluctuation) of the input was 5% of the free-stream velocity. For the corresponding results shown in Fig. 7 for the plane Poiseuille flow the Reynolds number (based on the center-line velocity and on one half channel width) was 10,000. The maximum disturbance input of the u'-fluctuation was 5% of the center-line velocity. These plots clearly show that no noticeable reflections of the disturbance waves at the outflow boundary do occur.

Finally in Fig. 8 instantaneous total vorticity and total stream function are shown for the boundary-layer flow with a step, again in perspective plotting. In this case the flow is disturbed at the inflow boundary by random perturbations (of Gaussian distribution). For this calculation the ω,ψ-formulation and the ω,u,v-formulation are employed in the same integration domain, shown in Fig. 3. The ω,ψ-formulation is used from the inflow boundary up to the line G-H (of Fig. 3) and the ω,u,v-formulation from there on to the outflow boundary. This is particularly convenient for this application because with the ω,ψ-formulation the incorporation of the wall vorticity calculation at the back-face wall E-F into the numerical scheme is somewhat easier than with the ω,u,v-formulation. On the other hand, the ω,u,v-formulation allows the use of less restrictive or softer conditions at the outflow boundary as discussed in Chapter IV. Thus the outflow boundary can be made more transparent to the large disturbance waves that eminate from behind the step as can be observed in Fig. 8.

The results shown in this chapter should give an impression of the dynamics involved in such flow simulations. Such calculations are discussed in more detail in [9] where their implications for investigations of laminar-turbulent transition are also elaborated upon.

REFERENCES

1 Thompson, J.F.: Numerical solution of flow problems using body-fitted coordinate systems. Lecture Notes on Computational Fluid Dynamics. Von Karman Institute for Fluid Dynamics, Brussels, 1978. Hemisphere Publishing Corporation (1979).

2 Roache, P.J.: Computational fluid dynamics. Hermosa Publishers, Albuquerque (1976).

3 Schubauer, G.B, Skramstad, H.K.: Laminar boundary-layer oscillations and transition on a flat plate. NACA Report No. 909 (1948).

4 Fasel H.F.: Investigation of the stability of boundary layers by a finite-difference model of the Navier-Stokes equations. J. Fluid Mech. 78, 355-383 (1976).

5 Fasel, H.F.: Reaktion von zweidimensionalen, laminaren, inkompressiblen Grenzschichten auf periodische Störungen in der Außenströmung. ZAMM 57, 180-183 (1977).

6 Briley, W.R.: A numerical study of laminar separation bubbles using the Navier-Stokes equations. J. Fluid Mech. 47, 713-736 (1971).

7 Wesseling, P., Sonneveld, P.: Numerical experiments with a multiple grid and a Lanczos type method. Proc. IUTAM-Symp. on Approximation Methods for Navier-Stokes Problems, Paderborn, Germany, 1979. Lecture Notes in Mathematics, Springer (1980).

8 Thomas, L.H.: Elliptic problems in linear difference equations over a network. Watson Scientific Computing Laboratory Report, Columbia University, New York (1949).

9 Fasel, H.F., Bestek, H., Schefenacker, R.: Numerical simulation studies of transition phenomena in incompressible two-dimensional flows. AGARD-CP-224 Paper No. 14 (1977).

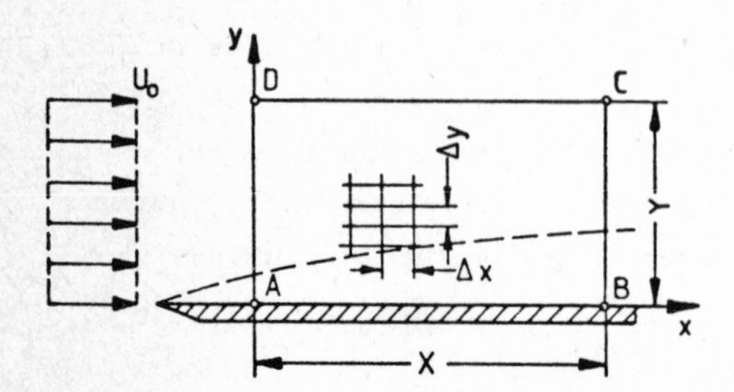

Figure 1. Integration domain for a boundary-layer flow on a flat plate.

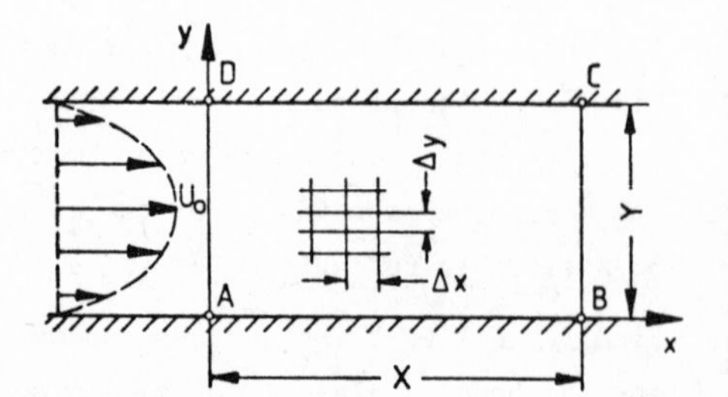

Figure 2. Integration domain for a plane Poiseuille flow.

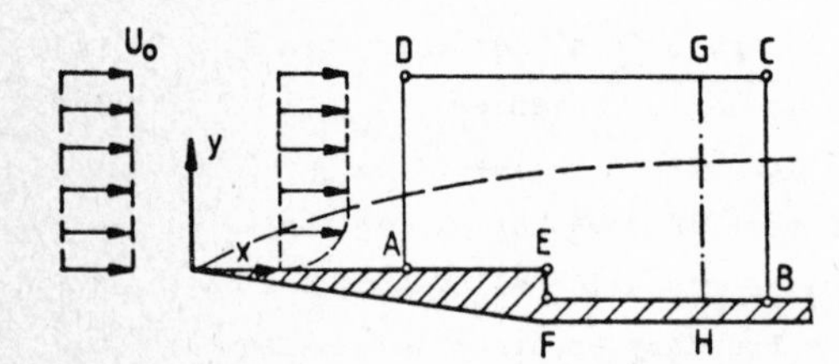

Figure 3. Integration domain for a boundary-layer flow with a backward-facing step.

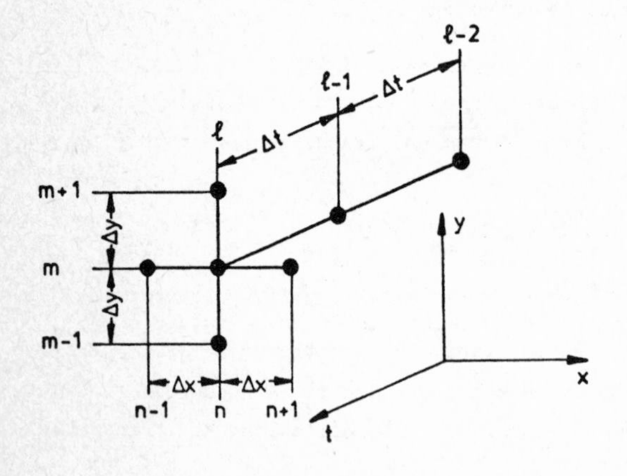

Figure 4. Difference molecule for fully implicit method.

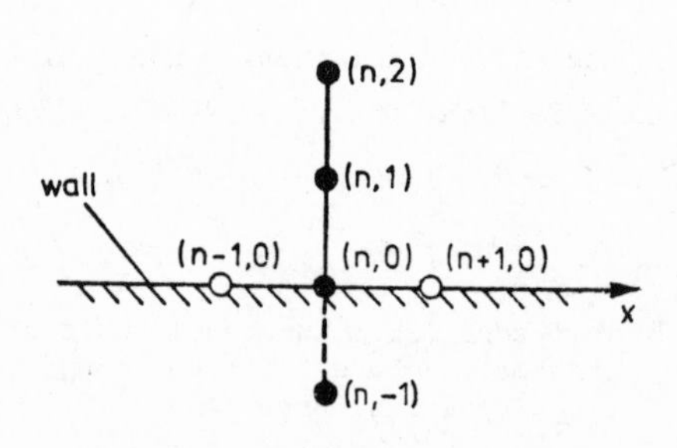

Figure 5. Arrangement of grid points for the calculation of vorticity at solid walls.

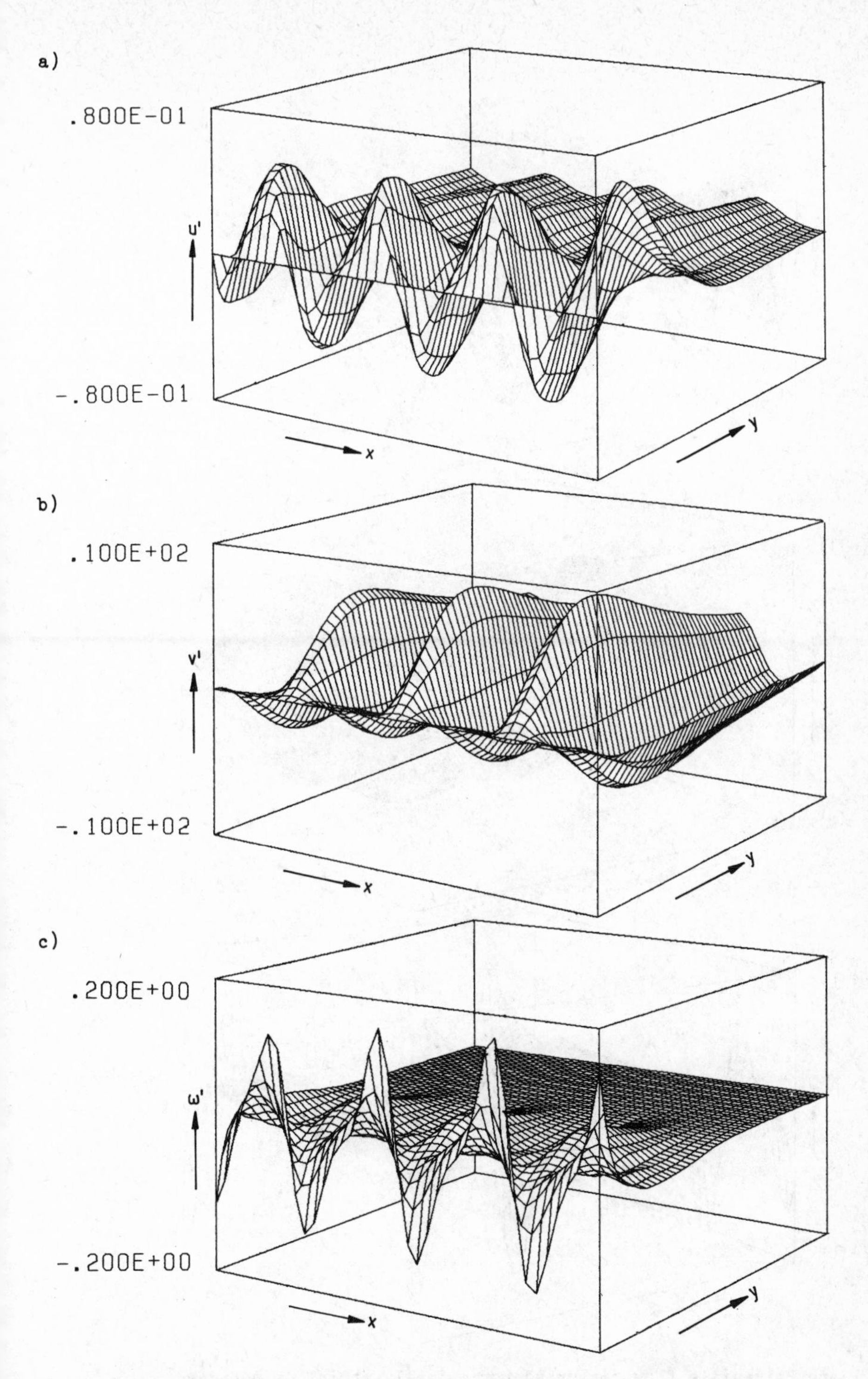

Figure 6. Boundary-layer flow disturbed periodically at inflow boundary. Instantaneous disturbance variables versus x and y (in perspective representation) at time $t/\Delta t = 180$ ($40\Delta t = 1$ period). a) u', b) v', c) ω'.

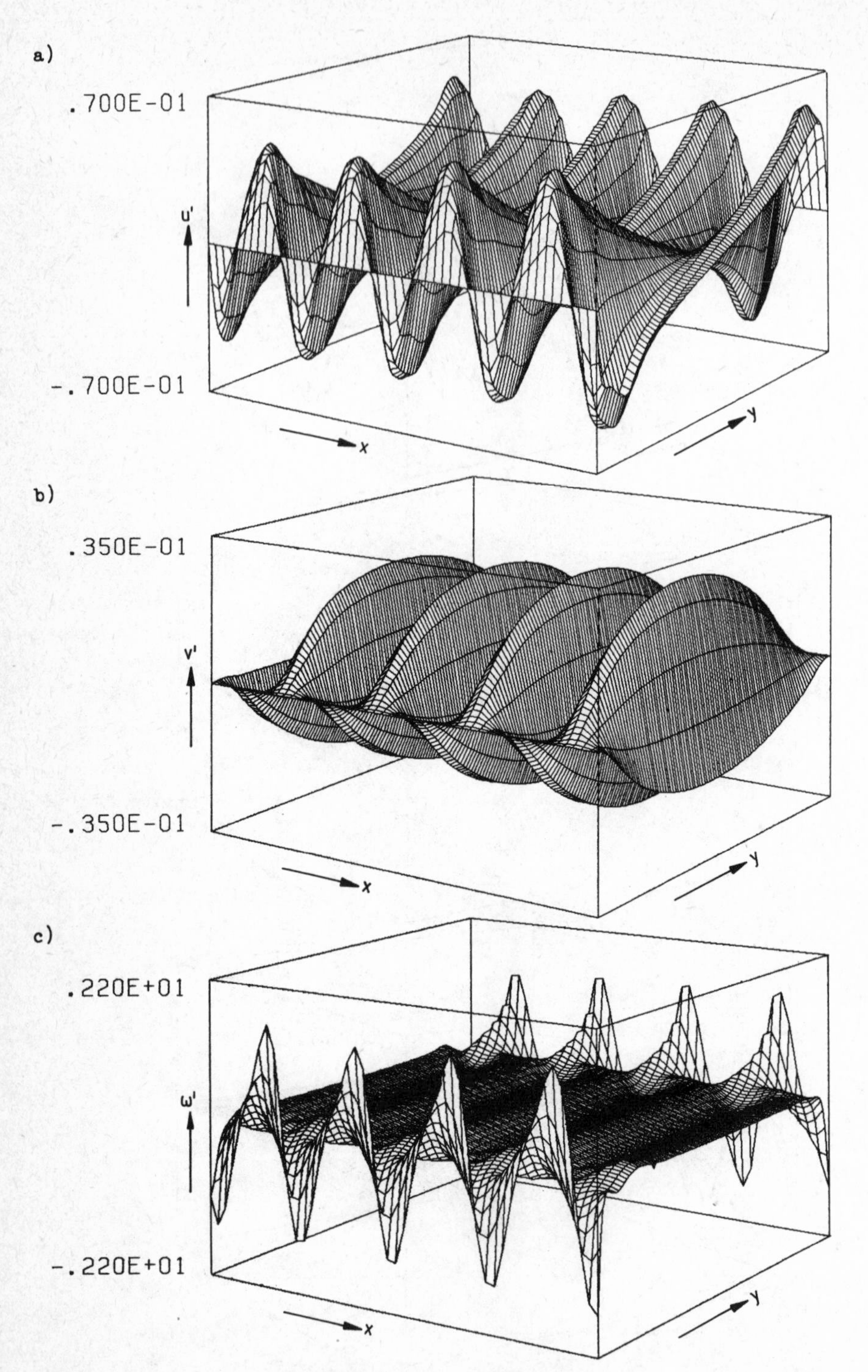

Figure 7. Plane Poiseuille flow disturbed periodically at inflow boundary. Instantaneous disturbance variables versus x and y (in perspective representation) at time $t/\Delta t = 216$ ($48\Delta t = 1$ period). a) u', b) v', c) ω'.

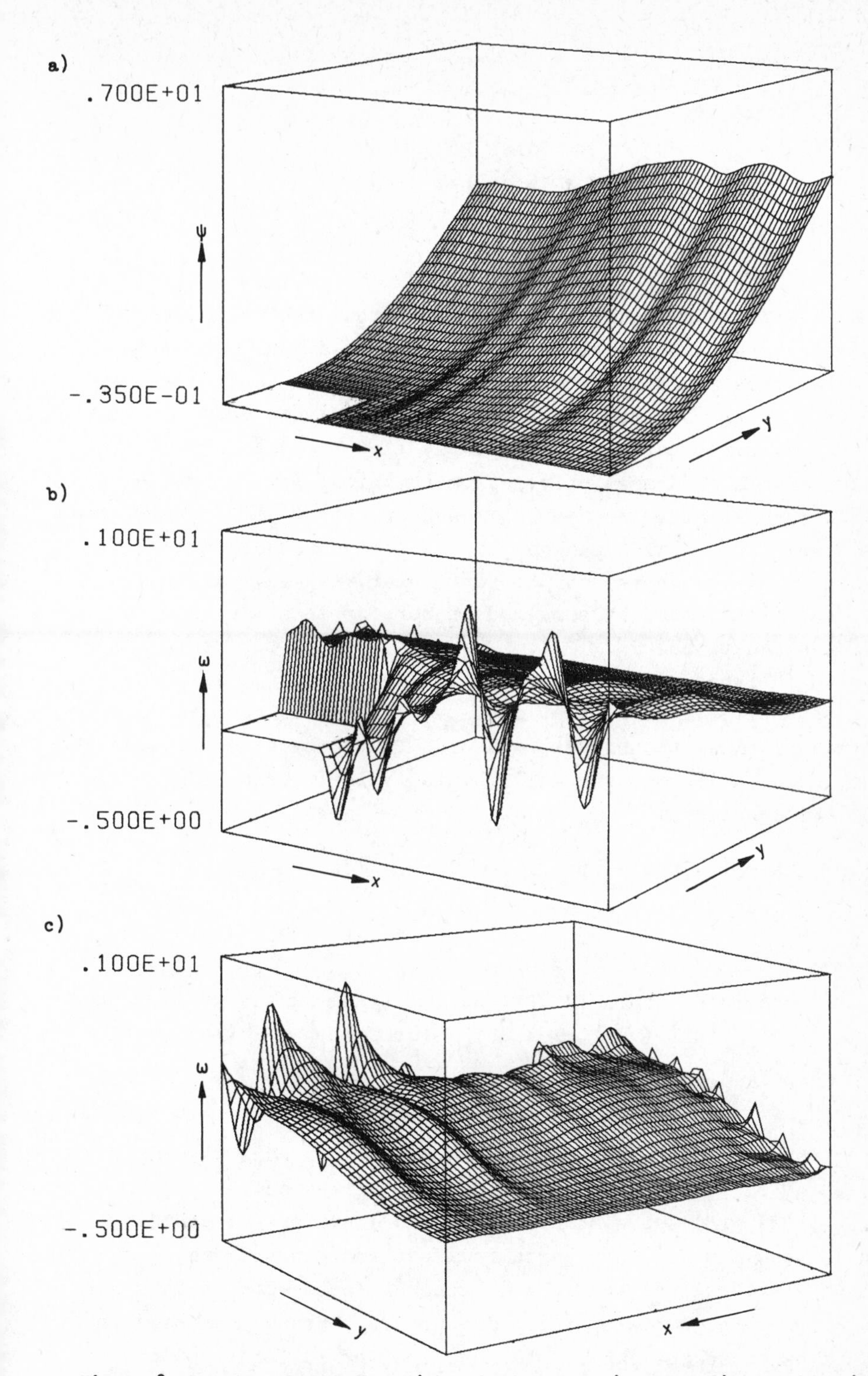

Figure 8. Boundary-layer flow with a backward-facing step disturbed at inflow boundary by random perturbations. Instantaneous variables versus x and y (in perspective representation). a) ψ , b) ω , c) ω (different view).

A SURVEY ON THE FUNCTIONAL DYNAMICAL SYSTEM GENERATED BY THE NAVIER-STOKES EQUATIONS

C. Foias
Université de Paris-Sud
91405 Orsay, France

Several approaches (Hopf [7], Landau [16], Ruelle-Takens [28]) to the mathematical understanding of the onset of turbulence are based on the assumption that the (partially hypothetical) dynamical system in the infinite dimensional space of divergence-free velocity fields, corresponding to the Navier-Stokes equations, has some very remarkable and rather peculiar behaviour, particularly that the complexity of its pattern depends in an almost universal monotonical way of some real parameters (for instance the Grasshoff number, the Taylor number, the Reynolds number, etc.) with sudden jumps at some treshold values of the parameters, which depend on the particular flow envisaged. Such a behaviour was illustrated by some models more or less far from the Navier-Stokes equations [7], [20], [27], [26], [24],.. ; it was also partially established for some remarkable flows [11], [33], [14], [25],.. . Unhappily it seems that it is still beyond the power of the nowadays mathematical techniques to decide if the same is true for the usual (not necessarily homogeneous) initial value problem for the Navier-Stokes equations:

(1) $\quad u_t - \nu \Delta u + (u.\nabla)u + \nabla p = f \ , \ \nabla . u = 0 \quad \text{for} \quad x \in \Omega \ , \ t > 0$

(2) $\quad u = \varphi \quad \text{for} \quad x \in \partial\Omega \ , \ t > 0$

(3) $\quad u = u_0 \quad \text{for} \quad x \in \Omega \ , \ t = 0,$

where Ω is a bounded domain in R^n $(n=2,3)$ lying on one side of its boundary $\partial\Omega \in C^2$, $f \in L^2(\Omega)^n$, $\varphi = \Phi|_{\partial\Omega}$ (for some $\Phi \in H^2(\Omega)^n$, $\nabla.\Phi = 0$ on Ω) and where

(4) $\quad u_0 \in H_\varphi := \{ u \in L^2(\Omega)^n : \nabla . u = 0 \text{ on } \Omega \ , \ u.n = \varphi.n \text{ on } \partial\Omega \},$

n denoting the unit outward normal on $\partial\Omega$. Moreover some general properties [4], [5] of the "dynamical system" in H_φ associated to the problem (1), (2), (3) give some contradictory evidence concerning the conjectured behaviour of this "dynamical system". In the sequel we shall present this evidence and the related general facts which have rigorous mathematical proofs, but skipping these proofs. Our presentation is essentially based on the papers [4] and [5] by R. Temam and the author (see also [3], [6], [32], [29]).

1. Let us recall some basic results concerning the problem (1), (2), (3) ([5],§§1-2; see also [18], pp. 102-105 and [31]). Thus for every $u_0 \in H_\varphi$ there exists at least one function

$$(1.1) \qquad u(t) := u(.,t) \in L^\infty([0,\infty);\, H_\varphi) \cap L^2_{loc}([0,\infty);\, H^1(\Omega)^n)$$

which (together with some adequate distribution p on $\Omega \times (0,\infty)$ is a (weak) solution of the problem (1), (2), (3); if moreover

$$(1.2) \qquad u_0 \in H^1(\Omega)^n \quad , \quad u_0|_{\partial\Omega} = \varphi \, ,$$

then there exists $t(u_0) > 0$ such that $u(t)$ is uniquely determined on $[0, t(u_0))$ and

$$(1.3) \qquad u(.) \in C([0,t(u_0));\, H^1(\Omega)^n).$$

Actually $u(.)$ is a $H^2(\Omega)^n$-valued analytic function on $(0,t(u_0))$ (see [5], § 3). In case $n=2$, the uniqueness of the function $u(.)$ is already assured on the whole time interval $[0,\infty)$ by the conditions (3), (4) and (1.1); moreover in this case $u(.)$ is a $H^2(\Omega)^2$-valued analytic function on $(0,\infty)$ such that

$$(1.4) \qquad \limsup_{t\to\infty} \|u(t)\|_{H^2(\Omega)^2} \leq c_1$$

where c_1 is independent of the solution $\{u(.),\, p\}$. In case $n=3$, one has the following well known estimates

$$(1.5) \qquad t(u_0) \geq c_2(1 + \|u_0\|_{H^1(\Omega)^3})^{-4}$$

$$(1.6) \qquad \limsup_{t\to\infty} \|u(t)\|_{L^2(\Omega)^3} \leq c_3$$

$$(1.7) \qquad \limsup_{t\to\infty} \frac{1}{t}\int_0^t \|u(\tau)\|^2_{H^1(\Omega)^3}\, d\tau \leq c_4$$

and also a new strange estimate (see [6])

$$(1.8) \qquad \limsup_{t\to\infty} \frac{1}{t}\int_0^t \|u(\tau)\|^{1/2}_{H^2(\Omega)^3}\, d\tau \leq c_5 \quad ,$$

where all positive constants $c_2, \ldots, c_5$ are independent of the envisaged solution $u(.)$. In this not yet mastered case $n=3$ there is a partial analyticity result, namely $u(.)$ is a $H^2(\Omega)^3$-valued analytic function on $(0,\infty)\setminus \mathfrak{S}$ where $\mathfrak{S}$ is a closed set (depending on the solution $u(.)$) of Hausdorff dimension $\leq 1/2$ ([30], [5], § 4 ; for the basic definitions see Sec. 5 below). Whether $\mathfrak{S}$ is always $= \emptyset$ or not, is essentially the tantalizing Leray's still open problem [17] . Therefore the "dynamical system" associated in H_φ to the problem (1), (2), (3) must, for the present time, be defined as follows: For $t_0 > 0$ define the map $S(t_0)$ on those $u_0 \in H_\varphi$ for which there exists a unique $u(.)$ satisfying (together with some adequate distribution p on $\Omega \times (0,\infty)$) the relations (1), (2), (3) and such that

$$(1.9) \qquad u(.) \in C((0,t_0] \,;\, H^1(\Omega)^n) \quad ,$$

by $S(t_o)u_o = u(t_o)$. The "dynamical system" is, by definition, the family of these maps $S(t_o)$ $(t_o \geq 0)$, $S(0)$ being the identical map; plainly

$$(1.10) \qquad S(t_1+t_2)u_o = S(t_1)(S(t_2)u_o)$$

whenever the right side of (1.10) makes sense. It is also obvious that $\{S(t)\}_{t\geq 0}$ is, in case $n=2$, a usual dynamical system. In case $n=3$, the domain $D(t)$ of $S(t)$ is open with respect to the $H^1(\Omega)^3$--norm and is certainly $\neq \emptyset$ for $t \in [0,c_2)$ (see (1.5)). For $n=2,3$ it is also known that $S(t)u$ is a $H^2(\Omega)^n$-valued analytic map on $\{\{t,u\} : u \in D(t)\}$ if $D(t)$ is endowed with the $H^1(\Omega)^n$-norm (see [2] , Ch.III).

2. The first question concerning the system $\{S(t)\}_{t\geq 0}$ is the study of its fixed points (i.e. the time independent solutions of the problem (1), (2), (3)). Let $S(\nu;f,\varphi)$ denote the set of all these points and let

$$(2.1) \qquad S(f,\varphi) = \bigcup_{\nu>0} S(\nu;f,\varphi) .$$

Then for fixed φ (resp. fixed f) there exists a dense G_δ -set G in $L^2(\Omega)^n$ (resp. in $\{\varphi = \Phi|_{\partial\Omega}: \Phi \in H^2(\Omega)^n , \nabla.\Phi = 0 \text{ on } \Omega\}$ endowed with the obvious quotient norm of the $H^2(\Omega)^n$-norm) such that for $f \in G$ (resp. $\varphi \in G$), the sets $S(\nu;f,\varphi)$ $(\nu>0)$ and $S(f,\varphi)$ are manifolds of dimension 0 and 1 , respectively (see [3], [4], [29], [31]); thus, in both cases, $S(\nu;f,\varphi)$ is generically finite [3], [29]. Moreover for any φ and f , $S(\nu;f,\varphi)$ is a $\mathbb{C}$-analytic set of finite dimension [4] .

It is a classical result (see for instance [15], [18] or [31]) that $S(\nu;f,\varphi)$ is, for ν large enough, a singleton. From the preceding statement it follows that "generically" the branch of $S(f,\varphi)$ on which $\nu \to \infty$ does not bifurcate. Since these results extend also to the boundary conditions involved in the Taylor flows [11], [33], [14](see also [13], § 3.1 and [31], Ch.II, § 4) we conclude that these real flows correspond to non generic boundary conditions. Therefore the study of the infinite dimensional generic case, although mathematically justified, does not have the à priori philisophical justification that "nature is well behaved".

3. In the study of $\{S(t)\}_{t\geq 0}$ a remarkable role is played by the orthonormal basis $\{w_m\}_{m=1}^{\infty}$ of H_0 (:= H_φ for $\varphi = 0$) formed by the eigenvectors of the following problem

$$(3.1) \qquad -\Delta w_m + \nabla q_m = \lambda_m w_m \quad , \quad \nabla.w_m = 0 \quad \text{on } \Omega$$

$$(3.2) \qquad w_m = 0 \quad \text{on } \partial\Omega ,$$

where q is some adequate distribution on Ω, $w_m \in H^1(\Omega)^n$ and the eigenvalues λ_m are counted in an increasing order:

(3.3) $$0 < \lambda_1 \leq \lambda_2 \leq \ldots \quad .$$

Actually we have a very precise estimate for the λ_m's , namely [23]

(3.4) $$\lim \frac{\lambda_m}{m^{2/n}} = \pi^2 ((n-1)B_n \text{ meas } \Omega)^{-2/n} \quad ,$$

where B_n denotes the measure of the unit ball in R^n .
We shall denote by P_m the orthogonal projection of H_0 onto $\mathbb{R}w_1 + \ldots + \mathbb{R}w_m$ (m=1,2,..). If φ is as in (3) we fix a $\Phi \in H^2(\Omega)^n$ such that $\varphi = \Phi|_{\partial\Omega}$, $\nabla\cdot\Phi = 0$ on Ω , and set $P_m u = P_m(u - \Phi) + \Phi$ for $u \in H_\varphi$.
It is easy to prove that, for m large enough, $P_m|_{S(\nu;f,\varphi)}$ is injective [3] , [4] . It is less simple to prove that the same is true for the larger set

(3.5) $$\{u_0 \in H_\varphi \ : \ S(t_0)u_0 = u_0\}$$

instead of $S(\nu;f,\varphi)$, where $t_0 > 0$ is fixed (see [5], §7). It is not yet known whether the same holds for the even larger set

(3.6) $$\{u_0 \in H_\varphi \ : \ S(t_0)u_0 = u_0 \text{ for some } t_0 > 0\} \ .$$

Also it is not known if these set (3.5) and (3.6) are, like $S(\nu;f,\varphi)$, generically finite. (See [32] for a related positive result.)
The proof of the statement concerning the set (3.5) is based on the following useful geometrical property (given in [5], §5).

4. With some adequate constants c_6 , c_7 , c_8 (depending only on Ω, ν , f and φ) we have for any m=1,2,.. , $u,v \in H_\varphi \cap H^1(\Omega)^n$ and

(4.1) $$0 < t \leq c_6 \ (1+r)^{-4} \quad (\text{where } r \leq \|u\|_{H^1(\Omega)^n} , \|v\|_{H^1(\Omega)^n}),$$

either

(4.2) $$\|S(t)u - S(t)v\|_{L^2(\Omega)^n} \leq \sqrt{2} \ \|P_m(S(t)u - S(t)v)\|_{L^2(\Omega)^n}$$

or

(4.3) $$\|S(t)u - S(t)v\|_{L^2(\Omega)^n} \leq c_7 \ e^{-c_8 m^{1/n} t^{5/4}} \|u - v\|_{L^2(\Omega)^n} \ .$$

In spite of its intricacy this property has many consequences, some of which will be given in the next sections.

5. Let us recall that if h is an increasing function from $(0,\infty)$ to $(0,\infty)$, X a metric space (which in the sequel will always be H_φ endowed with the L^2-norm) and Y a subset of X , then the Hausdorff measure of Y with respect to h is defined by

(5.1) $$\mu_h(Y) = \lim_{\varepsilon \to 0} \mu_{h,\varepsilon}(Y) = \sup_{\varepsilon > 0} \mu_{h,\varepsilon}(Y)$$

where

(5.2) $$\mu_{h,\varepsilon}(Y) = \inf \sum_j h(\text{diam } B_j) \quad ,$$

the infimum being taken over all coverings of Y by balls B_j of diameter $\operatorname{diam} B_j \leq \varepsilon$. Obviously $0 \leq \mu_h(Y) \leq \infty$; moreover μ_h is a Borel measure on X. If, for $D>0$, we set

$$(5.3) \qquad h_D(x) := x^D \quad \text{for} \quad x>0 ,$$

μ_{h_D} is called the D-dimensional Hausdorff measure on X. If $\mu_{h_D}(Y) < \infty$, then

$$\inf \{ D : h_D(Y) = 0 \} = \inf \{ D : h_D(Y) < \infty \}$$

and this last number is called the Hausdorff dimension of Y (see [1], [12]); Y is, in this case, homeomorphic to a set in an Euclidian (finite dimensional) space. If $\mu_{h_D}(Y) = \infty$ for all $D>0$, other function h instead of the functions h_D $(D>0)$ have to be considered. For instance, if $Y = H_\varphi \cap B \neq \emptyset$, where B is any open ball in $H^1(\Omega)^n$, we have for

$$(5.4) \qquad h(x) := e^{-\eta_1 x^{-\eta_2}} \qquad \text{for} \quad x>0 ,$$

(η_1, η_2 being fixed arbitrary positive constants)

$$(5.5) \qquad \mu_h(Y) = \begin{cases} 0 & \text{if } \eta_2 > n \\ \infty & \text{if } \eta_2 < n \end{cases} .$$

Therefore it is interesting to note that if Y is as above but if

$$(5.6) \qquad h(x) = O(e^{-\eta_1 (\log(1/x))^{\eta_2}}) \qquad \text{for } x \to 0 ,$$

where $\eta_1 > 0$, $\eta_2 > n+1$ are fixed, then (for $t>0$ small enough in order that $S(t)Y$ should make sense)

$$(5.7) \qquad \mu_h(S(t)Y) = 0 ;$$

thus with the choice (5.6) of h, we shall have

$$(5.8) \qquad \mu_h(S(t)Y) = \begin{cases} \infty & \text{if } t = 0 \\ 0 & \text{if } t>0 \text{ is small enough} . \end{cases}$$

Moreover for any θ, $0<\theta\leq 1$, and any D large enough we have

$$(5.9) \qquad \mu_{h_D}(S(t)Z) \leq \theta \mu_{h_D}(Z)$$

for all Borel subsets Z of Y. By (5.9) it is easy to construct (at least if $n=2$) some concrete finite Borel measures μ on H_φ enjoying the property

$$(5.10) \qquad \mu \circ S(t) \leq \theta \mu \qquad \text{for some } t>0 \text{ and } \theta,\ 0<\theta\leq 1 .$$

Thus instead of studying finite Borel measures on H invariant with respect to $\{S(t)\}_{t\geq 0}$, it is more tempting to study the finite Borel measures which satisfy (5.10).

6. A basic question concerning the system $\{S(t)\}_{t\geq 0}$ is the description

of its invariant sets, that is of the sets $Y \subset H_\varphi$ such that

$$(6.1) \qquad S(t)Y = Y \quad \text{for all} \quad t \geq 0 .$$

Whenever Y is bounded in $H^1(\Omega)^n$, Y is also bounded in $H^2(\Omega)^n$ and the Hausdorff dimension of Y is finite ([5], § 6; see also [21]). this always happens in case $n=2$; moreover in this case, there exists a maximal invariant set Y_{max} in H_φ such that :

$$(6.2) \qquad \text{distance}_{\text{in } H_\varphi}(u(t), Y_{max}) \to 0 \qquad (\text{for } t \to 0)$$

for every solution $u(.)$ of (1), (2), (3) (with $u_0 \in H_\varphi$ arbitrary). The Hausdorff dimension of Y_{max} is, for fixed f and φ , a remarkable function $D(\nu)$ of ν, $\nu > 0$. An upper estimate for $D(\nu)$ can be easily obtained from sec.4 above. This estimate is valid for a large class of equations related to the Navier-Stokes equations. In particular the same estimate holds for the maximal invariant set in $\mathbb{R}^n$ $\mathbb{R}w_1 + .. + \mathbb{R}w_m$ of the system obtained from the equation (1) (with the boundary conditions (2)) by cutting off the amplitudes of $u(.,t) - \Phi(.)$ corresponding to the eigenvectors w_{m+1} , w_{m+2} ,.. . Except these few facts nothing is really known about the nature of $D(\nu)$. Bifurcation theory yields cases in which $D(\nu)$ jumps from 0 to 1 or even to other larger integers ([8], [9], [11] ; see also [24]). The Landau-Hopf picture suggests that $D(\nu)$ is a non decreasing function of $1/\nu$ taking only integer values. But the Lorenz's model (see [20], [27], [34]) suggests that $D(\nu)$ might take also non integer values and be a non monotonic function in $1/\nu$. Anyway it is very plausible that for m large enough, P_m is injective on Y_{max} .

References:

[1] H. Federer, Geometric measure theory, Springer(New York,1969).

[2] C. Foias, Solutions statistiques des équations de Navier-Stokes, Cours au Collège de France(1974).

[3] C. Foias - R. Temam, Structure of the set of stationary solutions of the Navier-Stokes equations, Comm. Pure Appl. Math., 30(1977), 149-164.

[4] C. Foias - R. Temam, Remarques sur les équations de Navier-Stokes stationnaires et les problèmes succesifs de bifurcation, Ann. Sc. Norm. Sup. Pisa, (IV)5(1978), 29-63.

[5] C. Foias - R. Temam, Some analytic and geometric properties of the evolution Navier-Stokes equations, Journ. Math. Pures Appl., 58 (1979), 339-368.

[6] C. Foias - R. Temam, Sur certaines propriétés génériques des équations de Navier-Stokes, Orsay(1974); to appear.

[7] E. Hopf, A mathematical example displaying feature of turbulence, Comm. Pure Appl. Math., 1(1948), 303-322.

[8] E. Hopf, Abzweigung einer periodischen Lösung eines Differentialsystems, Berichten Math.-Phys. Kl. Wiss. Leipzig, 94(1942), 1-22.

[9] G. Iooss, Direct bifurcation of a steady solution of the Navier-Stokes equations into an invariant torus, Turbulence and Navier-Stokes equations, Orsay(1975), Lecture Notes in Math., N° 565, Springer(1976), 69-84.

[10] V.I. Iudovich, Secondary flows and fluid instability between rotating

cylinders, Prikl. Mat. Meh., 30(1966), 688-698.

[11] D.D. Joseph, Stability of fluid motions. (I&II), Springer(Berlin, 1976).

[12] J.P. Kahane, Mesures et dimensions, Turbulence and Navier-Stokes equations, Orsay(1975), Lecture Notes in Math., N° 565, Springer (1976), 94-103.

[13] K. Kirchgässner, Bifurcation in nonlinear hydrodynamic stability, SIAM Rev., 17(1975), 662-683.

[14] K. Kirchgässner - P. Sorger, Branching analysis for the Taylor problem, Quart. J. Mech. Appl. Anal., 32(1969), 183-209.

[15] O.A. Ladyzenskaya, The mathematical theory of viscous incompressible fluids, Gordon-Breach(New York, 1963).

[16] L.D. Landau - E.M. Lifshitz, Fluid Mechanics, Pergamon(Oxford, 1959).

[17] J. Leray, Sur le mouvement d'un liquide visqueux emlissant l'espace, Acta Math., 63(1934), 193-249.

[18] J.L. Lions, Quelques méthodes de résolution des problèmes aux limites non linéaires, Dunod(Paris, 1969).

[19] J.L. Lions - G. Prodi, Un théorème d'existence et unicité dans les équations de Navier-Stokes en dimension 2, C.R. Acad. Sci. Paris, 248(1959), 3519-3521.

[20] E.N. Lorenz, Deterministic nonperiodic flows, Journ. Atmos. Sci., 20(1963), 130-141.

[21] J. Mallet-Paret, Negatively invariant sets of compact maps and an extension of a theorem of Cartwright, Journ. Diff. Eq., 22(1976), 331-348.

[22] B. Mandelbrot, Inermittent turbulence and fractal kurtosis and the spectral exponent 5/3+D, Turbulence and the Navier-Stokes equations, Orsay(1975), Lecture Notes in Math., N° 565, Springer(1976), 121-145.

[23] G. Métivier, Etude asymptotique des valeurs propres et de la fonction spectrale de problèmes aux limites, Thèse, Univ. de Nice(1976).

[24] G. Minea, Remarques sur l'unicité de la solution d'une équation de type Navier-Stokes, Revue Roum. Math. P. Appl., 21(1976), 1071-1075.

[25] P.H. Rabinowitz, Existence and nonuniqueness of rectangular solutions of the Bénard problem, Arch. Rat. Mech. Anal., 29(1968), 32-57.

[26] D. Ruelle, Turbulence and Axiome A attractor, Proc. Intern. School math. Phys., Univ. di Camerino(1974), 162-181.

[27] D. Ruelle, A Lorenz attractor and the problem of turbulence, Turbulence and Navier-Stokes equations, Orsay(1975), N° 565, Springer (1976), 146-158.

[28] D. Ruelle - F. Takens, On the nature of turbulence, Comm. Math. Phys., 20(1971), 167-192; 23(1971), 343-344.

[29] J.C. Saut - R. Temam, Propriétés de l'ensemble des solutions stationnaires ou périodiques des équations de Navier-Stokes, C.R. Acad. Sci. Paris, 248(1977), A, 673-676; Generic properties of the Navier-Stokes equations, Orsay(1979).

[30] V. Sheffer, Turbulence and Hausdorff dimension, Turbulence and Navier-Stokes equations, Orsay(1975), Lecture Notes in Math., N° 565, Springer(1976), 174-183.

[31] R. Temam, Navier-Stokes equations. Theory and numerical analysis, North-Holland(New York, 1977).

[32] R. Temam, Une propriété générique de l'ensemble des solutions stationnaires ou périodiques des équations de Navier-Stokes, Funct. Anal. and Numerical Anal., Japan-France Seminar, Tokyo-Kyoto(1976), Japan Soc. for the Prom. of Sci.(1978), 483-493.

[33] W. Velte, Stabilität und Verzweigung stationäre Lösungen des Navier-Stokes Gleichungen beim Taylor Problem, Arch. Rat. Mech. Anal., 22(1966), 1-14.

[34] R.F. Williams, The structure of the Lorenz attractors, Turbulence Seminar, Berkeley(1976/77), Lecture Notes in Math., N° 615, Springer (1977), 94-112.

SOLUTION OF THE TIME-DEPENDENT INCOMPRESSIBLE NAVIER-STOKES AND BOUSSINESQ EQUATIONS USING THE GALERKIN FINITE ELEMENT METHOD

Philip M. Gresho, Robert L. Lee and Stevens T. Chan
Lawrence Livermore Laboratory, University of California
Livermore, California 94550/USA

and

Robert L. Sani
CIRES/NOAA
University of Colorado
Boulder, Colorado 80309/USA

I. INTRODUCTION

Our research is directed toward the generation of a time-dependent, three-dimensional model of the atmospheric boundary layer using the Galerkin finite element method (GFEM). Along the way, we have developed capabilities for solving the two-dimensional Navier-Stokes and Boussinesq equations, both steady and time-dependent. In this paper, we describe and demonstrate our innovative techniques for time-dependent flows.

The GFEM is applied to the primitive variable ($\underline{u}$,P,T) equations, thus generating, in the conventional manner, a coupled system of ordinary differential equations in time. We then discuss our time integration method which gives very accurate, and reasonably efficient solutions to these time-dependent problems; it is stable for any grid spacing and Reynolds number, is non-dissipative, and incorporates an automatic time step selection strategy.

The techniques are then demonstrated by means of two numerical examples — both starting with a motionless system: (1) a thermally driven square cavity which goes to a steady-state and (2) isothermal flow around a circular cylinder which leads to periodic vortex shedding.

II. BOUSSINESQ EQUATIONS AND SPATIAL DISCRETIZATION

A. CONTINUUM EQUATIONS

The equations of motion for a constant property (except for density in the buoyancy force term), incompressible fluid are the Boussinesq approximations to the Navier-Stokes equations,

$$\rho\left(\frac{\partial \underline{u}}{\partial t} + \underline{u} \cdot \nabla \underline{u}\right) = \nabla \cdot \underline{\underline{\tau}} - \rho\gamma \underline{g}(T-T_r) \tag{1a}$$

$$\nabla \cdot \underline{u} = 0 \quad , \tag{1b}$$

where $\underline{u} = (u,v)$ in two-dimensions (2-D) and $\underline{u} = (u,v,w)$ in three-dimensions (3-D), ρ is the (constant) density at the reference temperature (T_r), γ is the volumetric coefficient of thermal expansion, T is the temperature, $\underline{g}$ is the gravitational acceleration (directed opposite to the vertical coordinate), and

$$\tau_{ij} = -P\delta_{ij} + \mu\left(\frac{\partial u_i}{\partial x_j} + \frac{\partial u_j}{\partial x_i}\right) \tag{1c}$$

is the symmetric stress tensor in which μ is the viscosity, P is the pressure deviation from hydrostatic (at $T = T_r$), and δ_{ij} is the Kronecker delta. Finally, the energy equation for a constant property fluid, neglecting viscous dissipation, is

$$\frac{\partial T}{\partial t} + \underline{u} \cdot \nabla T = \kappa \nabla^2 T \quad , \tag{1d}$$

where κ is the thermal diffusivity. Eq. (1) can be used, given appropriate initial and boundary conditions (BC's), to obtain the velocity ($\underline{u}$), pressure (P), and temperature (T).

In an incompressible fluid the pressure is an intrinsic and independent variable of the motion — it is not a thermodynamic variable obtainable from an equation of state (Aris, 1962). Rather, it is an implicit variable which instantaneously 'adjusts itself' so that the incompressibility constraint (Eq. 1b) is always satisfied (for nonisothermal flows, the pressure field must also balance the 'hydrostatic' portion of the buoyancy force, a fact which makes the numerical simulation of Boussinesq flows more difficult than isothermal flows). This implicitness of the pressure is a characteristic of incompressible flow that invariably makes the problem difficult to solve; the remaining difficulties are caused by the nonlinear advection operator, $\underline{u} \cdot \nabla$.

B. INITIAL AND BOUNDARY CONDITIONS

1. Initial Conditions

The appropriate initial conditions for Eq. (1) are: (1) any velocity field, $\underline{u}_o(\underline{x})$, which is solenoidal (i.e., $\underline{u}_o$ must satisfy $\nabla \cdot \underline{u}_o = 0$) and (2) any temperature field, $T_o(\underline{x})$. Note that no initial conditions are required for pressure; the initial pressure field, $P_o(\underline{x})$, is contained implicitly in Eqs. (1a) and (1b), given $\underline{u}_o$ and T_o (it can be obtained explicitly from a Poisson equation formed by taking the divergence of Eq. (1a) using (1b)). We mention here, and will discuss in more detail later, that the system is not well-posed if the initial velocity field does not satisfy Eq. (1b).

2. Boundary Conditions

(a) Velocity

In general, the computational domain (Ω) is to be regarded as being bounded by a piecewise smooth boundary, Γ, along which the following BC's are permissible: specify u_n or

f_n and u_τ or f_τ, where u_n, u_τ are the normal and tangential components of $\underline{u}$ on Γ and f_n, f_τ are the respective components of the surface traction force; in particular,

$$f_n = -P + 2\mu \frac{\partial u_n}{\partial n} \quad , \tag{2a}$$

$$f_\tau = \mu \left(\frac{\partial u_n}{\partial \tau} + \frac{\partial u_\tau}{\partial n} \right) \quad , \tag{2b}$$

where $\underline{n}$ is the outward pointing unit normal vector and τ is the unit tangent vector.

(b) Temperature

The general BC appropriate to Eq. (1d) is

$$\kappa \frac{\partial T}{\partial n} + h(T - T_s) + q = 0 \quad , \tag{2c}$$

where h, T_s, and q are specified functions on Γ. The special case of specified temperature may be obtained from Eq. (2c) by letting $h \to \infty$; in practice, specified values are imposed more directly.

(c) Pressure

In general, the application of pressure BC's (i.e., specified pressure along a portion of Γ) is inconsistent with Eq. (1) and therefore 'illegal'. In the special case wherein u_n is specified on Γ (e.g., contained flow), a pressure datum is generally required since in this case the pressure is defined only up to an arbitrary additive constant. If, however, an explicit pressure equation is formed (the Poisson equation referred to above), then the requirement for pressure BC's does arise. In this case, however, they must be consistent with the original governing equations; hence, the appropriate BC's for the Poisson equation are the Neumann conditions obtained from the normal component of Eq. (1a) on Γ, using Eqs. (1b) and (1c). Even in this approach then, the application of Dirichlet BC's for pressure is generally not permissible. For further discussion of this point, and a numerical example, see Gresho et al. (1980a).

C. SPATIAL DISCRETIZATION VIA THE GALERKIN FINITE ELEMENT METHOD

The finite element spatial discretization of Eq. (1) is performed by the Galerkin method via the following expansions in the basis sets $\{\phi_i\}$ and $\{\psi_i\}$, where the piecewise-polynomial basis functions are endowed with the property that all but one are zero at a particular node, and the basis function for that node is unity; this conveniently identifies the amplitude coefficients of the expansions in the basis set with nodal values of the variables — thus,

$$\underline{u}^h(\underline{x},t) = \sum_{i=1}^{N} \underline{u}_i(t)\, \phi_i(\underline{x}) \quad , \tag{3a}$$

$$T^h(\underline{x},t) = \sum_{i=1}^{N} T_i(t)\, \phi_i(\underline{x}) \quad , \tag{3b}$$

and

$$P^h(\underline{x},t) = \sum_{i=1}^{M} P_i(t)\, \psi_i(\underline{x}) \quad , \tag{3c}$$

where, in the discretized domain, there are N nodes for velocity and temperature (neglecting boundary conditions) and M nodes for pressure; the superscript h indicates a finite dimensional approximation. For reasons too lengthy and digressionary to concern us in any detail here, the basis functions for pressure must be at least one order lower than those for velocity (see Olson, 1977; Sani et al., 1980); otherwise the final matrix of coefficients for the discretized dependent variables will be rank deficient and the solution (especially for the pressure) will be difficult or impossible to obtain. For convenience and higher accuracy, the basis functions for temperature are taken to be the same as those for velocity. Although our element "library" contains several workable combinations of approximation spaces, all based on the 'quadrilateral family' rather than the 'triangle family', in this paper we consider only one: it employs quadratic approximation for velocity and temperature and linear approximation for pressure. In 2-D, these approximations are manifest in an element which contains 9-nodes for $\underline{u}$ and T and 4-nodes (at the corners) for P; the 3-D counterpart would have 27-nodes for $\underline{u}$ and T with 8-nodes for P. Although capable of generating quite good results, our most recent experience has indicated that this element is probably not optimum, especially for strongly thermally-coupled flows. The reasons for this, which will be reported in detail elsewhere (Gresho et al., 1980b), relate to the approximation space used for the pressure; the space of continuous functions which we have been using is sometimes not large enough to adequately enforce incompressibility.

Inserting Eq. (3) into the weak (Galerkin) form of Eq. (1) (which reduces differentiability requirements — ϕ_i can be continuous with piecewise-discontinuous first derivatives and ψ_i can be piecewise discontinuous), leads to the following set of ordinary differential equations (ODE's), the Galerkin FEM equations, written in a compact matrix form,

$$M\dot{u} + [K + N(u)]\, u + CP + \beta M'T = f + \beta M'T_r \quad , \tag{4a}$$

$$C^T u = 0 \quad , \tag{4b}$$

and

$$M'\dot{T} + [K_T + N_T(u)]\, T = f_T \quad , \tag{4c}$$

where $\beta = \gamma g$. Now u is a global vector of length DN (D is the dimension of the physical space; 2 or 3) containing all nodal velocity components, P is a global M-vector of nodal pressures, and T is a global N-vector of nodal temperatures; f is a global vector (length DN) which incorporates any traction boundary conditions on velocity, and f_T is an N-vector which incorporates any of the 'mixed' boundary conditions of Eq. (2c). (Specified nodal values of velocity and temperature are imposed directly on the assembled system, typically by deleting the equation in question and transposing all coupling terms to the right hand side (RHS) or by replacing the equation by one with all zeros except for unity on the diagonal and placing the specified value on the RHS.) M is the DN x DN "mass" matrix, M' is an N x N subset of M (multiplied by the appropriate scalar), K is the DN x DN viscous matrix, K_T is the N x N thermal diffusion matrix, C is the DN x M pressure gradient matrix and its transpose, C^T is the M x DN divergence matrix, N(u) is the DN x DN nonlinear advection matrix, and $N_T(u)$ is a DN x N subset of N(u). M, N(u), and $N_T(u)$ are actually composed of blocks of smaller (N x N) matrices. For details regarding the matrix definitions, see Gresho et al. (1980a).

It is noteworthy that the Galerkin equations result in discretized approximations to the advection operator which are essentially centered difference-like and are basically nondissipative (actually, the advective form of the continuum equations leads to indefinite forms with respect to numerical diffusion; our recent experiments (Lee et al., 1980), in which this form was compared with conservation forms of the Galerkin equations which are truly non-dissipative, seem to indicate that the simpler advective form generally yields good results, displays little dissipation but can lead to nonlinear instability in long-time inviscid calculations). In particular, we do not subscribe to the notion of employing "modified" Galerkin or other ad hoc techniques which generate numerically diffusive "upwind approximations"; see Gresho and Lee (1979) and Leone and Gresho (1979).

III. TIME INTEGRATION METHOD

A. GENERAL CONSIDERATIONS

Because of the inherent implicitness of the pressure, it is not possible to solve Eqs. (4a) and (4b) by any purely explicit (time-marching) procedure. This fact, combined with the mass matrices which couple the time derivative vectors, suggests that implicit time integration techniques are appropriate. While they are mandatory for the treatment of the pressure and the associated continuity equation, the remaining terms in the momentum equation (4a) and all terms in the temperature equation could be treated explicitly, if desired. In our 2-D code development, we have thus far retained fully implicit methods for all terms in all equations, for the following 'standard' reason: explicit methods introduce time-step restrictions based on stability considerations, a disadvantage which could severely limit the use of good, graded meshes (an important inherent feature when using isoparametric finite element techniques) since time-step limits are then functions of the smallest mesh size. In a typical simulation, these stability restrictions would force the use of time steps which are much smaller (by perhaps several orders of magnitude) than would be required for temporal accuracy (of the ODE's). It is of course true that each time step is usually very much cheaper than that from implicit methods, and that there may be cases where explicit methods are more cost effective, since the greater stability of implicit methods exacts a significant computational price; viz, the nonlinear ODE's of Eq. (4) generate the requirement of solving a nonlinear algebraic system at each time step. While the implicit method we have developed and applied may not be optimally cost-effective in all situations, it is, we believe, robust and 'honest', and has worked quite well in a variety of situations. It is firmly based on the well-developed theory of ODE's from which we extracted a small, but useful portion.

The following sections will describe in some detail and demonstrate these techniques; here we provide an introductory, simplified overview. By an appropriate combination of two common integration techniques, the implicit trapezoid rule (TR) and an explicit Adams-Bashforth (AB) formula, we have developed a stable time integration scheme in which we are able to cheaply and automatically vary the step size based solely on temporal accuracy requirements by obtaining a good estimate of the local (single step) time truncation error. The nonlinear systems engendered by the TR algorithm are solved using a "one-step Newton" method, which requires the formulation and solution of one linear system per time step. The scheme has the useful additional property that some insight into the prevailing 'physics' is provided by the associated monitoring of its time scale. For example, the physics may require a small time step to follow the formation of a boundary layer or the shedding of vortices. Or a large time step may be sufficient to follow the slow growth of an attached eddy or a flow which is approaching steady state. In either case the algorithm will (usually) automatically select the appropriate time step, thus providing a cost-effective method in that the step size is increased whenever possible and decreased only when necessary. The overall scheme can be conveniently described by four steps (or stages) as discussed below.

B. INITIALIZATION

The discretized analog of the solenoidal constraint is Eq. (4b), which must be satisfied by the discretized equations in order that the ODE problem be well-posed. If $C^T u_o \neq 0$, it is easy to show that the following results will obtain (using TR) at the end of the first time step as $\Delta t_o \to 0$:

$$u_1 \to u_o - M^{-1}C(C^TM^{-1}C)^{-1}(C^Tu_o) \quad , \tag{5a}$$

$$\frac{P_1+P_o}{2} \approx (C^TM^{-1}C)^{-1}(C^Tu_o)/\Delta t_o \quad ; \tag{5b}$$

i.e., there is a finite jump in velocity and the pressure becomes arbitrarily large.

In order to obtain the initial compatible (with $C^Tu_o = 0$) pressure field, we solve Eq. (4a) and a time-differentiated form of Eq. (4b) (which is valid since $C^Tu = 0$ for all time); i.e., we must solve

$$\begin{bmatrix} M & C \\ C^T & 0 \end{bmatrix} \begin{Bmatrix} \dot{u}_o \\ P_o \end{Bmatrix} = \begin{Bmatrix} f - [K + N(u_o)]u_o + \beta M'(T_r - T_o) \\ 0 \end{Bmatrix} \tag{6}$$

simultaneously for P_o and the initial acceleration, $\dot{u}_o$. The extra cost associated with this procedure is amortized over the rest of the computation because our time step control strategy requires the acceleration vector; only at t = 0 must we solve a linear system to obtain it. For the same reason, Eq. (4c) is also solved for $\dot{T}_o$, given the initial temperature, T_o.

C. PREDICTOR EQUATION FOR VELOCITY AND TEMPERATURE

The variable step, second-order accurate AB formula applied to $\dot{y} = f$ is (Shampine and Gordon, 1975)

$$y^p_{n+1} = y_n + \frac{\Delta t_n}{2}\left[\left(2 + \frac{\Delta t_n}{\Delta t_{n-1}}\right)\dot{y}_n - \frac{\Delta t_n}{\Delta t_{n-1}}\dot{y}_{n-1}\right] \quad . \tag{7}$$

Since this is an explicit formula, it can only be used for velocity and temperature; i.e., y^p represents u and T from Eqs. (4a) and (4c). Note that two history vectors of 'acceleration' are required; these are obtained simply and recursively from the TR (corrector) results to be described below. This predictor step is cheap (relative to the TR step) and, since it performs two important functions, quite cost effective (it provides a very good starting guess and it permits local error control). Finally, since $\dot{y}_{n-1}$ is required, the AB formula cannot be applied until the second step, using $\dot{u}_o$ and $\dot{T}_o$ from the initialization procedure. Hence, error estimates commence at the end of the second step.

D. CORRECTOR STEP AND THE SOLUTION FOR PRESSURE

The TR formula, which is non-dissipative, completely stable (A-stable to be more precise), and also second-order accurate is, when applied to $\dot{y} = f$,

$$y_{n+1} = y_n + \frac{\Delta t_n}{2}(f_n + f_{n+1}) \quad . \tag{8}$$

The TR, being implicit, is applied to the full set of equations in Eq. (4), the solution of which yields the final (reported) values for u, T, and P. Application of the TR to Eq. (4) gives

$$\left[\begin{array}{c:c:c} \frac{2}{\Delta t_n}M+K+N(u_{n+1}) & \beta M' & C \\ \hdashline 0 & \frac{2}{\Delta t_n}M'+K_T+N'(u_{n+1}) & 0 \\ \hdashline C^T & 0 & 0 \end{array}\right] \left\{\begin{array}{c} u_{n+1} \\ \hdashline T_{n+1} \\ \hdashline P_{n+1} \end{array}\right\} = \left\{\begin{array}{c} M\left(\frac{2}{\Delta t_n}u_n+\dot{u}_n\right)+f_{n+1}+\beta M'T_r \\ \hdashline M\left(\frac{2}{\Delta t_n}T_n+\dot{T}_n\right)+f_{T_{n+1}} \\ \hdashline 0 \end{array}\right\}, \tag{9}$$

where we have utilized our knowledge of the time derivatives at t_n (see Eq. (10), below) to obtain the computationally efficient form of the RHS vector. We recognize Eq. (9) as a nonlinear algebraic system $A(x)x = b$ which, when solved for u, T, and P at t_{n+1} (the details of which are deferred), we can compute the 'acceleration' vector, $\dot{u}_{n+1}$ and $\dot{T}_{n+1}$, which is required for the next predictor step, by 'inverting' the TR in the form

$$\dot{y}_{n+1} = \frac{2}{\Delta t_n}(y_{n+1} - y_n) - \dot{y}_n \quad , \tag{10}$$

where again y represents u and T and $\dot{y}_n$ is available from the previous application of Eq. (10).

E. LOCAL TIME TRUNCATION ERROR AND TIME STEP SELECTION

The local time truncation error estimate begins with a Taylor series analysis of both predictor (AB) and corrector (TR) schemes, Eqs. (7) and (8). Denoting by $y(t_n)$ the exact solution at time t_n and assuming that $y_n = y(t_n)$ (i.e., local error estimates are based on the assumption that the exact solution is known at the beginning of a time step), we obtain the following error estimates:

$$y^p_{n+1} - y(t_{n+1}) = -\frac{1}{12}\left(2 + 3\frac{\Delta t_{n-1}}{\Delta t_n}\right)\Delta t_n^3\,\dddot{y}_n + O(\Delta t_n^4) \tag{11}$$

for AB and

$$y_{n+1} - y(t_{n+1}) \equiv d_{n+1} = \frac{1}{12}\Delta t_n^3\,\dddot{y}_n + O(\Delta t_n^4) \tag{12}$$

for TR; d_{n+1} is the local time truncation error of the actual (TR) solution. Since y^p_{n+1} and y_{n+1} are both available, Eqs. (11) and (12) may be used to eliminate the two unknowns, $y(t_{n+1})$ and $\dddot{y}_n$, which leads to

$$d_{n+1} = \frac{y_{n+1} - y^p_{n+1}}{3(1+\Delta t_{n-1}/\Delta t_n)} + O(\Delta t_n^4) \quad . \tag{13}$$

These results can be used to estimate the next step size based on the requirement that a (relative) norm of the error for the next step be equal to a pre-set input value, ε (we use a combined, weighted RMS norm on u and T based on the largest velocities and temperatures in the vector; the P part of y_{n+1} is ignored here). From Eq. (12) we have

$$\frac{|d_{n+2}|}{|d_{n+1}|} \cong \left(\frac{\Delta t_{n+1}}{\Delta t_n}\right)^3 \frac{|\dddot{y}_{n+1}|}{|\Delta\dddot{y}_n|} \quad , \tag{14}$$

where d_{n+1} is available. Since $\dddot{y}_{n+1} = \dddot{y}_n + O(\Delta t_n)$, and setting $|d_{n+2}| = \varepsilon$, we obtain

$$\Delta t_{n+1} = \Delta t_n\,(\varepsilon/|d_{n+1}|)^{1/3} \quad , \tag{15}$$

where higher order terms have been ignored. Equation (15) is used to compute the next step size. The only user-specified parameter is ε, which obviously can have a significant effect on the results. Too large an ε can cause (1) the theory behind the error estimate to be poor, (2) the well-known oscillatory behavior when large time steps are used in the TR, and (3) a more 'difficult' nonlinear system (Eq. (9); y_{n+1} is farther from y_n). Too small an ε, on the other hand, will cause excessively expensive computations. It has been our experience that $\varepsilon = .001$ (.1% relative error per step) is nearly optimum in that $\varepsilon = 10^{-4}$ (a factor of ~ 2 in total number of time steps) doesn't give significantly different results and $\varepsilon = .01$ will sometimes generate visible oscillatory solutions (the numerical oscillations generated by TR, whose amplitude is theoretically $0(\varepsilon)$ in our method, are only slowly damped, owing to physical diffusion, as $t \to \infty$).

Final details regarding time step selection are deferred until we discuss the solution of Eq. (9), since certain additional economic factors will arise.

IV. SOLUTION OF THE NONLINEAR SYSTEM

To solve Eq. (9), we employ a modified Newton-Raphson method. The modification consists of taking but one Newton iteration per time step, rather than iterating to some rather tight convergence criterion. Because the time integration error is usually quite small and by virtue of our good predictor value in the nonlinear terms (our AB result is close to the desired TR result), it turns out that the "one-step Newton" method works quite well and is obviously cost-effective. We converted to this scheme after some experimentation with both "full" Newton (which rarely required more than 1 or 2 iterations to reduce the iteration error to a value substantially less than the local time integration error, ε) and a chord method (see Gresho et al., 1980a), in which we employed an out-of-date Jacobian matrix. The results of these comparisons were that one-step Newton yields solutions (at least for $\varepsilon = .001$) which are virtually indistinguishable from full Newton and that the chord method is generally less cost-effective than either (the cost per iteration was higher than expected and several (4-6) iterations were typically required). Admittedly, our experience with the chord method was brief and it may still be a useful technique in some situations (and perhaps using a different linear equation 'solver', which will be described below). We have been so favorably impressed by the one-step Newton scheme, however, that we now use it exclusively.

The Newton method for solving $A(x)x = b$ leads to the following iterative sequence of linear systems:

$$J_m \, \delta X_{m+1} = b - A(X_m)X_m \quad , \quad m = 1,2,\ldots \quad ; \tag{16}$$

where $\delta X_{m+1} \equiv X_{m+1} - X_m$ is the change in X between iterations and $J_m \equiv \partial[A(X_m)X_m]/\partial X_m$ is the Jacobian matrix. For full Newton, Eq. (16) is solved repeatedly until $\delta X_{m+1} \to 0$. For one-step Newton, the (only) linear system is

$$J_p \, (X - X_p) = b - A(X_p)\, X_p \quad , \tag{17}$$

where X_p is a predicted value of X (first guess), J_p is the Jacobian matrix based on this predictor, and X is assumed to be the correct solution. Application of Eq. (17) to Eq. (9), with

$X = (u_{n+1}\ T_{n+1}\ P_{n+1})^T$ and $X_p = (u^p_{n+1}\ T^p_{n+1}\ 0)^T$ gives (there is no predictor equation for pressure — and none is needed since P appears linearly)

$$J_p = \begin{bmatrix} \frac{2}{\Delta t_n} M+K+N(u^p_{n+1})+N'(u^p_{n+1}) & \beta M' & C \\ N''(T^p_{n+1}) & \frac{2}{\Delta t_n} M'+K_T+N(u^p_{n+1}) & 0 \\ C^T & 0 & 0 \end{bmatrix} . \qquad (18)$$

$A(X_p)$ is the matrix on the left side of Eq. (9) where u^p_{n+1} is used to evaluate $N(u_{n+1})$, and b is the RHS vector of Eq. (9). Whereas N(u) corresponds to the operator $\underline{u} \cdot \nabla$, N'( ) and N''( ) correspond to terms like $\partial(\ )/\partial x$, etc. The nonlinear contributions to $A(X_p)$ and J_p are computed analytically, from the basis functions, with the resulting triply-subscripted arrays stored, element-by-element, on disk.

A summary of the entire algorithm for advancing the solution by one time step may be useful:

1. The predictors for velocity and temperature, u^p_{n+1} and T^p_{n+1}, are computed using Eq. (7).
2. The Jacobian matrix of Eq. (18) and the RHS of Eq. (17) are 'assembled' (by looping through the elements).
3. The final velocity, temperature and pressure are obtained by solving the linear system (17). We solve our linear algebraic systems using a disk-based unsymmetric frontal solver (Hood, 1976) without pivoting.
4. The 'acceleration' vector is updated via Eq. (10).
5. The next potential time step, Δt_{n+1}, is computed from Eqs. (13) and (15).

The final details of time step modification, which are mostly based on qualitative cost-effectiveness principles and are similar to those used in certain standard ODE software packages (Hindmarsh, 1972; Byrne et al., 1977) are these:

1. If $\Delta t_{n+1} \geq \Delta t_n$, we always accept the increase.
2. If $\alpha\Delta t_n \leq \Delta t_{n+1} < \Delta t_n$, where α is typically $\sim .8$, we accept the solution but do not change the time step.
3. If $\Delta t_{n+1} < \alpha\Delta t_n$, we reduce Δt and <u>repeat</u> the time step (i.e., we reject the solution at t_{n+1}).

A final potentially cost-effective modification of the algorithm, which we have not tested, would be to decouple the temperature equation from the momentum and continuity equations, thus solving two (uncoupled) smaller linear systems rather than one larger one. To effect this change, one must (effectively) omit the $\beta M'$ and $N''(T^p_{n+1})$ submatrices from J_p in Eq. (18), which in effect assumes that the vectors $\beta M'(T_{n+1} - T^p_{n+1})$ and $N''(T^p_{n+1})(u_{n+1} - u^p_{n+1})$ are negligibly small. An even better algorithm might result if the local time truncation errors are computed separately for velocity and temperature, which gives an idea of which variable is changing more rapidly (that with the larger truncation error — it has 'faster physics'), as follows: If $|d_{n+1}(u)| > |d_{n+1}(T)|$ from Eq. (13), the velocity is changing more rapidly; hence (1) neglect the $\beta M'$ submatrix and solve for $\delta u \equiv (u_{n+1} - u^p_{n+1})$ and P_{n+1}, (2) Solve for $\delta T \equiv T_{n+1} - T^p_{n+1}$ after evaluating the term $N''(T^p_{n+1})(u_{n+1} - u^p_{n+1})$ and transposing it to the RHS. On the other hand, if $|d_{n+1}(T)| > |d_{n+1}(u)|$, the velocity is changing more slowly than the temperature; hence (1) neglect the $N''(T^p_{n+1})$ submatrix and solve for δT, (2) solve for δu and P_{n+1} after evaluating the term $\beta M'(T_{n+1} - T^p_{n+1})$ and transposing it to the RHS.

V. NUMERICAL RESULTS

A. THERMALLY DRIVEN CAVITY

Initially, an isothermal ($T_o = 0$) fluid is at rest in a closed square cavity. The temperature of the left wall is suddenly increased by $\Delta T/2 = 0.5$ and that of the right wall is decreased by the same amount, while the top and bottom walls are insulated. The resulting, buoyancy-induced motion is governed by two dimensionless parameters: the Rayleigh number,

$$Ra = \frac{\gamma g \Delta T h^3}{\kappa \nu} , \tag{19}$$

where h is the cavity width and $\nu = \mu/\rho$ is the kinematic viscosity; and the Prandtl number,

$$Pr = \nu/\kappa , \tag{20}$$

which is a property of the fluid. In our simulation, we used $Ra = 10^5$ and $Pr = 1$.

A rather coarse, graded mesh (8 x 8) was employed for the simulation, following Marshall et al. (1978) so that we could compare our results with theirs. The mesh is shown in their paper and has nodal coordinates in both x and y at 0, .03125, .0625, .09375, .125, .1875, .25, .375, .5; and is symmetric about x = y = .5. There are 289 nodes and 948 equations in the total system. Figure 1 shows the time history of the step size (Δt) and the vertical velocity and temperature at a node located at x = .75, y = .5. The transient solution appears to be divisible into three stages: (1) the conduction stage, from 0 to $\sim$.01 during which linear effects (conduction, viscous momentum transfer, buoyancy) dominate and advection is small (i.e., the transient Stokes equations are applicable), (2) the "overshoot" stage, from t $\cong$ .01 to $\sim$.05, during which the inertial terms grow in size to such an extent that the temperature (buoyancy force) is advected 'too far' so that internal gravity wave oscillations are initiated, and (3) the recovery stage, from t $\cong$ .05 to .15 or so, during which the waves are damped by viscosity and heat conduction and a steady flow is attained. The time steps grow rapidly during the first stage, from $\Delta t_o = 6 \times 10^{-5}$ to $\sim$.0016, where the time integration is basically following a linear transient like $\sum_i a_i e^{-\lambda_i t}$, and more slowly during stages 2 and 3 where inertial effects are important, ultimately reaching $\sim$.026.

Figures 2 through 5 show a sequence of results at several interesting times. Figure 2 shows the evolution of two cells during phase 1, which start at the two vertical walls and

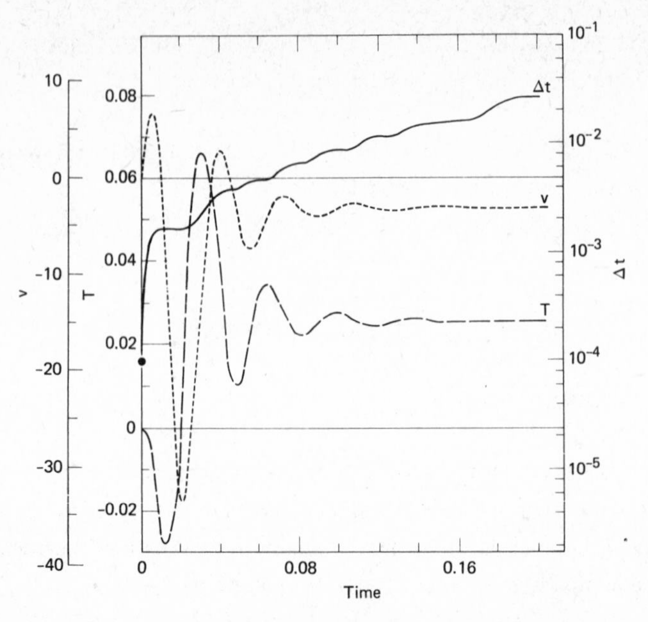

FIGURE 1. Time history of the step size (Δt), vertical velocity (v) and temperature (T) at x = .75, y = .5 for the thermally driven cavity.

grow outward into the fluid while the isotherms are still vertical. Figure 3 shows the onset of isotherm advection, two well-developed interior cells, and large pressure gradients in two corners (the flow is still accelerating). In Fig. 4 the maximum velocities have been attained, the isotherms have 'overshot', and, interestingly, the two cells have merged into one. This situation is short-lived, however, and during the final stage of damped oscillations, the flow returns to a two-cell configuration as shown in Fig. 5, which is essentially steady state (the grid is too coarse near the cavity center to obtain smooth streamline contours). At this time the value of the stream function at the cavity center is 9.56, in good agreement with that reported by Marshall et al. (9.54), who used the same grid but different basis functions (they employed the penalty function approach which corresponds to the same 9-node element for velocity and temperature, but a discontinuous bilinear pressure approximation — defined at the 2 x 2 Gauss points in each element). The entire simulation required 56 time steps and $\sim$6 minutes on a CDC-7600.

Finally, we should mention that this appears to be an 'easy' simulation in that the results 'look quite reasonable'. As mentioned earlier and reported in Gresho et al. (1980b), however, this element can generate unreasonable results for more difficult problems (e.g., a much higher Rayleigh number on the same grid).

B. FLOW PAST A CIRCULAR CYLINDER

The isothermal version of the code was employed to simulate the flow past a circular cylinder, starting from rest, and leading ultimately to periodic vortex shedding (Karman

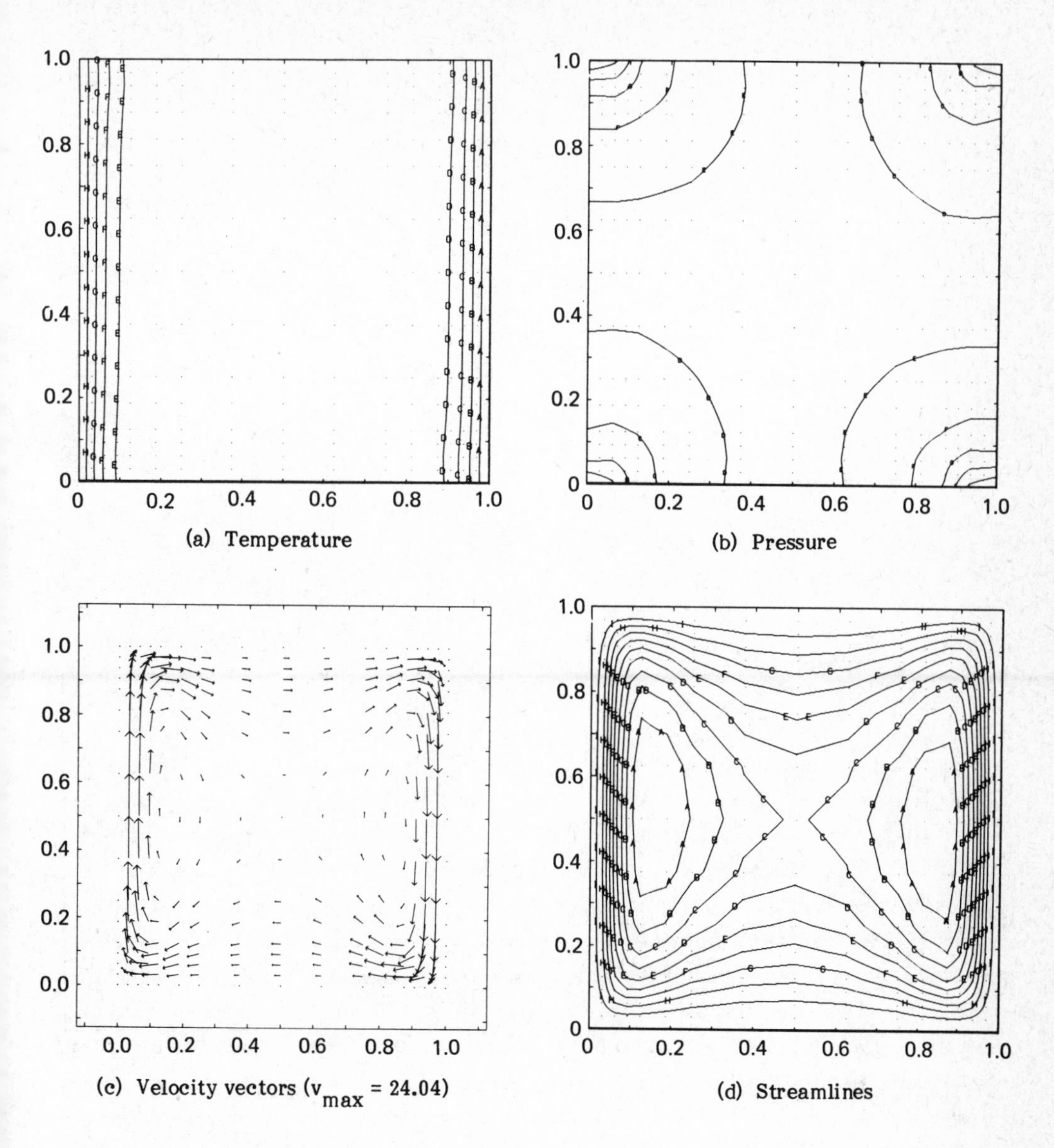

(a) Temperature

(b) Pressure

(c) Velocity vectors (v_{max} = 24.04)

(d) Streamlines

FIGURE 2. Thermally driven cavity (Ra = 10^5, Pr = 1); results at t = 0.00288.

vortices) at Re $\tilde{=}$ 100. The grid of 9-node isoparametric elements is shown in Fig. 6, which also depicts the BC's employed (the values of the applied normal traction force, f_n, which are unsymmetric owing to slight grid asymmetry, were taken to be the inlet pressures from the results of our steady code, at Re = 100, where an inlet BC of $\hat{u}$ = 1.0 was employed; this BC is, of course, illegal for the time-dependent calculation, since it would violate $C^T u_o = 0$). The grid contains 196 elements, 850 nodes, and leads to 1929 total equations (1700 are velocity, 229 are pressure).

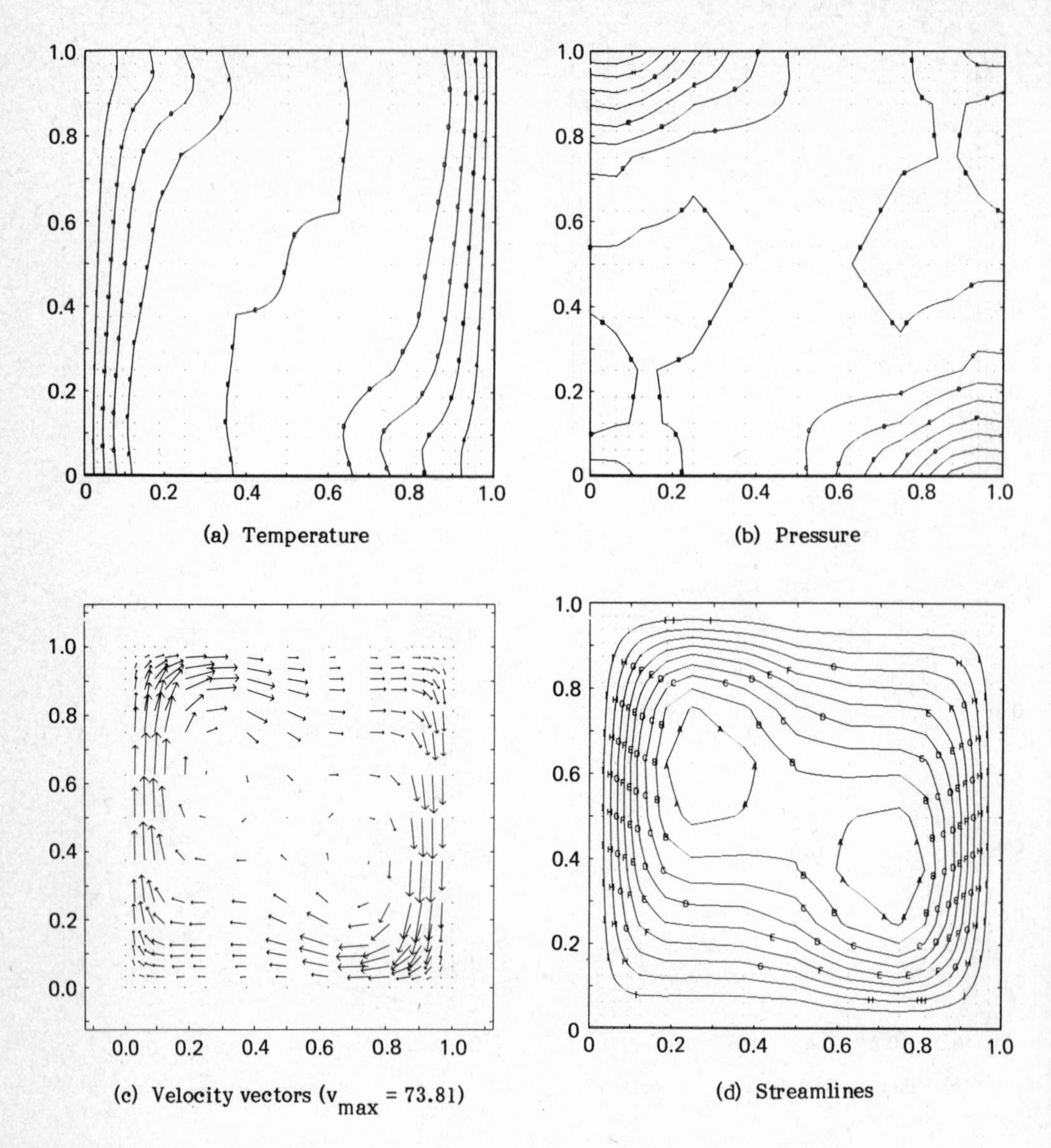

FIGURE 3. Thermally driven cavity (Ra = 10^5, Pr = 1), results at t = 0.01037.

In Fig. 7 is shown the time step history, which again provides some insight into the 'physics' of the flow. The rapid increase in Δt (from 2 to $\sim$5) up to t $\cong$ 20 is again typical of linear, viscous flow. At this time, some new 'physics' appears, however, since Δt stops increasing; the reversal of Δt followed by the slower growth from t $\cong$ 40 onward corresponds to the formation and growth of the separated flow regions behind the cylinder. The time step then grows monotonically while the (basically symmetric) eddies continue to grow in length; from a time of $\sim$150 to $\sim$350 or so, a constant Δt is sufficient to follow this growth. The continuous reduction in Δt, beginning at t $\cong$ 400, signals the beginning of a new dynamic

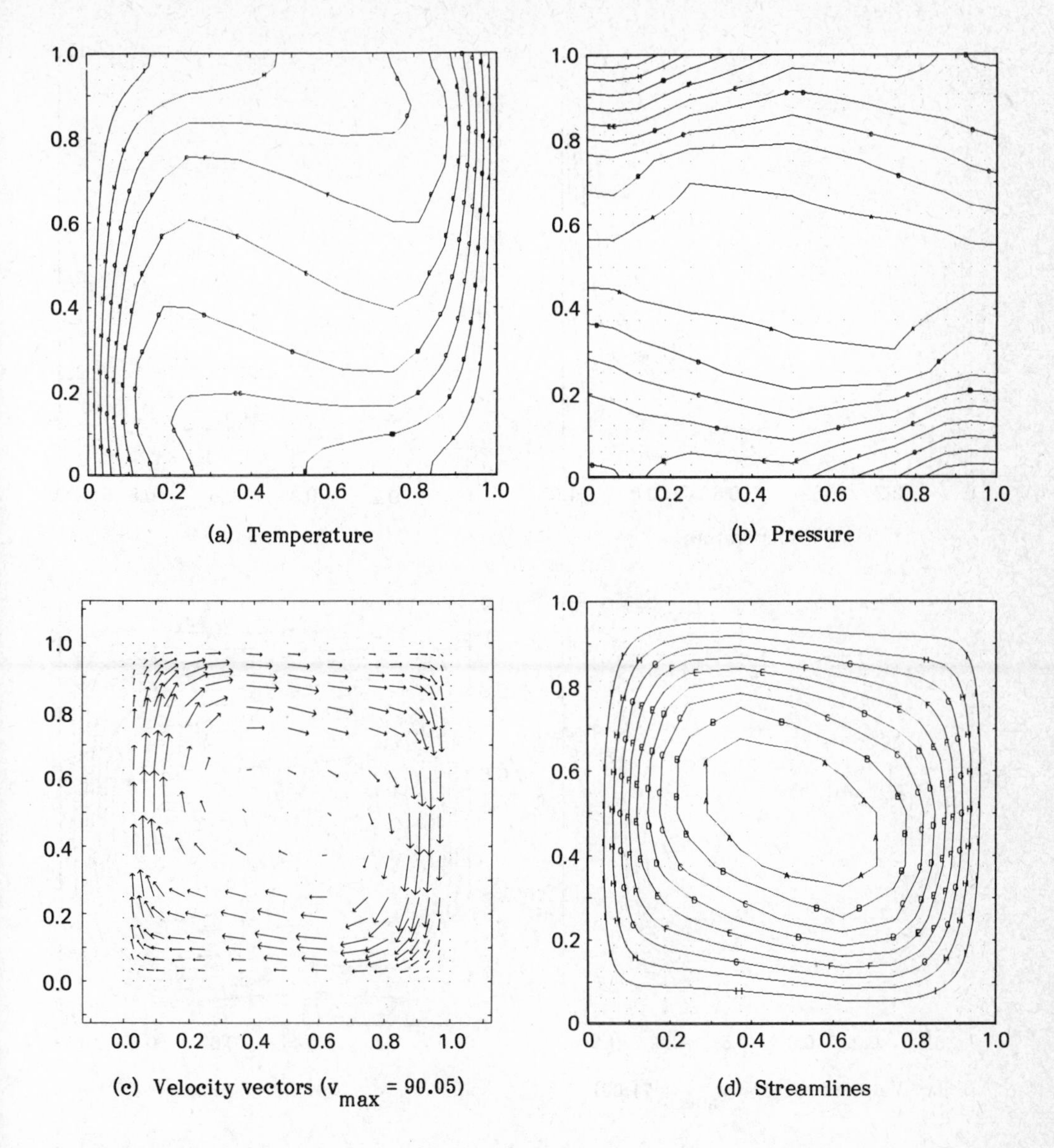

FIGURE 4. Thermally driven cavity (Ra = 10^5, Pr = 1), results at t = 0.02533.

phenomenon and corresponds, of course, to the oscillations which are caused by the inherent instability of the flow. The time step decreases from a maximum of ∿8.9 to ∿.37 as the Karman vortices gain in strength and set a final time scale for the flow. The number of time steps is also included in the figure; this curve is linear from t $\cong$ 495 to 520, and ends at ∿170 time steps. Using one-step Newton, each time step cost ∿20 seconds on a CDC-7600 with an early, 'research-version' of the code; current estimates from our more efficient code (although not yet truly optimised) are 10–12 seconds/step, which would give a cost of ∿ 3 minutes per shedding cycle.

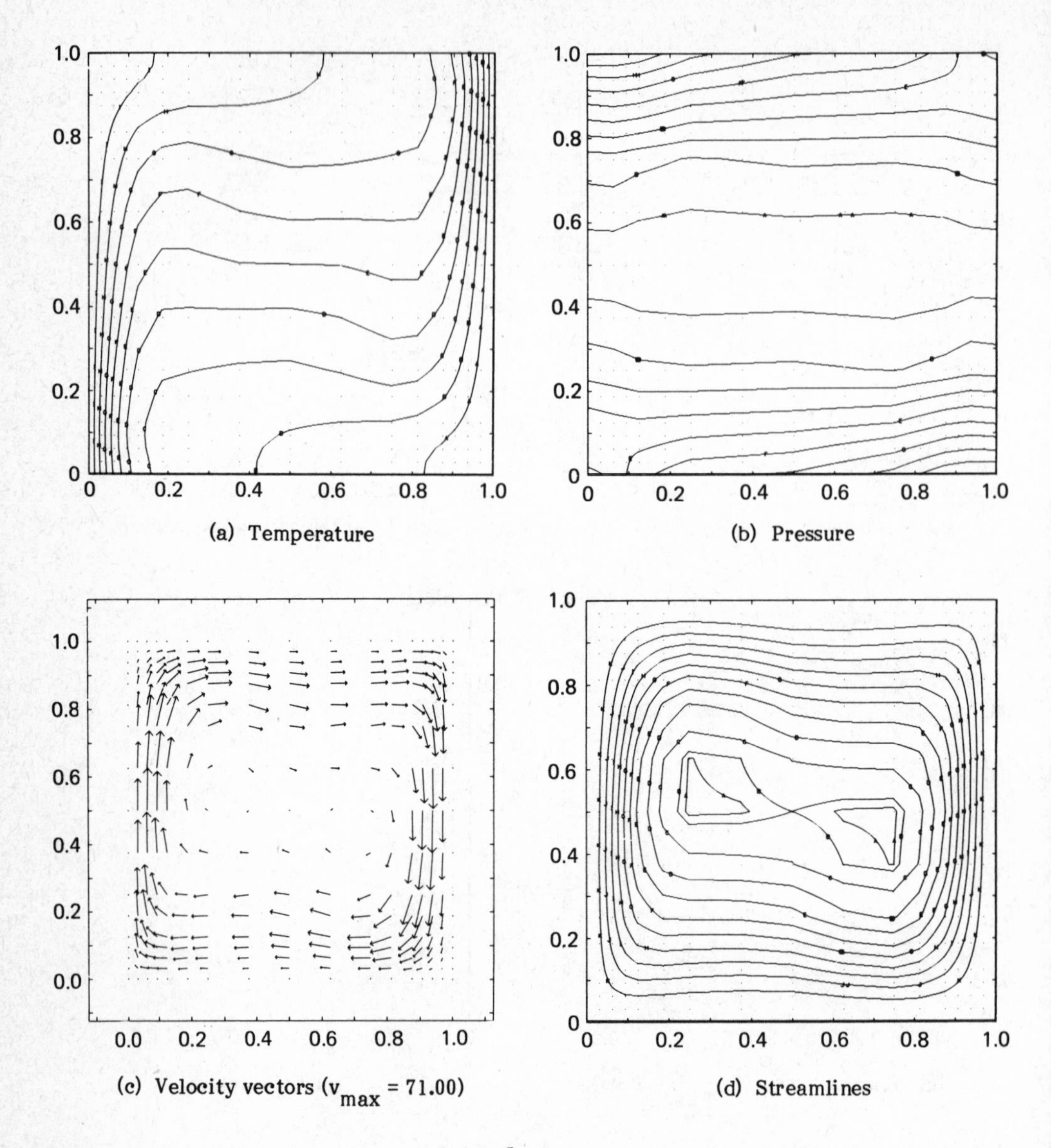

(a) Temperature

(b) Pressure

(c) Velocity vectors (v_{max} = 71.00)

(d) Streamlines

FIGURE 5. Thermally driven cavity (Ra = 10^5, Pr = 1), results at t = 0.21301.

Figure 8 shows a time history plot of the vertical velocity of a node located on-axis about 7.6 diameters downstream of the cylinder. The nearly explosive growth of the instability, beginning at t $\cong$ 450 or so, is clearly shown, as is the quasi-steady periodic oscillation beyond t $\cong$ 500; the final period of vortex shedding is 6.2-6.4, giving a dimensionless shedding frequency (Strouhal number, S = $fD/\hat{u}$) of about .16, which is in good agreement with experimental results (e.g. Schlichting, 1960).

A snapshot of the velocity vector field during vortex shedding is shown in Fig. 9a and the corresponding streamlines in Fig. 9b. Figure 10 shows a snapshot of 'relative' streamlines

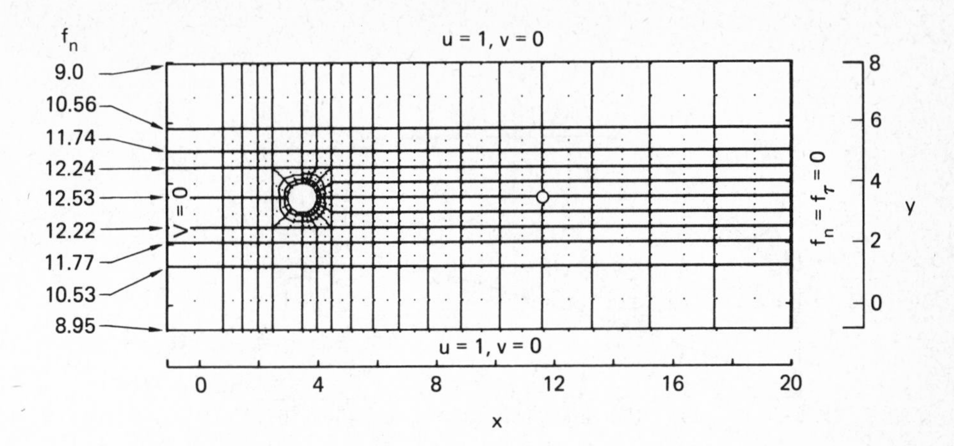

FIGURE 6a. The mesh of 9-node elements showing the boundary conditions for flow past a circular cylinder. (The open dot is the location of the node for which a time history plot is presented.)

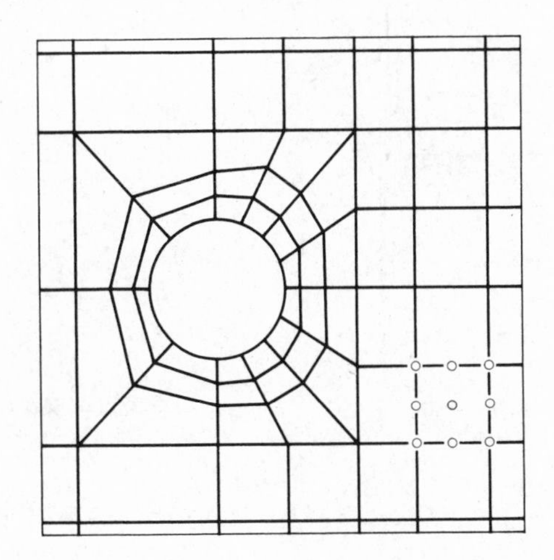

FIGURE 6b. Details of the mesh near the cylinder.

as seen by an observer moving with the nominal fluid velocity, and more clearly reveals the classic Karman vortex street. It is also noteworthy that the traction-free outflow BC works quite well in that this artificial boundary does not appear to degrade the quality of the solution as the vortices leave the grid.

Finally, a closeup snapshot of the vector field is presented in Fig. 11, showing some of the intricate details of the flow field in the near-wake region. For further discussion and more detailed results regarding this simulation, see Gresho et al. (1980a).

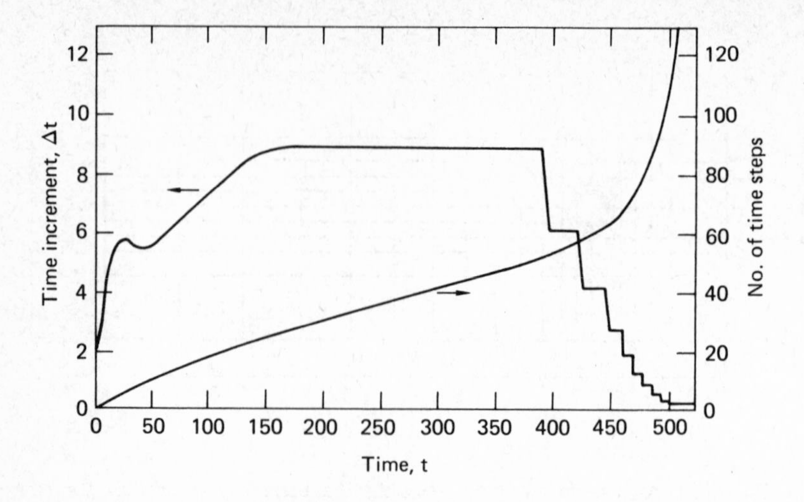

FIGURE 7. Time step history.

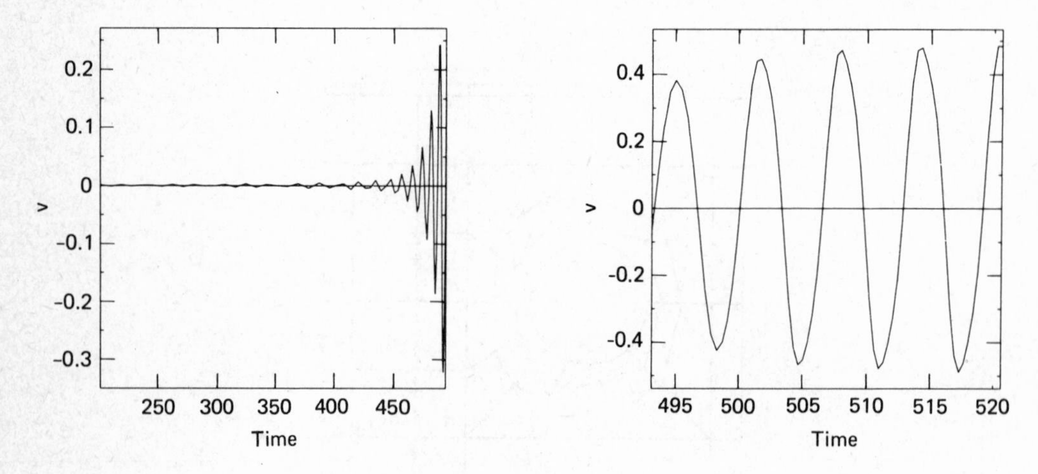

FIGURE 8. Time history of the vertical velocity at a node downstream of the cylinder (see Fig. 6a).

VI. SUMMARY AND CONCLUSIONS

We have described and demonstrated what we believe is a useful, accurate, and 'honest' technique for solving the time-dependent equations of motion for incompressible fluids in two-dimensions. While the computations are reasonably affordable in 2-D, they would be too expensive for significant 3-D simulations; our current research is directed toward the latter, and we are investigating various alternatives (Gresho et al., 1980a).

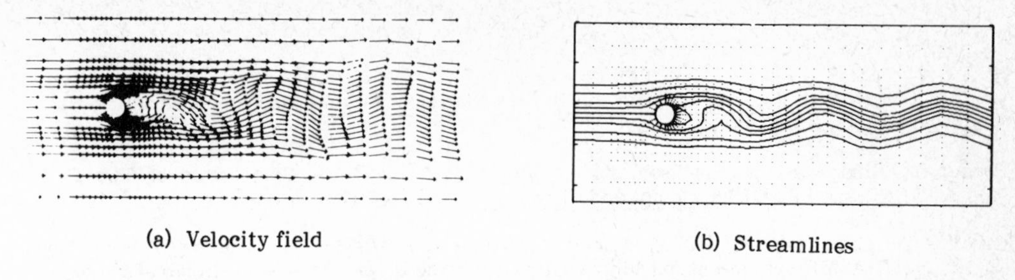

(a) Velocity field

(b) Streamlines

FIGURE 9. Vortex shedding at Re ≈ 100.

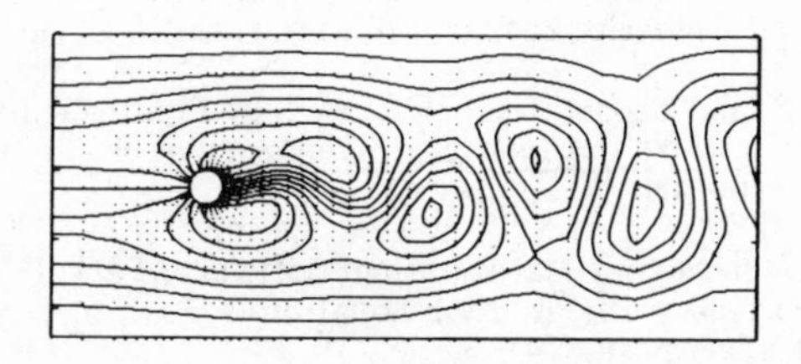

FIGURE 10. Relative streamlines as seen by an observer moving with the nominal fluid velocity.

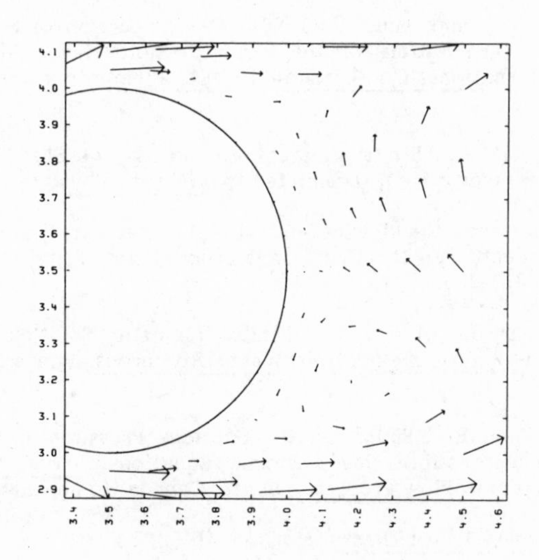

FIGURE 11. Details of flow field in the wake region.

VII. ACKNOWLEDGMENTS

The time integration scheme was developed with the assistance of Dr. A. C. Hindmarsh of LLL, whose contributions were crucial to the ultimate success of the method. This work was performed under the auspices of the U. S. Department of Energy by the Lawrence Livermore Laboratory under contract No. W-7405-Eng-48.

REFERENCES

Aris, R. (1962), Vectors, Tensors, and the Basic Equations of Fluid Mechanics, Prentice-Hall, Englewood Cliffs, NJ, p. 129.

Byrne, G., Hindmarsh, A., Jackson, K. and Brown, H. (1977), "A Comparison of Two ODE Codes: GEAR and EPISODE", Comp. and Chem. Eng., V. 1, pp. 133-147.

Gresho, P. and Lee, R. (1979), "Don't Suppress the Wiggles — They're Telling You Something!", ASME Symposium on Finite Element Methods for Convection-Dominated Flows, Winter Annual ASME Meeting, NY, December (1979).

Gresho, P., Lee, R., and Sani, R. (1980a), "On the Time-Dependent FEM Solution of the Incompressible Navier-Stokes Equations in Two- and Three-Dimensions," in Recent Advances in Numerical Methods in Fluids, Pineridge Press, Ltd., Swansea, U. K. (to appear).

Gresho, P., Lee, R., Chan, S. and Leone, J. (1980b), "A New Finite Element for Incompressible or Boussinesq Fluids," Proc. Third International Conference on Finite Elements in Flow Problems, Banff, Canada (to appear).

Hindmarsh, A. (1972), "Construction of Mathematical Software, Part III: The Control of Error in the Gear Package for Ordinary Differential Equations," Lawrence Livermore Laboratory Report UCID-30050, Part 3, Livermore, CA.

Hood, P. (1976), "Frontal Solution method for Unsymmetric Matrices," Int. J. for Num. Meth. Eng., 10, pp. 379-399.

Lee, R., Gresho, P., Chan, S., and Sani, R. (1980), "A Comparison of Several Conservative Forms for Finite Element Formulations of the Incompressible Navier-Stokes or Boussinesq Equations," Proc. Third International Conference on Finite Elements in Flow Problems, Banff, Canada (to appear).

Leone, J. and Gresho, P. (1979), "Finite Element Simulations of Steady, Two-Dimensional Viscous Incompressible Flow Over a Step," submitted to J. Comp. Phys.

Marshall, R., Heinrich, J., and Zienkiewicz, O. (1978), "Natural Convection in a Square Enclosure by a Finite-Element, Penalty Function Method Using Primitive Fluid Variables," Num. Heat Transfer, 1, pp. 315-330.

Olson, M. (1977), "Comparison of Various Finite Element Solution Methods for the Navier-Stokes Equations," in Finite Elements in Water Resources, Pentech Press, London, p. 4.185.

Sani, R., Gresho, P., and Lee, R. (1980), "On the Spurious Pressures Generated by Certain FEM Solutions of the Incompressible Navier-Stokes Equations," Proc. Third International Conference on Finite Elements in Flow Problems, Banff, Canada (to appear).

Schlichting, H. (1960), Boundary Layer Theory, McGraw-Hill, N. Y.

Shampine, L. and Gordon, M. (1975), Computer Solution of Ordinary Differential Equations: The Initial Value Problem, W. Freeman and Company, San Francisco, CA.

AUXILIARY FLUX AND PRESSURE CONDITIONS FOR NAVIER-STOKES PROBLEMS

John G. Heywood
University of British Columbia[1]
Vancouver, Canada

1. Introduction

Consider the matter of correctly posing problems for the Navier-Stokes equations in various types of domains. To be specific, consider first nonstationary problems and restrict attention to solutions which possess at least enough local regularity for the development of a uniqueness theory. For such solutions in bounded domains, physical intuition suggests both the fluid velocity and pressure (up to an additive function of t) should be determined by initial and boundary conditions for the velocity alone. Other non-Dirichlet type boundary conditions are also possible, but we will not go into that here. In unbounded domains, the picture is much less clear. A great variety of situations come to mind, few of which are well understood, either intuitively or mathematically. It turns out, boundary conditions for the velocity do not always suffice to determine a solution uniquely. Depending on the geometry of the domain and the particular class of solutions under consideration, further "auxiliary conditions" may be necessary. Our aim in this paper is to illustrate such general aspects of the existence-uniqueness theory by describing our results [2,4] for a particularly interesting special problem, that of flow through a hole in a wall.

2. The General Initial Boundary Value Problem

The initial boundary value problem traditionally studied in most of the mathematical literature is:

$$u_t + u\cdot\nabla u = -\nabla p + \Delta u\,, \qquad (x,t) \in \Omega \times (0,T), \tag{1a}$$

$$\nabla\cdot u = 0\,, \qquad (x,t) \in \Omega \times (0,T)\,, \tag{1b}$$

$$u(x,0) = 0\,, \qquad x \in \Omega\,, \tag{1c}$$

$$u(x,t) = 0\,, \qquad (x,t) \in \partial\Omega \times (0,T)\,, \tag{1d}$$

---

(1) This research was supported by the Natural Sciences and Engineering Research Council of Canada, grant no. A4150.

$$u(x,t) \to 0 \ , \quad \text{as} \quad |x| \to \infty, \quad t \in (0,T) \ , \tag{1e}$$

except that here, for simplicity, we have set all the usual prescribed data equal to zero. That is, the force and the initial and boundary values are zero. Ω is assumed to be three-dimensional. The first question to be asked, now, is whether the only possible solution is the trivial one, $u \equiv 0$. It turns out the answer depends on the geometry of the domain. However, even for domains in which the uniqueness of the trivial solution seems indicated, we will prove uniqueness only within a class of solutions satisfying certain integrability conditions. It is useful to make the following definition.

By a *class* J_1^* *solution of* (1), we mean a vector function $u(x,t)$ satisfying the integrability conditions (which imply, in particular, a weak form of (1e))

$$u, u_t, \nabla u \in L^2(\Omega \times (0,T)) \ ; \tag{2}$$

the initial and boundary conditions (1c) and (1d) in at least the generalized sense that the trace of u should vanish; the local regularity condition

$$\|u(t)\|_{L^4(\Omega)} < C \ , \qquad t \in (0,T) \ ; \tag{3}$$

the equation of incompressibility (1b); and the weak form of the Navier-Stokes equations (1a), i.e.,

$$\int_\Omega (u_t \cdot \phi + u \cdot \nabla u \cdot \phi + \nabla u : \nabla \phi) dx \ = \ 0 \ , \tag{4}$$

for all $\phi \in D(\Omega)$ and almost all $t \in (0,T)$, where

$$D(\Omega) \ = \ \{ \text{vectors } \phi : \phi \in C_o^\infty(\Omega) \text{ and } \nabla \cdot \phi = 0 \} \ .$$

It should be noted that a classical solution (or any bounded function) which satisfies the integrability conditions (2) automatically satisfies the regularity condition (3). This is not hard to show. Also, any classical solution automatically satisfies (4). For the uniqueness theorems which follow, the reader may, if he wishes, restrict consideration to classical J_1^* solutions, i.e., classical solutions satisfying the integrability conditions (2).

The uniqueness theory for the class of solutions just defined depends in a crucial way on the relation between the following two spaces of solenoidal functions:

$$J_1^*(\Omega) \ = \ \{ \text{vectors } \phi : \phi \in \overset{\circ}{W}{}_2^1(\Omega) \text{ and } \nabla \cdot \phi = 0 \} \ ,$$

$$J_1(\Omega) \quad = \quad \text{Completion of } D(\Omega) \text{ in } W_2^1(\Omega) .$$

As usual, $\overset{\circ}{W}{}_2^1(\Omega)$ is the completion of $C_o^\infty(\Omega)$ in $W_2^1(\Omega)$, i.e., in the norm

$$(\|\phi\|^2 + \|\nabla\phi\|^2)^{\frac{1}{2}} = \left(\sum_{i=1}^{3} \int_\Omega \phi_i^2 \, dx + \sum_{i,j=1}^{3} \int_\Omega (\frac{\partial \phi_i}{\partial x_j})^2 \, dx \right)^{\frac{1}{2}} .$$

The difference between these function spaces is a subtle one; the condition $\nabla \cdot \phi = 0$ is imposed after "taking the completion" in the definition of $J_1^*(\Omega)$, and before "taking the completion" in the definition of $J_1(\Omega)$. Clearly, one always has $J_1(\Omega) \subset J_1^*(\Omega)$. It can be shown, for any open set, whether bounded or not, that if ϕ is a function with $\phi, \nabla\phi \in L^2(\Omega)$ and if ϕ vanishes on $\partial\Omega$, then $\phi \in \overset{\circ}{W}{}_2^1(\Omega)$; see [2]. Thus, class J_1^* solutions of (1) belong to $J_1^*(\Omega)$ for almost every t.

It is instructive to consider for a moment the two definitions of "generalized" solution most commonly found in the literature prior to the writing of [2]. Except for some regularity conditions and initial conditions, they may be briefly characterized as follows.

Definition A. A "generalized" solution of (1) is a function u which, for every t, belongs to $J_1(\Omega)$ and satisfies (4) for all $\phi \in J_1(\Omega)$.

Definition B. A "generalized" solution of (1) is a function u which, for every t, belongs to $J_1^*(\Omega)$ and satisfies (4) for all $\phi \in J_1^*(\Omega)$.

Taking either one of these definitions, the uniqueness theorem (for sufficiently regular solutions) is proved almost trivially. However, if $J_1^*(\Omega) \not\subset J_1(\Omega)$, which is the case for some domains, one can prove at least the local existence of a non-trivial solution of (1) in each coset of, say, $C_o^\infty(0,T;J_1^*(\Omega))/C_o^\infty(0,T;J_1(\Omega))$. These solutions are class J_1^* solutions, according to our definition, and if $\partial\Omega$ is smooth they are fully classical. But, of course, they do not satisfy either of the definitions A or B, since the corresponding uniqueness theorems imply the trivial solution is the only one possible. What is wrong with these definitions? In the case of Definition A, if $J_1^*(\Omega) \not\subset J_1(\Omega)$, the condition $u \in J_1(\Omega)$ implies some further hidden constraint on u, beyond what is implied by the conditions of problem (1) and the integrability conditions (2). In the case of Definition B, if $J_1^*(\Omega) \not\subset J_1(\Omega)$, there is too large a space of test functions ϕ in the "weak form" (4) of equation (1a), resulting again in some hidden constraint on u. This will be clarified later.

We remark, the proper set of test functions in (4) is just $D(\Omega)$, as it is easily shown (for sufficiently regular functions u) that equation (1a) holds for some scalar function p, if and only if (4) holds for all $\phi \in D(\Omega)$. But, of course, if (4) holds for $\phi \in D(\Omega)$, then taking a limit, it also holds for $\phi \in J_1(\Omega)$; so the set of test functions in Definition A is not too large.

Before specializing our problem, we mention that one has uniqueness, as well as local existence, of class J_1^* solutions in each coset of $C_o^\infty(0,T;J_1^*(\Omega))/C_o^\infty(0,T;J_1(\Omega))$.

3. Relation between $J_1^*(\Omega)$ and $J_1(\Omega)$ in aperture domains

Through the rest of this paper, except Theorem 3, we assume Ω is an *aperture domain*, by which we mean all R^3 except for a wall with a hole in it. More precisely, we may take $\Omega = \{x = (x_1,x_2,x_3) : x_1 \neq 0 \text{ or } (x_2,x_3) \in S\}$, where S is a bounded open (not necessarily connected) set of the x_2,x_3-plane, or we may modify this domain slightly by giving the wall thickness and smoothing the edges of the hole at S. The normal to the x_1,x_2-plane is denoted by n.

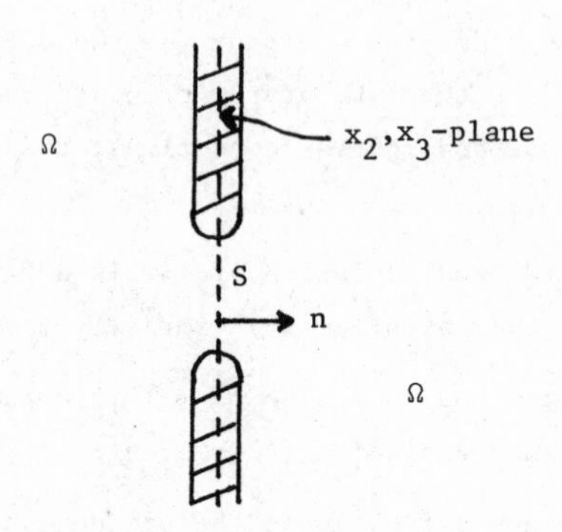

Figure 1. An aperture domain with smooth boundary $\partial\Omega$.

We can see immediately from the following two lemmas that, for aperture domains, $J_1^*(\Omega) \neq J_1(\Omega)$.

Lemma 1. If $u \in J_1(\Omega)$, then $\int_S u\cdot n\, ds = 0$.

This is obvious, as $\int_S \phi\cdot n\, ds = \int_{x_1<0} \nabla\cdot\phi\, dx = 0$, for $\phi \in D(\Omega)$, and the result holds in the limit for the trace of u on S. See Figure 2.

Lemma 2. There exists $b \in J_1^*(\Omega)$ such that $\int_S b\cdot n\, ds = 1$.

A suitable function $b(x)$ can be constructed as follows. Let θ be the angle between the positive x_1-axis and the ray joining a point x with the origin (it is assumed S contains the origin). Let

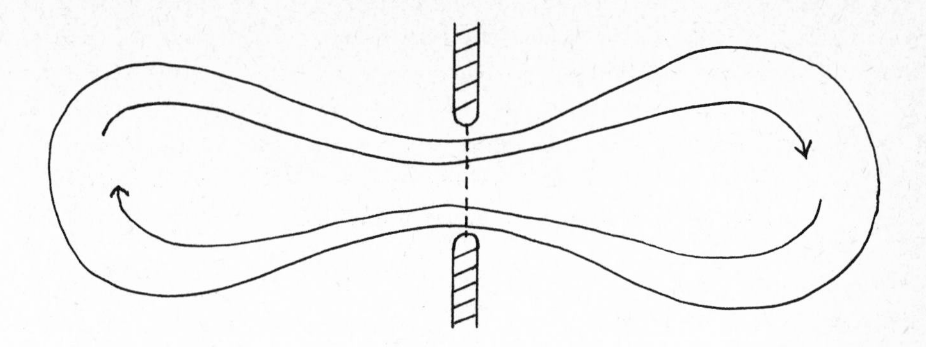

Figure 2. The support and velocity field of a function $\phi \in D(\Omega)$.

$$\hat{b}(x) = \begin{cases} (\cos 2\theta)^2 |x|^{-3} (x_1,x_2,x_3) & \text{for} \quad 0 \le \theta \le \pi/4 \\ 0 & \text{for} \quad \pi/4 \le \theta \le 3\pi/4 \\ -(\cos 2\theta)^2 |x|^{-3} (x_1,x_2,x_3) & \text{for} \quad 3\pi/4 \le \theta \le \pi . \end{cases}$$

Since the vector field $\hat{b}$ is purely radial and decays like $|x|^{-2}$, it is solenoidal. It remains solenoidal if we mollify it to remove the singularity at the origin. Clearly, the resulting vector field carries non-zero net flux through S, which can be adjusted to equal one by multiplying by a constant, giving the desired function b. Since $|b(x)| = O(|x|^{-1})$ and $|\nabla b(x)| = O(|x|^{-2})$, as $|x| \to \infty$, we have $b, \nabla b \in L^2(\Omega)$ and hence $b \in J_1^*(\Omega)$.

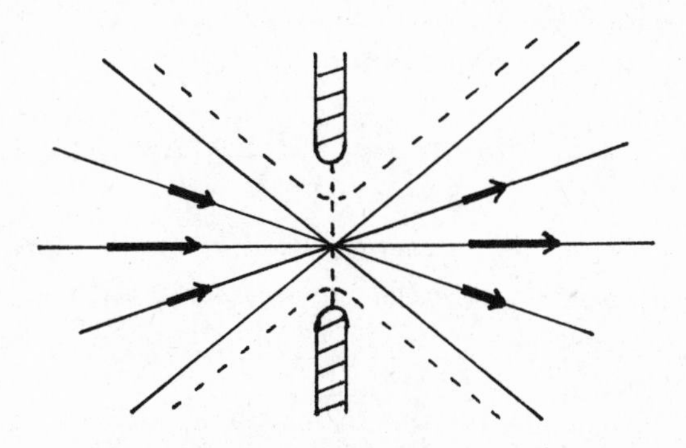

Figure 3. The vector field $\hat{b}$, with dotted line showing the support of b.

The following result is crucial for the existence-uniqueness theory in aperture domains.

Theorem 1. <u>For</u> $u \in J_1^*(\Omega)$, <u>there holds</u> $u \in J_1(\Omega)$ <u>if and only if</u> $\int_S u\cdot n\, ds = 0$.

Part of this is proved in Lemma 1. The other part, that $\int_S u\cdot n\,ds = 0$ implies $u \in J_1(\Omega)$, is relatively difficult and outside the scope of this paper; we refer the reader to [2] for the details. The reader may also wish to consult [7], where some of our results for the quotient space J_1^*/J_1 are extended to more complicated domains. We should point out though, that the authors of [7] argue (wrongly, we think) for the correctness of Definitions A and B of "generalized" solution.

4. Flux and Pressure Conditions

It seems evident, on the basis of physical intuition, that for an aperture domain, the conditions of problem (1) should be supplemented with either an *auxiliary flux condition*

$$\int_S u\cdot n\,ds \quad = \quad F(t) \;, \tag{5}$$

where F is a prescribed function of t (which vanishes for $t = 0$), or an *auxiliary pressure condition*

$$\lim_{\substack{|x|\to\infty \\ x_1<0}} p(x,t) \; - \; \lim_{\substack{|x|\to\infty \\ x_1>0}} p(x,t) \quad = \quad P(t) \;, \tag{6}$$

where P is a prescribed function of t. These intuitive expectations are borne out by the following mathematical theory, based on Theorem 1.

Theorem 2. <u>There is at most one class J_1^* solution of (1) satisfying the auxiliary flux condition (5)</u>.

This is shown by considering an energy identity for the difference $w = u - v$ of two solutions; the pressure term drops out because of the flux condition. From (4), we have

$$\int_\Omega (w_t\cdot\phi + w\cdot\nabla u\cdot\phi + v\cdot\nabla w\cdot\phi + \nabla w{:}\nabla\phi)\,dx \quad = \quad 0 \;, \tag{7}$$

for every $\phi \in D(\Omega)$, and almost all t. Since $\int_S w\cdot n\,ds = 0$, we have $w \in J_1(\Omega)$, for almost all t, by Theorem 1. Thus we may pass to the limit $\phi \to w$ in (7) obtaining, after several integrations by parts,

$$\frac{1}{2}\frac{d}{dt}\|w\|^2 + \|\nabla w\|^2 \quad = \quad \int_\Omega w\cdot\nabla w\cdot u\,dx \;. \tag{8}$$

If u and v are classical solutions, the right side is less than $\frac{1}{4}\sup_\Omega |u|^2 \cdot \|w\|^2 + \|\nabla w\|^2$, so that $\frac{1}{2}\frac{d}{dt}\|w\|^2 \le c\|w\|^2$. Hence $w \equiv 0$, because $\|w(0)\| = 0$. Assuming (3) rather than the classicality of the solutions, one still has

$$\begin{aligned}\int w\cdot\nabla w\cdot u\,dx &\le \|w\|_4 \cdot \|\nabla w\| \cdot \|u\|_4 \\ &\le c\|w\|^{\frac{1}{4}} \cdot \|\nabla w\|^{\frac{7}{4}} \cdot \|u\|_4 \\ &\le c\|u\|_4^8 \cdot \|w\|^2 + \|\nabla w\|^2 \\ &\le c\|w\|^2 + \|\nabla w\|^2 ,\end{aligned}$$

and hence $w \equiv 0$, as before. Above, we use the notation $\|\phi\|_4 = \|\phi\|_{L^4}$.

One might wonder at this point whether some kind of auxiliary conditions might also be necessary in exterior domains, or even bounded domains. We will only state the results for bounded and exterior domains which correspond to Theorems 1 and 2; the proofs are given in [2].

Theorem 3. If Ω is a bounded domain or an exterior domain (with some very mild regularity assumed of $\partial\Omega$), then $J_1^*(\Omega) = J_1(\Omega)$, and hence uniqueness holds for J_1^* solutions of (1), without auxiliary conditions.

The first question to be asked in connection with the pressure condition (6), is whether the limits of the pressure at infinity, in each half space, exist. This is assured in a generalized sense by the following theorem, in view of the divergence of the integral $\int |x|^{-2}\,dx$ over a half-space. To avoid a minor technical complication, we assume from now on that the wall has thickness and $\partial\Omega$ is smooth.

Theorem 4. If $p(x,t)$ is a pressure function associated with a class J_1^* solution of (1), then there exist functions $p_1(t)$ and $p_2(t)$ such that, for almost all t,

$$\int_{x_1<0} |p(x,t) - p_1(t)|^2 \cdot |x|^{-2}dx < \infty \quad \text{and} \quad \int_{x_1>0} |p(x,t) - p_2(t)|^2 \cdot |x|^{-2}dx < \infty . \tag{9}$$

To prove this, let t be fixed and consider u as a solution of the Stokes equations $\Delta u = \nabla p + f$, with $f = u_t + u\cdot\nabla u$. Since $f \in L^{3/2}_{loc}(\bar{\Omega})$, the estimates of Cattabriga [1] imply (after a second step) $u \in L^\infty(\Omega)$. Thus $f \in L^2(\Omega)$, and hence

$$\sum_{i,j,k=1}^{3} \left\|\frac{\partial^2 u_i}{\partial x_j \partial x_k}\right\|^2 \leq c(\|f\| + \|\nabla u\|) ,$$

by an estimate proved in [3]; Lemma 9 of [2] also suffices for our purpose, as regularity up to the boundary is not essential here; see [4]. Therefore $\nabla p \in L^2(\Omega)$. This implies the existence of limits p_1 and p_2 in the generalized sense of (9), in the left and right half spaces, respectively; see [2,p.95].

Theorem 5. <u>There is at most one class</u> J_1^* <u>solution of (1) satisfying the auxiliary flux condition</u> $p_1(t) - p_2(t) = P(t)$.

Consider, again, the difference $w = u - v$ of two solutions. Let $s = p - q$ be the difference of corresponding pressure functions. Since $w \in J_1^*(\Omega)$, for almost all t , Theorem 1 implies $w = \bar{w} + F(t)b(x)$ for some $\bar{w} \in L^2(0,T;J_1(\Omega))$ and some scalar function $F(t)$. Here, b is the function constructed in Lemma 2. Since $p_1(t) - p_2(t) = q_1(t) - q_2(t)$, we have $s_1(t) = p_1(t) - q_1(t) = p_2(t) - q_2(t) = s_2(t)$; i.e., $s(x,t)$ tends to the same limit at infinity in each half space. Without loss of generality, we may assume $s_1(t) = s_2(t) = 0$; then Theorem 4 implies

$$\int_\Omega |s(x,t)|^2 \cdot |x|^{-2} dx < \infty . \tag{10}$$

Since $\bar{w} \in J_1(\Omega)$, for almost all t , we can let $\phi \to \bar{w}$ in (7), giving

$$\int_\Omega (w_t \cdot \bar{w} + w \cdot \nabla u \cdot \bar{w} + v \cdot \nabla w \cdot \bar{w} + \nabla w : \nabla \bar{w}) dx = 0 . \tag{11}$$

Also, since u and v have enough regularity (as shown in the proof of Theorem 4) to satisfy (1a) strictly, we can multiply

$$w_t + w \cdot \nabla u + v \cdot \nabla w - \Delta w = -\nabla s$$

through by $F(t)b(x)$ and integrate over Ω to get

$$\int_\Omega (w_t \cdot Fb + w \cdot \nabla u \cdot Fb + v \cdot \nabla w \cdot Fb - \Delta w \cdot Fb) dx = -\int_\Omega \nabla s \cdot Fb \, dx . \tag{12}$$

Adding (11) and (12), and integrating several terms by parts, we obtain

$$\frac{1}{2}\frac{d}{dt}\|w\|^2 + \|\nabla w\|^2 = \int_\Omega w\cdot\nabla w\cdot u\,dx - \int_\Omega \nabla s\cdot Fb\,dx . \tag{13}$$

Now, for any sequence R_i tending to infinity,

$$\left|\int_\Omega \nabla s\cdot b\,dx\right| = \lim_{R_i\to\infty}\left|\int_{|x|=R_i} s\,b\cdot n\,ds\right|$$

$$\leq \lim_{R_i\to\infty}\left(\int_{|x|=R_i} |s|^2\cdot|x|^{-2}ds\right)^{\frac{1}{2}}\left(\int_{|x|=R_i} |x|^2|b|^2ds\right)^{\frac{1}{2}} . \tag{14}$$

In view of (10), there exists a sequence R_i tending to infinity, such that the first factor on the right of (14) tends to zero. The second factor remains bounded because $|b(x)| = O(|x|^{-2})$. We conclude that the pressure term on the right of (13) vanishes, so that (8) holds again and the uniqueness theorem can be completed as in Theorem 2.

Theorem 6. Let u be any class J_1^* solution of (1). Let $p_1(t)$ and $p_2(t)$ be the limits for the pressure found in Theorem 4, let b be the function constructed in Lemma 2, and let $F(t)$ be the flux defined by (5). Then there holds

$$\int (u_t\cdot b + u\cdot\nabla u\cdot b + \nabla u:\nabla b)dx = p_1(t) - p_2(t) , \tag{15}$$

and

$$\frac{1}{2}\frac{d}{dt}\|u\|^2 + \|\nabla u\|^2 = F(t)(p_1(t) - p_2(t)) . \tag{16}$$

Briefly, (15) is proved by multiplying (1a) through by b, integrating over Ω, and noting that $-\int_\Omega \nabla p\cdot b\,dx = p_1(t) - p_2(t)$. The techniques are the same as in Theorem 5. Finally, to obtain (16), multiply (15) by $F(t)$ and then add the result of letting $\phi \to \bar{u}$ in (7), where $\bar{u} = u - Fb$.

We can now identify, at least in the case of an aperture domain, the "hidden constraints" of the Definitions A and B of "generalized" solution. Clearly, Definition A incorporates the flux condition $F(t) \equiv 0$, and Definition B incorporates the pressure condition $p_1(t) - p_2(t) \equiv 0$. The latter follows from (15), since $b \in J_1^*(\Omega)$ is an admissible test function in Definition B.

The existence of a solution satisfying the flux condition (5) is proved in just the same way as one proves existence for a nonhomogeneous boundary value problem. The existence of a solution with a prescribed pressure drop is slightly more involved. We will outline the proof.

Theorem 7. There exists a class J_1^* solution of (1) satisfying the pressure condition (6), on some time interval $[0,T')$, provided $\int_0^T (P^2(t) + P_t^2(t))\,dt < \infty$.

We seek Galerkin approximations

$$u^n(x,t) = F_n(t)b(x) + \sum_{k=1}^{n} c_{kn}(t)a^k(x)$$

in terms of the function b of Lemma 2 and a system of functions $\{a^k\} \subset J_1(\Omega) \cap W_2^2(\Omega)$ which are complete in $J_1(\Omega)$ and orthonormal in $L^2(\Omega)$. The coefficients $F_n(t)$, $c_{kn}(t)$, $k=1,\dots,n$, are determined from the initial conditions $F_n(0) = 0$, $c_{kn}(0) = 0$, and the ordinary differential equations

$$\int_\Omega (u_t^n \cdot a^\ell + u^n \cdot \nabla u^n \cdot a^\ell + \nabla u^n : \nabla a^\ell)\,dx = 0, \quad \ell = 1,\dots,n, \tag{17}$$

and

$$\int_\Omega (u_t^n \cdot b + u^n \cdot \nabla u^n \cdot b + \nabla u^n : \nabla b)\,dx = P(t). \tag{18}$$

Taking linear combinations of these equations, one obtains

$$\frac{1}{2}\frac{d}{dt}\|u^n\|^2 + \|\nabla u^n\|^2 = F_n P,$$

and

$$\frac{1}{2}\frac{d}{dt}\|u_t^n\|^2 + \|\nabla u_t^n\|^2 = -(u_t^n \cdot \nabla u^n, u_t^n) + F_{nt}P_t.$$

Noting $|F_n| \le c\|\nabla u^n\|$ and $|F_{nt}| \le c\|\nabla u_t^n\|$, one obtains estimates parallel to those of Kiselev and Ladyzhenskaya [6]; see [4,5]. By a compactness argument, then, a subsequence of the approximations is seen to converge in various senses to a function u. Using (17), u is shown to satisfy (4), and thus to be a class J_1^* solution of (1). Using (18), it is seen that

$$\int_\Omega (u_t \cdot b + u \cdot \nabla u \cdot b + \nabla u : \nabla b)\,dx = P(t), \tag{19}$$

where $P(t)$ is the prescribed function we wish to show is the pressure drop. Indeed, this now follows from Theorems 4 and 6, since comparing (15) and (19) we have $p_1(t) - p_2(t) = P(t)$.

Open Problems

Our results on the quotient space $J_1^*(\Omega)/J_1(\Omega)$, for exterior domains and aperture domains, have been generalized to some more complicated types of domains by Ladyzhenskaya and Solonnikov [7], and Solonnikov [9]. Related results for the solvability of Navier-Stokes problems are given by these authors in [8] and [9]; we mention that a generalized form of the pressure condition is introduced by Solonnikov in [9]. In all domains which have been studied, if $J_1^*(\Omega) \neq J_1(\Omega)$, the cosets of J_1^*/J_1 are determined by flux conditions similar to that of Theorem 1. It is not known whether or not this is the case for all domains.

If Ω is a three-dimensional aperture domain, the existence of steady flows with any prescribed flux or pressure drop is known; see [2,4,8,9]. If the prescribed flux or pressure drop is small, uniqueness is expected. Arguing as in Theorems 2 and 5 we find

$$\|\nabla w\|^2 = \int w\cdot\nabla w\cdot u$$

$$\leq \begin{cases} \|u\|_3\cdot\|w\|_6\cdot\|\nabla w\| \leq c\|u\|_3\,\|\nabla w\|^2 \\ \qquad\text{or alternatively} \\ \|\nabla w\|\cdot\left(\int_\Omega |u|^2\cdot|w|^2 dx\right)^{\frac{1}{2}} \leq 2\sup_\Omega(|u(x)|\cdot|x|)\,\|\nabla w\|^2 . \end{cases}$$

This implies uniqueness if either $\|u\|_3$ or $\sup_\Omega|u(x)|\cdot|x|$ is sufficiently small. To date, it is not known whether these quantities are bounded for the solutions obtained in the existence theorems, though it seems highly probable.

The existence of two-dimensional nonstationary flow through an aperture is not yet proven. The methods of proving existence described in section 4 fail, because the analogue of the function b constructed in Lemma 2 decays like $|x|^{-1}$, and therefore does not belong to $L^2(r)$. For this problem, the methods of [3,5] may be useful.

REFERENCES

1. L. Cattabriga, *Su un problema al contorno relativo al sistema di equazioni di Stokes*, Rend. Sem. Mat. Univ. Padova 31 (1961), 308-340.

2. J.G. Heywood, *On uniqueness questions in the theory of viscous flow*, Acta Math. 136 (1976), 61-102.

3. J.G. Heywood, *The Navier-Stokes equations: on the existence, regularity and decay of solutions*, (to appear) Preprint, June 1978.

4. J.G. Heywood, *On the proper posing of problems for the Navier-Stokes equations*, (to appear), Canadian Math. Bulletin.

5. J.G. Heywood, *Classical solutions of the Navier-Stokes equations*, (these proceedings).

6. A.A. Kiselev and O.A. Ladyzhenskaya, *On the existence and uniqueness of the solution of the nonstationary problem for a viscous incompressible fluid*, Izv. Akad. Nauk SSSR Ser. Mat. 21 (1957), 655-680.

7. O.A. Ladyzhenskaya and V.A. Solonnikov, *Some problems of vector analysis and generalized formulations of boundary-value problems for the Navier-Stokes equations*, Journal of Soviet Math., Vol. 10, No. 2 (1978), 257-286.

8. O.A. Ladyzhenskaya and V.A. Solonnikov, *On the solution of boundary and initial value problems for the Navier-Stokes equations in domains with non-compact boundaries*, (in Russian) Leningrad Universitet Vestnik, No. 13 (1977), 39-47.

9. V.A. Solonnikov, *On the solvability of boundary and initial-boundary value problems for the Navier-Stokes system in domains with non-compact boundaries*, (to appear).

CLASSICAL SOLUTIONS OF THE NAVIER-STOKES EQUATIONS

John G. Heywood
University of British Columbia[1]
Vancouver, Canada

1. Introduction

The simplest, most elementary proofs of the existence of solutions of the Navier-Stokes equations are given via Galerkin approximation. The core of such proofs lies in obtaining estimates for the approximations from which one can infer their convergence (or at least the convergence of a subsequence of the approximations) as well as some degree of regularity of the resulting solution. The first to use this approach was Hopf [5], who based an existence theorem for the initial boundary value problem on an energy estimate for Galerkin approximations. However, based on this single estimate, Hopf's theorem provides very little regularity of the solution, in fact, insufficient regularity to prove the solution's uniqueness if the domain is three-dimensional. To remedy this situation, Kiselev and Ladyzhenskaya [7] introduced a second estimate for the approximations which yields enough further regularity for a uniqueness theorem. As is well known, this second estimate holds only locally in time unless the data are small or the domain is two-dimensional, a circumstance which has stimulated much speculation over the question of "unique solvability in the large". On the other hand, even during the time interval for which it holds, the estimate of Kiselev and Ladyzhenskaya provides far less than the full classical regularity of the solution.

An interesting varient of the Galerkin method, yielding a somewhat more regular solution, under weaker assumptions on the data, has been given by Prodi [11]. Prodi's existence theorem is based on an estimate, entirely different from those of Hopf and of Kiselev and Ladyzhenskaya, which is available when eigenfunctions of the Stokes operator are used as a basis for the approximations. This estimate is somewhat less elementary than those of Hopf and of Kiselev and Ladyzhenskaya, as it requires an L^2-theory of regularity for the steady Stokes equations. Still, like the estimate of Kiselev and Ladyzhenskaya, Prodi's holds only locally in time and yields only a generalized solution. Until now, the classical regularity of such generalized solutions has been proved only by resort to entirely different and more complicated methods, methods which have invariably depended on potential theoretic results for the Stokes equations. In this regard, we cite particularly the important

(1) This research was supported by the Natural Sciences and Engineering Research Council of Canada, grant no. A4150.

contributions of Ito [6], Fujita and Kato [3], Ladyzhenskaya [8,9], and Solonnikov [15,16].

The main point of the present paper is to show how the Galerkin approach to existence theorems can be pushed further, through additional estimates, to give the classical regularity of the solution directly and easily, with minimal reliance on the regularity theory for the Stokes equations. In fact, the only result which will be needed concerning the regularity of solutions of the Stokes equations is the L^2-estimate of the second derivatives of stationary solutions, which is already needed for Prodi's estimate of the Galerkin approximations, and which has been recently proved in a relatively simple way, independently of potential theory, by Solonnikov and Ščadilov [17].

Our procedure begins with the introduction of two infinite sequences of differential inequalities for the Galerkin approximations. Integration of the first inequality of one sequence, over a time interval $[0,T)$, gives Prodi's estimate. Integration of the first two inequalities of the other sequence, over $[0,T)$, gives, respectively, the estimates of Hopf and of Kiselev and Ladyzhenskaya. To proceed further, with the integration of the succeeding inequalities, it is necessary to work with both sequences simultaneously, using recursively the estimates obtained by integrating the inequalities of one sequence to linearize and integrate those of the other. To avoid the necessity of compatibility conditions for the data, which for the Navier-Stokes equations are of a very complicated non-local nature, these subsequent differential inequalities are integrated over time intervals $[\varepsilon,T)$, with $\varepsilon > 0$. For this, it is necessary to obtain "initial estimates" at $t = \varepsilon$, which we do utilizing yet another sequence of identities and inequalities for the Galerkin approximations.

Combining these estimates for the Galerkin approximations, one can infer the existence of a solution $u \in C^\infty(0,T; W_2^2(\Omega)) \cap L^\infty(0,T'; W_2^2(\Omega))$, where Ω is the spatial domain and T' is any number less than T. With this degree of regularity in hand, the solution's classical regularity follows by a standard argument, which is again based on only an L^2-estimate for the steady Stokes equations.

Although the procedure just described is simple, we will not attempt to give all the details here. The details are given in [4], along with a number of extensions and related results. One extension is the existence theorem, for classical solutions, in the case of unbounded three-dimensional domains with possibly non-compact boundaries. For such domains the result is new. Also in [4], the local existence theorem is proved for initial velocities merely required to possess a finite Dirichlet integral. This result is new in the case of unbounded domains, where, unless Poincaré's inequality holds, the initial velocity need not belong to L^2. One of the related topics studied in [4] is the decay of solutions, in unbounded domains, as $t \to \infty$. If the initial velocity is square-summable and the forces and boundary values are homogeneous, the decay is shown to be of order $t^{-\frac{1}{2}}$. The proof of this is outlined in the final part of the present paper.

2. Galerkin Approximations

Let $\Omega \subset R^3$ be a bounded domain with boundary $\partial\Omega$ of class C^3. We consider the initial boundary value problem

$$u_t + u\cdot\nabla u = -\nabla p + \Delta u \tag{1a}$$

$$\nabla\cdot u = 0 \tag{1b}$$

$$u(x,0) = u_o(x) \tag{1c}$$

$$u|_{\partial\Omega} = 0 \tag{1d}$$

for the vector velocity $u(x,t)$ and scalar pressure $p(x,t)$ of a viscous incompressible fluid. The problem has been normalized so that the density and viscosity are equal to one. It is required that the equations (1a), (1b) should be satisfied in a space-time cylinder $\Omega \times (0,T)$. The initial velocity is u_o. For simplicity, we have taken the external force and boundary values to be homogeneous; inhomogeneous boundary values and forces are considered in [4]. We call u,p a classical solution of problem (1) if $u \in C(\overline{\Omega} \times [0,T))$, if $u_t, \nabla u, \Delta u, \nabla p \in C(\Omega \times (0,T))$ and if the conditions of the problem are satisfied continuously.

Employing the Galerkin method, we consider approximate solutions

$$u^n(x,t) = \sum_{k=1}^{n} c_{kn}(t)\, a^k(x)$$

developed in terms of a system of functions $\{a^k\}$ which is complete in the space $J_1^*(\Omega)$ of divergence-free vector-valued functions from $\overset{\circ}{W}{}_2^1(\Omega)$. A special choice of the functions $\{a^k\}$ will be made shortly, but for now they are merely taken to be smooth and orthonormal in $L^2(\Omega)$. The coefficients $c_{kn}(t)$ are determined by the system of ordinary differential equations

$$(u_t^n, a^\ell) - (\Delta u^n, a^\ell) = -(u^n\cdot\nabla u^n, a^\ell), \tag{2}$$

$\ell = 1, \ldots, n$, with initial conditions $c_{kn}(0) = (a_k, u_o)$. Here, (ϕ,ψ) denotes the L^2 inner product $\int_\Omega \phi\cdot\psi\, dx$.

Hopf's energy identity for the Galerkin approximations is obtained by multiplying (2) by $c_{\ell n}(t)$, summing $\Sigma_{\ell=1}^n$, and integrating several terms by parts, noting in particular that $(u^n\cdot\nabla u^n, u^n) = 0$. The result is

$$\frac{1}{2}\frac{d}{dt}\|u^n\|^2 + \|\nabla u^n\|^2 = 0, \tag{3}$$

where $\|\cdot\|$ denotes the L^2-norm.

Clearly, if $u_o \in L^2(\Omega)$, one has a bound for the initial values

$$\|u^n(0)\| \leq \|u_o\| ,$$

which is uniform n. Hence (3) can be integrated from 0 to t, yielding the energy estimate

$$\frac{1}{2}\|u^n(t)\|^2 + \int_0^t \|\nabla u^n\|^2 d\tau \leq \frac{1}{2}\|u_o\|^2 \tag{4}$$

on which Hopf's existence theorem is based.

Kiselev and Ladyzhenskaya's estimate is based on an identity found by differentiating (2) with respect to t, multiplying by $d/dt\, c_{\ell n}$, summing, and integrating several terms by parts. The result is

$$\frac{1}{2}\frac{d}{dt}\|u_t^n\|^2 + \|\nabla u_t^n\|^2 = -(u_t^n\cdot\nabla u^n, u_t^n) . \tag{5}$$

The right side of (5) can be estimated by using successively Hölder's inequality, Sobolev's inequality, Young's inequality, and the inequality $\|\nabla u^n\|^2 \leq \|u^n\|\cdot\|u_t^n\|$, which follows from (3):

$$\begin{aligned}
|(u_t^n\cdot\nabla u^n, u_t^n)| &\leq \|\nabla u^n\|\cdot\|u_t^n\|_4^2 \\
&\leq c\|\nabla u^n\|\cdot\|u_t^n\|^{1/2}\cdot\|\nabla u_t^n\|^{3/2} \\
&\leq c\|\nabla u^n\|^4\cdot\|u_t^n\|^2 + \frac{1}{2}\|\nabla u_t^n\|^2 \\
&\leq c\|u_o\|^2\|u_t^n\|^4 + \frac{1}{2}\|\nabla u_t^n\|^2 .
\end{aligned}$$

Here, $\|\cdot\|_p$ denotes the L^p-norm. Using this estimate for its right side, (5) becomes

$$\frac{d}{dt}\|u_t^n\|^2 + \|\nabla u_t^n\|^2 \leq c\|u_o\|^2\|u_t^n\|^4 . \tag{6}$$

To obtain estimates for the Galerkin approximations by integrating (6), one needs a bound for the initial values $\|u_t^n(0)\|$, which is uniform in n. Using (2) one obtains

$$\|u_t^n(0)\| \leq \|\tilde{\Delta} u^n(0)\| + \|u^n(0)\cdot\nabla u^n(0)\| . \tag{7}$$

Here, $\tilde{\Delta} = P\Delta$, where P is the orthogonal projection of $L^2(\Omega)$ onto its subspace $J(\Omega)$, formed by completing the set of solenoidal test functions. A bound for the right side of (7) is found almost trivially if $u_o \in J_1^*(\Omega) \cap W_2^2(\Omega)$, provided the functions $\{a^k\}$ are chosen so that $a^1 = u_o/\|u_o\|$. However, we shall need to choose the $\{a^k\}$ differently, as eigenfunctions of the Stokes operator $\tilde{\Delta}$. In this case, a bound of the form $2\|\tilde{\Delta}u_o\| + c\|\nabla u_o\|^2 + c\|\nabla u_o\|^3$ is obtained for the right side of (7) using the inequality (18), below, and the orthogonality of the eigenfunctions $\{a^k\}$ in the inner-products $(\nabla\phi,\nabla\psi)$ and $(\tilde{\Delta}\phi,\tilde{\Delta}\psi)$. It follows, if $u_o \in J_1^*(\Omega) \cap W_2^2(\Omega)$, that by integrating (6) one obtains estimates of the form

$$\|u_t^n(t)\|, \int_0^t \|\nabla u_t^n\|^2 d\tau, \|\nabla u^n(t)\| \leq G(t), \tag{8}$$

for t in some interval $[0,T)$. Here, the inequality $\|\nabla u^n\|^2 \leq \|u_o\|\cdot\|u_t^n\|$ has been used again, in a final step, to get the estimate for $\|\nabla u^n(t)\|$. The estimates (8) are the ones on which the existence theorem of Kiselev and Ladyzhenskaya is based.

Prodi's estimate for the Galerkin approximations is based on the identity

$$\frac{1}{2}\frac{d}{dt}\|\nabla u^n\|^2 + \|\tilde{\Delta}u^n\|^2 = (u^n\cdot\nabla u^n, \tilde{\Delta}u^n), \tag{9}$$

which holds, simultaneously with (3) and (5), if the basis functions $\{a^k\}$ are taken to be the eigenfunctions of the eigenvalue problem

$$-\Delta a = \lambda a + \nabla p, \quad x \in \Omega \tag{10a}$$

$$\nabla\cdot a = 0, \quad x \in \Omega \tag{10b}$$

$$a\big|_{\partial\Omega} = 0. \tag{10c}$$

It follows from the regularity theory for the Stokes equations, discussed below, that the eigenfunctions a^k belong to $W_2^2(\Omega)$, so that one can write $\tilde{\Delta}a^k = -\lambda_k a^k$, where λ_k is the kth eigenvalue. Thus, multiplying (2) by λ_ℓ and summing $\Sigma_{\ell=1}^n$, one obtains

$$(u_t^n, -\tilde{\Delta}u^n) + (\Delta u^n, \tilde{\Delta}u^n) = (u^n\cdot\nabla u^n, \tilde{\Delta}u^n)$$

and hence (9).

The regularity theory needed above, and again below, consists of L^2-estimates of the general form

$$\|D^2u\|_{\Omega\cap G''} \leq c\|f\|_{\Omega\cap G'} + c\|\nabla u\|_{\Omega\cap G'} + c\|u\|_{\Omega\cap G'} \tag{11}$$

for solutions of Stokes' problem:

$$\Delta u = \nabla p - f, \quad x \in \Omega \tag{12a}$$

$$\nabla\cdot u = 0, \quad x \in \Omega \tag{12b}$$

$$u\big|_{\partial\Omega} = 0. \tag{12c}$$

Here, G'' and G' are bounded open subsets of R^3, with $\overline{G''} \subset G'$, and $\|D^2u\|^2 \equiv \Sigma_{i,j=1}^{3} \|\partial^2 u/\partial x_i \partial x_j\|^2$. For a simple proof of the "interior estimate", i.e., the case $G' \subset \Omega$, see Ladyzhenskaya [9 ,p.38]. For a relatively simple proof of the "estimate up to the boundary", i.e., the case that $G'' \cap \partial\Omega$ is nonempty, see Solonnikov and Ščadilov [17]. For a potential theoretic proof, giving general L^p-estimates, see Cattabriga [1].

To estimate the right side of (9) we shall need several consequences of (11). If Ω is a bounded domain, the global estimate

$$\|D^2 u\| \leq c(\|\tilde{\Delta} u\| + \|\nabla u\|) , \tag{13}$$

for solutions of (12), follows almost immediately from (11), setting $-f = \tilde{\Delta} u$. Using a slightly refined version of (11), we have also proved (13) for unbounded domains, even those with noncompact boundaries; see [4]. Of course, if Ω is unbounded, we require $u(x) \to 0$ as $|x| \to \infty$, in a generalized sense. Using, in addition to (13), the Sobolev inequality $\|\phi\|_3 \leq c\|\nabla\phi\|^{\frac{1}{2}} \cdot \|\phi\|^{\frac{1}{2}}$ for $\phi \in C_0^\infty(R^3)$, and estimates for the Sobolev norms of a function continued beyond its original domain of definition, one can show solutions of (12) satisfy

$$\|\nabla u\|_3 \leq c(\|\tilde{\Delta} u\|^{\frac{1}{2}} \cdot \|\nabla u\|^{\frac{1}{2}} + \|\nabla u\|). \tag{14}$$

This inequality, like (13), is valid for general unbounded domains; see [4].

The right side of (9) can now be estimated using successively Holder's inequality, Sobolev's inequality $\|\phi\|_6 \leq \|\nabla\phi\|$, the inequality (14), and Young's inequality:

$$\begin{aligned} |(u\cdot\nabla u , \tilde{\Delta} u)| &\leq \|u\|_6 \cdot \|\nabla u\|_3 \cdot \|\tilde{\Delta} u\| \\ &\leq c\|\nabla u\| (\|\tilde{\Delta} u\|^{\frac{1}{2}} \cdot \|\nabla u\|^{\frac{1}{2}} + \|\nabla u\|) \|\tilde{\Delta} u\| \\ &\leq c\|\nabla u\|^4 + c'\|\nabla u\|^6 + \tfrac{1}{2}\|\tilde{\Delta} u\|^2 . \end{aligned}$$

Thus we have

$$\frac{d}{dt} \|\nabla u^n\|^2 + \|\tilde{\Delta} u^n\|^2 \leq c\|\nabla u^n\|^4 + c'\|\nabla u^n\|^6 . \tag{15}$$

Also, if $u_0 \in J_1^*(\Omega)$, we have a bound for the initial values,

$$\|\nabla u^n(0)\| \leq \|\nabla u_0\| ,$$

because of the orthogonality of the eigenfunctions $\{a^k\}$ in the inner product $(\nabla\phi, \nabla\psi)$. Hence, (15) can be integrated, yielding estimates of the form

$$\|\nabla u^n(t)\|, \int_0^t \|\tilde{\Delta} u^n\|^2 d\tau \leq F(t) , \tag{16}$$

for t in some interval $[0,T)$.

An estimate for u_t^n is obtained by noting (2) implies

$$\|u_t^n\|^2 = (\tilde{\Delta}u^n , u_t^n) - (u^n\cdot\nabla u^n , u_t^n) , \tag{17}$$

so that, using (14),

$$\begin{aligned} \|u_t^n\| &\leq \|\tilde{\Delta}u^n\| + \|u^n\|_6\cdot\|\nabla u^n\|_3 \\ &\leq \|\tilde{\Delta}u^n\| + c\|\nabla u^n\|\cdot(\|\tilde{\Delta}u^n\|^{\frac{1}{2}}\cdot\|\nabla u_n\|^{\frac{1}{2}} + \|\nabla u^n\|) \\ &\leq 2\|\tilde{\Delta}u^n\| + c\|\nabla u^n\|^3 + c\|\nabla u^n\|^2 , \end{aligned} \tag{18}$$

and hence, using (16),

$$\int_0^t \|u_t^n\|^2 \, d\tau \leq \tilde{F}(t) , \tag{19}$$

for $t \in [0,T)$. The estimates (16) and (19) are the ones on which Prodi's existence theorem in [11] is based. Here, we have derived them in a manner independent of the "size" of either Ω or $\partial\Omega$. These estimates, that is, the functions F and $\tilde{F}$, depend only on $\|\nabla u_o\|$ and the regularity of $\partial\Omega$. In contrast, the estimates (8) depend on $\|u_o\|$, $\|\nabla u_o\|$ and $\|D^2 u_o\|$, but are independent of the regularity of $\partial\Omega$.

To obtain a classical solution, we need estimates of the solution's higher order derivatives. We work first to establish the regularity of u with respect to t, more precisely, to show $u \in C^\infty(0,T ; W_2^2(\Omega))$ by obtaining estimates

$$\|\nabla D_t^k u^n(t)\| , \int_\varepsilon^t \|\tilde{\Delta}D_t^k u^n\|^2 \, d\tau \leq F_k(t,\varepsilon) , \tag{20.k}$$

for $k = 0, 1, 2, \ldots$, and $0 < \varepsilon < t < T$. To this end, we write down three infinite sequences of identities for the Galerkin approximations (for brevity, the superscripts n are omitted):

$$\frac{1}{2}\frac{d}{dt}\|\nabla u\|^2 + \|\tilde{\Delta}u\|^2 = (u\cdot\nabla u , \tilde{\Delta}u) \tag{21.0}$$

$$\frac{1}{2}\frac{d}{dt}\|\nabla u_t\|^2 + \|\tilde{\Delta}u_t\|^2 = (u_t\cdot\nabla u , \tilde{\Delta}u_t) + (u\cdot\nabla u_t , \tilde{\Delta}u_t) \tag{21.1}$$

$$\frac{1}{2}\frac{d}{dt}\|\nabla u_{tt}\|^2 + \|\tilde{\Delta}u_{tt}\|^2 = (u_{tt}\cdot\nabla u , \tilde{\Delta}u_{tt}) + 2(u_t\cdot\nabla u_t\cdot\tilde{\Delta}u_{tt}) + (u\cdot\nabla u_{tt} , \tilde{\Delta}u_{tt}) \tag{21.2}$$

etc.

$$\|u_t\|^2 = (\tilde{\Delta}u , u_t) - (u\cdot\nabla u , u_t) \tag{22.1}$$

$$\|u_{tt}\|^2 = (\tilde{\Delta}u_t , u_{tt}) - (u_t\cdot\nabla u , u_{tt}) - (u\cdot\nabla u_t , u_{tt}) \tag{22.2}$$

$$\|u_{ttt}\|^2 = (\tilde{\Delta}u_{tt} , u_{ttt}) - (u_{tt}\cdot\nabla u , u_{ttt}) - 2(u_t\cdot\nabla u_t , u_{ttt}) - (u\cdot\nabla u_{tt} , u_{ttt}) \tag{22.3}$$

etc.

$$\frac{1}{2}\frac{d}{dt}\|u_t\|^2 + \|\nabla u_t\|^2 = -(u_t\cdot\nabla u , u_t) \tag{23.1}$$

$$\frac{1}{2}\frac{d}{dt}\|u_{tt}\|^2 + \|\nabla u_{tt}\|^2 = -(u_{tt}\cdot\nabla u , u_{tt}) - 2(u_t\cdot\nabla u_t , u_{tt}) \tag{23.2}$$

$$\frac{1}{2}\frac{d}{dt}\|u_{ttt}\|^2 + \|\nabla u_{ttt}\|^2 = -(u_{ttt}\cdot\nabla u , u_{ttt}) - 3(u_{tt}\cdot\nabla u_t , u_{ttt}) - 3(u_t\cdot\nabla u_{tt} , u_{ttt}) \tag{23.3}$$

etc.

Notice, we have already derived the leading identity of each sequence; (21.0) is just (9), (22.1) is (17), and (23.1) is (5). The succeeding identities are derived similarly, first differentiating (2) an appropriate number of times with respect to t.

Our basic plan now is to estimate the right sides of the identities (21.k), $k = 0, 1, 2, \ldots$, and integrate them with respect to t, obtaining the estimates (20.k). We have already done this for $k = 0$. To proceed, one thing which must be done, in order to be able to integrate each identity up to the same right limit T as (21.0), is to estimate the right side of (21.k), $k = 1, 2, 3, \ldots$, in such a way that it becomes a linear differential inequality when the estimates (20.ℓ), $\ell = 0, 1, \ldots, k-1$, are taken into account. This is easily done using the inequalities $\sup_\Omega|u| \le c(\|\tilde{\Delta}u\| + \|\nabla u\|)$, $\|\nabla u\|_3 \le c(\|\tilde{\Delta}u\| + \|\nabla u\|)$ and $\|u\|_6 \le c\|\nabla u\|$, which are derived from (13) and various Sobolev inequalities. For instance, for the right side of (21.1) we have

$$\begin{aligned}|(u_t\cdot\nabla u , \tilde{\Delta}u_t) + (u\cdot\nabla u_t , \tilde{\Delta}u_t)| &\le \|u_t\|_6\cdot\|\nabla u\|_3\cdot\|\tilde{\Delta}u_t\| + \sup_\Omega|u|\cdot\|\nabla u_t\|\cdot\|\tilde{\Delta}u_t\| \\ &\le c(\|\tilde{\Delta}u\|^2 + \|\nabla u\|^2)\|\nabla u_t\|^2 + \tfrac{1}{2}\|\tilde{\Delta}u_t\|^2,\end{aligned}$$

so that (21.1) becomes

$$\frac{d}{dt}\|\nabla u_t\|^2 + \|\tilde{\Delta}u_t\|^2 \le c(\|\tilde{\Delta}u\|^2 + \|\nabla u\|^2)\|\nabla u_t\|^2 . \tag{24}$$

The right sides of (21.k), $k \ge 2$, are estimated similarly; we omit the details.

Finally, we must find estimates, independent of n, for the "initial values" of

$\|\nabla D_t^k u^n\|$. This presents a rather severe difficulty; it is to deal with it that we have introduced the identities (22.k) and (23.k), k = 1, 2, ... The difficulty is that for initial values $u_o \in J_1^*(\Omega)$, or even for $u_o \in D(\Omega)$, one must generally expect, for the solution u, that $\|\nabla u_t(t)\| \to \infty$ as $t \to 0^+$, and hence, for the Galerkin approximations u^n, that $\|\nabla u_t^n(0)\| \to \infty$ as $n \to \infty$. To see this, suppose $u_o \in D(\Omega)$, i.e., $u_o \in C_o^\infty(\Omega)$ and $\nabla \cdot u_o = 0$. Then conditions (1) imply $\Delta p = -\nabla \cdot (u_o \cdot \nabla u_o)$ and $\nabla p|_{\partial\Omega} = 0$, at $t = 0$. This is an overdetermined Neumann problem for the initial pressure. In general, the tangential components of ∇p will not vanish on $\partial\Omega$ at $t = 0$, and yet, initially, $u_t = \nabla p$ in a neighborhood of $\partial\Omega$, if $u_o \in D(\Omega)$. Thus, one can not expect $\lim_{t\to 0^+} u_t(t) \in \overset{\circ}{W}{}_2^1(\Omega)$. The compatibility condition which would be needed is a non-local one, expressible as an integral identity involving the Neumann function, and virtually uncheckable in practice.

Instead of imposing such conditions on the initial velocity, we use (22.k) and (23.k) to obtain estimates for $\|\nabla D_t^k u^n(\varepsilon)\|$, at arbitrarily small values of ε. For this, we only need to assume $u_o \in J_1^*(\Omega)$. The procedure is as follows. First, (21.0) is integrated giving (20.0), that is, (15) is integrated giving (16), which we have already done. Then, (22.1), i.e., (18), is integrated giving (19), which we have already done. Now, from (19), we see that for every $\varepsilon > 0$ and every positive integer n, there exists a number τ_n, $0 < \tau_n < \varepsilon$, such that

$$\|u_t^n(\tau_n)\|^2 \leq \tilde{F}(\varepsilon)/\varepsilon. \tag{25}$$

Also, the right side of (23.1) can be estimated, using part of the derivation of (6), giving

$$\frac{d}{dt}\|u_t^n\|^2 + \|\nabla u_t^n\|^2 \leq c\|\nabla u^n\|^4\|u_t^n\|^2 , \tag{26}$$

which is a linear differential inequality when (20.0) is taken into account. Thus, using (25), we can integrate (26) over the interval $[\tau_n, t]$, obtaining

$$\int_\varepsilon^t \|\nabla u_t^n\|^2 d\tau \leq \int_{\tau_n}^t \|\nabla u_t^n\|^2 d\tau \leq G(t;\varepsilon) , \tag{27}$$

for $\varepsilon < t < T$. Now, from (27), we see that for every $\varepsilon > 0$ and every positive integer n, there exists a number σ_n, $\varepsilon < \sigma_n < 2\varepsilon$, such that

$$\|\nabla u_t^n(\sigma_n)\|^2 \leq G(2\varepsilon;\varepsilon)/\varepsilon . \tag{28}$$

This provides the estimate of the "initial values" needed for integrating (21.1), or rather the corresponding differential inequality (24). Integrating (24) over

$[\sigma_n, t]$, one obtains estimates of the form

$$\|\nabla u^n_t(t)\|^2, \quad \int_{2\varepsilon}^{t} \|\tilde{\Delta} u^n_t\|^2 d\tau \leq F_1(t;2\varepsilon) , \tag{29}$$

for $2\varepsilon < t < T$. This is just (20.1).

We have come full cycle. One can continue by integrating (22.2), using (20.1), to obtain estimates of $\|u^n_{tt}(\rho_n)\|$, with $2\varepsilon < \rho_n < 3\varepsilon$. Then, one can integrate (23.2) to obtain estimates of $\|\nabla u^n_{tt}(\delta_n)\|$, with $3\varepsilon < \delta_n < 4\varepsilon$. And then, one can integrate (21.2) obtaining (20.2), etc., etc.. The full argument is given by induction in [4].

To prove the continuous assumption of the initial values, one last estimate for the Galerkin approximations will be needed. We derive it here assuming $u_o \in W^2_2(\Omega)$, though the condition $u_o \in L^2(\Omega)$ is not necessary; see [4]. Multiplying (2) by λ_ℓ and summing, we obtain

$$(\Delta u^n , \tilde{\Delta} u^n) = (u^n_t , \tilde{\Delta} u^n) + (u^n \cdot \nabla u^n , \tilde{\Delta} u^n) ,$$

and hence, using (14) as in the derivation of (18),

$$\|\tilde{\Delta} u^n\| \leq 2\|u^n_t\| + c\|\nabla u^n\|^3 + c\|\nabla u^n\|^2 . \tag{30}$$

Thus, (8) and (16) imply

$$\|\tilde{\Delta} u^n(t)\| \leq \hat{G}(t) , \tag{31}$$

for $t \in [0,T']$, for some $T' > 0$.

3. Passage from the Approximations to a Classical Solution

Using only the estimates (16) and (19), one can show the Galerkin approximations converge to a generalized solution $u \in L^\infty(0,T' ; J_1^*(\Omega))$ with u_t, $D^2_x u$, $\nabla p \in L^2(0,T';L^2(\Omega))$, for $0 < T' < T$. Here, we only need the convergence of a subsequence of the approximations, which is proved by a compactness argument in [4]; in fact, the whole sequence of approximations is known to converge; see Rautmann [12,13] and Foias [2]. Using (13) and Sobolev's inequality, the estimates (20.k) imply $u \in C^\infty(0,T;W^2_2(\Omega))$. It follows, of course, that $u \in C(\bar{\Omega} \times (0,T))$. In passing to the limit on the basis of estimates (16) and (19), the solution is only shown to satisfy the initial condition in a generalized sense: $u(t) \to u_o$ in $\overset{\circ}{W}{}^1_2(\Omega)$, as $t \to 0^+$. However, the estimate (31) implies $u \in L^\infty(0,T';W^2_2(\Omega))$, and hence that $u(t) \to u_o$

weakly in $W_2^2(\Omega)$. So, from the compactness of the imbedding $W_2^2(\Omega) \subset C(\Omega)$, it follows that $u(\bar{x},t) \to u_o(x)$ continuously as $(\bar{x},t) \to (x,0)$. Thus $u \in C(\bar{\Omega} \times [0,T))$, if $u_o \in W_2^2(\Omega)$.

We noted above, the estimates (20.k) imply $u \in C^\infty(0,T;W_2^2(\Omega))$. To establish further interior regularity with respect to the spatial variables, one observes, for any fixed t, and for $k = 0, 1, 2, \ldots$, that $D_t^k u$ is a solution of (12), with force

$$f_k = -D_t^{k+1}u - \sum_{\beta=0}^{k} c(D_t^{k-\beta}u)\cdot\nabla(D_t^\beta u) .$$

From the known regularity of u, it is clear $f_k \in C^\infty(0,T;W_2^1(\Omega))$. Thus, viewing $D_x^1 D_t^k u$ as a solution of (12a), (12b) with force $D_x^1 f_k \in C^\infty(0,T;L^2(\Omega))$, the interior estimate (11) implies $u \in C^\infty(0,T;W_2^3(G))$, for every $G \subset\subset \Omega$, i.e., for every bounded set G with closure $\bar{G} \subset \Omega$. This, in turn, implies $f_k \in C^\infty(0,T;W_2^2(G))$, for every $G \subset\subset \Omega$. And thus, viewing $D_x^2 D_t^k u$ as a solution of (12a), (12b) with force $D_x^2 f_k \in C^\infty(0,T;L^2(\Omega))$, (11) implies $u \in C^\infty(0,T;W_2^4(G))$, for every $G \subset\subset \Omega$. By induction, one sees $u \in C^\infty(0,T;W_2^\ell(G))$, for every $G \subset\subset \Omega$, and for $\ell = 3, 4, \ldots$. It follows that $u \in C^\infty(\Omega \times (0,T))$.

4. Decay, as $t \to \infty$, in Unbounded Domains

Using Poincaré's inequality, i.e., the inequality $\|\phi\| \leq C_\Omega \|\nabla\phi\|$ for $\phi \in \mathring{W}_2^1(\Omega)$, once can show the Galerkin approximations of section 2 decay exponentially as $t \to \infty$. More precisely, there exists a number T^* dependent on C_Ω and $\|u_o\|$, such that for every $\gamma < C_\Omega^{-2}$, one obtains an estimate of the form

$$\sup_{x\in\Omega} |u^n(x,t)| \leq c(\|\tilde{\Delta} u^n(t)\| + \|\nabla u^n(t)\|) \leq Ce^{-\gamma t} , \tag{32}$$

for $t \geq T^*$. From this follows the global existence and exponential decay of the classical solution constructed in section 3, provided $T^* < T$, with T as in either (8) or (16). One can show $T^* < T$ under various hypotheses, for instance, if $\|u_o\|_{W_2^1(\Omega)}$ and C_Ω are sufficiently small. In general, if $T^* > T$, one still has the global existence of Hopf's generalized solution, and its classical regularity can be proved for $t \in (0,T) \cup (T^*,\infty)$. During the interval $[T,T^*]$, it is classical, except perhaps, for values of t belonging to a set of t-measure zero, whose complement consists of intervals. These results for bounded domains are rather standard, see [4 ,14]. Instead of giving the details, we will describe some analogous results for unbounded domains.

All the estimates given in section 2 are independent of the size of Ω and $\partial\Omega$, though some of them depend on the C^3-regularity of $\partial\Omega$. This makes possible the construction of a solution of problem (1) in any three-dimensional domain with uniformly C^3 boundary, by considering an expanding sequence of subdomains. All the estimates of section 2 remain valid for the eventual solution. Also, assuming $u_o \in J_1(\Omega)$, the solution which is constructed belongs to $J_1(\Omega)$, for almost every t.

Of course, the estimate (32) generally fails in unbounded domains. Still, an explicit estimate for the solution's rate of decay can be obtained from the energy estimate (4) and the differential inequality (15), i.e., from

$$\int_o^\infty \|\nabla u^n\|^2 dt \leq \frac{1}{2}\|u_o\|^2 \equiv E_o \tag{33}$$

and

$$\frac{d}{dt}\|\nabla u^n\|^2 \leq c\|\nabla u^n\|^4 + c'\|\nabla u^n\|^6 . \tag{34}$$

If $\|\nabla u^n(t)\|$ were known to be monotonically decreasing, (33) would clearly imply $\|\nabla u^n(t)\|^2 \leq E_o t^{-1}$. In fact, (34) implies such a slow rate of growth of $\|\nabla u^n\|$, when $\|\nabla u^n\|$ is small, that one gets a similar result for large t, namely

$$\|\nabla u^n(t)\|^2 \leq H(t) \leq \left(\frac{\exp(cE+1) - 1}{c + E^{-1}}\right) t^{-1} \tag{35}$$

for $t \geq c'E^2 \exp(cE+1) \geq T^*$. Here c and c' are the same as in (34). This estimate is proved by comparison with solutions of the differential equation $\phi' = \alpha\phi^2$, which are of the form $\phi = \alpha^{-1}(t_o - t)^{-1}$. It is easily checked that if such a function ϕ is defined for $t \in [0,\tau]$ and satisfies $\int_o^\tau \phi\, dt < E$, then $\phi(\tau) < (\exp \alpha E - 1)/\alpha\tau$. The more complicated form of the estimate (35) is due to the presence of the term $c'\|\nabla u^n\|^6$ in (34); the details are given in [4]. Once (35) is proven, (15) can be integrated to give

$$\int_t^\infty \|\tilde{\Delta} u^n\|^2 d\tau \leq C\, t^{-1} ,$$

for $t \geq T^*$. Then, integration of (18) gives

$$\int_t^\infty \|u_t^n\|^2 d\tau \leq C\, t^{-1} ,$$

for $t \geq T^*$. Since (6) is a differential inequality of the form $\phi' \leq \alpha\phi^2$, this implies

$$\|u_t^n(t)\|^2 \leq C_\delta t^{-1} , \tag{36}$$

for $t > T^* + \delta$, for any $\delta > 0$. The estimates (35) and (36), together with (30), imply

$$\|\tilde{\Delta} u^n(t)\|^2 \leq C_\delta t^{-1} , \tag{37}$$

for $t > T^* + \delta$. Finally, (35) and (37) imply

$$\sup_{x\in\Omega} |u(x,t)| \leq c(\|\tilde{\Delta} u^n\| + \|\nabla u^n\|) \leq c_\delta t^{-\frac{1}{2}} , \tag{38}$$

for $t \geq T^* + \delta$. This estimate is a variant of one proved by Masuda [10] for exterior domains. Masuda's estimate is based essentially on (6) rather than (15); while it gives a slower rate of decay, it remains valid in the case of nonhomogeneous boundary values.

The relation between the estimates (16) and (35) is shown in Figure 1. For every n, $\|\nabla u^n(t)\|^2$ would be represented by a smooth curve defined for all $t \geq 0$ and bounded by the graphs of $F(t)$ and $H(t)$. T depends on $\|\nabla u_o\|$, and T^* on $\|u_o\|$.

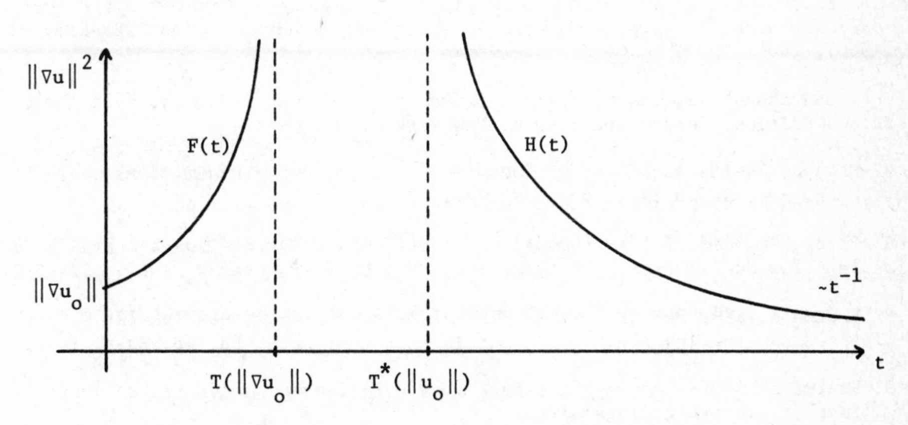

Figure 1. Estimates (16) and (35) for the Galerkin approximations.

If $T \leq T^*$, we lack a bound for $\|\nabla u^n(t)\|$, uniform in n, on the interval $[T,T^*]$; so, during this time interval, the regularity of solutions constructed from the Galerkin approximations may break down. If $T^* < T$, there is only one solution and it is regular in the classical sense for all $t \geq 0$. It is shown in [4] that $T^* < T$, if, for some number β,

$$\frac{1}{2}\|u_o\|^2 \leq \frac{\log(\beta/\|\nabla u_o\|^2)}{c + c'\beta} .$$

REFERENCES

1. L. Cattabriga, *Su un problema al contorno relativo al sistema di equazioni di Stokes,* Rend. Sem. Mat. Univ. Padova 31 (1961), 308-340.

2. C. Foias, *Statistical study of the Navier-Stokes equations I,* Rend. Sem. Math. Un. Padova 48 (1973), 219-348.

3. H. Fujita and T. Kato, *On the Navier-Stokes initial value problem, I,* Arch. Rational Mech. Anal. 16 (1964), 269-315.

4. J.G. Heywood, *The Navier-Stokes equations: on the existence, regularity and decay of solutions,* (to appear) Preprint, June 1978.

5. E. Hopf, *Über die Anfangswertaufgabe für die hydrodynamischen Grundgleichungen,* Math. Nachr. 4 (1951), 213-231.

6. S. Ito, *The existence and the uniqueness of regular solution of nonstationary Navier-Stokes equation,* J. Fac. Sci. Univ. Tokyo Sect. I A, 9 (1961), 103-140.

7. A.A. Kiselev and O.A. Ladyzhenskaya, *On the existence and uniqueness of the solution of the nonstationary problem for a viscous incompressible fluid,* Izv. Akad. Nauk SSSR Ser. Mat. 21 (1957), 655-680.

8. O.A. Ladyzhenskaya, *On the classicality of generalized solutions of the general nonlinear nonstationary Navier-Stokes equations,* Trudy Mat. Inst. Steklov 92 (1966), 100-115.

9. O.A. Ladyzhenskaya, *The mathematical theory of viscous incompressible flow,* Second Edition, Gordon and Breach, New York, 1969.

10. K. Masuda, *On the stability of incompressible viscous fluid motions past objects,* J. Math. Soc. Japan 27 (1975), 294-327.

11. G. Prodi, *Teoremi di tipo locale per il sistema di Navier-Stokes e stabilità delle soluzione stazionarie,* Rend. Sem. Mat. Univ. Padova 32 (1962), 374-397.

12. R. Rautmann, *Eine Fehlerschranke für Galerkinapproximationen lokaler Navier-Stokes-Lösungen,* in: Int. Schriftenreihe zur num. Math. Bd. 48, Basel 1979.

13. R. Rautmann, *On the convergence-rate of nonstationary Navier-Stokes approximations* (these proceedings).

14. M. Shinbrot and S. Kaniel, *The initial value problem for the Navier-Stokes equations,* Arch. Rational Mech. Anal. 21 (1966), 270-285.

15. V.A. Solonnikov, *Estimates of solutions of nonstationary linearized systems of Navier-Stokes equations,* Trudy Mat. Inst. Steklov 70 (1964), 213-317, Amer. Math. Soc. Transl. 75 (1968), 1-116.

16. V.A. Solonnikov, *On differential properties of the solutions of the first boundary-value problem for nonstationary systems of Navier-Stokes equations,* Trudy Mat. Inst. Steklov 73 (1964), 221-291.

17. V.A. Solonnikov and V.E. Ščadilov, *On a boundary value problem for a stationary system of Navier-Stokes equations,* Trudy Mat. Inst. Steklov 125 (1973), 196-210; Proc. Seklov Inst. Math. 125 (1973), 186-199.

DIRECT AND REPEATED BIFURCATION INTO TURBULENCE

Daniel D. Joseph
Department of Aerospace Engineering and Mechanics
University of Minnesota
Minneapolis, MN 55455

Lecture Given at the IUTAM Symposium on
Approximation Methods for the Navier-Stokes Equations

This lecture is a review of the applications of the theory of bifurcation to the problem of transition to turbulence. Most of the material in this lecture can be found in detail in my recent review [11], in other reviews in the same volume and in the monographs [12]. We shall discuss some new results having to do with frequency-locked solutions and bifurcation into higher dimensional tori in the transition to turbulence which were not discussed in [11] and [12]. Some of these results are derived in the new book on bifurcation theory by Iooss and Joseph [9]. To keep the lecture and this written report of it discursive, I am not going to do much citing and attributing of old results; complete citations for the older work can be found in [11] and [12].

I will confine my remarks to a discussion of the bifurcation of solutions of the Navier-Stokes equations for an incompressible fluid when the velocity $\underline{V}_B(\underline{x},t)$ of the boundary B of the region $\mathcal{V}$ occupied by the fluid is prescribed together with field forces $\underline{G}(\underline{x},t)$:

$$\left.\begin{aligned}&\frac{\partial \underline{V}}{\partial t} + \underline{V}\cdot\nabla\underline{V} = -\nabla p + \frac{1}{R}\nabla^2\underline{V} + \underline{G}(\underline{x},t),\\ &\qquad \operatorname{div}\underline{V} = 0\ ,\end{aligned}\right\}\ \underline{x}\in\mathcal{V}$$

$$\underline{V} = \underline{V}_B(\underline{x},t),\quad \underline{x}\in B\ .$$

We call the prescribed values $\underline{V}_B(\underline{x},t)$ and $\underline{G}(\underline{x},t)$, the <u>data</u>. R is the Reynolds number, a dimensionless parameter composed of the product of a velocity times a length divided by the kinematic viscosity. We can think of it as a dimensionless speed.

The motion of the fluid must ultimately be determined by the data. When the Reynolds number is small the motion is uniquely determined by the <u>data</u>. The meaning of this is as follows: given an initial condition

$$\underline{V}_0(\underline{x}) = \underline{V}(\underline{x},0),\ \underline{x}\in\mathcal{V} \tag{1.2}$$

we may suppose that the initial-boundary-value problem (1.1) and (1.2) have a unique solution. When R is small, each of these different solutions belonging to different $\underline{V}_0$, tend to a single one determined ultimately by $\underline{G}$ and $\underline{V}_B$ and not by $\underline{V}_0$. So when R small we ultimately get solutions which reproduce the symmetries of the data. Steady data gives rise to steady solutions, periodic data to periodic solutions.

When R is large solutions are not uniquely determined by the data. The relation between the data and solutions is subtle and elusive.

Let us consider what happens when the data is steady as we increase the Reynolds number. For technical reasons we suppose now and hereafter that V is a bounded domain, or it can be made bounded by devices such as restricting solutions to spatially periodic ones which can be confined to a period cell. We suppose $U(R)$ is a steady solution which is the continuation of the unique steady solution which exists when R is small. When the equation satisfied by this solution is subtracted from Navier-Stokes equation, we get equations for the disturbance $\underline{u}$ of $U(R)$ which has zero data and nonzero initial conditions

$$(1.3)\qquad \begin{aligned} &\frac{\partial \underline{u}}{\partial t} + \underline{U}\cdot\nabla\underline{u} + \underline{u}\cdot\nabla\underline{U} + \underline{u}\cdot\nabla\underline{u} = -\nabla p + \frac{1}{R}\nabla^2\underline{u} \quad \text{in } V \\ &\operatorname{div} \underline{u} = 0 \\ &\underline{u} = 0, \ \underline{x} \in B \end{aligned}$$

$$(1.4)\qquad \underline{u}(\underline{x},0) \neq 0, \ \underline{x} \in V$$

If the null solution $\underline{u} = 0$ of (1.3) is stable, then $\underline{U}(R)$ is stable. It is stable when R is small. We want to catalogue what can happen when $\underline{u} = 0$ loses stability as R is increased.

For simplicity we first write (1.3) as an evolution equation in some space, say a Banach space

$$(1.5)\qquad \frac{d\underline{u}}{dt} = \underline{F}(R,\underline{u}), \quad \underline{F}(R,0) = 0 \ .$$

It does no harm to think of (1.5) as a system of ordinary differential equations in $\mathbb{R}^n$.

To study the stability of $\underline{u} = 0$ we linearize (1.5) and introduce exponential solutions in order to derive the associated spectral problem:

$$\frac{d\underline{v}}{dt} = \underline{F}_u(R|\underline{v}) \ ,$$

$$\underline{v} = e^{\sigma t}\underline{\zeta} \ ,$$

$$\sigma = \xi(R) + i\eta(R) \in \textstyle\sum \underline{F}_u(R|\cdot) \ ,$$

$$\sigma\underline{\zeta} = \underline{F}_u(R|\underline{\zeta})$$

where $\sum \underline{F}_u$ means the spectrum of $\underline{F}_u$. If σ is in the spectrum so is $\bar{\sigma}$. When V is a bounded domain the spectrum of $\underline{F}_u$ is all of eigenvalues and when R is small all of the eigenvalues are bounded by a parabola on the left hand side of the complex σ plane. As R is increased past its first critical value some eigenvalues cross into the right side of the complex σ plane. In the usual case a single eigenvalue or a complex conjugate pair of eigenvalues cross over.

We state the foregoing conditions, which are sufficient for bifurcation, in a precise mathematical sense as follows. $R = R_c$ is the first critical value of R such that $\xi(R) < 0$ for all eigenvalues belonging to $\underline{F}_u(R|\cdot)$ when $R < R_c$, $\xi(R_c) = 0$, $\sigma(R_c) = i\omega_0$ (where $\omega_0 = \eta(R_c)$) is an algebraically simple eigenvalue of $\underline{F}_u(R_c|\cdot)$ and the loss of stability of $\underline{u} = 0$ at R_c is strict; that is,

$$\xi'(R_c) > 0 \ .$$

Given the assumptions made in the last paragraph there are two possibilities:

(I) $\omega_0 = 0$ and one real eigenvalue crosses at critically. A steady solution which breaks the spatial symmetry of the data, bifurcates.

It is usually enough to consider three possible types of bifurcation into steady solutions (see Figure 1). Transcritical bifurcation occurs when the projection of the quadratic part of the nonlinear terms into the null space of $F_u(R_c|\cdot)$ is nonvanishing. When this projection does vanish bifurcation is controlled by cubic terms. When these terms don't vanish there are two possibilities: bifurcation to the right (supercritical) and bifurcation to the left (subcritical). Solutions which bifurcate supercritically are stable; subcritical solutions are unstable

Figure 1: Bifurcation of Steady Solutions. Dotted lines indicate unstable solution.

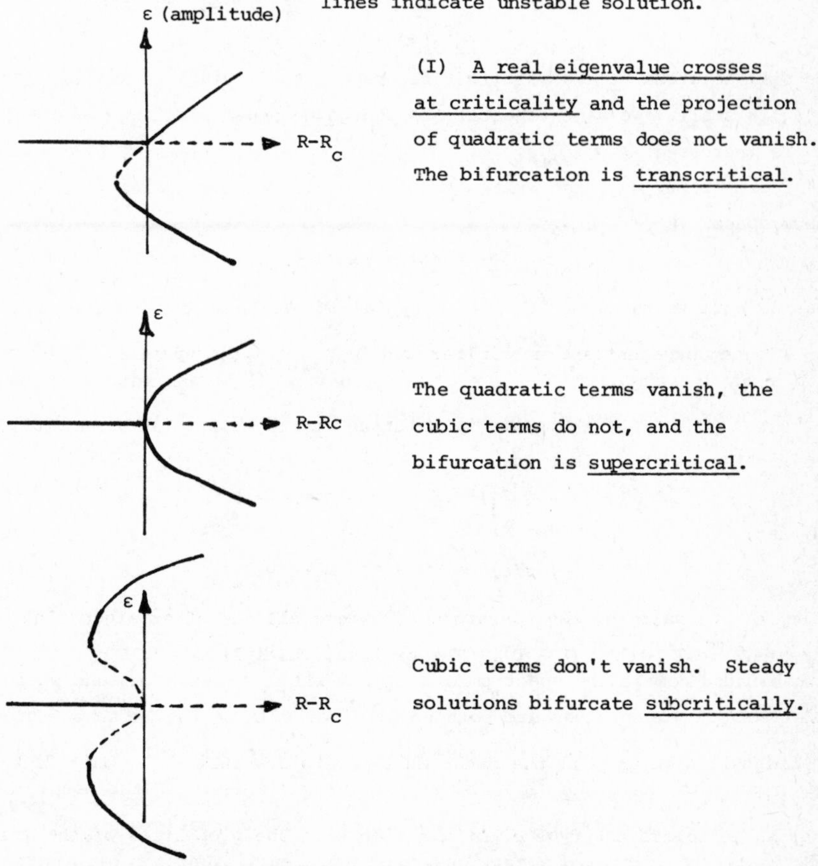

(II) A complex pair crosses. The quadratic projection vanishes automatically and we never get the transcritical case

$$R(\varepsilon) = R(-\varepsilon)$$
$$\omega(\varepsilon) = \omega(-\varepsilon)$$

As R is increased, new steady solutions, with different patterns of symmetry, may bifurcate. After some number of these steady bifurcations a periodic solution will typically bifurcate.

Now we ask what happens when a periodic solution bifurcates? Suppose we have a stable periodic solution with velocity given by

$$\underline{V}(\omega(\varepsilon)(t + \delta_1), \varepsilon)$$

where ε is the amplitude, $\omega(\varepsilon) = \omega(-\varepsilon)$ and δ_1 is an arbitrary phase which may be set to zero by a suitable choice of the origin of time. A small disturbance $\underline{q}$ of $\underline{V}$ satisfies the linearized equation

$$\frac{d\underline{q}}{dt} = \underline{F}_u(R(\varepsilon), \underline{V}(\omega(\varepsilon)t, \varepsilon)|\underline{q}) .$$

We can derive a spectral problem for $\underline{F}_u(R(\varepsilon), \underline{V}(\omega(\varepsilon)t, \varepsilon)|\cdot)$ by the method of Floquet.

$$q = e^{\sigma t}\zeta(t), \quad \underline{\zeta} \in \mathbb{P}_{\frac{2\pi}{\omega}}$$

$$\sigma\zeta = -\frac{d\zeta}{dt} + \underline{F}_u(R, \underline{V}|\underline{\zeta}) \stackrel{\text{def}}{=} \Pi\zeta$$

$$\text{dom } \Pi = \mathbb{P}_{\frac{2\pi}{\omega}}$$

where

$$\sigma = \xi(\varepsilon) + i\Omega(\varepsilon) \in \textstyle\sum\Pi$$

and

$$\lambda = e^{2\pi\sigma/\omega}$$

are the Floquet exponent and multiplier and $\mathbb{P}_{2\pi/\omega}$ is the space of $2\pi/\omega$ periodic functions.

We now suppose that the periodic solution $\underline{u}$ loses stability strictly when $\varepsilon = \varepsilon_1$, $R = R(\varepsilon_1) = R_1$; that is,

$$\xi(\varepsilon_1) = \xi_1 = 0, \quad \xi'(\varepsilon_1) > 0$$

and further,

$$\sigma(\varepsilon_1) = \sigma_1 = i\Omega_1, \Omega_1 = \Omega(\varepsilon_1)$$

is a simple eigenvalue of the operator Π, where all the other eigenvalues of Π have negative real parts $(\xi < 0)$. The critical multipliers

$$\lambda_1 = e^{2\pi i\Omega_1/\omega(\varepsilon_1)}$$

are on the unit disk and all the other multipliers are inside the unit disk (see Figure 2).

We can correlate the type of bifurcation with the properties of the multipliers which pass out of the unit disk at criticality. These properties are determined by the values of the frequency ratio Ω_1/ω_1. There are two possibilities: Ω_1/ω_1 is irrational or Ω_1/ω_1 is rational. We get all the rational points on the Floquet circle shown in Figure 3 if we take

$$\frac{\Omega_1}{\omega_1} = \frac{m}{n}, \quad 0 \leq \frac{m}{n} < 1 .$$

The rational points are called points of resonance and the irrational points are called quasiperiodic points. The resonant points are roots of unity, $\lambda_1^n = 1$. We further divide the resonant points into

(i) <u>points of strong resonance</u>: $n = 1,2,3,4$ and,

(ii) <u>points of weak resonance</u>: $n > 4$

Under the assumptions we have made we get bifurcation of periodic solutions. At points of strong resonance a subharmonic solution (a periodic solution) with a new period $2\pi/\Omega(\varepsilon)$. $(\Omega(\varepsilon_1) = \Omega_1)$ approximately $n(n = 1,2,3,4)$ times the old

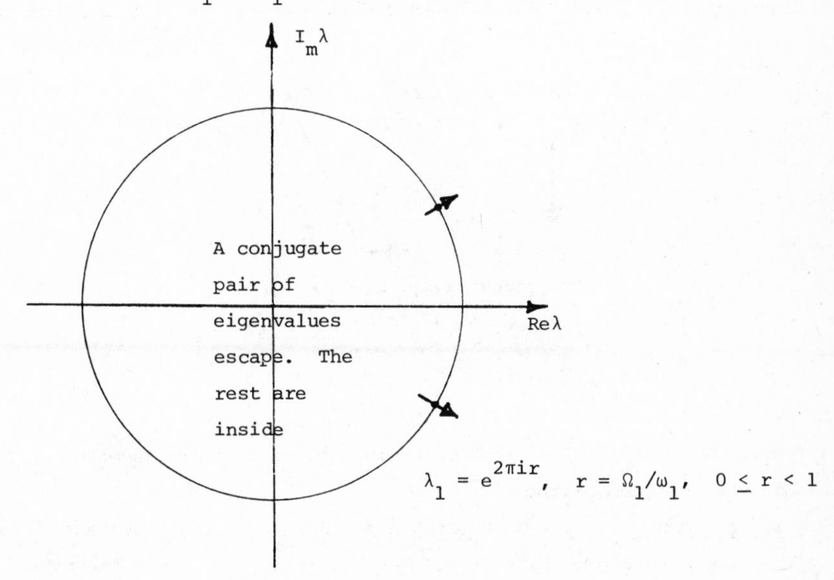

Figure 2: Floquet circle at criticality. r is irrational at quasiperiodic points. The resonant points are fractions

$$r = \frac{m}{n} < 1, \ \lambda_1^n = 1$$

At points of strong resonance $(n = 1,2,3,4)$ we get subharmonic solutions with periods nearly n times $2\pi/\omega$. In the other cases a torus of asymptotically doubly periodic flows bifurcate.

one, bifurcates when ε is near to ε_1. If the period $2\pi/\omega$ of the old solution is independent of the amplitude as in the case of periodic forcing, the period of the new solution is also independent of the amplitude and the ratio Ω/ω of frequencies is exactly m/n. For ease of description we will confine our remarks immediately below to the forced periodic case. The results for bifurcation of periodic solutions in autonomous problems are slightly different [9].

At quasiperiodic points we get bifurcation into asymptotically doubly-periodic solutions on a torus. The asymptotic expressions have two frequencies $\omega = 2\pi/T$ and $\Omega(\varepsilon)$ and $\Omega(\varepsilon)$ varies continuously, so that for almost all ε in the range of these solutions the doubly periodic solution is quasiperiodic, with two frequencies, whilst for a dense set of rational points the two periods of the doubly periodic solution fit into a common period (see [9]).

Subharmonic solutions can bifurcate at points of weak resonance ($n \geq 5$) only when exceptional conditions hold. In the usual case a torus of asymptotically quasi-periodic solutions bifurcates, even at points of weak resonance.

In Figures 3 and 4 we summarize the various possibilities for bifurcation which have been discussed so far.

Figure 3: Bifurcation of steady solutions into periodic ones and the bifurcation of periodic solutions

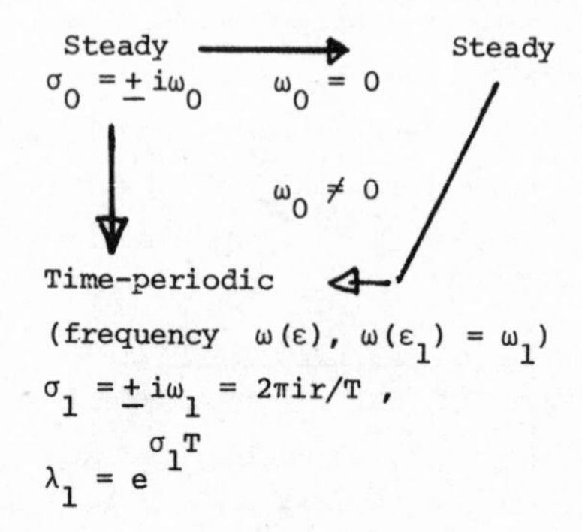

r is irrational. A torus T^2 bifurcates. Solutions on T^2 are asymptotically doubly-periodic with two frequencies.
r is rational, $r = m/n$, $\lambda_1^n = 1$. Case (i): $n = 1,2,3,4$, strong resonance. Subharmonic, nT-periodic solutions bifurcate. Case (ii): $n \geq 5$, weak resonance. A torus T^2 of asymptotically doubly periodic bifurcates. Case (iii): $n \geq 5$, weak resonance. Subharmonic solutions on T^2 will bifurcate if certain coefficients vanish.

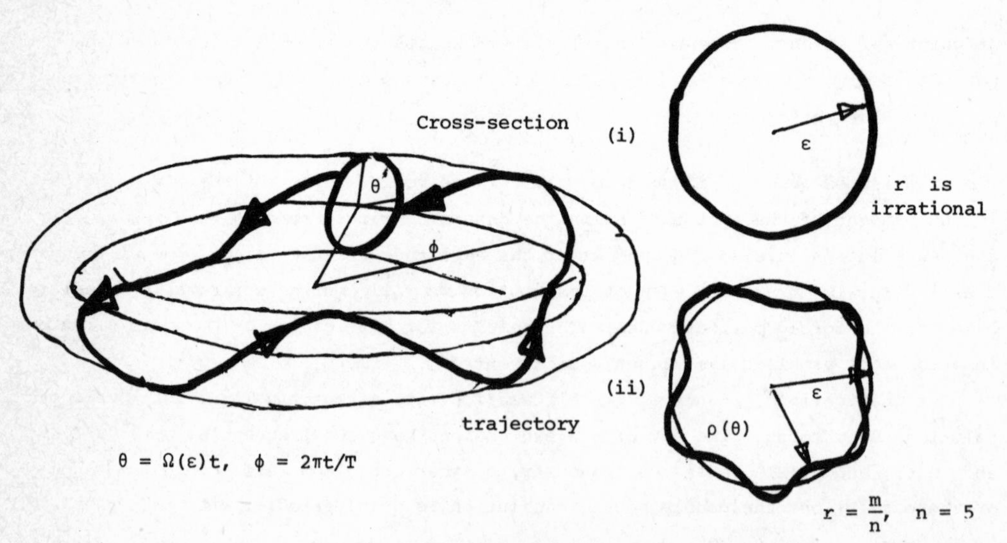

Figure 4: The bifurcating torus [9]. The amplitude ε is the mean radius of the torus. The cross-section is circular when the Floquet multipliers escape at quasi-periodic points. At points of weak resonance $(n \geq 5)$ a torus, with n lobes bifurcates unless special conditions hold. Whenever there is a torus, the solutions on it are asymptotically doubly-periodic. When the special conditions hold for $n \geq 5$, subharmonic, nT-periodic solutions on the torus will bifurcate. When the Floquet multiplier crosses at a point of strong resonance, $n = 1,2,3,4$ an nT periodic solution, not on a torus, will bifurcate.

The circumstances under which the solutions on T^2 are exactly and not just asymptotically quasiperiodic are presently unknown. In our asymptotic result, and in the experiments, there are two frequencies $\omega(\varepsilon)$ and $\Omega(\varepsilon)$ which appear to vary smoothly with the amplitude ε, when ε is not too large, and the asymptotic solutions are doubly-periodic, of the form $f(\omega t, \Omega t)$, where f is 2π-periodic in both arguments. Higher dimensional tori T^n may be associated with n frequencies in the same way.

A good way to determine the properties of solutions in experiments to get the Fourier transform of measured data, say the power spectrum of the fluctuating values of velocity at a point, see Figure 5. A periodic solution T^1 shows sharp peaks in the power spectrum and these peaks are harmonics of one frequency ω. The power spectrum of a doubly periodic solution also has sharp peaks and all of them may be matched to linear combinations of two frequencies. And for multiperiodic solutions the same thing goes for n frequencies.

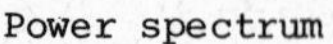

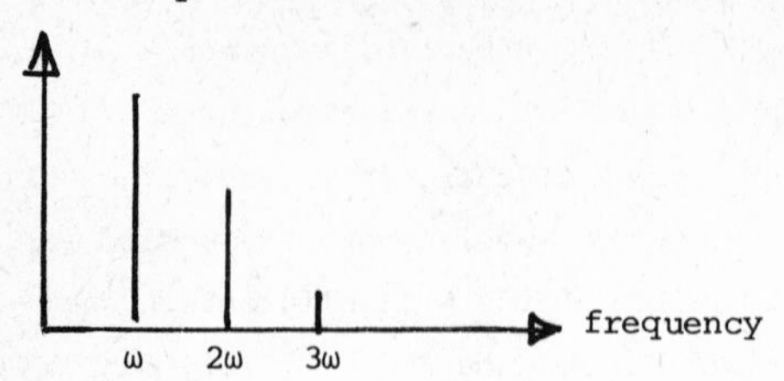

T^1: There is one frequency and harmonics.

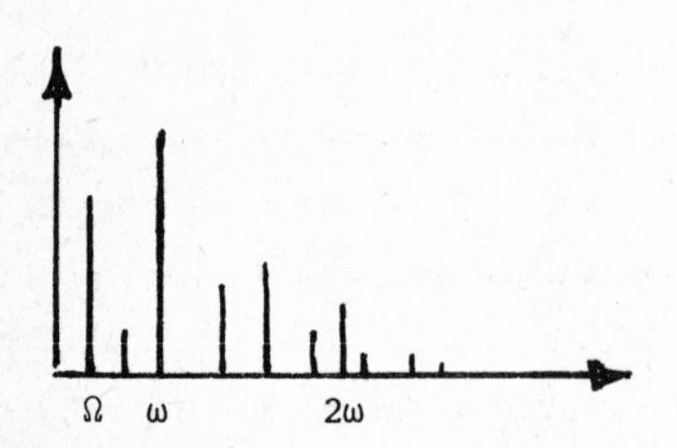

T^2: All spectral lines are of the form $2\left(\frac{n}{\Omega}+\frac{m}{\omega}\right)$. If Ω/ω is irrational the solution is quasiperiodic.

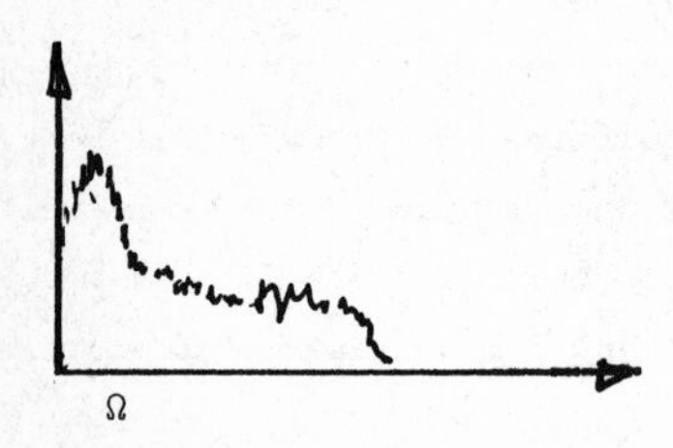

Nonperiodic (strange) attractor (dynamical noise) centered at Ω.

Figure 5: Power spectrum for periodic, doubly periodic and nonperiodic solution.

As long as there is a periodic solution we can study its stability by Floquet theory. But the study of the stability and bifurcation of quasiperiodic solutions on a two-dimensional torus as R is increased is more complicated.

Three types of changes of T^2 are observed in experiments when the Reynolds number is increased.

(1) The solutions lock frequencies. A locked solution is a periodic solution on the torus in which the ratio of frequencies is rational. Locked in solutions are subharmonic; the time taken by N times one cycle is the same as M times the other cycle

$$\tau = \frac{2\pi N}{\omega} = \frac{2\pi M}{\omega}$$

The locked in solutions appear to be related to those which bifurcate at points of weak resonance. A <u>locked in</u> solution has the property that the ratio of frequencies on T^2 remains constant even when R varies.

(2) T^2 bifurcates directly into nonperiodic attractor. This is not well understood theoretically.

(3) T^2 bifurcates into T^3 (Gollub and Benson, unpublished). The bifurcation of T^2 into T^3 (and T^n into T^{n+1}) was discussed by Landau and Hopf. Mathematical conditions which are sufficient to guarantee bifurcation of the torus T^n into the torus T^{n+1} have recently been given by Chenciner and Iooss [3], Haken [8] and Sell [18]. If the fluid systems satisfied the conditions set out by these authors we would get turbulence of the type proposed by Landau and Hopf.

According to Landau and Hopf we get turbulence by adding new frequencies through bifurcation as the Reynolds number is increased. With each frequency we have an associated arbitrary phase so the motion looks chaotic. In the triply periodic case we have a velocity at a point in the form

$$\underline{u}(t,\varepsilon) = \hat{\underline{u}}(\omega_1(\varepsilon)(t - \delta_1), \omega_2(\varepsilon)(t - \delta_2), \omega_3(\varepsilon)(t - \delta_3), \varepsilon)$$

where the ω_i are the frequencies and δ_i the phases. Turbulence then is always quasiperiodic with a finite number of discrete frequenceies.

Real turbulence is phase mixing, quasiperiodic turbulence is not phase mixing. If $\underline{u}(t)$ is a fluctuation with mean value zero and is almost periodic, then

$$\underline{u}(t) \sim \sum_{-\infty}^{\infty} \underline{u}_n e^{-i\lambda_n t}, \quad \lambda_n \neq 0 .$$

The autocorrelation for this is

$$g(\tau) = \lim_{T\to\infty} \frac{1}{T} \int_0^T \underline{u}(t + \tau)\underline{u}(t)\,dt = \sum_{-\infty}^{\infty} |\underline{u}_n|^2 e^{-i\lambda_n \tau}$$

and $g(\tau)$ does not vanish for solutions of the Landau-Hopf type, as it must for true turbulence. In true turbulence events at distant times are presumably uncorrelated. In some experiments [20] a noisy part of spectrum coexists with a peaked part. In these cases the autocorrelation function will decay as the noisy part of power spectrum grows larger, but it will not decay to zero.

Lorenz [14] and Ruelle-Takens [17] suggested that turbulence could occur after a finite number of bifurcations. Then there would be an attracting set of lower dimensionality in phase space in which solutions are:

(1) Sensitive to initial conditions. Two velocity fields which are initially close evolve into very different fields.

(2) Mixing, with a decaying autocorrelation function.

(3) Noisy with broad band components as well as sharp peaks in the spectrum.

Attracting sets of this type are sometimes called nonperiodic or strange.

Experiments favor Lorenz-Ruelle-Takens rather than Landau-Hopf. But it seems like there is no universal sequence of bifurcation into turbulence. We turn next to experiments.

In comparing bifurcation theory to experiments it is necessary to remember that bifurcation results are local and therefore do not cover all the possibilities in experiments. To make this point more strongly we consider the equilibrium solutions $F(R,\varepsilon) = 0$ of the evolution equation $\overset{\circ}{u} = F(R,u)$ in $\mathbb{R}$. We may imagine

$$F(R,u) = uF_1(R,u)F_2(R,u)\cdots F_n(R,u)$$

Each vanishing factor $F_\ell(R,\varepsilon) = 0$ gives a different solution as in Figure 6. Only the intersecting ones could in principle be studied by local bifurcation theory. Isolated solutions would escape analysis.

In experiments we frequently see the early sequence of bifurcating solutions predicted by bifurcation theory:

(1) Steady solutions bifurcate into time-periodic ones.

(2) Time-periodic ones bifurcate into subharmonic ones.

(3) Time-periodic ones bifurcate into doubly-periodic ones. But we also see the bifurcation of

(4) Steady solutions into nonperiodic ones as in the examples given below. This property holds for solutions of the Lorenz [14] equations as the Rayleigh number is increased.

(5) Doubly periodic ones (T^2) into nonperiodic ones [1].

(6) Doubly periodic ones into triply periodic ones (T^3) (Gollub and Benson, unpublished)

(7) Frequency locking followed by bifurcation into nonperiodic solutions [5], [6], [7], [13].

(8) Frequency locking followed by a cascade of repeated bifurcation of periodic doubling solution into turbulence, [13].

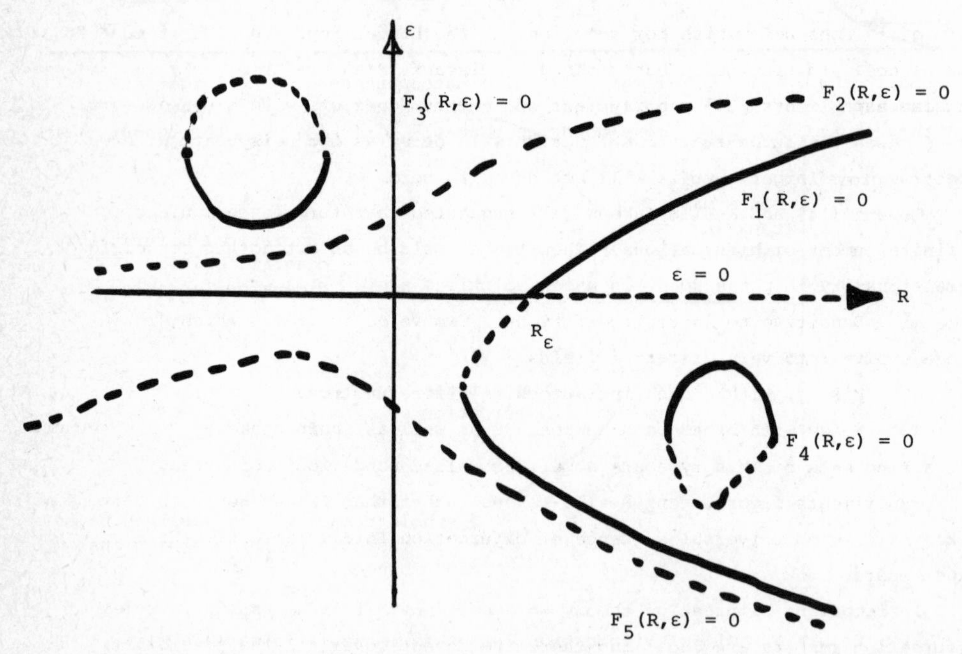

Figure 6: Bifurcation and stability of steady solutions (R,ε) of $\overset{\circ}{u} = uF_1F_2F_3F_4F_5$

In convection experiments in boxes the sequence of observed bifurcating states is very dependent on the aspect ratio of the box and on the spatial form of the convection [15]. The observations are so varied and so recent that it is not possible or desirable to systematize them. Instead we can report on some interesting cases.

We start first by giving an account of some direct transitions to turbulence. Consider flow induced by a pressured drop ΔP down a plane channel. When ΔP is small the flow is laminar, and the velocity is unidirectional and varies across the cross-section like a parabola. At larger pressure drops there are alternate patches of laminar and turbulent flow, at still larger ΔP the flow is turbulent throughout. The bifurcation diagram of Figure 7 is for an idealized two-dimensional problem in which disturbances are assumed to be spatially periodic. The bifurcation is subcritical. On general theoretical grounds we expect the bifurcation diagram to recover stability when the amplitude is large (see "Factorization theorems" in [9]). In fact Orszag and Kelm [16] calculate a curve like shown in Figure 7. They integrate the initial-value problem by brute force, using interesting spectral methods. They show that the large amplitude branch which is stable for two dimensional disturbances is unstable to three dimensional disturbances. Their numerical results are in agreement with experimental observations.

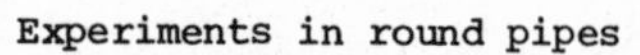

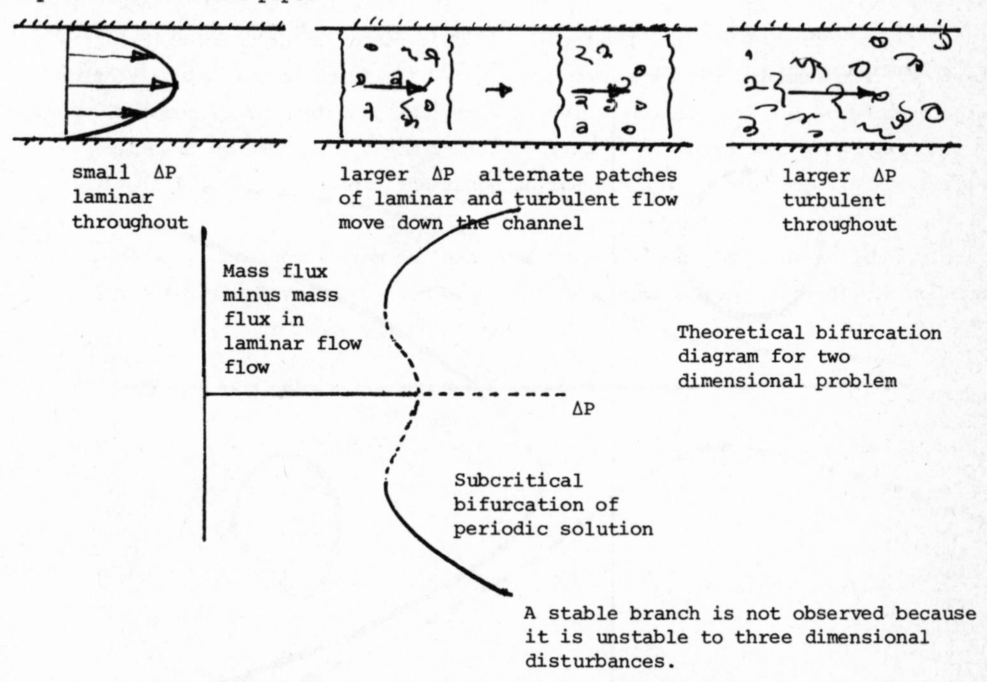

Figure 7: Direct bifurcation into turbulence in Poiseuille flow.

The same type of direct transition to turbulence occurs in Couette flow with the inner cylinder at rest and the outer in steady rotation with angular velocity Ω

Figure 8: Spiral turbulence

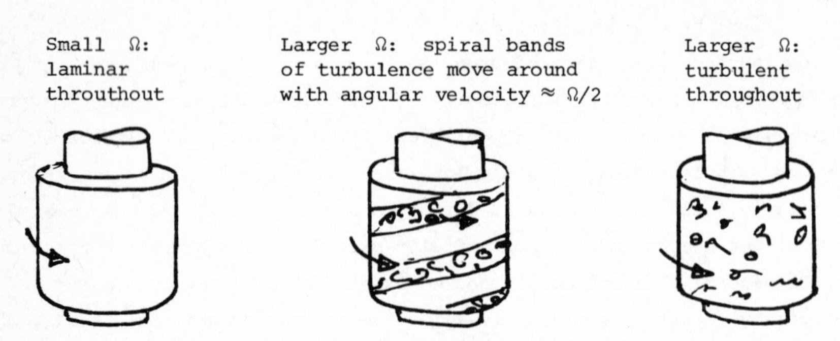

We turn next to experiments which exhibit repeated supercritical branching leading to turbulence after a finite number of bifurcations. The experiments which seem most interesting are in relatively small, enclosed volumes of fluid in which the eigenvalues in the spectrum of the linearized operator are widely separated. This separation seems to be associated with the fact that the dynamics of these fluid systems behave very nearly as if they were governed by ODE's in $\mathbb{R}^n$ with small values of n. This feature is of very great interest because it suggests that some features of turbulence are governed by a small number of ordinary differential equations. For example, the complicated sequence of bifurcations in a box of fluid heated from below [6] are well simulated in numerical solutions of 14 coupled ODE's [4] which arise by truncation into spatial Fourier modes with unknown time-dependent coefficients of solutions of the convection equations used by Lorenz [14].

The first experiments to report frequency data for different bifurcations leading to turbulence were done by Swinney-Gollub and the most recent and comprehensive report of developments coming from that work have been given by Fenstermacher, Swinney and Gollub. A summary of their observations are shown in Figure 9.

Figure 9: Couette flow, outer cylinder stationary, increasing $\Omega(R) \rightarrow$

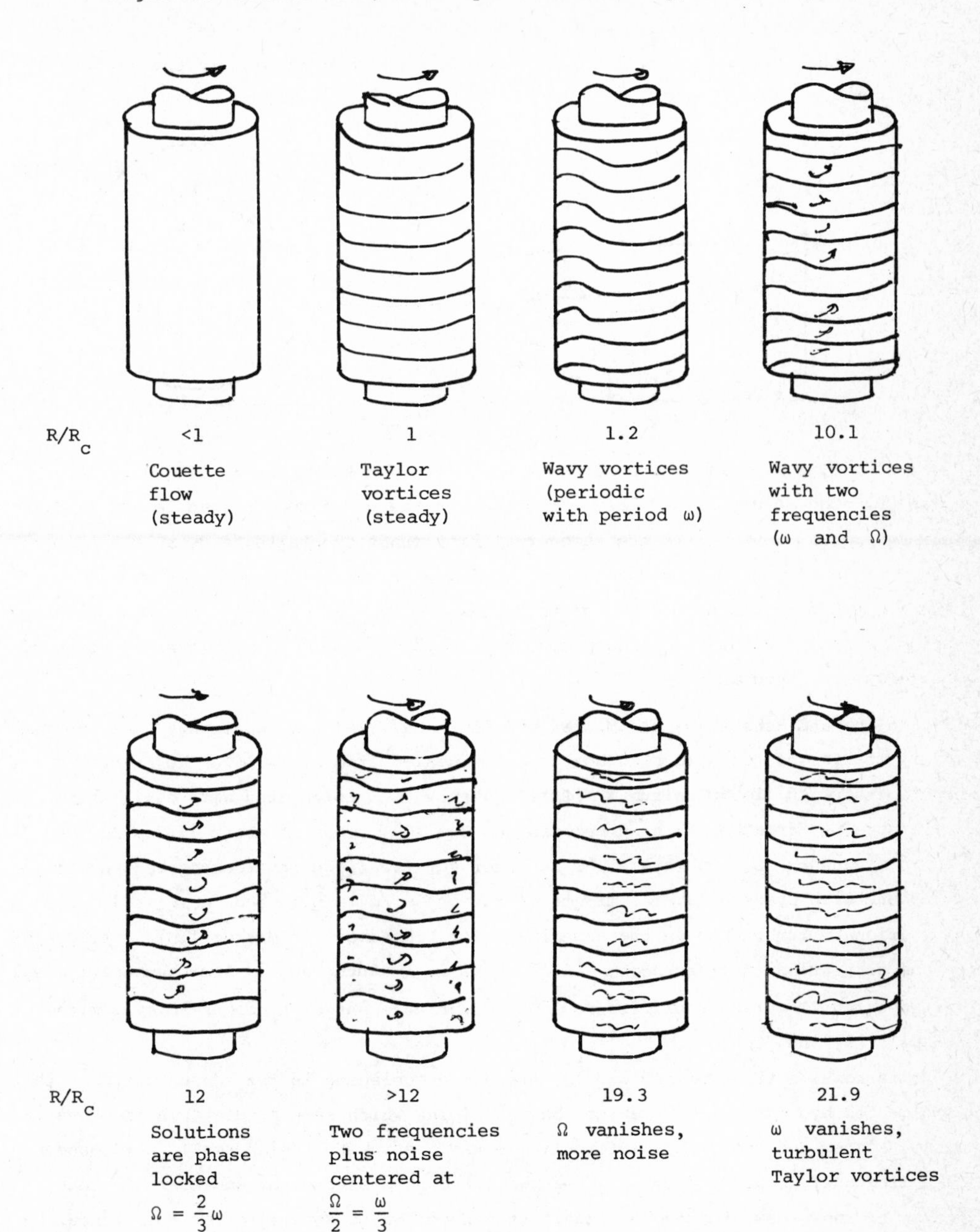

Yavorskaya, Beleyaev, Monakov and Scherbakov [20] have carried out bifurcation experiments for the problem of flow between rotating spheres when the inner sphere rotates and the gap is wide. In Figure 10 I have sketched the frequency versus Reynolds number graph given as Figure 1 of their paper. They get their results by monitoring the fluctuating velocity at a point and they also measure the autocorrelation function.

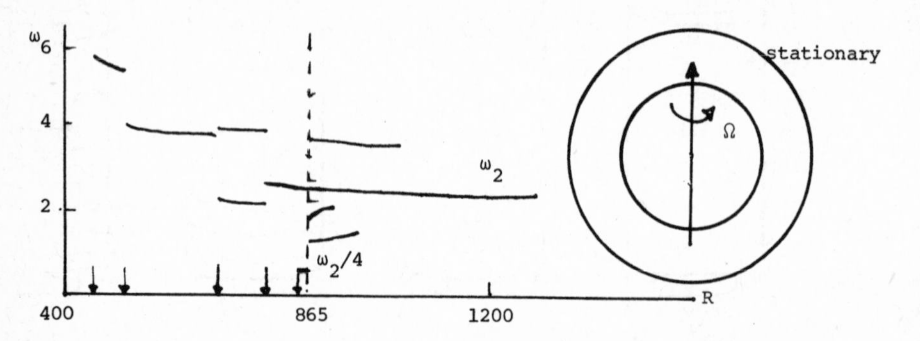

Figure 10: The flow between spheres is periodic when there is one frequency at a given R. The solution is doubly periodic when there are two frequencies present. Just before $R = 895$ where the autocorrelation function starts to decay there is $4(2\pi/\omega_2)$ subharmonic solution. The first decay of the autocorrelation at $R = 895$ is accompanied by the appearance of three new frequencies. The autocorrelation never does decay fully because the sharp spectral component coexists with dynamic noise for the range of R considered.

Gollub and Benson [7] and Maurer and Libchaber [15] have done many experiments on bifurcation of convection in box of fluid heated from below. In the French experiments with liquid helium a first frequency ω_1 associated with oscillating rolls appears for a Rayleigh number around 2×10^4, then at about 2.7×10^4 a second frequency ω_0, much smaller is observed, two frequency locking regimes are observed, with hysteresis, for frequency ratios $\omega_1/\omega_2 = 6.5$ and $\omega_1/\omega_2 = 7$. The transition to turbulence in the experiments of Libchaber and Maurer [13] is triggered by the generation of frequencies $\frac{\omega_2}{2}, \frac{\omega_2}{4}, \frac{\omega_2}{8}, \frac{\omega_2}{16} \to$ turbulence. A mathematical model for repeated 2T-periodic bifurcation into turbulence has been discussed by Tomita and Kai [19] and Ito [10].

The results reviewed in this lecture are astonishing in the sense that show that complicated hydrodynamical problems have dynamics which seem predictable from analysis of systems of nonlinear differential equations in $\mathbb{R}^n$, with small n. Concepts like frequency locking which have been well known to electrical engineers for many years, are now known to have an important connection to some types of turbulence. The idea of nonperiodic or strange attractors is a very major advance in the subject. On the other hand, experiments do not seem to suggest the transitor to turbulence can be characterized in any simple way.

[1] Ahlers, G. and Behringer, R. P. Evolution of turbulence from Rayleigh-Benard instability. Phys. Rev. Lett. 40, 712-716 (1978).

[2] Bowen, Rufus. A model for Couette flow data. Turbulence Seminar. Springer lecture notes in mathematics 615, 117-134, 1977.

[3] Chenciner, A. and Iooss, G. Bifurcation de tores invariants. Arch. Rational Mech. Anal. 69, 109-198 (1979).

[4] Curry, J. H. A generalized Lorenz system. Commun. Math. Phys. 60, 193-204 (1978).

[5] Fenstermacher, P. R., Swinney, H. L. and Gollub, J. P. Dynamical instabilities and transition to chaotic Taylor vortex flow. J. Fluid Mech. (to appear).

[6] Gollub, J. P. and Benson, S. V. Phase locking in the oscillations leading to turbulence. To appear in Pattern formation (approximate title), edited by H. Haken, Springer-Verlag, Berlin, 1979

[7] Gollub, J. P. and Benson, S. V. Chaotic response to a periodic perturbation of a convecting fluid. Phys. Rev. Letters, 41, 948-950 (1978).

[8] Haken, H. Nonequilibrium phase transitons of limit cycles and multiperiodic flows. Z. Physik. B. 29, 61-66 (1978) and Nonequilibrium phase transitions of limit cycles and multiperiodic flow in continuous media. Z. Physik. B. 30, 423-428 (1978).

[9] Iooss, G. and Joseph, D. D. Elementary stability and bifurcation theory (to appear).

[10] Ito, A. Perturbation theory of self-oscillating system with a periodic perturbation. Prog. Theor. Phys. 61, 45 (1979)
Successive subharmonic bifurcations and chaos in a nonlinear Mathieu equation. Prog. Theor. Phys. 61, 815 (1979).

[11] Joseph, D. D. Hydrodynamic stability and bifurcation. In hydrodynamic instabilities and the transition to turbulence. Springer Topics in Current Physics. Eds. Swinney H. L. and Gollub, J. P.

[12] Joseph, D. D. Stability of fluid motions Vols. I and II. Springer tracts in Nat. Phil. Vols. 27 and 28, 1976.

[13] Libchaber, A. and Maurer, J. An experiment of Rayleigh-Benard in small domains; multiplation, locking and division of frequencies (to appear).

[14] Lorenz, E. N. Deterministic nonperiodic flow. J. Atmos. Sci. 20, 130 (1963).

[15] Maurer, J. and Libchaber, A. Rayleigh-Benard experiment in liquid helium; frequency locking and the onset of turbulence. J. Phys. Letters (to appear in July 1979).

[16] Orszag, S. A. and Kelms, L. C. Transition to turbulence in Plane Poiseuille and Plane Couette Flow. J. Fluid Mech. (to appear).

[17] Ruelle D. and Takens, F. On the nature of turbulence. Comm. Math. Phys. 20, 167-192 (1971).

[18] Sell, G. R. Bifurcation of higher dimensional tori. Arch. Rational Mech. Anal. 69, 199-230 (1979).

[19] Tomita K. and Kai, T. Phys. Letters 66A, 91 (1978).

[20] Yavorskaya, I. M., Beleyayev, J. N., Monakov, A. A., Scherbakov, N. M. Generation of turbulence in a rotating visious fluid. JETP, 29, 329-334

Time-periodic motion of the drop of climbing STP. Rod radius, 0.635; rotational speed, 13.3 rev s^{-1}; frequency of periodic motion, 0.4 cycles s^{-1}.

(From "The Rotating Rod Viscometer", by G. S. Beavers and D. D. Joseph, Journal of Fluid Mechanics, 69, 1975, pp. 475-511.)

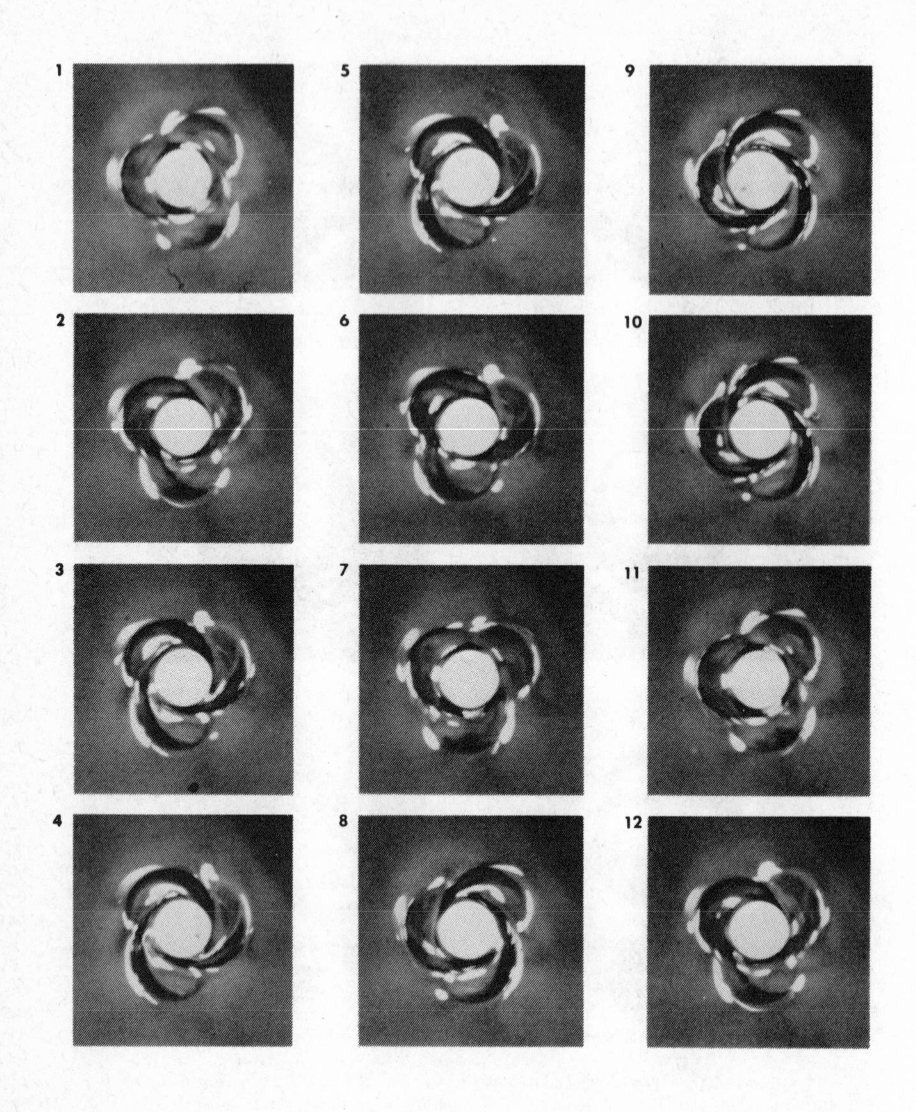

The motion of the fully-developed three-lobe flower instability as the rod goes through one complete cycle. The time between frames is approximately 0.015 sec.

(From "Experiments on Free Surface Phenomena" by G. S. Beavers and D. D. Joseph, Journal of Non-Newtonian Fluid Mechanics, 5, 1979, pp. 323-352.)

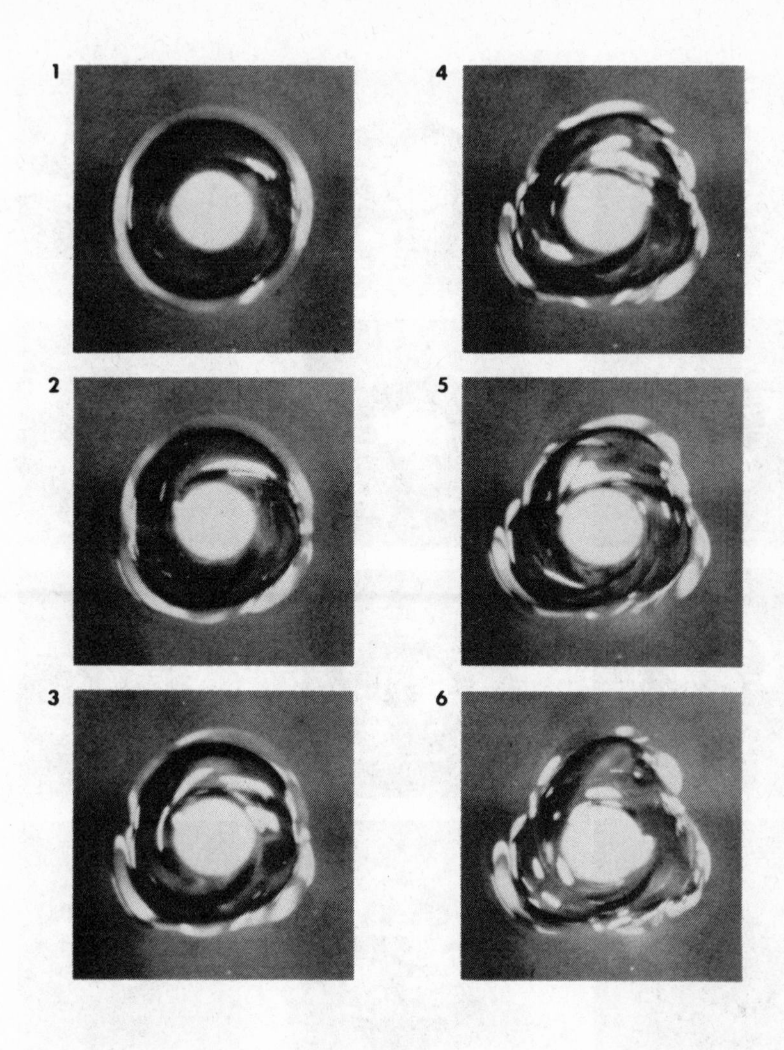

The growth of the flower instability in TLA-227 as viewed from above. The angle of twist is 4 radians and the frequency of oscillation is 7 cycles per sec. The photographs show the rod at approximately the same position in its cycle. The time between photographs is approximately 1.1 sec.

(From "Experiments on Free Surface Phenomena" by G. S. Beavers and D. D. Joseph, Journal of Non-Newtonian Fluid Mechanics, 5, 1979, pp. 323-352.)

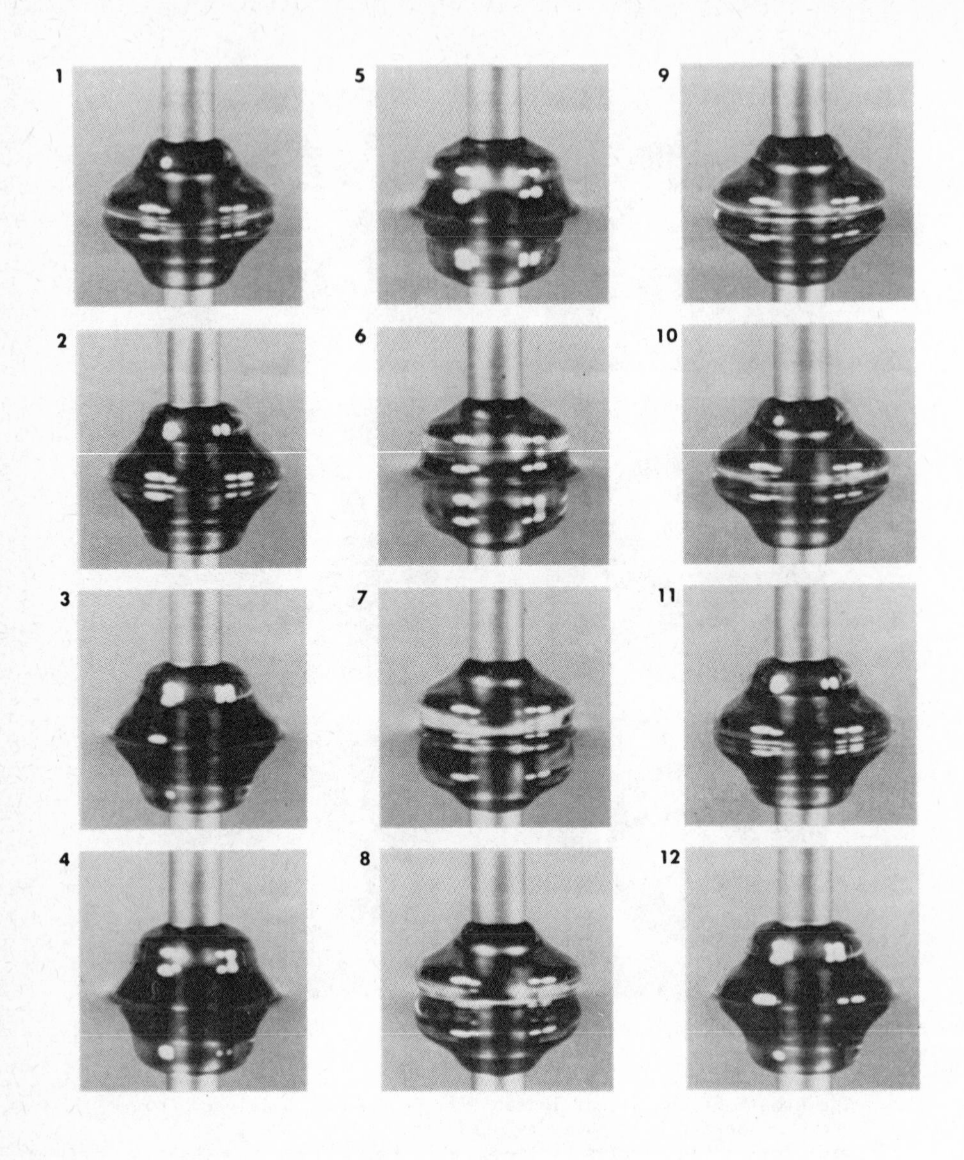

The breathing instability of a bubble of STP on a rod of radius 0.476 cm rotating at 19 rev s^{-1}. Frames 1-10 represent one complete cycle. Time between frames = 0.15 sec.

(From "Experiments on Free Surface Phenomena" by G. S. Beavers and D. D. Joseph, Journal of Non-Newtonian Fluid Mechanics, 5, 1979, pp. 323-352.)

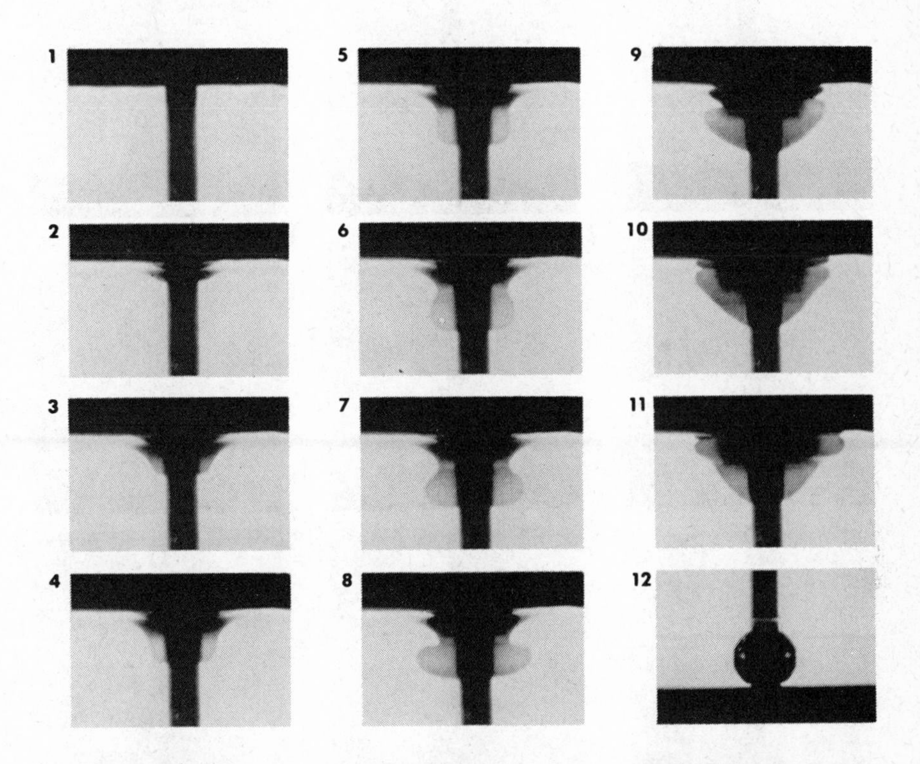

The motion of the bubble of TLA-227 on a rod which rotates at 14.5 rev s^{-1} in TLA-227 floating on water. The periodic motion of the bubble is controlled by the competing effects of normal stresses, inertia, surface tension and buoyancy forces. The bubble on the upper surface of the TLA-227 (frame 12) is steady and stable. The time between frames is 6 sec.

(From "Experiments on Free Surface Phenomena" by G. S. Beavers and D. D. Joseph, Journal of Non-Newtonian Fluid Mechanics, 5, 1979, pp. 323-352.

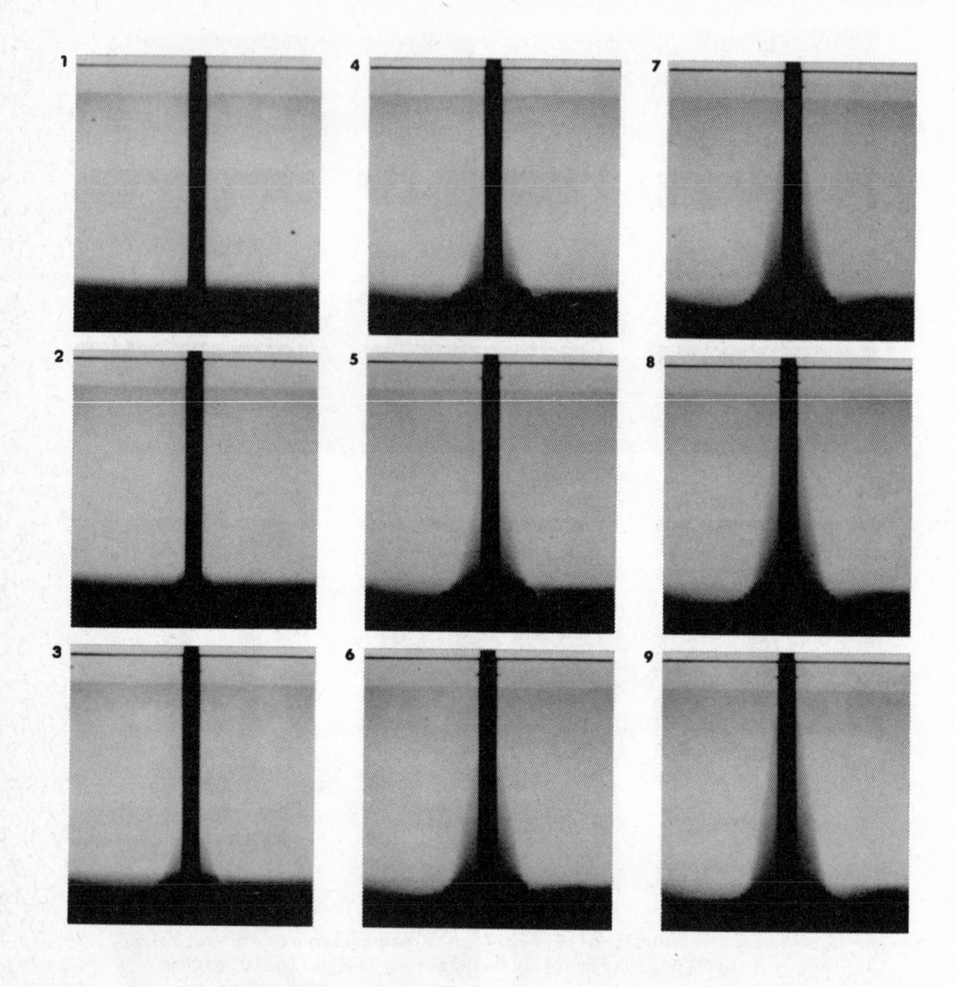

The climb of TLA-227 on a rod of radius 0.620 cm for a configuration in which STP floats on TLA-227. The steady rotational speed is 3.2 rev s^{-1}. The density difference at the STP/TLA-227 interface is 0.0005 g cm^{-3} and the difference in climbing constants is approximately 19 g cm^{-1}. The TLA-227 climbs without bound through the STP because the normal stresses have essentially no gravity forces to oppose them. Time between frames is approximately 7.5 sec.

(From "Experiments on Free Surface Phenomena" by G. S. Beavers and D. D. Joseph, Journal of Non-Newtonian Fluid Mechanics, 5, 1979, pp. 323-352.)

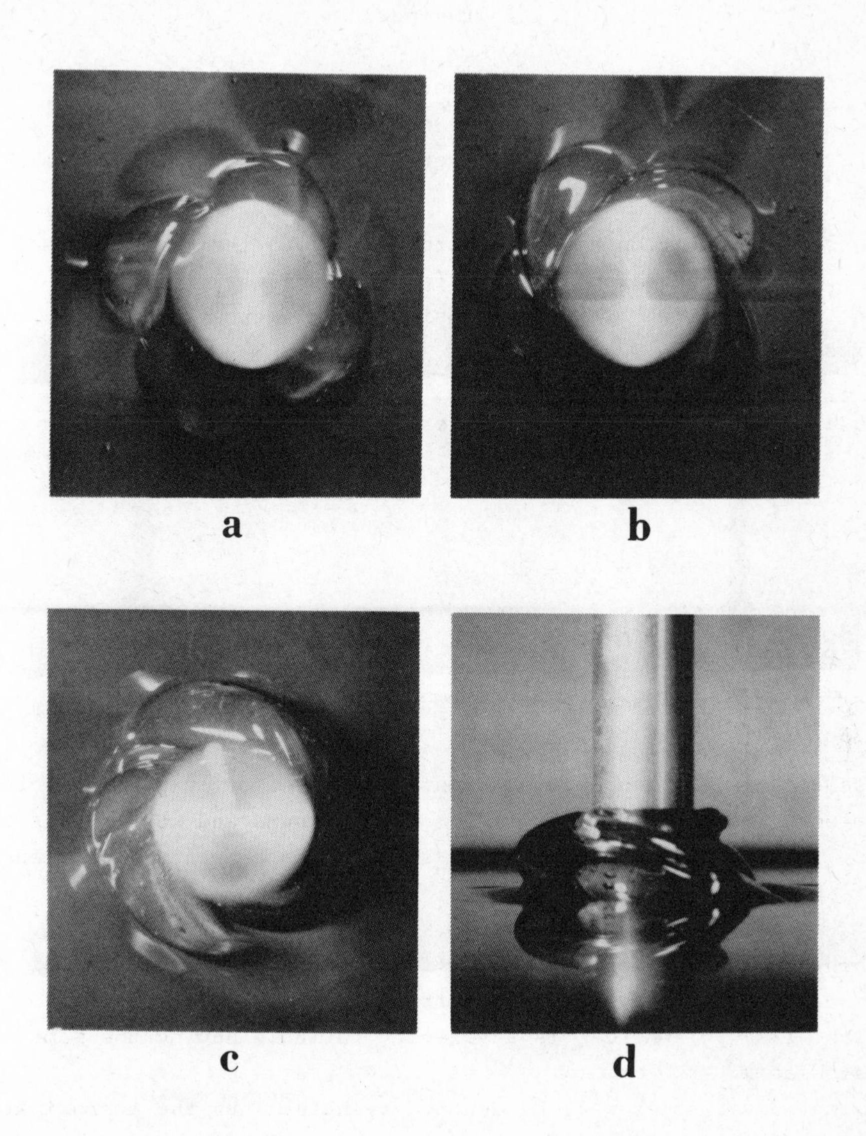

(a), (b) Top views of the four-petal configurations bifurcating from an axisymmetric time-periodic flow. The two views are photographs at two different instants during a cycle: $\omega = 9.5$ cycles/s, $\Theta = 200°$. (c) Top view and (d) side view of the three-petal configuration bifurcating from an axisymmetric time-periodic flow: $\omega = 9.2$ cycles/s, $\Theta = 235°$.

(From "Novel Weissenberg Effects", by G. S. Beavers and D. D. Joseph, Journal of Fluid Mechanics, 81, 1977, pp. 265-272.)

Approximation of the Hydrodynamic Equations by a Transport Process

Shmuel Kaniel
Department of Mathematics
The Hebrew University of Jerusalem

1. This is a report on a novel approach towards the numerical solution of the compressible flow equations. It will be stated, for an inviscid flow. Work on extension to viscous flow is in progress..

The equations governing the motion of inviscid compressible flow are:

$$(1.1) \qquad \frac{\partial \rho}{\partial t} + \nabla \cdot \rho u = 0$$

$$(1.2) \qquad \frac{\partial u}{\partial t} + u \cdot \nabla u + \frac{1}{\rho} \nabla p = 0$$

$$(1.3) \qquad \frac{\partial s}{\partial t} + u \cdot \nabla S = 0$$

$$(1.4) \qquad p = F(\rho, S)$$

The scalar $\rho(x,t)$ denotes the density, the vector $u(x,t)$ denotes the velocity, the scalar $p(x,t)$ in the pressure and $S(x,t)$ is the entropy. F is a general equation of state. x is a point in n dimensional space, $n = 1,2,3$.

The main idea is the approximation for a short time interval, of the system (1.1)-(1.4) by the transport of moments of a distribution function $f(x,\xi,t)$ where ξ is a velocity variable having the same dimension as x.
Approximated variables will be denoted by hats. Thus the approximate density and momentum are defined by

$$(1.5) \qquad \hat{\rho}(x,t) = \int f(x,\xi,t)\,d\xi$$

$$(1.6) \qquad \hat{m}(x,t) = \hat{\rho}(x,t) \cdot \hat{u}(x,t) = \int \xi f(x,\xi,t)\,d\xi$$

The evolution of $f(x,\xi,t)$ is taken to be free streaming i.e.

$$(1.7) \qquad f(x,\xi,t) = f(x-\xi t,\xi,0)$$

The corresponding differential equation is

$$(1.8) \qquad \frac{\partial f}{\partial t} + \sum_{i=1}^{n} \xi_i \frac{\partial f}{\partial x_i} = 0.$$

Since (1.7) is well defined for discontinuous functions, it satisfies (1.8) in a "weak sense". We are concerned here with the derivation of approximating system. No proof of convergence will be exhibited. So we will use, rather freely, equation (1.8). Take equation (1.8) and integrate $d\xi$.

$$(1.9) \qquad \frac{\partial}{\partial t} \int f d\xi + \sum_{i=1}^{n} \frac{\partial}{\partial x_i} \int \xi_i f = 0 \quad .$$

By (1.5) and (1.6) one gets

$$(1.10) \qquad \frac{\partial}{\partial t} \hat{\rho} + \sum_{i=1}^{n} \frac{\partial}{\partial x_i} \hat{m}_i = 0$$

which is the equation of continuity (1.1) in the hatted variables. Let us try to satisfy (1.2). Multiply (1.8) by ξ_k and integrate $d\xi$.

$$(1.11) \qquad \frac{\partial}{\partial t} \int \xi_k f d\xi + \sum_{i=1}^{n} \frac{\partial}{\partial x_i} \int \xi_i \xi_k f d\xi = 0 \ .$$

This equation should approximate the momentum equation

$$(1.12) \qquad \frac{\partial m_k}{\partial t} + \sum_{i=1}^{n} \frac{\partial}{\partial x_i} \frac{m_i m_k}{\rho} + \frac{\partial p}{\partial x_i} = 0 \ .$$

The time derivative in (1.11) is, indeed, equal to $\frac{\partial \hat{m}_k}{\partial t}$, but the second term does not look like the spatial derivatives in (1.12). On the other hand, $f(x,\xi,t)$ is constructed only for the approximation of (1.1)-(1.4). Thus only few moments of f are needed to be known. The freedom in the definition of f will be utilized in order to construct $f(x,\xi,0)$ in such a manner that the momentum equation, for $t=0$, will be satisfied.

2. $f(x,\xi,0)$ will be constructed by a generating function $g_S(|\xi|)$. This generating function has to be viewed as a thermodynamic function i.e. it is functionally dependent on the variables ρ and S. f, then, will be constructed from g by specifying the values of $\rho(x)$, $u(x)$ and $S(x)$.

$g_S(|\xi|)$ is defined, implicitly, as follows

$$(2.1) \qquad \rho(\alpha) = \int_{|\xi|\leq\alpha} g_S(|\xi|)\,d\xi$$

$$(2.2) \qquad p(\alpha) = \frac{1}{n} \int_{|\xi|\leq\alpha} |\xi|^2 g_S(|\xi|)\,d\xi \qquad\qquad n = \text{dimension}$$

where

$$p(\alpha) = F(\rho(\alpha),S)$$

Let us solve the equations in the 3 dimensional space. Using polar coordinates.

$$(2.3) \qquad \rho(\alpha) = 4\pi \int_{r\leq\alpha} r^2 g_S(r)\,dr$$

$$(2.4) \qquad p(\alpha) = \frac{4\pi}{3} \int r^4 g_S(r)\,dr$$

Differentiate:

$$(2.5) \qquad \frac{\partial\rho}{\partial\alpha} = 4\pi\cdot\alpha^2 g_S(\alpha)$$

$$(2.6) \qquad \frac{\partial p}{\partial\alpha} = \frac{4\pi}{3}\cdot\alpha^4 g_S(\alpha)$$

Thus

$$\frac{\partial p}{\partial\alpha} = \frac{\alpha^2}{3}\cdot\frac{\partial\rho}{\partial\alpha} \; .$$

On the other hand $\frac{\partial p}{\partial\alpha} = \frac{\partial p}{\partial\rho}\cdot\frac{\partial\rho}{\partial\alpha}$ where $\frac{\partial p}{\partial\rho} = c^2(\rho,S)$, c is the local speed of sound, which is known from the equation of state (1.4). Hence

$$(2.7) \qquad \alpha^2 = 3c^2(\rho,S) \; .$$

By (2.7), for fixed S

$$\frac{\partial\rho}{\partial\alpha} = \{\frac{\partial\alpha}{\partial\rho}\}^{-1} = \frac{\partial}{\partial\rho}\{\sqrt{3}\,\frac{\partial p}{\partial\rho}^{1/2}\}^{-1} = \frac{2}{\sqrt{3}}\,\frac{\partial^2 p}{\partial\rho^2}^{-1} \cdot \frac{\partial p}{\partial\rho}^{1/2}$$

Thus

$$(2.8)\qquad g_S(\alpha) = \frac{1}{2\pi\cdot\sqrt{3}} \cdot \frac{1}{\alpha^2} \cdot \frac{\partial^2 p}{\partial\rho^2}^{-1} \cdot \frac{\partial p}{\partial\rho}^{1/2}$$

It follows that if $\frac{\partial^2 p}{\partial\rho^2} > 0$, (a condition satisfied by a "reasonable" equation of state), $g_S(\alpha)$ is well defined and, moreover, it is positive. The derivation for other dimensions is similar. For $n = 1$ it follows that $\alpha = c$.

Consider, for example, the equation of state of an ideal gas $p = A(S)\rho^\gamma$.
In this case

$$\frac{\partial p}{\partial\rho} = A\gamma\rho^{\gamma-1}$$

$$\alpha = \sqrt{3A\gamma}\cdot\rho^{\frac{\gamma-1}{2}}$$

$$(2.9)\qquad \frac{\partial\rho}{\partial\alpha} = (3A\gamma)^{-\frac{1}{\gamma-1}} \cdot \frac{2}{\gamma-1} \cdot \alpha^{\frac{3-\gamma}{\gamma-1}}$$

$$(2.10)\qquad g_S(\alpha) = \frac{1}{4\pi}(A(S)\cdot 3\cdot\gamma)^{-\frac{1}{\gamma-1}} \cdot \frac{2}{\gamma-1} \cdot \alpha^{\frac{5-3\gamma}{\gamma-1}}$$

In particular, for $\gamma = \frac{5}{3}$, $g_S(\alpha)$ is a constant depending on S. The analysis for one and two dimensional distributions is similar. For two dimensional distributions the result is

$$(2.11)\qquad g_S(\alpha) = \frac{1}{2\pi}(A(S)\cdot 2\cdot\gamma)^{-\frac{1}{\gamma-1}} \cdot \frac{2}{\gamma-1} \cdot \alpha^{\frac{4-2\gamma}{\gamma-1}} \qquad n = 2$$

For one dimensional distribution

$$(2.12)\qquad g_S(\alpha) = \frac{1}{\gamma-1} \cdot (A(S)\cdot\gamma)^{-\frac{1}{\gamma-1}} \cdot \alpha^{\frac{3-\gamma}{\gamma-1}} \qquad n = 1\,.$$

For $n = 2$, $\gamma = 2$ or $n = 1$, $\gamma = 3$ it turns out that $g_S(\alpha) = K(S)$

3. $f(x,\xi,0)$ is defined by

$$(3.1) \qquad f(x,\xi,0) = g_{S(x)}(|\xi-u(x)|) \qquad |\xi-u(x)| \leq \sqrt{n}\, c(x)$$

$$(3.2) \qquad f(x,\xi,0) = 0 \qquad |\xi-u(x)| > \sqrt{n}\, c(x)$$

Theorem 3.1 If $f(x,\xi,0)$ is defined by (3.1) and (3.2) then, for $t=0$, the momentum equation is satisfied, i.e.

$$(3.3) \qquad \frac{\partial \hat{m}_k}{\partial t} + \sum_{i=1}^{n} \frac{\partial}{\partial x_i}\left(\frac{m_i m_k}{\rho}\right) + \frac{\partial p}{\partial x_k} = 0$$

Proof: Let us evaluate the integrals in (1.11).

$$\int \xi_i \xi_k f(x,\xi,0)\,d\xi = \int_{|\xi-u(x)|\leq\sqrt{n}\,c(x)} \xi_i \xi_k \cdot g_{S(x)}(|\xi-u(x)|)\,d\xi$$

Change variables $\eta = \xi - u(x)$. The last integral is equal to:

$$\int_{|\eta|\leq\sqrt{n}c(x)} (\eta_i + u_i(x))(\eta_k + u_k(x))\, g_{S(x)}(|\eta|)\,d\eta =$$

$$= u_i(x)\cdot u_k(x)\cdot \int_{|\eta|\leq\alpha(x)} g_{S(x)}(|\eta|)\,d\eta + u_i(x) \int_{|\eta|\leq\alpha(x)} \eta_k g_{S(x)}(|\eta|)\,d\eta +$$

$$+ u_k(x) \int_{|\eta|\leq\alpha(x)} \eta_i g_{S(x)}(|\eta|)\,d\eta + \int_{|\eta|\leq\alpha(x)} \eta_i \eta_k g_{S(x)}(|\eta|)\,d\eta \quad .$$

The first integral is equal, by (2.1) to $u_i(x)\cdot u_k(x)\cdot\rho(x) =$ $= \frac{m_i(x)\cdot m_k(x)}{\rho(x)}$. Since $g_{S(x)}(|\eta|)$ is spherically symmetric and the domain of integration is a ball it follows that the second as well as the third integrals vanish. As for the fourth integral, if $i \neq k$ then, again for symmetry, it vanishes. If $i=k$ it is equal to

$$\int_{|\eta|\leq\alpha(x)} \eta_k^2\, g_{S(x)}(|\eta|)\,d\eta = \frac{1}{n} \int_{|\eta|\leq\alpha(x)} |\eta|^2 g_S(|\eta|)\,d\eta = p(x)$$

by (2.2).

Remark: In the usual derivation of the equations of compressible flow one uses a local Maxwellian distribution in ξ. Then, by an incomplete argument, the hyperbolic system (1.1) - (1.4) is reached.
For hyperbolic system the domain of dependence propagates with finite speed. Thus, in a transport process the speed of propagation (which

is ξ) has also to be finite. Observe that $g_S(|\xi|)$ is unique, provided that a finite speed condition is imposed.

4. The construction in section 3 can already be used for the computation of inviscid isentropic flow. In this case S is considered to be fixed and (1.3) is omitted.
The computation may be conveniently done in a conservation law form. The region is divided into cells and in each cell the total mass and momentum is recorded.

The first step in the computation is the definition of $\rho(x)$ and $m(x)$ for all x in the region. Since the transport mechanism is well defined for discontinuous distributions one may consider, for example, $\rho(x)$ and $m(x)$ to be constant in each cell or to take piecewise linear distributions with possible discontinuities at the cells boundaries.

Now $f(x,\xi,0)$ is constructed by (3.1) and (3.2), a time interval $(0,\Delta t)$ is chosen, f is advanced to Δt by (1.7), then the flux of the appropriate moments of f is being computed.

For example, the integrated flux of a typical moment $\int \xi_k f(x,\xi,t)d\xi$, in a time interval $(0,t)$ across a flat boundary $x_j = x_j^{(\ell)}$, from the left is:

$$\int_{x_j \le x_j^{(\ell)}} \int_{x_j-\xi_j t \ge x_j^{(\ell)}} \xi_k f(x,\xi,0)\,d\xi dx =$$

$$= \int_{x_j \le x_j^{(\ell)}} \int_{\substack{x_j-\xi_j t \ge x_j^{(\ell)} \\ |\xi-u(x)|^2 \le 3c^2(x)}} g_S(\xi-u(x))\,d\xi dx =$$

$$= \int_{x_j \le x_j^{(\ell)}} \int_{\substack{|\eta|^2 \le 3c^2(x) \\ x_j - u_j(x)t - \eta_j t \ge x_j^{(\ell)}}} (\eta_k + u_k(x))\, g_S(|\eta|)\,d\eta dx =$$

$$= \int_{x_j \le x_j^{(\ell)}} \Big\{ \int_{\substack{|\eta|^2 \le 3c^2(x) \\ \eta_j \le \frac{x_j^{(\ell)} - x_j}{t} + u_j}} \eta_k g_S(|\eta|)\,d\eta + u_k(x) \int_{\substack{|\eta|^2 \le 3c^2(x) \\ \eta_j \le \frac{x_j^{(\ell)} - x_j}{t} + u_j}} g_S(|\eta|)\,d\eta \Big\}\, dx$$

The η-integrals are defined on the intersection of a ball with a half-space. Recall that g is a function of $|\eta|$ only. Thus the η-integrals can be precomputed either by analytical or by numerical methods to furnish a known function which is dependent only on the quantity $d = \frac{x_j^{(\ell)} - x_j}{t} + u_j(x)$ and on the density at x.
The precomputation can be performed for all moments involved. Thus the variable is eliminated so that the flux is dependent on x-integrals of known functions.

Likewise, in a two dimensional domain, the integrated flux of the moment above, across a corner, $x_j = x_j^{(\ell)}$, $x_i = x_i^{(s)}$, is done by adding the inequalities $x_i \leq x_i^{(s)}$ and $\eta_i \leq \frac{x_i^{(\ell)} - x_i}{t} + u_i$ under the last integral. Again, the η-integrals can be precomputed as functions of three variables: ρ, $d_j = \frac{x_i^{(\ell)} - x_j}{t} + u_j$ and $d_i = \frac{x_i^{(\ell)} - x_i}{t} + u_i$.

One still has to perform the x-integration. The construction of quadrature formulas for these is done in a way similar to the quadrature formulas in the Finite Element Method.

One dimensional computations have been performed ("shock tube" computations). Two dimensional computations are in progress. They will be reported in the 4th International Congress on Numerical Methods in Physics and Engineering, sponsored by I.R.I.A.

5. The approximation of equations (1.1) and (1.2) was derived for the general, variable entropy, case. In this case equation (1.3) has also to be integrated. In principle it can be done. Some methods, like the F.C.T. method of Boris and Book, are geared to the integration of such an equation. The basic transport approach suggests, however, to consider the pressure as the third thermodynamic variable.

Define, accordingly, pseudo pressure $\hat{p}(x,t)$ by

$$(5.1) \qquad \hat{p}(x,t) = \frac{1}{n} \int \sum_{i=1}^{n} (\xi_i - \hat{u}_i(x,t))^2 f(x,\xi,t)\, d\xi \qquad .$$

$\hat{p}$ is well defined in the approximate system and $\hat{p}(x,0) = p(x,0)$ by the definition of $f(x,\xi,0)$ and $g_{S(x)}(|\xi|)$.

Unfortunately $\frac{\partial\hat{p}}{\partial t} \neq \frac{\partial p}{\partial t}$ even for $t=0$. In fact the following holds.

Theorem 5.1: for $t=0$

$$\frac{\partial\hat{p}}{\partial t} - \frac{\partial p}{\partial t} = -\frac{1}{n}((n+2)p - n\rho\frac{\partial p}{\partial\rho})\sum_{i=1}^{n}\frac{\partial u_i}{\partial x_i} \tag{5.2}$$

Proof: Multiply (1.8) by $\xi_i\xi_j$ and integrate:

$$\frac{\partial}{\partial t}\int\xi_i\xi_j f(x,\xi,t)d\xi = -\sum_{k=1}^{n}\frac{\partial}{\partial x_k}\int\xi_i\xi_j\xi_k\, f(x,\xi,t)d\xi$$

For $t = 0$, the right hand side is equal to

$$-\sum_{k=1}^{n}\frac{\partial}{\partial x_k}\int_{|\eta|\leq\sqrt{n}c(x)}(\eta_i+u_i)(\eta_j+u_j)(\eta_k+u_k)g_{S(x)}(|\eta|)d\eta$$

This term is easily evaluated to yield for $t = 0$ and $i \neq j$

$$\frac{\partial}{\partial t}\int\xi_i\xi_j f(x,\xi,t)d\xi = -\sum_k\frac{\partial}{\partial x_k}(\rho u_i u_j u_k) - \frac{\partial}{\partial x_i}(u_j p) - \frac{\partial}{\partial x_j}(u_i p) \tag{5.3}$$

Whereas

$$\frac{\partial}{\partial t}\int\xi_i^2 f(x,\xi,t) = -\sum_k\frac{\partial}{\partial x_k}(u_k p + \rho u_k u_i^2) - 2\frac{\partial}{\partial x_i}(u_i p)\ . \tag{5.4}$$

Thus, for $t = 0$

$$n\cdot\frac{\partial}{\partial t}\hat{p} = \frac{\partial}{\partial t}\sum_i(\xi_i - \hat{u}_i(x))^2 f(x,\xi,t)d\xi =$$

$$= \frac{\partial}{\partial t}\sum_i\int\xi_i^2 f(x,\xi,t)d\xi - \hat{u}_i^2\int f(x,\xi,t)d\xi$$

$$= -2\sum_i\frac{\partial}{\partial x_i}(u_i p) - \sum_{i,k}\frac{\partial}{\partial x_k}(u_k p + \rho u_k u_i^2) - \sum_i\frac{\partial}{\partial t}\,\hat{}\,\hat{u}_i^2$$

Since, for $t = 0$, $\frac{\partial\hat{\rho}}{\partial t} = \frac{\partial\rho}{\partial t}$ and $\frac{\partial\hat{m}}{\partial t} = \frac{\partial m}{\partial t}$ it follows that

$$-\sum_i\frac{\partial}{\partial t}\hat{\rho}\,\hat{u}_i^2 = \sum_{i,k}\frac{\partial}{\partial x_k}(\rho u_k)\cdot u_i^2 + 2\sum_{i,k}\rho u_k u_i\frac{\partial u_i}{\partial x_k} + 2\sum_i u_i\frac{\partial p}{\partial x_i}$$

Thus

$$\frac{\partial \hat{p}}{\partial t} = - \sum_i u_i \frac{\partial p}{\partial x_i} - \frac{n+2}{n} p \sum_i \frac{\partial u_i}{\partial x_i} \tag{5.5}$$

On the other hand

$$\begin{aligned} \frac{\partial p}{\partial t} &= \frac{\partial p}{\partial \rho}\frac{\partial \rho}{\partial t} + \frac{\partial p}{\partial S}\frac{\partial S}{\partial t} = - \frac{\partial p}{\partial \rho} \sum_i \frac{\partial}{\partial x_i}(\rho u_i) \\ &- \frac{\partial p}{\partial S} \sum_i u_i \frac{\partial S}{\partial x_i} = - \sum_i u_i \frac{\partial p}{\partial x_i} - \frac{\partial p}{\partial \rho} \sum_i \frac{\partial u_i}{\partial x_i} \end{aligned} \tag{5.6}$$

The comparison of (5.5) and (5.6) establishes (5.1).

Such a result had to be expected. The "thermodynamic argument" that led to the construction of $g_S(|\xi|)$ did not take into account any deformation in the flow. Indeed, the difference turns out to be the divergence multiplied by a thermodynamic function.

Let us compute the order of the approximation of the continuity and momentum equations. It is easy to see that
$\frac{\partial^2 \hat{\rho}}{\partial t^2} = \frac{\partial^2 \rho}{\partial t^2}$ while, for the momentum equation the following holds.

<u>Theorem 5.2</u>: for $t = 0$

$$\frac{\partial^2 \hat{m}_k}{\partial t^2} - \frac{\partial^2 m_k}{\partial t^2} = \sum_i \frac{\partial}{\partial x_i} \{(\frac{\partial u_i}{\partial x_i} + \frac{\partial u_k}{\partial x_i}) p\} + \frac{\partial}{\partial x_k} \{(p - \frac{\partial p}{\partial \rho} \cdot \rho) \sum_i \frac{\partial u_i}{\partial x_i}\} \tag{5.7}$$

<u>Proof</u>: Second derivatives of the momenta, at $t = 0$, are computed by differentiating equation (1.8)

$$\begin{aligned} \frac{\partial^2}{\partial t^2} \hat{m}_k &= \sum_{i,j} \frac{\partial^2}{\partial x_i \partial x_j} \int \xi_i \xi_j \xi_k f(x,\xi,t) d\xi = \\ &= \sum_{i,j} \frac{\partial^2}{\partial x_i \partial x_j} \int (\eta_i + u_i)(\eta_j + u_j)(\eta_k + u_k) g_S(|\eta|) d\eta \\ &= \sum_{i,j} \frac{\partial^2}{\partial x_i \partial x_j} (\rho u_i u_j u_k) + (\sum_i \frac{\partial^2}{\partial x_i^2})(u_k p) + 2 \sum_i \frac{\partial^2}{\partial x_i \partial x_k} (u_i p). \end{aligned} \tag{5.8}$$

$$\frac{\partial^2 m_k}{\partial t^2} = -\frac{\partial}{\partial t}\{\sum_i \frac{\partial}{\partial x_i}(\rho u_i u_k) + \frac{\partial p}{\partial x_k}\} =$$

$$(5.9) \quad = \sum_{i,j} \frac{\partial^2}{\partial x_i \partial x_j}(\rho u_i u_j u_k) + \sum_i \frac{\partial}{\partial x_i}(\frac{\partial p}{\partial x_i} u_k + \frac{\partial p}{\partial x_k} u_i)$$

$$+ \frac{\partial}{\partial x_k}\{\sum_i u_i \frac{\partial p}{\partial x_i} + \rho \frac{\partial p}{\partial \rho} \sum_i \frac{\partial u_i}{\partial x_i}\}$$

The comparison of (5.8) and (5.9) yields (5.7).

Observe that in any neighborhood where the flow is a pure rotation it follows that $\frac{\partial \hat{p}}{\partial t} = \frac{\partial p}{\partial t}$ and $\frac{\partial^2 \hat{m}}{\partial t^2} = \frac{\partial^2 m}{\partial t^2}$. $\frac{\partial \hat{p}}{\partial t} = \frac{\partial p}{\partial t}$ also whenever $\operatorname{div} u = 0$ or for the particular equation of state $p = A(S)\rho^{5/3}$ (in 3 dimensions). The latter case can be characterized by the condition $g_S(|\xi|) = K(S)$.

It turns out that there is a way to get second order correction to the equation (1.8). It can be done, again, by a transport process.

<u>Theorem 5.3</u>: Let a function $h(x,\xi,0)$ be given. Let $f^*(x,\xi,t)$ be defined by

$$(5.10) \qquad f^*(x,\xi,t) = f(x-\xi t,\xi,0) + th(x-\frac{\xi}{2}t,\xi,0) \ .$$

Let the variables $\hat{\rho}, \hat{m}$ and $\hat{p}$ be defined as the corresponding moments of f^*, then by an appropriate choice of $h(x,\xi,0)$, for $t = 0$ it follows that

$$\frac{\partial^3 \hat{\rho}}{\partial t^3} = \frac{\partial^3 \rho}{\partial t^3}, \quad \frac{\partial^2 \hat{m}}{\partial t^2} = \frac{\partial^2 m}{\partial t^2}, \quad \frac{\partial \hat{p}}{\partial t} = \frac{\partial p}{\partial t} \ .$$

The construction of $h(x,\xi,0)$ will be exhibited in section 6. Note the factor $\frac{1}{2}$ in the argument under h. It is a must if one has to consider the first order derivative in (5.2) and the second order derivative in (5.7).

6. The second order correction is constructed in a way similar to the construction $g_S(\xi)$. Consider the 3 dimensional case. Let $q(\xi)$ be a function of $|\xi|$.

Define, for $|\xi - u(x)| \le \varphi(x)$

$$h(x,\xi,0) = \sum_{i\neq j} \frac{1}{2}\left(\frac{\partial u_i}{\partial x_j} + \frac{\partial u_j}{\partial x_i}\right)(\xi_i - u_i)(\xi_j - u_j)q(\xi - u(x))$$

(6.1)

$$+ \sum_i \frac{\partial u_i}{\partial x_i}\{(\xi_i - u_i)^2 - \lambda^2(x)\}q(\xi - u(x))$$

Otherwise define

$$h(x,\xi,0) = 0 \qquad\qquad |\xi - u(x)| > \varphi(x)$$

The function $q(|\xi|)$ and the constants $\lambda(x)$ and $\varphi(x)$ should be thermodynamic variables, i.e., they should be determined by the equation of state and the dependence on x should be through the variables $\rho(x)$ and $S(x)$.

The particular form of $h(x,\xi,0)$ is chosen by the requirement that, for $t = 0$, the term $th(x - \frac{\xi}{2}t,\xi,0)$ should not affect $\frac{\partial\hat{\rho}}{\partial t}$, $\frac{\partial^2\hat{\rho}}{\partial t^2}$ or $\frac{\partial\hat{m}}{\partial t}$. Thus we require that

$$\int h(x,\xi,0)d\xi = 0 \tag{6.2}$$

$$\int \xi_k h(x,\xi,0)d\xi = 0 \tag{6.3}$$

Equation (6.2) is satisfied if λ is defined by

$$\lambda^2 \cdot 3 \int_{|\eta|\le\varphi} q(|\eta|)d\eta = \int_{|\eta|\le\varphi} |\eta|^2 q(|\eta|)d\eta \tag{6.4}$$

Since q depends only on $|\xi|$ and h is a multiple of q by a polynomial of degree 2 it follows that (6.2) and (6.4) imply (6.3). Multiply $h(x,\xi,0)$ by $\xi_k = (\xi_k - u_k) + u_k$ and use the symmetry arguments to show that each of the resulting integrals vanishes.

Let f^* be defined by (5.10). Then

$$\frac{\partial f^*}{\partial t} = -\sum_\ell \xi_\ell \frac{\partial f}{\partial x_\ell} + h - \frac{t}{2}\sum_\ell \xi_\ell \frac{\partial h}{\partial x_\ell} \tag{6.5}$$

$$\frac{\partial^2 f^*}{\partial t^2} = \sum_{\ell,m} \xi_\ell\xi_m \frac{\partial^2 f}{\partial x_\ell \partial x_m} - \sum_\ell \xi_\ell \frac{\partial h}{\partial x_\ell} + \frac{t^2}{4}\sum_{\ell,m} \xi_\ell\xi_m \frac{\partial^2 h}{\partial x_\ell \partial x_m} \tag{6.6}$$

In order to compute, for $t = 0$, the correction terms for $\frac{\partial^2 m_k}{\partial t^2}$, we have to multiply equation (6.6) by ξ_k and integrate $d\xi$. the contribution of the correction term will be

$$(6.7) \qquad -\sum_{\ell} \frac{\partial}{\partial x_\ell} \int \xi_k \xi_\ell h(x,\xi,0)\,d\xi$$

For the evaluation of (6.7) we need to compute the following integrals.

For $i \neq j$

$$(6.8) \qquad \int_{|\xi-u(x)|\leq\varphi(x)} (\xi_i - u_i)(\xi_j - u_j)\xi_k\xi_\ell \cdot q(\xi - u(x))\,d\xi =$$

$$= \int_{|\eta|\leq\varphi} \eta_i\eta_j(\eta_k + u_k)(\eta_\ell + u_\ell)q(|\eta|)\,d\eta$$

$$= \int_{|\eta|\leq\varphi} \eta_\ell^2\eta_k^2 q(|\eta|)\,d\eta$$

If $i = \ell$, $j = k$ or $i = k$, $j = \ell$. Otherwise the integrals vanish.

$$\int_{|\xi-u(x)|\leq\varphi(x)} \{(\xi_i - u_i)^2 - \lambda^2\}\ \xi_k\xi_\ell q(\xi - u(x))\,d\xi =$$

$$(6.9) \qquad = \int_{|\eta|\leq\varphi} (\eta_i^2 - \lambda^2)(\eta_k + u_k)(\eta_\ell + u_\ell)q(|\eta|)\,d\eta =$$

$$= \int_{|\eta|\leq\varphi} (\eta_i^2\eta_k^2 - \lambda^2\eta_k^2)q(|\eta|)\,d\eta$$

If $k = \ell$. Otherwise the integrals vanish. (We used here equation(6.4)).

Suppose now that $q(|\eta|)$ satisfies the following equations

$$(6.10) \qquad \int_{|\eta|\leq\varphi} \eta_i^2\eta_k^2 q(|\eta|)\,d\eta = p \qquad\qquad i \neq k$$

$$(6.11) \qquad \int_{|\eta|\leq\varphi} \eta_i^4 q(|\eta|)\,d\eta = 3p$$

$$(6.12) \qquad \lambda^2 \int_{|\eta|\leq\varphi} \eta_i^2 q(|\eta|)\,d\eta = \rho\,\frac{\partial p}{\partial \rho}$$

Then, by straight forward computation, it follows that (6.7) cancels the right-hand side of (6.3).

Equations (6.10) and (6.11) are consistent for any function $q(|\eta|)$. Indeed, using spherical coordinates:

$$\int \eta_i^4 q(|\eta|)d\eta = \int r^6 \cos^4\theta \; q(r) \sin\theta \; d\theta d\varphi \; dr = \frac{4\pi}{5} \int r^6 q(r)dr$$

$$\int \eta_i^2\eta_k^2 q(|\eta|)d\eta = \int r^6 \cos^2\theta \sin^2\theta \cos^2\varphi q(r) \sin\theta \; d\theta d\varphi dr =$$

$$= \pi(\frac{2}{3}-\frac{2}{5}) \int r^6 q(r)dr = \frac{4\pi}{15} \int r^6 q(r)dr \quad .$$

The correction term for $\frac{\partial p}{\partial t}$ and $t = 0$ is

$$\frac{1}{3} \int \sum_k (\xi_k - u_k)^2 h(x,\xi,0)d\xi \; .$$

The evaluation of the last integral, by the use of (6.8) - (6.12) cancels, again, the right hand side of (5.2).

The equations (6.4), (6.10) and (6.12) should determine the three functions q, φ and λ. In the general case the computation is quite complex. Some important particular cases can be easily solved.

For $p = A(S)\rho^\gamma$ try to set up $q(|\eta|) = A(S)B(S)|\eta|^\beta$.

Then (6.10) reduces to

$$p = AB \cdot \frac{4\pi}{15} \int_{r\le\varphi} r^{6+\beta}dr = \frac{AB \cdot 4\pi}{15(7+\beta)} \varphi^{7+\beta} \tag{6.13}$$

(6.4) reduces to

$$\lambda^2 = \varphi^2 \frac{3+\beta}{3(5+\beta)} \tag{6.14}$$

(6.12) reduces to

$$\rho \frac{\partial p}{\partial \rho} = \gamma p = \frac{\gamma \cdot AB \cdot 4\pi}{15(7+\beta)}\varphi^{7+\beta} = \tag{6.15}$$

$$= \{\frac{3+\beta}{3(5+\beta)} \cdot \varphi^2\}\{\frac{AB \cdot 4\pi}{3(5+\beta)} \varphi^{5+\beta}\}$$

Thus

$$\frac{\gamma}{5(7+\beta)} = \frac{3+\beta}{3(5+\beta)^2}$$

So

$$(6.16) \qquad \beta^2 + 10\beta + 25 - \frac{20}{5 - 3\gamma} = 0$$

For $\gamma < 5/3$ the equation has two real roots. B is a free parameter. If $\gamma \geq 5/3$ then $q(|\eta|)$ cannot be a power of $|\eta|$.

Take $\gamma = 1.4$ (the equation of state of air). It turns out that, in this case, we may choose $\beta = 0$. Moreover, if $\gamma = 1.4$ then (6.13) implies $\rho = C \cdot \varphi^5$ and

$$\frac{\partial p}{\partial \rho} = \frac{\partial p}{\partial \varphi} / \frac{\partial \rho}{\partial \varphi} = C_1 \cdot \varphi^2$$

So φ is a multiple of the speed of sound! B may be chosen in order to adjust the multiplier.

7. This work is devoted to the derivation of the approximate transport process and to prove consistency of various approximations. For the isentropic case the approximation is already in divergence form. Thus the approximation is valid, in a weak sense, also for discontinuous distributions. Still, for efficient numerical procedures, the space as well as time discretization will have to be analysed.

As for the general, variable entropy, case. The definition of f^* in (5.10) leads to equations in divergence form, but the definition of $h(x,\xi,0)$ involves the deformations $\frac{\partial u_i}{\partial x_j} + \frac{\partial u_j}{\partial x_i}$ which are infinite near a discontinuity. The singularity may be "integrated out" when appropriate approximation based on discrete cell data is used.

Extension to viscous flows poses a very interesting problem. There is no known derivation of viscosity terms in the compressible case which is convincing enough (at least not to the knowledge of the author). Since the terms in (5.7) are of "viscosity type" the derivation of the transport approximation may have a bearing also on the derivation of viscous flow models.

Finally, with proper definition of viscosity one may attempt to use this method for the approximation of incompressible flow by "hardly compressible flow" using, in turn, the transport approach for the latter.

These problems, as well as others, will have to be analysed in order to fully understand the principle of approximation by transport, introduced here.

SOME DECAY PROPERTIES OF SOLUTIONS OF THE NAVIER-STOKES EQUATIONS

George H. Knightly*
Department of Mathematics
University of Massachusetts
Amherst, MA 01003 / USA

I. INTRODUCTION

In this paper we investigate the rates of decay, in space and time, of classical solutions of the Navier-Stokes equations describing the flow of a viscous, incompressible fluid past a finite body. In a coordinate system attached to the body and moving with prescribed velocity $-b_\infty(t)$, the equations become

(1a) $w_t - \Delta w + w\cdot\nabla w + \nabla p = b_\infty'$, $x \in \Omega$, $t > 0$,

(1b) $\nabla\cdot w = 0$, $x \in \Omega$, $t > 0$,

where the vector $w(x,t)$ denotes the velocity relative to the body, the scalar $p(x,t)$ denotes the pressure and the flow region Ω is the exterior of a bounded domain in $\mathbb{R}^3$. Here we take unit kinematic viscosity and we assume there are only conservative external forces, which are then included in the pressure term. In addition, w satisfies prescribed initial and boundary values:

(1c) $w(x,0) = b_\infty(0) + \alpha(x)$, $x \in \Omega$

(1d) $w(x,t) = \gamma(x,t)$, $x \in \partial\Omega$, $t > 0$,

(1e) $w(x,t) \to b_\infty(t)$ as $|x| \to \infty$, $t > 0$.

Some of the properties we seek are exhibited by solutions of the Cauchy problem associated with (1), in which $\Omega = \mathbb{R}^3$ and (1d) is omitted. When $b_\infty(t) \equiv 0$ and α is smooth, divergence-free and satisfies

(2) $|\alpha(x)| \le A(1+|x|)^{-s}$, $x \in \mathbb{R}^3$,

for some $s \in [1,3)$, it is shown in [10] (see also [11;12]), for A sufficiently small, that problem (1) has a unique global solution, w_0, p_0, which satisfies

(3) $|w_0(x,t)| \le C[1+t+|x|^2]^{-s/2}$, $x \in \mathbb{R}^3$, $t > 0$,

for some constant $C > 0$. For the same α and general smooth $b_\infty(t)$, if we set

(4) $x_0(t) = \int_0^t b_\infty(\tau)d\tau$, $t > 0$,

then

(5) $w(x,t) = b_\infty(t) + w_0(x - x_0(t),t)$, $p(x,t) = p_0(x - x_0(t),t)$

---

*This research was supported by NSF Grants No. MCS77-04927 and MCS79-03555.

solves problem (1) and there holds, for some constant $C > 0$,

(6) $$|w(x,t) - b_\infty(t)| \le C[1 + t + |x - x_0(t)|^2]^{-s/2}, \quad x \in \mathbb{R}^3,\ t \ge 0.$$

The estimate (6) shows that the "disturbance" $w - b_\infty$ decays in time as $t^{-s/2}$, uniformly on $\mathbb{R}^3$, and decays in space as $|x|^{-s}$, pointwise for $t \ge 0$. In addition the center, $x_0(t)$, of the disturbance is swept downstream at the rate $b_\infty(t)$.

Another property we investigate is the wake. If we set $\gamma^*(x) = \lim_{t\to\infty} \gamma(x,t)$ and $\beta^* V = \lim_{t\to\infty} b_\infty(t)$, where β^* is a scalar and V a unit vector, then the stationary problem associated with (1)

(7a) $$-\Delta w + w\cdot\nabla w + \nabla p = 0, \quad x \in \Omega,$$

(7b) $$\nabla\cdot w = 0, \quad x \in \Omega,$$

(7c) $$w(x) = \gamma^*(x), \quad x \in \partial\Omega,$$

(7d) $$w(x) \to \beta^* V \ \text{ as } \ |x| \to \infty,$$

has been studied extensively by Finn [2;3;4]. When $\gamma^*(x) - \beta^* V$ is small in a suitable sense, Finn showed that (7) has a Physically Reasonable solution, $w(x) = \beta^* V + b(x)$, satisfying, for an appropriate constant $C > 0$,

(8) $$|b(x)| \le C|x|^{-1}, \quad x \in \Omega.$$

If $\beta^* \ne 0$ and $|\gamma^*(x) - \beta^* V| \le k\beta^*$ for some fixed constant k, then the solution possesses a wake in the direction V. One aspect of the wake is the inequality

(9) $$|b(x)| \le \beta^* C Q_0(x,0), \quad x \in \Omega,$$

where the constant C is independent of β^*. Here $Q_\varepsilon(x,t)$, $0 \le \varepsilon \le \frac{1}{2}$, is defined for $x \in \mathbb{R}^3$ and $t > 0$ by

(10) $$Q_\varepsilon(x,t) = \begin{cases} (1 + t + |x|^2)^{-1/2} & \text{in the paraboloid } \xi > 0,\ \eta^2 \le \xi, \\ (1 + t + |x|^2)^{-1/2-\sigma(1-2\varepsilon)} & \text{along the surfaces } \xi > 0,\ \xi^{1+2\sigma} = \eta^2, \text{ for } 0 \le \sigma \le \frac{1}{2}, \\ (1 + t + |x|^2)^{-1+\varepsilon} & \text{otherwise, i.e., outside the cone } \xi \ge 0,\ 0 \le \eta \le \xi, \end{cases}$$

where $\xi = |x| \cos\theta$, $\eta = |x| \sin\theta$ for $\theta = \theta(x) \in [0,\pi]$ defined by $\cos\theta = (x\cdot V)/|x|$.

We shall study solutions of (1) in terms of the disturbance velocity u and pressure q defined by

(11) $$w(x,t) = b_\infty(t) + b(x) + u(x,t), \quad p(x,t) = p(x) + q(x,t)$$

where $w(x) = \beta^* V + b(x)$, $p(x)$ is the corresponding solution of (7). We will derive for u estimates of the form

(12a) $$|u(x,t)| \le C[Q_{\frac{1}{2}}(x,t)]^s, \quad x \in \Omega,\ t > 0,$$

when $b \equiv \beta^* = 0$ and

(12b) $$|u(x,t)| \le C Q_\varepsilon(x - \beta^* tV, t) \quad x \in \Omega,\ t > 0$$

when $b_\infty(t) \equiv \beta^*V$ for $t \geq 1$. Both inequalities (12a,b) establish the type of decay property we seek. In addition, (12b) demonstrates, because of the β^*tV term, the convection of the disturbance in the V (i.e., downstream) direction with speed β^*. When $\varepsilon < \frac{1}{2}$ (12b) suggests that u possesses a paraboloidal wake region in the V-direction since a slower rate of decay is indicated within a paraboloid in the V direction than in other directions.

Our derivation of (12a,b) requires small data, a decay rate of the form (2) and some knowledge of the decay in time of u and the associated stress on $\partial\Omega$. When $b_\infty(t)$ and $\gamma(x,t)$ vanish, the assumed decay rates are possessed by certain global smooth solutions obtained by Heywood [9]; this application is discussed in section 4. In section 5 our results and related work in [1;7;8;9;14] are summarized.

The estimates (12) are obtained from a representation of u as solution of a nonlinear integral equation

$$(13) \qquad u = u_0 + F(u)$$

in terms of fundamental solution tensors of the linearized systems (Stokes or Oseen equations). The representation (13) is derived in section 2; u_0 is a sum of integrals depending on the initial and boundary data, on $b(x)$ and on the stress Tu on $\partial\Omega$, while $F(u)$ contains terms linear and quadratic in u. In section 3 we prove the inequalities (12) for solutions of (13) when u_0 is small in a suitable class.

The work reported here is a continuation of that in [13]. In particular, Theorem 2 and the associated lemmas are improved versions of the results in [13].

2. THE INTEGRAL REPRESENTATIONS

If we study the solution of (1) in the form (11), then we find that u,q solves the following problem, where $U(t) = b_\infty(t) - \beta^*V$,

$$(14a) \qquad u_t - \Delta u + \beta^*V\cdot\nabla u + \nabla q = -\{U\cdot\nabla b + U\cdot\nabla u + b\cdot\nabla u + u\cdot\nabla b + u\cdot\nabla u\}, \quad x \in \Omega,\ t > 0,$$

$$(14b) \qquad \nabla\cdot u = 0, \quad x \in \Omega,\ t > 0$$

$$(14c) \qquad u = -U + \gamma - \gamma^*, \quad x \in \partial\Omega,\ t > 0$$

$$(14d) \qquad u(x,t) \to 0 \text{ as } |x| \to \infty,\ t > 0, \text{ uniformly on bounded time intervals}$$

$$(14e) \qquad u(x,0) = \alpha(x) - b(x), \quad x \in \Omega.$$

The representation (13) for a solution u of (14) involves a fundamental solution tensor E of the Oseen equations

$$(15) \qquad u_t - \Delta u + \beta^*V\cdot\nabla u + \nabla q = 0, \quad \nabla\cdot u = 0.$$

The components of E are obtained from the following definitions (see Oseen [15] or [10;11;13]). Here δ_{ij} is the Kronecker delta symbol:

$$(16) \qquad \Phi_0(r,t) = (16\pi^3 t)^{-1/2}\int_0^1 \exp[-s^2r^2/(4t)]ds,$$

(17) $\Phi(x,t;\beta^*) = \Phi_0(|x - \beta^* tV|,t)$,

(18) $E_{ij} = -\Delta\Phi\delta_{ij} + \dfrac{\partial^2\Phi}{\partial x_i \partial x_j}$, for $i=1,2,3$,

(19) $E = (E_{ij}(x - y,t - \tau;\beta^*))$.

If E_j denotes the jth column of $E(x-y,t-\tau;\beta^*)$, then for $t > \tau$ the pair $(u,q) = (E_j,0)$ satisfies (15) in the (x,t)-variables and the adjoint system

(20) $u_t + \Delta u + \beta^* Vu + \nabla q = 0, \quad \nabla\cdot u = 0$

in the (y,τ)-variables. Moreover, E becomes singular at $(y,\tau) = (x,t)$ in such a way that for any smooth divergence-free vector field $u(x,t)$ we have

(21) $$\lim_{\tau\to t^-}\int_G u(y,\tau)\cdot E(x-y,t-\tau;\beta^*)dy = u(x,t) + \int_S \frac{y - x}{4\pi|y-x|^3}u(y,t)\cdot n d\sigma_y,$$

where G is any bounded region in $\mathbb{R}^3$, S its boundary, $x \in G$ and n is the unit exterior normal on S.

Let $G = \{y\in\Omega: |x-y| < R\}$ for large R, let u,q be a smooth solution of (14) such that

(22) $\nabla u(x,t),\ q(x,t) = O(|x|)$ as $|x| \to \infty$,

locally uniformly in t, and let $b(x)$ be bounded. Then the identity

(23) $$u\cdot[E_\tau + \Delta E + \beta^* V\cdot\nabla_y E] + E\cdot[u_\tau - \Delta u + \beta^* V\cdot\nabla u + \nabla q] = -E\cdot[U\cdot\nabla b + U\cdot\nabla u + b\cdot\nabla u + u\cdot\nabla b + u\cdot\nabla u]$$

may be integrated by parts over $G\times(\delta,t)$, using (21), (22) and (14), letting $R \to \infty$ then $\delta \to 0^+$ to obtain equation (13) with

(24) $$u_0 = \sum_{i=1}^{7} F_i(x,t), \quad F(u)(x,t) = \sum_{i=8}^{10} F_i(x,t),$$

where

(25) $$F_1(x,t) = \int_\Omega [b(y) - \alpha(y)]\Delta\Phi(x-y,t)dy,$$

$$F_2(x,t) = \int_{\partial\Omega} [-U(0) - \gamma^*(y) + \gamma(y,0)]\cdot n\nabla\Phi(x-y,t)d\sigma_y,$$

$$F_3(x,t) = \int_{\partial\Omega} [U(t)-\gamma(y,t)+\gamma^*(y)]\cdot n(y-x)[4\pi|y-x|^3]^{-1}d\sigma_y,$$

$$F_4(x,t) = \int_0^t\int_{\partial\Omega} [\gamma(y,\tau)\cdot nE(x-y,t-\tau)\cdot(U(\tau)-\gamma(y,\tau)+\gamma^*(y)) + (\gamma(y,\tau)-\gamma^*(y))\cdot nE(x-y,t-\tau)\cdot(\beta^* V-\gamma^*(y))]d\sigma_y d\tau,$$

$$F_5(x,t) = \int_0^t\int_{\partial\Omega} n\cdot TE(x-y,t-\tau)\cdot[U(\tau)-\gamma(y,\tau)+\gamma^*(y)]d\sigma_y d\tau,$$

$$F_6(x,t) = \int_0^t\int_{\partial\Omega} n\cdot Tu(y,\tau)\cdot E(x-y,t-\tau)d\sigma_y d\tau,$$

$$F_7(x,t) = \int_0^t\int_\Omega U(\tau)\cdot\nabla E(x-y,t-\tau)\cdot b(y)dyd\tau,$$

$$F_8(x,t) = \int_0^t\int_\Omega U(\tau)\cdot\nabla E(x-y,t-\tau)\cdot u(y,\tau)dyd\tau,$$

$$F_9(x,t) = \int_0^t\int_\Omega [b(y)\cdot\nabla E(x-y,t-\tau)\cdot u(y,\tau) + u(y,\tau)\cdot\nabla E(x-y,t-\tau)\cdot b(y)]dyd\tau,$$

$$F_{10}(x,t) = \int_0^t\int_\Omega u(y,\tau)\cdot\nabla E(x-y,t-\tau)\cdot u(y,\tau)dyd\tau.$$

Here the components of the stress tensor, Tu, are given by

$$(26)\qquad (Tu)_{ij} = -q\delta_{ij} + \left(\frac{\partial u_i}{\partial x_j} + \frac{\partial u_j}{\partial x_i}\right), \quad i,j=1,2,3,$$

and TE is formed using (26) with $q = 0$.

Remark. When dealing with F_6 in (25) we may replace $q(y,\tau)$ by $q(y,\tau) - q(y_0,\tau)$ in (26) for any fixed y_0, since the resulting contribution of the $q(y_0,\tau)$ term to the integral over $\partial\Omega$ in F_6 is zero:

$$\int_{\partial\Omega} q(y_0,\tau)n\cdot Ed\sigma_y = -q(y_0,\tau)\int_{\tilde\Omega}\nabla\cdot Edy = 0,$$

where $\tilde\Omega$ is the complement of Ω.

We assume throughout the paper that $\partial\Omega$ and the functions $U,\gamma,\gamma^*,\alpha,b$ are sufficiently smooth (e.g. C^3) functions of their arguments.

3. DECAY ESTIMATES

In this section we obtain properties of the solution u of (13) based on various assumptions concerning u_0; we regard u_0 as given data and $F(u)$ defined by (24). The following uniqueness statement ensures that we are always studying the vector field u of interest, and not some other solution of (13). The method of proof follows that of Theorem 3 in [12].

Lemma 1. If $U(t)$ and $b(x)$ are bounded, then, given $u_0(x,t)$, equation (13) has at most one solution $u(x,t)$ that is bounded and continuous on $\Omega\times[0,T]$.

In the next result we obtain an estimate of the form (12a).

Theorem 1. In the definition (24) of $F(u)$ let $b \equiv \beta^* = 0$ and let U satisfy

$$(27)\qquad |U(t)| \le \sigma(1+t)^{-1/2}, \quad t \ge 0,$$

for some constant $\sigma \ge 0$. Suppose the vector field u_0 is continuous and satisfies

$$(28)\qquad |u_0(x,t)| \le \delta[Q_{\frac{1}{2}}(x,t)]^s, \quad x\in\Omega,\ t>0,$$

for some $\delta > 0$ and some $s \in [1,2]$. Then there are positive numbers δ_0 and σ_0 depending only on s, such that if $0 \le \delta < \delta_0$ and $0 \le \sigma < \sigma_0$ then there exists a continuous solution u, of (13), which satisfies

(29) $|u(x,t)| \le C\delta[Q_{\frac{1}{2}}(x,t)]^s, \quad x \in \Omega,\ t > 0,$

with positive constant C depending only on σ_0.

In Theorem 2 we suppose that the body reaches its steady velocity $-\beta^* V$ in finite time, i.e., that

(30) $|U(t)| \le \begin{cases} \sigma, & \text{if } t \le 1 \\ 0, & \text{if } t \ge 1, \end{cases}$

for some $\sigma > 0$.

Theorem 2. In the definition (24) of $F(u)$ let $b(x)$ satisfy (9) and let $U(t)$ satisfy (30). Suppose the vector field $u_0(x,t)$ is continuous and satisfies

(31) $|u_0(x,t)| \le \delta Q_\varepsilon(x - \beta^* tV, t), \quad x \in \Omega,\ t > 0,$

for some $\varepsilon \in (0,\frac{1}{2}]$. There are positive numbers β_0, δ_0 and σ_0 depending only on ε such that if $0 < \beta^* < \beta_0$, $0 < \delta < \delta_0$ and $0 < \sigma < \sigma_0$ then equation (13) has a unique solution, which satisfies

(32) $|u(x,t)| \le 2\delta Q_\varepsilon(x - \beta^* tV, t), \quad x \in \Omega,\ t > 0.$

Theorems 1 and 2 are proved using properties of the linear integral operator $Lu \equiv F_8(u) + F_9(u)$ and the bilinear form

$$N(u,v) = \int_0^t\int_\Omega u\cdot\nabla E\cdot v \, dy d\tau.$$

In this notation $F_{10}(u) = N(u,u)$ and (13) becomes

(33) $u = u_0 + Lu + N(u,u).$

The proofs of Theorems 1 and 2 require the next two lemmas, which establish mapping properties of the operators L and N.

Lemma 2. Suppose that $b(x) \equiv \beta^* = 0$ and U satisfies (27). Let v^1 and v^2 be vector fields satisfying

(34) $|v^i(x,t)| \le k_i[Q_{\frac{1}{2}}(x,t)]^s, \quad x \in \Omega,\ t > 0,$

for $i=1,2$, some $s \in [1,2]$ and some constants k_1 and k_2. Then there are constants A_1 and H_1 depending only on s, such that for $x \in \Omega$, $t > 0$

(35) $|Lv^i(x,t)| \le k_i\delta A_1[Q_{\frac{1}{2}}(x,t)]^s, \quad i=1,2$

and

(36) $|N(v^1,v^2)(x,t)| \le k_1k_2H_1[Q_{\frac{1}{2}}(x,t)]^s.$

Lemma 3. Suppose b satisfies (9) and U satisfies (30). Let v^1 and v^2 be vector fields satisfying

(37) $|v^i(x,t)| \le k_iQ_\varepsilon(x - \beta^* tV, t), \quad x \in \Omega,\ t > 0,$

for $i=1,2$, some $\varepsilon \in (0,\frac{1}{2}]$ and some constants k_1, k_2. Then there are constants A_2 and H_2, independent of $\varepsilon, \beta^*, k_1$ and k_2, and a constant B depending only on an upper bound for β^*, such that for $x \in \Omega$ and $t > 0$,

(38) $\quad |Lv^i(x,t)| \le k_i\varepsilon^{-1}[\delta A_2 + \beta^{*\frac{1}{2}}B]Q_\varepsilon(x - \beta^* tV,t), \quad i=1,2,$

and

(39) $\quad |N(v^1,v^2)| \le k_1 k_2 \varepsilon^{-1} H_2 Q_\varepsilon(x - \beta^* tV,t).$

Lemma 3 is close to Lemma 3 in [13] and the proof differs only in a minor way; the proof of Lemma 2 is similar and will not be given here.

The following proof of Theorem 2 is based on Lemma 3; similar steps involving Lemma 2 yield Theorem 1. We insert a parameter λ in equation (33) to get

(40) $\quad u = u_0 + \lambda Lu + \lambda N(u,u)$

and we seek a solution of (40) in the form

(41) $\quad u(x,t) = \sum_{n=0}^{\infty} \lambda^n w_n(x,t).$

If we substitute (41) in (40) and equate coefficients of like powers of λ, we get the following formulas:

(42) $\quad w_0 = u_0, \quad w_{n+1} = Lw_n + \sum_{j=0}^{n} N(w_j,w_{n-j}), \quad n=0,1,2,\ldots.$

Setting $W_0 = 1$ and using (31) we have

$$|w_0(x,t)| \le \delta W_0 Q_\varepsilon(x - \beta^* tV,t), \quad x \in \Omega,\ t > 0.$$

Suppose that $|w_j(x,t)| \le \delta W_j Q_\varepsilon(x - \beta^* tV,t)$ for $j=0,1,\ldots n$. Then (42) and Lemma 3 imply that

$$|w_{n+1}(x,t)| \le \delta W_{n+1} Q_\varepsilon(x - \beta^* tV,t), \quad x \in \Omega,\ t > 0,$$

where

(43) $\quad W_{n+1} = \varepsilon^{-1}(\delta A_2 + \beta^{*1/2}B)W_n + \varepsilon^{-1}\delta H_2 \sum_{j=0}^{n} W_j W_{n-j}.$

Now the equation

(44) $\quad W = 1 + \lambda\varepsilon^{-1}(\delta A_2 + \beta^{*1/2}B)W + \lambda\varepsilon^{-1}\delta H_2 W^2$

has a solution

(45) $\quad W(\lambda) = \sum_{n=0}^{\infty} \lambda^n W_n$

analytic in a neighborhood of $\lambda = 0$, with $W_0 = 1$ and W_n given by (43) for $n=1,2,\ldots$. A calculation based on (44) shows that the circle of convergence of the series in (45) includes $\lambda = 1$ wherever

(46) $\quad (\sigma A_2 + \beta^{*1/2}B + 2\delta H_2) + [(\sigma A_2 + \beta^{*1/2}B + 2\delta H_2)^2 - (\sigma A_2 + \beta^{*1/2}B)^2]^{\frac{1}{2}} < \varepsilon$

If (46) holds, then convergence of the majorizing series (45) at $\lambda = 1$ implies convergence of (41) at $\lambda = 1$ to a solution of (40), such that

(47) $\quad |u(x,t)| \le \sum_{n=0}^{\infty} |w_n(x,t)| \le \delta W(1) Q_\varepsilon(x - \beta^* tV,t), \quad x \in \Omega,\ t > 0.$

The inequality (46) holds, e.g., if $0 \le \sigma < \sigma_0$, $0 \le \beta^* < \beta_0$ and $0 \le \delta < \delta_0$, where

σ_0, β_0 and δ_0 are defined by

$$\sigma_0 A_2 = \beta_0^{1/2} B = \delta_0 H_2 = \varepsilon/(4 + 2\sqrt{3}).$$

From these latter restrictions it follows that $W(1) \le 2$, which with (47) implies (32).

4. APPLICATION

To apply Theorems 1 and 2 to a solution of (14) we must show that the representation (13) is valid and verify the hypotheses (28) or (31) on u_0. We have shown how to obtain (13) for classical solutions of (14) in the class (22) but (13) can be obtained for other classes of solutions as well (e.g., related representations of generalized solutions are obtained in [6]).

Next we state conditions that imply the bounds (28) and (31). The proofs of Lemmas 4 and 5 are omitted; they are similar to the proof of Lemma 2 in [13].

Lemma 4. Let $b(x) \equiv \beta^* = 0$ and suppose for some $s \in [1,2]$, $\mu \in (0,1]$ that (2) and the following estimates hold:

(48) $|U(t)| \le U_0(1+t)^{-s/2}$, $t > 0$,

(49) $|U(t) - U(\tau)| \le U_0(1+\tau)^{-s/2}|t-\tau|^{\mu}$, $t \ge \tau > 0$,

(50) $|\gamma(x,t)| \le \gamma_0(1+t)^{-s/2}$, $x \in \partial\Omega$, $t > 0$,

(51) $|\gamma(x,t) - \gamma(y,\tau)| \le \gamma_0(1+\tau)^{-s/2}[|t-\tau| + |x-y|^2]^{\mu/2}\tau^{-\mu/2}$, $x,y \in \partial\Omega$, $t \ge \tau > 0$.

Let u,q be a solution of (14) for which the stress on $\partial\Omega$ satisfies

(52) $|Tu(x,t)| \le T_0(1+t)^{-s/2}$, $x \in \partial\Omega$, $t > 0$.

Then u_0, defined by (24), satisfies (28) with constant δ that vanishes with $|(A,U_0,\gamma_0,T_0)|$.

In the next result $|D^m\gamma(x,t)|$ and $|D^m\gamma^*(x)|$ denote bounds for the tangential mth derivatives on $\partial\Omega$.

Lemma 5. Suppose $b(x)$ satisfies (9), $U(t)$ satisfies (30), $\alpha(x)$ satisfies (2) with $s = 1$ and suppose that $\gamma(x,t) \equiv \gamma^*(x)$ for $t \ge 1$ and

(53) $|D^m\gamma(x,t)| \le \gamma_0$, $|D^m\gamma^*(x)| \le \gamma_0$,

for $m = 0,1$, $x \in \partial\Omega$ and $t > 0$. Let u,q be a solution of (14) for which the stress on $\partial\Omega$ satisfies

(54) $|Tu(x,t)| \le T_0(1+t)^{-1}$, $x \in \Omega$, $t > 0$.

Then u_0, defined by (24), satisfies (31) with $\varepsilon = \frac{1}{2}$ and constant δ that vanishes with $|(A,\sigma,\gamma_0,T_0,\beta^*)|$. If, in addition, $\alpha(x) \equiv 0$ and for some $\varepsilon \in [0,\frac{1}{2})$

(55) $|Tu(x,t)| \le T_0(1+t)^{2\varepsilon-2}$, $x \in \Omega$, $t > 0$,

then u_0 satisfies (31) with this same value of ε and constant δ that vanishes with $|(\sigma,\gamma_0,T_0,\beta^*)|$.

While most of the hypotheses of Lemmas 4 and 5 restrict the data of problem (14), the conditions (52), (54), (55) require knowledge of the solution. Such conditions seem reasonable in that certain global solutions of the Cauchy problem possess the decay rates in (52), (54) or (55) as $t \to \infty$, uniformly for $x \in \mathbb{R}^3$ (see [10]).

Now we show that Theorem 1 applies to some of the global solutions obtained by Heywood [9]. Suppose U, b, β^*, γ and γ^* are all zero and α lies in the closure of smooth solenoidal vector fields of compact support in Ω, under the norm

$$\|\alpha\|_{1,2} = [\|\alpha\|^2 + \|\nabla\alpha\|^2]^{\frac{1}{2}},$$

where $\|\cdot\|$ denotes the usual norm in $L^2(\Omega)$. If $\|\alpha\|_{1,2}$ is sufficiently small, then Heywood proves the existence of a global solution of (14). For large t this solution satisfies, in particular, $|u(x,t)| \le M_1 t^{-\frac{1}{2}}$ uniformly Ω, and $\|u_t(\cdot,t)\| \le M_2 t^{-1}$, with constants M_1 and M_2 that are small with $\|\alpha\|_{1,2}$. If also the second derivatives $D^2\alpha$ lie in $L^2(\Omega)$, then the solution is classical and satisfies the inequalities

$$(56) \quad |u(x,t)| \le M_3(1+t)^{-\frac{1}{2}}, \quad x \in \Omega,\ t \ge 0,$$

$$(57) \quad \|u_t(\cdot,t)\| \le M_4(1+t)^{-1}, \quad t \ge 0,$$

with constants M_3 and M_4 that vanish with $\|\alpha\|_{1,2} + \|D^2\alpha\|$. If we make the additional assumptions on α that (2) holds with $s = 1$ and $|\nabla\alpha(x)| \le A_1$ holds in a neighborhood of $\partial\Omega$, then the next lemma applies with $s = 1$ and Lemma 4 and Theorem 1 apply in turn to give for u an estimate on

$$(58) \quad \Omega_\nu = \{x \in \Omega\colon \text{distance from } x \text{ to } \partial\Omega \text{ exceeds } \nu\}$$

of the form (29) with $s = 1$, provided that $A, A_1, \|\alpha\|_{1,2}$ and $\|D^2\alpha\|$ are sufficiency small. This estimate, together with (56) implies that (29) with $s = 1$ holds for the full set Ω.

We use estimates such as (56), (57) to verify the hypotheses of Lemma 4, not on $\partial\Omega$ but on $\partial\Omega_\nu$, where Ω_ν is given by (58). Let A_ν denote the annular region between $\partial\Omega$ and $\partial\Omega_{2\nu}$. The proof of the following lemma again involves representing $u(x,t)$ in terms of integrals, but now a truncated fundamental solution is used as in Fujita and Kato [6]. From such representations, estimates of the form

$$(59) \quad |\nabla u(x,t)| \le M(1+t)^{-\frac{1}{2}},\ |\nabla q(x,t)| \le M(1+t)^{-\frac{1}{2}}$$

are derived for $x \in \partial\Omega_\nu$, $t > 0$. Using the mean value theorem and the Remark following (26) one sees that the bounds (59) imply (52), with $s = 1$, for dealing with the term F_6 in (25).

<u>Lemma 6</u>. Suppose $U, b, \beta^*, \gamma, \gamma^*$ are zero in (14) and α satisfies $|\alpha(x)| \le A_1$ for $x \in A_\nu$, some $\nu > 0$. If u, q is a classical solution of (14) such that

$$(60) \quad |u(x,t)| \le M(1+t)^{-s/2}, \quad x \in A_\nu,\ t > 0,$$

and

(61) $\quad [\int_{A_\nu} |u_t(x,t)|^2 dx]^{\frac{1}{2}} \le M(1+t)^{-s/2}, \quad t > 0,$

then on $\partial\Omega_\nu$ the hypotheses (50), (51) and (52) of Lemma 4 are satisfied with constants γ_0 and T_0 that vanish with $|(A_0,A_1,M)|$.

5. SUMMARY

Here we summarize the foregoing work and indicate other possible ways to deduce decay properties similar to (12) for solutions of problem (1).

When the steady velocity is zero, $w(x) \equiv 0$, and $\alpha(x)$, $b_\infty(t)$ and $\gamma(x,t)$ satisfy the hypotheses of Lemma 4, then Theorem 1 together with Lemma 4 shows for small data that a solution, $w = b_\infty + u$, of problem (1) satisfying (52) with small T_0 also possesses an estimate

(61) $\quad |w(x,t) - b_\infty(t)| \le C[Q_{\frac{1}{2}}(x,t)]^s \quad x \in \Omega, t > 0.$

On the other hand, if $w(x) \equiv 0$, $b_\infty(t) \equiv 0$, $\gamma(x,t) \equiv 0$, $s = 1$ and α satisfies (2) and the hypotheses of Heywood leading to (56), then the extra condition (52) is not needed and from Theorem 1, Lemma 4 and Lemma 6 it follows for small data that the global solutions of problem (1) obtained in [9] also satisfy

(62) $\quad |w(x,t)| \le CQ_{\frac{1}{2}}(x,t) \quad x \in \Omega, t > 0.$

One way to obtain (62) in some cases, without appealing to Theorem 1, is suggested by results of Bemelmans [1] for problem (1) with $\gamma \equiv 0$ and $b_\infty(t) \equiv \beta^*V$ (for $t \ge 1$, say). If α satisfies (2) and lies in the domain of an appropriate fractional power of a certain linear operator then for small data Bemelmans shows there exists a solution $w(x,t)$, global in time and satisfying

(63) $\quad |w(x,t) - \beta^*V| \le C(1+|x|)^{-1}, \quad x \in \Omega, t > 0.$

Thus, if $\beta^* = 0$, if α also satisfies the hypotheses of Heywood leading to the estimate (56) and if one can show under these conditions that the Heywood and Bemelmans global solutions are the same, then (62) is an immediate consequence of (56) and (63).

Next, we give a brief discussion of the starting problem concerning the acceleration to steady velocity in a finite time of a finite body initially at rest in a fluid occupying its entire three dimensional exterior Ω. This corresponds to problem (1) or problem (14) with $\alpha = 0$; we also suppose that (30) holds, $\gamma(x,t) \equiv \gamma^*(x)$ for $t \ge 1$, $U(0) = 0$ and $\gamma(x,0) = 0$. Then the initial value in (14e) is $u(x,0) = -b(x)$ and we speak of solutions of (14) as "finite energy" solutions if $b(x) \in L^2(\Omega)$ (this is of course in addition to (9)). Finn [5] has shown that $b(x) \in L^2(\Omega)$ if and only if the net force exerted by the fluid on the body is balanced by a certain momentum flux across the boundary. In particular, if $\beta^* \ne 0$ and $\gamma^*(x) \equiv 0$ on $\partial\Omega$ then

$b(x) \notin L^2(\Omega)$. Thus both finite energy solutions and non-finite energy solutions are of interest.

In [7] Heywood proved the existence of global finite energy solutions of the starting problem for small data. (See also [8] where the linear starting problem is treated.) Masuda [14] showed that these solutions decay to steady state uniformly as $t^{-1/8}$, while Heywood [9] obtained an improved decay rate $t^{-1/4}$. Thus

$$(64) \qquad |w(x,t) - w(x)| \le C(1+t)^{-1/4} \qquad x \in \Omega, t > 0.$$

On the other hand, Bemelmans (see [1] and also his paper in these Proceedings) has obtained non-finite energy solutions in the case $\gamma = \gamma^* = 0$ and $b_\infty(t)$ small. These solutions satisfy

$$(65) \qquad |w(x,t) - b_\infty(t)| \le C(1+|x|)^{-1}, \qquad x \in \Omega, t > 0.$$

Since $\beta^* \equiv b(x) = 0$ is the only steady finite energy solution when $\gamma^* = 0$, the Heywood and Bemelmans results (64), (65) can overlap only when $w(x) \equiv 0$ and $b_\infty(t) = 0$ for $t > 1$. But in this case the results of [9] leading to (56) apply. So the exponent 1/4 in (64) can be replaced by 1/2 and again the estimate (62) is obtained.

Finally, we observe that Theorem 2 and Lemma 5, taken together, offer the prospect of the better estimate (32) for a solution of the starting problem provided (54) or (55) can be verified. If (32) holds then the disturbance decays uniformly as $t^{-1/2}$ is carried downstream with velocity $\beta^* V$ and, if $\varepsilon < \frac{1}{2}$, also possesses a paraboloidal wake in the direction V.

REFERENCES

1. Bemelmans, J., Eine Aussenraumaufgabe für die instationären Navier-Stokes-Gleichungen, *Math. Z. 162*(1978), 145-173.

2. Finn, R., Estimates at infinity for stationary solutions of the Navier-Stokes equations, *Bull. Math. Soc. Sci., Math. Phys. R.P. Roumaine, 3*(51) (1959), 387-418.

3. Finn, R., On the exterior stationary problem for the Navier-Stokes equations, and associated perturbation problems, *Arch. Rational Mech. Anal. 19*(1965), 363-406.

4. Finn, R., Mathematical questions relating to viscous fluid flow in an exterior domain, *Rocky Mt. J. Math. 3*(1973), 107-140.

5. Finn, R., An energy theorem for viscous fluid motions, *Arch. Rational Mech. Anal. 6*(1960), 371-381.

6. Fujita, H. and T. Kato, On the Navier-Stokes initial value problem, I, Stanford University Technical Report #131, 1963.

7. Heywood, J., The exterior nonstationary problem for the Navier-Stokes equations, *Acta. Math. 129*(1972), 11-34.

8. Heywood, J., On nonstationary Stokes flow past an obstacle, *Indiana Univ. Math. J. 24*(1974), 271-284.

9. Heywood, J., The Navier-Stokes equations: on the existence, regularity and decay of solutions, preprint, University of British Columbia, 1978.

10. Knightly, G., On a class of global solutions of the Navier-Stokes equations, *Arch. Rational Mech. Anal.* *21*(1966), 211-245.

11. Knightly, G., Stability of uniform solutions of the Navier-Stokes equations in n-dimensions, (Tech. Summary Rep. 1085, Mathematics Research Center, United States Army, University of Wisconsin, Madison, 1970).

12. Knightly, G., A Cauchy problem for the Navier-Stokes equations in R^n. *SIAM J. Math. Anal.* *3*(1972), 506-511.

13. Knightly, G., Some asymptotic properties of solutions of the Navier-Stokes equations, *Dynamical Systems*, A.R. Bednarek and L. Cesari, eds., Academic Press, New York, 1977, 139-155.

14. Masuda, K., On the stability of incompressible viscous fluid motions past objects, *J. Math. Soc. Japan* *27*(1975), 294-327.

15. Oseen, C. W., *Neuere Methoden und Ergebnisse in der Hydrodynamik*, Akademische Verlagsgesllschaft, Leipsig, 1927.

THE IMPLICIT DIFFERENCE SCHEMES FOR NUMERICAL SOLVING THE NAVIER-STOKES EQUATIONS

V.M.Kovenya, N.N.Yanenko
Institute of Theoretical & Applied Mechanics
USSR Academy of Sciences, Novosibirsk 630090

The increase in the dimensionality and the complication of computed region geometries impose certain requirements on the methods employed: the method should be economical, possess a sufficient accuracy and be simple in operating. Explicit difference scheme can prove to be ineconomical in view of the rigid restrictions imposed on the stability, especially when solving the problems at moderate and low Reynolds numbers. The use of non-uniform grids makes in fact impossible their application when solving the multidimensional problems. Therefore lately the researchers' main efforts are directed at the development of implicit or hybrid difference schemes. The methods review of solving the Navier-Stokes equations up to 1975 is presented in [1] .

The difficulties in constructing and operating the implicit difference schemes grow with the increase in the equation dimensionality. The splitting-up [2] and factorization [3]methods are practised on a large scale for constructing the economical schemes. These methods allow to reduce the solution of multidimensional problems to a set of their one-dimensional analogues. Nowadays a considerable number of implicit difference schemes for solving the multidimensional equations (see, for example,[4-15]) using the ideas of the both methods of factorization and splitting-up is developed.

As known [16]the application of difference schemes based on the approximation of differential equations in the divergent form permits to increase the calculation accuracy as the conservation laws both for each element of the computed cell and for the whole computed region for the present schemes are implemented. The divergent difference schemes obtained on the approximation basis are as a rule nonlinear with respect to the upper time layer and for their realization either iteration methods or linearization of initial nonlinear schemes are employed. While using the first approach the schemes are conservative both for stationary and nonstationary equations. While using the second approach the difference schemes are conservative only in the stationary case.

A considerable number of problems is devoted to the solution of stationary are weakly changing in time Navier-Stokes equations for a compressible gas. For this class of solutions the second approach is widely used. The increase in the calculation accuracy may be achieved by using the schemes of raised approximation order. However the operation of these schemes is more complex in comparison with the schemes of the first or second approximation order. The second approach is based on the difference schemes application on the non-uniform grids which condense in a region of large gradients and are solved in a region of small gradients. There are some approaches of constructing the adapted to the solution grids, for example, [17-19] which allow to increase essentially the calculation accuracy.

The present paper is devoted to constructing the implicit absolutely stable difference schemes for numerical solution of Navier-Stokes equations for a compressible gas. The schemes proposed are based on the splitting of the differential operators in the physical processes and the space variables, that enables one to construct the economical schemes that are operated by scalar sweeps. To increase the calculation accuracy, a moving difference grid [19] which adapts itself automatically to the solution may be used.

Some means of constructing the difference grids are discussed in the first part of the paper and the economical difference schemes are proposed for numerical solution of one-dimensional Navier-Stokes equations for compressible gas. In the second part of the paper the generalization of the one-dimensional case is given. In the third paragraph the constructing of the moving grid automatically adapting to the solution (when considering the one-dimensional grid condensation) is described.

1. One-dimensional case

1. Let us present the system of Navier-Stokes equations in the vector form

$$\frac{\partial \vec{F}}{\partial t} + \frac{\partial}{\partial x}\vec{W} = 0 \qquad (1)$$

where

$$\vec{F} = \begin{pmatrix} \rho \\ \rho u \\ E \end{pmatrix}, \quad \vec{W} = \begin{pmatrix} \rho u \\ \rho u^2 + p - \frac{4}{3}\mu\frac{\partial u}{\partial x} \\ u(E+p) - \frac{4}{3}\mu u\frac{\partial u}{\partial x} - \lambda\frac{\partial T}{\partial x} \end{pmatrix}, \quad E = \rho\left(e + \frac{u^2}{2}\right)$$

To close the system of equations (1), we set the equation of state $p = p(\rho, e)$, the law of dependence of viscosity coefficients and thermal conductivity, for instance, as a function of temperature, and the connection between internal energy and temperature $e = e(\rho, T)$.

Assume that the numerical solution of the system of equations (1) is sought in the domain $D\{0 \le t \le t^1, 0 \le x \le x^1\}$ with appropriate boundary conditions. Let us introduce the difference grid $D_{\tau h}$ with the steps τ and h. We shall determine the difference functions f_j^n in the mesh points n, j. The second derivatives in the operator $\partial \vec{W}/\partial x$ we approximate by symmetrical differences with the second order $(\frac{\partial}{\partial x} a \frac{\partial}{\partial x} \approx \bar{\Lambda} a \Lambda)$, and the first derivatives we approximate either by central differences or by non-symmetrical operators with regard for the velocity sign $\Lambda^k_\mp$ with the order k using the formulae (at h =Const.) :

$$\Lambda^1_\mp f_j = \pm \frac{1}{h}(f_j - f_{j\mp 1}) \quad , \quad \Lambda^2_\mp f_j = \pm \frac{1}{2h}(3f_j - 4f_{j\mp 1} + f_{j\mp 2}) \tag{2}$$

$$\Lambda^2 f_j = \pm \frac{1}{2h}(f_{j+1} - f_{j-1}) \quad , \quad \bar{\Lambda} a \Lambda = \frac{1}{h}[a_{j+1/2}\Lambda^1_+ - a_{j-1/2}\Lambda^1_-] ,$$

$$a_{j\pm 1/2} = \frac{1}{2}(a_j + a_{j\pm 1})$$

In what follows, for simplicity of the designation we shall designate the operator $\partial \vec{W}/\partial x$ approximation by $\Lambda^k \vec{W}$ realizing that the first and second derivatives are approximated with the help of the formulae (2).

The difference scheme

$$\frac{\vec{F}^{n+1} - \vec{F}^n}{\tau} + \Lambda^k[\alpha \vec{W}^{n+1} + (1-\alpha)\vec{W}^n] = 0 \tag{3}$$

approximates the system of equations (1) with the order $O(\tau^z + h^k)$, where $z=1$ at $\alpha \neq 0,5$ and $z=2$ at $\alpha=0,5$; $k=1,2$ and is conservative (the difference laws of the mass, impulse and total energy conservation are implemented). The solution of difference equations may be obtained by means of iterations.

To obtain the iterationless scheme, we shall linearize the vectors $\vec{F}^{n+1}$ and $\vec{W}^{n+1}$ with respect to the vector

$$\vec{F}^{n+1} = \vec{F}^n + \tau A^n \frac{\vec{f}^{n+1} - \vec{f}^n}{\tau} + O(\tau^2)$$
$$\vec{W}^{n+1} = \vec{W}^n + \tau B^n \frac{\vec{f}^{n+1} - \vec{f}^n}{\tau} + O(\tau^2) \tag{4}$$

where $A = \partial \vec{F}/\partial \vec{f}$, $B = \partial \vec{W}/\partial \vec{f}$. An analogous linearization was used apparently in the paper [4] for the first time. With regard for the relations (4) the scheme (3) may be presented in the form:

$$(A^n + \tau\alpha \Lambda^k B^n)\frac{\vec{f}^{n+1} - \vec{f}^n}{\tau} = -\Lambda^k \vec{W}^n \tag{5}$$

The scheme of the form (5) at $\vec{f} = \vec{F}$ is proposed in the papers [7,14], at $\vec{f} \neq \vec{F}$ - in the paper [8]. The difference scheme (5) approximates the system of equations (1) with the same order in space, at $\vec{F} = \vec{f}$ - in the time is conservative in the case of obtaing the stationary solution, but it does not require the iteration procedure. In the case of Navier-Stokes equations the matrix B has a complex structure on account of linearizing the terms which contain the second derivatives. The scheme is operated by the vector sweeps.

To obtain the solution of stationary equations by means of the relaxation method or the solution weakly depending on the time, the economical difference schemes operated by scalar sweeps are proposed in the paper [12] . Let us present the system of equations (1) in the form

$$\frac{\partial \vec{f}}{\partial t} = - A^{-1} \frac{\partial \vec{W}}{\partial x} \tag{6}$$

where as $\vec{f}$ the vector with the components ρ , u , e either ρ , u p , or ρ , u , T may be chosen. The matrix $A = \partial \vec{F} / \partial \vec{f}$ is a non-degenerate. For numerical solution of equations (6) we shall consider the difference scheme

$$C \frac{\vec{f}^{n+1} - \vec{f}^{n}}{\tau} = -(A^n)^{-1} \Lambda^k \vec{W}^n \tag{7}$$

A stabilizing operator $C = \prod\limits^{2} (I + \tau \alpha B_j^n)$ is chosen from the conditions of scheme's absolute stability and economy. In [12,13] as C the operator is chosen obtained when splitting the equations in terms of the physical processes, and in the multidimensional case when splitting the equations also in terms of the space variables. For instance, if the vector $\vec{f}$ has the components ρ , u , e , then the matrices B_1 and B_2 where $B = B_1 + B_2$ have the form:

$$B_1^n = \begin{pmatrix} u^n \Lambda_{\mp}^k & 0 & 0 \\ 0 & u^n \Lambda_{\mp}^k - \frac{4}{3\rho^n} \bar{\Lambda} \mu^n \Lambda & 0 \\ 0 & 0 & u^n \Lambda_{\mp}^k - \frac{1}{\rho^n} \bar{\Lambda} \lambda^n \Lambda \end{pmatrix}, \quad B_2^n = \begin{pmatrix} 0 & \rho^n \Lambda_{\mp}^k & 0 \\ (a^2)^n \Lambda_{\pm}^k & 0 & (b^2)^n \Lambda_{\pm}^k \\ 0 & (c^2)^n \Lambda_{\mp}^k & 0 \end{pmatrix}.$$

Note for obtaining an absolute stable scheme the pressure terms in the operator B_2 are approximated with respect to conjugated to the convective terms formulae. The right-hand sides of the difference scheme (7) are approximated with the order k in concord with the approximation of the operators B_1 and B_2 . The difference scheme (7) approximates the governing equations (1) with the order $O(\tau + h^k)$ and is absolutely stable at $\alpha \geqslant 0{,}5$ and in the case of obtaining the stationary solution is conservative. $a^2 = \frac{1}{\rho} \frac{\partial p}{\partial \rho}$, $b^2 = \frac{1}{\rho} \frac{\partial p}{\partial e}$, $c^2 = \frac{p}{\rho}$.
An equivalent scheme in fractional steps (I-is a single matrix) is employed for numerical operating:

$$(I + \tau \alpha B_1^n) \vec{\xi}^{n+1/2} = -(A^n)^{-1} \Lambda^k \vec{W}^n$$

$$(I + \tau \alpha B_2^n) \vec{\xi}^{n+1} = \vec{\xi}^{n+1/2} \tag{8}$$

$$\vec{f}^{n+1} = \vec{f}^{n} + \tau \vec{\xi}^{n+1}$$

At each fractional step the difference equation can be solved by scalar sweeps.

A shortcoming of the scheme of the type (5) is the operating complicasy and non-economy - the matrix sweeps, and the one of the scheme of the type (7) is the non-divergent form of the stabilizing operator. One can expect the usage of factorization schemes that are operated by scalar sweeps enables one firstly to increase the calculation accuracy, secondly, decrease the number of iterations up to the convergence as the stabilizing operator is chosen in the divergent form (for the terms containing the derivatives in space variables). Some difference schemes will be considered below that are constructed on the bases of the operator $A^n + \tau\alpha\Lambda^k B^n$ factorization so that the schemes possessed the property of absolute stability but were operated by scalar sweeps.

II. It seems to be obvious that both the scheme operating and the calculating accuracy may essentially depend on the choice of unknown functions. In the case of Navier-Stokes equations it seems to be expedient to choose the vector $\vec{f}$ with components ρ , $m = \rho u$ and p as unknown functions. This choice is stipulated by the following circumstances.

1. Navier-Stokes equations in the variables ρ , m , p have a sufficiently simple form, and hence can be simply operated.

2. The continuity and motion equations are written in the divergent form that allows to increase the calculation accuracy as the difference laws of the mass and impulse conservation are precisely implemented.

3. The equation of energy may be used both in the divergent and non-divergent form. In the second case one can follow the implementation of the internal energy conservation law, whereas the divergent form of the energy equations gives the implementation only of the difference law of the total energy conservation.

Let us consider the difference schemes for the different presentation of the energy equation.

Present the system of equations (1) in the form

$$\frac{\partial \vec{f}}{\partial t} = -A^{-1}\frac{\partial \vec{W}}{\partial x} = -\vec{W}_0 , \tag{9}$$

where

$$\vec{f} = \begin{pmatrix} \rho \\ m \\ p \end{pmatrix}, \quad A = \begin{pmatrix} 1 & 0 & 0 \\ 0 & 1 & 0 \\ -\frac{u^2}{2} & u & e \end{pmatrix}, \quad \vec{W}_0 = \begin{pmatrix} \partial m/\partial x \\ \frac{\partial}{\partial x}\left[p + m^2/\rho - \frac{4}{3}\mu\frac{\partial}{\partial x} m/\rho\right] \\ u\frac{\partial p}{\partial x} + c^2\left(\frac{\partial m}{\partial x} - u\frac{\partial \rho}{\partial x}\right) - \frac{\partial}{\partial x}\lambda\frac{\partial}{\partial x}\frac{p}{\rho} - \frac{4}{3}u\frac{\partial}{\partial x}\frac{m}{\rho} \end{pmatrix}$$

Here for simplicity the equation of state is taken in the form $p = \rho e/\ell$ and $c^2 = P/\rho$, $u = m/\rho$. $\ell = 1/(\gamma - 1)$

For numerical solving the equations (9) we shall consider the difference schemes of the universal algorithm type

$$C\,\frac{\vec{f}^{n+1} - \vec{f}^{n}}{\tau} = -\vec{W}_0^{\,n} \tag{10}$$

or

$$A^n C\,\frac{\vec{f}^{n+1} - \vec{f}^{n}}{\tau} = -\Lambda^k \vec{W}^{n} \tag{11}$$

where $\vec{W}_0^{\,n}$ and $\Lambda^k \vec{W}^n$ is the difference approximation of appropriate difference operators with the order k . In the case of obtaining the stationary solution the difference scheme (11) is conservative. The stabilizing operator C is chosen so that the difference schemes (10), (11) were a) absolutely stable at the appropriate parameter value α, b) were operated by scalar sweeps, c) required a minimum number of computations by operating. Present the operator C in the form,

$$C = (I + \tau\alpha B_1^n)(I + \tau\alpha B_2^n) \tag{12}$$

where $B = B_1 + B_2$. The difference operator $B = \partial\vec{W}_0/\partial\vec{f}$ is obtained by linearizing the operator W_0^{n+1} with the help of the formulae (4) under assumption that the viscosity and thermal conductivity coefficients are taken into account at n-time layer. When approximating the viscous terms $1/\rho$ is taken into account also at n-layer.

There exist four splittings of the operator B at B_1 and B_2 for which the conditions a), b) are implemented. The latter condition satisfying leads to the following form of the operators

$$B_1^n = \begin{pmatrix} 0 & 0 & 0 \\ -\Lambda^k_{\mp}(u^2)^n & \Lambda^k_{\mp} u^n - \bar{\Lambda}\mu^n \Lambda \frac{1}{\rho^n} & 0 \\ (uc^2)^n \Lambda^k_{\mp} & 0 & \Lambda^k_{\mp} u^n - \bar{\Lambda}\lambda^n \Lambda \frac{1}{\rho^n} \end{pmatrix}, \quad B_2^n = \begin{pmatrix} 0 & \Lambda^k_{\mp} & 0 \\ 0 & \Lambda^k_{\mp} u^n & \Lambda^k_{\pm} \\ 0 & (c^2)^n \Lambda^k_{\mp} & 0 \end{pmatrix}$$

To obtain a stable scheme (10) or (11) the term with pressure in the operator B_2 is approximated with respect to conjugated to the convective terms formulae. For numerical scheme (10) operating it is employed either fractional steps of the form (8) or the scheme of the stabilizing correction type

$$\begin{aligned} &\frac{\vec{f}^{n+1/2} - \vec{f}^{n}}{\tau} + \alpha B_1^n (\vec{f}^{n+1/2} - \vec{f}^{n}) = -\vec{W}_0^{\,n} \\ &\frac{f^{n+1} - f^{n+1/2}}{\tau} + \alpha B_2^n (\vec{f}^{n+1} - \vec{f}^{n}) = 0 \end{aligned} \tag{13}$$

Analogously to (13) the scheme in fractional steps

$$(I + \tau\alpha B_1^n)\vec{\xi}^{n+1/2} = -(A^n)^{-1}\Lambda^k \vec{W}^n \tag{14}$$

$$(I + \tau\alpha B_2^n)\vec{\xi}^{n+1} = \vec{\xi}^{n+1/2}$$

$$\vec{f}^{n+1} = \vec{f}^n + \tau\vec{\xi}^{n+1}$$

is equivalent to the scheme (11).

As follows from the form of operators B_1 and B_2 the difference schemes (10), (11) or equivalent schemes (13,14) are operated by scalar sweeps. In fact, all the difference equations can be solved with respect to $\vec{f}^{n+1/2}$ or $\vec{\xi}^{n+1/2}$ independently one from another at the first fractional step as the matrix is lower triangular. At the second fractional step, for instance, for the schemes (14) eliminating the pressure from the equation of motion, we obtain for ξ_m^{n+1} the difference equation

$$[I + \tau\alpha\Lambda_{\mp}^k u^n - \tau^2\alpha^2\Lambda_{\pm}^k (c^2)^n \Lambda_{\mp}^k]\,\xi_m^{n+1} = \xi_m^{n+1/2} - \tau\alpha\Lambda_{\pm}^k \xi_p^{n+1/2}$$

which is solved by scalar sweep, after this procedure ξ_ρ^{n+1} and ξ_p^{n+1} are explicitly computed. Using the Fourier technique for linearized equations one can easy show that the difference schemes (10), (11) are absolutely stable at $\alpha \geqslant 0,5$.

Consider the difference scheme for solving the equations (11) in the form

$$C\,\frac{\vec{\xi}^{n+1} - \vec{\xi}^n}{\tau} = -\Lambda^k \vec{W}^n \tag{15}$$

where the stabilizing operator C is used in the divergent form. Present C in the form

$$C = (I + \tau\alpha\Lambda^k B_1^n)(A^n)^{-1}(I + \tau\alpha\Lambda^k B_2^n)$$

Here

$$B_1 = \begin{pmatrix} 0 & 0 & 0 \\ -u^2 & u - \frac{4}{3}\mu\Lambda\frac{1}{\rho} & 0 \\ -u(u^2+s^2) & u^2 - \frac{4}{3}\mu u\Lambda\frac{1}{\rho} & \gamma e u - \lambda\Lambda\frac{1}{\rho} \end{pmatrix}, \quad B_2 = \begin{pmatrix} 0 & 1 & 0 \\ 0 & u^n & 1 \\ 0 & s^2 + \frac{u^2}{2} & 0 \end{pmatrix}$$

(the matrix B_1 and B_2 are written for the equation of state $p = \rho e/e$). The operator A is determined in (9).

The difference scheme

$$(A^n + \tau\alpha \Lambda^k B_1^n)\vec{\xi}^{\,n+1/2} = -\Lambda^k \vec{W}^n$$

$$(A^n + \tau\alpha \Lambda^k B_2^n)\vec{\xi}^{\,n+1} = A^n \vec{\xi}^{\,n+1/2} \tag{16}$$

$$\vec{f}^{\,n+1} = \vec{f}^{\,n} + \tau \vec{\xi}^{\,n+1}$$

is equivalent in whole steps to the scheme (15), approximates the governing system of equations with the order $O(\tau + h^k)$, is absolutely stable at $\alpha \geq 0,5$ and conservative in the case of obtaining the stationary solution.

Notion I. In the above operators B , B_1 and B_2 the non-symmetrical approximation of the first derivatives with the order k $(k=1,2)$ was employed.By using the symmetrical approximation of the second order the solution of the difference equations can be obtained by scalar three-point sweeps and k =2 if the term $\Lambda^2 a \Lambda^2$ approximate on the three-point (and not on the five-point one) in the equation of motion after eliminating the pressure. A shortcoming of the symmetric approximation is a poor conditionality of the sweep technique at $\tau u/h \geq 1$ that requires employing the special methods for solving poorly stipulated systems of the difference equations.

Notion 2. For the gas dynamic equation ($\mu = \lambda$ =0) the presentation of the operator B in the form of sum of upper and lower triangular operators

$$B_1 = \begin{pmatrix} 0 & 0 & 0 \\ -\Lambda^k_{\mp} u^2 & \Lambda^k_{\mp} u & 0 \\ uc\Lambda^k_{\mp} & c^2\Lambda^k_{\mp} & u\Lambda^k_{\mp} \end{pmatrix}, \quad B_2 = \begin{pmatrix} 0 & \Lambda^k_{\mp} & 0 \\ 0 & \Lambda^k_{\mp} u & \Lambda^k_{\pm} \\ 0 & 0 & 0 \end{pmatrix}$$

enables one to construct the difference scheme equivalent to the implicit scheme of "running" calculation at each fractional step and absolutely stable at $|u| \geq c$.

2. Multidimensional case

By constructing the difference schemes for the multidimensional Navier-Stokes equations the splitting of equations (or the factorization of operators) in the terms of the physical processes and the space variables. We shall consider the difference schemes for two-dimensional equations. The generalization on the space case can be carried out similarly to the two-dimensional case.

Consider the system of Navier-Stokes equations in cartesian coordinates in the following form:

$$\frac{\partial \vec{f}}{\partial t} = -\sum_{j=1}^{2} \vec{W}_{0j} = -A^{-1}\sum_{j=1}^{2}\frac{\partial \vec{W}_j}{\partial x_j} ,$$

where

$$\vec{f} = \begin{pmatrix} \rho \\ u\rho \\ v\rho \\ p \end{pmatrix} = \begin{pmatrix} \rho \\ m \\ n \\ p \end{pmatrix}, \quad \vec{W}_1 = \begin{pmatrix} m \\ m^2/\rho + p - G_{11} \\ mn/\rho - G_{12} \\ mH/\rho - \lambda\frac{\partial T}{\partial x_1} - uG_{11} - vG_{12} \end{pmatrix}$$

$$\vec{W}_2 = \begin{pmatrix} n \\ mn/\rho - G_{12} \\ n^2/\rho + p - G_{22} \\ nH/\rho - \lambda\frac{\partial T}{\partial x_2} - uG_{12} - vG_{22} \end{pmatrix}, \quad E = ep + \frac{m^2+n^2}{2\rho}, \quad H = E + p.$$

$$G_{11} = 2\mu\frac{\partial u}{\partial x_1} + (\zeta' - \tfrac{2}{3}\mu)\,\mathrm{div}\,\vec{v}, \quad G_{22} = 2\mu\frac{\partial v}{\partial x_2} + (\zeta' - \tfrac{2}{3}\mu)\,\mathrm{div}\,\vec{v}$$

$$G_{12} = \mu\left(\frac{\partial u}{\partial x_2} + \frac{\partial v}{\partial x_1}\right), \quad \mathrm{div}\,\vec{v} = \frac{\partial u}{\partial x_1} + \frac{\partial v}{\partial x_2}.$$

and the equation of state is taken in the form $p = e\rho/e$. By analogy with the one-dimensional case, let us consider the difference scheme of the universal algorithm type

$$C\,\frac{\vec{f}^{n+1} - \vec{f}^{n}}{\tau} = -(A^n)^{-1}\sum_{j=1}^{2}\Lambda_j^k\,\vec{W}_j \tag{17}$$

approximating the system of equations (1) with the order $O(\tau + h^k)$ and conservative in the case of obtaining the stationary solution. The stabilizing operator $C = \prod_{j=1}^{2}(I + \tau\alpha B_{j1}^n)(I + \tau\alpha B_{j2}^n)$ is chosen from the conditions of scalar solvability of the scheme

$$\begin{aligned} (I + \tau\alpha B_{11}^n)\,\vec{\xi}^{n+1/4} &= -(A^n)^{-1}\sum_{j=1}^{2}\Lambda_j^k\,\vec{W}_j^n \\ (I + \tau\alpha B_{12}^n)\,\vec{\xi}^{n+1/2} &= \vec{\xi}^{n+1/4} \\ (I + \tau\alpha B_{21}^n)\,\vec{\xi}^{n+3/4} &= \vec{\xi}^{n+1/2} \\ (I + \tau\alpha B_{22}^n)\,\vec{\xi}^{n+1} &= \vec{\xi}^{n+3/4} \\ \vec{f}^{n+1} &= \vec{f}^{n} + \tau\,\vec{\xi}^{n+1} \end{aligned} \tag{18}$$

equivalent in whole steps to the difference scheme (17). The operators B_{ij} $(i,j = 1,2)$ are presented in the form:

$$B_{11} = \begin{pmatrix} 0 & 0 & 0 & 0 \\ -\Lambda^k_{\mp 1} u^2 & \Lambda^k_{\mp 1} u - \frac{4}{3}\bar{\Lambda}_1 \mu \Lambda_1 \frac{1}{\rho} & 0 & 0 \\ -\Lambda^k_{\mp 1} uv & \Lambda^k_{\mp 1} v & \Lambda^k_{\mp 1} u - \bar{\Lambda}_1 \mu \Lambda_1 \frac{1}{\rho} & 0 \\ uc^2 \Lambda^k_{\mp 1} & 0 & 0 & \Lambda^k_{\mp 1} u - \bar{\Lambda}_1 \lambda \Lambda_1 \frac{1}{\rho} \end{pmatrix}$$

$$B_{12} = \begin{pmatrix} 0 & 0 & 0 & 0 \\ -\Lambda^k_{\mp 2} uv & \Lambda^k_{\mp 2} v - \Lambda_2 \mu \Lambda_2 \frac{1}{\rho} & \Lambda^k_{\mp 2} u & 0 \\ -\Lambda^k_{\mp 2} v^2 & 0 & \Lambda^k_{\mp 2} v - \frac{4}{3\rho} \Lambda_2 \mu \Lambda_2 \frac{1}{\rho} & 0 \\ vc^2 \Lambda^k_{\mp 2} & 0 & 0 & \Lambda^k_{\mp 2} v - \bar{\Lambda}_2 \lambda \Lambda_2 \frac{1}{\rho} \end{pmatrix}$$

$$B_{21} = \begin{pmatrix} 0 & \Lambda^k_{\mp 1} & 0 & 0 \\ 0 & \Lambda^k_{\mp 1} u & 0 & \Lambda^k_{\pm 1} \\ 0 & 0 & 0 & 0 \\ 0 & c^2 \Lambda^k_{\mp 1} & 0 & 0 \end{pmatrix}$$

$$B_{22} = \begin{pmatrix} 0 & 0 & \Lambda^k_{\mp 2} & 0 \\ 0 & 0 & 0 & 0 \\ 0 & 0 & \Lambda^k_{\mp 2} v & \Lambda^k_{\pm 2} \\ 0 & 0 & c^2 \Lambda^k_{\mp 2} & 0 \end{pmatrix}$$

$\bar{\Lambda}_j a \Lambda_j$, $\Lambda^k_{\pm j}$ $(k = 1, 2)$ is the approximation of appropriate derivatives in the j-direction. Such a splitting of the operator B allows to operate the difference scheme (18) by scalar sweeps. As shows the analysis of stability for the linearized equations the difference scheme (17) is absolutely stable at $\alpha \geq 0{,}5$.

For the system of equations written in the divergent form,

$$\frac{\partial \vec{F}}{\partial t} = -\sum_{j=1}^{2} \frac{\partial \vec{W}_j}{\partial x_j} \tag{19}$$

the difference scheme

$$C \frac{\vec{f}^{n+1} - \vec{f}^{n}}{\tau} = -\sum_{j=1}^{2} \Lambda^k_j \vec{W}_j^{n} \tag{20}$$

can be constructed approximating the governing equations with the order $O(\tau + h^k)$. The stabilizing operator $C = (A^n + \tau\alpha \Lambda^k_1 B^n_{11}) \cdot (A^n)^{-1}(A^n + \tau\alpha \Lambda^k_1 B^n_{12})(A^n)^{-1}(A^n + \tau\alpha \Lambda^k_2 B^n_{21})(A^n)^{-1}(A + \tau\alpha \Lambda^k_2 B^n_{22})$ is obtained by the factorization of the operator B where $B = \partial \vec{W} / \partial \vec{f}$. By analogy with one-dimensional case for numerical operating the scheme

(20), we shall consider the equivalent difference scheme in fractional steps

$$(A^n + \tau\alpha \Lambda_1^k B_{11}^n)\vec{\xi}^{n+1/4} = -\sum_{j=1}^{2} \Lambda_j^k \vec{W}_j^n$$

$$(A^n + \tau\alpha \Lambda_1^k B_{12}^n)\vec{\xi}^{n+1/2} = A^n \vec{\xi}^{n+1/4}$$

$$(A^n + \tau\alpha \Lambda_2^k B_{21}^n)\vec{\xi}^{n+3/4} = A^n \vec{\xi}^{n+1/2}$$

$$(A^n + \tau\alpha \Lambda_2^k B_{22}^n)\vec{\xi}^{n+1} = A^n \vec{\xi}^{n+3/4}$$

$$\vec{f}^{n+1} = \vec{f}^n + \tau\vec{\xi}^{n+1} \quad .$$

The matrices B_{ij} are chosen from the condition of the scheme's solvability at each fractional step by scalar sweeps and the scheme's stability at the whole step. The matrix B_{ij} form is similar to (19) except the last lines. Note that the mixed derivatives are taken explicitly into account at the lower time layer, however the scheme's stability is conserved.

The schemes generalizing for the space Navier-Stokes equations is of no difficulties. It is instructive to note that the property of absolute stability of the schemes proposed in the papers [7,8] is not conserved in the case of space equations. Apparently it holds also for the splitting schemes of type (17), (20). For proving it is sufficient to consider the difference schemes for the three-dimensional model transfer equation.

3. The method of moving grids.

For increasing the calculation accuracy and obtaining a detailed pattern of the flow, it is necessary to employ the difference grids with steps in the space. As a rule, the regions of a sharp change of functions occupy a small part of computed region. Therefore, the non-uniform grids which condence in separate subregions are employed in practical calculations, for example, by solving the problems of outer flow-part in the boundary layer region or in the shock wave zone. Thus, when setting a priori the grid condensation the information on the solution is employed. In the case of complex flows the information on the solution can be absent. Therefore, it arises the problem of constructing the grids which would depend on the solution, i.e. were condensed in the regions of large gradients and solved in the region of small gradients. The problems of creating the adapting to the solution grids were considered by various authors, f.e., [17,19] . Taking as the basis the paper [19] , let us consider the method of constructing the moving grids. For simplicity we shall consider only the case of one-dimensional condensation, i.e. assume that the functions change sharply along one of the coordinate direction. Introduce the non-degenerate transformation of coordinates

$$\xi = \xi(x_1)\ , \quad \eta = \eta(x_1, x_2)$$

transforming the computed region into a standard one, f.e., a square. The reverse transformation

$$x_1 = x_1(\xi)\ , \quad x_2 = x_2(\xi, \eta) \tag{21}$$

serves for finding the coordinates of the points. It is known [20] that the system of Navier-Stokes equations written in the divergent form in new variables ξ and η may be presented in the divergent form. For numerical solving the Navier-Stokes equations in the variables ξ and η the difference schemes of the type (17), (20) operated by scalar sweeps can be considered. The nonuniform grid set by the transformation (21) corresponds to the uniform one in the region $R\{0 \le \xi \le 1, 0 \le \eta \le 1\}$. The mesh points in the x_1 direction are set from the first equations (21). The coordinates of the mesh points in the x_2 direction we shall obtain from the solution of the equation

$$\left(\left| \frac{\partial f}{\partial x_2} \right|^\alpha + \delta \right) \frac{\partial x_2}{\partial \eta} = \mathrm{Const} \tag{22}$$

where as f one of the unknown functions of solutions or their combination is chosen. The condition (22) implementation signifies that in the region of large gradients of the function f in the x_2 direction the grid steps will be small A necessary condensation of the mesh points is achieved by the choice of coefficients α and δ . It is more convenient to find the coordinate value from the solution of parabolic equation with the stationary boundary-value conditions

$$\frac{\partial x_2}{\partial t} = \frac{\partial}{\partial \eta} \left(\left| \frac{\partial f}{\partial x_2} \right|^\alpha + \delta \right) \frac{\partial x_2}{\partial \eta}$$

The present approach enables one to increase essentially the calculation accuracy at a relatively small number of mesh points (see [13, 21]) and a negligible increase in the computer time required.

References:

1. P.Peyret, H.Viviand. Computatian of Viscous Compressible Flows Based on the Navier-Stokes. AGARD-AG-212, 1975.
2. N.N.Yanenko. The Method of Fractional Steps. New-York, 1971.
3. J.Douglas and J.E.Gunn. A General Formulation of Alternating Direction Method. Numerische Mathematik , vol.6, 1964, pp.428--453.
4. N.N.Yanenko, V.D.Frolov, V.E.Neuvazhaev. O primenenii metoda rasshepleniya dlya chislennogo rascheta dvizheniya teploprovodnogo gaza v krivolineynykh koordinatakh. Izvestiya SO AN SSSR ser.techn.nauk, t.8, N°2, 1967.
5. V.I.Polezhaev. Chislennoye resheniye sistemy dvymernykh nestatsionarnykh uravneniy Navier-Stoksa dlya szhimaemogo gaza. Izvestiya AN SSSR. Mekh.Zhidk. i Gaza, N°2, 1967.
6. V.M.Kovenya. Application of Implicit Difference Schemes to the Solution of Aerodynamic Problems. Lecture Notes in Physics, vol.58, 1976.
7. R.M.Beam, R.F.Warming. An Implicit Finite-Difference Algorithm for Hyperbolic System in Conservation Law Form. Journal of Computational Physics , vol.22, Sept. 1976.
8. W.R.Briley, H.McDonald.Solution of the Multidimensional Compressible Navier-Stokes Equations by a Generalized Implicit Method. Journal of Computational Physics, vol.224, N°4, 1977.
9. J.S.Shang, W.L.Hankey. Numerical Solution of the Navier-Stokes for a Three-Dimensional Corner. AIAA Journal, vol.15, N°11, 1977.
10. A.I.Tolstykh. O neyavnykh raznostnykh skhemakh tret'yego poryadka tochnosti dlya mnogomernykh zadach.ZhVM, t.16, N°5,1976.
11. L.I.Severinov. Sposob resheniya nelinejnykh raznostnykh kraevykh zadach mekhaniki sploshnoj sredy, ZhVM i MF, t.8, N°4, 1978.
12. N.N.Yanenko, V.M.Kovenya. A Difference Method for Solving the Multidimensional Equations of Gas Dynamics. Sovjet.Math.Dokl., vol.18, N°1, 1977.
13. V.M.Kovenya, N.N.Yanenko. Raznostnaya skhema na podvizhnykh setkakh dlya resheniya uravnenij vyazkogo gaza. ZnVM i MF, t.19, N°1, 1979.
14. R.M.Beam, R.F.Warming. An Implicit Factored Scheme for the Compressible Navier-Stokes Equations, AIAA Journal, vol.16, N°4, 1978.

15. J.T.Steger. Implicit Finite-Difference Simulation of Flow about Arbitrart Two-Dimensional Geometries. AIAA Journal, vol.16, N°7, 1978.

16. A.A.Samarsky, Yu.P.Popov. Raznostnye Skhemy Gazovoi Dinamiki. M., "Nauka", 1975.

17. J.F.Thompson, F.C.Thames and C.M.Msstin. Automatic Numerical Generation of Body-Fitted Curvilinear Coordinate System for Field Containing a Number of Arbitrary Two-Dimensional Bodies. Journal of Computational Physics, vol.15, July 1974.

18. A.I.Tolstykh. O metode chislennogo resheniya uravnenij Navier--Stoksa szhimaemogo gaza v shirokom diapazone chisel Reynoldsa. Dokl. AN SSSR, t.210, N°1, 1973.

19. N.N.Yanenko, N.T.Danaev, V.D.Lisejkin. O Variatsionnom Metode Postroyeniya Setok. In: Chislennye metody Mekhaniki Sploshnoi Sredy, t.8, N°4, Novosibirsk, 1977.

20. M.Vinokur. Conservation Equations of Gas Dynamics in Curvilinear System. Journal of Computational Physics, vol.14, N°2, 1974, pp.105-125.

21. N.N.Yanenko, V.M.Kovenya, V.D.Fomin, E.F.Vorozhtsov, V.D.Liseikin. On Some Methods for the Numerical Simulation of Flow with Complex Structure, Lecture Notes in Physics,vol.90, 1979.

Finite-Difference Solutions of the Navier-Stokes Equations For Axially Symmetric Flows in Spherical Gaps

Egon Krause and Fritz Bartels*
Aerodynamisches Institut, RWTH Aachen
Aachen, Germany

Summary

Finite-difference solutions of the Navier-Stokes equations are adapted to the calculation of flows in spherical gaps. The equations of motion are restricted to axially symmetric flows so that the stream-function-vorticity formulation can be employed. Thereby the rather difficult calculation of the pressure field is avoided. Only the velocity and vorticity distributions can be determined. Since Taylor-Görtler vortices can be formed in such flows the solution is time dependent and may not reach a steady state, not even for large times. For that reason the truncation errors must be kept sufficiently small and the damping introduced in the discretization cannot exceed certain critical values. Since the solution is almost periodic in the colateral direction, error bounds can be derived for the spatial resolution. The influence of the time-step is analysed with a model equation. The results indicate that for the reasons mentioned the spacing is never arbitrary not even for implicit formulations. Several sample calculations confirm the validity of the error bounds derived.

Introduction

Taylor-Görtler vortices can be generated in cylindrical and spherical gaps by rotating the inner cylinder or sphere and holding the outer one at rest.

This problem was investigated experimentally by Sawatzki and Zierep who first [1] demonstrated that several flow modes can exist due to the influence of the centrifugal and Coriolis forces during the acceleration phase. Continuing their work, Wimmer [2] was able to show in his experiments that the number of the vortices depends on the gap width, on the angular acceleration of the inner sphere and on the Reynolds number. For a gap width, nondimensionalized with the radius of the outer sphere of 0.15 [2] , three steady axially symmetric and two unsteady asymmetric flow modes were found.

These flow modes are the simplest ones observed. The vortices are steady and do not change after the final angular velocity has been reached. Other two-and three-dimensional steady and

---

* Postdoctoral research scientist, presently supported by the Deutsche Forschungsgemeinschaft. On leave of absence at the David W. Taylor Naval Ship Research and Development Center, Bethesda, USA.

unsteady modes exist. With decreasing gap width and ancreasing Reynolds numbers, the vortices become smaller in diameter and do not reach a steady state for large times. Finally, transition to turbulent flow occurs.

Since the Reynolds number, based on the gap width is of the order of a few hundred, this problem is an excellent test case for the application of finite approximations to the Navier-Stokes equations. In [3] an explicit and an implicit solution were constructed to solve the vorticity-transport equation for the vorticity component in the meridional plane, the Poisson-equation for the stream-function and the momentum equation for the azimuthal direction. With this formulation only the symmetric modes can be described. The restriction was necessary as it was to be expected that the computational effort necessary for solving the problem was relatively large. The major aim of the investigation [3] was to find out whether a consistent and stable formulation would be sufficient for a correct prediction of the various flow modes mentioned. It was soon discovered that the spatial resolution could not be chosen arbitrarily.

Certain step sizes yielded a flow field, which did not belong to the prescribed boundary conditions, although the solution was smooth and did not exhibit any oscillations observed when the step sizes are too large. The influence of the error can be demonstrated by reducing the difference equations to consistent differential expressions. The major development of the error analysis given in [3] will be briefly reviewed here.

Formulation of the problem

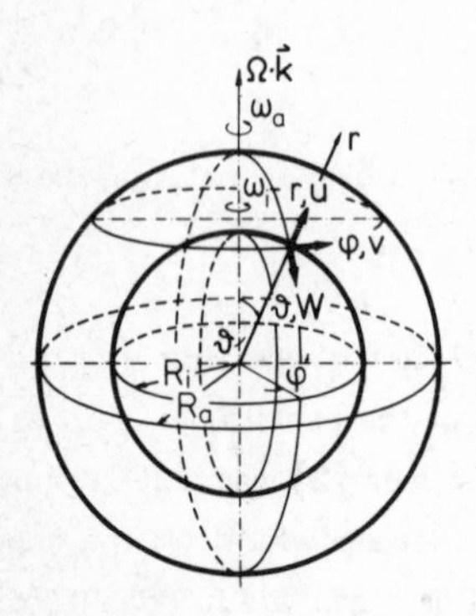

Fig. 1 Coordinates and velocity components of the flow in a spherical gap. After [3].

The coordinates and velocity components are shown in Fig. 1. If it is assumed that the flow is symmetric with respect to the k-axis, the stream-function-vorticity formulation can be employed. The vorticity transport equation then advances the vorticity component of the meridional plane in time, while for each time step the stream function ψ has to be determined by solving the Poisson equation. The three velocity components are finally computed by numerically differentiating the stream function and solving the circumferential momentum equation. The differential problem may then be formulated as follows:

Velocity components:

$$u = -\frac{1}{r^2 \sin\vartheta}\frac{\partial\psi}{\partial\vartheta} \; ; \; v = \frac{\phi}{r\sin\vartheta} \; ; \; w = \frac{1}{r\sin\vartheta}\frac{\partial\psi}{\partial r} \tag{1}$$

Circumferential momentum:

$$\frac{\partial \phi}{\partial t} + u\frac{\partial \phi}{\partial r} + \frac{w}{r}\frac{\partial \phi}{\partial \vartheta} = \nu D^2 \phi \tag{2}$$

Vorticity equation:

$$\begin{aligned}&\frac{\partial \zeta}{\partial t} + u\frac{\partial \zeta}{\partial r} + \frac{w}{r}\frac{\partial \zeta}{\partial \vartheta} - \frac{2\zeta}{r}(u + w\cot\vartheta) \\ &\quad + \frac{2}{r}\frac{\phi}{r^2 \sin^2\vartheta}\left(\frac{\partial \phi}{\partial \vartheta}\sin\vartheta - \frac{\partial \phi}{\partial r}r\cos\vartheta\right) = \nu D^2 \zeta\end{aligned} \tag{3}$$

Poisson equation for the stream function:

$$D^2 \psi = \zeta \,, \tag{4}$$

where the differential operator is defined as

$$D^2 = \frac{\partial^2}{\partial r^2} + \frac{1}{r^2}\frac{\partial^2}{\partial \vartheta^2} - \frac{\cot\vartheta}{r^2}\frac{\partial}{\partial \vartheta} \tag{5}$$

The boundary conditions are:

Circumferential velocity component:

$$R_i \le r \le R_a \begin{cases} \vartheta = 0 & \phi = 0\,; \\ \vartheta = \frac{\pi}{2} & \frac{\partial \phi}{\partial \vartheta} = 0\,; \end{cases} \quad 0 \le \vartheta \le \frac{\pi}{2}\,,\ r = R_{i,a}\,,\ \phi = \omega_{i,a}(t)\,.\,R_{i,a}^2 \sin^2\vartheta \tag{6}$$

Stream function:

$$R_i \le r \le R_a \begin{cases} \vartheta = 0 & \psi = 0\,; \\ \vartheta = \frac{\pi}{2} & \psi = 0\,; \end{cases} \quad 0 \le \vartheta \le \frac{\pi}{2}\,,\ r = R_{i,a}\,,\ \psi = 0 \tag{7}$$

The boundary conditions for the vorticity component have to be determined from the discretized Poisson equation. Since the radial and colateral velocity components vanish for $r = R_{i,a}$ the Poisson equation contains only normal derivatives on the boundaries. Along the axis of rotation and in the equatorial plane the vorticity component has to vanish for reasons of symmetry. The boundary conditions for the vorticity component then read

$$R_i \le r \le R_a \begin{cases} \vartheta = 0 & \zeta = 0\,; \\ \vartheta = \frac{\pi}{2} & \zeta = 0\,; \end{cases} \quad 0 \le \vartheta \le \frac{\pi}{2}\,,\ r = R_{i,a}\,,\ \zeta = \frac{\partial^2 \psi}{\partial r^2} \tag{8}$$

For the solution it was assumed that the flow is initially at rest, i.e.

$$t = 0 \qquad \phi(r,\vartheta,0) = \psi(r,\vartheta,0) = \zeta(r,\vartheta,0) = 0 \tag{9}$$

Standard discretization procedures are available for the above problem. Explicit as well as implicit methods can be employed to solve the vorticity-transport equation and the circumferential momentum equation; several methods are also available for solving the Poisson equation. Details need not to be reported here as the solution of the difference problem corresponding to the differential problem (1) - (9) is described in [3]. After a few test calculations it was found that the finite-difference approximation converged to a solution, which in comparison to the experimental data did not correspond to the prescribed boundary conditions. A change of step sizes resulted in completely different flow fields (Fig. 2). The distributions of the flow variables were smooth indicating that the calculation was stable.

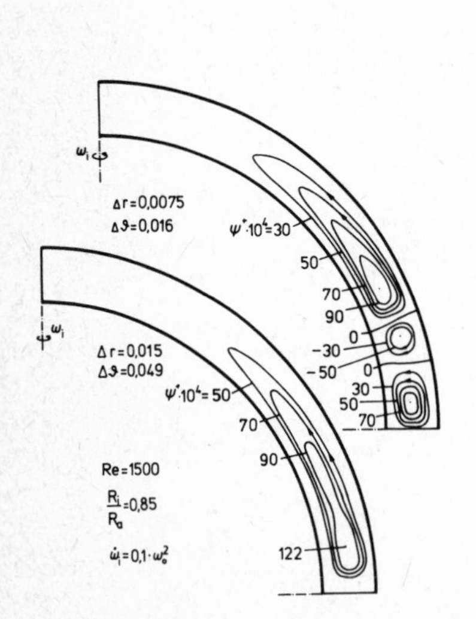

Fig. 2 Influence of the spatial resolution. After [3].

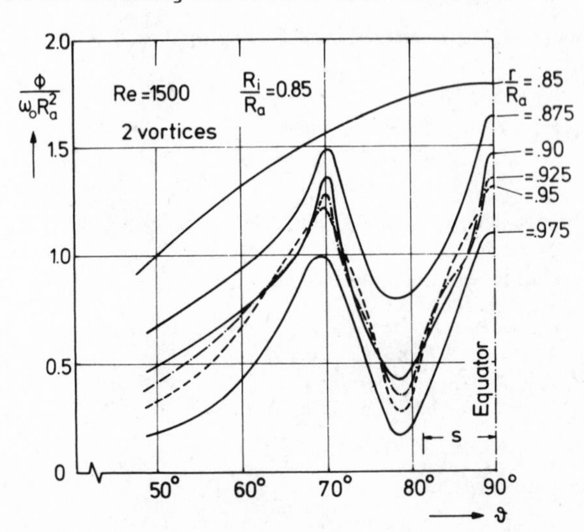

Fig. 3 Azimuthal velocity distribution in a spherical gap with two Taylor-Görtler-vortices. After [3].

If one or more vortices exist, the flow variables exhibit an almost periodic behaviour in the meridional direction. This is demonstrated in Fig. 3, where the normalized azimuthal velocity distribution is plotted versus the meridional coordinate ϑ for a flow containing two Taylor-Görtler vortices. It is seen that the influence of the vortices extends over a distance of about two gap width. Close to the equator, Φ can be expanded in a Fourier series.

$$\phi(r,\vartheta) = \sum_{m=0}^{\infty} \hat{\phi}_m(r) \exp\left[im \frac{\pi r}{s} \left(\frac{\pi}{2} - \vartheta\right)\right] \tag{10}$$

If only the first two terms are retained, the relative error with which Φ_ϑ is approximated is

$$\varepsilon_\vartheta = \frac{1}{\phi_\vartheta} \sum_{l=1}^{\infty} \Delta\vartheta^{2l} \phi_\vartheta^{(2l+1)} = -\frac{1}{3!}\left(\frac{\Delta\vartheta\pi r}{s}\right)\left[1 - \frac{1}{20}\left(\frac{\Delta\vartheta\pi r}{s}\right)^2 + \ldots\right] \tag{11}$$

For $\varepsilon_\vartheta << 1$, it follows that

$$\Delta\vartheta \leq \frac{\sqrt{6}\,s}{\pi r} k_\vartheta , \tag{12}$$

where k_ϑ is the positive value of the square root of the maximum permissible error in the meridional direction. If k_ϑ is known eqn. (12) can be used to determine the step size $\Delta\vartheta$. It is clear from this relationship that the ratio $r\Delta\vartheta/s$ is a constant, so that the spacing must be decreased if the gap width is made smaller. A similar estimate can be obtained for the radial direction. The wave length is in this case proportional to one half of the gap width, so that

$$\frac{\Delta r}{s} \leq \frac{\sqrt{6}}{4\pi} k_r \tag{13}$$

The error bounds in the radial and meridional direction have been determined by decreasing the step sizes and comparing the flow modes obtained to those of the experiments in [2]. It was found that the flow fields were correctly predicted if the bounds were chosen $k_\vartheta^2 \leq 0.05$ and $k_r^2 \leq 0.1$. At the present time there does not seem to exist a method with which these bounds can be determined without making use of flow observations. The influence of the step size Δt on the solution cannot be analysed in this manner. Instead a simplified model equation was used to determine the major parameters. Consider the linearized Burger's equation

$$\frac{\partial u}{\partial t} = -a\frac{\partial u}{\partial x} + b\frac{\partial^2 u}{\partial x^2} \tag{14}$$

where a is a constant wave speed and b the viscosity, which is also assumed to be constant. If the initial conditions are periodic, i.e.

$$u(0,x) = \sum_{m=0}^{\infty} A_m \exp(imx) \tag{15}$$

the solution of eqn. (14) is

$$u(x,t) = \sum_{m=0}^{\infty} A_m \exp(-bm^2 t)\exp\left[im(x-at)\right] \tag{16}$$

With the Fourier-coefficients A_m the above solution describes the timedependent exponential damping of the initial wave. It is clear from eqn. (16) that the damping of the various Fourier-Components is increased with increasing wave number. An explicit difference-approximation to eqn. (16) is

$$v_i^{n+1} = v_i^n - \frac{a\Delta t}{2\Delta x}\left(v_{i+1}^n - v_{i-1}^n\right) + \frac{b\Delta t}{\Delta x^2}\left(v_{i+1}^n - 2v_i^n + v_{i-1}^n\right) \tag{17}$$

Through Taylor-series development one obtains the corresponding differential equation

$$v_t + \sum_{l=2}^{\infty} \frac{\Delta t^{l-1}}{l!} v_t^{(l)} = -a v_x - a \sum_{l=1}^{\infty} \frac{\Delta x^{2l}}{(2l+1)!} v_x^{(2l+1)} + b v_x^{(2)} + b \sum_{l=2}^{\infty} \frac{2\Delta x^{(2l-2)}}{(2l)!} \tag{18}$$

which serves to establish the solution to the difference equation (17). The initial condition $v(0, x) = u(0, x)$ finally yields

$$v(x,t) = \sum_{m=0}^{\infty} A_m \exp(\alpha_r t) \exp\left[i(\alpha_i t + mx)\right] \tag{19}$$

where the damping factor α_r is given by

$$\alpha_r = \frac{b \ln\left[\left(1 - 4d \sin^2\left(\frac{m\Delta x}{2}\right)\right) + c^2 \sin^2(m\Delta x)\right]}{2d\,\Delta x^2} \tag{20}$$

and the wave speed is

$$\alpha_i = -\frac{a \tan^{-1}\left[\frac{c \sin(m\Delta x)}{1-4d \sin^2(m\Delta x/2)}\right]}{C\,\Delta x} \tag{21}$$

The quantities $c = a\,\Delta t/\,\Delta x$ and $d = b\,\Delta t/\,\Delta x^2$ stand for the Courant number and the diffusion number. Eqn. (21) reduces to the relationship between phase error and step size Δx for a hyperbolic differential equation with periodic initial conditions, derived by Kreiss and Oliger in [4] , if the viscosity is set equal to zero.

The terms α_r/bm^2 and α_i/am represent the ratios of the damping factors and the wave speeds of the Fourier-components in the differential equation and the difference equation. The numerical damping falsifies the physical damping, and, due to the change in the wave speed, introduces phase errors, which depend on the step sizes Δx, Δt, the wave number m and the constants a and b. It is seen that v converges to u for $\Delta x \to 0$, and $\Delta t \to 0$. The stability in the von-Neumann-sense requires $\alpha_r < 0$. This means that

$$g = \left(1 - 4d \sin^2\left(\frac{m\Delta x}{2}\right)\right)^2 + C^2 \sin^2(m\Delta x) < 1 \tag{22}$$

The corresponding components are damped out instantaneously, if $g = 0$, i.e. $m\Delta x = \frac{2k-1}{2}$ ($k = 0,1,2...$) and $d = 1/4$, and $c = 0$ for $m\Delta x = 2 \sin^{-1}\sqrt{\frac{1}{4}\ d}$. All components for which $g < \exp(-2d\,(m\Delta x)^2$ are damped out faster and those for which $g > \exp(-2d\,(m\Delta x)^2$ are damped out slower than in the corresponding solution for u. If $m\Delta x = 2k\pi$ the damping is zero and the Fourier-components remain as errors in the asymptotic solution for $t \to \infty$.

The change in the wave speed, which is different for every wave number in the finite-difference solution causes dispersion of the various components. If $d > 1/4$, the components for which $\sin^2(\frac{m\,\Delta x}{2}) > 1/4$ or $\sin(m\,\Delta x) < 0$, experience a phase shift by π. This means that for each time step there is a change in sign in the corresponding Fourier coefficient. No phase shift is observed for $d = 1/2$ and $c = 1$. If $m\,\Delta x << 1$, as required already earlier for the spatial resolution, equns (20) and (21) can be expanded to yield

$$\frac{\alpha_r}{bm^2} = -1 + \frac{c^2}{2d} - \frac{d}{2}(m\Delta x)^2 + ... \tag{23}$$

and

$$\frac{\alpha_i}{am} = \frac{-1}{1 - 4d\sin^2(m\Delta x/2)} \tag{24}$$

If the physical damping is not to be falsified, it is necessary to require that $c^2/2\,d << 1$, or $\Delta t << \frac{2\,b}{\alpha^2}$. For the flow problem discussed here this result implies that the finite-difference solution is a meaningful approximation only, if the Fourier-components with large wave numbers do not influence the solution. In actual flow calculations the choice of the time step may act as an additional angular acceleration $\dot{\omega}_i$ of the inner spere. This has been demonstrated with an implicit solution of [3]. With increasing Δt the angular acceleration had to be decreased if the same mode of the flow was to be predicted for $t \to \infty$. Fig. 4 shows the limit of the angular acceleration, as a function of Δt for which the flow changes over from no and two vortices in the gap.

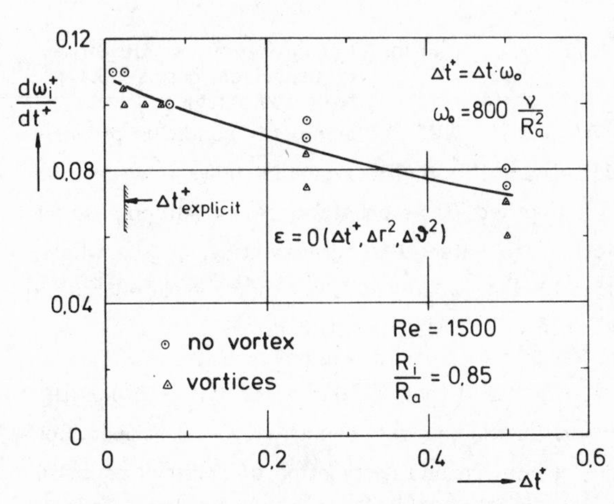

Fig. 4 Influence of the time step Δt on the finite difference solution. After [3].

Results

Some flow fields determined with the solution described in [3] are shown in the next few figures. The bounds for the spatial resolution and the time step were obtained with the analysis just described. Comparison to the experiments of [1] and [2] asserted that vortex formation can accurately be predicted as long as the flow does not become turbulent. Fig. 5 shows the calculated stream-line pattern for a gap width of 0.15 and a Reynolds number of 1500 for two angular accelerations. If the ratio of $\dot{\omega}_i / \omega_o^2$ is changed from 0.12 to 0.10, the flow develops two vortices, while for $\dot{\omega}_i / \omega_o^2 = 0.12$ no vortex appears. All attempts to determine the flow mode containing only one vortex for $s = 0.15$ and Re = 625 failed until the symmetry was slightly disturbed near the equatorial plane. By changing ϑ_{max} only by less than one permill from 90 to 90.7 degrees and imposing the symmetry condition in the new position, the finite difference solution predicted the one-vortex mode at a Reynolds number of about Re = 650. It was also

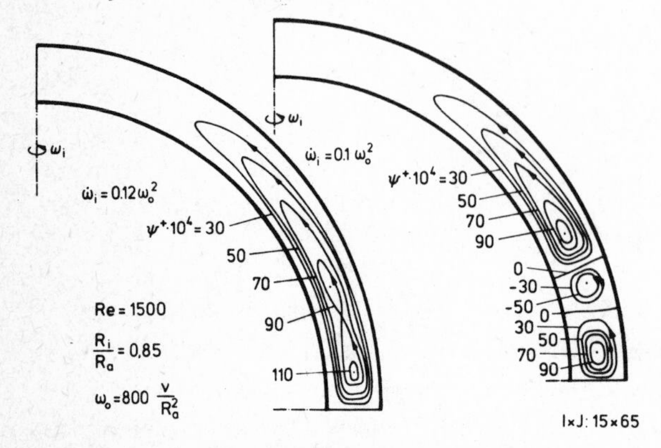

Fig. 5 Influence of the angular acceleration $\dot{\omega}_i$ on the flow in a spherical gap. After [3].

Fig. 6 Formation of a single vortex in a spherical gap; s = 0.15; Re = 700. After [3].

found that the vortex remained stable above Re = 700 with the symmetry condition properly imposed. The streamlines of the flow are depicted in Fig. 6. This example shows how sensitive the solution reacts to the boundary conditions imposed in the equatoral plane. Although slight differences between experiment and prediction are noted, the agreement is, on the whole, rather satisfactory. In some cases the differences between calculated and measured torque coefficient in Fig. 7 are only of the order of magnitude of the truncation error.

Among other interesting results predicted, it was found that for gap width $0.07 \leq s \leq 0.09$ the solution did not reach a steady state for $t \rightarrow \infty$ as for the boundary conditions just described. For example, for s = 0.08 and Re = 2000 periodic generation and destruction of vortices causes a time dependent variation of the torque coefficient (Fig. 8). The extremum values (1 - 4) indicated in the figure correspond to the flow modes shown in Fig. 9.

As pointed out before the vortices are formed near the equatior. First disturbances are observed at $\vartheta \approx 70^0$; they travel towards the equator and form vortex cells, which in general for small gap widths ($s \leq 0.09$) and certain Reynolds numbers are destroyed at the equator. This can be seen in Fig. 10, where the torque coefficient is plotted versus time for several Reynolds numbers and gap widths. The relationship between the characteristic periods, the Reynolds

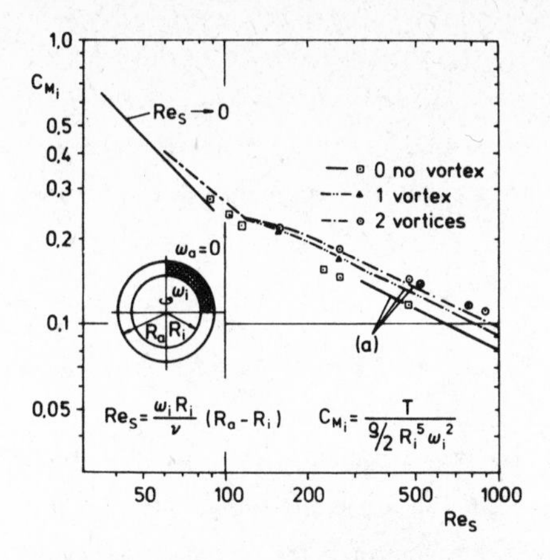

Fig. 7 Calculated and measured torque coefficient. After [3]. Experimental data are taken from [2].

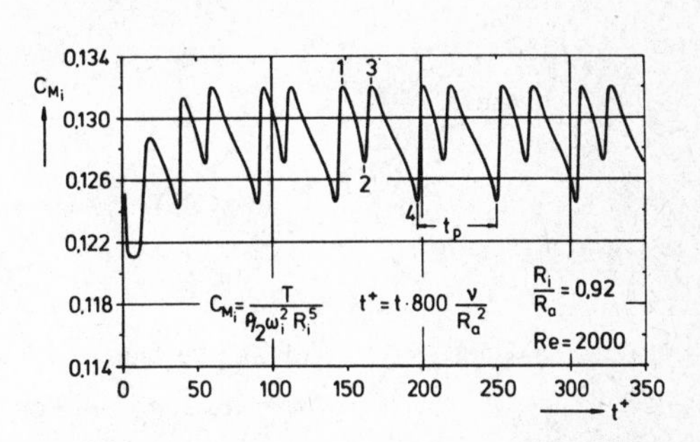

Fig. 8 Time-dependent torque coefficient; s = 0.08; Re = 2000. After [3].

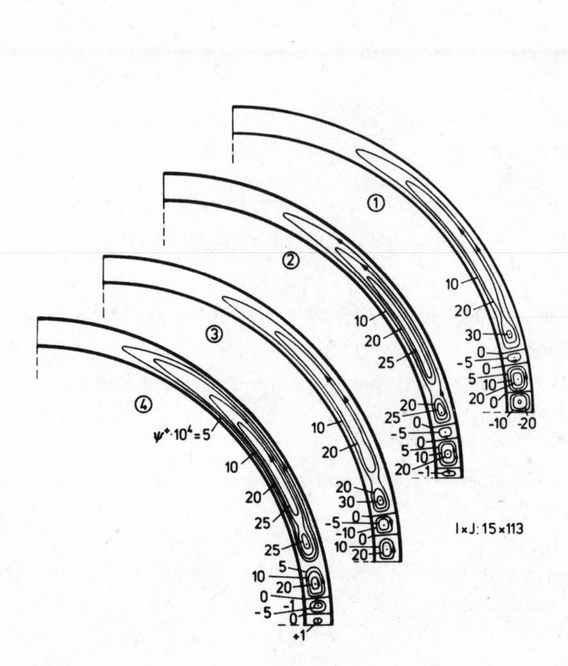

Fig. 9 Streamlines of unsteady flow motion in a spherical gap. After [3].

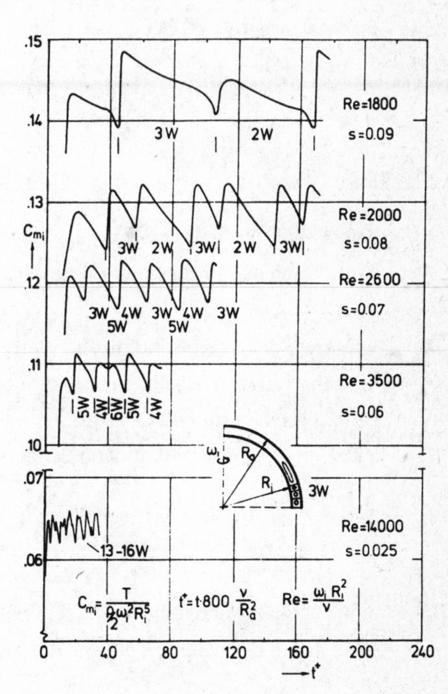

Fig. 10 Time-dependent torque coefficient for several Reynolds-numbers and gap widths. After [3].

number and the gap width has not been explored. Comparison of these predictions to experimental data has also not been possible as the measurements of [1] and [2] were carried out only for s = 0.15. Extension to truely three-dimensional flow modes seems, at the present time, difficult, since for the accuracy required, large storage capacities and computation speeds will be necessary.

Conclusions

In spherical gaps, Taylor-Görtler vortices may be generated if the inner sphere is set in motion with a certain angular acceleration. An error analysis of this problem shows that finite-difference approximations to the Navier-Stokes equations can correctly predict the flow behaviour, if accuracy conditions, derived in the analysis, are met. Although for most problems sufficient, consistent and stable finite-difference formulation of the locally linearized conservation equations does not ensure uniqueness of the solution. Only with proper spatial and time resolution can the various flow modes be predicted. Torque coefficients determined for some of the steady modes investigated show excellent agreement with experimentally determined values.

References

[1] Sawatzki, O., and Zierep, J., Das Stromfeld im Spalt zwischen zwei konzentrischen Kugelflächen,von denen die innere rotiert. Acta Mechanica, 9, 1970, pp. 13-35

[2] Wimmer, M., Experimentelle Untersuchungen der Strömung im Spalt zwischen zwei konzentrischen Kugeln, die beide um einen gemeinsamen Durchmesser rotieren. Dissertation Techn. Universität Karlsruhe, 1974. See also: Experiments on a Viscous Fluid Flow between Concentric Rotating Spheres. J. Fluid Mech. 78, 1976, pp. 317-335

[3] Bartels, F., Rotationssymmetrische Strömungen im Spalt konzentrischer Kugeln. Dissertation RWTH Aachen, 1978

[4] Kreiss, H.O., and Oliger, J., Comparison of Accurate Methods for the Integration of Hyperbolic Equations. Tellus XXIV, 3, 1972, 199-215

[5] Krause, E., and Bartels, F., Rotationssymmetrische Strömung im Kugelspalt. Reprint from Recent Developments in Theoretical and Experimental Fluid Mechanics. Springer-Verlag Berlin, Heidelberg, 1979.

NUMERICAL INVESTIGATION OF UNSTEADY VISCOUS INCOMPRESSIBLE FLOW ABOUT BODIES FOR VARYING CONDITIONS OF THEIR MOTION

V.I.Kravchenko, Yu.D.Shevelev, V.V.Shchennikov
Institute of Mechanical Problems,
Academy of Sciences of the USSR,
Moscow, USSR

Numerical solutions for unsteady viscous incompressible flows occuring due of impulsive and gradual variation of the velocity of a body using as examples flow about a cylinder, a sphere and ellipsoids are presented.

The following conditions have been considered as the conditions of a body moving in a stream:

a) flow past an abruptly started body;
b) sudden impulsive acceleration of the body which was previously moving steadily;
c) a start with finite acceleration;
d) gradual acceleration of a body which was previously moving steadily;
e) gradual deceleration of a body to a stop.

Consideration of the impulsive conditions of velocity variations is carried out using the transformation of the stream function, the vorticity and independent variables to eliminate surface singularity of vorticity. By dividing the computation region into two subregions followed by matching of numerical solutions obtained in these subregions, the interaction is studied of the new vortex layer adjacent to the body surface with the vortex field which existed prior to the change of conditions.

Numerical solution of the problems in question is sought after using the monotone finite-difference schemes proposed in this paper, which approximate the initial equations to an accuracy of the second order. The numerical results are given, the corresponding experimental data are compared with computations performed by other authors.

— · — · —

The present paper is devoted to mathematical modelling of certain unsteady flows of a viscous incompressible fluid about a body of finite dimensions.

As a object of this study is a flow about a body moving in a viscous fluid with a considerably varying velocity, it is naturally interesting to consider the most "unfavorable" cases of varying conditions of a flow about a body. (Fig. 1).

In the present paper, the bodies shaped as a sphere, a cylinder and oblate and prolate ellipsoids are considered in numerical experiments pertinent to the investigation of the above mentioned conditions of flow about a body.
In part I, a method of numerical investigation of unsteady flow about oblate and prolate ellipsoids, instanteneously accelerated from the state of rest, is presented. To eliminate initial singularity of the vorticity upon the body surface, the transformation of the vorticity, of the stream function and independent variables is used. So, the transformed initial equations are essentially simplified when $t \to 0$ and become self-similar.

The existing initial conditions for solving the problem on an impulsively accelerated ellipsoid are obtained by integration of this self-similar system of equations. The transformation of time coordinate used here involves increasing automatically the time step from a necessary sufficiently small value at the initial moment. Numerical integration of the transformed initial equations is performed according to the same algorithm, including also the initial moment, the departure from the origin of the time coordinate being effected with an extremely small step which then gradually increases.

The approach suggested here permits one to obtain solutions both for small and later times and in a wide range of the Reynolds numbers. Note, that the method used in this part is a natural modification of the numerical investigation technique for unsteady flows about a circular cylinder and sphere instanteneously set into motion ([1] , [2]).

When a viscous flow developes about a body of finite dimensions and when the body is set into motion from the state of rest by a jerk, the flow at the initial moment is irrotational. The vorticity in fluid flow has an infinitely large value and is concentrated in a limited vinicity of the body. As time passes, the vorticity diffuses into the external flow and a viscous region is formed around a body. The phenomenon of "flow separation" occurs which begins at the rear critical point, and the "separation" point shifts upstream reaching at later times the values corresponding to the stationary solution. After some time, a pair of vorticies are formed in the vicinity of the body that grow with time. When time is small, the length of vortices is little affected by the viscosity. As the viscosity decreases, this phenomenon becomes more tangible. The vortex region grows with time, the vortices may become unstable and be carried away by the external flow. When a body is accelerated by a jerk from one stationary state to another, the flow in the initial moment is irrotational and then the vortex formation begins. The vortex region grows similarly as in the case of motion from the state of rest. When a body is abruptly decelerated from one stationary state to another, the character of the flow becomes essentially more complex. (Part 2). This is due the fact that the vortex region moves forward and the disturbance at the rear part of the body is transferred to its front part, increasing with the bulk of the fluid flow. Numerical modelling of the flow makes it possible to study the principal regularities in case of deceleration at small Reynolds numbers.

In Part 2 the problem of a viscous incompressible fluid flow about a cylinder which was initially moving with a constant velocity and then begins to be impulsively accelerated or decelerated in the direction perpendicular to the cylinder axis (Fig. 1, modes b,c) is discussed. The first results pertaining to the solution of the problem of unsteady flow about a body of finite dimensions, when the motion is abruptly disturbed, have been obtained within the boundary layer framework. It has been proved [3] , that at the initial moment, following any sudden disturbance of the external flow, the fluid flow field is a superposition of the velocity field, which existed prior to the disturbance, and the velocity field which characterizes the disturbance of the external flow, with the exception of the infinitely thin vortex layer adjacent to the body.

It should be noted, that the study of the flow in question is much complicated by the fact that at the moment of a sudden change in the flow conditions a new, infinitely thin, secondary vortex layer begins to develop adjacent to the body inside the original layer, these two layers interact with each other to finally form one quasi-stationary vortex layer.

In the boundary layer approximation, the question of the interaction of the nonstationary secondary vortex layer with the original one which is stationary in the neighborhood of the frontal critical point, has been discussed in [4] .

In the present paper, for the purpose of investigating the interaction of the new vortex layer, arising at the cylinder surface at the moment of impulsive velocity change, with the former vortex field, the region of computation is divided into two subregions. In each of these subregions numerical integration of the initial system of equations is conducted with regard to the corresponding transformations of the vorticity, of the stream function and the independent variables and with a subsequent matching of the obtained solutions.

Necessary initial conditions are found by integration of the initial system of equations, which in the internal subregion adjacent to the surface of the cylinder, has with regard to the proposed transformation a self-similar character at the initial moment of time. In the outer subregion, the initial parameter distribution corresponding to the initial stationary flow about a body is used as the initial condition.

The approach presented here allows one to eliminate the initial singularity of the vorticity on the surface of a body in a stream and to consider the influence of the initial vortex field both in case of acceleration and deceleration of the body.

The numerical solution of the problem in question is realized using implicit second-order-accurate difference schemes that make it possible to carry out computations starting directly from impulsive velocity change.

The treatment used here allows one to compute the flow about a body without the boundary condition for the vorticity upon the body surface. The results of numerical computations are given for the cases of impulsive acceleration from the steady-state flow conditions when $Re_1 = 36.5$ (where $Re_1 = 2V_1R/\nu$) and the states corresponding to $Re_2 = 40, 100, 550$ ($Re_2 = 2V_2R/\nu$) and impulsive deceleration from $Re_1 = 36.5$ to $Re_2 = 31$.

I. Numerical Investigation of Unsteady Viscous Fluid Flow about an Ellipsoid Impulsively Set into Motion

I. Let an ellipsoid being initially at rest, be abruptly set into motion in the direction of the OZ axis of the Cartesian coordinate system attached to the ellipsoid. (Fig. 2).

Consider a generalized Helmholtz equation describing the vorticity transportation in the unsteady viscous incompressible fluid flow

$$\frac{\partial \omega}{\partial t} + rot(\vec{\omega} \times \vec{v}) = -\nu \, rot \, rot \, \vec{\omega}, \qquad (I.I)$$

where ν is the kinematic viscosity.

With the assumption of the axial flow symmetry in the ellipsoidal coordinate system (τ , η , φ) we have

$$\vec{\omega} = \{0, 0, \omega\}, \quad \vec{v} = \{v_\tau, v_\eta, 0\}, \tag{1.2}$$

where

$$\omega = (rot\,\vec{v})_\varphi = \frac{1}{H_\tau H_\varphi}\left\{\frac{\partial(H_\eta v_\eta)}{\partial \tau} - \frac{\partial(H_\tau v_\tau)}{\partial \eta}\right\},$$

$$v_\tau = \frac{1}{H_\eta H_\varphi}\frac{\partial \psi}{\partial \eta}, \quad v_\eta = -\frac{1}{H_\tau H_\varphi}\frac{\partial \psi}{\partial \tau}, \quad H_\tau = H_\eta, \quad H_\varphi \; -$$

are the Lame coefficients.

Converting into the scalar form of Eq. (1.1) in the ellipsoidal coordinate system we have after reduction to the dimensionless form

$$\frac{\partial^2 \omega}{\partial \tau^2} + \frac{\partial^2 \omega}{\partial \eta^2} + A(\tau)\frac{\partial \omega}{\partial \tau} + ctg\,\eta\,\frac{\partial \omega}{\partial \eta} + B(\tau,\eta)\omega = \frac{Re}{2}H_\tau\Big\{H_\tau\frac{\partial \omega}{\partial t} +$$

$$+ v_\tau\frac{\partial \omega}{\partial \tau} + v_\eta\frac{\partial \omega}{\partial \eta} - v_\tau A(\tau)\omega - v_\eta\,ctg\,\eta\,\omega\Big\}, \tag{1.3}$$

$$H_\tau^2 H_\varphi\,\omega = \frac{\partial^2 \psi}{\partial \tau^2} - A(\tau)\frac{\partial \psi}{\partial \tau} - ctg\,\eta\,\frac{\partial \psi}{\partial \eta} + \frac{\partial^2 \psi}{\partial \eta^2},$$

where ψ and ω are the dimensionless stream function and vorticity, $Re = 2fV/\nu$ is the Reynolds number, V is the initial velocity of ellipsoid motion, f is the focal distance.

Besides, for an prolate ellipsoid

$$A(\tau) = cth\,\tau, \quad B(\tau,\eta) = 1/sh^2\tau + 1/sin^2\eta,$$

$$H_\tau^2 = H_\eta^2 = sh^2\tau + sin^2\eta, \quad H_\varphi = sh\,\tau\,sin\,\eta,$$

for an oblate ellipsoid

$$A(\tau) = th\,\tau, \quad B(\tau,\eta) = 1/sin^2\eta - 1/ch^2\tau,$$

$$H_\tau^2 = H_\eta^2 = ch^2\tau - sin^2\eta, \quad H_\varphi = ch\,\tau\,sin\,\eta$$

Let a surface $\tau = \tau_0 - const$ correspond to the solid surface of the body, then

$$\tau_0 = 1/2\,ln\{(1+\alpha)/(1-\alpha)\},$$

where $\alpha = b/a$, a , b are the half-axes of the ellipsoid, $a > b$,

$$a = ch\,\tau_0, \quad b = sh\,\tau_0$$

Boundary conditions for the problem in question have the form

$$\psi = 0,\quad \partial\psi/\partial r = 0 \quad \text{when} \quad t \geq 0,\ r = r_0;$$
$$\psi = 0,\quad \omega = 0 \quad \text{when} \quad t \geq 0,\ \eta = 0, \pi; \tag{1.4}$$
$$\psi = 0,\quad \omega = 0 \quad \text{when} \quad t = 0,\ r \to \infty$$
$$\omega = 0,\quad \partial\psi/\partial r \to \frac{1}{2}\, sh\, 2r \sin^2\eta \quad \text{when} \quad t > 0,\ r \to \infty.$$

Consider the transformation of independent variables

$$r = r_0 + \varkappa(\tau)\xi,\quad t = \mathcal{T}(\tau), \tag{1.5}$$

where $\varkappa(\tau) = 2r_0\sqrt{2/Re}\,\mathcal{Y}(\tau)$. Here $\mathcal{Y}(\tau)$ and $\mathcal{T}(\tau)$ are present smooth functions of the following form

$$\mathcal{Y}(\tau) = \gamma\{1 - \exp(-\tau)\} \quad \text{when} \quad \tau \geq 0, \tag{1.6}$$
$$\mathcal{T}(\tau) = \varepsilon\tau^2/(1+\tau) \quad \text{when} \quad \tau \geq 0.$$

ε, γ, $-\ const > 0$ - are the constant parameters.

Such choice of the functions $\mathcal{Y}(\tau)$ and $\mathcal{T}(\tau)$ in transformation (1.5) allows to consider uniformly with respect to η the dynamics of the vortex layer evolution on the surface of ellipsoid at different moments of time, that is of major importance in an investigation of the flow about very oblate or prolate ellipsoids.

Let introduce a transformation of the stream function and vorticity

$$\psi = \varkappa(\tau)\,\Psi(\eta,\xi,\tau),\quad \omega = W(\eta,\xi,\tau)/\varkappa(\tau), \tag{1.7}$$

which warrants elimination of singularity of the initial vortex field on the body surface.

Eqs (1.3) with (1.5)-(1.7) take the form

$$\frac{\partial^2 W}{\partial\xi^2} + Q_1(\eta,\xi,\tau)\frac{\partial W}{\partial\xi} + Q_2(\eta,\xi,\tau)W + Q_3(\tau)\left\{\frac{\partial^2 W}{\partial\eta^2} + ctg\,\eta\frac{\partial W}{\partial\eta} - \right.$$
$$\left. - Q_4(\eta,\xi,\tau)W\right\} - P_1(\eta,\xi,\tau)\frac{\partial W}{\partial\tau} + P_2(\eta,\xi,\tau)\left\{\frac{\partial\Psi}{\partial\eta}\frac{\partial W}{\partial\xi} - \right.$$
$$\left. - \frac{\partial\Psi}{\partial\xi}\frac{\partial W}{\partial\eta} - P_3(\xi,\tau)W\frac{\partial\Psi}{\partial\eta} + P_4(\eta)W\frac{\partial\Psi}{\partial\xi}\right\}, \tag{1.8}$$
$$\frac{\partial^2\Psi}{\partial\xi^2} - L_1(\xi,\tau)\frac{\partial\Psi}{\partial\xi} - L_2(\eta,\tau)\frac{\partial\Psi}{\partial\eta} + \varkappa^2(\tau)\frac{\partial^2\Psi}{\partial\eta^2} = H_r^2 H_\varphi W,$$

Here, the following designations are used:

a) in the case of an prolate ellipsoid

$$Q_1(\eta,\xi,\tau) = \mathcal{H}(\tau)\,cth\,\tau + Re\,H_\tau^2\,\xi\,\mathcal{H}(\tau)\mathcal{H}'(\tau)/2\beta,$$

$$Q_2(\eta,\xi,\tau) = Re\,H_\tau^2\,\mathcal{H}(\tau)\mathcal{H}'(\tau)/2\beta, \quad Q_3(\tau) = \mathcal{H}^2(\tau),$$

$$Q_4(\eta,\xi,\tau) = 1/sh^2\tau + 1/sin^2\eta, \quad P_1(\eta,\xi,\tau) = \mathcal{H}^2(\tau)H_\tau^2\,Re/2\beta,$$

$$P_2(\eta,\xi,\tau) = Re\,\mathcal{H}^2(\tau)/2H_\varphi, \quad P_3(\xi,\tau) = \mathcal{H}(\tau)\,cth\,\tau, \quad P_4(\eta) = ctg\,\eta,$$

$$L_1(\xi,\tau) = \mathcal{H}(\tau)\,cth\,\tau, \quad L_2(\eta,\tau) = \mathcal{H}^2(\tau)\,ctg\,\eta, \quad \beta = \mathcal{T}'(\tau)$$

Boundary conditions (1.4) with (1.5)-(1.7) are rewritten in the form

$$\begin{aligned} &\Psi = 0, \quad \frac{\partial\Psi}{\partial\xi} = 0 && \text{when } \tau \geq 0, \ \xi = 0; \\ &\Psi = 0, \quad W = 0 && \text{when } \tau \geq 0, \ \eta = 0, \pi; \\ &W = 0, \quad \partial\Psi/\partial\xi \to \frac{1}{2}\,sh\,2r(\xi,\tau)\,sin^2\eta && \text{when } \tau \geq 0, \ \xi \to \infty \end{aligned} \tag{1.9}$$

Hence, Eqs (1.8) with conditions (1.9) form a closed system for determination $\Psi(\eta,\xi,\tau)$ and $W(\eta,\xi,\tau)$ when $\tau \geq 0$.

2. It is easy to obtain the initial distributions of the given stream function and vorticity by integrating system (1.8) when τ =0 with conditions (1.9).

Really, assuming τ = 0 in (1.8), we obtain

$$\begin{aligned} &\frac{\partial^2 W}{\partial\xi^2} + 2H_\tau^2(0)\,\xi\,\frac{\partial W}{\partial\xi} + 2H_\tau^2(0)W = 0, \\ &H_\tau^2(0)H_\varphi(0)W = \frac{\partial^2\Psi}{\partial\xi^2}, \end{aligned} \tag{1.10}$$

where $H_\tau(0) = H_\tau(\eta,\xi,0)$, $H_\varphi(0) = H_\varphi(\eta,\xi,0)$,

$$(\eta,\xi) \in \{0 \leq \eta \leq \pi, \quad 0 \leq \xi < \infty\}.$$

Solving (1.10) with conditions (1.9) when τ = 0, we obtain the initial conditions sought after in order to solve the differential problem (1.8), (1.9):

$$W(\eta,\xi,0)=\frac{2c}{\sqrt{\pi}\,H_\tau(0)}\,e^{-(H_\tau(0)\xi)^2}\sin\eta,$$

(1.11)

$$\Psi(\eta,\xi,0)=\frac{1}{2}\,sh\,2\tau_0\left\{\xi\,erf(\xi)+\frac{1}{\sqrt{\pi}\,H_\tau(0)}\left(e^{-(H_\tau(0)\xi)^2}-1\right)\right\}\sin^2\eta,$$

where

$$erf(\xi)=\frac{2H_\tau(0)}{\sqrt{\pi}}\int_0^\xi e^{-(H_\tau(0)\xi)^2}\,d\xi,$$

$$c=\begin{cases} ch\,\tau_0 & \text{- for the prolate ellipsoid,}\\ sh\,\tau_0 & \text{- for the oblate ellipsoid.}\end{cases}$$

3. Let the orthogonal grid $\bar{\omega}_{h_1 h_2 \delta\tau}=\{(ih_1, jh_2, \delta\tau\cdot m)$, $i = 0,1,\ldots,N_1$; $j = 0,1,\ldots,N_2$; $m = 0,1,\ldots,M\}$ with increments $\delta\eta=h_1=\pi/N_1$, $\delta\xi=h_2=\xi_{max}/N_2$, $\delta\tau$ be defined in the domain $\Omega=\{(\eta,\xi,\tau)$, $0\le\eta\le\pi$, $0\le\xi\le\xi_{max}$, $0\le\tau\le T\}$.

Denote as $\omega_{h_1,h_2}=\{(ih_1, jh_2)$, $i = 1,2,\ldots,N_1-1$; $j = 1,2,\ldots,N_2-1\}$ the famely of the internal points of computational grid in the (η,ξ) plane.

Then for $(i,j)\in\omega_{h_1,h_2}$ the following finite-difference equations are valid that approximate differential system (1.10) with an error of order $O(\delta\tau^2, h^2)$

$$\frac{W_{ij+1}^{n-\frac{1}{2}}-2W_{ij}^{n-\frac{1}{2}}+W_{ij-1}^{n-\frac{1}{2}}}{h_2^2}+2H_\tau^2(i)\xi_j\,\frac{W_{ij+1}^{n-\frac{1}{2}}-W_{ij-1}^{n-\frac{1}{2}}}{2h_2}+$$

(1.12)

$$+2H_\tau^2(i)W_{ij}^{n-1}=\frac{W_{ij}^{n-\frac{1}{2}}-W_{ij}^{n-1}}{\sigma_1}$$

or $$FW_{ij+1}^{n-\frac{1}{2}}-GW_{ij}^{n-\frac{1}{2}}+HW_{ij-1}^{n-\frac{1}{2}}=Z,$$

where

$$F=\frac{\sigma_1}{h_2^2}+\frac{\sigma_1 H_2^2(i)}{h_2}\xi_j,\quad H=\frac{\sigma_1}{h_2^2}-\frac{\sigma_1 H_2^2(i)}{h_2}\xi_j,$$

$$G = \frac{2\sigma_1}{h_2^2} + 1 \,, \quad Z = \left\{ 2\sigma_1 H_\eta^2(i) + 1 \right\} W_{ij}^{n-1} \,.$$

If conditions (1.15) are applied to the coefficients of Eq. (1.16), the restriction on a grid step with respect to ξ follows

$$h_2 < \min_i \left\{ 1/\xi_{max} \, H_\eta^2(i) \right\} \,, \tag{1.13}$$

where $H_\eta(i) = H_\eta(\eta_i)\,, \quad i = 1, 2, \dots, N_1 - 1$

The finite-difference equations can be presented in the form

$$F^{n-1m}(i,j) W_{ij+1}^{n-\frac{1}{2}m} - G^{n-1m}(i,j) W_{ij}^{n-\frac{1}{2}m} + H^{n-1m}(i,j) W_{ij-1}^{n-\frac{1}{2}m} = Z^{n-1m}(i,j) \,,$$

$$K^{m}(j) \Psi_{ij+1}^{n-\frac{1}{2}m} - L^{m}(i,j) \Psi_{ij}^{n-\frac{1}{2}m} + M^{m}(j) \Psi_{ij-1}^{n-\frac{1}{2}m} = R^{n-1m}(i,j) \,,$$

$$W_{ij}^{nm} = \sigma_2 W_{ij}^{n-\frac{1}{2}m} + (1 - \sigma_2) W_{ij}^{n-1m} \,, \tag{1.14}$$

$$\Psi_{ij}^{nm} = \sigma_3 \Psi_{ij}^{n-\frac{1}{2}m} + (1 - \sigma_3) \Psi_{ij}^{n-1m} \,,$$

at $(i,j) \in \omega_{h_1 h_2}$

where n is the iteration subscript, $\sigma_1 > 0$ is regularization parameter, $\sigma_2 < 2$, $\sigma_3 > 0$ are relaxation parameters, the numerical values of the subscripts r , s , p , q are chosen depending on the sign of values:

$$V_\eta(i,j) = -\frac{1}{H_\eta H_\varphi} \frac{\Psi_{ij+1}^{n-1m} - \Psi_{ij-1}^{n-1m}}{2h_2}, \quad V_\eta(i,j) = \frac{\mathcal{H}_m}{H_\eta H_\varphi} \frac{\Psi_{i+1j}^{n-1m} - \Psi_{i-1j}^{nm}}{2h_1}$$

at every point $(i,j) \in \omega_{h_1 h_2}$ according to the following rule

a) $s = n - 1/2,\ r = n-1,\ q = n - 1/2,\ p = n-1$ when $V_\eta \geq 0$, $V_r \geq 0$;

b) $s = n-1,\ r = n - 1/2,\ q = n-1,\ p = n - 1/2$ when $V_\eta < 0$, $V_r < 0$;

c) $s = n - 1/2,\ r = n-1,\ q = n-1,\ p = n - 1/2$ when $V_\eta < 0$, $V_r \geq 0$; (1.13)

d) $s = n-1,\ r = n - 1/2,\ q = n - 1/2,\ p = n-1$ when $V_\eta \geq 0$, $V_r < 0$;

Thus, the system of finite-difference equations (1.12) reduces to the system of linear algebraic equations (1.14), the matrix of which has tridiagonal form. An effective method of solving systems of this kind is the factorization method. It is well-known ([5]) that for monotonous and stable computation when a system like (1.14) is to be solved by the factorization method, it is sufficient to fulfil the conditions:

$$F(i,j)>0,\ G(i,j)>0,\ H(i,j)>0,\ G(i,j)\geq F(i,j)+H(i,j)$$
$$K(j)>0,\ L(i,j)>0,\ M(j)>0,\ L(i,j)\geq K(j)+M(j) \tag{1.15}$$

It is easy to show that the choice of the subscripts s , r , p , q is system (1.12) according to the rule (1.13) ensures the necessary form of the system coefficients to satisfy conditions (1.15). Besides, conditions (1.15) imposes restrictions on the computational grid step.

4. Let us choose the boundary condition for the vorticity on the surface of a body in flow and the computational procedure.

Let the contour $j = 0$ correspond to the solid body surface. As a boundary condition for the vorticity on the solid surface which, generally speaking, is absent in the physical formulation of the problem, the Thom's condition [6] is usually used, or the Wood's condition [7] , or that of the higher order of accuracy with respect to h [8] . The presence of an additional iterational process connected with the boundary condition for the vorticity on the solid surface of the body in a stream, may essentially restrict the convergence rate of a numerical method.

The method of boundary condition approximation presented in the paper [9] is based on solving of the vorticity transportation equation in the auxiliary domain disposed within the main computational domain. The boundary conditions for the vorticity on the surface of the auxiliary domain are found from the second equation of system (1.8).

The stream function field is corrected at each iterational step so that the prescribed boundary conditions are satisfied.

As it is pointed out in the above work, it is possible to successfully apply this approach to the solution of nonstationary problems, and in the case of implicit difference schemes, it makes possible to considerably shorten the required computation time.

In the present paper the stream function and vorticity fields are computed at every time step by means of iterations. The approach presented in [9] is used to obtain a correlated pattern of these fields.

The corresponding computational formulas by which the distributions of values sought for are iterated in the grid nodes adjacent the boundary, have the form

$$\psi_{i1}^{nm} = 1/2\, \psi_{i2}^{n-1m} - 1/9\, \psi_{i3}^{n-1m} + O(h_2^4),$$

$$W_{i1}^{nm} = \Big\{ \frac{1}{h_2^2}\left(\psi_{i2}^{n-1m} - 2\psi_{i1}^{n-1m}\right) + \frac{\mathcal{H}_m^2}{h_1^2}\Big(\psi_{i+11}^{n-1m} - 2\psi_{i1}^{n-1m} +$$

$$+ \psi_{i-11}^{nm}\Big) - L_1(1)\psi_{i2}^{n-1m}/2h_2 - L_2^m(i)\frac{\psi_{i+11}^{n-1m} - \psi_{i-11}^{n-1m}}{2h_1}\Big\} H_r^2 H_\varphi(i,1,m) + O(h^2),$$

where $h = max(h_1, h_2)$.

After the convergence of iterational process on the corresponding time step is achieved, the vorticity distribution on a body surface is once for all computed using the second-order-accurate formula (an analog of the Wood's presentation)

$$W_{i0}^{nm} = \left\{ \frac{3\psi_{i1}^{nm}}{h_2^2 L_4(0)} - \frac{W_{i1}^{nm}}{2} L_4(1) \right\} \Big/ \left\{ 1 + \frac{h_2}{2} L_1(1) \right\} + O(h^2), \quad (1.16)$$

where

$$L_4(0) = H_r^2(i,0,m) H_\varphi(i,0,m), \quad L_4(1) = H_r^2 H_\varphi(i,1,m),$$

$$L_1(1) = L_1(1,m)$$

As pressure is excepted from consideration when a problem is solved in the (ψ , ω) formulation, we shall treat it separately.

From the Navier-Stokes equation we have

$$\vec{\nabla p} = -\frac{2}{Re}\nabla \times \vec{\omega} - \frac{\partial \vec{v}}{\partial t} - \nabla\left(\frac{\vec{v}}{2}\right)^2 + \vec{v} \times \vec{\omega}, \quad (1.17)$$

where p is the pressure reduced to dimensionless form with respect to ρv^2 , ρ is the fluid density.

Consider the projection of equation (1.17) upon the unit vector $O\bar{pm}$ $\vec{i_\eta}$ (Fig. 2), then, considering the conditions $v_\eta = 0$, $v_r = 0$

on the surface of a body in a flow, we obtain after integration with respect to η an expression for determination of pressure distribution on an ellipsoid surface

$$\frac{P(\eta)-P(0)}{1/2\,\rho V^2} = -\frac{4}{\varkappa^2(\tau)Re}\int_0^{\eta}\left(\frac{\partial W}{\partial \xi}+\varkappa(\tau)\lambda W\right)_{\xi=0} d\eta \tag{1.18}$$

In a similar manner we obtain a formula for determining the pressure in the rear critical point

$$\frac{P(0)-P_\infty}{1/2\,\rho V^2} = 1+2\int_{\tau_0}^{\infty}\left(H_\tau \frac{\partial V_\tau}{\partial t}+\frac{4}{Re}\frac{\partial \omega}{\partial \eta}\right)_{\eta=0} d\tau\,, \tag{1.19}$$

where P_∞ is the pressure at infinite distance from the body.

To compute the skin friction drag C_{D_f} and pressure drag C_{D_p} coefficients, we can easily obtain the following formulas

$$C_{D_f} = \frac{D_f}{1/2\,\rho V^2 \pi K^2} = \frac{8\lambda}{\varkappa(\tau)Re}\int_0^{\pi} W(\eta,0,\tau)\sin^2\eta\, d\eta\,,$$

$$C_{D_p} = \frac{D_p}{1/2\,\rho V^2 \pi K^2} = -\int_0^{\pi} P(\eta)\sin 2\eta\, d\eta\,, \tag{1.20}$$

where $\lambda = cth\,\tau_0$, $K = sh\,\tau_0$ - in case of elongated ellipsoid, $\lambda = th\,\tau_0$, $K = ch\,\tau_0$ - in case of oblate ellipsoid.

5. Computations of the unsteady flow about ellipsoids of revolution have been performed in the present work with the purpose of assessing the proposed numerical method. It has been suggested, that the flow about the body was axisymmetric. Body shape is varied.

Oblate ellipsoids with the half-axes ratio α = 0.15, 0.25, 0.5 and the elongated ellipsoids with α = 0.25, 0.5 have been selected for testing. Besides, for the purpose of comparing results, the unsteady flow about a sphere instantly set into motion has been computed by this method.

Some numerical results obtained in the range of $Re = 1 \div 100$ are presented in Fig. 3 - 9. The total number of the computational grid nodes as chosen equal 61×61, the value of a step along the radial coordinate was varied depending on the Reynolds number.

Displacement of the separation point with time is shown in Fig. 3a, 3b, where for comparison the corresponding data for a sphere are given. It is evident from the figures that the shape of the body in a flow does influence the separation. In case of a more oblate ellipsoid, the separation occurs earlier in time.

The dynamics of the reverse-circulation flow zone growth in time is illustrated in Fig. 4. The corresponding results for the sphere are also given there.

The effect of the Reynolds number on the position of the separation point in case of an almost steady-state flow at large times after an impulsive start of an oblate ellipsoid is shown in Fig. 5.

In Fig. 6,7 distributions of the vorticity and reduced pressure are presented, computed according to formulas (1.16) and (1.18), respectively. The data given in these figures correspond to an almost steady fluid flow about ellipsoids of different shape.

In Fig. 8 and 9 the results obtained in the present paper are compared with the data of [11,12]. In Fig. 8 the comparison is given with respect to the frontal pressure $P(\pi)$, and in Fig. 9 with respect to the reverse-circulation flow zone length.

2. Numerical Investigation of Impulsive Velocity Variations of a Body Moving in a Viscous Fluid Flow

1. Let a circular cylinder, which initially moved with the velocity V_1, abruptly change its velocity and begin to move in the former direction with velocity V_2. The flow about the cylinder is assumed to be symmetric.

The Navier-Stokes equations that describe the unsteady viscous incompressible fluid flow, written in the polar coordinate system (θ, r) and reduced to the dimensionless form, are as follows:

$$\frac{\partial \omega}{\partial t} + \frac{1}{r}\frac{D(\psi,\omega)}{D(\theta,r)} = \frac{2}{Re}\left\{\frac{\partial^2 \omega}{\partial r^2} + \frac{1}{r}\frac{\partial \omega}{\partial r} + \frac{1}{r^2}\frac{\partial^2 \omega}{\partial \theta^2}\right\}, \quad (2.1a)$$

$$\omega = \frac{\partial^2 \psi}{\partial r^2} + \frac{1}{r}\frac{\partial \psi}{\partial r} + \frac{1}{r^2}\frac{\partial^2 \psi}{\partial \theta}, \quad (2.1b)$$

where $Re = 2VR/\nu$ is the Reynolds number, V is the cylinder velocity relative to the undisturbed fluid, R is the radius of the cylinder, t is the dimensionless time counting from the moment of abrupt change of velocity.

Consider the transformation of independent variables

$$\begin{aligned} r &= \exp\{\alpha\pi K(\tau)\xi\} && \text{when } 0 \le \xi < \xi_1; \\ r &= \exp\{\alpha\pi(\xi - \xi_1 + K(\tau)\xi_1)\} && \text{when } \xi_1 \le \xi \le \xi_{max}; \\ \theta &= \pi\eta && \text{when } 0 \le \eta \le 1; \\ t &= \mathcal{T}(\tau) && \text{when } \tau \ge 0, \end{aligned} \quad (2.2)$$

where $K(\tau) = 2\sqrt{2}/Re_2\, \mathcal{Y}(\tau)$, $Re_2 = 2V_2R/\nu$, $\alpha > 0$ is the constant parameter, $\mathcal{Y}(\tau)$ and $\mathcal{T}(\tau)$ are given smooth functions which are presented in Part I - (1.6).

The character of transformation (2.2) is illustrated in Fig. 10, where γ is the cylinder surface, Γ_1 is the intermediate contour, Γ is a contour sufficiently remote from the body, are the half-axes of symmetry, $\Omega_1 = \{(\eta,\xi),\ 0 \le \eta \le 1,\ 0 \le \xi \le \xi_1\}$ is the inner subdomain, $\Omega_2 = \{(\eta,\xi),\ 0 \le \eta \le 1;\ \xi_1 \le \xi \le \xi_{max}\}$ is the outer subdomain, Γ_1 being equal to $\Omega_1 \cap \Omega_2$, $\xi_1 \in (0, \xi_{max})$ will be chosen below.

The boundary conditions of the problem in question for Eqs. (2.1) in Ω_1 have the form

$$\begin{aligned} \gamma&: \psi = 0,\ \frac{\partial \psi}{\partial r} = 0 && \text{when } t \ge 0,\ r = 1; \\ C_\pm&: \psi = 0,\ \omega = 0 && \text{when } t \ge 0,\ \theta = 0, \pi; \end{aligned}$$

$$\Gamma_1: \frac{\partial \psi}{\partial r} = \left(1 - \frac{v_1}{v_2}\right)\sin\theta, \quad \omega = \omega_s(1,\theta)\frac{v_1}{v_2} \qquad \text{when } t = 0,\ r = 1;$$

$$\Gamma_1: \psi = \psi_{\Gamma_1}(\theta,t), \quad \omega = \omega_{\Gamma_1}(\theta,t) \qquad \text{when } t > 0,\ r \in \Gamma_1;$$

in the Ω_2 subdomain:

$$\begin{aligned}
&\Gamma_1: \psi = \psi_{\Gamma_1}(\theta,t), \quad \omega = \omega_{\Gamma_1}(\theta,t) && \text{when } t > 0,\ r \in \Gamma_1; \\
&C_\pm: \psi = 0, \quad \omega = 0 && \text{when } t \geq 0,\ \theta = 0, \pi; \\
&\Gamma: \omega = 0, \quad \psi = r\sin\theta && \text{when } t > 0,\ r \in \Gamma,
\end{aligned} \tag{2.4}$$

where $\psi_{\Gamma_1}(\theta,t)$, $\omega(\theta,t)$ are the distributions of the stream function and vorticity along the contour Γ_1 , $\omega_s(1,\theta)$ is the stationary distribution of the vorticity over the cylinder surface, which corresponds to a flow about the cylinder moving with velocity v_1 when $t < 0$.

In the internal subdomain $(\eta, \xi) \in \Omega_1$ we introduce the transformation

$$\psi = K(\tau)\Psi(\eta,\xi,\tau), \qquad \omega = W(\eta,\xi,\tau)/K(\tau) \tag{2.5}$$

to eliminate the initial singularity of the vorticity field. By virtue of transformations (2.2), (2.5) the boundary conditions (2.3) take the form

$$\begin{aligned}
&\gamma: \Psi = 0, \quad \frac{\partial \Psi}{\partial \xi} = 0 && \text{when } \tau \geq 0,\ \xi = 0; \\
&C_\pm: \Psi = 0, \quad W = 0 && \text{when } \tau \geq 0,\ \eta = 0, 1; \\
&\Gamma_1: W = 0, \quad \frac{\partial \Psi}{\partial \xi} = \alpha\pi\left(1 - \frac{v_1}{v_2}\right)\sin\pi\eta && \text{when } \tau = 0,\ \xi = \xi_1; \\
&\Gamma_1: W = W_{\Gamma_1}(\eta,\tau), \quad \Psi = \Psi_{\Gamma_1}(\eta,\tau) && \text{when } \tau > 0,\ \xi = \xi_1,
\end{aligned} \tag{2.6}$$

where $W_{\Gamma_1}(\eta,\tau) = K(\tau)\omega_{\Gamma_1}(\theta,t)$, $\Psi_{\Gamma_1}(\eta,\tau) = \psi_{\Gamma_1}(\theta,t)/K(\tau)$.

Eqs (2.1a), (2.1b) with transformations (2.2), (2.5) may be rewritten in the following form when $(\eta,\xi,\tau) \in \Omega_1 \times \mathcal{T}$, where

$$\mathcal{T} = \{0 \leq \tau \leq \tau_M\}$$

$$P(\tau)=\frac{\partial W}{\partial \tau}+\frac{K^2(\tau)}{2\alpha E^2}\frac{D(\Psi,W)}{D(\eta,\xi)}=\frac{2}{Re}\left\{\frac{1}{2\alpha^2 E^2}\frac{\partial^2 W}{\partial \xi^2}+\right.$$

$$\left.+Q(\tau)\xi\frac{\partial W}{\partial \xi}+Q(\tau)W+\frac{K^2(\tau)}{2E^2}\frac{\partial^2 W}{\partial \eta}\right\}, \tag{2.7}$$

$$E^2 W=\frac{1}{\alpha^2}\frac{\partial^2 \Psi}{\partial \xi^2}+K^2(\tau)\frac{\partial^2 \Psi}{\partial \eta^2},$$

where

$$P(\tau)=K^2(\tau)/2\beta,\quad \beta=\mathcal{T}'(\tau),\quad Q(\tau)=Re_2 K(\tau)K'(\tau)/4\beta,$$

$$E=\pi\exp(\alpha\pi K\xi)$$

Solving system (2.7) when $\tau = 0$ with conditions (2.6) it is not difficult to obtain expressions for initial distributions of the reduced stream function and the reduced vorticity when $(\eta,\xi)\in\Omega_1$, namely

$$W(\eta,\xi,0)=\frac{2}{\sqrt{\pi}}\left(1-\frac{V_1}{V_2}\right)\exp\left\{-(\alpha\pi\xi)^2\right\}\sin\pi\eta,$$

$$\Psi(\eta,\xi,0)=\alpha\pi\left(1-\frac{V_1}{V_2}\right)\left\{\xi\,erf(\xi)+\frac{1}{\alpha\pi^{3/2}}\left(e^{-(\alpha\pi\xi)^2}-1\right)\right\}\sin\pi\eta, \tag{2.8}$$

where $erf(\xi)=2\alpha\sqrt{\pi}\int_0^{\xi}\exp\left\{-(\alpha\pi\xi)^2\right\}d\xi$

In the external subdomain $(\eta,\xi)\in\Omega_2$ the system of Eqs (2.1) with transformations (2.2) when $\tau > 0$ takes the form

$$\frac{\partial\omega}{\partial t}+\frac{\beta}{\alpha H^2}\frac{D(\psi,\omega)}{D(\eta,\xi)}=\frac{2\beta}{Re_2 H^2}\left\{\frac{1}{\alpha^2}\frac{\partial^2\omega}{\partial\xi^2}+\frac{\partial^2\omega}{\partial\eta^2}\right\}+$$

$$+\xi_1 K'(\tau)\frac{\partial\omega}{\partial\xi},\qquad H^2\omega=\frac{1}{\alpha^2}\frac{\partial^2\psi}{\partial\xi^2}+\frac{\partial^2\psi}{\partial\eta^2} \tag{2.9}$$

with the boundary conditions

$$\Gamma_1: \psi = \psi_{\Gamma_1}(\eta,\tau), \quad \omega = \omega_{\Gamma_1}(\eta,\tau) \quad \text{when } \tau > 0, \ \xi = \xi_1 ;$$

$$C_\pm : \psi = 0, \ \omega = 0 \quad \text{when } \tau \geq 0, \ \eta = 0,1 ; \tag{2.10}$$

$$\Gamma : \omega = 0, \ \psi = exp\{\alpha\pi(\xi - \xi_1 + K\xi_1)\} sin\pi\eta \ \text{when } \tau \geq 0, \ \xi = \xi_{max},$$

Here

$$H = \pi \, exp\{\alpha\pi(\xi - \xi_1 + K(\tau)\xi_1)\}.$$

Note the fact, that the initial conditions for Eqs (2.9) in Ω_2 are the result of superposition of the initial steady flow velocity field corresponding to the flow about a body when $V = V_1$, and of the velocity field characterizing a disturbance, i.e. fluid velocity variation at infinity and at $\Delta V = V_2 - V_1$ [3].

Since all dimensionless characteristics of the initial undisturbed flow about a body when $Re_1 = 2V_1R/\nu$ have been computed beforehand by reducing to a dimensionless form with respect to the velocity V_1 scale, to obtain dimensionless initial conditions in the V_2 scale, we shall change the scale. As a result of this we finally obtain when $\tau = 0$

$$\omega(\eta,\xi,0) = \omega_s(\eta,\xi)\frac{V_1}{V_2}, \tag{2.11}$$

$$\psi(\eta,\xi,0) = \psi_s(\eta,\xi)\frac{V_1}{V_2} + \left(1 - \frac{V_1}{V_2}\right)\left\{e^{\alpha\pi(\xi-\xi_1)} - 1\right\} sin\pi\eta,$$

at $(\eta,\xi) \in \Omega_2$,

here, $\omega_s(\eta,\xi)$, $\psi_s(\eta,\xi)$ are the dimensionless vorticity and stream function fields corresponding to the initial steady flow about a body when the undisturbed fluid flow velocity is V_1.

2. Let the orthogonal uniform grid $\omega_{h_1 h_2 \delta\tau} = \{(ih_1, jh_2, m\delta\tau), \ i = 0,1,\dots,N_1; \ j = 0,1,\dots,N_2; \ m = 0,1,\dots,M\}$ be defined in the domain $\overline{D} = \{\Omega_1 \cup \Omega_2\} \times \mathcal{T}$, the grid being self-correlated on the contour Γ_1:

$$\xi_1 = N_3 h_2, \quad 0 < N_3 < N_2$$

In the internal subdomain $(\eta_i, \xi_j) \in \Omega_1$ the finite-difference scheme proposed in Part. I has been used for numerical integration of Eqs (2.7), therefore we shall not dwell upon it.

In the external subdomain $(\eta_i, \xi_j) \in \Omega_2$ the implicit finite-difference scheme with asymmetric approximation of convective terms is constructed to solve numerically the differential problem (2.9), (2.10) ([13]).

It is not difficult to show that scheme (2.12a,b) approximates (2.9) with an error of order $O(h^2, \delta\tau)$, where $h = max(h_1, h_2)$.

Difference Eqs (2.12a,b) are reduced to the relationships, which make it possible to determine the values of the stream function and vorticity in every internal nodal point of the grid in Ω_2.

These relationships have the form

$$
\begin{aligned}
&A_1^{n-1m} \psi_{ij}^{n-\frac{1}{2}m} + B_1^{n-1m} \omega_{ij}^{n-\frac{1}{2}m} = C_1^{n-1m}, \\
&A_2^{n-1m} \psi_{ij}^{n-\frac{1}{2}m} + B_2^{n-1m} \omega_{ij}^{n-\frac{1}{2}m} = C_2^{n-1m}, \\
&\omega_{ij}^{nm} = \sigma_1 \omega^{n-\frac{1}{2}m} + (1-\sigma_1)\omega^{n-1m}, \\
&\psi_{ij}^{nm} = \sigma_2 \psi_{ij}^{n-\frac{1}{2}m} + (1-\sigma_2)\psi_{ij}^{n-1m}
\end{aligned}
\tag{2.13}
$$

where $\sigma_1 < 2$, $\sigma_2 > 0$ 0 are the relaxation parameters, n is the iterational number.

The coefficients A_1, A_2, B_1, B_2, C_1, C_2 in system (2.13) are chosen so that the following conditions are satisfied: the constant sign of the system determinant, its distinction from zero and reality of the eigenvalue matrix of the same system. The fulfilment of these conditions ensures computation stability and monotonous character of the numerical solution [13] .

The numerical solution of the problem in question is found by successive integration of solutions obtained in Ω_1 and Ω_2 followed by their matching on the contour Γ_1, such matching the solutions being carried out during every iteration.

As condition of exit from the iterational process, the following condition has been used

$$
max\{\varepsilon_1, \varepsilon_2\} < 10^{-4}, \tag{2.14}
$$

where

$$
\varepsilon_1 = \max_{(i,j)\in\Omega_1} \left\{ \left| \frac{\Delta \psi_{ij}^{nm}}{\psi_{ij}^{nm}} \right|, \left| \frac{\Delta W_{ij}^{nm}}{W_{ij}^{nm}} \right| \right\}
$$

$$\mathcal{E}_2 = \max_{(i,j)\in\Omega_2}\left\{\left|\frac{\Delta\psi_{ij}^{nm}}{\psi_{ij}^{nm}}\right|, \left|\frac{\Delta\omega_{ij}^{nm}}{\omega_{ij}^{nm}}\right|\right\},$$

$\Delta F_{ij}^{nm} = F_{ij}^{nm} - F_{ij}^{n-1m}$ is difference between two successive approximations of the grid function F_{ij}^{m}.

Correlation the stream function and vorticity fields has been achieved in the iterational process without resort to the boundary condition for the vorticity on the body surface which is also absent in the physical formulation of the problem.

After conditions (2.14) are satisfied, the reduced vorticity on the surface of a body in a stream is computed using the expression

$$W_{i0}^{nm} = \frac{3\psi_{i1}^{nm}}{(\alpha\pi h_2)^2} - \frac{W_{i1}^{nm}}{2}\exp(2\alpha\pi K_m h_2) + O(h_2^2),$$

which is an analog of the Wood's condition [7].

3. It should be noted, that the crucial point of the method set forth in this section is the choice of the value ξ_1, and, consequently, of the position of the contour Γ_1.

According to (2.2) we have

$$\Gamma_1: \quad z = \exp\{\alpha\pi K(\tau)\xi_1\}$$

i.e. the position of the contour Γ_1 in the physical plane is varying with time.

Besides, said contour at every instant must overlap the boundary layer of intensive vorticity due to the abrupt variation of velocity of the body moving in a stream, which expands while interacting with the "old" vortex field.

On the other hand, the value ξ_1 is limited from above due to the monotony and stability condition of the numerical method used in Ω_1, i.e.

$$\xi_1 < 2/h_2$$

Considering the abovesaid the value ξ_1 has been experimentally chosen in the present work through a number of preliminary computations.

The numerical method proposed in the present paper allows to perform by one general algorithm the numerical investigation of the unsteady flow about a cylinder both in case of impulsive acceleration and deceleration.

Numerical computations have been conducted for cases where the cylinder undergoes abrupt acceleration from the state with $Re_1 = 36.5$ to $Re_2 = 40, 100, 550$ and abrupt deceleration from $Re_1 = 36.5$ to $Re_2 = 31$.

Some of the computation results are shown in Fig. 11-17.

In Fig. 11 relationships between the angle of flow separation on the cylinder and time are given for different conditions of velocity variation. It should be noted, that the relationship $\theta_s(t)$ has non-monotonous character in case of instant deceleration.

The character of the interaction of the vorticity layer being formed on a body, with the initial vortex field are presented in Fig. 12 and 13 for the cases of the body accelerated or decelerated in a stream, respectively. As it follows from the given data, the vorticity layer formed on the body has a small thickness in the initial stage of flow evolution both in the case of acceleration and deceleration.

The skin friction drag coefficient C_{Df} versus time relationships are given in Fig. 14.

The vorticity and reduced pressure distributions on a body in the conditions of acceleration and deceleration and in different moments of time are illustrated in Fig. 15-16. Behavior of curves in Fig. 15 apparently indicates the fact, that the flow initially formed in the neighborhood of an abruptly decelerated body has the mode of a reverse flow about the body with respect to its surface. It is evident, from that the reverse-circulation zone penetrates upstream to the regions adjacent to a body surface.

It should also be noted, that the stagnation zone considerably increases in length as compared to its initial value that is characteristic of deceleration.

Reference

1. Кравченко В.И., Шевелев Ю.Д., Щенников В.В. Численное последование нестационарного обтекания тел конечных размеров потоком вязкой несжимаемой жидкости при различных режимах разгона и торможения. Препринт Института проблем механики, № 84, 1977.
2. Кравченко В.И., Шевелев Ю.Д., Щенников В.В. Численное исследование течения вязкой несжимаемой жидкости около цилиндра, мгновен-

но приведенного в движение. Журнал прикладной механики и техн. физики, № 5, 1976.

3. Watson J. The two-dimensional laminar flow near the stagnation point of a cylinder which has an arbitrary transverse motion. Quart. J. of Mech. and Appl.Math., vol.12, 150-190, 1959.

4. Nanbu K. Flow near the stagnation point of a body which undergoes a sudden change in a steady stream. "Trans.ASME", E40, No.1, 1973.

5. Годунов С.К., Рябенький В.С. Введение в теорию разностных схем. М., Физматгиз, 1962.

6. Том А., Эйплт С. Числовые расчеты полей в технике и физике. Изд-во "Энергия", 1964.

7. Woods L. Note on the numerical solution of a fourth order differential equation. Aer. Quart., No. 5, 1954.

8. Кускова Т.В., Чудов Л.А. О приближенных граничных условиях для вихря при расчете течений вязкой несжимаемой жидкости. В Сб. "Вычислительные методы и программирование", М., Изд-во МГУ, вып. XI, 1968.

9. Грязнов В.Л., Полежаев В.И. Исследование некоторых разностных схем и аппроксимация граничных условий для численного решения уравнений тепловой конвекции. Препринт Института проблем механики № 40, Москва.

10. Отрощенко И.В., Федоренко Р.П. О приближенном решении стационарных уравнений Навье-Стокса. Препринт № 6 за 1976г., Институт прикладной математики АН СССР, Москва, 1975.

11. Masliyah J.H., Epstein N. Numerical study of steady flow past spheroids. J.Fluid Mech., vol. 44, pp. 493-512, 1970.

12. Masliyah J.H. Steady wakes behind oblate spheroids: flow visualisation. Phys.Fluids, vol.15, No.6, pp. 1144-1146, 1972.

13. Люлька В.А., Щенников В.В. Численное решение уравнений Навье-Стокса. В сб. "Сборник теоретических работ по гидромеханике". Тр. ВЦ АН СССР, М., 1970, 107-149.

14. Гущин В.А., Щенников В.В. Об одной монотонной разностной схеме второго порядка точности. Ж. выч. мат. и матем. физ., 1974, 14, № 3, 789-792.

15. Шлихтинг Г. Теория пограничного слоя. Изд-во "Наука", 1969.

16. Collins W.M., Dennis S.C.R. Symmetrical flow past a uniformly accelerated circular cylinder. J.Fluid Mech., v.65, p.3, 1974.

17. Taneda S. Visualization experiments on unsteady viscous flows around cylinders and plates. IUTAM Symp. on Unsteady Boundary Layers Laval. Univ. Queb. Canada, 1971.

18. Lugt H.J., Haussling H.J. Laminar flow past an abruptly accelerat-

ed elliptic cylinder at 45° incidence. J. Fluid Mech., vol. 65, p.4, pp. 711-734, 1974.

19. Bar-lev M., Yang H.T. Initial flow field over an impulsively started circular cylinder. J.F.M., vol. 72, p.4, pp. 625-647, 1975.

Fig. 1

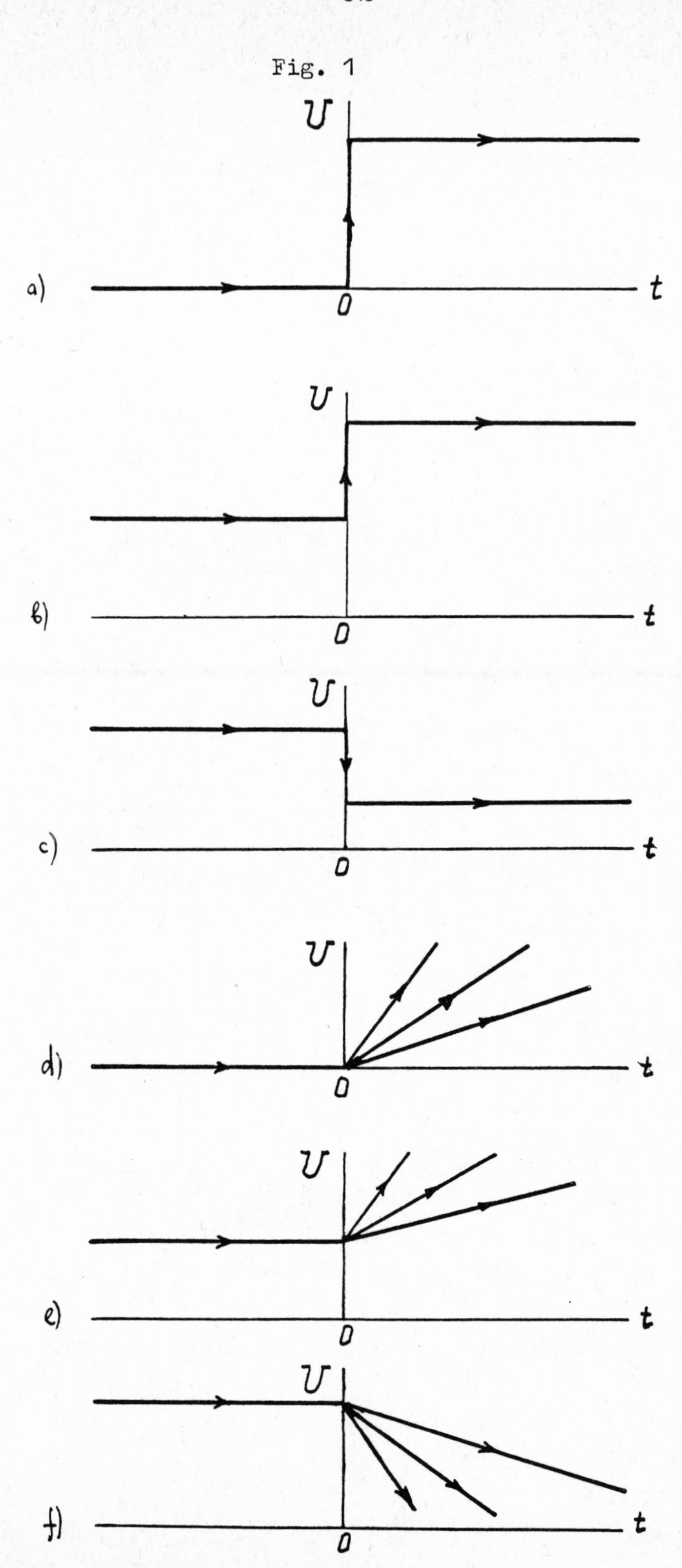

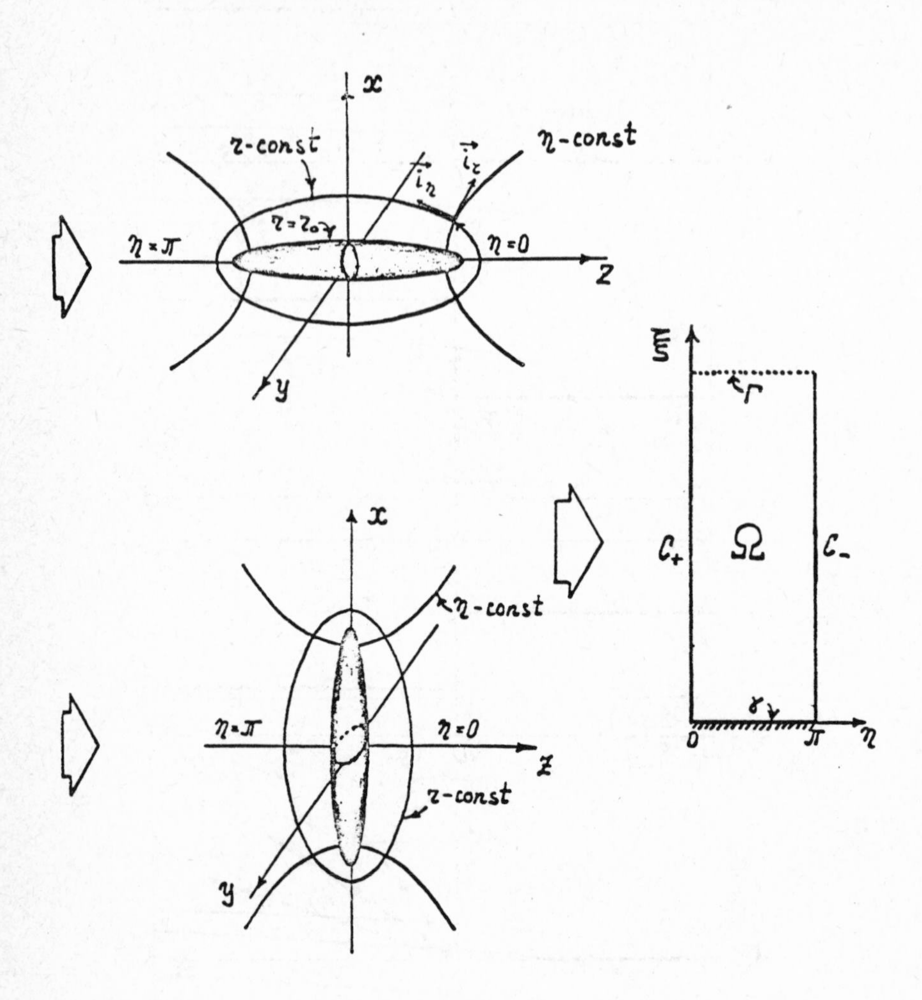
x
τ-const
η-const
η=π
τ=τ₀
η=0
z
y
ξ
Γ
C+
Ω
C-
x
η-const
η=π
η=0
z
τ-const
y
γ
0
π
η

Fig. 2

Fig. 3a

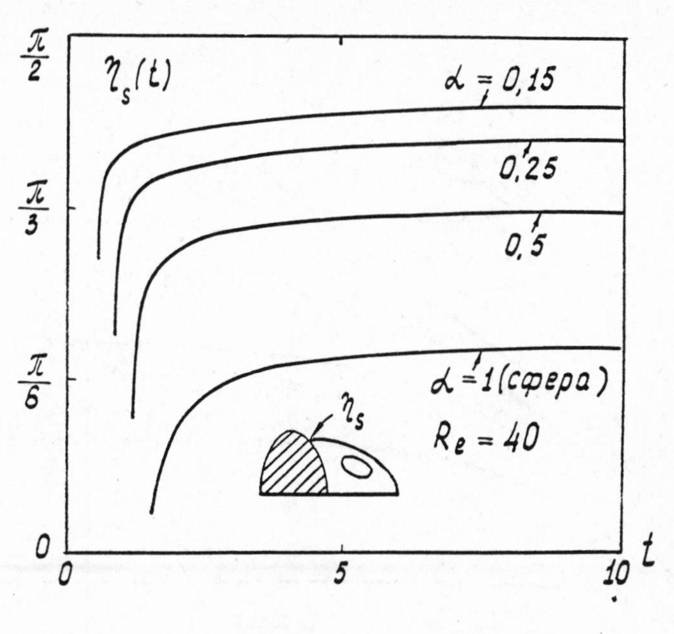
$\eta_s(t)$
$\alpha = 0,15$
0,25
0,5
$\alpha = 1$ (сфера)
$R_e = 40$
η_s
$\frac{\pi}{2}$
$\frac{\pi}{3}$
$\frac{\pi}{6}$
0
0
5
10
t

Fig. 3b

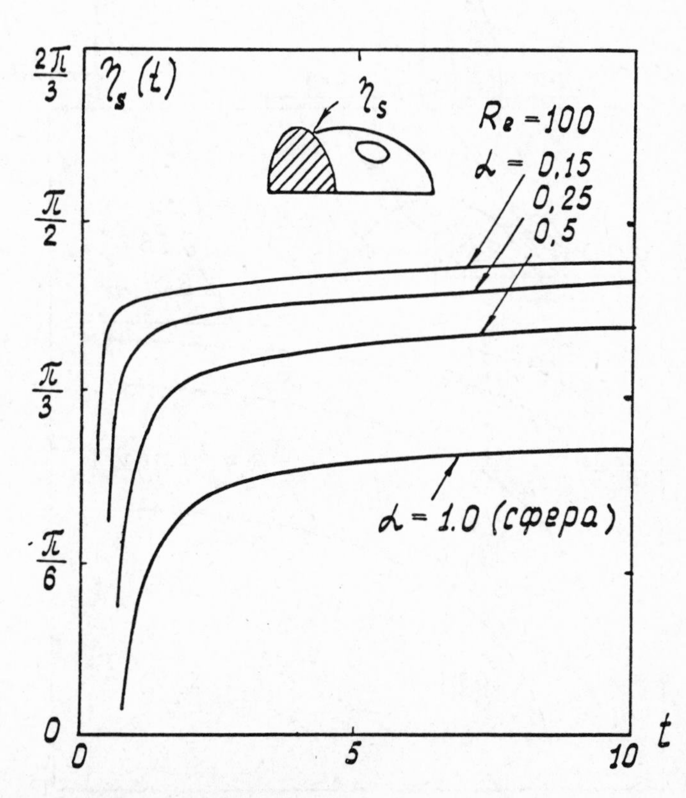
$\eta_s(t)$
η_s
$R_e = 100$
$\alpha = 0,15$
0,25
0,5
$\alpha = 1.0$ (сфера)
$\frac{2\pi}{3}$
$\frac{\pi}{2}$
$\frac{\pi}{3}$
$\frac{\pi}{6}$
0
0
5
10
t

Fig. 4

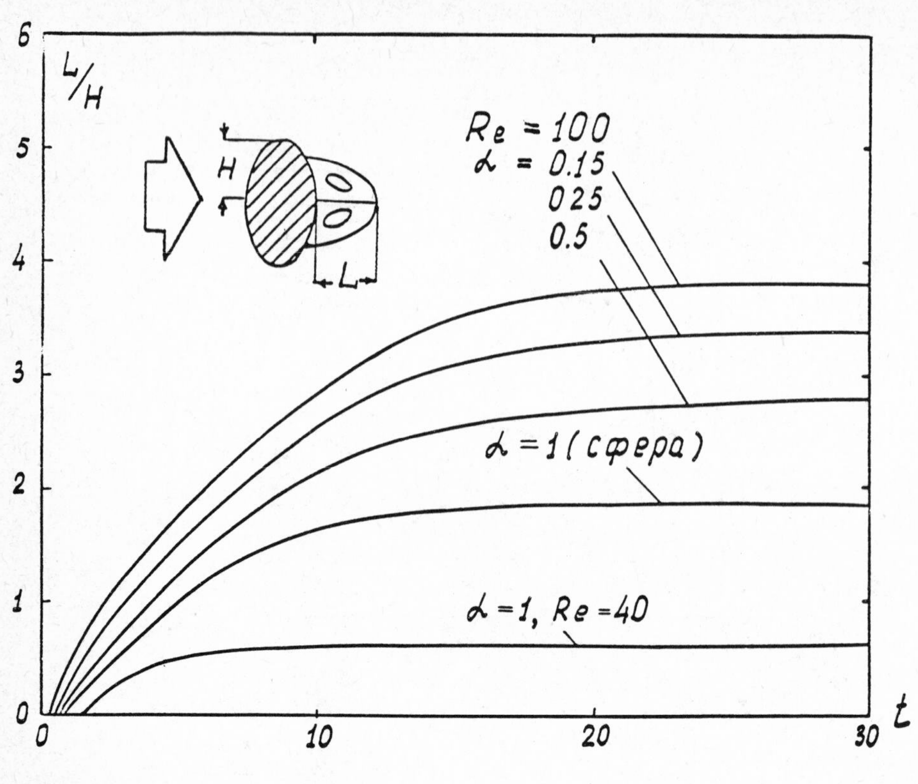
L/H
H
L
Re = 100
α = 0.15
0.25
0.5
α = 1 (сфера)
α = 1, Re = 40
t
0
1
2
3
4
5
6
0
10
20
30

Fig. 5

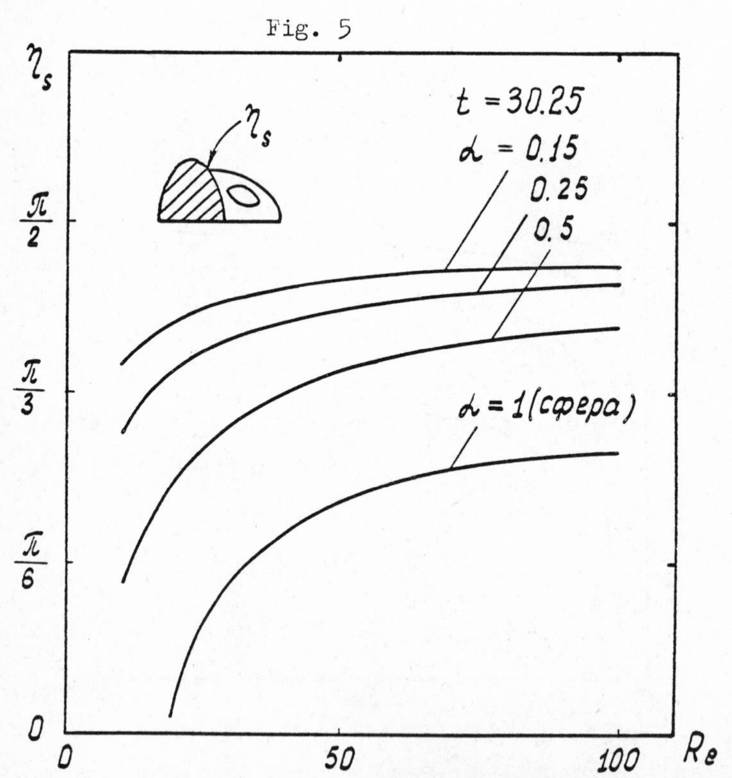
η_s
η_s
t = 30.25
α = 0.15
0.25
0.5
α = 1 (сфера)
π/2
π/3
π/6
0
0
50
100
Re

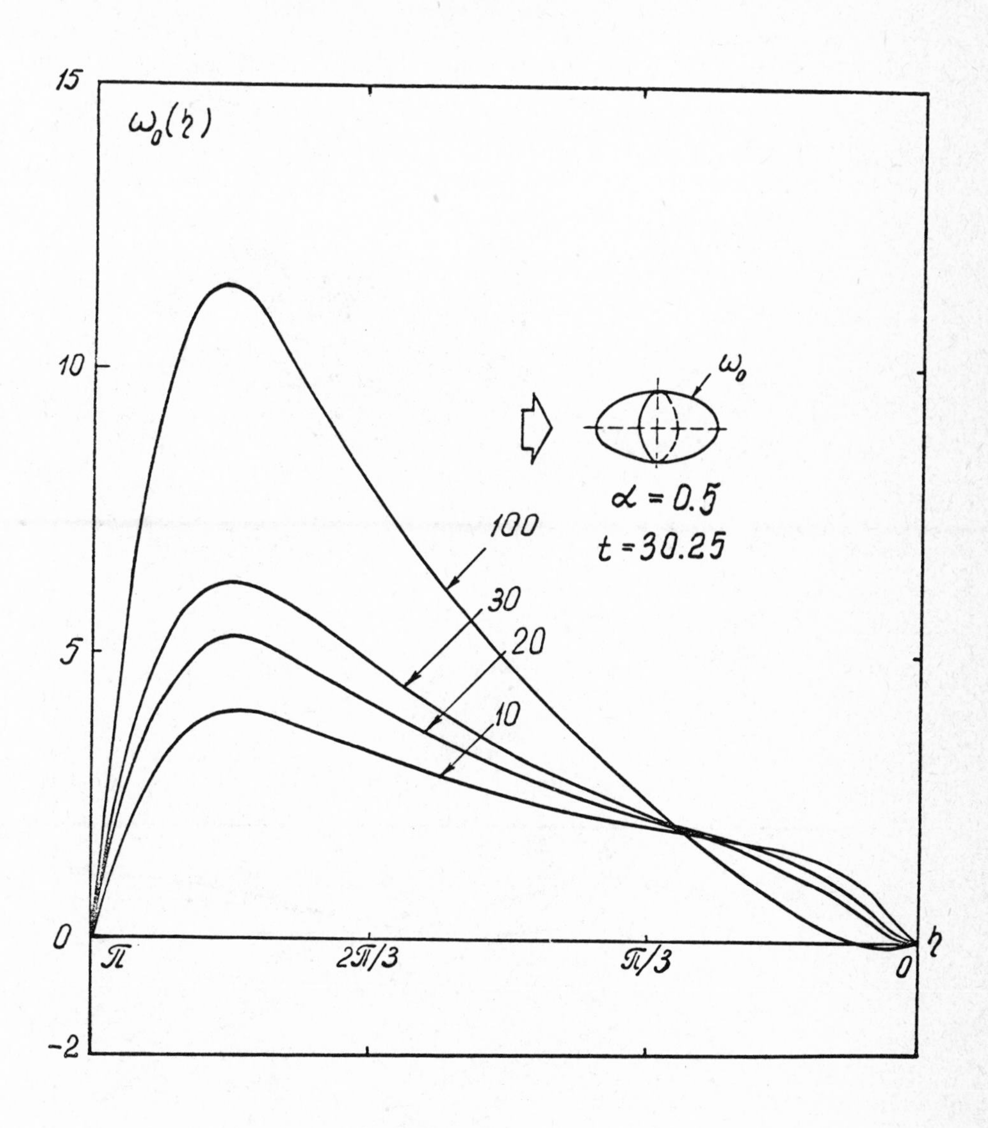
15
10
5
0
-2
$\omega_0(\eta)$
ω_0
$\alpha = 0.5$
$t = 30.25$
100
30
20
10
π
$2\pi/3$
$\pi/3$
0
η

Fig. 6

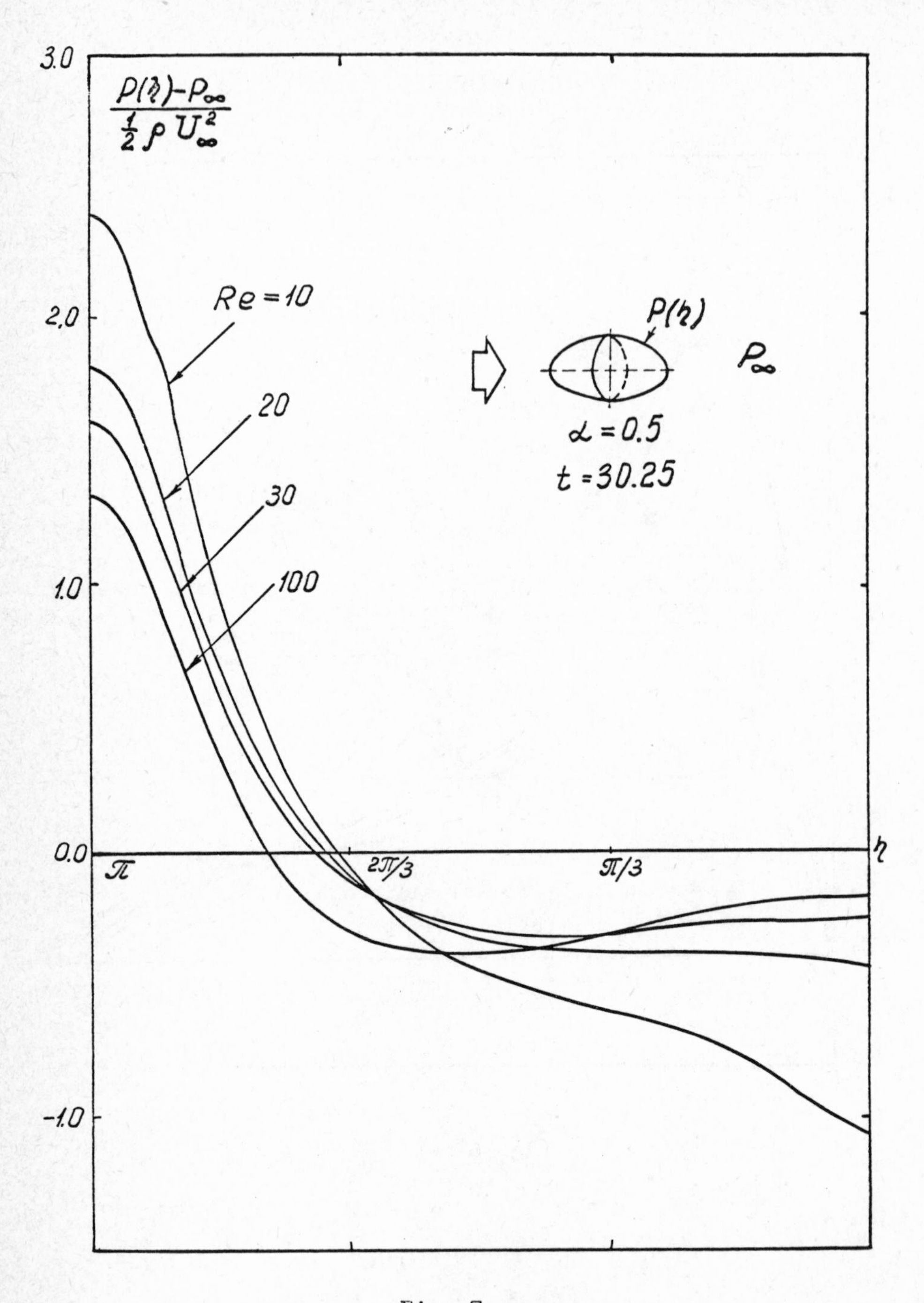
3.0
$\frac{P(\eta)-P_\infty}{\frac{1}{2}\rho U_\infty^2}$
2.0
1.0
0.0
-1.0
Re = 10
20
30
100
P(η)
P_∞
α = 0.5
t = 30.25
π
2π/3
π/3
η

Fig. 7

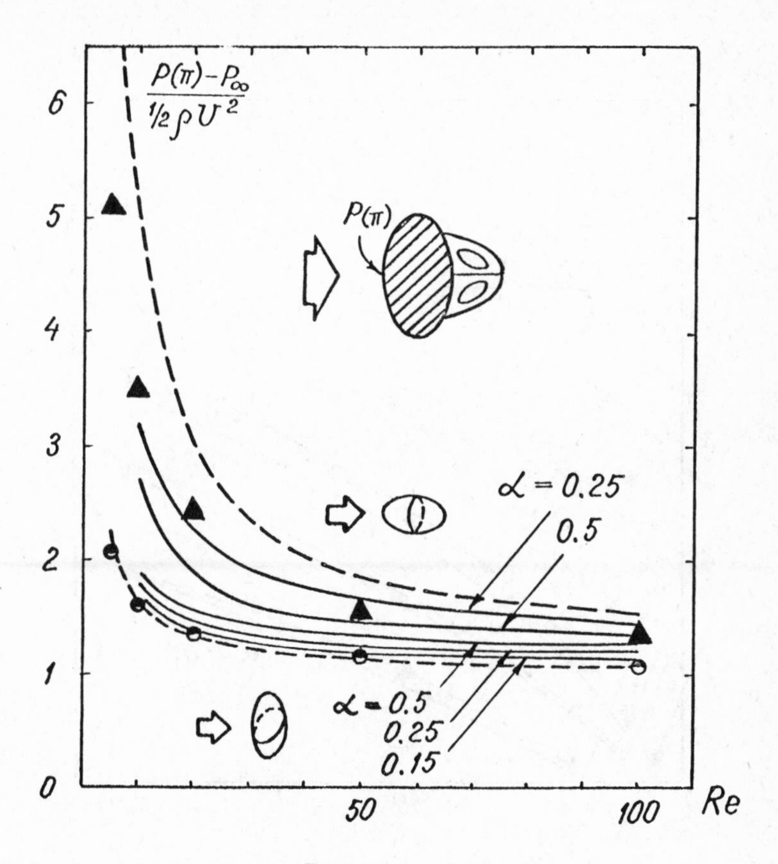

Fig. 8

▲ – α = 0.2 (prolate ellipsoid),

● – α = 0.2 (oblate ellipsoid)

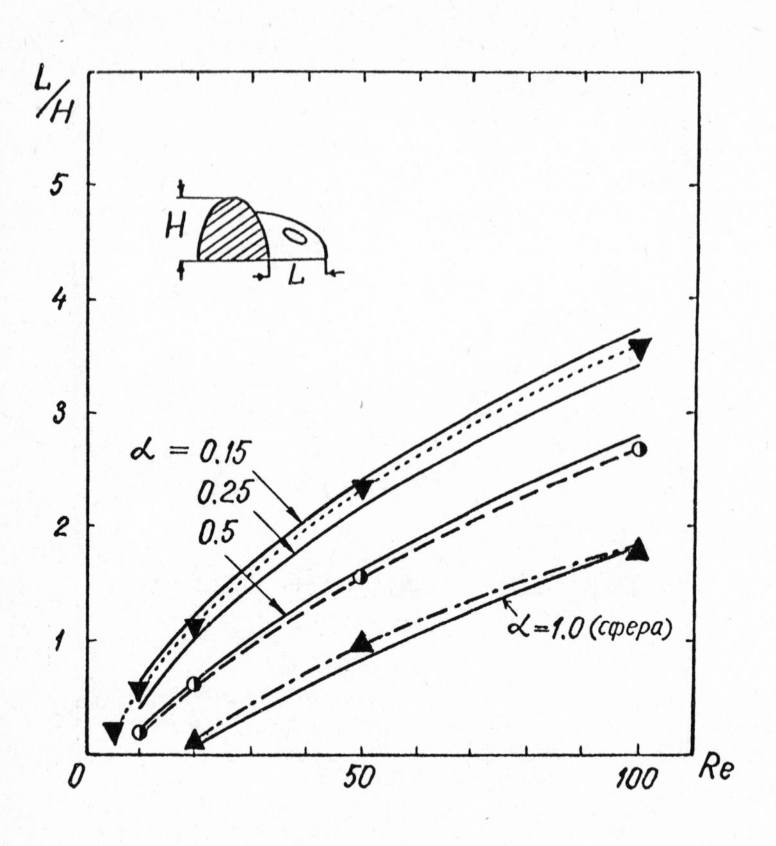

Fig. 9

---◐--- α = 0.5, ...▼..... α = 0.2,
.-.▲-.--α = 0.9

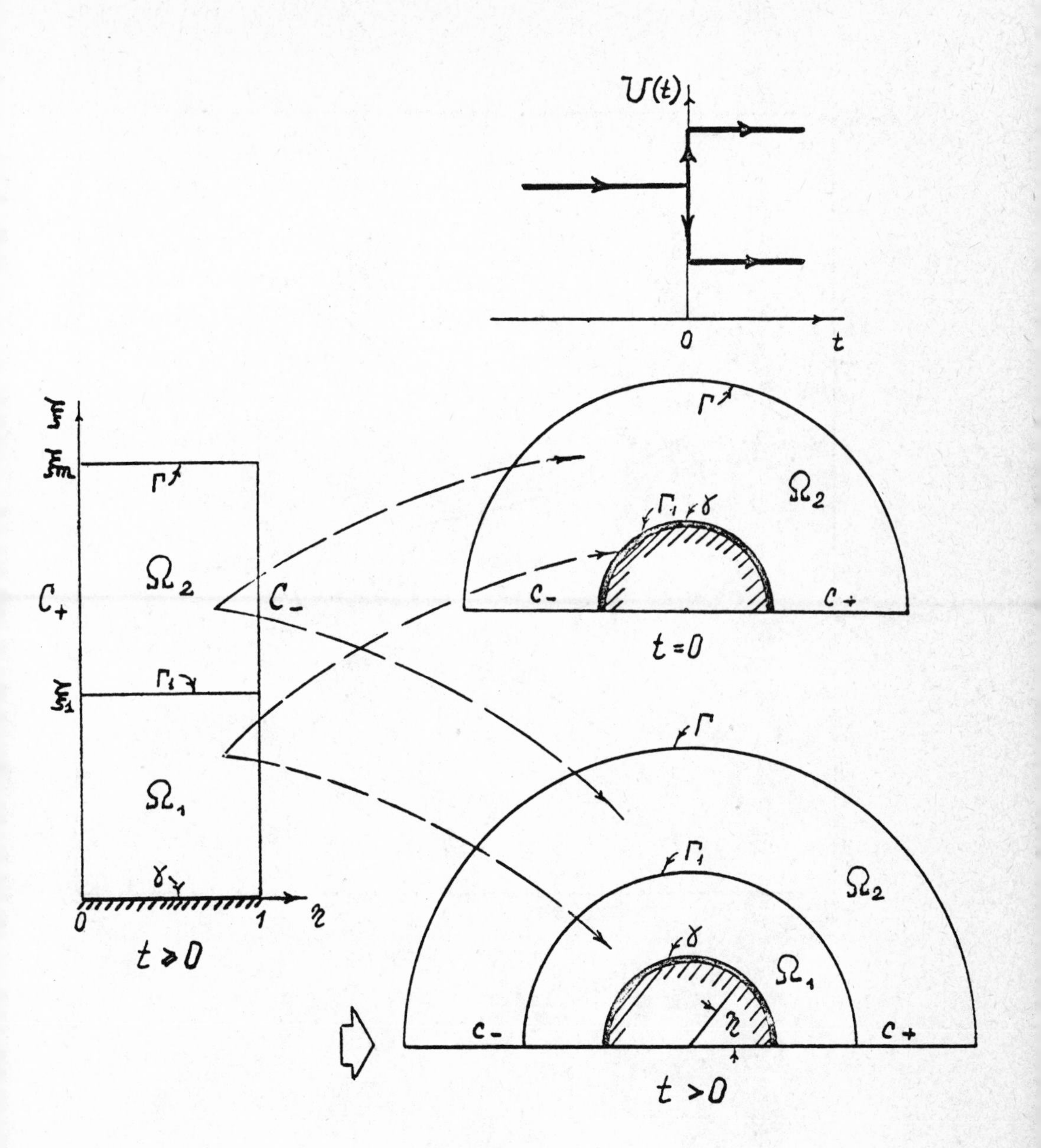

Fig. 10

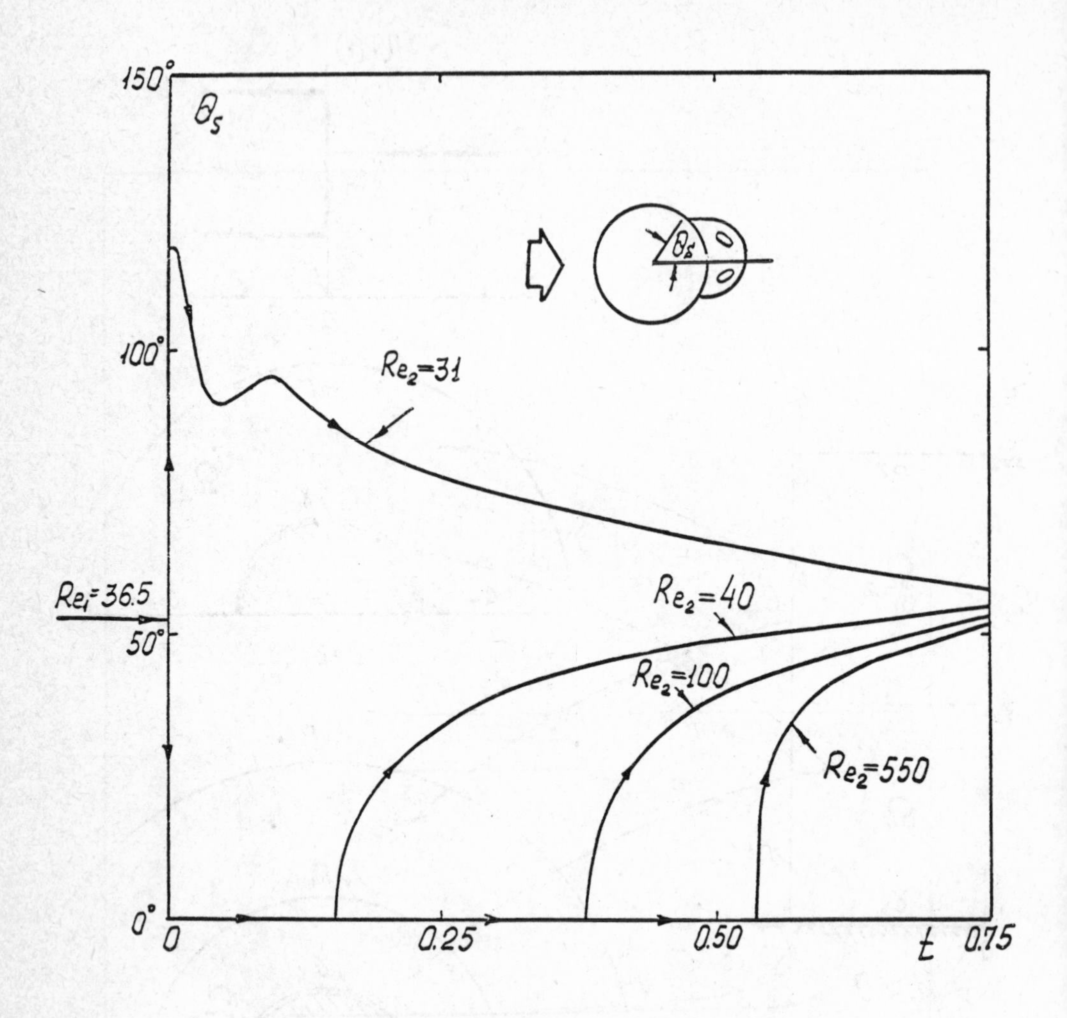
150°
θ_s
θ_s
100°
$Re_2=31$
$Re_1=36.5$
$Re_2=40$
50°
$Re_2=100$
$Re_2=550$
0°
0
0.25
0.50
t
0.75

Fig. 11

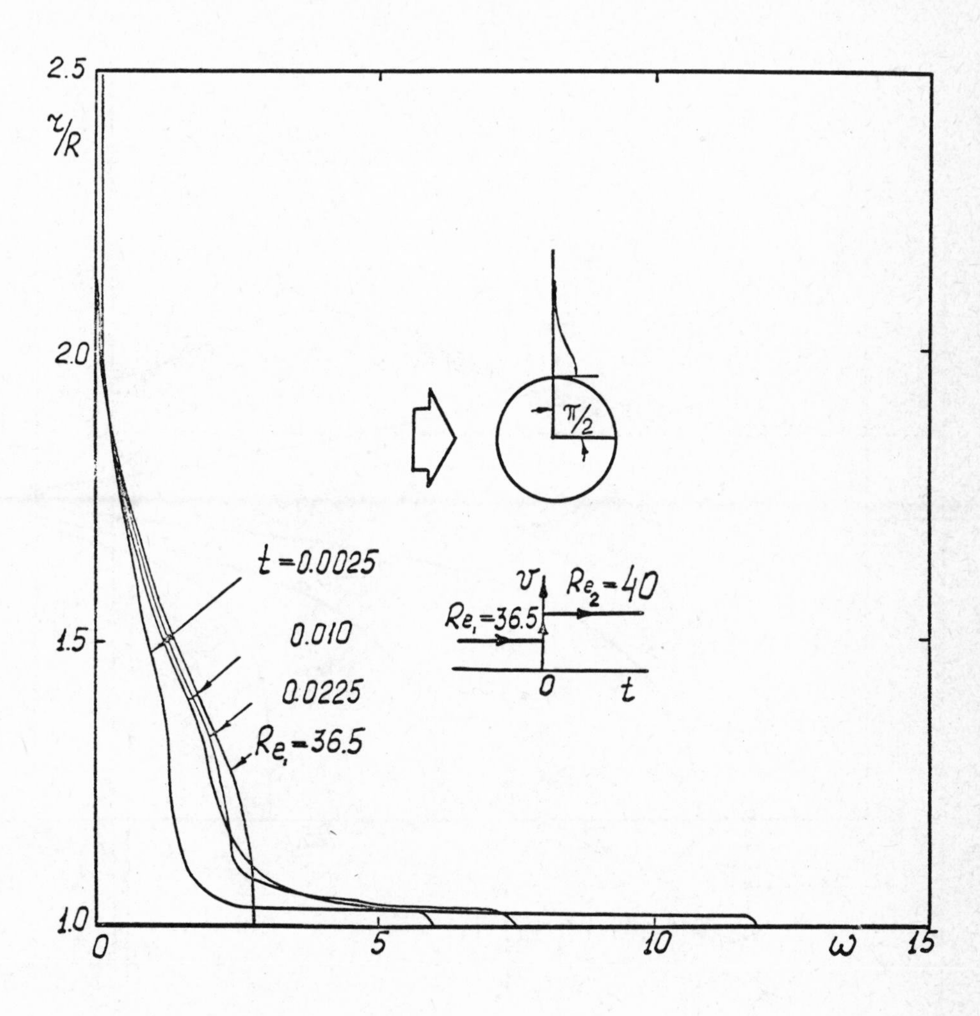
2.5
r/R
2.0
1.5
1.0
0
5
10
ω
15
t = 0.0025
0.010
0.0225
Re₁ = 36.5
π/2
v
Re₂ = 40
Re₁ = 36.5
0
t

Fig. 12

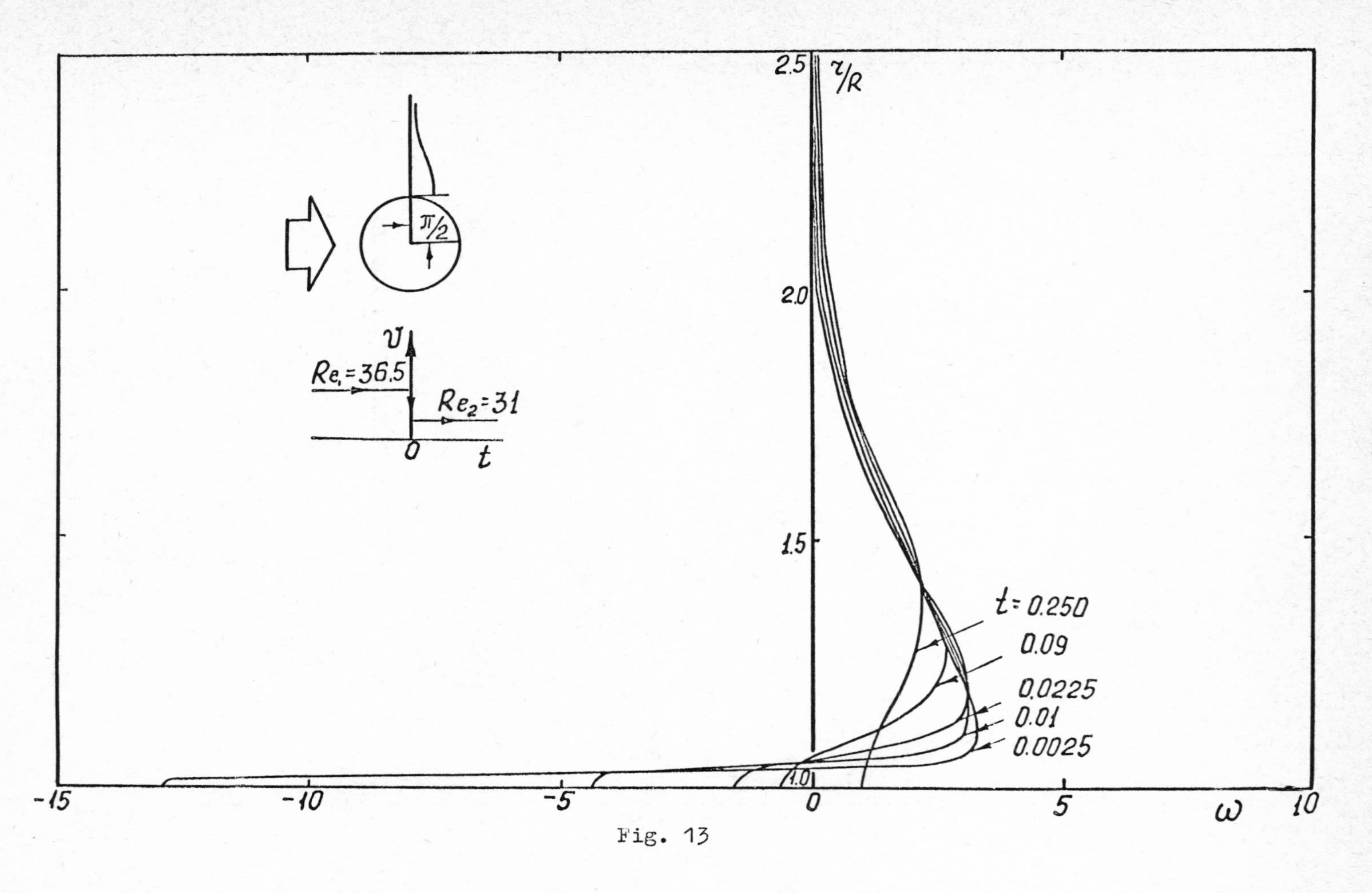

π/2
U
Re₁ = 36.5
Re₂ = 31
0
t
2.5
r/R
2.0
1.5
1.0
t = 0.250
0.09
0.0225
0.01
0.0025
-15
-10
-5
0
5
ω
10

Fig. 13

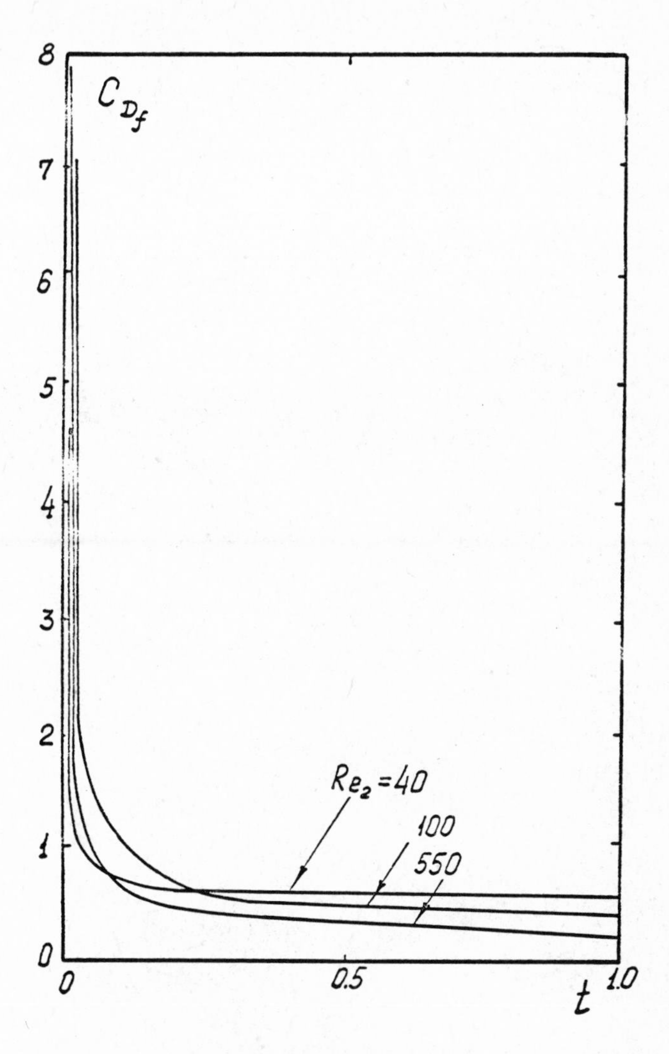

Fig. 14

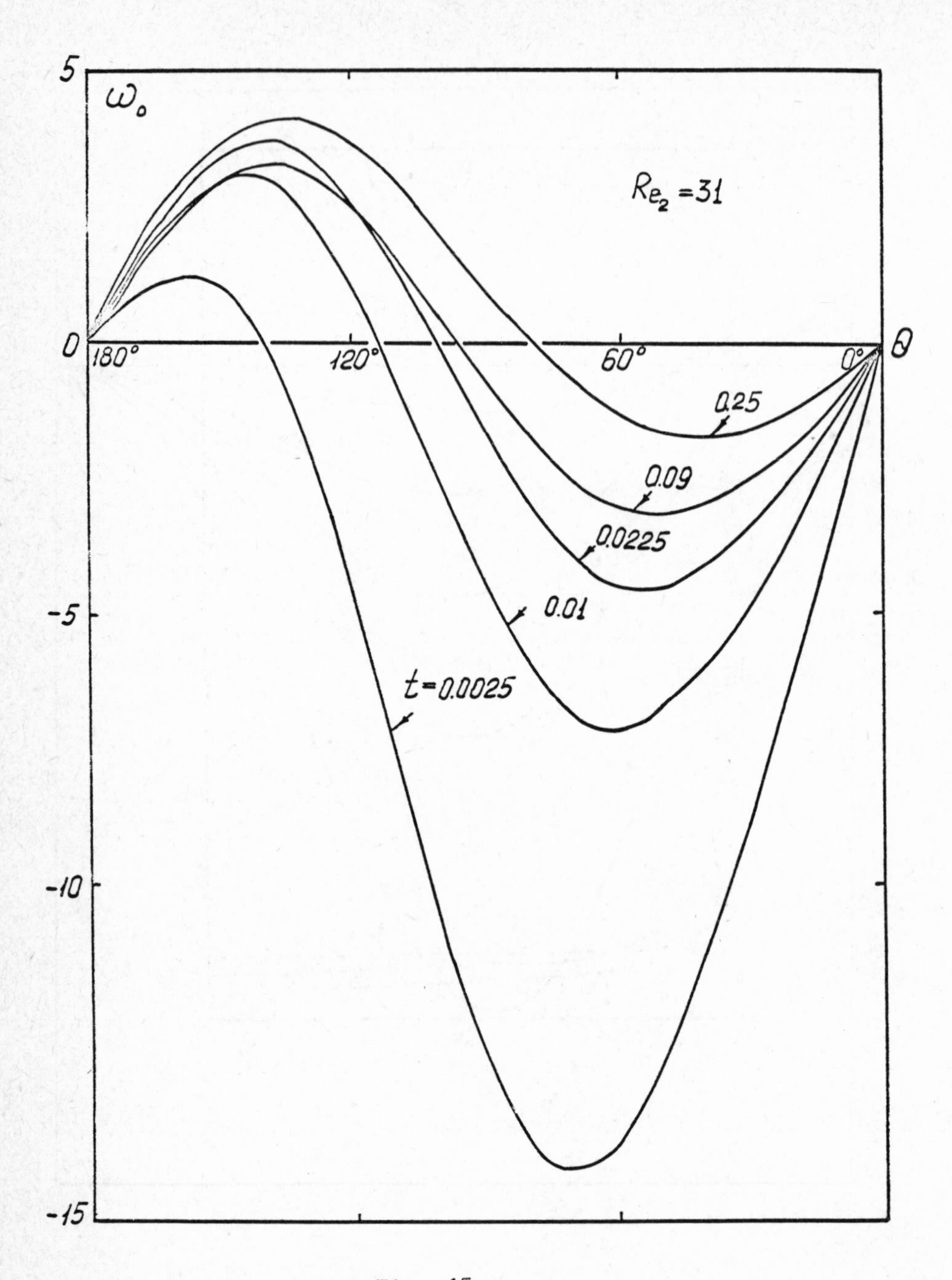

Fig. 15

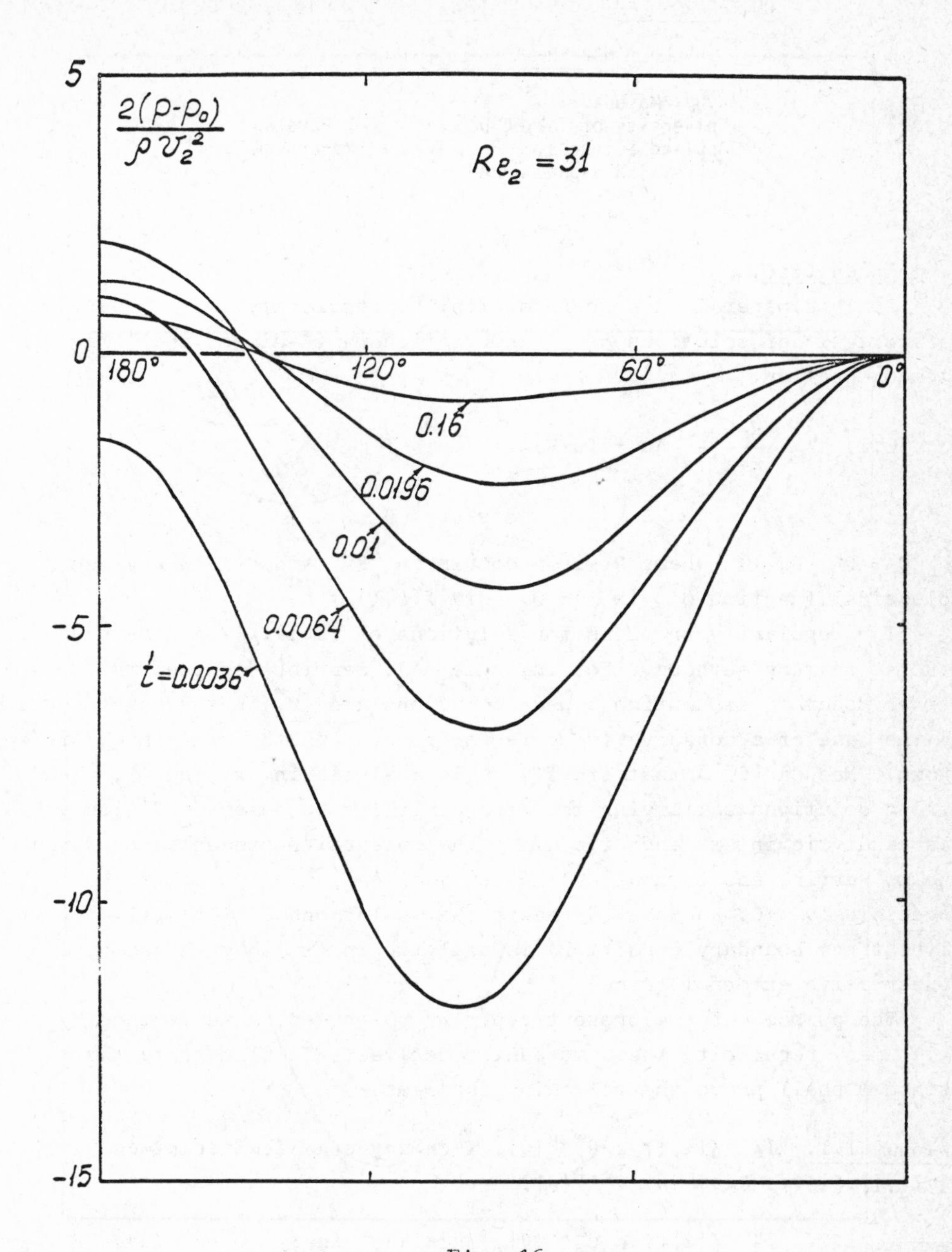
5
$\frac{2(p-p_0)}{\rho v_2^2}$
$Re_2 = 31$
0
180°
120°
60°
0°
0.16
0.0196
0.01
0.0064
t=0.0036
-5
-10
-15

Fig. 16

ON THE REGULARITY OF SOLUTIONS OF THE NONSTATIONARY NAVIER-STOKES EQUATIONS

Kyûya Masuda
University of Tokyo, Department of Pure and Applied Sciences, 3-8-1, Komaba, Meguro-ku, Tokyo, 153 Japan, JAPAN

1. Introduction.

In this paper I am concerned with the regularity in the space variable x of solutions of the nonstationary (3-dimensional) Navier-Stokes equation:

$$(1.1) \qquad \frac{\partial u}{\partial t} - \Delta u + (u\cdot\nabla)u + \nabla p = f ;$$

$$(1.2) \qquad \operatorname{div} u = 0$$

in $Q \equiv \Omega\times (0, T)$ where Ω is a domain in R^3, and f is a given solenoidal function of $(x,t)\varepsilon\, Q$; $\operatorname{div}_x f(x,t) = 0$.

The regularity problems for solutions of (1.1),(1.2) have been studied by many authors. For instance, J. Serrin[1] showed that under rather moderate assumptions, weak solutions are C^∞ in the space variable in the case of a conservative external force. On the other hand, it was shown (Masuda [2]) that if $f(x,t)$ is analytic in x and t, then strong solutions satisfying the zero Dirichlet boundary condition are also analytic in x and t; using the integral representation formula due to Serrin, and estimating the derivatives in x of solutions successively, C. Kahane [3] showed that solutions of (1.1),(1.2) (without any boundary condition) is analytic in x in the case of a conservative external force.

The purpose of the present paper is to present a new method by which many regularity theorems can be derived. To illustrate the method I shall prove the following theorems.

Theorem 1.1. *If $f(x,t)\ \varepsilon\ C^{\infty,0}_\sigma(Q)$, then any classical solution u of (1.1),(1.2) is also in $C^{\infty,0}_\sigma(Q)$.*

Theorem 1.2. *If $f(x,t)\ \varepsilon\ C^{\omega,0}_\sigma(Q)$, then any classical solution u of (1.1),(1.2) is also in $C^{\omega,0}_\sigma(Q)$.*

Here by the function spaces $C^{\infty,0}_\sigma(Q)$, $C^{\omega,0}_\sigma(Q)$, I mean:

$C^{\infty,0}_\sigma(Q)$ is the set of all continuous solenoidal (in x)vector functions on Q such that for any multi-index α, partial derivative $\partial_x^\alpha f(x,t)$ exists and is continuous in Q.

$C_\sigma^{\omega,o}(Q)$ is the set of all continuous solenoidal vector functions u such that $u(x,t)$ has an analytic extension in x for each fixed t ; there is a continuous function $\tilde{u}(z,t)$ on $\tilde{\Omega} \times (0,T)$ with $\tilde{u}(x,t) = u(x,t)$ for $(x,t) \varepsilon Q$ and such that $\tilde{u}(z,t)$ is analytic in $z \varepsilon \tilde{\Omega}$ for each fixed t, $\tilde{\Omega}$ being some n-dimensional complex neighborhood of Ω.

I shall also fix some notations.

$H^s(\Omega)$ (s;integer): the usual Sobolev space.
$C_{o,\sigma}^\infty(\Omega)$: the set of all 3-dimensional vector functions $\phi \varepsilon C_o^\infty(\Omega)$ with $\text{div}\, \phi = 0$.
$C_{o,\sigma}^\infty(Q)$: the set of all 3-dimensional vector functions $\phi \varepsilon C_o^\infty(Q)$ with $\text{div}_x\, \phi = 0$.
$H_{o,\sigma}^1(\Omega)$: the closure of $C_{o,\sigma}^\infty(\Omega)$ with respect to the norm of H^1.
$L_\sigma^2(\Omega)$: the closure of $C_{o,\sigma}^\infty(\Omega)$ with respect to the norm of $L^2(\Omega)$.
P: the orthogonal projection from $L^2(\Omega)$ onto $L_\sigma^2(\Omega)$.

2. Proof of Theorem 1.2.

1st step. I first fix an R^3-valued function $\bar{\phi} = (\bar{\phi}_1, \bar{\phi}_2, \bar{\phi}_3) \varepsilon C_o^\infty(Q)$, and define $\phi = (\phi_1, \phi_2, \phi_3)$ by $y_j \equiv \phi_j(x,t,\lambda) = x_j + \lambda_j \bar{\phi}_j(x,t)$ for $(x,t) \varepsilon \bar{Q}$ and $\lambda \varepsilon \mathbb{C}^3$(3-dimensional complex space). Set

$$\alpha_{j,k}(x,t,\lambda) = \partial\phi_j(x,t,\lambda)/\partial x_k, \qquad (j,k = 1,2,3)$$

and let $\tilde{\alpha}_{j,k}$ be the (j,k)-cofactor of the matrix $(\alpha_{j,k})$. Take δ so small that the Jacobian of $\phi \equiv \det(\alpha_{j,k}) \neq 0 \quad ((x,t)\varepsilon\bar{Q}; |\lambda| \leq \delta)$. Then;

i) $\alpha_{j,k}(x,t,\lambda) = \delta_{j,k}$ for $(x,t,\lambda) \varepsilon K$,

ii) $\phi(x,t,\lambda)$ is a diffeomorphism of class C^∞ on Ω for each fixed real λ ($|\lambda| \leq \delta$) and t,

where

$$K = \{ (x,t,\lambda) \varepsilon Q\times\mathbb{C}^3;\ (x,t) \notin \text{supp}(\bar{\phi}), |\lambda| \leq \delta \} \cup \{(x,t,0); (x,t)\varepsilon Q\} .$$

Put

$$v_j(x,t,\lambda) = \tilde{\alpha}_{k,j}(x,t,\lambda)\, u_k(\phi(x,t,\lambda),t) - u_j(x,t)$$

$$q(x,t,\lambda) = p(\phi(x,t,\lambda),t) - p(x,t)$$

for real λ and $(x,t) \in Q$; here a repeated index is summed from 1 to 3. Then one can easily see that v is in $C^2_{o,\sigma}(Q)$ and satisfies the equation of the form:

$$(2.1)\quad B(x,t,\lambda)\frac{\partial v}{\partial t} = C_{jk}(x,t,\lambda)\frac{\partial^2 v}{\partial x_j \partial x_k} + C_j(x,t,\lambda)\frac{\partial v}{\partial x_j} + C(x,t,\lambda)v$$

$$+ N_j(x,t,\lambda)[v,\frac{\partial v}{\partial x_j}] + N(x,t,\lambda)[v,v] + g(x,t,\lambda)$$

$$+ \nabla q.$$

(B, C_{jk}, C_j, C are matrix functions of x,t,λ ; N_j, N are bilinear forms with the coefficient depending on x,t,λ ; g is a vector function of x,t,λ)

We may assume without loss of generality that $\partial_x^\alpha u$ ($|\alpha| \le 2$), $\partial_t u$ exist and are continuous on $\overline{Q}$; consider u in a slightly small subdomain of Q from the beginning. Then, by a direct calculation one can easily prove:

<u>Lemma 2.1.</u>

i) $B(x,t,\lambda)$, $C_{jk}(x,t,\lambda)$, $C_j(x,t,\lambda)$, $C(x,t,\lambda)$, $g(x,t,\lambda)$ *are* $C^1(\overline{Q})$-*valued holomorphic functions of* λ .

ii) $N_j(x,t,\lambda)[\xi,\eta]$, $N(x,t,\lambda)[\xi,\eta]$ *are bilinear forms of* ξ *and* η *with the coefficients being* $C^1(\overline{Q})$-*valued holomorphic functions of* λ .

iii) for $\lambda = 0$, *we have*

$$B(x,t,\lambda) = E ;$$

$$C_{jk}(x,t,0) = \delta_{jk} E, \quad C_j(x,t,0)\frac{\partial w}{\partial x_j} = (u(x,t)\ \nabla)w,$$

$$C(x,t,0)w = (w\ \nabla)u(x,t) ;$$

$$N_j(x,t,0)[w,\frac{\partial w}{\partial x_j}] = (w\ \nabla)w , \quad N(x,t,0)[w,w] = 0 ;$$

$$g(x,t,0) = 0,$$

where E *is the unit matrix.*

<u>2nd step.</u>

To study how the solution v of (2.1) depends on λ , we shall rewrite the differential equation (2.1) in the form of operator equation in $L^2_\sigma(\Omega)$, and make use of the abstract theory of evolution equations. For this purpose we define the bounded operator $D(t,\lambda)$ by:

$$D(t,\lambda)w = P\mathcal{B}(t,\lambda)w , \qquad w \in L^2_\sigma(\Omega)$$

; here and in what follows we shall delete the letter x from $\mathcal{B}(x,t,\lambda)$ etc when considered as multiplication operators in $L^2_\sigma(\Omega)$.

Then from Lemma 2.1 the following lemma easily follows.

Lemma 2.2. *We have*

i) $D(t,0)$ = *the identity operator.*

ii) $D(t,\lambda)$ *is an* $L(L^2_\sigma(\Omega), L^2_\sigma(\Omega))$*-valued continuously differentiable function of* t *and* λ $(|\lambda| \leq \delta)$*; and so for each fixed* t, $D(t,\lambda)$ *depends analytically on* λ *in the operator (uniform) norm.*

iii) If δ_1 *is sufficiently small, then for any* t *and* λ *with* $|\lambda| \leq \delta_1$, $D(t,\lambda)^{-1}$ *exists as a bounded operator in* $L^2_\sigma(\Omega)$*, and enjoys the same properties i), ii) as* $D(t,\lambda)$.

($L(L^2_\sigma(\Omega), L^2_\sigma(\Omega))$ is the Banach space consisting of all bounded linear operators in $L^2_\sigma(\Omega)$, and with the uniform operator norm.)

Now operating the projection P first, and then the inverse operator $D(t,\lambda)^{-1}$ to both sides of (2.1), one obtains

$$\frac{dv}{dt} = - A(t,\lambda)v + F(t,\lambda)[v] + h(t,\lambda) \tag{2.2}$$

for $0< t <T$ and real λ $(|\lambda| \leq \delta_1)$, and

$$v = 0 \quad (t = 0) \tag{2.3}$$

where

$A(t,\lambda)$ is the operator defined by

the domain of $A(t,\lambda) = H^2(\Omega)\cap H^1_{o,\sigma}(\Omega)$

$$A(t,\lambda)w = - D(t,\lambda)^{-1}P\{ C_{jk}(t,\lambda)\partial^2 w/\partial x_j\partial x_k + C_j(t,\lambda)\, \partial w/\partial x_j + C(t,\lambda)w \} ;$$

$F(t,\lambda)$ is the nonlinear operator defined by

the domain of $F(t,\lambda) = L^\infty(\Omega)\cap H^1_{o,\sigma}(\Omega)$,

$$F(t,\lambda)[w]= D(t,\lambda)^{-1}P\, \{N_j(t,\lambda)[w,\partial w/\partial x_j] + N(t,\lambda)[w,w]\} ;$$

and

$$h(t,\lambda) = D(t,\lambda)^{-1}Pg(t,\lambda).$$

3rd step.

Let us rewrite the equation (2.2) in the integral form. To this end we shall show that a family of operators $\{-A(t,\lambda)\}$ generates the family of evolution operators $\{U(t,s;\lambda)\}$, $0 \leq s \leq t \leq T$, of "holomorphic type", by applying Theorem A.1 in the appendix. We first recall the following

Propositon 2.3. (T.Kato-H.Fujita [4]) *Let* A_o *be the Stokes operator defined by*

$$\begin{cases} D(A_o) = H^2(\Omega) \cap H^1_{o,\sigma}(\Omega) \\ A_o w = - P\Delta w. \end{cases}$$

Then A_o *is a positive self-adjoint operator in* $L^2_\sigma(\Omega)$, *and* $D(A_o^{3/4}) \subset L^\infty(\Omega)$. *Furthermore,*

$$\|w\|_{H^2} \leq M \|A_o w\| \tag{2.4}$$

M *being a positive constant.* ($\|\cdot\|$; L^2*-norm*).

Remark 2.4. We may assume that the boundary of Ω is sufficiently smooth.

Remark 2.5. The fractional power $A_o^{3/4}$ of A_o is defined through the spectral representation for the self-adjoint operator A_o.

Since $A(t,0)w = A_o w - (u\cdot\nabla)w - (w\cdot\nabla)u$, it easily follows from the above proposition that the resolvent set of the operator $A(t,0)$ contains a sectorial domain

$$\{\zeta \,;\, |\arg(\zeta-\xi_o)| \leq \tfrac{1}{2}\pi + \omega_o\} \qquad (\equiv \Sigma(\omega_o,\xi_o))$$

with some ω_o ($0<\omega_o<\frac{1}{2}\pi$) and some real ξ_o , and the estimate

$$\|(\zeta - A(t,0))^{-1}\| \leq \frac{M}{|\zeta-\xi_o|} \tag{2.5}$$

holds for $\zeta \in \Sigma(\omega_o,\xi_o)$, M being a positive constant.
On the other hand, by Lemma 2.1, Lemma 2.2, and (2.4),

$$\|(A(t,\lambda)- A(t,0))(\xi_o + A(t,0))^{-1}\| \leq M|\lambda| \tag{2.6}$$

and

(2.7) $$\left\| (A(t,\lambda) - A(s,\lambda))(\xi_o + A(t,0))^{-1} \right\| \le M\,|t-s|$$

for $0 \le s,t \le T$ and $|\lambda| \le \delta$, M being a positive constant. Hence we see from (2.5), (2.6) that if λ is sufficiently small, say $|\lambda| \le \delta_2$, the assumption A.2 is satisfied for ω_o and ξ_o . The assumption A.3 follows from (2.6),(2.7) and the relation :

$$(A(t,\lambda) - A(s,\lambda))(\xi_o + A(s,\lambda))^{-1}$$

$$= (A(t,\lambda)-A(s,\lambda))(\xi_o+A(s,0))^{-1}(1+(A(s,\lambda)-A(s,0))(\xi_o+A(s,0))^{-1})^{-1}$$

$$= (A(t,\lambda)-A(s,\lambda))(\xi_o + A(s,0))^{-1} \sum_{j=o}^{\infty} (-1)^j V(s,\lambda)^j$$

where

$$V(s,\lambda) = (A(s,\lambda) - A(s,0))(\xi_o + A(s,0))^{-1}$$

Since all the other assumptions are clearly satisfied, $-A(t,\lambda)$ generates the family of evolution operators $\{U(t,s;\lambda)\}$ with the properties stated in Theorem A.1. for $|\lambda| \le \delta_2$. By the well-known property of the evolution operator, the solution v of (2.2) can be expressed in the form

(2. 8) $$v(t,\lambda) = \int_o^t U(t,s;\lambda)F(s,\lambda)[v(s,\lambda)]\,ds + \int_o^t U(t,s;\lambda)h(s,\lambda)\,ds$$

for $0 \le t \le T$ and <u>real</u> λ with $|\lambda| \le \delta_2$.

<u>4th step</u>.

Apart from the fact that $v(t,\lambda)$ is a solution of (2.8) , we shall try to solve the equation (2.8) for $0 \le t \le T$ and <u>complex</u> λ with $|\lambda| \le \delta_2$. We define some function spaces.

$D_{A_o^{3/4}}$; the Banach space consisting of all (complex valued) functions u in $D(A_o^{3/4})$ with the norm $\| A_o^{3/4} u \|$

Y ; the Banach space consisting of all $D_{A_o^{3/4}}$-valued continuous functions of t ($0 \le t \le T$), and with the uniform norm; $\|u\|_Y = \max_t \| A_o^{3/4} u \|$ ($0 \le t \le T$).

Z ; the Banach space consisting of all $L^2_\sigma(\Omega)$-valued continuous functions u of t $(0 \le t \le T)$ and with the uniform norm; $\|u\|_Z = \max_t \|u\|$, $(0 \le t \le T)$.

We next define the mapping $J(\lambda,w)$ of $\{\lambda \varepsilon \mathbb{C}^3; |\lambda| \le \delta_2\} \times Y$ into Y ;

$$J(\lambda,w) = w(t) - \int_0^t U(t,s;\lambda)F(s,\lambda)[w(s)]\,ds - \int_0^t U(t,s;\lambda)h(s,\lambda)ds$$

$(0 \le t \le T$ and $|\lambda| \le \delta_2)$.

In the next section, we shall prove;

<u>Lemma 2.6</u>. $J(\lambda,w)$ *is an analytic mapping.*

Clearly, $J(0,0) = 0$, and the Fréchet derivative in w of $J(\lambda,w)$ at $\lambda=0$, $w=0$ is the identity operator in Y. Hence the equation

$$(2.9) \qquad J(\lambda, w(\lambda)) = 0; \quad w(0) = 0$$

has the unique analytic solution $w(\lambda)$ in the neighborhood of $\lambda = 0$, by the implicit function theorem. By a standard argument, it can be shown that the function satisfying (2.9) for real λ is unique . Since v satisfies (2.9) for real λ by (2.8). we have

$$v(t,\lambda) = w(t,\lambda)$$

for $0 \le t \le T$ and real λ with $|\lambda| \le \delta_3$, δ_3 being a sufficiently small positive number. Hence $v(t,\lambda)$ and so $u(\phi(x,t,\lambda),t)$ has an analytic extension in $|\lambda| \le \delta_3$ as a Y-valued(and so Z-valued)function, since

$$(2.10) \qquad u_j(\phi(x,t,\lambda),t) = \frac{\alpha_{jk}(x,t,\lambda)}{\det(\alpha_{jk}(x,t,\lambda))}\{v_k(x,t,\lambda) + u_k(x,t)\}$$

Here $v(x,t,\lambda)$ means the function $w(\lambda)$ considered as a function of x, t,λ . Let $Q_0 = \Omega_0 \times (\varepsilon, T-\varepsilon)$ be a subdomain of Q with the closure in Q , and let $\bar{\phi}$ be a $C_0^\infty(Q)$-function with $\bar{\phi} = 1$ on Q_0 . Then, by (2.10), $u(x+\lambda,t)$ has an analytic extension in λ as an $L^2(\Omega_0)$-valued function uniformly in t. From the Sobolev imbedding theorem it follows that u(x,t) has an analytic extension in x , uniformly in t. This proves Theorem 1.2.

3. <u>Proof of Lemma 2.6.</u>

Since the mapping $(\lambda,w)\varepsilon \{\lambda; |\lambda| \le \delta_2\} \times Y \longrightarrow F(x,t,\lambda)[w(x,t)]\ \varepsilon Z$

is analytic as can be easily seen, it follows from Theorem A.1 that for $0<\alpha<1$

$$(3.1)\qquad (t-s)^{\alpha}(\xi_0+A(t,\lambda))^{\alpha}\,U(t,s;\lambda)F(s,\lambda)[w(s)]$$

is analytic in λ and w as a $C(\Delta;\ L^2_\sigma(\Omega))$-valued function where $\Delta=\{(s,t);\ 0\leq s\leq t\leq T\}$. We note the fractional power $(\xi_0+A(t,\lambda))^{-\alpha}$ of $(\xi_0+A(t,\lambda))^{-1}$ is defined by

$$S^{-\alpha}=\frac{\sin\pi\alpha}{\pi}\int_0^\infty\eta^{-\alpha}(\eta+S)^{-1}d\eta$$

$$(S\equiv\xi_0+A(t,\lambda))$$

and $(\xi_0+A(t,\lambda))^{\alpha}$ is defined as the inverse of $(\xi_0+A(t,\lambda))^{-\alpha}$. (For more detailed properties of the fractional power of linear operators, see K. Yosida [5].)

Using $\|\nabla w\|=\|A_0^{1/2}w\|\leq\|A_0w\|^{1/2}\|w\|^{1/2}$, we get

$$\|A_0w\|\leq\|A(t,0)w\|+M\|w\|.$$

Hence, in the same way as in the third step, we have

$$\|A_0(\eta+S)^{-1}\|\leq\|A(t,0)(\eta+S)^{-1}\|+M\|(\eta+S)^{-1}\|\leq M'$$

$(S\equiv\xi_0+A(t,\lambda))$, M and M' being positive constants. Hence, by the interpolation theorem,

$$(3.2)\qquad \|A_0^{\alpha}(\eta+\xi_0+A(t,\lambda))^{-1}\|\leq M\,|\eta|^{\alpha-1}\ .$$

On the other hand, by Lemma 2.1, Lemma 2.2, and Proposition 2.3, for $w\in D(A(t,0))=D(A_0)$, $A(t,\lambda)w$ is differentiable in λ and satisfies

$$(3.3)\qquad \|\partial_\lambda A(t,\lambda)w\|\leq M\|A_0w\|\ .$$

Moreover, $(\eta+\xi_0+A(t,\lambda))^{-1}w$ is also differentiable, and

$$\partial_\lambda(\eta+\xi_0+A(t,\lambda))^{-1}w=-(\eta+\xi_0+A(t,\lambda))^{-1}[\partial_{\lambda'}A(t,\lambda')(\eta+\xi_0+A(t,\lambda))^{-1}]w\Big|_{\lambda'=\lambda}$$

from which it follows by (3.2) and (3.3) that $A_0^{3/4}(\eta+\xi_0+A(t,\lambda))^{-1}w$ is differentiable in λ and satisfies the estimate

$$\|\partial_\lambda A_0^{3/4}(\eta+\xi_0+A(t,\lambda))^{-1}\|\leq M\,|\eta|^{-1/4}$$

Hence, it follows from the definition of the fractional power that for $3/4 < \alpha < 1$, the function

$$(3.4) \qquad A_o^{3/4}(\xi_o + A(t,\lambda))^{-\alpha} w$$

is differentiable in λ uniformly in t. By (3.1) and (3.4), the function

$$(t-s)^{\alpha} A_o^{3/4} U(t,s;\lambda) F(s,\lambda)[w(s)]$$

is analytic in λ and w as a $C(\Delta; L_\sigma^2(\Omega))$-valued function. Hence, the integral

$$\int_0^t A_o^{3/4} U(t,s;\lambda) F(s,\lambda)[w(s)] ds$$

is a Z-valued holomorphic function of λ and w ; to see this, note $A_o^{3/4} U(t,s;\lambda) = (t-s)^{-\alpha}(t-s)^{\alpha} A_o^{3/4} U(t,s;\lambda)$. This implies $\int_0^t U(t,s;\lambda) F(s,\lambda)[w(s)] ds$ is a Y-valued holomorphic function. Similarly we can show $\int_0^t U(t,s;\lambda) h(s,\lambda) ds$ is a Y-valued holomorphic function of λ. This proves Lemma 2.6.

4. Proof Theorem 1.1.

To prove Theorem 1.1, we have only to restrict ourselves to real λ and to make use of the corresponding abstract theory for evolution operators.

Remark 4.1. Without using the theory of evolution operators, we can show the regularity in λ of solution v of (2.1) (see Masuda [6]).

5. Appendix.

Let $\omega_\delta = \{ \lambda \in \mathbb{C}^n; |\lambda| < \delta \}$. Let $\{A(t,\lambda)\}$, $0 \le t \le T, \lambda \in \omega_\delta$, be a family of linear operators in a Banach space X with the norm $\| \cdot \|$, satisfying the following assumptions:

Assumption A.1. The domain of $A(t,\lambda)$ is independent of t and λ.

Assumption A.2. The resolvent sets of all $A(t,\lambda)$ contain some fixed sector

$$\{\zeta \; ; \; |\arg(\zeta - \xi)| \le \tfrac{1}{2}\pi + \theta\} \; (\equiv \Sigma(\theta,\xi))$$

($0 < \theta < \frac{\pi}{2}$ and ξ ; real) and the inequality

$$\|(\zeta + A(t,\lambda))^{-1}\| \leq \frac{M}{|\zeta - \xi|}$$

holds for $0 \leq t \leq T$, $\lambda \in \omega_\delta$ and $\zeta \in \Sigma(\theta,\xi)$, M being a positive constant.

Assumption A.3. For some γ ($0 < \gamma \leq 1$), the inequality

$$\|(\xi + A(t,\lambda))(\xi + A(s,\lambda))^{-1} - I\| \leq M'|t-s|^\gamma$$

holds for $0 \leq s,t \leq T$, $\lambda \in \omega_\delta$, M' being a positive constant.

Assumption A.4. For any $x \in D(A(t,\lambda))$, $A(t,\lambda)x$ is a $C([0.T];X)$-valued holomorphic function of λ .

Then we have;

Theorem A.1. *Under the above assumptions, there is a family* $\{U(t,s;\lambda)\}$ ($0 \leq s \leq t \leq T$; $\lambda \in \omega_\delta$) *of bounded linear operators in* X *with the following properties;*

a) $U(t,s;\lambda)U(s,r;\lambda)= U(t,r;\lambda)$, $U(t,t;\lambda) = I$

$$(0 \leq r \leq s \leq t \leq T \ ; \ \lambda \in \omega_\delta)$$

b) $\frac{\partial}{\partial t} U(t,s;\lambda) = - A(t,\lambda)U(t,s;\lambda)$, $(s,t) \in \Delta$, $\lambda \in \omega_\delta$, $s \neq t$,

c) *For any* α ($0 \leq \alpha \leq 1$) *and* $x \in X$,

$$(t-s)^\alpha(\xi + A(t,\lambda))^\alpha U(t,s;\lambda)x$$

is a $C(\Delta; X)$*-valued holomorphic function of* λ

Remark A.2. The proof of the above theorem is similar to H, Tanabe [7], and will be published elsewhere under more general situations.

References

[1] J.Serrin - On the interior regularity of weak solutions of the Navier-Stokes equations. Arch.Rational Mech.Anal.9 (1962),p.187-195.

[2] K.Masuda - On the analyticity and the unique continuation theorem for solutions of the Na vier-Stokes equations. Proc Japan Acad., 43(1967), p.827-832.

[3] C.Kahane - On the spatial analyticity of solutions of the Navier-

Stokes equations. Arch.Rational Mech.Anal.33 (1969),p.386 -405.

[4] H.Fujita and T.Kato - On the Navier-Stokes initial value problem, I. Arch.Rational Mech.Anal., 16 (1964),p.269-315.

[5] K. Yosida - *Functional Analysis*, Grundlehren Band 123 ,Springer, (1965).

[6] K. Masuda - On the regularity of nonlinear elliptic and parabolic systems of partial differential equations. (to appear).

[7] H. Tanabe - On the equation of evolution in a Banach space. Osaka Math.J. 12 (1960),p.363-376.

THE ASYMPTOTIC BEHAVIOUR OF SOLUTIONS OF THE NAVIER-STOKES EQUATIONS NEAR SHARP CORNERS

H.K. Moffatt
School of Mathematics
University of Bristol
Bristol BS8 2XL, U.K.

When the boundary of a fluid domain has any sharp corners, an understanding of the asymptotic structure of the flow near the corners is invariably helpful as a preliminary to determining the overall flow structure, and as a means of checking subsequent numerical calculations. The simplest example is that of Poiseuille flow along a sharp corner; a less simple example is the low Reynolds number flow between two hinged plates in relative angular motion. These problems are discussed here in §§1 and 2 (full details are given in Moffatt & Duffy 1979) and it is shown that the local similarity solution is not invariably asymptotically valid near the corner, but that it may (for certain ranges of corner angle) be dominated by 'eigenfunction' contributions which are sensitive to conditions remote from the corner. For critical angles, a transitional behaviour occurs.

In §§3-5, further examples are discussed of problems for which, despite the existence of a local similarity solution, the flow may (again for certain angle ranges) be dominated by eigenfunction contributions which may imply the presence of an infinite sequence of eddies in the corner (Moffatt 1964). These examples include the curved duct problem of Collins & Dennis (1976), the free convection problems of Liu & Joseph (1977), and the secondary flow problem in a cone-and-plate viscometer (Hynes 1978).

1. Poiseuille flow near a sharp corner

The following simple problem exemplifies a conflict between 'forced' solutions of similarity form and 'free' solutions of eigenfunction form which arises in the more complicated problems treated in subsequent sections. Consider steady pressure-driven flow along a straight duct whose cross-section $\mathcal{D}_\alpha$ is the sector of a circle illustrated in figure 1a.

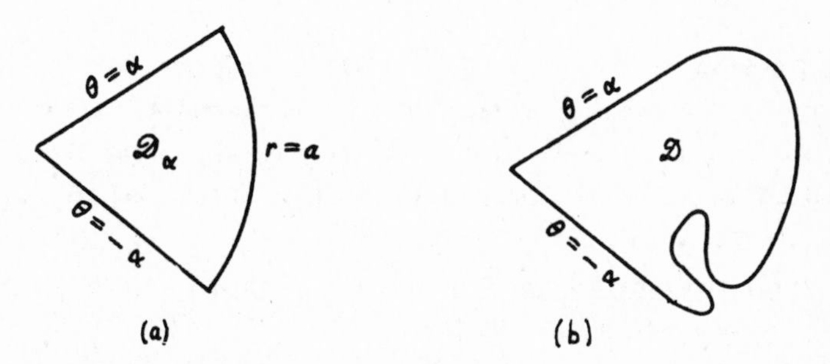

Figure 1. Cross-section of ducts; when $2\alpha < \pi/2$, the flow near O has the similarity form (1.2a); when $2\alpha > \pi/2$, it has the form (1.5) where A_0 depends on conditions far from the corner.

Taking polar coordinates (r,θ), with origin at the corner, and axis $\theta = 0$ along the bisector of the angle, the velocity $w(r,\theta)$ satisfies

$$\nabla^2 w = -G/\mu \quad \text{in } \mathcal{D}_\alpha, \tag{1.1a}$$

$$w(r, \pm\alpha) = 0, \qquad w(a,\theta) = 0, \tag{1.1b,c}$$

where G is the pressure gradient, μ the fluid viscosity, and 2α the angle of the corner. The (unique) solution to this problem is easily obtained in the form $w = w_1 + w_2$, where

$$w_1 = \frac{Gr^2}{4\mu}\left(\frac{\cos 2\theta}{\cos 2\alpha} - 1\right), \qquad w_2 = \sum_{n=0}^{\infty} A_n r^{\lambda_n} \cos\lambda_n\theta, \tag{1.2a,b}$$

with

$$\lambda_n = \frac{(2n+1)\pi}{2\alpha}, \qquad A_n = \frac{(-1)^{n+1}2G\,a^{2-\lambda_n}}{\alpha\mu\lambda_n(\lambda_n^2 - 4)}. \tag{1.3a,b}$$

The contribution w_1 is a particular integral satisfying (1.1a,b) but not (1.1c); the contribution w_2 is a sum of eigenfunctions of the associated homogeneous problem

$$\nabla^2 w_2 = 0 \quad \text{in } \mathcal{D}_\alpha, \qquad w_2(r, \pm\alpha) = 0, \tag{1.4}$$

with coefficients chosen so that $w = w_1 + w_2$ satisfies $w(a,\theta) = 0$ also.

Near the corner, the asymptotic form of w depends on the angle 2α. If $2\alpha < \pi/2$ (so that $\lambda_0 > 2$), then $w \sim w_1$ as $r \to 0$, i.e. the solution has a similarity form which does not depend on the radius a of the sector. This behaviour is insensitive to the 'remote geometry' of the cross-section, and we must evidently expect the same asymptotic behaviour if the remote boundary is distorted in any manner as in figure 1b.

If $2\alpha > \pi/2$ however, then $\lambda_0 < 2$, and so the leading term of the series for w_2 dominates as $r \to 0$, i.e.

$$w \sim A_0 r^{\pi/2\alpha} \cos(\pi\theta/2\alpha). \tag{1.5}$$

Here, A_0 <u>does</u> depend on the radius a (eqn. 1.3b), so that the solution cannot be regarded as locally determinate; indeed, although an asymptotic behaviour of the form (1.5) may be expected for the general geometry of figure 1b, the coefficient A_0 will in general depend on the shape of the portion of the boundary that is remote from the corner.

When $2\alpha = \pi/2$, the behaviour is evidently transitional. Putting $2\alpha = \pi/2 + \varepsilon$ and letting $\varepsilon \to 0$, we find

$$w \sim -\frac{Gr^2}{\pi\mu}\left[\frac{\pi}{4} + \left(\log\frac{r}{a}\right)\cos 2\theta - \theta\sin 2\theta\right]. \tag{1.6}$$

The singularity in w_1 when $2\alpha = \pi/2$ is compensated by an equal and opposite singularity in the first term of the series for w_2. There is a similar cancellation with the second term when $2\alpha = 3\pi/2$.

2. The hinged plate and Jeffery-Hamel paradoxes

The apparent breakdown of a corner similarity solution at a critical value of the corner angle can occur in other circumstances, when the means of resolution may not be so obvious. Consider the problem depicted in figure 2a: fluid is contained in the angle between the two hinged plates which are rotated towards each other with angular velocities $\pm\omega$.

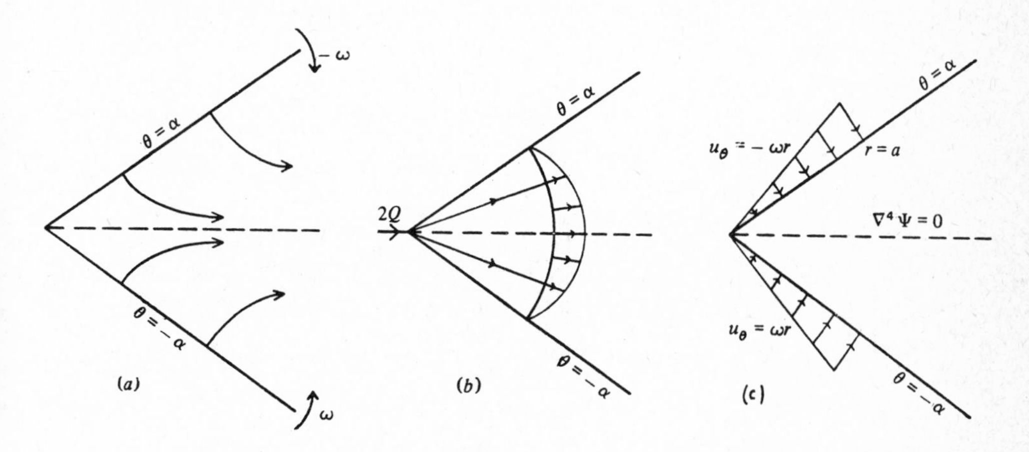

Figure 2. (a) The hinged plate problem; (b) the Jeffery-Hamel problem; (c) a hybrid problem which reduces to the hinged plate problem when $a \to \infty$ and to the Jeffery-Hamel problem when $a \to 0$, $\omega \to \infty$ with ωa^2 fixed.

Near the corner, inertia forces are negligible, and the stream function $\psi(r,\theta)$ satisfies the biharmonic equation $\nabla^4 \psi = 0$. The relevant similarity solution (Moffatt 1964) is

$$\psi_1 = \tfrac{1}{2}\omega r^2 f(\theta), \qquad f(\theta) = \frac{\sin 2\theta - 2\theta\cos 2\alpha}{\sin 2\alpha - 2\alpha\cos 2\alpha}, \tag{2.1a,b}$$

which is singular when $\tan 2\alpha = 2\alpha$, i.e. when $2\alpha \simeq 257^{\circ}$ (more accurately 257.45°). A similar singularity occurs in the low Reynolds number version of the Jeffery-Hamel problem (figure 2b) : if fluid is introduced at a volume rate $2Q$ per unit length of intersection, then the relevant solution is $\psi = Qf(\theta)$ where $f(\theta)$ is as defined by (2.1b) with the same singularity when $2\alpha \simeq 257^{\circ}$.

The same type of paradox was identified and resolved, in an elasticity context, by Sternberg and Koiter (1958). We may resolve the difficulty here in a similar manner, by considering the problem sketched in figure 2c, in which (in effect) the source is distributed over a segment $0 < r < a$ of the two planes. This problem may be solved by means of the Mellin transform (for full details, see Moffatt and Duffy 1979). For $r < a$, the solution is given by a series

$$\psi = \omega a^2 \sum_{n=1}^{\infty} \left(\frac{r}{a}\right)^{p_n+2} f_n(\theta), \tag{2.2a}$$

where

$$f_n(\theta) = \frac{(p_n+2)^{-1}\cos p_n\alpha \sin(p_n+2)\theta - p_n^{-1}\cos(p_n+2)\alpha \sin p_n\theta}{\sin 2\alpha - 2\alpha\cos 2(p_n+1)\alpha}\ ; \tag{2.2b}$$

here, the p_n are the (generally complex) roots of the equation

$$(p+1)\sin 2\alpha = \sin 2(p+1)\alpha, \tag{2.2c}$$

with Re $p_n > -1$, and ordered so that

$$-1 < p_1 \leqslant \operatorname{Re} p_2 \leqslant \operatorname{Re} p_3 \leqslant \cdots . \tag{2.2d}$$

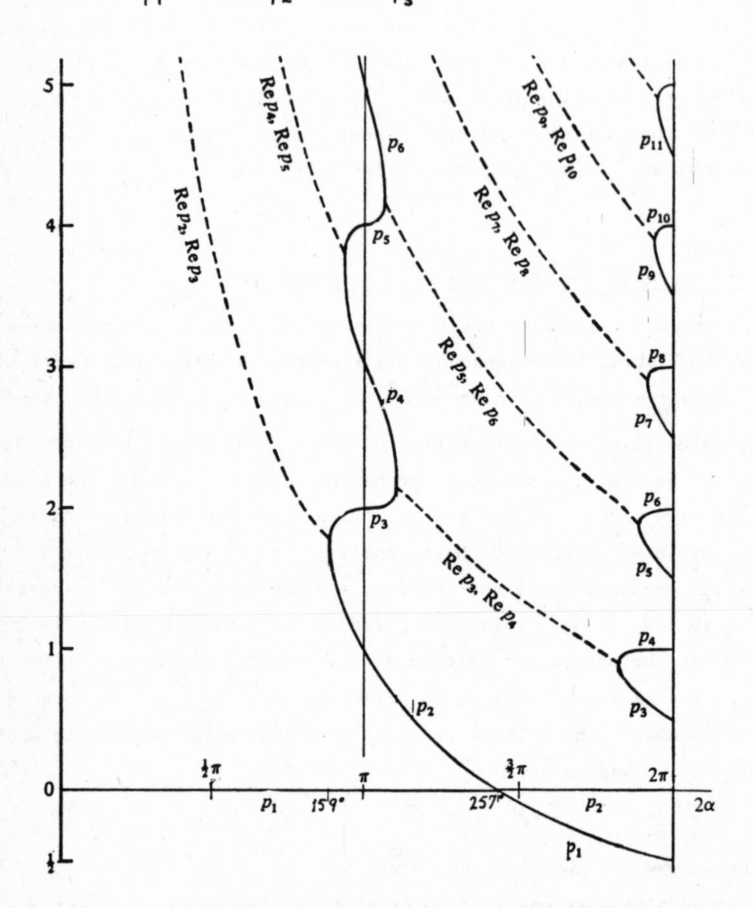

Figure 3. The real part of the roots p_n of equation (2.2c), as functions of the corner angle 2α. Where the curves are solid, p_n is real; where they are dashed, p_n is complex, and only its real part is shown. Note that $p_1 = 0$ for $2\alpha < 257^\circ$, $p_1 < 0$ for $2\alpha > 257^\circ$.

Figure 3 shows the rather interesting behaviour of Re p_n $(n = 1,2\ldots)$ as functions of α : note that p_1 is always real. For $r > a$, there is a reciprocity relation

$$\psi(r,\theta) = \left(\frac{r}{a}\right)^2 \psi\left(\frac{a^2}{r},\theta\right), \tag{2.2e}$$

relating the solution to the expression (2.2a).

As $r/a \to 0$, (2.2a) generally gives the asymptotic behaviour

$$\psi \sim \omega a^2 \left(\frac{r}{a}\right)^{p_1+2} f_1(\theta). \tag{2.3}$$

For $2\alpha < 257^\circ$ (the root of $\tan 2\alpha = 2\alpha$), $p_1 = 0$ and (2.3) gives $\psi \sim \psi_1$ where ψ_1 is given by (2.1), i.e. the 'expected' similarity solution is indeed valid near the corner. Similarly, from (2.2e), for $r/a \to \infty$, $\psi \sim \omega a^2 f_1(\theta)$ which again has the expected similarity form. However, when $2\alpha > 257^\circ$, p_1 is negative (actually $-\frac{1}{2} < p_1 < 0$), and we do not recover the usual similarity solution; as in the simpler problem of §1, the solution for $r \to 0$ retains a dependence on the outer scale a, while for $r \to \infty$, the solution similarly retains a dependence on the inner scale a. Again, as we pass through the critical angle 257° (more accurately $257{\cdot}45^\circ$), there is a transitional behaviour, the stream-function involving logarithmic terms.

3. Flow through curved ducts of triangular cross-section

In the examples discussed above, the simple similarity solution emerges as the particular integral of an inhomogeneous boundary-value problem, and its validity as an asymptotic solution breaks down whenever there is a solution of the associated homogeneous problem (i.e. a complementary function) which dominates over this particular integral. We now consider some further examples. Consider the problem of pressure-driven flow along a curved duct of triangular cross-section, as sketched in figure 4. It is supposed that the triangle is isosceles, the centre of curvature being on the line of symmetry of the section, and the angles of the triangle being β, β and 2α, with $\beta = \pi/2-\alpha$. This configuration was studied numerically by Collins & Dennis (1976) in the particular case $2\alpha = \pi/2$, and with particular attention to the behaviour of the secondary flow in the cross-section and near the corners. When the curvature κ is small, the primary velocity component w in the duct is unaffected by the secondary flow, and is given by

$$\left.\begin{aligned} \nabla^2 w &= -G/\mu \quad \text{in } \mathcal{D}, \\ w &= 0 \quad \text{on } \partial\mathcal{D}. \end{aligned}\right\} \tag{3.1}$$

The stream function of the secondary flow is then determined by

$$\left.\begin{aligned} \nabla^4 \psi &= -2\kappa\nu^{-1} w\, \partial w/\partial y \quad \text{in } \mathcal{D} \\ \psi &= 0,\ \partial\psi/\partial n = 0, \quad \text{on } \partial\mathcal{D} \end{aligned}\right\} \tag{3.2}$$

where the coordinate Oy is as indicated in figure 4, and ν is the kinematic viscosity of the fluid. Consider first the behaviour near the corner A. Let (r,θ) be local polar coordinates; then $w = O(r^2)$ and $w\partial w/\partial y = O(r^3)$ near A, and, as pointed out by

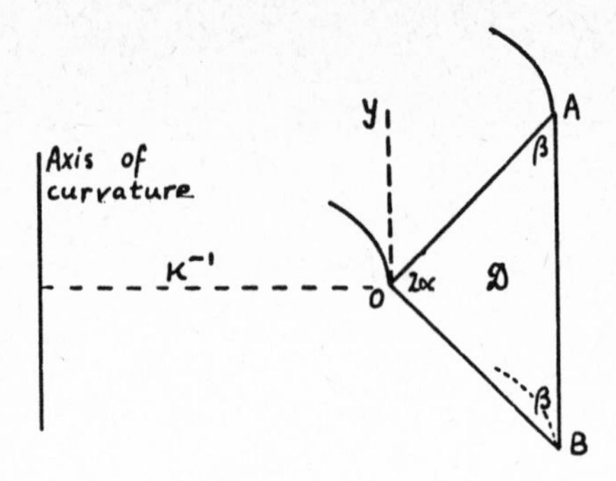

Figure 4. The curved duct configuration of Collins & Dennis (1976). When the flow is pressure-driven, eddies form at A and B if $\beta > 40.4^{\circ}$, and at O if $71.9^{\circ} < 2\alpha < 159.1^{\circ}$. When the flow is driven by rotation of the boundary AB about the axis of curvature (§4), eddies do not form at A and B, but they do form at O if $35.0^{\circ} < 2\alpha < 159.1^{\circ}$.

Collins & Dennis, the particular integral of (3.2), ψ_1 say, is $O(r^7)$ near A. The dominant part of the complementary function has the form $r^{\lambda} f(\theta)$, where λ is complex. Near A, the flow certainly includes an ingredient whose r-component of velocity is antisymmetric about the corner bisector (we refer to this as an 'antisymmetric mode'), and the dependence of $\mathrm{Re}\lambda$ on the corner angle (from tables given by Moffatt 1964) is as indicated in figure 5. If $\mathrm{Re}\lambda < 7$, then the flow near A (and likewise of course near B) is characterised by an infinite sequence of corner eddies. From figure 5, this is the case for $\beta > 40.4^{\circ}$ (and in particular for the value $\beta = 45^{\circ}$ chosen by Collins & Dennis). If $\beta < 40.4^{\circ}$, then corner eddies will not form near A.

Consider now the situation near O. If $2\alpha < \pi/2$, then $w = O(r^2)$ where r is now distance from O, and again $\psi_1 = O(r^7)$. The secondary flow is symmetric about the bisector of the angle at O, and only symmetric eigenfunctions of the homogeneous problem are relevant. Again from figure 5, we find that, for the dominant symmetric mode, $\mathrm{Re}\lambda < 7$ if $2\alpha > 71.9^{\circ}$. Hence corner eddies do not form at O if $2\alpha < 71.9^{\circ}$.

If $2\alpha > \pi/2$, then $w = O(r^{\pi/2\alpha})$ and so $\psi_1 = O(r^{\pi/\alpha + 3})$. It is easily ascertained that $\mathrm{Re}\,\lambda < \pi/\alpha + 3$ for $\pi/2 < 2\alpha < \pi$, so that the complementary function dominates throughout this range; however λ is real for $2\alpha > 159.1^{\circ}$, and so corner eddies form at O only if $71.9^{\circ} < 2\alpha < 159.1^{\circ}$.

Ducts of other polygonal cross-sections could clearly be treated similarly.

4. Secondary eddies in circular Couette flow

Suppose now that the flow in the curved duct of figure 4 is generated not by a

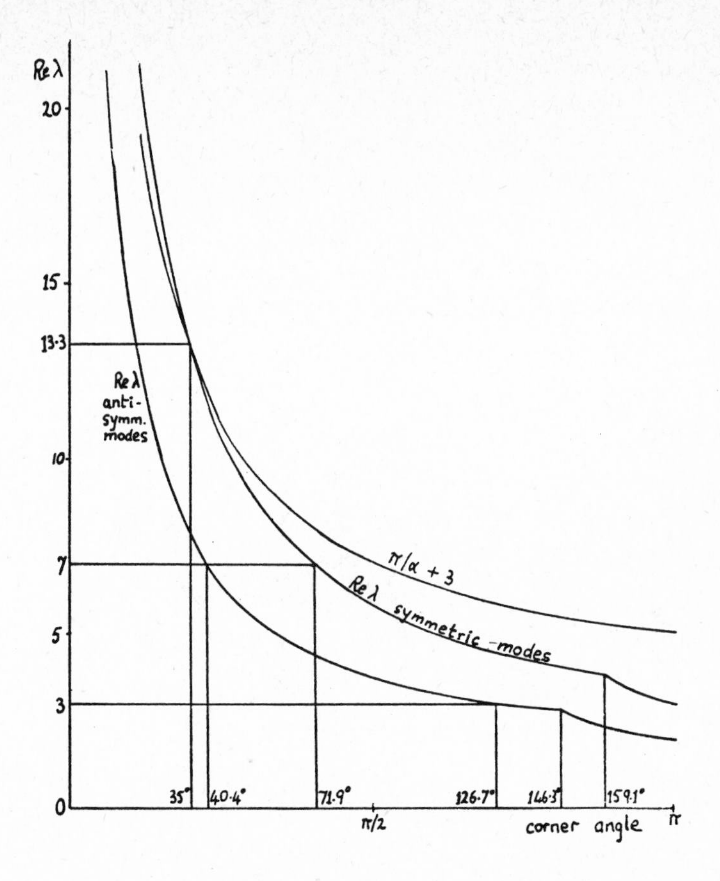

Figure 5. Variation of the real part of the exponent λ for symmetric and antisymmetric corner modes; note the jump in gradient at the critical angles (146.3° and 159.1°) where Imλ = 0 and corner eddies disappear.

pressure gradient, but by rotation of the cylindrical boundary AB about its axis of curvature, i.e. $w = w_0$ say on AB, while $w = 0$ on OA and OB. The primary flow is then of the Couette class, and $w = O(1)$ as $r \to 0$ (near A). Hence $w\partial w/\partial y = O(r^{-1})$ and so the particular integral of (3.2), ψ_1, is $O(r^3)$ near A. The relevant exponent λ in the complementary function satisfies $\mathrm{Re}\lambda > 3$ for all $\beta < \pi/2$, and so eddies do not form at A in this situation.

Near O, since there is no pressure gradient, we have $w = O(r^{\pi/2\alpha})$ and $\psi_1 = O(r^{\pi/\alpha+3})$, and eddies form if λ is complex and $\mathrm{Re}\lambda < \pi/\alpha + 3$; from figure 5, this condition is satisfied when $2\alpha > 35.0^\circ$, (but it will be noticed that we have here a situation where the particular integral and complementary function have nearly the same exponent over a wide range of corner angles). The range of angles over which corner eddies will form is

$$35{\cdot}0^\circ < 2\alpha < 159{\cdot}1^\circ. \tag{4.1}$$

There is a related behaviour in a situation studied by Hynes (1978), viz. the flow in the space between two cones (angles α, β with $\alpha < \beta$) with common vertex and axis of symmetry, when one of the cones is rotated about the axis of symmetry, the other being fixed (figure 6a). If $\beta = \pi/2$ and the inner cone is rotated, we have

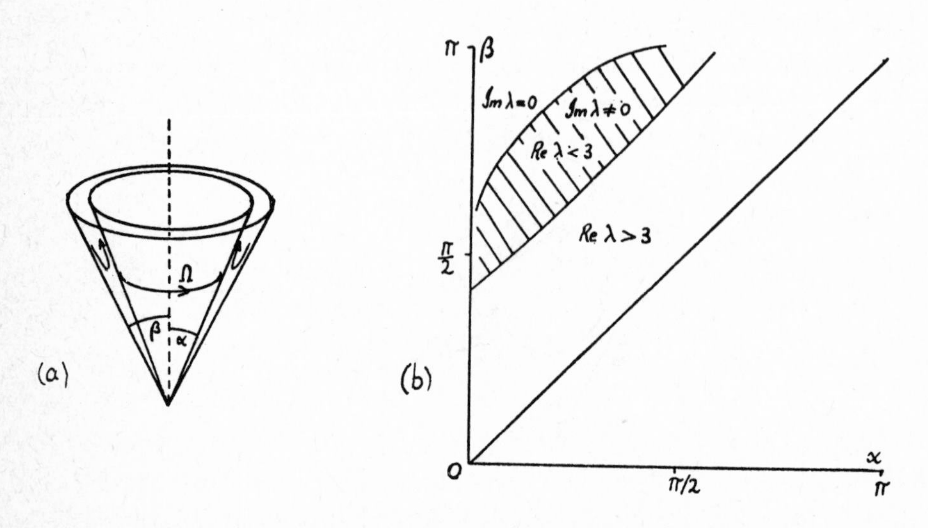

Figure 6. (a) Secondary flow between two cones, the inner one rotating. (b) The shaded area corresponds to values of α and β for which an infinite sequence of toroidal vortices occurs near the vertex.

here the familiar configuration of the cone-and-plate viscometer. The appearance of eddies (or rather toroidal vortices) in the secondary flow is often interpreted as a manifestation of non-Newtonian behaviour. However, Hynes has shown that eddies will occur even in a Newtonian fluid for a certain range of values of α and β. The primary flow component is azimuthal and has the form $w = rf(\theta)$. The corresponding forced secondary flow is $O(r^3)$ as $r \to 0$. The Stokes stream-function ψ consists again of two parts, the forced ingredient $\psi_1 = O(r^5)$, and the complementary function ψ_2 which satisfies

$$\left.\begin{aligned} D^4\psi_2 &= 0 \qquad && (\alpha < \theta < \beta) \\ \partial\psi_2/\partial\theta = \psi_2 &= 0 \qquad && (\theta = \alpha,\ \theta = \beta) \end{aligned}\right\} \tag{4.2}$$

where D^2 is the Stokes operator. This admits solutions of the form

$$\psi_2 = r^{\lambda+2} f_2(\theta), \tag{4.3}$$

and this dominates over ψ_1 near $r = 0$ if $\mathrm{Re}\lambda < 3$. Toroidal vortices therefore occur near 0 if $\mathrm{Re}\lambda < 3$ and $\mathrm{Im}\lambda \neq 0$. Figure 6b (inferred from Hynes 1978) shows the region of the (α,β) plane in which these conditions are simultaneously satisfied. In the cone and plate viscometer situation ($\beta = \pi/2$), eddies will not form if $\alpha > 15^{\circ}$. If eddies do appear for $\alpha > 15^{\circ}$, then this must indeed be due to non-Newtonian effects (see, for example, Rivlin 1976).

5. Corner flows driven by heating

Finally, let us apply similar considerations to flows of the type studied by Liu & Joseph (1977), as depicted in figure 7. The applied temperature difference ΔT induces a temperature field $T(r,\theta)$ in the fluid, and the resulting buoyancy force

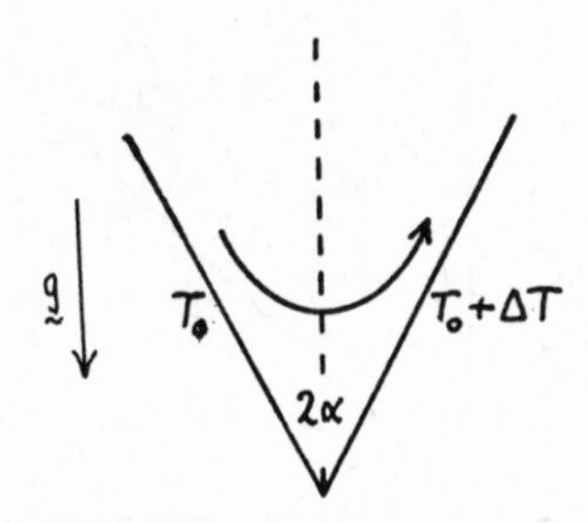

Figure 7. Corner flow driven by side-wall heating (Liu & Joseph 1977). Corner eddies occur if $126.7^{\circ} < 2\alpha < 146.3^{\circ}$.

drives the motion. Sufficiently near the corner, T satisfies $\nabla^2 T = 0$, and so $T \sim T_0 + \Delta T\, f(\theta)$ as $r \to 0$. Hence $\nabla T = O(r^{-1})$, and the resulting forced ingredient ψ_1 of the stream-function is $O(r^3)$. This is like the situation studied in §4, and we know that corner eddies will not occur if $2\alpha < \pi/2$. Liu & Joseph gave full details for the cases $2\alpha = 10^{\circ}$, 60° and found that the forced ingredient indeed dominated in these cases.

When $2\alpha > \pi/2$, the condition $\mathrm{Re}\lambda < 3$ is satisfied if $2\alpha > 126.7^{\circ}$ (figure 5), and λ is real if $2\alpha < 146.3^{\circ}$; hence corner eddies may be expected to form for

$$126.7^{\circ} < 2\alpha < 146.3^{\circ}, \tag{5.1}$$

although again, the forced and free solutions are finely balanced in this range of angles ($\mathrm{Re}\lambda \simeq 2.8$ when $2\alpha = 146.3^{\circ}$)

Acknowledgements

The work described in §§1 and 2 was carried out in collaboration with Dr. B.R. Duffy, and is supported by S.R.C. Research Grant No. GR/A/5593.4.

References

Collins, W.M. & Dennis, S.C.R. 1976. Viscous eddies near a 90° and a 45° corner in flow through a curved tube of triangular cross-section. J.Fluid Mech. 76, 417-432.

Hynes, T. 1978. Slow viscous flow between cones. Ph.D. thesis, Cambridge University.

Liu, C.H. & Joseph, D.D. 1977. Stokes flow in wedge-shaped trenches. J.Fluid Mech. 80, 443-463.

Moffatt, H.K. 1964. Viscous and resistive eddies near a sharp corner. J.Fluid Mech. 18, 1-18.

Moffatt, H.K. & Duffy, B.R. 1979. Local similarity solutions and their limitations. J.Fluid Mech. (to appear).

Rivlin, R.S. 1977. Secondary flows in viscoelastic fluids. in Theoretical and Applied Mechanics, Proc. 14th IUTAM Congress, Delft 1976, Ed. W. Koiter, North Holland Publ.Co., 221-232.

HIGH RESOLUTION SPECTRAL CALCULATIONS OF INVISCID COMPRESSIBLE FLOWS

Steven A. Orszag
Department of Mathematics
M.I.T.
Cambridge, MA 02139 USA

and

David Gottlieb
Department of Applied Mathematics
Tel-Aviv University
Tel-Aviv, ISRAEL

Summary

The extension of spectral methods to inviscid compressible flows is considered. Techniques for high resolution treatment of shocks and contact discontinuities are introduced. Model problems that demonstrate resolution of moderate-strength shocks and contact discontinuities over one effective grid interval are given.

Introduction

Spectral methods have been very successfully applied to the numerical simulation of a variety of viscous flows, including the numerical simulation of turbulence (Orszag & Patterson 1972) and the numerical simulation of transition to turbulence (Orszag & Kells 1979) in incompressible fluids. Spectral methods (Gottlieb & Orszag 1977) are based on representing the solution to a problem as a truncated series of smooth functions of the dependent variables. It may seem that these methods are limited to problems where Fourier series are appropriate in rectangular geometries and spherical harmonic series are appropriate in spherical geometries. In these cases, smooth solutions can be represented as rapidly converging spectral series, leading to great economies over discrete approximations that lead to finite difference methods. For example, it has been estimated that three-dimensional spectral calculations of homogeneous turbulence are between one and two orders of magnitude more efficient as regards both memory and speed than fourth-order finite-difference calculations.

Spectral methods have now developed as a useful tool in areas far removed from their original, and perhaps obvious, applications. Transform methods (Orszag 1971) have allowed their application to problems with general nonlinear and nonconstant coefficients. Orthogonal polynomial expansions (Gottlieb & Orszag 1977) have expanded widely the kinds of boundary conditions amenable to spectral treatment. New extensions of the fast Fourier transform to nearly arbitrary Sturm-Liouville eigenfunction bases (Orszag 1979a) may improve the efficiency of spectral methods based on exotic function bases. A fast, general iteration method has improved the efficiency of solving general spectral equations so that spectral solution of problems in general complicated geometries requires little more work than that required to solve the lowest-order finite-difference approximation to the problem in the complex geometry (Orszag 1979b).

One kind of problem has not yet received much attention for treatment by spectral methods, namely, the approximation of discontinuous solutions by spectral methods. Some early partial results were encouraging. It was shown (Orszag & Jayne 1974) that continuous solutions with discontinuous derivatives are well represented spectrally and that the accuracy advantages of spectral methods over finite difference methods survive for such solutions.

In the present paper, we provide an initial glimpse into the extension of spectral methods to treat discontinuous solutions with shocks and contact discontinuities. The results are very encouraging. It seems possible that spectral methods allow the resolution of shock fronts and contact discontinuities within shock-capturing techniques with only one grid interval across the discontinuity. Of course, spectral methods may also be used with shock-fitting techniques to represent a perfectly sharp discontinuity.

Linear Hyperbolic Problems

It has been shown by Majda, McDonough & Osher (1978) that by pre- and post-processing discontinuous data it is possible to achieve high accuracy with spectral approximations to discontinuous solutions of linear problems.

As a model problem, let us consider the solution to the one-dimensional wave equation

$$u_t(x,t) + u_x(x,t) = 0 \tag{1}$$

with periodic boundary conditions on the interval $0 \le x \le 2\pi$. With nonsmooth initial data $u(x,0) = f(x)$, the solution $u(x,t) = f(x-t)$ (extended periodically) is nonsmooth for all t and a Fourier-spectral method should be expected to converge slowly. Nevertheless, results obtained by pseudospectral Fourier solution of (1) are spectacular (at least in comparison with those obtained by more standard techniques). In Fig. 1, we plot the solution to (1) at $t = 4\pi$ (after two full propagation periods over the spatial domain) with $f(x) = \exp(-(x-\pi)^2/4\Delta x^2)$ so the solution is a Gaussian with width $2\Delta x$, where $\Delta x = 2\pi/64$ is the effective grid resolution with 64 Fourier modes. The results plotted in Fig. 1 were obtained using a weak low-pass filter to pre- and post-process the results.

In Fig. 2, we give similar plots for the solution to (1) for $f(x)$ a top-hat function. Here a stronger post-processing was necessary to remove large oscillations, due to the Gibbs phenomenon, near the discontinuities. The particular post-processing does not appear to be too important for this problem; we used a one-sided average in the neighborhood of rapid changes of the solution. The motivation for this choice is similar to that of the flux-corrected transport (FCT) algorithm (Boris & Book 1976). The results plotted in Figs. 1 and 2 are obtained using 64 Fourier modes.

Compressible Flow Problems

In this Section, we study the application of spectral methods to one-dimensional compressible flow problems and compare the results with those obtained by more conventional finite-difference methods. The one-dimensional Eulerian equations of motion in a finite shock tube are, in conservation form,

$$\frac{\partial \vec{w}}{\partial t} + \frac{\partial}{\partial x} \vec{F}(\vec{w}) = 0 \quad (-1 \leq x \leq 1) \tag{2}$$

$$\vec{w} = (\rho, m, E)^T, \; \vec{F}(\vec{w}) = u\,\vec{w} + (0, p, pu)^T \tag{3}$$

where ρ is the mass density, m is the momentum density, E is the total energy density, p is the pressure and $u = m/\rho$ is the velocity.

We assume the equation of state to be

$$p = (\gamma-1)\;[E - \tfrac{1}{2}\rho u^2]; \tag{4}$$

usually, we take $\gamma = 1.4$, as is approximate for a diatomic gas.

For simple shock tube problems, that involve only constant states separated by shocks and contact discontinuities , there is no immediate need for high-order accurate numerical methods. However, for more complicated flow problems, especially in higher dimensions, where there may be interacting shocks, rarefaction waves, contact discontinuities as well as the interaction of shocks with boundary layers and interfaces, it is necessary to have both high accuracy in the interior of the flow, in the boundary layers, near the interfaces, as well as good representations of discontinuities. From this point of view it seems appropriate to use the Chebyshev spectral method since it provides both high interior accuracy and very high resolution in the boundary layer region (Gottlieb & Orszag 1977). We regard the classical simple shock tube problems as extremely severe tests of spectral methods. After all, spectral methods are designed to give good resolution of complicated flow structures throughout the flow domain and, since they are based on expansions in orthogonal functions, it would seem that isolated local jumps at shocks and contact discontinuities would be most inhospitable for them. On the other hand, it would seem that low-order finite-difference methods would be best for treating shock discontinuities because of their localized character. One of the important conclusions of this paper is that the accuracy (or rather, the resolution) advantages of spectral methods holds up in the neighborhood of (at least, moderately strong) shock discontinuities, not just in regions of smooth flow and boundary layers.

There are three ways to apply Chebyshev-spectral methods to these problems, namely, collocation, Galerkin, and tau approximation. We choose to use the collocation (or pseudospectral) technique here for two reasons. First, collocation is the easiest and most efficient method to apply for complicated problems. Also, since collocation involves solving the equations in physical space rather than in transform space with transforms used only to evaluate derivatives, boundary conditions are also easier to apply.

Let us give a brief description of the collocation method. At each time step, we evaluate the components of $\vec{F}$ in (2) at the points

$$x_j = \cos \pi j/N \quad (0 \le j \le N) \tag{5}$$

Next, the Chebyshev expansion coefficients $\underset{\sim}{a}_n$ of $\vec{F}$ are found from

$$\vec{F}(x_j) = \sum_{n=0}^{N} \vec{a}_n T_n(x_j) \quad (0 \le j \le N), \tag{6}$$

where $T_n(x)$ is the Chebyshev polynomial of degree n defined by $T_n(x) = \cos(n \cos^{-1} x)$. Since $T_n(x_j) = \cos \frac{\pi j n}{N}$ it follows that

$$\vec{a}_n = \frac{2}{N\bar{c}_n} \sum_{p=0}^{N} \frac{1}{\bar{c}_p} \vec{F}(x_p) \cos \frac{\pi p n}{N} \quad (0 \le n \le N), \tag{7}$$

where $\bar{c}_0 = \bar{c}_N = 2$, and $\bar{c}_n = 1$ for $0<n<N$. Differentiating (6) gives

$$\frac{\partial \vec{F}}{\partial x} = \sum_{n=0}^{N} \vec{a}_n T'_n = \sum_{n=0}^{N} \vec{s}_n T_n \tag{8}$$

where

$$\vec{s}_n = \frac{2}{\bar{c}_n} \sum_{\substack{p=n+1 \\ p+n \text{ odd}}}^{N} p\vec{a}_p. \tag{9}$$

Eqs. (7)-(8) are implemented using the fast Fourier transform. In order to avoid having to perform order N^2 operations to evaluate (9) we observe that $\vec{s}_N = 0, \vec{s}_{N-1} = 2N\vec{a}_N$ and $\vec{s}_n$ $(n \le N-2)$ satisfies the recurrence relation

$$\bar{c}_n \; \vec{s}_n = \vec{s}_{n+2} + 2(n+1)\; \vec{a}_{n+1} \quad (n \leq N-2) \tag{10}$$

Using (10), only $0(N)$ operations are necessary to obtain $\vec{s}_n$ from $\vec{a}_n$. Thus, using the fast Fourier transform, evaluation of $\partial\vec{F}/\partial x$ from $\vec{F}$ requires only order $N \log N$ operations.

It should be noted at this point that we do not make use of the fact that the collocation points x_j are crowded in the neighborhood of $x = 1$ and $x = -1$. [x_1 and x_{N-1} are located at distances of $\pi^2/2N^2$ from $x = 1$, respectively]. For $N = 128$, there are 40 points located between $0.9 < |x| \leq 1$. This high boundary resolution can be of great value when boundary layers or interfaces are also present, but is irrelevant for the simple test problems of the present paper.

Time differencing

In the last Section, we described briefly spectral methods for space discretization. Here we describe some time marching techniques. Let us denote the discrete approximation for $\vec{w}$ at collocation point x_j and time step $n\Delta t$ by $\vec{w}_j^n$ the discrete approximation for $\vec{F}$ by $\vec{F}_j^n$. Then we advance in time using the two-step (modified Euler) method

$$\begin{aligned} \vec{w}_j^{\,n+\frac{1}{2}} &= \vec{w}_j^{\,n} - \frac{1}{2}\,\Delta t\left(\frac{\partial\vec{F}}{\partial x}\right)_j^n \\ \vec{w}_j^{\,n+1} &= \vec{w}_j^{\,n} - \Delta t\,\left(\frac{\partial\vec{F}}{\partial x}\right)_j^{n+\frac{1}{2}} \end{aligned} \tag{11}$$

This scheme results in a linear stability condition of the form

$$\Delta t \leq \frac{8}{N^2 \max(|u|+c)} \tag{12}$$

where c is the sound speed. This condition is very severe; using finite difference methods with N points leads to stability restrictions like $\Delta t = 0(1/N)$. An alternative approach (Gottlieb & Turkel 1979) that avoids the latter difficulty is described below.

From (9) it follows that

$$(s_0,\ldots,s_N)^T = A\;(a_0,\ldots,a_N)^T \tag{13}$$

where A is an $(N+1)\times(N+1)$ matrix. The stability condition (12) results from the fact that $||A|| = O(N^2)$. However, if we replace the formula

$$\Delta t\ s_n = \frac{1}{\bar{c}_n} \sum 2\Delta t\ p\ a_p \tag{14}$$

by

$$\Delta t\ s_n = \frac{1}{\bar{c}_n} \sum \frac{f(\Delta t p^2)}{p}\ a_p \tag{15}$$

where $f(z)$ satisfies

$$\lim_{z\to 0} f(z)/z = 2, \tag{16}$$

$$|f(z)| \leq 1 \quad (0<z<\infty), \tag{17}$$

then the time step restriction (12) is avoided. With spectral derivatives evaluated by (15) instead of (14), time integrations are unconditionally stable.

Gottlieb & Turkel (1979) show that a good choice for $f(z)$ is

$$f(z) = \frac{1}{3\alpha}\,[11-18e^{-\alpha z} + 9e^{-2\alpha z} - 2e^{-3\alpha z}] \tag{18}$$

where α is a suitable constant.

A systematic way a priori of choosing α has not yet been found. Our computational experience is that the best α is given roughly by

$$\alpha = \max(|u|+c). \tag{19}$$

We note that when Δt satisfies the explicit stability condition (12), the results are nearly identical using either (14) or (15).

Boundary conditions

Boundary conditions play a crucial role in the application of spectral methods. Incorrect boundary treatment may give strong instabilities whereas, with finite difference methods, instabilities due to boundaries usually appear as relatively weak oscillations. On the other hand, in contrast to high-order finite-difference methods, spectral methods normally do not require numerical boundary conditions in addition to the physical boundary conditions required by the partial differential equation.

For shock tube problems, we must specify all the flow variables at supersonic inflow points, two flow variables at subsonic inflow points, and one flow variable at subsonic outflow points. If we overspecify or underspecify the boundary conditions, the spectral calculation is usually unstable after a few time steps.

At subsonic outflow points, it is not satisfactory to specify arbitrarily any one of the flow variables m, ρ, or E. With arbitrary outflow boundary conditions, one can obtain oscillations coming originating at the boundary. This phenomenon was analyzed by Gottlieb, Gunzburger & Turkel (1979); the actual outflow boundary conditions used here were obtained following their analysis. We advance one time step without imposing the boundary conditions and denote the calculated quantities by w_c. We observe that if $x = +1$ is a subsonic outflow point, the incoming characteristic quantity at $x = +1$ is

$$v_1 = p - (\rho c)\, u$$

whereas the outgoing characteristic quantities are

$$v_2 = p+(\rho c)u, \quad v_3 = p-c^2\rho.$$

Let us assume that one flow variable is given on the inflow characteristic at $x = +1$. Then we solve the system

$$v_2 = (v_2)_c\,, \quad v_3 = (v_3)_c \tag{20}$$

for the two remaining flow variables at $x = 1$. This procedure yields a stable scheme with no oscillations emanating from the boundaries.

In contrast to inflow-outflow boundaries whose treatment is quite systematic by the above procedure, material boundaries evidently do require boundary conditions in addition to those required by the mathematical theory of characteristic initial value problems. At characteristic surfaces, like material boundaries, the specification of one flow variable, like $u = 0$ should suffice. However, we found it necessary to supplement this boundary condition at only <u>one</u> characteristic boundary by <u>one</u> additional condition, like p given. An analysis of these boundary conditions will be published elsewhere.

<u>Smoothing and Filtering</u>

The approximation of discontinuous functions by truncated Chebyshev polynomial expansions exhibits the Gibbs phenomenon near the jump and has two point oscillations over the whole region. Similar oscillations are observed when approximating shock waves in inviscid flows. It is essential to use some damping or filtering to eliminate these oscillations. Several methods have been used including pseudo-viscosity, Shuman filtering, and a new spectral filtering method. The best results have, so far, been obtained by the latter two methods that will now be described.

Shuman filtering involves applying the filter

$$\bar{w}_j^n = w_j^n + \beta[\alpha_1(w_{j+1}^n - w_j^n) + \alpha_2(w_j^n - w_{j-1}^n)] \tag{21}$$

to a dynamical field w_j^n. Here α_1 and α_2 are chosen such that the last term approximates the second derivative of w on the non-uniform grid x_j. We choose the constant β to localize the effect of dissipation in the neighborhood of the shock. The surprising thing is that β is a function of Δt; for smaller Δt we need smaller β (in contrast to finite difference methods that usually utilize constant β to achieve dissipation). Also surprising is the result that one step of Shuman filtering every 200 or so time steps suffices to remove oscillations for shocks of moderate strength (pressure jumps of 10 to 1, say). Shocks of width 1 1/2 - 2 grid intervals normally result from this kind of filtering (see below).

A much more intriguing kind of filtering is based on the following idea. If a low-pass spectral filter just strong enough to remove those high frequency waves that lead to numerical instabilities is applied to the inviscid compressible flow equations, the spectral equations will give bad oscillations near shock fronts and other discontinuities. However, as recently pointed out by Lax (1978), these oscillatory solutions obtained by a high-order method like a spectral method should contain enough information to be able to reconstruct the proper non-oscillatory discontinuous solution by a post-processing filtration. The idea is that very weak filtering or damping to stabilize together with a final 'cosmetic' filter to present the results should be able to give great improvements in resolution.

In practice, we use a low-pass filter for stabilization purposes that is of the form

$$f(k) = \begin{cases} 1 & k < k_0 \\ e^{-a(k-k_0)^4} & k > k_0 \end{cases} \tag{22}$$

where k is a spectral (wavenumber) index, k_0 is typically 1/2-2/3 the maximum wavenumber and a is a constant chosen so that the highest modes are damped by a factor of roughly e^{-1} to e^{-10}. We have found that, for moderate shocks, the results are insensitive to the detailed form of (22). However, it is important that $f(k)$ be low-pass so $f(k) = 1$ for $k < k_0$, in order that large-scales be treated very accurately (without phase error) by the spectral method.

The post-processing cosmetic filter is usually chosen to be of the form of the Shuman filter (21) with $\beta = (\frac{\Delta x}{2})^2$, except in the neighborhood of large jumps where a one-sided smoothing is used. Further work will likely concentrate on more sophisticated filtering methods like those employed in signal analysis applications.

Results

The first model problem is a shock tube problem with a diatomic gas ($\alpha = 1.4$) satisfying supersonic inflow at $x = -1$ and subsonic outflow at $x = +1$. The initial conditions are

$$p = 1, \; \rho = 1, \; u = 1 \quad (-1<x\leq 1), \tag{23}$$

while the boundary conditions applied at the inflow point $x = -1$ are

$$p = p_1, \rho = \frac{1+6p_1}{p_1+6}, \quad u = 1+\sqrt{5}(p_1-1)/\sqrt{1+6p_1}, \tag{24}$$

where p_1 is a constant. With these initial and boundary conditions, a pure shock propagates from $x = -1$ toward $x = +1$ at a speed

$$v_s = 1+\left(\frac{1+6p_1}{5}\right)^{1/2} \tag{25}$$

In Figs. 3,4, we plot the structure of the resulting shock wave at $t = 0.1, 0.2$, respectively, determined by the Chebyshev spectral method with $N = 128$ polynomials and $p_1 = 5$ in (24). The exact pressure jump across this shock wave is 5 and the density jump is 2.81818. The shock speed is 3.48997. In these calculations, Shuman filtering is applied. The results plotted in Figs. 3,4 demonstrate that this method achieves high resolution shocks (1 1/2 grid intervals) without significant oscillations.

The second model problem is a shock tube problem with $x = \pm 1$ as material boundaries. The initial conditions at $t = 0$ are:

$$p = \begin{cases} 1.0 & x<0 \\ 0.55 & x=0 \\ 0.1 & x>0 \end{cases}$$

$$= \begin{cases} 1.0 & x<0 \\ 0.5625 & x=0 \\ 0.125 & x>0 \end{cases} \tag{26}$$

$$u = 0$$

while the boundary conditions are

$$u = 0 \quad x = \pm 1 \tag{27}$$

$$\rho = 1 \quad x = -1$$

For moderate t, the solution to this problem consists of one shock, one contact discontinuity and a rarefaction wave.

A variety of difference methods for solution of the flow that evolves from (26)-(27) have been compared by Sod (1978). We have solved this problem using the spectral filtering method with $N = 64$ Chebyshev polynomials to represent the flow (in contrast to Sod's 100 point grid). The results are plotted in Fig. 5 at $t = 0.3$. Both the shock and contact discontinuity are resolved over only one grid point. The overall solution is in good agreement with the exact solution to the problem. However, without the 'cosmetic' final filter, the results are highly oscillatory. Evidently, the highly oscillatory, but stable, spectral solutions do contain enough information to reconstruct sharp discontinuities.

This work was supported by the Office of Naval Research under Contract N00014-77-C-0138 and the Air Force Office of Scientific Research under Grants 77-3405 and 78-3651.

References

Boris, J. P. & Book, D. L. 1976 Methods in Computational Physics, Academic, Vol. 16, Chap. 11.

Gottlieb, D., Gunzburger, M. & Turkel, E. 1979 SIAM J. Num. Anal., in press.

Gottlieb, D. & Orszag, S. A. 1977 Numerical Analysis of Spectral Methods: Theory and Applications, NSF-CBMS Monograph No. 26, Soc. Ind. Appl. Math.

Gottlieb, D. & Turkel, E. 1979 Stud. in Appl. Math., in press.

Lax, P. 1978 in Proc. Symp. Mathematics Research Center (Univ. of Wisconsin), Academic, p. 107.

Majda, A., McDonough, J. & Osher, S. 1978 Math. Comp. 32, 1041.

Orszag, S. A. 1971 Stud. in Appl. Math. 50, 395.

Orszag, S. A. 1979a 'Fast' eigenfunction transforms, to be published.

Orszag, S. A. 1979b J. Comp. Phys., in press.

Orszag, S. A., & Jayne, L. W. 1974 J. Comp. Phys. 14, 93.

Orszag, S. A. & Kells, L. C. 1979 J. Fluid Mech., in press.

Orszag, S. A. & Patterson, G. S. 1972 Phys. Rev. Letters 28, 76.

Sod, G. A. 1978 J. Comp. Phys. 27, 1.

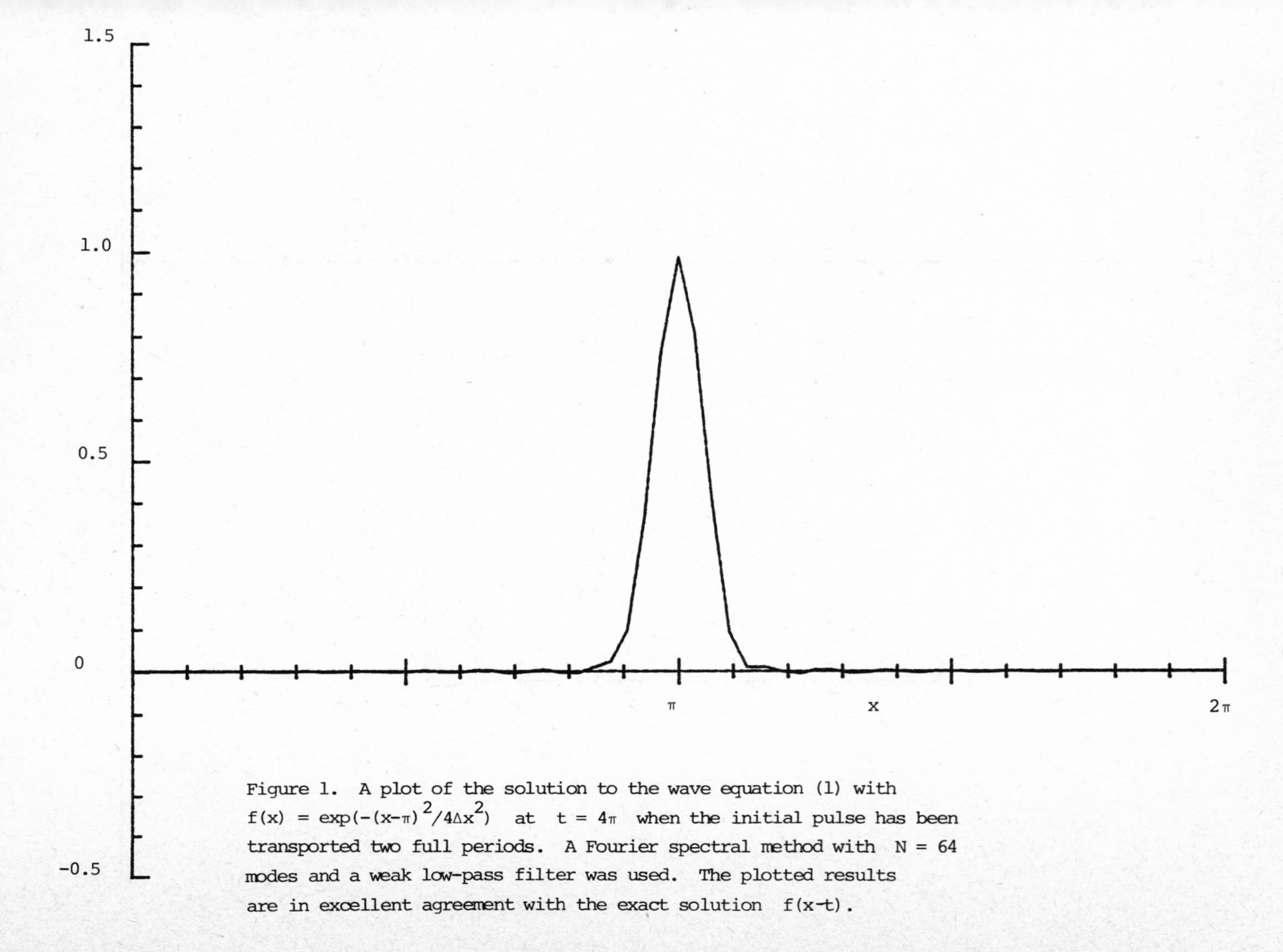

Figure 1. A plot of the solution to the wave equation (1) with $f(x) = \exp(-(x-\pi)^2/4\Delta x^2)$ at $t = 4\pi$ when the initial pulse has been transported two full periods. A Fourier spectral method with $N = 64$ modes and a weak low-pass filter was used. The plotted results are in excellent agreement with the exact solution $f(x-t)$.

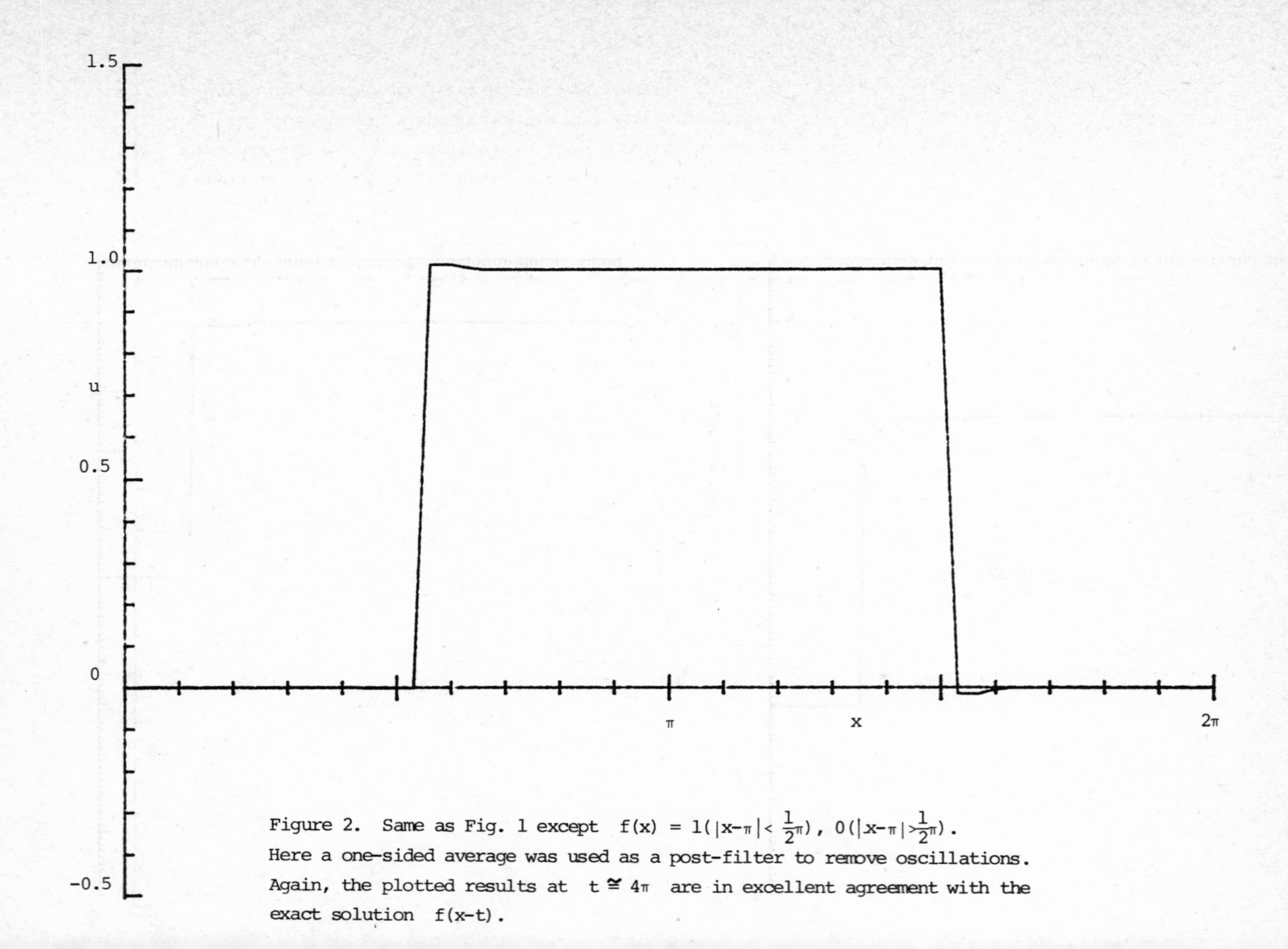

Figure 2. Same as Fig. 1 except $f(x) = 1(|x-\pi| < \frac{1}{2}\pi)$, $0(|x-\pi| > \frac{1}{2}\pi)$. Here a one-sided average was used as a post-filter to remove oscillations. Again, the plotted results at $t \cong 4\pi$ are in excellent agreement with the exact solution $f(x-t)$.

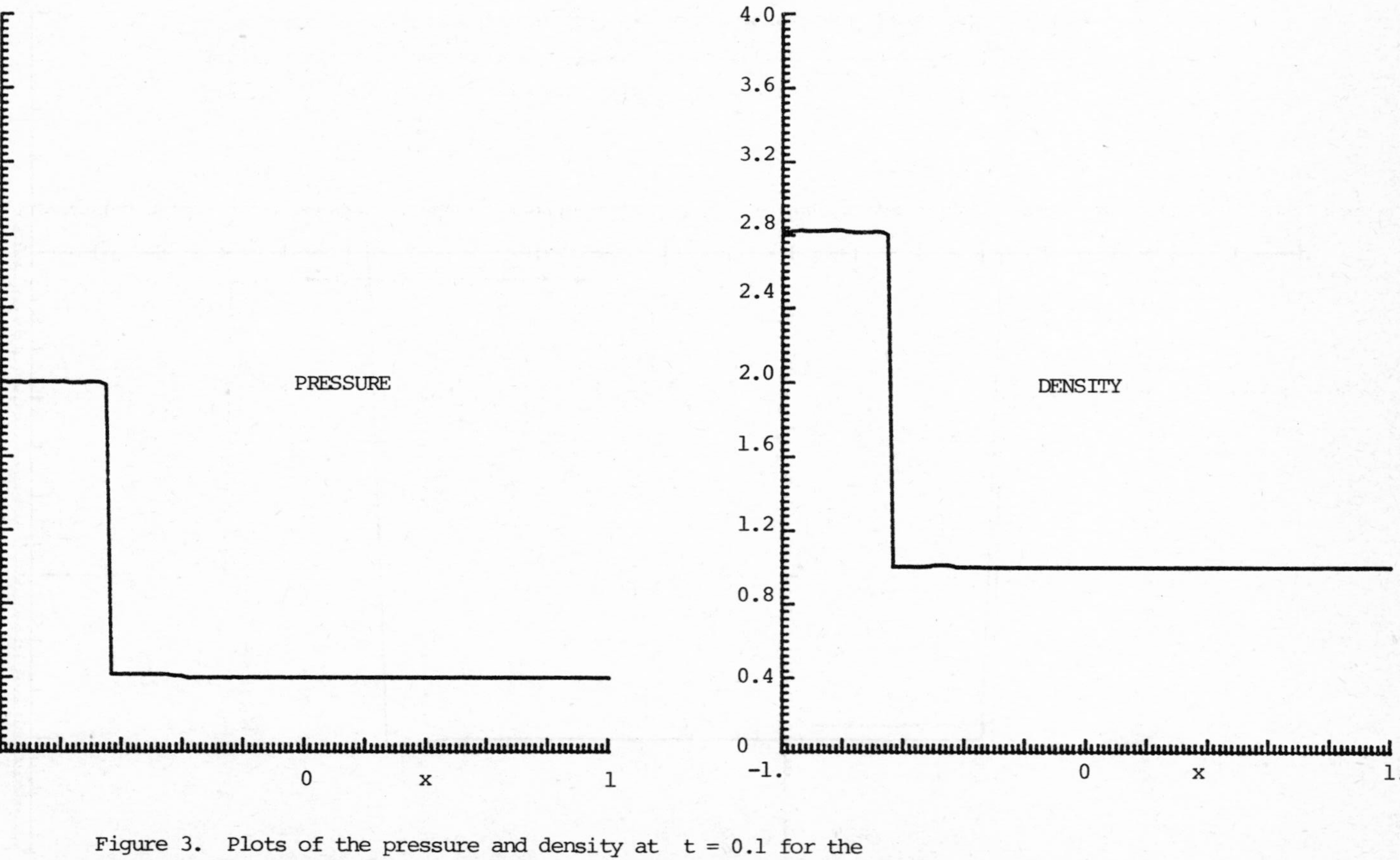

Figure 3. Plots of the pressure and density at $t = 0.1$ for the shock tube problem with initial-boundary conditions (23)-(24) with $p_1 = 5$. The Chebyshev spectral equations were truncated at $N = 128$ polynomials and Shuman filtering was applied.

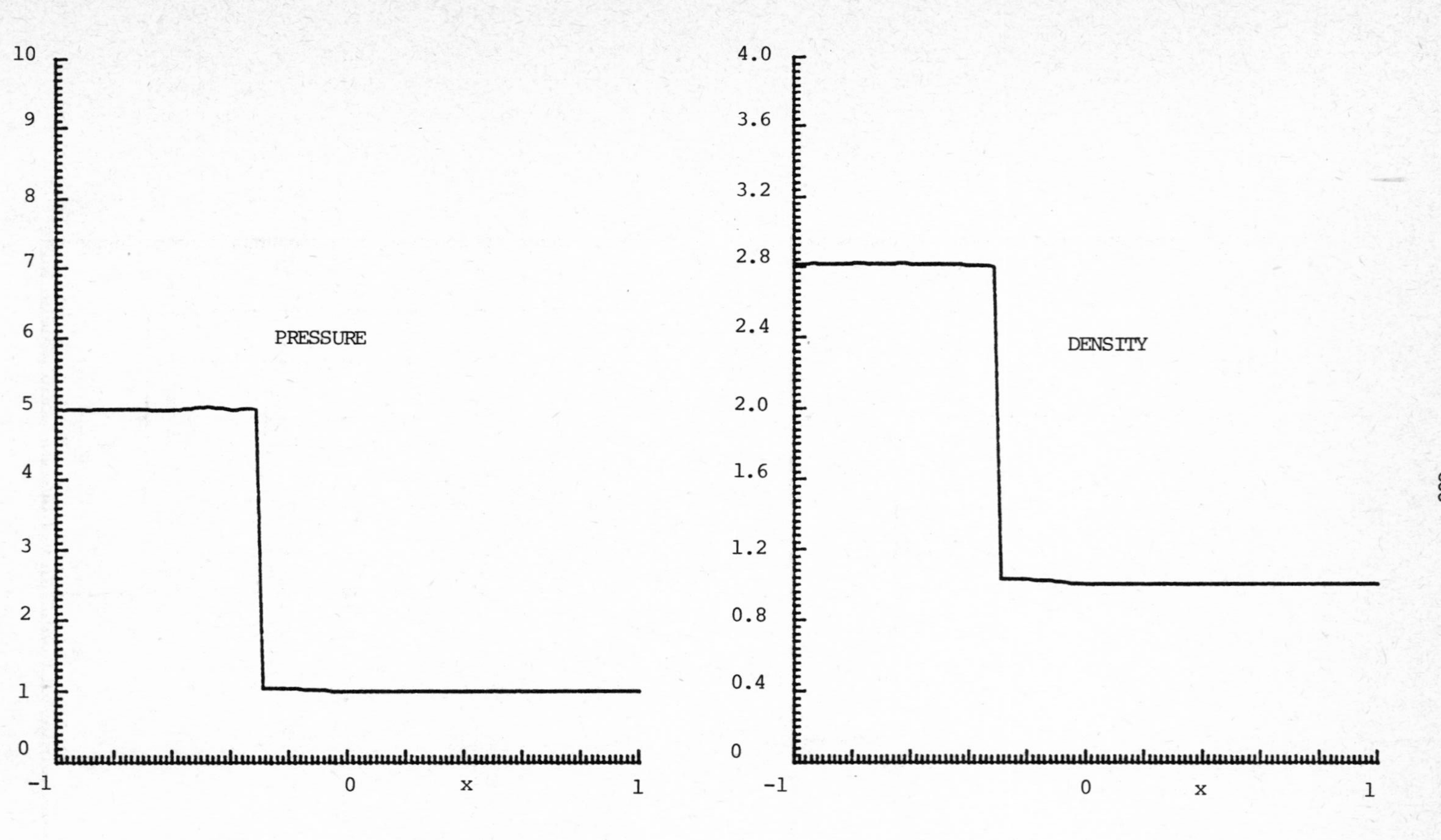

Fig. 4. Same as Fig. 3 except $t = 0.2$.

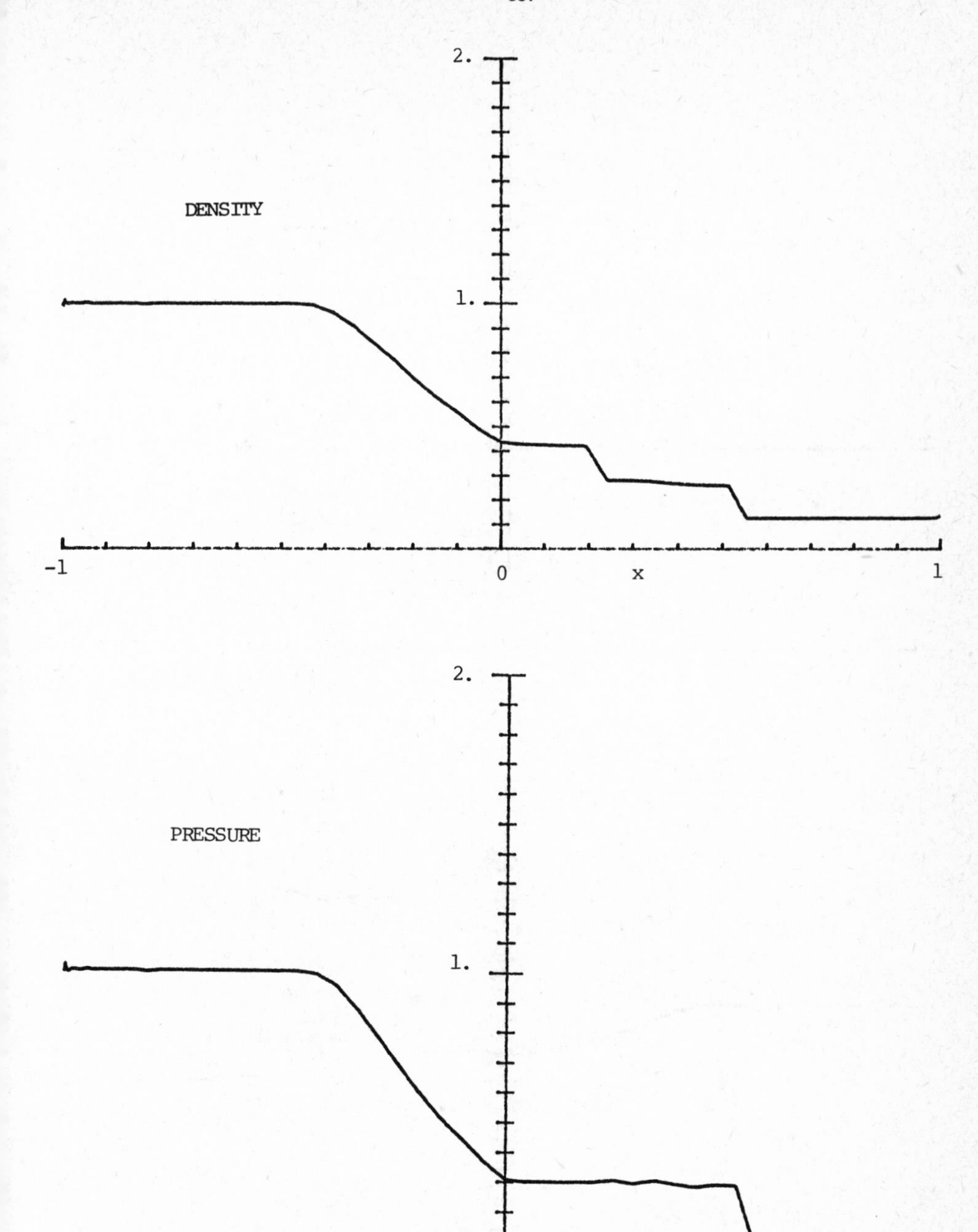

Fig. 5. Plots of the density, pressure, velocity, and energy for the shock tube problem initial-boundary conditions (26)-(27) at $t = 0.3$. The Chebyshev spectral equations were truncated at $N = 64$ polynomials and the spectral filtering method was applied to smooth the final results.

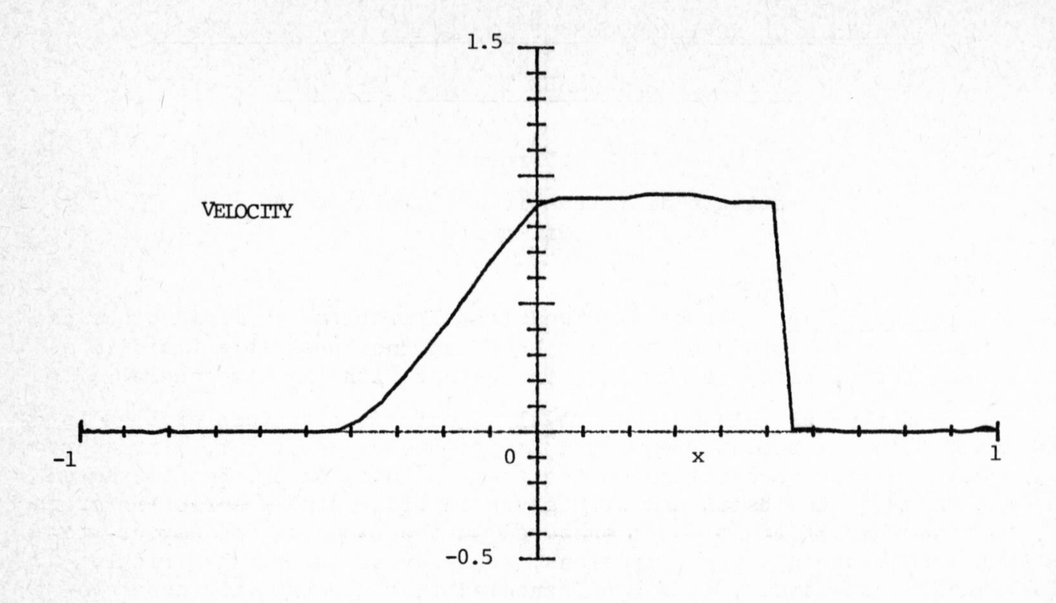

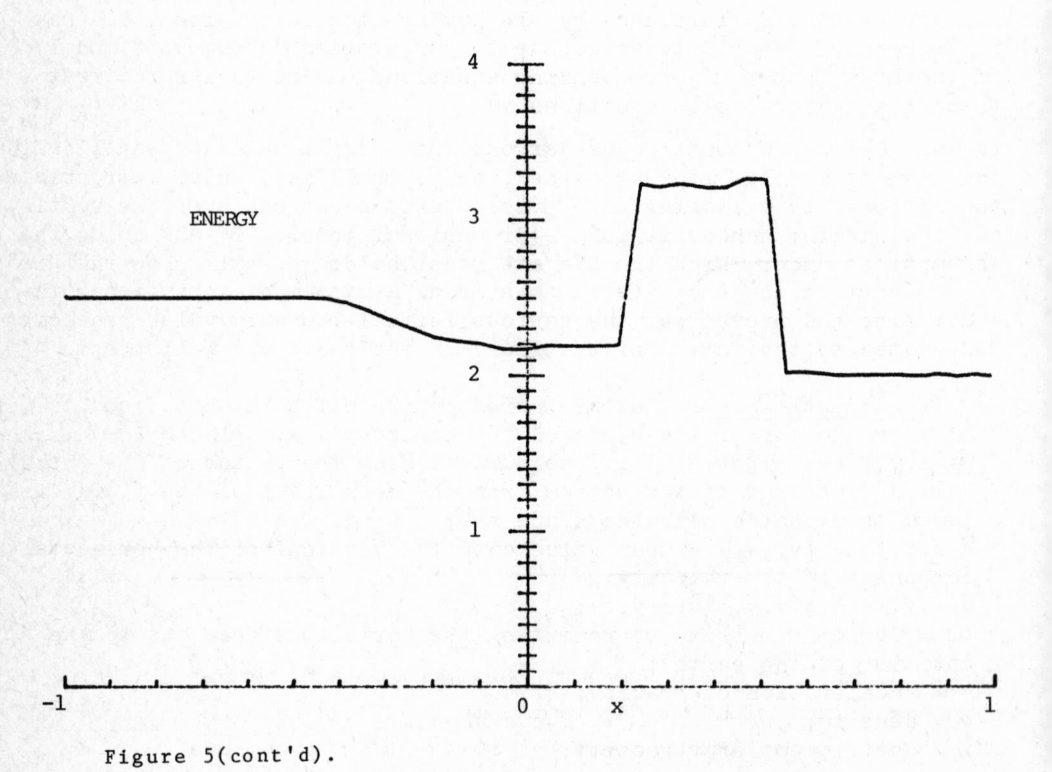

Figure 5(cont'd).

ANALYSIS OF NAVIER-STOKES TYPE EQUATIONS ASSOCIATED TO MATHEMATICAL MODELS IN FLUID DYNAMICS

G. Prouse

Istituto di Matematica del Politecnico
Milano

I - Introduction. One of the fundamental problems in fluid dynamics consists in the study of the motion of an incompressible fluid in a closed basin, with free surface in contact with the atmosphere.

The most general mathematical model associated with this problem is constituted by the 3-dimensional Navier-Stokes equations, with appropriate boundary conditions on the free surface, on the bottom and on the sides of the basin and with given initial data. Bearing however in mind the difficulties encountered in the study of the Navier-Stokes equations in three dimensions, even for the simplest Dirichlet boundary conditions, it may be expected that a relatively complete analysis of the solutions of our problem (global existence, uniqueness and continuous dependence theorems, etc.) is not at present possible. We can observe, on the other hand, that, in general, the information required for practical applications is less detailed than the one eventually furnished by the Navier-Stokes equations, so that it appears reasonable to associate to our problem a "coarser" model, in the hope that the corresponding equations may be easier to treat than the Navier-Stokes equations.

In what follows we shall consider one such model, which we shall call the Liggett model. It is obtained, as we shall see, under appropriate assumptions, by avaraging the Navier-Stokes equations over the vertical coordinate; hence, it is a 2-dimensional model. By examining the assumptions under which the Liggett model holds we shall show that it is appropriate to associate to this model a system of variational inequalities and prove that the various initial-boundary value problems associated to the motion of the fluid in the basin are well posed.

2 - The Liggett model. Let us introduce the following notations.

Ω = projection of the basin on the horizontal x,y plane; $\Gamma = \partial\Omega$;

$\xi(x,y,t)$ = height of the free surface from the bottom of the basin;

$-h(x,y)$ = height of the bottom from the mean level of the fluid, assumed to coincide with the plane $z=0$;

$\vec{v}(x,y,t) = \{v_1, v_2\}$ = mean value over the vertical of the horizontal component of the velocity;

$\vec{u}(x,y,t) = \xi(x,y,t)\vec{v}(x,y,t)$;

ω = Coriolis matrix (representing the Coriolis effect due to the rotation of the earth);

g = acceleration of gravity;

ρ = density;

μ = coefficient of viscosity;

p = atmospheric pressure on the free surface;

$\vec{\chi}_s(x,y,t) = \{\chi_{1,s}, \chi_{2,s}\}$ = tangential stress on the free surface due to the action of the wind.

We make, moreover, the following assumptions, which appear reasonable for a wide class of applications.

a) The fluid is homogeneous;
b) The stress tensor is given by

$$\tau_{11} = 2\mu \frac{\partial v_1}{\partial x} \;;\; \tau_{22} = 2\mu \frac{\partial v_2}{\partial y} \,,\; \tau_{12} = \tau_{21} = \mu\left(\frac{\partial v_1}{\partial y} + \frac{\partial v_2}{\partial x}\right) \,,\; \tau_{13} = \tau_{31} = \tau_{23} = \tau_{32} = \tau_{33} = 0 \,;$$

c) The particles on the free surface have velocity tangent to the free surface itself;
d) The resistance exercised by the bottom of the basin is given by

$$\vec{\chi}_b = \alpha \vec{v} + \beta \vec{v} |\vec{v}| \quad , \quad \alpha, \beta \geq 0 \;;$$

e) The motion is fairly uniform, with respect to both space and time;
f) The depth ξ is strictly positive;
g) $|\vec{v}|$ does not approach the speed of light.

From assumptions e), f), g), it follows that there exist suitable positive constants M_1, M_2, M_3, σ such that

$$(2.1) \qquad \left|\frac{\partial \xi}{\partial t}\right| < M_1 \quad , \quad |\mathrm{grad}\, \xi| < M_2 \quad , \quad \xi > \sigma \quad ,$$

$$(2.2) \qquad |\vec{u}| \leq M_3 \;.$$

Consider now the 3-dimensional Navier-Stokes equations and calculate the avarages of the first two equations and of the equation of continuity over the vertical coordinate z from -h to ξ-h; we obtain then, under the assumptions given above, the equations (see [1], [2], where an extensive list of references regarding hydrodynamical models can also be found)

$$\frac{\partial \xi}{\partial t} + \mathrm{div}\, \vec{u} = 0$$

$$(2.3) \qquad \frac{\partial \vec{u}}{\partial t} - \mu \Delta \vec{u} + (\vec{u} \cdot \mathrm{grad}) \frac{\vec{u}}{\xi} + \frac{\vec{u}}{\xi}\, \mathrm{div}\, \vec{u} + g\xi\, \mathrm{grad}\, \xi =$$

$$= \omega \vec{u} - \frac{\xi}{\rho} \mathrm{grad}\, p + g\xi\, \mathrm{grad}\, h + \frac{1}{\rho} \vec{\chi}_s - \frac{1}{\rho}\left(\alpha \frac{\vec{u}}{\xi} + \beta \frac{\vec{u}|\vec{u}|}{\xi^2}\right) =$$

$$= F(x, y, t, \xi, \vec{u}) \,.$$

Observe that assumptions a), b), c), d) appear directly in the formulation of (2.3), whereas e), f), g) are simply given in order to justify approximations introduced in the deduction of (2.3) and have therefore to be taken into account separately. We shall then call <u>Liggett model</u> the hydrodynamical model <u>associated to relations</u> (2.I), (2.2), (2.3).

It is obvious that it is not possible to know "a priori" if the Liggett model is applicable to a certain problem, since there are no physical conditions that impose on ξ, $\vec{u}$ to satisfy (2.I), (2.2); the validity of the model can only be established "a posteriori", by checking if the eventual solutions of (2.3) satisfy (2.I), (2.2).

Observation. Assumption c) implies that a particle that , at a certain time $t=\bar{t}$, is on the free surface, remains there $\forall\, t>\bar{t}$; the formation of vortices which are not parallel to the free surface is therefore excluded. Regarding d), which is one example of the many empyrical formulas that can be adopted, it must be noted that, since the unknown functions are avarage velocities, the effect on the motion of the fluid produced by the resistance at the bottom has to be expressed as a function of such velocities.

3 - Initial-boundary value problems. To the relations corresponding to the Liggett model are associated initial-boundary value problems which consist in assigning the values of $\vec{u}$ and ξ in Ω at the initial time t=0

$$(3.1)\qquad \vec{u}(x,y,0)=\bar{\vec{u}}(x,y)\ ,\quad \xi(x,y,0)=\bar{\xi}(x,y)\qquad (x,y)\in\Omega$$

and boundary conditions on Γ corresponding to various physical problems.

Denoting by $\vec{\tau}$ and $\vec{\nu}$ respectively the unit vectors tangent and normal to Γ and setting $u_\tau=\vec{u}\cdot\vec{\tau}$, $u_\nu=\vec{u}\cdot\vec{\nu}$, we shall consider the following three boundary conditions, which are the most common in practical applications:

α) $\vec{u}(x,y,t)=0$ when $(x,y)\in\Gamma$;

β) $\vec{u}(x,y,t)=0$ when $(x,y)\in\Gamma_1$
$u_\tau(x,y,t)=0$, $\xi(x,y,t)=q$ when $(x,y)\in\Gamma_2$ $(\Gamma_1\cup\Gamma_2=\Gamma)$;

γ) $\vec{u}(x,y,t)=0$ when $(x,y)\in\Gamma_1$,
$u_\nu(x,y,t)=0$, $\dfrac{\partial u_\tau(x,y,t)}{\partial\nu}=0$ when $(x,y)\in\Gamma_2$ $(\Gamma_1\cup\Gamma_2=\Gamma)$.

Condition α) occurs when Γ represents a solid surface; in this case, by the boundary layer theory, the velocity on Γ vanishes.

Suppose that Γ is constituted by two parts, Γ_1 and Γ_2, the first of which solid, while Γ_2 represents a separation from a fluid inlet or outlet (river, canal, etc.). Condition β) imposes that $\vec{u}$ vanishes on Γ_1, while, on Γ_2 , the velocity is directed normally to Γ_2 and the depth is assigned.

Finally, γ) represents a "symmetricity" condition and holds if we can assume that Ω is "half" of a domain Ω' in which the motion is symmetrical with respect to Γ_2 .

We shall therefore associate to the Liggett model the initial-boundary conditions (3.I)-α, (3.I)-β, (3.I)-γ.

4 - Weak formulation of the problems. We shall now give a weak formulation of equations (2.3).

Let $\varphi(x,y,t)$, $\vec{\Psi}(x,y,t)$ be two "test functions" which we shall assume, for the time being, to be sufficiently smooth and set

$$\vec{v}(t)=\{\vec{v}(x,y,t);(x,y)\in\Omega\}\ ,\quad \zeta(t)=\{\zeta(x,y,t)\,;\,(x,y)\in\Omega\}\ ,$$

$$\vec{v}'(t)=\left\{\frac{\partial\vec{v}(x,y,t)}{\partial t};\ (x,y)\in\Omega\right\},\quad \zeta'(t)=\left\{\frac{\partial\zeta(x,y,t)}{\partial t};\ (x,y)\in\Omega\right\},$$

$$(\zeta,\varphi)_{L^2}=\int_\Omega \zeta(x,y,t)\,\varphi(x,y,t)\,d\Omega,$$

$$(\vec{v},\vec{\psi})_{L^2}=\int_\Omega \sum_{i=1}^{2} v_i(x,y,t)\,\psi_i(x,y,t)\,d\Omega,$$

$$(\vec{v},\vec{\psi})_{H_0^1}=\int_\Omega \left(\frac{\partial\vec{v}}{\partial x}\cdot\frac{\partial\vec{\psi}}{\partial x}+\frac{\partial\vec{v}}{\partial y}\cdot\frac{\partial\vec{\psi}}{\partial y}\right)d\Omega,$$

$$b(\vec{u},\vec{v},\vec{w})=\int_\Omega (\vec{u}\cdot\mathrm{grad})\,\vec{v}\cdot\vec{w}\,d\Omega .$$

Multiplying the first of (2.3) by φ , the second by $\vec{\psi}$, integrating over $\Omega\times[0,T]$ $(T>0)$ and applying Green's formula, we obtain, by a standard procedure, bearing in mind (3.I),

$$\int_0^T (\xi'+\mathrm{div}\,\vec{u},\varphi)_{L^2}\,dt=0$$

$$(4.1)\quad \int_0^T\Big\{-(\vec{u},\vec{\psi}')_{L^2}+\mu(\vec{u},\vec{\psi})_{H_0^1}-b(\vec{u},\vec{\psi},\tfrac{\vec{u}}{\xi})+(g\xi\,\mathrm{grad}\,\xi,\vec{\psi})_{L^2}-$$
$$-(\vec{F},\vec{\psi})_{L^2}\Big\}\,dt+\int_0^T\!\!\int_\Gamma\Big\{-\mu\frac{\partial\vec{u}}{\partial\nu}\cdot\vec{\psi}+(\vec{u}\cdot\vec{\psi})\frac{u_\nu}{\xi}\Big\}\,d\Gamma\,dt+$$
$$+(\vec{u}(T),\vec{\psi}(T))_{L^2}-(\vec{u},\vec{\psi}(0))_{L^2}.$$

The various boundary conditions correspond to different choices of the test functions and to different forms of the boundary integral

$$B(T;\vec{u},\xi,\vec{\psi})=\int_0^T\!\!\int_\Gamma\Big\{-\mu\frac{\partial\vec{u}}{\partial\nu}\cdot\vec{\psi}+(\vec{u}\cdot\vec{\psi})\frac{u_\nu}{\xi}\,d\Gamma\,dt.$$

Considering separately the three boundary conditions introduced in the preceding §, we obtain:

Condition α): We choose $\vec{\psi}$ such that $\vec{\psi}(x,y,t)=0$ when $(x,y)\in\Gamma$; it follows directly that

$$(4.2)\qquad B(T,\vec{u},\xi,\vec{\psi})=0 .$$

Condition β) : Assume that $\vec{\psi}(x,y,t)=0$ when $(x,y)\in\Gamma_1$, $\psi_\tau(x,y,t)=0$ when $(x,y)\in\Gamma_2$; we obtain then

$$(4.3)\quad B(T,\vec{u},\xi,\vec{\psi})=\int_0^T\!\!\int_{\Gamma_2}\Big\{\mu\Big(u_\nu\,\mathrm{div}\,\vec{\nu}+\frac{\partial q}{\partial t}\Big)\psi_\nu+\frac{u_\nu^2}{q}\psi_\nu\Big\}\,d\Gamma\,dt .$$

Observe, in fact, that, if $\vec{n}(x,y)=a(x,y)\vec{i}+b(x,y)\vec{j}$ is a unit vector $\in C^1$, with $\vec{n}=\vec{\nu}$ on Γ_2 , we have

$$\mathrm{div}\,\vec{\nu}\big|_{\Gamma_2}=\frac{\partial a}{\partial\nu}\cos\nu x+\frac{\partial a}{\partial\tau}\sin\nu x+\frac{\partial b}{\partial\nu}\cos\nu y+\frac{\partial b}{\partial\tau}\sin\nu y\Big|_{\Gamma_2}=$$
$$=\frac{\partial a}{\partial\tau}\sin\nu x+\frac{\partial b}{\partial\tau}\sin\nu y\Big|_{\Gamma_2},$$

being, by construction,

$$\frac{\partial}{\partial\nu}\cos n\nu\Big|_{\Gamma_2} = \frac{\partial a}{\partial\nu}\cos\nu x + \frac{\partial b}{\partial\nu}\cos\nu y\Big|_{\Gamma_2} = 0 .$$

Hence, in (4.3), the value of div $\vec{\nu}$ on Γ_2 can be explicitely calculated. Moreover, by the conditions imposed on $\vec{u}$ and $\vec{\psi}$,

$$\operatorname{div}\vec{u}\big|_{\Gamma_2} = \frac{\partial u_\nu}{\partial\nu} + u_\nu \operatorname{div}\vec{\nu} + \frac{\partial u_\tau}{\partial\tau} + u_\tau \operatorname{div}\tau\big|_{\Gamma_2} = \frac{\partial u_\nu}{\partial\nu} + u_\nu \operatorname{div}\vec{\nu}\big|_{\Gamma_2} ,$$

while, bearing in mind the first of (2.3),

$$-\operatorname{div}\vec{u}\big|_{\Gamma_2} = \frac{\partial\xi}{\partial t}\Big|_{\Gamma_2} = \frac{\partial q}{\partial t}\Big|_{\Gamma_2} .$$

Hence,

$$\frac{\partial u_\nu}{\partial\nu}\Big|_{\Gamma_2} = -u_\nu \operatorname{div}\vec{\nu} - \frac{\partial q}{\partial t}\Big|_{\Gamma_2} .$$

We have, on the other hand,

$$\frac{\partial\vec{u}}{\partial\nu}\cdot\vec{\psi}\Big|_{\Gamma_2} = \left(\frac{\partial\vec{u}}{\partial\nu}\cdot\vec{\nu}\right)\psi_\nu\Big|_{\Gamma_2} =$$

$$= \left(\frac{\partial u_\nu}{\partial\nu}\vec{\nu} + u_\nu\frac{\partial\vec{\nu}}{\partial\nu} + \frac{\partial u_\tau}{\partial\nu}\vec{\tau} + u_\tau\frac{\partial\vec{\tau}}{\partial\nu}\right)\cdot\vec{\nu}\,\psi_\nu\Big|_{\Gamma_2} =$$

$$= \frac{\partial u_\nu}{\partial\nu}\psi_\nu + \frac{1}{2}\frac{\partial}{\partial\nu}|\vec{\nu}|^2 u_\nu\psi_\nu\Big|_{\Gamma_2} = \frac{\partial u_\nu}{\partial\nu}\psi_\nu\Big|_{\Gamma_2}$$

and, consequently,

$$\int_{\Gamma_2}\frac{\partial\vec{u}}{\partial\nu}\cdot\vec{\psi}\,d\Gamma_2 = \int_{\Gamma_2}\frac{\partial u_\nu}{\partial\nu}\psi_\nu\,d\Gamma_2 = -\int_{\Gamma_2}\left(u_\nu\operatorname{div}\vec{\nu} + \frac{\partial q}{\partial t}\right)\psi_\nu\,d\Gamma_2 .$$

Substituting the above relation into the expression of B, we obtain immediately (4.3).

Condition γ): We choose $\vec{\psi}(x,y,t)=0$ when $(x,y)\in\Gamma_1$, $\psi_\nu(x,y,t)=0$ when $(x,y)\in\Gamma_2$; we obtain then directly

$$B(T,\vec{u},\xi,\vec{\psi}) = 0 . \tag{4.4}$$

5 - Analysis of the Liggett model. According to what was stated in the preceding §§ , the Liggett model is associated to the relations

$$\int_0^T (\xi' + \operatorname{div}\vec{u}, \varphi)_{L^2}\,dt = 0 \tag{5.1}$$

$$\begin{aligned}\int_0^T \Big\{ &-(\vec{u},\vec{\psi}')_{L^2} + \mu(\vec{u},\vec{\psi})_{H_0^1} - b\left(\vec{u},\vec{\psi},\frac{\vec{u}}{\xi}\right) + (g\,\xi\operatorname{grad}\xi,\vec{\psi})_{L^2} - \\ &-(\vec{F},\vec{\psi})_{L^2}\Big\}\,dt + B(T,\vec{u},\xi,\vec{\psi}) + \\ &+(\vec{u}(T),\vec{\psi}(T))_{L^2} - (\bar{\vec{u}},\vec{\psi}(0))_{L^2}\end{aligned} \tag{5.2}$$

$$\left|\frac{\partial\xi}{\partial t}\right| < M_1 ,\ |\operatorname{grad}\xi| < M_2 ,\ \xi > \sigma ,\ |\vec{u}| \le M_3 \text{ a.e. in } \Omega\times[0,T], \tag{5.3}$$

where $\xi(x,y,0)=\bar{\xi}(x,y)$ $(x,y)\in\Omega$, $B(T,\vec{u},\xi,\vec{\psi})$ is given respectively by (4.2), (4.3), (4.4) according to the different boundary conditions considered and ξ, $\vec{u}$, φ, $\vec{\psi}$ belong to appropriate functional spaces.

Observe now that it is possible to substitute, in system (5.I), (5.2), (5.3), variational inequalities in place of the first two equations. Let, in fact, K_1, K_2, K_3 be three closed convex sets defined by

$$K_1=\{\xi(t)\in L^2(0,T;L^2):|\frac{\partial\xi}{\partial t}|\leq M_1,|\mathrm{grad}\,\xi|\leq M_2,\ \xi\geq\sigma \text{ a.e. in } \Omega\times[0,T]\}$$

$$K_2=\{\vec{u}(t)\in L^2(0,T;L^2):|\vec{u}|\leq M_3' \text{ a.e. in } \Omega\times[0,T]\}$$

$$K_3=\{\vec{u}(t)\in L^2(0,T;L^2):|\vec{u}|\leq M_3' \text{ a.e. in } \Gamma_2\times[0,T]\},$$

where $M_3'>M_3$ and M_i, σ are defined by (2.I), (2.2); assume, moreover, that $\{\xi,\vec{u}\}$ satisfy the following conditions:

i_α) $\xi(t)\in H^1(0,T;L^2)\cap K_1=K_1$, $\xi(0)=\bar{\xi}$;

ii_α) $\vec{u}(t)\in L^2(0,T;H_0^1)\cap C^0(0,T;L^2)\cap K_2$;

iii_α) $\{\xi,\vec{u}\}$ are solutions of the inequalities

$$\int_0^T(\xi'+\mathrm{div}\,\vec{u},\xi-\varphi)_{L^2}\,dt\leq 0$$

$$\int_0^T\{(\vec{\psi}',\vec{u}-\vec{\psi})_{L^2}+\mu(\vec{u},\vec{u}-\vec{\psi})_{H_0^1}-b(\vec{u},\vec{u}-\vec{\psi},\frac{\vec{u}}{\xi})+$$
$$+(g\xi\,\mathrm{grad}\,\xi,\vec{u}-\vec{\psi})_{L^2}-(\vec{F},\vec{u}-\vec{\psi})_{L^2}\}\,dt+$$
$$+\frac{1}{2}\|\vec{u}(T)-\vec{\psi}(T)\|_{L^2}^2-\frac{1}{2}\|\vec{\bar{u}}-\vec{\psi}(0)\|_{L^2}^2\leq 0$$

$\forall\ \varphi(t)\in K_1$, $\vec{\psi}(t)\in L^2(0,T;H_0^1)\cap H^1(0,T;L^2)\cap K_2$.

The relationship between such functions and the eventual solutions of (5.I), (5.2), (5.3), (4.2) is given by the following theorem (for the proof of which, see [2], [3]).

Theorem 5.I: *If* $\{\xi,\vec{u}\}$ *satisfy* i_α), ii_α), iii_α) *and* (5.3), *then they are also solutions of* (5.I), (5.2), (4.2).

The converse is also easily shown to be true: if $\{\xi,\vec{u}\}$ satisfy (5.I), (5.2), (5.3), (4.2), then they will also satisfy i_α), ii_α), iii_α).

Assume now that $\{\xi,\vec{u}\}$ satisfy i_α), ii_α), iii_α) and that

$$(5.4)\qquad |\frac{\partial\xi}{\partial t}|<M_1,\quad |\mathrm{grad}\,\xi|<M_2,\quad \xi>\sigma,\quad |\vec{u}|\leq M_3$$

a.e. in $\Omega\times[0,t']$, with

$$t'=\sup\{t:0\leq t\leq T,|\frac{\partial\xi}{\partial t}|<M_1,|\mathrm{grad}\,\xi|<M_2,\xi>\sigma,|\vec{u}|\leq M_3 \text{ a.e. in } \Omega\times[0,t]\}.$$

Then, by theorem 5.I, $\{\xi,\vec{u}\}$ satisfies (5.I), (5.2), (5.3), (4.2) only when $t\leq t'$, while, in general, if $t'<T$, they differ from the solutions of this system when $t>t'$. Observe, however, that, in this case, the Liggett model is no longer applicable for $t > t'$, since some of the assumptions made when deducing it from the Navier-Stokes equations are not verified.

Hence, i_α), ii_α), iii_α) correspond to problem (3.I)-α for the Liggett model wherever this model is physically consistent; the functions $\{\xi, \vec{u}\}$ satisfying i_α), ii_α), iii_α) will therefore be considered solutions of problem (3.I)-α associated to the Liggett model.

Following the same procedure, we shall say that $\{\xi, \vec{u}\}$ are solutions of problem (3.I)-β (resp. (3.I)-γ) associated to the Liggett model if conditions i_β), ii_β), iii_β) (resp. i_γ), ii_γ), iii_γ)) given below are satisfied:

i_β) $\xi(t) \in H^1(0,T;L^2) \cap K_1 = K_1,\ \xi(0) = \bar{\xi}$;

ii_β) $\vec{u}(t) \in L^2(0,T;H^1) \cap C^0(0,T;L^2) \cap K_2 \cap K_3$, $\vec{u}=0$ on Γ_1, $u_\tau = 0$ on Γ_2;

iii_β) $\{\xi, \vec{u}\}$ are solutions of the inequalities

$$\int_0^T (\xi' + \operatorname{div}\vec{u}, \xi - \varphi)_{L^2}\, dt \le 0$$

$$\int_0^T \{(\vec{\psi}', \vec{u} - \vec{\psi})_{L^2} + \mu(\vec{u}, \vec{u} - \vec{\psi})_{H_0^1} - b(\vec{u}, \vec{u} - \vec{\psi}, \frac{\vec{u}}{\xi}) +$$

$$+ (g\xi \operatorname{grad}\xi, \vec{u} - \vec{\psi})_{L^2} - (\vec{F}, \vec{u} - \vec{\psi})_{L^2}\}\, dt +$$

$$+ \int_0^T \int_{\Gamma_2} \{[\mu(u_\nu \operatorname{div}\vec{\nu} + \frac{\partial q}{\partial t}) + \frac{u_\nu^2}{q}](u_\nu - \psi_\nu)\, d\Gamma_2\, dt +$$

$$+ \frac{1}{2}\|\vec{u}(T) - \vec{\psi}(T)\|_{L^2}^2 - \frac{1}{2}\|\vec{\bar{u}} - \vec{\psi}(0)\|_{L^2}^2 \le 0$$

$\forall\ \varphi(t) \in K_1$, $\vec{\psi}(t) \in L^2(0,T;H^1) \cap H^1(0,T;L^2) \cap K_2 \cap K_3$, with $\vec{\psi}=0$ on Γ_1, $\psi_\tau = 0$ on Γ_2.

i_γ) $\xi(t) \in H^1(0,T;L^2) \cap K_1 = K_1,\ \xi(0) = \bar{\xi}$;

ii_γ) $\vec{u}(t) \in L^2(0,T;H^1) \cap C^0(0,T;L^2) \cap K_2$, $\vec{u}=0$ on Γ_1, $u_\nu = 0$ on Γ_2 ;

iii_γ) $\{\xi, \vec{u}\}$ are solutions of the inequalities

$$\int_0^T (\xi' + \operatorname{div}\vec{u}, \xi - \varphi)_{L^2}\, dt \le 0$$

$$\int_0^T \{(\vec{\psi}', \vec{u} - \vec{\psi})_{L^2} + \mu(\vec{u}, \vec{u} - \vec{\psi})_{H_0^1} - b(\vec{u}, \vec{u} - \vec{\psi}, \frac{\vec{u}}{\xi}) +$$

$$+ (g\xi \operatorname{grad}\xi, \vec{u} - \vec{\psi})_{L^2} - (\vec{F}, \vec{u} - \vec{\psi})_{L^2}\}\, dt +$$

$$+ \frac{1}{2}\|\vec{u}(T) - \vec{\psi}(T)\|_{L^2}^2 - \frac{1}{2}\|\vec{\bar{u}} - \vec{\psi}(0)\|_{L^2}^2 \le 0$$

$\forall\ \varphi(t) \in K_1$, $\vec{\psi}(t) \in L^2(0,T;H^1) \cap H^1(0,T;L^2) \cap K_2$, with $\vec{\psi}=0$ on Γ_1, $\psi_\nu = 0$ on Γ_2.

We now want to establish if problems (3.I)-α, (3.I)-β, (3.I)-γ are well posed for the Liggett model. Taking, at first, into consideration problem (3.I)-α , the following theorems can be proved (see [2], [3]).

Theorem 5.2: Assume that

$$p(t)\in L^2(0,T;H^1)\ ,\quad \vec{\chi}_s(t)\in L^2(0,T;L^2)\ ,$$
$$\bar{\xi}\in L^2\ ,\quad |\mathrm{grad}\,\bar{\xi}|\leq M_2\ ,\quad \bar{\xi}\geq\sigma \qquad \text{a.e. in } \Omega\ ,$$
$$\vec{u}\in H^1_0\ ,\quad |\vec{u}|\leq M_3' \qquad \text{a.e. in } \Omega\ .$$

There exists then $\{\xi,\vec{u}\}$ satisfying i_α), ii_α), iii_α).

Theorem 5.3: Assume that $\{\xi^{(1)},\vec{u}^{(1)}\}$, $\{\xi^{(2)},\vec{u}^{(2)}\}$ satisfy i_α), ii_α), iii_α) and (5.4) a.e. in $\Omega\times[0,t']$. Then $\xi^{(1)}=\xi^{(2)}$, $\vec{u}^{(1)}=\vec{u}^{(2)}$. Moreover, the unique solution depends continuously on the data.

We can therefore conclude that problem (3.I)-α is well posed for the Liggett model.

Analogous results hold for the other two problems considered.

Observation. In a similar way other hydrodynamical models (Welander, Saint-Venant) can be studied. The results obtained are comparable to the ones given above.

6 - Final remarks. Let us make a few general remarks on the results obtained and the methods used, since these can be applied to many other problems which arise in Mathematical Physics.

In general, a "mathematical model" of a physical problem consists of:

(a) A system of partial differential equations;

(b) Initial and boundary value conditions;

(c) Consistency conditions (under which equations (a) have been deduced).

In the study of a mathematical model, (c) is generally overlooked and the model itself is associated only to (a), (b); consequently, the eventual solutions of (a), (b) may not have any physical significance. The fact, in some cases, it may not be possible to prove that problem (a), (b) is well posed, even if the original physical problem is well posed, should not therefore be too unexpected.

Associating, as we have done, to the mathematical model all three conditions (a), (b), (c), it can be seen, in first place, that (a) can be substituted by

(a') A system of partial differential inequalities.

If then, considering (a'), (b), we can prove that such a problem is well posed (which can be very often accomplished more easily than for (a), (b)), it is possible to conclude that the mathematical model is well posed, provided it is physically consistent. This is the best result we can hope to obtain since, as already observed at §2, there appears to be no possibility of determining this consistency "a priori".

References

[1] M. Verri: Analisi teorica di modelli matematici di idrodinamica relativi a canali aperti e bacini. To appear on Pubbl. IAC, Roma.

[2] F. Rolandi and M. Verri: Su un modello idrodinamico bidimensionale di tipo Navier-Stokes. To appear on Pubbl. IAC, Roma.

[3] F. Rolandi and M. Verri: On the variational inequalitie associated to a hydrodynamical model. To appear.

ON THE FINITE ELEMENT APPROXIMATION OF THE NONSTATIONARY

NAVIER-STOKES PROBLEM

Rolf Rannacher
Institut für Angewandte Mathematik
der Universität Bonn, Beringstr. 4-6
5300 Bonn 1, GERMANY

In this note we report some basic convergence results for the semi-discrete finite element Galerkin approximation of the nonstationary Navier-Stokes problem. Asymptotic error estimates are established for a wide class of so-called conforming and nonconforming elements as described in the literature for modelling incompressible flows. Since the proofs are lengthy and very technical the present contribution concentrates on a precise statement of the results and only gives some of the key ideas of the argument for proving them. Complete proofs for the case of conforming finite elements may be found in a joint paper of J. Heywood, R. Rautmann and the author [5], whereas the nonconforming case will be treated in detail elsewhere.

1. The Navier-Stokes problem

We consider the nonstationary Navier-Stokes problem

$$(1)\qquad \left.\begin{aligned} u_t - \nu\Delta u + u\cdot\nabla u - \nabla p &= f \\ \nabla\cdot u &= o \end{aligned}\right\} \quad \text{in} \quad \Omega \times (o,\infty)$$
$$u_{|\partial\Omega} = o\ , \quad u_{|t=o} = a\ ,$$

where $\Omega \subset R^n$, $n = 2,3$, is a bounded domain, $u = u(x,t)$ is the velocity field in R^n and $p = p(x,t)$ the corresponding pressure function. For simplicity we assume homogeneous boundary data and the domain Ω to be convex polyhedral.

Throughout the paper $L^p(\Omega)^n$ denotes the Lebesgue space of n-vector functions with components being to the p-th power integrabel over Ω. $H^m(\Omega)^n$ is the m-th order Sobolev space of L^2-functions having generalized derivatives up to order m in $L^2(\Omega)$. The corresponding norms are

$$\|u\|_{L^p} = \Big(\int_\Omega |u|^p\,dx\Big)^{1/p}\ , \quad \|u\|_{H^m} = \Big(\sum_{k=o}^{m} \|\nabla^k u\|^2_{L^2}\Big)^{1/2}\ ,$$

$$\|u\|_{L^\infty} = \operatorname*{ess\,sup}_{x\in\Omega} |u(x)|\ ,$$

where $\nabla^k u$ is the tensor of all k-th order derivatives of u. In the case $p = 2$ we set for convenience

$$(u,v) = \int_\Omega u\cdot v \, dx \ , \quad \|u\| = \|u\|_{L^2} = (u,u)^{1/2} \ .$$

$H_o^1(\Omega)^n$ denotes the closure of the space $C_o^\infty(\Omega)^n$ of C^∞-vector functions having compact support in Ω and J_o is the subspace of all solenoidal functions in $H_o^1(\Omega)^n$:

$$J_o = \{v \in H_o^1(\Omega)^n : \quad \nabla\cdot v = o \ \text{ a.e. in } \ \Omega\} \ .$$

For time dependent functions into some Banach space X we use the notation

$$L^p(o,T;X) = \{u = u(t) : (o,T) \to X \ \text{ measurable} : \int_o^T \|u(\tau)\|_X^p \, d\tau < \infty\}$$

with the usual modification for $p = \infty$.
Finally we introduce the bilinear and trilinear forms, respectively,

$$a(u,v) = \nu \int_\Omega \nabla u\cdot \nabla v \, dx \ , \quad b(u,v,w) = \int_\Omega u\cdot\nabla v\cdot w \, dx \ ,$$

where the dot "$\cdot$" denotes the usual tensor multiplication.

Using these notation the weak formulation of problem (1) is as follows

(2) Find some $u = u(t) \in J_o$ such that $u(o) = a$ and

$$(u_t,\phi) + a(u,\phi) + b(u,u,\phi) = (f,\phi) \ , \quad \forall\phi \in J_o \ .$$

It is well known (see Ladyzhenskaya [5] and Heywood [3]) that under the assumptions

$$a \in J_o \ , \quad f \in L^2(o,\infty;L^2(\Omega)^n)$$

there is an unique solution $u \in L^\infty(o,T;J_o)$ of (2) on some time interval $[o,T)$ where $T > o$. Furthermore,

$$u \in L^2(o,T;H^2(\Omega)^n) \quad , \ u_t \in L^2(o,T;L^2(\Omega)^n)$$

and the corresponding pressure satisfying equation (1) is

$$p \in L^2(o,T;H^1(\Omega)).$$

Under the additional stronger conditions

$$a \in H^2(\Omega)^n \quad , \quad f_t \in L^2(o,\infty;L^2(\Omega)^n)$$

one even has

$$u \in L^\infty(o,T;H^2(\Omega)^n)\ ,\ u_t \in L^\infty(o,T;L^2(\Omega)^n) \cap L^2(o,T;H^1(\Omega)^n)$$
$$p \in L^\infty(o,T;H^1(\Omega)).$$

In the following we shall study the convergence behavior of the finite element Galerkin method for approximating the weak solution u of problem (1) under the above minimal assumptions.

2. Finite element Galerkin method

Let $\Pi_h = \{K\}$ be finite "triangulations" of the polyhedral domain Ω which satisfy the usual regularity conditions (see [1] and [2]) for mesh size h tending to zero, namely that each $K \in \Pi_h$ contains a n-ball of radius κh and is contained in a n-ball of radius $\kappa^{-1}h$.

We consider finite element spaces $J_{o,h}$ consisting of piecewise polynomial functions which are proper approximations of the basic space J_o in the following sense:

(3) *Each function* $v_h \in J_{o,h}$ *satisfies*

(i) $\displaystyle\int_{\partial K \cap \partial K'} \{v_{h|K} - v_{h|K'}\}\, ds = o, \quad \forall\, K,K' \in \Pi_h$,

(ii) $\displaystyle\int_{\partial K \cap \partial\Omega} v_{h|K}\, ds = o, \quad \forall\, K \in \Pi_h$,

(iii) $\displaystyle\int_K \nabla\cdot v_h\, dx = o, \quad \forall\, K \in \Pi_h$.

Furthermore, there are operators $r_h : \{J_o \cap H^2(\Omega)^n\} \oplus J_{o,h} \to J_{o,h}$ *such that for* $1 \le p \le \infty$

(iv) $r_h v_h = v_h \quad , \quad \forall v_h \in J_{o,h}$,

(v) $\| v - r_h v\|_{L^p} \le c\, h^{2+n/p-n/2} \| v\|_{H^2} \quad , \quad \forall v \in J_o \cap H^2(\Omega)^n$.

By these conditions the spaces $J_{o,h}$ are approximations of J_o of order $m = 1$. They include a wide class of conforming and even nonconforming finite elements for modelling incompressible flows as studied for instance by Crouxeiz and Raviart [2] and Fortin [3]. By (3i, ii) it is guaranteed that the functions in $J_{o,h}$ are at least approximately H^1_o-functions, i.e. their jumps along the element boundaries ∂K and their boundary values on $\partial\Omega$ are in some sense small. Condition (3iii) means that even the divergence of functions in $J_{o,h}$ is in some sense small. Hence the spaces $J_{o,h}$ are approximately admissible with respect to the space J_o. The conditions (3iv,v) ensure that each function in J_o can be approximated arbitrarily close by a sequence of functions in $J_{o,h}$ for h tending to zero.

As examples of elements satisfying all the conditions (3i-v) we mention the <u>nonconforming linear element</u> in two or even three dimensions with the corresponding nodal values:

values $v_h(b_i)$ at the centers b_i of (n-1)-faces of all $K \in \Pi_h$

and the <u>conforming quadratic element</u> in two dimensions with the corresponding nodal values:

values $v_h(a_i)$ at the vertices a_i and line integrals $\int_\Gamma v_h \, ds$ over the sides of all triangles $K \in \Pi_h$.

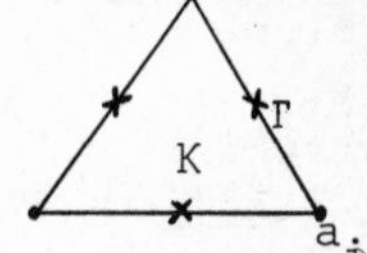

Using a so-called "bulb-function" the order of the quadratic element can be raised to m = 2 without increasing the dimension of the space $J_{o,h}$. In all these cases the spaces $J_{o,h}$ are of the type

$$J_{o,h} = \{ \phi \in L^2(\Omega)^n : \text{i) } \phi|_K \in P_m \text{ \underline{for some} } m \geq 1, \; K \in \Pi_h,$$

ii) ϕ <u>is continuous with respect to the prescribed nodal values</u>;

iii) ϕ <u>has vanishing nodal values along the boundary</u> $\partial\Omega$;

iv) $\int_K \nabla \cdot \phi \, dx = o \, , \quad \forall K \in \Pi_h \}$.

As operator r_h one can choose the usual interpolation operator with respect to the prescribed nodal values which maps the space $J_o \cap H^2(\Omega)^n$ into $J_{o,h}$ and leaves the spaces $J_{o,h}$ invariant by its special construction. For a detailed description of these finite element spaces and for further examples of even higher order m > 1 we refer to the literature [2] , [3] and [8].

For functions $\phi \in J_o \oplus J_{o,h}$ the discrete gradient $\nabla_h \phi$ is defined piecewise with respect to the patchs $K \in \Pi_h$. In this sense we can define the following bilinear and trilinear forms, respectively, on the direct sum $J_o \oplus J_{o,h}$ of the spaces J_o and $J_{o,h}$:

$$a_h(v,w) = \nu(\nabla_h v, \nabla_h w) = \sum_{K \in \Pi_h} \int_K \nabla v \cdot \nabla w \, dx$$

$$b_h(u,v,w) = \frac{1}{2} \sum_{K \in \Pi_h} \int_K \{u \cdot \nabla v \cdot w - u \cdot \nabla w \cdot v\} \, dx .$$

Obviously the forms b_h are compatible with b in the sence

$$b_h(u,v,w) = b(u,v,w) \quad , \quad u,v,w \in J_o \; ,$$

and they even satisfy

$$b_h(u,v,v) = o \; , \quad u,v \in J_o \oplus J_{o,h} \; .$$

Using these notations the semi-discrete analogues of problem (2) are

(4) <u>Find some</u> $u_h = u_h(t) \in J_{o,h}$ <u>such that</u> $u_h(o) = a_h$ <u>and</u>

$$(u_{ht}, \phi_h) + a_h(u_h, \phi_h) + b_h(u_h, u_h, \phi_h) = (f, \phi_h) \; , \quad \forall \phi_h \in J_{o,h} \; ,$$

where $a_h \in J_{o,h}$ are appropriate approximations to the initial data $a \in J_o$ satisfying uniformly for $h \to o$

$$\| a - a_h \| \le c\, h^k \, \| a \|_{H^k} \quad , \; k = 1,2 \; ,$$

provided that $a \in H^k(\Omega)^m$.

If $\{\phi_h^{(i)}, \, i = 1,\ldots,N_h = \dim(J_{o,h})\}$ is a basis of $J_{o,h}$, then problem (4) is equivalent to a system of first-order ordinary differential equations for the coefficient functions $\xi_i(t)$ in the representation

$$u_h(t) = \sum_{i=1}^{N_h} \xi_i(t) \phi_h^{(i)} \quad .$$

For some of the spaces satisfying the conditions (3i-v) quasi-local bases are known at least in two dimensions (see [3]). This problem which is crucial for the numerical realization of the scheme (4) will be discussed in a subsequent paper.

Setting $\phi_h = u_h(t)$ in (4), we obtain the discrete energy inequality

$$\frac{1}{2}\frac{d}{dt}\|u_h(t)\|^2+\|\nabla_h u_h(t)\|^2 \le \|f(t)\|\,\|u_h(t)\| \quad , \quad t \ge o,$$

and from that via Gronwall's inequality the bound

$$\|u_h(t)\|_{L^\infty} \le c(h)\|u_h(t)\| \le c(t) \quad , t \ge o \ .$$

This guarantees the existence of unique discrete solutions u_h of problem (4) which are in $L^\infty(o,T';J_{o,h})$ for all times $T' > o$.

For these approximate solutions u_h we have the following basic convergence results:

Theorem. Assume that the finite element spaces $J_{o,h}$ *are first-order approximations of the space* J_o *in the sense described above. Further assume that the data of problem* (1) *satisfy*

$$a \in J_o \ , \quad f \in L^2(o,\infty;L^2(\Omega)^n),$$

and let $u \in L^\infty(o,T;J_o)$ *and* $u_h \in L^\infty(o,T;J_{o,h})$ *for some* $T > o$ *be the corresponding unique solutions of problem* (2) *and* (4), *respectively. Then the error function* $e = u - u_h$ *satisfies the estimate*

$$(5) \qquad \|e(t)\| + \Big(\int_o^t \|\nabla_h e\|^2\, d\tau\Big)^{1/2} \le c(t)\, h \ , \ t \in [o,T),$$

and, if additionally

$$a \in H^2(\Omega)^n \quad , \quad f_t \in L^2(o,\infty;L^2(\Omega)^n) \ ,$$

even the pointwise estimate

$$(6) \qquad \|e(t)\|_{L^\infty} \le c(t) \begin{cases} h^{1/4} & \textit{for} \quad n = 3 \\ h^{1/2}|\ln h|^{1/2} & \textit{for} \quad n = 2 \end{cases} , \quad t \in [o,T) \ .$$

Moreover, if the solution u *also satisfies*

$$u_t \in L^2(o,T;H^2(\Omega)^n) \ , \quad p_t \in L^2(o,T;H^1(\Omega)) \ ,$$

then we have

$$(7) \qquad \|\nabla_h e(t)\| + \Big(\int_o^t \|e_t\|^2\, d\tau\Big)^{1/2} \le c(t)\, h \ , \quad t \in [o,T),$$

and finally the improved pointwise estimate

$$(8) \qquad \|e(t)\|_{L^\infty} \le c(t) \begin{cases} h^{1/2} & \textit{for } n = 3 \\ h|\ln h|^{1/2} & \textit{for } n = 2 \end{cases} , \quad t \in [o,T).$$

All the constants $c(t)$ *depend continuously on the specified data but are independent on the mesh size* h.

In order to illustrate the statements of the theorem we add the following remarks.

1. Note that the estimates (5) - (8) hold for any time $T > o$ such that the solution of problem (2) is known to be $u \in L^{\infty}(o,T;J_o)$, regardless where this information comes from. In the case $n = 2$ or, if the data of problem (1) are in a certain sense small enough, even in the case $n = 3$ one has $T = \infty$ and the constants $c(t)$ remain bounded for $t \to \infty$ (see [7] for the behaviour of the solution u of problem (1) and [5] for the behaviour of the discrete solutions u_h for time t tending to infinity).

2. The estimates (5) and (6) obviously hold under the above mentioned minimal assumptions on the data a and f. The sharper results (7) and (8) are a little problematic since the corresponding assumptions on u_t and p_t seem to be realistic only if certain global compatibility conditions for the initial data a are satisfied (see [3]). These unnatural strong assumptions on u_t, p_t are essentially forced by the allowed nonconformity of the spaces $J_{o,h}$. If the spaces $J_{o,h}$ are fully conforming, i.e. $J_{o,h} \subset J_o$, then the estimates (7), (8) even hold under the same natural assumptions as made for the estimate (6).(This corresponds with the estimates given by Rautmann [8] for the approximation of problem (1) by means of eigenfunctions of the Stokes operator.) It would be desirable to remove this weak point in our results.

3. For finite element spaces $J_{o,h}$ of higher order satisfying the conditions (3i-v) in a stronger sense corresponding higher order error estimates hold provided the solution u is regular enough (see[4]). But again this a priori assumption might be problematic unless certain compatibility conditions are satisfied by the initial data (see [3]).

4. Having already computed the approximation u_h to the velocity vector u one can generate approximations p_h to the corresponding pressure p by a suitable Galerkin Ansatz. Since this procedure is just the same as for the simplest steady state Stokes problem we only refer to the literature [2], [9] for a further discussion of pressure computation.

5. All the above results remain valid if the usual techniques for

approximating curved boundary and nonhomogeneous boundary data are used. The additionally required technical argument is again just the same as for the well known steady state Stokes problem or even as for any of the usual elliptic model problems (see [1]).

6. For time discretization several of the methods known for parabolic problems may also be used for the Navier-Stokes problem (see [9]). For instance the (nonlinear) Crank-Nicolson scheme is unconditionally stable and of second order convergent. It has the form

$$\frac{1}{\Delta t}(U_h^k - U_h^{k-1}, \phi_h) + a_h(\tilde{U}_h^k, \phi_h) + b_h(\tilde{U}_h^k, \tilde{U}_h^k, \phi_h) = (f^k, \phi_h) \;, \quad \forall \phi_h \in J_{o,h} \;,$$

where U_h^k is the discrete solution for the time level $k\cdot\Delta t$ and

$$\tilde{U}_h^k = \tfrac{1}{2}(U_h^k + U_h^{k-1}) \;, \quad U_h^o = a_h \;, \quad f^k = \frac{1}{\Delta t}\int_{(k-1)\cdot\Delta t}^{k\cdot\Delta t} f(\tau)\,d\tau \;.$$

The error estimate

$$\|u_h(k\cdot\Delta t) - U_h^k\| \le c(t)\,\Delta t^2 \;, \quad t = k\cdot\Delta t \in [o,T) \;,$$

holds provided that $u_{htt} \in L^2(o,T;L^2(\Omega)^n)$. This may be shown by standard techniques for time discretization of parabolic problems. The linearized C.-N. scheme using the term $b_h(U_h^{k-1}, \tilde{U}_h^k, \phi_h)$ instead of $b_h(\tilde{U}_h^k, \tilde{U}_h^k, \phi_h)$ is also stable (see [9]) but only of first order convergent.

3. Proof of the theorem

In the following we present the key ideas of the argument for proving the error estimates (5) - (8). Most of the technical complications arise from the allowed nonconformity of the spaces $J_{o,h}$ with respect to the divergence condition "$\nabla\cdot v = o$" as well as with respect to the continuity requirement "$v \in H_o^1(\Omega)^n$". Since the techniques for overcoming these problems are essentially the same as for the steady state case (see [2] and [6]) we mainly concentrate on the argument concerning the nonstationarity of the problem.

A) Technical preliminaries

At first we provide some more technical tools which will be used below. Let the L^2-projections $L_h : L^2(\Omega)^n \to J_{o,h}$ be defined by

$$(v-L_h v,\phi_h) = o , \quad \forall\phi_h \in J_{o,h} ; \; v \in L^2(\Omega)^n ,$$

and the (generalized) Stokes projections $S_h : J_o \oplus J_{o,h} \to J_{o,h}$ by

$$a_h(v-S_h v,\phi_h) = o , \quad \forall\phi_h \in J_{o,h} ; \; v \in J_o \oplus J_{o,h} ,$$

where again $J_o \oplus J_{o,h}$ denotes the direct sum of the vector spaces J_o and $J_{o,h}$. Both of the projection operators commute with time differentiation:

$$L_h v_t = (L_h v)_t \quad , \quad S_h v_t = (S_h v)_t .$$

Furthermore, under our assumptions on the spaces $J_{o,h}$, one may prove the following estimates (see for instance [2] and [5])(+)

$$\text{(9)} \quad \begin{array}{l} \| L_h v\| + \| S_h v\| \le c\|v\| + ch\| \nabla_h v\| \\ \| \nabla_h L_h v\| + \| \nabla_h S_h v\| \le c \|\nabla_h v\| \end{array} , \quad v \in J_o \oplus J_{o,h} ,$$

and

$$\text{(1o)} \quad \begin{array}{l} \| v - L_h v\| + h\|\nabla_h (v-L_h v)\| \le c h^2 \|v\|_{H^2} \\ \| v - S_h v\| + h\|\nabla_h (v-S_h v)\| \le c h^2 \|v\|_{H^2} \end{array} , \; v \in J_o \cap H^2(\Omega)^n.$$

Below we shall frequently use the Sobolev inequality for $n \le 3$

$$\|v\|_{L^6} \le c \|v\|_{H^1} \quad , \quad v \in H^1(\Omega)^n ,$$

the Poincare inequality

$$\| v\|_{H^1} \le c\| \nabla v\| \quad , \quad v \in H_o^1(\Omega)^n ,$$

and the a priori estimate

$$\| v\|_{H^2} \le c \|\Delta v\| \quad , \; v \in H_o^1(\Omega)^n \cap H^2(\Omega)^n ,$$

where the latter holds on any bounded convex domain. This leads together with the Hölder inequality

$$\| v\|_{L^3} \le \| v \|^{1/2} \| v\|_{L^6}^{1/2}$$

to the estimates

$$\| v\|_{L^3} \le c \|\nabla v\|^{1/2} \|\Delta v\|^{1/2} , \; v \in H_o^1(\Omega)^n \cap H^2(\Omega)^n ,$$

$$\| v\|_{L^3} \le c\| v\|^{1/2} \| \nabla v\|^{1/2} , \; v \in H_o^1(\Omega)^n ,$$

which are useful for handling the nonlinear term $u \cdot \nabla u$ in equation (1).

(+) "c" always denotes a generic constant which may change with the context

Similar inequalities as stated above even hold for the functions in the discrete spaces $J_{o,h}$, namely for $n \leq 3$:

$$(11) \qquad \|v_h\|_{L^6} \leq c\|\nabla_h v_h\| \quad , \quad v_h \in J_{o,h} ,$$

and

$$(12) \qquad \|\nabla_h v\|_{L^6} \leq c\|\Delta_h v_h\| \quad , \quad v_h \in J_{o,h} .$$

Here Δ_h denotes a discrete analogue of the Laplacian Δ which is defined by means of the eigenvalues $\{\lambda_h^{(i)}\}$ and the corresponding eigenvector systems $\{w_h^{(i)}\} \subset J_{o,h}$, $i = 1,\ldots,N_h = \dim(J_{o,h})$, of the discrete Stokes operator:

$$a_h(w_h^{(i)}, \phi_h) = \lambda_h^{(i)}(w_h^{(i)},\phi_h) \quad , \quad \forall\phi_h \in J_{o,h} .$$

Using this notation we set for $v_h \in J_{o,h}$

$$(13) \qquad \Delta_h v_h := \sum_{i=1}^{N_h} \lambda_h^{(i)}(v_h,w_h^{(i)})w_h^{(i)} .$$

By definition we have the discrete Green's formula

$$(14) \qquad (\nabla_h v_h,\nabla_h w_h) = -(\Delta_h v_h,w_h) \quad , \quad v_h,w_h \in J_{o,h} .$$

The estimates (11) and (12) will be proved for the conforming case, $J_{o,h} \subset H_o^1(\Omega)^n$, in [4] and for the more complicated nonconforming case in a forthcoming paper of the author (see also some similar estimates derived in [6] and [9]). A sketch of the proof of the presumably most surprising estimate (12) will be given below in step (C).

B) Proof of the estimates (5) and (7)

Because of their nonconformity the functions $\phi_h \in J_{o,h}$ may not be used directly as test functions in the weak formulation (2) of problem (1). To overcome this complication we note that under our assumptions equation (1) even holds in the strong sense

$$u_t - \nu\Delta u + u\cdot\nabla u + \nabla p = f \quad \text{a.e. in } \Omega \times (o,T).$$

Then multiplying this identity with $\phi_h \in J_{o,h}$ and integrating then by parts leads to

$$(u_t,\phi_h) + a_h(u,\phi_h) + b_h(u,u,\phi_h) = (f,\phi_h) + (p,\nabla_h\cdot\phi_n) + r_h(u,u,\phi_h) ,$$

where

$$r_h(u,u,\phi_h) = \sum_{K \in \Pi_h} \int_{\partial K} \{\nu u_n + \tfrac{1}{2}(u\cdot n)u - pn\} \cdot \phi_h \, ds$$

(n = outer normal unit vector to ∂K, $u_n = \frac{\partial u}{\partial n}$ normal derivative).

Combining this with the corresponding relation (4) for the discrete solutions u_h, we get for the error $e = u-u_h$ the identity

$$(15) \quad (e_t,\phi_h) + a_h(e,\phi_h) = b_h(u_h,u_h,\phi_h) - b_h(u,u,\phi_h) + (p,\nabla_h\cdot\phi_h) + r_h(u,u,\phi_h) \ , \ \forall\ \phi_h \in J_{o,h} \ .$$

The two terms on the right hand side coming from the nonconformity of $J_{o,h}$ will be estimated as follows:

By assumption (3iii) we have

$$(p,\nabla_h\cdot\phi_h) = (p-q_h,\nabla_h\cdot\phi_h) \le \|p-q_h\| \ \|\nabla_h\cdot\phi_h\| \quad ,$$

where q_h is any piecewise constant approximation to the pressure p, and hence

$$(16) \qquad (p,\nabla_h\cdot\phi_h) \le c\ h\ \|\nabla_h\phi_h\| \ \|\nabla p\| \quad .$$

The boundary term may be rewritten as

$$r_h(u,u,\phi_h) = \sum_\Gamma \int_\Gamma \{\nu u_n + \tfrac{1}{2}(u\cdot n)u - pn\} \cdot [\phi_h] ds \quad ,$$

where the summation is taken over all $(n-1)$-faces of the $K \in \Pi_h$ and $[\phi_h]$ is the jump of ϕ_h along such a face Γ . By assumption (3i,ii) these jumps $[\phi_h]$ have vanishing mean value on Γ and hence allow us to insert appropriate mean values ω_Γ of the sum set in brackets:

$$r_h(u,u,\phi_h) = \sum_\Gamma \int_\Gamma \{\nu u_n + \tfrac{1}{2}(u\cdot n)u - pn - \omega_\Gamma\} \cdot [\phi_h] ds \quad .$$

Applying a Poincare type inequality to the term in brackets as well as to the jumps $[\phi_h]$ we conclude

$$r_h(u,u,\phi_h) \le c \sum_{K \in \Pi_h} h^{1/2} \ \|\nabla_h\phi_h\| \ h^{1/2}(\|\nabla^2 u\| + \|\nabla u^2\| + \|\nabla p\|) .$$

Finally the estimates provided in step (A) lead us to

$$(17) \quad r_h(u,u,\phi_h) \le c\ h\ \|\nabla_h\phi_h\| \ (\ \|\Delta u\| + \|\nabla u\|^3 + \|\nabla p\|) .$$

Now in order to prove the estimate (5) we insert $\phi_h = L_h e \in J_{o,h}$ into the identity (12) and get by a simple rearrangement of terms

$$\frac{1}{2}\frac{d}{dt}\|e\|^2 + \nu\|\nabla_h e\|^2 = (e_t, u-L_h u) + a_h(e, u-L_h u) + b_h(u_h, u_h, L_h e) -$$

$$- b_h(u,u,L_h e) + (p, \nabla_h \cdot L_h e) + \Gamma_h(u,u,L_h e)$$

and furthermore

$$b_h(u_h,u_h,L_h e) - b_h(u,u,L_h e) = b_h(e,e,L_h e) - b_h(e,u,L_h e) - b_h(u,e,L_h e).$$

Now applying the estimates preserved in part (A), one concludes by a somewhat lengthy but straightforward calculation that for any $\varepsilon \in (0,1]$.

$$(20) \qquad b_h(u_h,u_h,L_h e) - b_h(u,u,L_h e) \le \varepsilon\|\nabla_h e\|^2 + c\|\nabla u\|^4\|e\|^2 .$$

From the estimates (16) and (17) we have

$$(p, \nabla_h \cdot L_h e) + \Gamma_h(u,u,L_h e) \le \varepsilon\|\nabla_h e\|^2 + c\,h^2(\|\Delta u\|^2 + \|\nabla u\|^6 + \|\nabla p\|^2).$$

Choosing ε sufficiently small we arrive at

$$\frac{d}{dt}\|e\|^2 + \|\nabla_h e\|^2 \le c\,h^2\frac{d}{dt}\|\nabla u\|^2 + c\,h^2(\|\Delta u\|^2 + \|\nabla u\|^6 + \|\nabla p\|^2).$$

From that the estimate (5) immediately follows by applying Gronwall's inequality.

To prove the estimate (7) we insert $\phi_h = S_h e_t$ into the identity (15) and get again by a simple rearranging of terms

$$\|e_t\|^2 + \frac{1}{2}\frac{d}{dt}\|\nabla_h e\|^2 = (e_t, u_t - S_h u_t) + a_h(e, u_t - S_h u_t) + b_h(u_h, u_h, S_h e_t) -$$

$$- b_h(u,u,S_h e_t) + (p, \nabla_h \cdot S_h e_t) + \Gamma_h(u,u,S_h e_t) .$$

By the definition of the Stokes projection we have

$$(21) \qquad a_h(e, u_t - S_h u_t) = a_h(u - S_h u, u_t - S_h u_t) = \frac{1}{2}\frac{d}{dt}\|\nabla_h(u - S_h u)\|^2$$

$$= h^2\frac{d}{dt}[c_h\|\Delta u\|^2] , \quad c_h := \frac{\|\nabla_h(u-S_h u)\|^2}{h^2\|\Delta u\|^2} \le c.$$

Furthermore, with any $\varepsilon \in (0,1]$,

$$(22) \qquad (e_t, u_t - S_h u_t) \le \varepsilon\|e_t\|^2 + c\,h^2\|\nabla u_t\|^2 .$$

The terms representing the nonlinearity may be rearranged to

$$b_h(u_h,u_h,S_he_t)-b_h(u,u,S_he_t) = b_h(e,e,S_he_t)-b_h(e,u,S_he_t)-b_h(u,e,S_he_t)$$

$$= \frac{d}{dt}[b_h(e,e,S_he_t)-b_h(e,u,S_he_t)-b_h(u,e,S_he_t)] - b_h(e_t,e,S_he_t) -$$

$$- b_h(e,e_t,S_he_t)+b_h(e_t,u,S_he_t)+b_h(e,u_t,S_he_t)+b_h(u_t,e,S_he_t) +$$

$$+ b_h(u,e_t,S_he_t) .$$

Now again a rather lenghty calculation using the estimates provided in part (A) and in addition the already proved basic error estimate (5) leads us to ($\varepsilon \in (o,1]$)

$$(23) \quad b_h(u_h,u_h,S_he_t)-b_h(u,u,S_he_t) \leq \frac{d}{dt}[c_o(t)\|\nabla_h e\|^2] + \varepsilon\|e_t\|^2 + $$
$$+ c_\varepsilon(t)\|\nabla_h e\|^2 + c_\varepsilon(t)\,h^2 ,$$

where the constants $c_o(t)$ and $c_\varepsilon(t)$ depend on $\|\Delta u(t)\|$ and $\|\Delta u_t\|$, respectively, but are bounded with respect to h.

The two terms coming from the nonconformity of the spaces $J_{o,h}$ are estimated in a similar way using the estimates (16) and (17) as follows

$$(24) \quad (p,\nabla_h\cdot S_he_t) = \frac{d}{dt}(p,\nabla_h\cdot S_he_t) - (p_t,\nabla_h\cdot S_he)$$
$$\leq c\,\frac{d}{dt}(h^2\|\nabla p\|^2 + \|\nabla_h e\|^2) + c\,h^2\|\nabla p_t\|^2 + c\,\|\nabla_h e\|^2$$

and

$$(25) \quad r_h(u,u,S_he_t) = \frac{d}{dt}r_h(u,u,S_he) - r_h(u_t,u,S_he) - r_h(u,u_t,S_he)$$
$$\leq \frac{d}{dt}[c_o(t)h^2 + \|\nabla_h e\|^2] + c(t)h^2 + c\,\|\nabla_h e\|^2 ,$$

where again the constants $c_o(t)$ and $c(t)$ depend on the norms $\|\Delta u_t\|$ and $\|\nabla p_t\|$, respectively, but are bounded with respect to h.

Collecting all the estimates (21) - (25) we obtain for ε, sufficiently small,

$$\|e_t\|^2 + \frac{d}{dt}\|\nabla_h e\|^2 \leq \frac{d}{dt}[c_o(t)\|\nabla_h e\|^2 + c_1(t)h^2] + c_2(t)\|\nabla_h e\|^2 .$$

From that one gets the desired estimate (7) again by applying the Gronwall inequality.

C) Proof of the pointwise estimates (6) and (8)

At first we give a short proof of the estimate (12) for the discrete Laplacian Δ_h.

To any given $v_h \in J_{o,h}$, which clearly has a representation of the form

$$v_h = \sum_{i=1}^{N_h} \alpha_i w_h^{(i)} \quad , \quad \alpha_i \in R ,$$

we attach a function $v \in J_o$ and a corresponding pressure q by solving the Stokes problem

$$(26) \qquad \left.\begin{aligned} -\nu\Delta v + \nabla q &= -\Delta_h v_h \\ v &= o \end{aligned}\right\} \quad \text{in } \Omega , \quad v = o \text{ on } \partial\Omega.$$

Then v_h turns out to be just the finite element approximation to v which is defined by the discretized Stokes problem

$$(27) \qquad a_h(v_h,\phi_h) = -(\Delta_h v_h,\phi_h) \quad , \quad \forall\, \phi_h \in J_{o,h} \quad .$$

For that there are the following error estimates available (see [2])

$$(28) \quad \| v-v_h \| + h\| \nabla_h(v-v_h)\| \leq ch(\| \Delta v\| + \|\nabla q\|) .$$

Furthermore, we have

$$\| \nabla_h v_h\|_{L^6} \leq \| \nabla_h(v_h-r_h v)\|_{L^6} + \|\nabla_h(r_h v-v)\|_{L^6} + \| \nabla v\|_{L^6} ,$$

and so-called "inverse" property of finite elements (see [1]) gives us

$$\begin{aligned} \|\nabla_h(v_h-r_h v)\|_{L^6} &\leq c\, h^{-1} \|v_h-r_h v \|_{L^6} \\ &\leq c\, h^{-1}\| v_h-v\|_{L^6} + c\, h^{-1}\|v-r_h v\|_{L^6} \quad . \end{aligned}$$

Applying Sobolev's inequality

$$\| \nabla v \|_{L^6} \leq c\| v\|_{H^2}$$

and the estimate in assumption (3v)

$$\| v-r_h v\|_{L^6} + h\| \nabla_h(v-r_h v)\|_{L^6} \leq c\, h \|v \|_{H^2} ,$$

we find

$$\| \nabla_h v_h\|_{L^6} \leq c \|v\|_{H^2} \quad .$$

Now the usual a priori estimate for the Stokes problem (see [7]) yields

$$\| v \|_{H^2} + \| \nabla q \| \leq c \| \Delta_h v_h \|$$

and hence the desired estimate

$$\| \nabla_h v_h \|_{L^6} \leq c \| \Delta_h v_h \| \quad , \qquad v_h \in J_{o,h} .$$

For proving the pointwise error estimates we shall use the following set of inequalities for functions $v_h \in J_{o,h}$

$$(29) \qquad \| v_h \|_{L^\infty} \leq \begin{cases} h^{1/4} \| \nabla_h v_h \|_{L^6} + c h^{-3/4} \| v_h \| & (n=3) \\ h^{1/2} |\ln h|^{1/2} \| \nabla_h v_h \|_{L^6} + c\, h^{-1/2} \| v_h \| & (n=2) \end{cases}$$

and

$$(3o) \qquad \| v_h \|_{L^\infty} \leq c \begin{cases} h^{-1/2} \| \nabla_h v_h \| & (n=3) \\ |\ln h|^{1/2} \| \nabla_h v_h \| & (n=2) \end{cases} .$$

Proofs of these estimates will be given for the conforming case in [5] and for the again more troublesome nonconforming case in a subsequent paper of the author already announced above.

Now we use the discrete Sobolev inequalities (11) and (12) from part (A) and conclude for the error $e = u - u_h$ in the case $n = 3$

$$\begin{aligned} \| e \|_{L^\infty} &\leq \| u - r_h u \|_{L^\infty} + \| r_h e \|_{L^\infty} + h^{1/4} \| \nabla_h r_h e \|_{L^6} + c\, h^{-3/4} \| r_h e \| \\ &\leq \| u - r_h u \|_{L^\infty} + h^{1/4} (\| \Delta u \| + \| \Delta_h u_h \|) + \\ &\qquad + c\, h^{-3/4} (\| u - r_h u \| + \| e \|) \end{aligned}$$

and even

$$\| e \|_{L^\infty} \leq \| u - r_h u \|_{L^\infty} + c\, h^{-1/2} (\| \nabla_h (u - r_h u) \| + \| \nabla_h e \|) .$$

Then assumption (3v), combined with the already proved error estimates (5) and (7), lead to

$$\| e \|_{L^\infty} \leq c\, h^{1/4} (\| \Delta u \| + \| \Delta_h u_h \|) + c(t) h^{1/4}$$

and to the desired estimate (8)

$$\| e \|_{L^\infty} \leq c(t) h^{1/2} .$$

In two dimensions, n = 2, one proceeds analoguously. So, in order to prove also the estimate (6), we have to bound the norm $\|\Delta_h u_h(t)\|$ for all times $t \in [o,T)$. This may be done in a similar way as for the norm $\|\Delta u(t)\|$ by inserting $\phi_h = -\Delta_h u_{ht}$ as test function into equation (4) (see [4] for the conforming case). This gives

$$\|\nabla_h u_{ht}\|^2 + \frac{1}{2}\frac{d}{dt}\|\Delta_h u_h\|^2 = \frac{d}{dt}\, b_h(u_h,u_h,\Delta_h u_h) + \frac{d}{dt}(f,\Delta_h u_h) -$$

$$- (f_t,\Delta_h u_h) - b_h(u_{ht},u_h,\Delta_h u_h) - b_h(u_h,u_{ht},\Delta_h u_h) .$$

By a somewhat complicated calculation the terms on the right hand side may be estimated in terms of $\|\nabla_h u_{ht}\|$ and $\|\Delta_h u_h\|$ such that again Gronwall's inequality yields the desired bound

$$\|\Delta_h u_h(t)\| \le c(t) \quad , t \in [o,T) .$$

References

[1] Ciarlet, P.G.: The Finite Element Method for Elliptic Problems. North-Holland: Amsterdam 1978.

[2] Crouxeiz, M.; Raviart, P.-A.: Conforming and nonconforming finite element methods for solving the stationary Stokes equation I. R.A.I.R.O. Anal. Numer. 3 (1973), 33-76.

[3] Fortin, M.: Résolution des équations des fluides incompressibles par la méthode des éléments finis. In Proceedings of the Third International Conference on the Numerical Methods in Fluid Mechanics, Springer: Berlin-Heidelberg-New York 1972.

[4] Heywood, J.G.: Classical solutions of the Navier-Stokes equations. In this Proceedings.

[5] Heywood, J.G.; Rannacher, R.; Rautmann, R.: Semidiscrete finite element Galerkin approximation of the nonstationary Navier-Stokes problem. To appear.

[6] Jamet, P.,; Raviart, P.-A.: Numerical solution of the stationary Navier-Stokes equation by finite element methods. In Computing Methods in Applied Sciences and Engineering, Part 1, Lecture Notes in Computer Sciences, Vol. 1o, Springer: Berlin-Heidelberg-New York 1974.

[7] Ladyžhenskaya, O.A.: The Mathematical Theory of Viscous Incompressible Flow. Gordon and Breach: New York 1962.

[8] Rautmann, R.: On the convergence-rate of nonstationary Navier-Stokes approximations. In this Proceedings.

[9] Temam, R.: Navier Stokes Equations. North-Holland: Amsterdam 1978.

Address: The University of Michigan
Department of Mathematics
Ann Arbor, Michigan 481o9 (USA)

Institut für Angewandte Mathematik
der Universität Bonn
Beringstraße 4-6
D-53oo Bonn 1 (West-Germany)

ON THE CONVERGENCE RATE OF NONSTATIONARY NAVIER-STOKES APPROXIMATIONS.

Reimund Rautmann

Fachbereich Mathematik-Informatik der
Gesamthochschule Paderborn
Warburger Straße 100
D 479o Paderborn
Germany

Recently in many papers the solution of Navier-Stokes boundary value problems has been investigated with the help of Galerkin approximations which are built upon the basis of the eigenfunctions of corresponding linear Stokes boundary value problems. Since the general properties of the Stokes boundary value problems and their eigenfunctions are relatively well known, one may ask, how "well" an instationary Navier-Stokes solution can be approximated on the basis of the Stokes eigenfunctions.

We will prove that the error bounds which hold for eigenfunction expansions in self-adjoint elliptic boundary value problems (and wich are reciprocal to the eigenvalues) can be carried over to the Galerkin approximations of instationary Navier-Stokes solutions. Since asymptotic estimates of the eigenvalues in general cases are known [15], asymptotic error estimates for the Galerkin approximations will result.

Generally spoken the error bounds for eigenfunction expansions depend essentially on L^2-bounds for the first and second partial derivatives of the function to be expanded. As John Heywood has shown in 1978, for a Navier-Stokes solution, its Galerkin approximations and its time derivatives of any order, norm bounds of this kind can be deduced very elegantly, if the Galerkin approximations are formed with the eigenfunctions of the corresponding Stokes boundary value problem. The error estimates, which are the aim of this paper, result then from a differential inequality for the respective norm of the difference of two Galerkin approximations and a subsequent limiting process. The result shows that for any of Heywood's *norm* estimates for Navier-Stokes solutions in L^2 and H_1 there is a corresponding *error* estimate of the same kind for the respective Galerkin approximations. As we will see the same holds true for the time derivatives of any order, too.

In section 1 we will state well known properties of the Stokes boundary value problem as far as we will use them. The error estimates for eigenfunction expansions, the starting point of our investigation, are formulated together with short proofs in section 2. Using the norm estimates from [7] and [16] cited in sections 3 and 4, we will prove in sections 4 - 6 the error estimates stated in the Theorems 4.1 , 5.1 , 6.1 and Corollary 5.1.

1. Stokes' boundary value problem. Notations.

Let Ω be an open bounded set in the three-dimensional space R^3 with points $x = (x^1,x^2,x^3)$. For short we assume the boundary $\partial\Omega$ is a C_3-submanifold of R^3. For the precise formulation of Stokes' boundary value problem

$$(1.1) \qquad -\Delta v + \nabla q = h, \quad \nabla\cdot v = o \text{ in } \Omega, \qquad v = o \quad \text{on } \Omega$$

with the given vector function h [1)] we use the Hilbert space H_m of the vector functions $f = (f^1,f^2,f^3)$ (defined almost everywhere) on Ω which are in $L^2(\Omega)$ together with their generalized derivatives up to the order m (m included).

The norm on H_m is as usual

$$|f|_{H_m} = \Big(\sum_{o \le \alpha_1+\alpha_2+\alpha_3 \le m} \int_\Omega |\partial_x^\alpha f(x)|^2 dx \Big)^{1/2}$$

with $\partial_x^\alpha f = \dfrac{\partial^{\alpha_1+\alpha_2+\alpha_3}}{(\partial x^1)^{\alpha_1}(\partial x^2)^{\alpha_2}(\partial x^3)^{\alpha_3}} f$, $\alpha = (\alpha_1,\alpha_2,\alpha_3)$, and the Euclidean norm $|\cdot|$ in R^3. In addition let D be the linear space of all divergence-free C_∞-vector-functions having compact support in Ω. The closures of D in $L^2 = H_o$ or H_1 are denoted with $\mathcal{H}_o$ or $\mathcal{H}_1$, respectively.

According to Weyl's theorem [24], the orthogonal projection

$$P : L^2 \to \mathcal{H}_o$$

sends into zero exactly the generalized gradients on Ω. Then the Stokes boundary value problem takes the following shape: For a given $h \in \mathcal{H}_o$ we are looking for a solution $v \in \mathcal{H}_1 \cap H_2$ of the equation

$$(1.2) \qquad -P\Delta v = h \quad .$$

We state the fundamental properties of the operator $-P\Delta$ in

---

1) and $\Delta = \dfrac{\partial^2}{(\partial x^1)^2} + \dfrac{\partial^2}{(\partial x^2)^2} + \dfrac{\partial^2}{(\partial x^3)^2}$, $\nabla = \left(\dfrac{\partial}{\partial x^1}, \dfrac{\partial}{\partial x^2}, \dfrac{\partial}{\partial x^3}\right)$

Lemma 1.1.: The map $-P\Delta : \mathcal{H}_1 \cap H_2 \to \mathcal{H}_0$ defines on $\mathcal{H}_1 \subset \mathcal{H}_0$ a symmetric positive-definite operator having the compact inverse $(-P\Delta)^{-1} : \mathcal{H}_0 \to \mathcal{H}_0$,

c.p. [12, p. 44-45], [9, p. 115-116], [7, p. 17], [22, p. 2o following pages].

The proof of Lemma 1.1 results from the following remarks, which will be useful also for the next sections: The linear operator $P\Delta$ is symmetric in $\mathcal{H}_0$, since its domain $\mathcal{H}_1 \cap H_2$ is dense in $\mathcal{H}_0$ and the equation

$$(1.3) \qquad \int_\Omega (-P\Delta f) g = \int_\Omega (\nabla f)\cdot(\nabla g)$$ [2)]

holds for any $f \in \mathcal{H}_1 \cap H_2$, $g \in \mathcal{H}_1$. Therefore we have $\int_\Omega (-P\Delta f)\cdot g = \int_\Omega f\cdot(-P\Delta g)$ if g belongs to H_2, too.

With $f_k \in C_2(\bar{\Omega})$ instead of f and $g_k \in D$ instead of g the equation (1.3) follows by application of Gauss' divergence theorem. Passing to the limit $f_k \to f$ in H_2 and $g_k \to g$ in H_1 we get (1.3); for (due to the smoothness of $\partial\Omega$, c.p. [5, p. 18] and more generally [1, p. 54]) $C_2(\bar{\Omega})$ is dense in H_2. Obviously the equation (1.3) defines the operator $-P\Delta$ on $\mathcal{H}_1$.

The positive definiteness of the operator $-P\Delta$ is equivalent to Poincaré's inequality

$$(1.4) \qquad |f|_{L^2} \leq c_0 \cdot |\nabla f|_{L^2} ,$$

which holds on the bounded open set Ω with a constant $c_0 > o$ [3)] for all $f \in \mathcal{H}_1$. From (1.4) results too, that for any fixed $h \in \mathcal{H}_0$ at most one solution $v \in \mathcal{H}_1$ of the equation

$$(1.5) \qquad \int \nabla v \cdot \nabla g = \int h \cdot g$$

exists, (1.5) holding for any $g \in \mathcal{H}_1$.

The existence of at least one solution of (1.5) in $\mathcal{H}_1$ follows in the well-known manner by means of a Galerkin-ansatz using Poincaré's inequality and the compactness

---

2) For short, here and in the following we omit the volume differential under the integral sign.

3) In the following, we denote by $c_0, c_1, \ldots$ positive real numbers, which do not depend on the functions in question.

of the imbedding of $\mathcal{H}_1$ in $\mathcal{H}_o$. Thus on $\mathcal{H}_o$ the operator $-P\Delta$ has the inverse $(-P\Delta)^{-1} : \mathcal{H}_o \to \mathcal{H}_1$. This inverse is compact in $\mathcal{H}_o$, because of the estimate $|v|_{H_1} \le c_1 |h|_{L^2}$ resulting from (1.5) with a constant $c_1 > o$. A direct consequence of Lemma 1.1 is

Lemma 1.2.: The operator $-P\Delta$ has the sequence (λ_i) of positive eigenvalues $\lambda_i > o$, $\lambda_1 \le \lambda_2 \le \ldots, \lambda_i \to \infty$, and the corresponding eigenfunctions (e_i) form a complete orthonormal system in $\mathcal{H}_o$.

The proof follows from Lemma 1.1 by well-known functional analytic facts, [21, p. 343].

Let $\mathcal{H}_{o,m}$ be the m-dimensional subspace of $\mathcal{H}_o$, which is spanned by the first m eigenfunctions $e_1,\ldots,e_m$ of (1.1), taken in any fixed order corresponding to the eigenvalues $\lambda_1 \le \ldots \le \lambda_m$. We denote by P_m the orthogonal projection of L^2 on $\mathcal{H}_{o,m}$, i.e. we have

$$P_m f = \sum_{i=1}^{m} e_i \cdot \int_\Omega (f \cdot e_i)$$

for any $f \in L^2$. On the space $\mathcal{H}_1$ equipped with the inner product $\int_\Omega (\nabla f)(\nabla g)$ we have

Lemma 1.3.: The functions $\left(\frac{e_i}{\lambda_i^{1/2}}\right)$ form a complete orthonormal set in $\mathcal{H}_1$. For any $f \in \mathcal{H}_1$ the sequence $(P_i f)$ converges to f in $\mathcal{H}_1$.

For the proof, using the equation

$$(1.6) \qquad -P\Delta e_i = \lambda_i e_i$$

for the eigenfunctions e_i, $i = 1,2,\ldots$, and (1.3) we verify the asserted orthonormality

$$(1.7) \qquad \int_\Omega \nabla \frac{e_i}{\lambda_i^{1/2}} \cdot \nabla \frac{e_k}{\lambda_k^{1/2}} = \begin{cases} 1 & \text{for } i = k, \\ o & \text{for } i \neq k. \end{cases}$$

Then the completeness of the $\frac{e_i}{\lambda_i^{1/2}}$ in $\mathcal{H}_1$ and the second statement of the Lemma follow by standard argumentation, c.p. [14, p. 144].

We state the regularity of the eigenfunctions in

Lemma 1.4.: Let $\partial\Omega$ be a C_m-submanifold of R^3, $m \ge 2$. Then the eigenfunctions e_i of (1.2) belong to $H_m(\Omega)$.

The proof is given in [22, p.39].

2. Error bounds for expansions in terms of the eigenfunctions of Stokes' boundary value problem.

If we approximate points of $\mathcal{H}_1$ by their projections on the finite dimensional space $\mathcal{H}_{o,m}$, we can estimate the error according to

Lemma 2.1.: Assume $f \in \mathcal{H}_1$. Then the error estimate

(2.1) $$|f-P_m f|^2_{L^2} \le \frac{1}{\lambda_{m+1}} |\nabla f|^2_{L^2}$$

holds, [1o, p. 39].

For the proof, using (1.3),(1.6) and $\lambda_i \ge \lambda_{m+1}$ for any $i \ge m+1$ we get the estimate

$$(\int_\Omega fe_i)^2 = \frac{1}{\lambda_i^2} (\int_\Omega f\cdot(-P\Delta e_i))^2 = \frac{1}{\lambda_i^2} (\int_\Omega (\nabla f)(\nabla e_i))^2$$

$$\le \frac{1}{\lambda_{m+1}} (\int_\Omega (\nabla f)(\nabla \frac{e_i}{\lambda_i 1/2}))^2.$$

Inserting this in Parseval's equality

(2.2) $$|f_m|^2_{L^2} = \sum_{i=m+1}^{\infty} (\int_\Omega fe_i)^2$$

for the function $f_m = f-P_m f$, we obtain

$$|f_m|^2_{L^2} \le \frac{1}{\lambda_{m+1}} \sum_{i=m+1}^{\infty} (\int_\Omega (\nabla f)(\nabla \frac{e_i}{\lambda_i 1/2}))^2$$

and hence (2.1) by application of Bessel's inequality, for the $\frac{e_i}{\lambda_i^{1/2}}$ form an orthonormal system in $\mathcal{H}_1$ with respect to the inner product $\int_\Omega (\nabla f)(\nabla g)$.

Lemma 2.2.: Assume $f \in \mathcal{H}_1 \cap H_2$. Then the error estimate

(2.3) $$|f-P_m f|^2_{L^2} \le \frac{1}{\lambda_{m+1}^2} |P\Delta f|^2_{L^2}$$

holds, [1o, p. 38].

For the proof we note, that from $-P\Delta e_i = \lambda_i e_i$ and $\lambda_i \ge \lambda_{m+1}$ for any $i \ge m+1$ we obtain the estimate

$$(\int_\Omega fe_i)^2 = \frac{1}{\lambda_i^2} (\int_\Omega (-P\Delta f)e_i)^2$$

$$\le \frac{1}{\lambda_{m+1}^2} (\int_\Omega (-P\Delta f)e_i)^2$$

because of the symmetry of $P\Delta$. Inserting this in Parseval's equality (2.2) for the function $f_m = f-P_m f$, we find

$$|f_m|^2_{L^2} \le \frac{1}{\lambda^2_{m+1}} \sum_{i=m+1}^{\infty} (\int_\Omega (-P\Delta f)e_i)^2$$

and hence (2.3) by means of Bessel's inequality for the function $-P\Delta f \in \mathcal{H}_o$.

Lemma 2.3.: Assume $f \in \mathcal{H}_1 \cap H_2$. Then the error estimate

(2.4) $$|\nabla f - \nabla P_m f|^2_{L^2} \le \frac{1}{\lambda_{m+1}} |P\Delta f|^2_{L^2} \qquad \text{holds.}$$

For the proof, we get from (1.3), (1.6) and $\lambda_i \ge \lambda_{m+1}$ for any $i \ge m+1$ the estimate

(2.5) $$(\int_\Omega \nabla f \, \nabla \frac{e_i}{\lambda_i^{1/2}})^2 = \left(\int_\Omega (-P\Delta f) \frac{e_i}{\lambda_i^{1/2}}\right)^2 \le \frac{1}{\lambda_{m+1}} (\int_\Omega (-P\Delta f)e_i)^2 .$$

Since the expansion $f_m = f - P_m f = \sum_{i=m+1}^{\infty} e_i \int_\Omega f e_i$ converges in $\mathcal{H}_1$, we have

$$\nabla f_m = \sum_{i=m+1}^{\infty} \nabla e_i \int_\Omega f e_i = \sum_{i=m+1}^{\infty} \nabla \frac{e_i}{\lambda_i^{1/2}} \int_\Omega (\nabla f)(\nabla \frac{e_i}{\lambda_i^{1/2}}) \quad ,$$

the latter equation following again by (1.3), (1.6). Thus f_m being orthogonal to $e_1,\dots,e_m$ also in $\mathcal{H}_1$ with the inner product $\int_\Omega \nabla f \nabla g$, Parseval's equation with respect to the complete orthonormal system $(\frac{e_i}{\lambda_i^{1/2}})$ takes the shape

$$|\nabla f_m|^2_{L^2} = \sum_{i=m+1}^{\infty} (\int_\Omega \nabla f \, \nabla \frac{e_i}{\lambda_i^{1/2}})^2 \quad .$$

Substituting on the right side the bounds from (2.5) we obtain

$$|\nabla f_m|^2_{L^2} \le \frac{1}{\lambda_{m+1}} \sum_{i=m+1}^{\infty} (\int_\Omega (-P\Delta f)e_i)^2$$

and hence (2.4) by application of Bessel's inequality for the function $P\Delta f \in \mathcal{H}_o$.

3. Bounds for local Navier-Stokes solutions and their Galerkin approximations.

The velocity field $u(t,x) = (u^1,u^2,u^3)$ of a viscous flow in Ω at time t with constant mass-density satisfies the Navier-Stokes equations

(3.1) $$\begin{aligned} \partial_t u - P\Delta u &= -P(u\nabla u), \ \nabla \cdot u = o \text{ in } \Omega, \ t > o, \\ u &= o \quad \text{on } \partial\Omega \text{ for } t \ge o, \\ u &= u_o \quad \text{for } t = o, \end{aligned}$$

if we assume the condition of adherence on $\partial\Omega$ and for short consider external potential forces only, and if finally distance and time are measured in the appropriate units.

Local strong solution $u \in L^\infty([o,T], \mathcal{H}_1)$ of (3.1) on a (possibly small) time interval $[o,T]$, which - by definition - have derivatives $\partial_t u, \partial_x^\alpha u$ with $\alpha_1+\alpha_2+\alpha_3 \le 2$ in

$L^2((o,T], L^2(\Omega))$ and satisfy $|\nabla u(t,\cdot) - \nabla u_o|_{L^2(\Omega)} \to o$ as $t \to o$, are constructed by means of the Galerkin-ansatz

$$u_k(t,x) = \sum_{i=1}^{k} a_{ki}(t)e_i(x) \tag{3.2}$$

on the basis of the eigenfunctions e_i of the linear Stokes problem (1.1). For the k unknown coefficients

$$a_{ki}(t) = \int_\Omega u_k(t,\cdot)\cdot e_i,$$

the inner products with e_i in L^2, $i = 1,\ldots,k$ of the approximate equations

$$\begin{aligned} \partial_t u_k - P\Delta u_k &= -P_k(u_k\cdot\nabla u_k) \quad \text{for } t > o, \\ u_k &= P_k u_o \qquad\qquad \text{for } t = o \end{aligned} \tag{3.3}$$

constitute a system of k ordinary differential equations and initial conditions. The differential equations are quadratic in the unknown functions a_{ki}, thus locally Lipschitz-continuous. According to a wellknown theorem on ordinary differential equations, the global existence and uniqueness of the $a_{ki} \in C_\infty[o,\infty)$ results from the energy equation

$$\partial_t |u_k(t,\cdot)|^2_{L^2(\Omega)} + 2|\nabla u_k|^2_{L^2(\Omega)} = o \tag{3.4}$$

which follows by taking the inner product in L^2 of (3.3) with $a_{ki}e_i$ and summing up about $i = 1,\ldots,k$, [8, p. 226]. The nonlinear term drops out because of

Lemma 3.1.: On $\mathcal{H}_1$, the integral $\int_\Omega (f\cdot\nabla g)\cdot h$ is a bounded trilinearform and skewsymmetric in g and h, [22, p. 162-163].

The error estimates in the next sections are based on

Theorem 3.1. [J. Heywood 1978]: Let the initial value u_o be in $\mathcal{H}_1$. Then on a (possibly small) time interval [o,T] the Navier-Stokes problem (3.1) has a unique strong solution u.
The partial derivatives of u and its Galerkin approximations u_k satisfy the estimates

$$|\nabla u(t,\cdot)|^2_{L^2(\Omega)} + \int_o^t |\partial_\tau u(\tau,\cdot)|^2_{L^2(\Omega)}\, d\tau \le F_o(t) \le F_{o,o}, \tag{3.5}$$

$$\int_o^t |P\Delta u(\tau,\cdot)|^2_{L^2(\Omega)}\, d\tau \le h_o(t) \quad \text{on } [o,T], \tag{3.6}$$

$$(3.7)\quad |\nabla\partial_t^n u(t,\cdot)|^2_{L^2(\Omega)} + \int_\varepsilon^t |\partial_\tau^{n+1} u(\tau,\cdot)|^2_{L^2(\Omega)}\, d\tau \le F_n(t,\varepsilon),$$

$$(3.8)\quad \int_\varepsilon^t |P\Delta\partial_\tau^n u(\tau,\cdot)|^2_{L^2(\Omega)}\, d\tau \le h_n(t,\varepsilon),$$

$$(3.9)\quad |\partial_t^n u(t,\cdot)|^2_{L^2(\Omega)} + \int_\varepsilon^t |\nabla\partial_\tau^n u(\tau,\cdot)|^2_{L^2(\Omega)}\, d\tau \le G_n(t,\varepsilon)$$

$$(3.1o)\quad |P\Delta\partial_t^n u(t,\cdot)|^2_{L^2(\Omega)} \le g_n(t,\varepsilon) \le g_{n,\varepsilon}$$

on $[\varepsilon,T]$ for any ε in the open interval (o,T) and any $n = o,1, \ldots$

Additionally in the case $u_o \in \mathcal{H}_1 \cap H_2$ we have

$$(3.11)\quad |\partial_t u(t,\cdot)|^2_{L^2(\Omega)} + \int_o^t |\nabla\partial_\tau u(\tau,\cdot)|^2_{L^2(\Omega)} \le G_1(t)\quad ,$$

$$(3.12)\quad |P\Delta u(t,\cdot)|^2_{L^2(\Omega)} \le g_o(t) \le g_{o,o}$$

on $[o,T]$. The functions on the right sides depend on their argument t or (t,ε) and in addition on n, T and the Dirichlet norm $|\nabla u_o|^2_{L^2(\Omega)}$ only, in the latter case on $|u_o|_{H_2}$ too. On the interval in question these functions are continuous in the variable t, the functions h_o, F_n, G_n being continuously differentiable with respect to t by their definition, [7 ,p. 29-3o, 63].

In the following we assume the bounds on the right sides are chosen in such a way that they are monotonously increasing in n for any fixed values of t and ε coming into question.

Kiselev and Ladyženskaja have proved the local existence of unique strong solutions of (3.1) with the assumption $u_o \in \mathcal{H}_1 \cap H_2$, [12, p. 143, 161]. Existence-proofs for local strong solutions with initial value $u_o \in \mathcal{H}_1$ are given by Prodi [16] and Kaniel and Shinbrot [11] , [20].

As an immediate consequence we obtain

Corollary 3.1.: For any natural number $n = o,1, \ldots$ and any $\varepsilon \in (o,T)$ there exists a subsequence (u_{k^*}) of the u_k from (3.2), (3.3), $k = 1,2, \ldots$ which converges in H_m uniformly on $[\varepsilon,T]$ to the Navier-Stokes solution u, $m = o,1$. This convergence holds for the time derivatives up to the order n, too. In the cases $u_o \in \mathcal{H}_1$ and $n = m = o$ or $u_o \in \mathcal{H}_1 \cap H_2$ and $n = o$, $m = o,1$, the convergence is uniform even on $[o,T]$.

For the proof of the Corollary we remark first of all, that a sequence of L^2-differentiable maps $f_k: [\varepsilon,T]\rightarrow \mathcal{H}_m \cap H_{m+1}$, for which the terms $|f_k(t,\cdot)|_{H_{m+1}}$ and $\int_\varepsilon^t |\partial_\tau f_k(\tau,\cdot)|^2_{H_m}\, d\tau$ are uniformly (with respect to t and k) bounded, contains a subsequence $(f_{k'}(t,\cdot))$, which converges uniformly on $[\varepsilon,T]$ in $\mathcal{H}_m$. This remark follows according to the theorem of Ascoli and Arzelá from the compactness of the natural imbedding $\mathcal{H}_m \cap H_{m+1} \subset \mathcal{H}_m$, the $f_k(t,\cdot)$ being equicontinuous in $\mathcal{H}_m$; namely from the Cauchy-Schwarz inequality we get for $t,s \in [\varepsilon,T]$ the estimate

$$(f_k(t,x)-f_k(s,x))^2 = \left(\int_s^t \partial_\tau f_k(\tau,x)d\tau\right)^2 \leq |t-s|\left|\int_s^t (\partial_\tau f_k(\tau,x))^2 d\tau\right|$$

almost everywhere on Ω and hence after integrating

$|f_k(t,\cdot)-f_k(s,\cdot)|^2_{L^2(\Omega)} \leq |t-s|\cdot$ constant value by our assumption. In the case m=1 the last two formulas hold with ∇f_k instead of f_k, too, thus also showing the equicontinuity in the norm of $\mathcal{H}_1$.

We keep in mind the estimates of Theorem 3.1. From (3.1o) or (3.12) we obtain bounds in H_2 by means of Cattabriga's inequality $|f|_{H_2} \leq c|P\Delta f|_{L^2}$, which holds on $\mathcal{H}_1 \cap H_2$ with a constant $c > o$, [2, p. 311]. Taking firstly $f_k = u_k$, then $f_k = \partial_t u_k$, and so on, at each step selecting a suitable subsequence we get a subsequence (u_{k^*}) of (u_k), which converges in $\mathcal{H}_m$ uniformly on $[\varepsilon,T]$ together with the sequences of all its time derivatives up to the order n. By a wellknown theorem in Banach spaces, cp. [13, p. 117], the limit $\partial_t^j \lim_{k^*\to\infty} u_{k^*} = \lim_{k^*\to\infty} \partial_t^j u_k$ exists in $C_o([\varepsilon,T],\mathcal{H}_m)$ for $j = 0,\ldots,n$. As (3.5) or (3.12) show, in the case $u_o \in \mathcal{H}_1$ and $n = m = o$ or $u_o \in \mathcal{H}_1 \cap H_2$ and $n = o$, $m = o,1$, respectively, our proof works also with $\varepsilon = o$.

Let $u^* \in C_o([o,T],\mathcal{H}_o)$ be the limit of the u_{k^*}. From Theorem 3.1 and the rules for weak convergence in the spaces H_m [5, p. 16] follows as in [7, p. 31], that u^* and even any accumulation point of the u_k in $L^2([o,T],\mathcal{H}_o)$ (with respect to the uniform strong convergence) is a solution of (3.1), which has the properties listed in Theorem 3.1. Because of the uniqueness of such a solution we have $u^*(t,\cdot)=u(t,\cdot)$ in $L^2(\Omega)$, and the sequence (u_k) can have at most one accumulation point in $L^2([o,T],\mathcal{H}_o)$. Therefore, the whole sequence $(u_k(t,\cdot))$ converges to $u(t,\cdot)$ in $L^2(\Omega)$ uniformly with respect to t.

Thus the sequences $(\partial_t^j u_k(t,\cdot))$ converge strongly in $\mathcal{H}_m$ to the respective limit $\partial_t^j u(t,\cdot)$, uniformly on $[\varepsilon,T]$ and even on $[o,T]$ in the special cases mentioned above.

4. Error bounds in $L^2(\Omega)$ for Galerkin approximations of Navier-Stokes solutions.

Let $[o,T]$ be a time interval as in Theorem 3.1, and (e_i) be the system of eigenfunctions corresponding to the eigenvalues (λ_i) of the linear Stokes problem (1.2) as in Lemma 1.2.

We will prove

Theorem 4.1: Suppose the initial velocity u_o belongs to $\mathcal{H}_1$. Then the Galerkin appoximation u_k from (3.2), (3.3) of the solution u of (3.1) satisfies the error estimate

$$(4.1)\qquad |u(t,\cdot) - u_k(t,\cdot)|^2_{L^2(\Omega)} + \int_0^t |\nabla(u-u_k)|^2_{L^2(\Omega)}\, d\tau \le \frac{F_o^*(t)}{\lambda_{k+1}}$$

for any $t \in [o,T]$. The continuous function F_o^* of the variable t depends on T and the Dirichlet norm of u_o only. (4.1) holds also with any u_l instead of u, $l > k$.

For the proof, let u_k and u_l with $l > k$ be two Galerkin approximations (3.2) of the solution u of (3.1). Subtracting from the equations (3.3) for u_l these equations for u_k, we get for the difference $w = u_l - u_k$ the basic equations

$$(4.2)\qquad \begin{aligned} &\partial_t w - P\Delta w = -(P_l-P_k)(u_l\nabla u_l) - P_k(u_l\nabla w + w\nabla u_k)\ , \\ &w(o,\cdot) = (P_l-P_k)u_o . \end{aligned}$$

Because of (1.3), the inner product of (4.2) with w in $L^2(\Omega)$ results in

$$(4.3)\qquad \begin{aligned} &\partial_t |w|^2_{L^2(\Omega)} + 2|\nabla w|^2_{L^2(\Omega)} = -2\int_\Omega \{(P_l-P_k)(u_l\nabla u_l) + P_k(u_l\nabla w + w\nabla u_k)\}\cdot w\ , \\ &|w(o,\cdot)|^2_{L^2(\Omega)} = |(P_l-P_k)u_o|^2_{L^2(\Omega)} . \end{aligned}$$

We estimate the single terms on the right sides one after the other.

From Lemma 2.1 and Theorem 3.1, (3.5) we get for the initial value

$$(4.4)\qquad |w(o,\cdot)|^2_{L^2} \le \frac{|\nabla u_l(o,\cdot)|^2_{L^2}}{\lambda_{k+1}} \le \frac{F_o(o)}{\lambda_{k+1}}$$

For rewriting the first term of the integral in (4.3), we note, that the equation

$$(4.5)\qquad \int_\Omega ((P_l-P_k)f)w = \int_\Omega f\,(1-P_k)u_l$$

holds for any $f \in L^2$ and $l > k$, since P_l, P_k and P_l-P_k are orthogonal projections of L^2 and the equation $(P_l-P_k)(u_l-u_k) = (1-P_k)u_l$ follows from $P_l u_k = P_k u_k = u_k$, $P_l u_l = u_l$, 1 denoting the identical map.

In the resulting equation

$$(4.6)\qquad \int_\Omega ((P_l-P_k)(u_l\nabla u_l))w = \int_\Omega (u_l\nabla u_l)(1-P_k)u_l$$

we estimate the first factor on the right side in L^∞ by means of the inequality

$$(4.7)\qquad |f|_{L^\infty} \le c_2 |P\Delta f|_{L^2(\Omega)}\ ,$$

which follows for any $f \in \mathcal{H}_1 \cap H_2$ from Sobolev's imbedding theorem on 3-dimensional regions and Cattabriga's inequality

$$(4.8) \qquad |f|_{H_2} \leq c_3 |P\Delta f|_{L^2(\Omega)} , \qquad [2, p.311].$$

Then for arbitrary $g \in H_1$, $h \in L^2$ with the Cauchy-Schwarz inequality we obtain the estimate

$$(4.9) \qquad |\int_\Omega (f\nabla g)h| \leq |f\nabla g|_{L^2} |h|_{L^2} \leq c_2 |P\Delta f|_{L^2} |\nabla g|_{L^2} |h|_{L^2} ,$$

from which the inequalities

$$(4.1o) \qquad |\int_\Omega (f\nabla g)h| \leq \delta |\nabla g|^2_{L^2} + c_\delta |P\Delta f|^2_{L^2} \; |h|^2_{L^2}$$

and

$$(4.11) \qquad |\int_\Omega (f\nabla g)h| \leq \delta |h|^2_{L^2} + c_\delta |P\Delta f|^2_{L^2} \; |\nabla g|^2_{L^2}$$

follow according to Young's inequality. In any case the coefficient $c_\delta > o$ depends on the arbitrary value $\delta > o$ only. (4.1o) and (4.11) will be used below. We estimate the right side of (4.6) by means of (4.9). Substituting the bounds of Lemma 2.2 and Theorem 3.1, (3.5) yields the estimate

$$(4.12) \qquad |\int_\Omega ((P_1 - P_k)(u_1 \nabla u_1))w| \leq c_3 \frac{|P\Delta u_1(t,\cdot)|^2_{L^2(\Omega)} \cdot F^{1/2}_{o,o}}{\lambda_{k+1}}$$

for the first term of the integral in (4.3).

For estimating the second term of this integral we note, that we have

$$(4.13) \qquad |P_k F|_{L^2} \leq |F|_{L^2}$$

for any $F \in L^2$, P_k being an orthogonal projection in L^2. Then with the help of (4.9), (4.1o) we obtain

$$(4.14) \qquad |\int_\Omega (P_k(u_1 \nabla w))w| \leq \delta |\nabla w|^2_{L^2(\Omega)} + c_\delta |w|^2_{L^2(\Omega)} \; |P\Delta u_1|^2_{L^2(\Omega)} .$$

For the third term $\int_\Omega (w\nabla u_k)w$ of the integral in (4.3) we need a bound containing first order derivatives of the first factor w only, which are bounded by the second term on the left side of (4.3).

Thus in order for later use to estimate [4)] the more general triple

---

4) as in [16, p. 38o] and [7, p. 14, 15]

product $\int_\Omega (f\cdot\nabla g)h$, first of all we apply the Cauchy-Schwarz inequality and the Hölder inequality

$$(4.15)\qquad |f\nabla g|_{L^2} \le |f|_{L^6}\ |\nabla g|_{L^3}$$

for $f \in L^6$, $\nabla g \in L^3$. Secondly we use the inequality

$$(4.16)\qquad |\nabla g|_{L^3} \le c_4 |P\Delta g|_{L^2} \quad ,$$

which holds for any $g \in \mathcal{H}_1 \cap H_2$. Namely, if in the general Sobolev inequality

$$|G|_{L^3} \le c_5 \{ |\nabla G|^{1/2}_{L^2}\ |G|^{1/2}_{L^2} + |G|_{L^2} \}$$

[5, p. 27] for functions $G \in H_1$ we substitute ∇g in the place of G and use Cattabriga's bound (4.8) we obtain (4.16), [7, p. 14].

Thirdly we use the Sobolev inequality

$$(4.17)\qquad |f|_{L^6} \le c_6\ |\nabla f|_{L^2} \quad ,$$

which holds for any $f \in \mathcal{H}_1$, [12, p. 1o].

The estimates (4.13), (4.15)-(4.17) yield the inequality

$$(4.18)\qquad |\int_\Omega (P_k(f\nabla g))h| \le c_7 |\nabla f|_{L^2}\ |P\Delta g|_{L^2}\ |h|_{L^2} \quad ,$$

which implies

$$(4.19)\qquad |\int_\Omega (P_k(f\nabla g))h| \le \delta |\nabla f|^2_{L^2} + c_\delta |P\Delta g|^2_{L^2} |h|^2_{L^2}$$

and

$$(4.2o)\qquad |\int_\Omega (P_k(f\nabla g))h| \le \delta |h|^2_{L^2} + c_\delta |P\Delta g|^2_{L^2}\ |\nabla f|^2_{L^2}$$

for any $f \in \mathcal{H}_1$, $g \in \mathcal{H}_1 \cap H_2$, $h \in L^2$, according to Young's inequality. Again the coefficient $c_\delta > o$ depends on the arbitrary value $\delta > o$ only.
The estimates (4.19) and (4.2o) will be used in the next sections, too. From (4.19) we get for the third term of the integral in (4.3)

$$(4.21)\qquad |\int_\Omega (P_k(w\nabla u_k))w| \le \delta |\nabla w|^2_{L^2(\Omega)} + c_\delta |P\Delta u_k|^2_{L^2(\Omega)}\ |w|^2_{L^2(\Omega)} \quad .$$

Now we assume $\delta \in (o,\frac{1}{4}]$. With the bounds from (4.4), (4.12), (4.14), (4.21) , the integration of (4.3) with respect to the variable t yields the integral inequality

$$(4.22)\qquad |w(t,\cdot)|^2_{L^2(\Omega)} + \int_0^t |\nabla w(\tau,\cdot)|^2_{L^2(\Omega)} d\tau \le \frac{a(t)}{\lambda} + \int_0^t b(\tau)\, |w(\tau,\cdot)|^2_{L^2(\Omega)}\, d\tau$$

containing the continuous functions

$$(4.23)\qquad a(t) = F_o(o) + 2c_3\, F_{o,o}^{1/2}\, h_o(t)\ ,$$

$$b(t) = 2c_\delta \cdot \{|P\Delta u_l(t,\cdot)|^2_{L^2(\Omega)} + |P\Delta u_k(t,\cdot)|^2_{L^2(\Omega)}\},$$

$$\int_0^t b(\tau)d\tau \le 4c_\delta\, h_o(t) \text{ by Theorem 3.1 , (3.6).}$$

a is even continuously differentiable. Thus Theorem 4.1 is an immediate consequence of the following variant of Gronwall's Lemma:

Corollary 4.1.: Let the function $a(t) \ge o$ be absolutely continuous with $a'(t) \ge o$ and $b(t) \ge o$ summable in $[o,T]$. Assume the integral inequality

$$(4.24)\qquad \varphi(t) + \int_0^t \varphi^*(\tau)d\tau \le \frac{a(t)}{\lambda} + \int_0^t b(\tau)\varphi(\tau)d\tau$$

holds for the positive continuous functions φ and φ^* on $[o,T]$ with a constant $\lambda > o$.

Then we have

$$(4.25)\qquad \varphi(t) + \int_0^t \varphi^*(\tau)d\tau \le \frac{A(t)}{\lambda}$$

with the function

$$(4.26)\qquad A(t) = (1+\int_0^t b(\tau)d\tau)\psi(t)\ ,\qquad \psi(t) = a(t)\, e^{\int_0^t b(\tau)d\tau}\ .$$

For the proof of the Corollary first of all we neglect the integral on the left side in (4.24). Then according to Gronwall's Lemma [23, p. 15], keeping in mind $a' \ge o$ and $b \ge o$ we obtain the inequality $\varphi \le \frac{\psi}{\lambda}$ with the ψ in (4.26). Hence (4.25) follows from (4.24) with A from (4.26).

In the case of (4.22), (4.23) the Corollary 4.1 yields the inequality

$$(4.27)\qquad |u_l(t,\cdot)-u_k(t,\cdot)|^2_{L^2(\Omega)} + \int_0^t |\nabla(u_l(\tau,\cdot)-u_k(\tau,\cdot))|^2_{L^2(\Omega)} d\tau \le \frac{F_o^*(t)}{\lambda_{k+1}}\ ,$$

with $F_o^* = A$ from (4.26) and a,b from (4.23), the right side being independent of the special Galerkin approximation u_l, $l > k$. By Corollary 3.1 we can find a subsequence $(u_{l'}(t,\cdot))$ converging in $L^2(\Omega)$ to $u(t,\cdot)$ uniformly on $[o,T]$ and hence pass to the limit $l' \to \infty$ in the first term on the left side.

Because of (4.27) the values $\int_0^t |\nabla(u_l-u_k)|^2_{L^2(\Omega)} d\tau$ being uniformly bounded, the convergence rules in the spaces H_m [5, p.16] show, that the $\nabla(u_l-u_k)$ converge weakly in $L^2([o,t] \times \Omega)$ to $\nabla(u-u_k)$ for any $t \in (o,T)$. Thus we have

$$\int_0^t |\nabla(u-u_k)|^2_{L^2(\Omega)} d\tau \leq \limsup_{l'\to\infty} \int_0^t |\nabla(u_l-u_k)|^2_{L^2(\Omega)} d\tau \quad .$$

This additional remark shows, that passing to the limit $l' \to \infty$ we obtain (4.1) from (4.27), the function F_o^* being continuous and depending on T and $|\nabla u_o|_{L^2(\Omega)}$ only by our construction and Theorem 3.1 .
An <u>open problem</u> is, wether in the case $u_o \in \mathcal{H}_1 \cap H_2$ the left side of (4.1) has a bound being proportional to $\frac{1}{\lambda_{k+1}^2}$ as suggested by Lemma 2.2.

5. Error bounds in the Dirichlet norm for Galerkin approximations of Navier-Stokes solutions.

As before we denote by $[o,T]$ a time interval, on which the assertions of Theorem 3.1 hold. A counterpart of Lemma 2.3 in the instationary case is established in

<u>Theorem 5.1.:</u> Suppose the initial velocity u_o belongs to $\mathcal{H}_1 \cap H_2$. Then the Galerkin approximation u_k from (3.2), (3.3) of the solution u of (3.1) satisfies the error estimate

$$(5.1) \qquad |\nabla u(t,\cdot)-\nabla u_k(t,\cdot)|^2_{L^2(\Omega)} + \int_0^t |\partial_\tau u(\tau,\cdot)- \partial_\tau u_k(\tau,\cdot)|^2_{L^2(\Omega)} \leq \frac{G^*(t)}{\lambda_{k+1}}$$

for any $t \in [o,T]$. The continuous function G^* of the variable t depends on T and the norm $|u_o|_{H_2}$ only. (5.1) holds also with any u_l instead of u, $l > k$.

For the <u>proof</u>, let $w=u_l-u_k$ be the difference of two Galerkin approximations (3.2) with $l > k$ of the solution u of (3.1). Taking the inner product of (4.2) with $\partial_t w \in \mathcal{H}_1$ in $L^2(\Omega)$ and using (1.3), we obtain

$$(5.2) \qquad \partial_t |\nabla w(t,\cdot)|^2_{L^2(\Omega)} + 2 |\partial_t w(t,\cdot)|^2_{L^2(\Omega)} = -2 \int_\Omega \{(P_l-P_k)(u_l \nabla u_l)+P_k(u_l \nabla w+w\nabla u_k)\}\partial_t w \quad .$$

Additionally we have the initial condition

$$(5.3) \qquad |\nabla w(o,\cdot)|^2_{L^2} = |\nabla(P_l-P_k)u_o|^2_{L^2} \quad .$$

We estimate the single terms on the right sides one after the other.

Lemma 2.3 and Theorem 3.1 , (3.12) yield the inequality

$$(5.4)\qquad |\nabla w(o,\cdot)|^2_{L^2} \leq \frac{g_o(o)}{\lambda_{k+1}} \quad .$$

The orthogonal projections P_1 and P_k commute with the time derivation ∂_t in this case. By (4.6) we have

$$(5.5)\qquad \int_\Omega \{(P_1-P_k)(u_1\nabla u_1)\}\, \partial_t w = \int_\Omega (u_1\nabla u_1)\, \partial_t (1-P_k)u_1 \quad .$$

Integration by parts with respect to $t \in [o,T]$ results in

$$(5.6)\qquad \int_\varepsilon^t \int_\Omega \{(P_1-P_k)(u_1\nabla u_1)\}\, \partial_\tau w \, d\tau = \int_\Omega (u_1\nabla u_1)(1-P_k)u_1 \Big|_\varepsilon^t - \int_\varepsilon^t\int_\Omega (u_1\nabla u_1)_\tau (1-P_k)u_1 d\tau$$

for any $\varepsilon \in [o,T]$.[5] Using (4.9) or (4.18) and Lemma 2.2 we get

$$(5.7)\qquad |\int_\Omega (u_1\nabla u_1)_t (1-P_k)u_1| \leq \frac{c_2+c_7}{\lambda_{k+1}} |P\Delta u_1(t,\cdot)|^2_{L^2(\Omega)} |\nabla \partial_t u_1(t,\cdot)|_{L^2(\Omega)} \quad .$$

On the right side we substitute the constant bound $g_{o,\varepsilon}$ from Theorem 3.1, (3.1o) (or (3.12) in the case $\varepsilon = o$). Using Young's inequality and subsequently integrating (5.7) with respect to t we obtain

$$(5.8)\qquad |\int_\varepsilon^t \int_\Omega (u_1\nabla u_1)_\tau (1-P_k)u_1 d\tau| \leq \frac{c_8}{\lambda_{k+1}}\; g_{o,\varepsilon}\{(t-\varepsilon)+G_1(t,\varepsilon)\}$$

with G_1 from Theorem 3.1 , (3.9) or (3.11), respectively.

Again with (4.9) and Lemma 2.2 , using Young's inequality we obtain immediately

$$(5.9)\qquad |\int_\Omega (u_1\nabla u_1)(1-P_k)u_1 \Big|_{t_o}^t | \leq \frac{2c_2}{\lambda_{k+1}} \{|P\Delta u_1(t,\cdot)|^4_{L^2(\Omega)} + |\nabla u_1(t,\cdot)|^2_{L^2(\Omega)}\}$$

for any $t_o \in [o,t]$.

For both remaining terms of the integral in (5.2) bearing in mind (4.13) we get

$$(5.1o)\qquad |\int_\Omega (P_k(u_1\nabla w))\partial_t w| \leq \delta |\partial_t w|^2_{L^2(\Omega)} + c_\delta |P\Delta u_1|^2_{L^2(\Omega)} |\nabla w|^2_{L^2(\Omega)}$$

---

[5] We will use these formulas for the case $\varepsilon > o$ in the later proof of Corollary 5.1.

from (4.11) and further

$$(5.11)\qquad |\int_\Omega (P_k(w\nabla u_k))\partial_t w| \le \delta|\partial_t w|^2_{L^2(\Omega)} + c_\delta |P\Delta u_k|^2_{L^2(\Omega)} |\nabla w|^2_{L^2(\Omega)}$$

from (4.2o), the coefficient c_δ depending on the arbitrary value $\delta > o$ only.

Now we assume $\delta \in (o,\frac{1}{4}]$, $\varepsilon = o$, the latter being possible according to the last part of Theorem 3.1 and our requirement $u_o \in \mathcal{H}_1 \cap H_2$. On the right sides of the inequalities (5.9)-(5.11) we substitute the bounds of Theorem 3.1 , (3.5), (3.11) and (3.12). Then these inequalities together with (5.4) and (5.8) show, that the initial value problem (5.2) and (5.3) leads to the integral inequality

$$(5.12)\qquad |\nabla w(t,\cdot)|^2_{L^2(\Omega)} + \int_o^t |\partial_\tau w(\tau,\cdot)|^2_{L^2(\Omega)} d\tau \le \frac{a(t)}{\lambda_{k+1}} + b\int_o^t |\nabla w(\tau,\cdot)|^2_{L^2(\Omega)} d\tau$$

on [o,T] with the continuously differentiable function

$$(5.13)\qquad \begin{aligned} a(t) &= g_o(o) + 2c_8 g_{o,o}\,(t+G_1(t)) + 4c_2\cdot(g^2_{o,o} + F_o(t)) \text{ and} \\ b &= 4c_\delta g_{o,o}\,. \end{aligned}$$

Thus from Corollary 4.1 the equation

$$(5.14)\qquad |\nabla u_1(t,\cdot)-\nabla u_k(t,\cdot)|^2_{L^2(\Omega)} + \int_o^t |\partial_\tau u_1(\tau,\cdot)-\partial_\tau u_k(\tau,\cdot)|^2_{L^2(\Omega)} \le \frac{G^*(t)}{\lambda_{k+1}}$$

follows with $G^*(t) = A(t)$ from (4.26), a,b from (5.13).

The right side of (5.14) being independent of the special u_1, by Corollary 3.1 (with $u_o \in \mathcal{H}_1 \cap H_2$) we can use a subsequence $(u_{1'}(t,\cdot))$ converging in H_1 to $u(t,\cdot)$ uniformly on [o,T]. The last remark in the proof of Theorem 4.1 holds for the second term on the left side of (5.14), too (with ∂_τ instead of ∇). Hence passing in (5.14) to the limit $1'\to\infty$ we obtain (5.1), the function G^* being continuous and depending on T and $|u_o|_{H_2}$ only, as our construction and Theorem 3.1 show.

Without the restriction $u_o \in H_2$ we have no initial bound for $|\nabla w|^2_{L^2(\Omega)}$ in t = o. Using the first part of Theorem 3.1 only, we prove

<u>Corollary 5.1.:</u> Suppose the initial velocity u_o belongs to $\mathcal{H}_1$. Then the Galerkin approximation u_k from (3.2), (3.3) of the solution u of (3.1) satisfies

$$(5.15)\qquad |\nabla u(t,\cdot)-\nabla u_k(t,\cdot)|^2_{L^2(\Omega)} + \int_\varepsilon^t |\partial_\tau u(\tau,\cdot)-\partial_\tau u_k(\tau,\cdot)|^2_{L^2(\Omega)} d\tau \le \frac{G^*_o(t,\varepsilon)}{\lambda_{k+1}}$$

for any $t \in [\varepsilon,T]$ and $\varepsilon > 0$. Being continuous with respect to t, the function G_0^* depends on t,ε,T and the Dirichlet norm of u_0 only. (5.15) holds also with any u_l instead of u, $l > k$.

Firstly for the proof we establish in the manner of [7, p. 36] an equivalent to the initial condition (5.3). Let $w = u_l - u_k$ with $l > k$ be the difference of two Galerkin approximations. From Theorem 4.1 we get

$$\int_{\varepsilon}^{2\varepsilon} |\nabla w|^2_{L^2(\Omega)} \, d\tau \le \frac{F_0^*(2\varepsilon)}{\lambda_{k+1}} ,$$

the integrant being continuous. Hence the mean value theorem of integral calculus insures the existence of at least one $t_* \in (\varepsilon, 2\varepsilon)$ with

$$\varepsilon \, |\nabla w(t_*,\cdot)|^2_{L^2(\Omega)} = \int_{\varepsilon}^{2\varepsilon} |\nabla w|^2_{L^2(\Omega)} \, d\tau .$$

Therefore we have

$$(5.16) \qquad |w(t_*,\cdot)|^2_{L^2(\Omega)} \le \frac{F_0^*(2\varepsilon)}{\varepsilon \cdot \lambda_{k+1}} .$$

The inequalities (5.7) - (5.11) have been proved without the restriction $u_0 \in H_2$. Assuming $\delta \in (0,\frac{1}{4}]$, $\varepsilon > 0$ and substituting the bounds of Theorem 3.1 , (3.5), (3.9) and (3.10) we see from the inequalities (5.16) and (5.7)-(5.11), that the differential equality (5.2) together with the initial condition (5.16) yield the integral inequality

$$(5.17) \qquad |\nabla w(t,\cdot)|^2_{L^2(\Omega)} + \int_{t_*}^{t} |\partial_\tau w(\tau,\cdot)|^2_{L^2(\Omega)} \, d\tau \le \frac{a(t)}{\lambda_{k+1}} + b \int_{t_*}^{t} |\nabla w(\tau,\cdot)|^2_{L^2(\Omega)} \, d\tau$$

on $[t_*,T]$ with the continuously differentiable function

$$(5.18) \qquad a(t) = \frac{F_0^*(2\varepsilon)}{\varepsilon} + 2c_8 g_{0,\varepsilon} \cdot \{(t-\varepsilon)+G_1(t,\varepsilon)\} + 4c_2 \cdot (g_{0,\varepsilon}^2 + F_0(t)) \quad \text{and}$$
$$b = 4c_\delta g_{0,\varepsilon} .$$

For the calculation of $a(t)$ we have used, that the integral of the right side of (5.7) is a monotonously decreasing function of its lower limit.

Thus using Corollary 4.1 on $[t_*,T]$ we get

$$(5.19) \qquad |\nabla w(t,\cdot)|^2_{L^2(\Omega)} + \int_{2\varepsilon}^{t} |\partial_\tau w(\tau,\cdot)|^2_{L^2(\Omega)} \, d\tau \le \frac{G_0^*(t,2\varepsilon)}{\lambda_{k+1}}$$

with the upper bounds

$$G_0^*(t,2\varepsilon) = (1+(t-\varepsilon)b)\,\psi^*(t,\varepsilon) \ge (1+(t-t_*)b)\psi(t) = A(t) \text{ and}$$
$$\psi^*(t,\varepsilon) = a(t)\, e^{(t-\varepsilon)b} \ge a(t)\, e^{(t-t_*)b} = \psi(t)$$

of the functions $A(t)$ and $\psi(t)$ from (4.26), respectively, a,b from (5.18). Then Corollary 5.1 follows by passing to the limit $l \to \infty$ as in the proof of Theorem 5.1.

6. Error bounds in H_1 for the time derivatives of any order of the Galerkin approximations.

On the basis of John Heywood's method and in correspondance to his *norm* estimates in [7], in this final section we will construct an infinite system of *error* estimates for the Galerkin approximations u_k and their time derivatives of any order in the norm of $L^2(\Omega)$ and H_1. We prove

Theorem 6.1.: Assume the initial velocity u_o belongs to $\mathcal{H}_1$. Then the Galerkin approximation u_k from (3.2), (3.3) of the Navier-Stokes solution u in (3.1) satisfies the infinite system of estimates

$$(6.1.\,m)\qquad |\partial_t^m(u(t,\cdot))|^2_{L^2(\Omega)} + \int_\varepsilon^t |\nabla\partial_\tau^m(u(\tau,\cdot)-u_k(\tau,\cdot))|^2_{L^2(\Omega)}\,d\tau \le \frac{G_m^*(t,\varepsilon)}{\lambda_{k+1}} ,$$

$$(6.2.\,m)\qquad |\nabla\partial_t^m(u(t,\cdot)-u_k(t,\cdot))|^2_{L^2(\Omega)} + \int_\varepsilon^t |\partial_\tau^{m+1}(u(\tau,\cdot)-u_k(\tau,\cdot))|^2_{L^2(\Omega)}\,d\tau \le \frac{F_m^*(t,\varepsilon)}{\lambda_{k+1}}$$

on $[\varepsilon,T]$ for any $\varepsilon > o$. Being continuous with respect to t and monotonously increasing in t and m = o,1,..., the functions G_m^* and F_m^* depend on t,ε,m,T and $|\nabla u_o|_{L^2(\Omega)}$ only. (6.1.m) and (6.2.m) hold with any u_l instead of $u, l > k$, too.

The Theorem will be proved by induction with respect to m. For m = o, the assertion is true by Theorem 4.1 ((6.1.o) holds even for any $\varepsilon \in [o,T]$) and Corollary 5.1 . Now let the statement be true for all m = o,...,n - 1. We will prove it for m = n.

Since the solutions $a_{ki}(t)$ of the differential equation (3.3) have derivatives of any order, we can differentiate the equation (4.2) for the difference $w = u_l - u_k$ (of two Galerkin approximations (3.2) with $l > k$) n times with respect to t. According to Leibniz' rule, we get the basic differential equation

$$(6.3)\qquad (\partial_t^n w)_t - P\Delta\partial_t^n w = -\sum_{j=o}^{n}\binom{n}{j}\left\{\begin{array}{l}(P_l-P_k)((\partial_t^j u_l)\nabla\partial_t^{n-j}u_l) + \\ + P_k((\partial_t^j u_l)\nabla\partial_t^{n-j}w) + \\ + P_k((\partial_t^j w)\nabla\partial_t^{n-j}u_k)\end{array}\right\} ,$$

because P_k and ∂_t^n commute. Using (1.3), the inner product of (6.3) with

$\partial_t^n w \in \mathcal{H}_1$ in L^2 results in the differential equality

$$(6.4)\qquad \partial_t |\partial_t^n w|^2_{L^2(\Omega)} + 2|\nabla \partial_t^n w|^2_{L^2(\Omega)} = -2 \sum_{j=o}^{n} \binom{n}{j} \int_\Omega \left\{ \begin{array}{l} (P_1-P_k)((\partial_t^j u_1)\nabla\partial_t^{n-j} u_1) + \\ + P_k((\partial_t^j u_1)\nabla\partial_t^{n-j} w) + \\ + P_k((\partial_t^j w)\nabla\partial_t^{n-j} u_k) \end{array} \right\} \cdot \partial_t^n w \, .$$

In a first step, we establish a bound for the initial value of the function $|\partial_t^n w(t,\cdot)|^2_{L^2(\Omega)}$ in a point $t_* \in (\varepsilon, 2\varepsilon)$. Then using bounds for the three single terms of the integral in (6.4), which follow from the general inequalities stated in section 4, we obtain a linear differential inequality and thus solving it we get (6.1.n).

Inequality (6.2. n-1) being true by assumption, we have

$$\int_\varepsilon^{2\varepsilon} |\partial_\tau^n w|^2_{L^2(\Omega)} \, d\tau \le \frac{F^*_{n-1}(2\varepsilon,\varepsilon)}{\lambda_{k+1}} ,$$

the integrant being a continuous function by its definition. Thus the mean value theorem of integral calculus ensures us that the equation

$$\varepsilon \cdot |(\partial_t^n w)(t^*,\cdot)|^2_{L^2(\Omega)} = \int_\varepsilon^{2\varepsilon} |\partial_\tau^n w|^2_{L^2(\Omega)} \, d\tau$$

and therefore the inequality

$$(6.5)\qquad |(\partial_t^n w)(t^*,\cdot)|^2_{L^2(\Omega)} \le \frac{F^*_{n-1}(2\varepsilon,\varepsilon)}{\varepsilon \cdot \lambda_{k+1}}$$

holds for at least one $t_* \in (\varepsilon, 2\varepsilon)$.

Now we rewrite the first term of the integral in (6.4) according to (4.6)[6]. Then using (4.9) we get

$$\left|\int_\Omega \{(P_1-P_k)((\partial_t^j u_1)\ \nabla \partial_t^{n-j} u_1)\} \partial_t^n w \right| \le c_2 |P\Delta\partial_t^j u_1|_{L^2(\Omega)} |\nabla\partial_t^{n-j} u_1|_{L^2(\Omega)} |(1-P_k)\partial_t^n u_1|_{L^2(\Omega)} .$$

Thus by Theorem 3.1 , (3.1o) and Lemma 2.2 we have

$$(6.6)\qquad \left|\int_\Omega \{(P_1-P_k)((\partial_t^j u_1)\nabla\partial_t^{n-j} u_1)\} \partial_t^n u_1 \right| \le \frac{c_2}{\lambda_{k+1}} F_n^{1/2}(t,\varepsilon) g_{n,\varepsilon}$$

using the monotonicity of F_m and $g_{m,\varepsilon}$ with respect to m.

For the second and third term of the integral on the right side of (6.4) we note, that P_k is an orthogonal projection in L^2. Then by means of (4.1o) we obtain

---

6) As (3.2) shows, equation (4.6) also holds with $\partial_t^n w$ and $(1-P_k)\partial_t^n u_1$ instead of w and $(1-P_k)u_1$.

$$|\int_\Omega \{P_k((\partial_t^j u_1)\nabla\partial_t^{n-j}w)\}\partial_t^n w| \le \delta|\nabla\partial_t^{n-j}w|^2_{L^2(\Omega)} + c_\delta|P\Delta\partial_t^j u_1|^2_{L^2(\Omega)}|\partial_t^n w|^2_{L^2(\Omega)}$$

with $c_\delta > o$ depending on the arbitrary value $\delta > o$ only. Therefore using the bound from Theorem 3.1 , (3.1o) and in the case $j > o$ the bound from (6.2. n-j) for the first term on the right side too, we get

$$(6.7)\quad |\int_\Omega \{P_k(u_1\nabla\partial_t^n w)\}\partial_t^n w| \le \delta|\nabla\partial_t^n w|^2_{L^2(\Omega)} + c_\delta g_{o,\varepsilon}|\partial_t^n w|^2_{L^2(\Omega)}$$

and

$$(6.8)\quad |\int_\Omega \{P_k((\partial_t^j u_1)\nabla\partial_t^{n-j}w)\}\partial_t^n w| \le \frac{\delta}{\lambda_{k+1}} F^*_{n-1}(t,\varepsilon) + c_\delta\cdot g_{n,\varepsilon}|\partial_t^n w|^2_{L^2(\Omega)}$$

for $j > o$, keeping in mind the monotonicity of F^*_m and $g_{m,\varepsilon}$ in m.

Using (4.19) for the third term of the integral in (6.4) we obtain

$$|\int_\Omega \{P_k((\partial_t^j w)\nabla\partial_t^{n-j}u_k)\}\partial_t^n w| \le \delta|\nabla\partial_t^j w|^2_{L^2(\Omega)} + c_\delta|P\Delta\partial_t^{n-j}u_k|^2_{L^2(\Omega)}|\partial_t^n w|^2_{L^2(\Omega)}$$

with c_δ depending on the arbitrary value $\delta > o$ only. Thus using the bound from Theorem 3.1 , (3.1o) and the bound from (6.2. j) in the case $j < n$, too, we arrive at

$$(6.9)\quad |\int_\Omega \{P_k((\partial_t^n w)\nabla u_k)\}\partial_t^n w| \le \delta|\nabla\partial_t^n w|^2_{L^2(\Omega)} + c_\delta g_{o,\varepsilon}|\partial_t^n w|^2_{L^2(\Omega)}$$

and

$$(6.1o)\quad |\int_\Omega \{P_k((\partial_t^j w)\nabla\partial_t^{n-j}u_k)\}\partial_t^n w| \le \frac{\delta}{\lambda_{k+1}} F^*_{n-1}(t,\varepsilon) + c_\delta g_{n,\varepsilon}|\partial_t^n w|^2_{L^2(\Omega)}$$

for $j < n$. The coefficient c_δ depends on the arbitrary value $\delta > o$ only.

Now we assume $\delta \in (o,\frac{1}{4}]$. Substituting the bounds from (6.6) - (6.1o) we transform (6.4) into the differential inequality

$$(6.11)\quad \partial_t|\partial_t^n w|^2_{L^2} + |\nabla\partial_t^n w|^2_{L^2} \le \frac{a}{\lambda_{k+1}} + b|\partial_t^n w|^2_{L^2}$$

for the continuous differentiable function $|\partial_t^n w|^2_{L^2(\Omega)}$ with continuous coefficients

$$(6.12)\quad \begin{aligned} a(t) &= 2^{n+1}(c_2 F_n^{1/2}(t,\varepsilon)g_{n,\varepsilon} + 2\,\delta F^*_{n-1}(t,\varepsilon)) \quad \text{and} \\ b &= 4\,c_\delta\{g_{o,\varepsilon} + 2^n g_{n,\varepsilon}\}. \end{aligned}$$

The differential inequality (6.11) together with the initial bound (6.5) yields the assertion (6.1.n) with the help of

Corollary 6.1.: Let $a(t) \geq o$, $b(t) \geq o$, $\varphi^*(t) \geq o$ be continuous functions, $\varphi(t)$ be continuously differentiable on $[\varepsilon_0,T]$ and assume the inequalities

$$\varphi' + \varphi^* \leq \frac{a}{\lambda} + b\varphi \quad \text{on} \quad [\varepsilon_0,T] \quad \text{and} \tag{6.13}$$

$$\varphi(t^*) \leq \frac{a_0}{\lambda}$$

hold with constants $a_0 \geq o$, $\lambda > o$ and a value $t_* \in [\varepsilon_0, \varepsilon_0 + \varepsilon]$. Then we have

$$\varphi(t) + \int_{\varepsilon_0+\varepsilon}^{t} \varphi^* \, d\tau \leq \frac{A(t,\varepsilon_0+\varepsilon)}{\lambda} \quad \text{on} \quad [\varepsilon_0+\varepsilon,T] \tag{6.14}$$

with the continuous functions

$$A(t, \varepsilon_0+\varepsilon) = a_0 + \int_{\varepsilon_0}^{t} \{a(\tau)+b(\tau)\psi(\tau,\varepsilon_0+\varepsilon)\} \, d\tau \quad \text{and} \tag{6.15}$$

$$\psi(t,\varepsilon_0+\varepsilon) = \{a_0 + \int_{\varepsilon_0}^{t} a(\tau) e^{-\int_{\varepsilon_0+\varepsilon}^{t} b(\tau)d\tau}\} e^{\int_{\varepsilon_0}^{t} b(\tau)d\tau}$$

being monotonously increasing in the variable t.

The proof of the Corollary follows with the methods of differential inequalities from the fact that our assumptions on φ and the definition of ψ imply

$$\varphi' - (\frac{a}{\lambda} + b\varphi) \leq o \leq (\frac{\psi}{\lambda})' - (\frac{a}{\lambda} + b\cdot(\frac{\psi}{\lambda})) = \frac{a}{\lambda} \cdot (e^{\int_{\varepsilon_0}^{\varepsilon_0+\varepsilon} b(\tau)d\tau} - 1)$$

as well as $\varphi(t_*) \leq \frac{a_0}{\lambda} \leq \frac{\psi(t*)}{\lambda}$, hence $\varphi \leq \frac{\psi}{\lambda}$ on $[t_*,T]$. Substituting $\frac{\psi}{\lambda}$ in (6.13) and integrating as indicated yield the assertion.

This Corollary with $\varepsilon_0 = \varepsilon$ shows, that the differential inequality (6.11) together with the initial condition (6.5) result in

$$|\partial_t^n(u_l(t,\cdot)-u_k(t,\cdot))|^2_{L^2(\Omega)} + \int_{2\varepsilon}^{t} |\nabla\partial_\tau^n(u_l(\tau,\cdot)-u_k(\tau,\cdot))|^2_{L^2(\Omega)} d\tau \leq \frac{A(t,2\varepsilon)}{\lambda_{k+1}} \tag{6.16}$$

on $[2\varepsilon,T]$ with the functions A from (6.15), a, b from (6.12) which are independent of the special Galerkin approximation u_l, $l > k$. Now we write ε instead of 2ε. According to Corollary 3.1 we can find a subsequence $(u_{l'}(t,\cdot))$ converging together with all $\partial_t^j u_{l'}(t,\cdot)$, $j = o, \ldots ,n$ in H_1 to the respective limit $\partial_t^j u(t,\cdot)$. The convergence being uniform on $[\varepsilon,T]$, in (6.16) we can pass to the limit $l' \to \infty$ thus obtaining (6.1.n) with the right side $\frac{A(t,\varepsilon)}{\lambda_{k+1}}$, which is monotonously increasing in t. Defining $G_n^*(t,\varepsilon) = \max\{A(t,\varepsilon), G_{n-1}^*(t,\varepsilon)\}$, the sequence (G_m^*) becomes monotonously increasing in $m = o,\ldots,n$ too.

In order to prove (6.2. n) we take the inner product of (6.3) in $L^2(\Omega)$ with the vector function $\partial_t^{n+1} w \in \mathcal{X}_1$. With (1.3) we obtain the differential equation

$$(6.17)\quad \partial_t|\nabla\partial_t^n w|^2_{L^2(\Omega)} + 2|\partial_t^{n+1}w|^2_{L^2(\Omega)} = -\sum_{j=0}^{n} 2\binom{n}{j}\int_\Omega \left\{\begin{array}{l}(P_1-P_k)((\partial_t^j u_1)\nabla\partial_t^{n-j}u_1) + \\ P_k((\partial_t^j u_1)\nabla\partial_t^{n-j}w) + \\ P_k((\partial_t^j w)\nabla\partial_t^{n-j}u_k)\end{array}\right\}\cdot\partial_t^{n+1}w$$

for the function $|\nabla\partial_t^n w(t,\cdot)|^2_{L^2(\Omega)}$. A bound for its value in a point $t^* \in (2\varepsilon,3\varepsilon)$ follows from (6.1.n). Namely by this inequality we have

$$\int_{2\varepsilon}^{3\varepsilon} |\nabla\partial_t^n w|^2_{L^2(\Omega)}\, d\tau \le \frac{G_n^*(3\varepsilon,2\varepsilon)}{\lambda_{k+1}} .$$

The integrant being a continuous function we conclude with the mean value theorem of integral calculus that the equation

$$\varepsilon\cdot|(\nabla\partial_t^n w)(t^*,\cdot)|^2_{L^2(\Omega)} = \int_{2\varepsilon}^{3\varepsilon} |\nabla\partial_t^n w|^2_{L^2(\Omega)}\, d\tau$$

and hence the inequality

$$(6.18)\quad |(\nabla\partial_t^n w)(t^*,\cdot)|^2_{L^2(\Omega)} \le \frac{G_n^*(3\varepsilon,2\varepsilon)}{\varepsilon\cdot\lambda_{k+1}}$$

holds for at least one $t^* \in (2\varepsilon,3\varepsilon)$.

The first term of the integral in (6.17) we rewrite according to (4.6)[6]. Then using (4.9) we get

$$(6.19)\quad |\int_\Omega\{(P_1-P_k)((\partial_t^j u_1)\nabla\partial_t^{n-j}u_1)\}\,\partial_t^{n+1}w| \le c_2|P\Delta\partial_t^j u_1|_{L^2(\Omega)}|\nabla\partial_t^{n-j}u_1|_{L^2(\Omega)}\cdot$$

$$\cdot\,|(1-P_k)\partial_t^{n+1}u_1|_{L^2(\Omega)}$$

$$\le \frac{c_2}{\lambda_{k+1}}\, g_{n+1,\varepsilon}\, F_n^{1/2}(t,\varepsilon) .$$

The last inequality follows by Lemma 2.2 and Theorem 3.1 ,(3.7), (3.1o) because of the monotonicity of $g_{m,\varepsilon}$ and F_m with respect to m.

For the second and third term of the integral in (6.17) we keep in mind (4.13). Then by means of (4.11) we obtain

$$|\int_\Omega\{P_k((\partial_t^j u_1)\nabla\partial_t^{n-j}w)\}\partial_t^{n+1}w| \le \delta|\partial_t^{n+1}w|^2_{L^2(\Omega)} + c_\delta|P\Delta\partial_t^j u_1|^2_{L^2(\Omega)}|\nabla\partial_t^{n-j}w|^2_{L^2(\Omega)} .$$

Therefore using the bound of Theorem 3.1 , (3.1o) and in the case $j > o$ the bound from (6.2. n-j) too, we get

$$(6.2o)\quad |\int_\Omega \{P_k(u_1\nabla\partial_t^n w)\}\partial_t^{n+1}w| \le \delta|\partial_t^{n+1}w|^2_{L^2(\Omega)} + c_\delta g_{o,\varepsilon}|\nabla\partial_t^n w|^2_{L^2(\Omega)}$$

and

$$(6.21)\quad |\int_\Omega \{P_k((\partial_t^j u_1)\nabla\partial_t^{n-j}w)\}\partial_t^{n+1}w| \le \delta|\partial_t^{n+1}w|^2_{L^2(\Omega)} + c_\delta g_{n,\varepsilon}\cdot F^*_{n-1}(t,\varepsilon)$$

for $j > o$ due to the monotonicity of $g_{m,\varepsilon}$ and F^*_m in m.

With the help of (4.2o) for the third term of the integral in (6.17) we get

$$|\int_\Omega \{P_k((\partial_t^j w)\nabla\partial_t^{n-j}u_k)\}\partial_t^{n+1}w| \le \delta|\partial_t^{n+1}w|^2_{L^2(\Omega)} + c_\delta|\nabla\partial_t^j w|^2_{L^2(\Omega)}|P\Delta\ \partial_t^{n-j}u_k|^2_{L^2(\Omega)}$$

with the coefficient $c_\delta > o$ depending on the arbitrary value $\delta > o$ only. Substituting the bound from Theorem 3.1 , (3.1o) and in the cases $j < n$ the bounds from (6.2.j) too, results in

$$(6.22)\quad |\int_\Omega \{P_k((\partial_t^n w)\nabla u_k)\}\partial_t^{n+1}w| \le \delta\cdot|\partial_t^{n+1}w|^2_{L^2(\Omega)} + c_\delta g_{o,\varepsilon}|\nabla\partial_t^n w|^2_{L^2(\Omega)}$$

and

$$(6.23)\quad |\int_\Omega \{P_k((\partial_t^j w)\nabla\partial_t^{n-j}u_k)\}\partial_t^{n+1}w| \le \delta|\partial_t^{n+1}w|^2_{L^2(\Omega)} + \frac{c\delta}{\lambda_{k+1}}\, g_{n,\varepsilon}\cdot F^*_{n-1}(t,\varepsilon) \quad \text{for } j < n.$$

Now assume $\delta \in (o,2^{-(n+3)})$. Collecting the bounds from (6.18)-(6.23) we obtain from (6.17) the differential inequality

$$(6.24)\quad \partial_t|\nabla\partial_t^n w|^2_{L^2(\Omega)} + |\partial_t^{n+1}w|^2_{L^2(\Omega)} \le \frac{a}{\lambda_{k+1}} + b|\nabla\partial_t^n w|^2_{L^2(\Omega)}$$

on $[\varepsilon,T]$ for the continuously differentiable function $|\nabla\partial_t^n w(t,\cdot)|^2_{L^2(\Omega)}$ with the continuous coefficients

$$(6.25)\quad \begin{aligned} a(t) &= 2^{n+1}(c_2F_n^{1/2}(t,\varepsilon)g_{n+1,\varepsilon} + 2c_\delta\ g_{n,\varepsilon}\ F^*_{n-1}(t,\varepsilon)) \quad \text{and} \\ b &= 4\ c_\delta g_{o,\varepsilon} \quad . \end{aligned}$$

According to Corollary 6.1 (with $\varepsilon_o = 2\varepsilon$) this differential inequality together with the initial condition (6.18) yields

$$(6.26)\quad |\nabla\partial_t^n(u_l(t,\cdot)-u_k(t,\cdot))|^2_{L^2(\Omega)} + \int_{3\varepsilon}^{t} |\partial_\tau^{n+1}(u_l(\tau,\cdot)-u_k(\tau,\cdot))|^2_{L^2(\Omega)}\, d\tau \le \frac{A(t,3\varepsilon)}{\lambda_{k+1}} ,$$

the functions $A(t,3\varepsilon)$ from (6.15) and a, b from (6.25) being independent of the special Galerkin approximation u_l, $l > k$. Further on we write ε instead of 3ε. As Corollary 3.1 shows, we can take a subsequence $(u_{l'}(t,\cdot))$ converging with all time-derivatives $\partial_t^m u_{l'}(t,\cdot)$, $m = 0,\dots,n+1$ in H_1 to $\partial_t^m u(t,\cdot)$, respectively. This convergence being uniform on $[\varepsilon,T]$, in (6.26) we can pass to the limit $l' \to \infty$ thus obtaining (6.2.n) with the right side $\frac{A(t,\varepsilon)}{\lambda_{k+1}}$, which is monotonously increasing in t. Defining $F_n^*(t,\varepsilon) = \max\{A(t,\varepsilon), F_{n-1}^*(t,\varepsilon)\}$ the sequence (F_m^*) becomes monotonously increasing in $m = 0,1,..,n$, too.

References:

1. Adams, R.A., Sobolev Spaces, Academic Press, New York 1975.

2. Cattabriga, L., Su un problema al contorno relativo al sistema di equazioni di Stokes, Rend. Mat. Sem. Univ. Padova 31 (1961) 3o8-34o.

3. Foias, C., Statistical study of Navier-Stokes equations I, Rend. Mat. Sem. Univ. Padova 48 (1972) 219-348.

4. Foias, C., Temam, R., Some analytic and geometric properties of the solutions of the evolution Navier-Stokes equations, Prepublication Univ. de Paris-Sud, Orsay, 1979.

5. Friedman, A., Partial Differential Equations, Holt, Rinehart and Winston, New York 1969.

6. Heywood, J.G., On nonstationary Stokes flow past an obstacle, Indiana Univ. Math. J. 24 (1974) 271-284.

7. Heywood, J.G., The Navier-Stokes equations: On the existence, regularity and decay of solutions, Preprint, Univ. of British Columbia, Vancouver 1978.

7a. Heywood, J.G., Classical Solutions of the Navier-Stokes Equations (these proceedings).

8. Hopf, E., Über die Anfangswertaufgabe für die hydrodynamischen Grundgleichungen, Math. Nachr. 4 (1951) 213-231.

9. Ito, S., The existence and the uniqueness of regular solution of non-stationary Navier-Stokes equation, J. Fac. Sci. Univ. Tokyo Sect. I A, 9 (1961) 1o3-14o.

1o. Jörgens, K., Rellich, F., Eigenwerttheorie gewöhnlicher Differentialgleichungen Springer, Berlin 1976.

11. Kaniel, S., Shinbrot, M., The initial value problem for the Navier-Stokes equations, Arch. Rat. Mech. An. 21 (1966) 27o-285.

12. Ladyženskaja, O.A., The Mathematical Theory of Viscous Incompressible Flow, Second Edition, Gordon and Breach, New York 1969.

13. Lang, S., Real Analysis, Addison-Wesley Publ., Reading (Mass.) 1973.

14. Leis, R., Vorlesungen über partielle Differentialgleichungen zweiter Ordnung, Bibliographisches Institut, Mannheim 1967 (BI 165/165 a).

15. Métivier, G., Etude asymptotique des valeurs propres et de la fonction spectrale des problèmes aux limites. Thèse, Univ. de Nice 1976.

16. Prodi, G., Teoremi di tipo locale per il sistema di Navier-Stokes e stabilità delle soluzione stazionarie Rend. Mat. Sem. Univ. Padova 32 (1962) 374-397.

17. Rautmann, R., On the convergence of a Galerkin method to solve the initial value problem of a stabilized Navier-Stokes equation, Numerische Behandlung von Differentialgleichungen, edited by R. Ansorge, L. Collatz, G. Hämmerlin, W. Törnig, ISNM 27, Birkhäuser, Basel (1975) 255-264.

18. Rautmann, R., Eine Fehlerschranke für Galerkinapproximationen lokaler Navier-Stokes-Lösungen, Constructive Methods for Nonlinear Boundary Value Problems and Nonlinear Oscillations, edited by J. Albrecht, L. Collatz, K. Kirchgässner, ISNM 48, Birkhäuser, Basel (1979) 11o-125.

19. Serrin, J., The initial value problem for the Navier-Stokes equations, Nonlinear Problems, edited by R.E. Langer, The University of Wisconsin Press, Madison, 1963.

2o. Shinbrot, M., Lectures on Fluid Mechanics, Gordon and Breach, New York 1973.

21. Taylor, A.E., Introduction to Functional Analysis, Wiley, New York 1958.

22. Temam, R., Navier-Stokes Equations, North-Holland Publ. Comp., Amsterdam 1977.

23. Walter, W., Differential and Integral Inequalities, Springer, Berlin 197o.

24. Weyl, H., The method of orthogonal projection, Duke Math. J. 7 (194o) 411-444.

OPTIMISATION OF HERMITIAN METHODS FOR NAVIER-STOKES EQUATIONS IN THE VORTICITY AND STREAM-FUNCTION FORMULATION

by B. ROUX and P. BONTOUX

Institut de Mécanique des Fluides de Marseille
1, rue Honnorat, 13003 MARSEILLE

and by TA PHUOC LOC and O. DAUBE

Laboratoire d'Informatique pour la Mécanique et les Sciences
de l'Ingénieur - B.P. 30 Campus d'Orsay
91406 ORSAY CEDEX

*

ABSTRACT

The aim of this paper is to show the efficiency of a "combined" method proposed to solve the 2D incompressible Navier-Stokes equations in the vorticity and stream-function formulation. A compact hermitian scheme is used for the stream function equation while a classical second order accurate scheme is taken for the vorticity equation. Comparisons have been made with purely hermitian or purely second order accurate methods. Numerical experiments have been carried out for a large variety of steady and unsteady, internal or external flows. For these problems, comparisons have been made with available experimental or computed data.

1. INTRODUCTION

The main difficulty in solving the Navier-Stokes equations in the vorticity and stream function formulation, for the numerical simulation of viscous incompressible flows is related to the boundary condition for the vorticity on the rigid walls. This condition, namely :

$$\zeta_w = (\nabla^2\psi)_w$$

couples the vorticity, ζ, and the streamfunction, ψ. This coupling is the source of limitations for stability and accuracy of a numerical solution of NS equations by an ADI (alternating - direction implicit) method, as shown by Pearson (1965), Roache (1972),Orszag and Israeli (1974) and Hirsh (1975), among others. More recently an heuristic extension of the Von Neuman - Fourier analysis, by Bontoux et al. (1979) has clarified the role of ζ_w on the numerical stability.

To partly avoid the problem involved by the vorticity at the wall, and mainly to ease the use of a divergence form of the NS equations, the so-called "combined" method has been proposed by Ta Phuoc Loc and Daube (1977). In such a method a high order hermitian technique is used, in the compact form proposed by Hirsh (1975), for the streamfunction equation ; but the transport equation of the vorticity which involves transport terms is discretized with a second order accurate scheme.

The aim of this paper is to show the efficiency of this combined method, compared to purely hermitian (compact) and to purely second order accurate methods. This discussion is made mainly in the frame of two model problems. The first one, proposed by Pearson (1965), admits analytical solutions. The second is the well known driven cavity problem studied by numerous authors.

The flexibility of the method allowed its use for several others problems as steady natural convection in cavities, oscillatory flow in constricted channel, unsteady flows around elliptic cylinders. In these problems, special attention has been devoted to trying situations characterized by high values of Reynolds or Rayleigh numbers, by using the divergence form of the NS equations. The results are compared with the available experimental or computed data.

2. GOVERNING EQUATIONS

The study is devoted to the simulation of a number of viscous flow problems, which are governed by the 2D - NS equations. The NS equations are considered with the vorticity, the streamfunction and the local temperature as dependent variables. Under the convective form, they are written as follows :

$$(1) \qquad \zeta_t + u\,\zeta_x + v\,\zeta_y = a_1\,(\zeta_{xx} + \zeta_{yy}) + a_2\,(\sin\Omega T_y + \cos\Omega\, T_x)$$

$$(2) \qquad \psi_{xx} + \psi_{yy} = \zeta$$

$$(3) \qquad T_t + u\,T_x + v\,T_y = a_3\,(T_{xx} + T_{yy})$$

where $u = \psi_y$ and $v = -\psi_x$

The coefficients a_1, a_3 and a_2 of the diffusive and source terms depend on the physical parameters of the particular problems considered : the Reynolds number, Re, the Prandtl number, Pr, and the Rayleigh number, Ra. The different problems studied herein are summarized in Table 1. They concern model problems (I,II), internal flows (III,IV) and external flows (V, VI, VII). In problem III devoted to the natural convection problem in a differentially heated cavity, the NS equations are coupled with the energy equation through buoyancy forces (Roux et al, 1978). Problem IV is related to the influence of an oscillating pressure gradient on the flow through a stenotic obstruction (Forestier et al, 1977). Problems V, VI and VII concern unsteady flows around bodies, the purpose is the prediction of the time-dependent flow characteristic and structure. The mechanism of the creation and the transport of Von Karman vortices behind an elliptic cylinder is studied (Daube and Ta Phuoc Loc, 1978). Secondary vortices observed experimentally behind an impulsively started circular cylinder can be now simulated numerically (Ta Phuoc Loc, 1979). Finally unsteady flow properties of a heaving elliptic cylinder are for the first time analyzed (Daube and Ta Phuoc Loc, 1979).

The boundary conditions on the rigid walls are given by the no-slip and no-permeability conditions for the streamfunction, and by equation (2) for the vorticity :

$$\psi_x = 0 \qquad\qquad (\psi_\eta)_w = 0 \qquad\qquad \psi_w = (\psi_{\eta\eta})_w$$

where η is the normal to the rigid wall.

		GOVERNING EQUATIONS	a_1	a_2	a_3	CHARACTERISTICS	REFERENCES
I	PEARSON'S PROBLEM	(1), (2)	1	0		$\psi = \exp(-2\pi t^2) \sin \pi y \cos \pi x$	PEARSON (1965)
II	DRIVEN CAVITY	(1), (2)	$\frac{1}{Re}$	0		V_0	BONTOUX et al (1978)
III	NATURAL CONVECTION	(1),(2),(3)	Pr	RaPr	1	T_1 Ω $T_2>T_1$ g	ROUX et al (1978)
IV	STENOSIS OSCILLATORY FLOW ($p_z = A \sin wt$)	(1), (2)	$\frac{1}{Re}$	0			FORESTIER et al (1977)
	UNSTEADY FLOWS AROUND CYLINDERS:	(1), (2)	$\frac{1}{Re}$	0			
V	ELLIPTIC CYL.					V_0	DAUBE et TA PHUOC LOC (1978)
VI	IMPULSIVELY STARTED CIRCULAR CYL.					V_0	TA PHUOC LOC (1979)
VII	HEAVING ELLIPTIC CYL.					V_0	DAUBE et TA PHUOC LOC (1979)

Table 1

3. NUMERICAL METHODS

The numerical solution of the NS equations is performed by using the ADI scheme proposed by Peaceman and Rachford (1955) for the diffusion equation and which has been extended by Pearson (1965) to the case of the transport equations. The scheme is known as one of the most efficient (Isaacson et Keller, 1966).

3.1. Temporal discretization

The advancement is accomplished over two time steps, $\Delta t/2$, for the vorticity and energy equation (1) and (3). The streamfunction equation (2) is similarly solved under a pseudo-unsteady formulation.

According to Roache (1972), the ADI scheme can be written as :

$$\frac{2}{\Delta t}(\zeta^{*}-\zeta^{n}) + u^{\alpha}\zeta_x^{n} + v^{\alpha}\zeta_y^{*} = a_1(\zeta_{xx}^{n} + \zeta_{yy}^{*}) + a_2(\sin\Omega\, T_y^{\beta} + \cos\Omega\, T_x^{\beta}) \quad (4a)$$

$$\frac{2}{\Delta t}(\zeta^{n+1}-\zeta^{*}) + u^{\alpha}\zeta_x^{n+1} + v^{\alpha}\zeta_y^{*} = a_1(\zeta_{xx}^{n+1} + \zeta_{yy}^{*}) + a_2(\sin\Omega\, T_y^{\beta} + \cos\Omega\, T_x^{\beta}) \quad (4b)$$

$$\lambda_{\ell 1}(\psi^{n+1,*} - \psi^{n+1,\ell}) = \psi_{xx}^{n+1,\ell} + \psi_{yy}^{n+1,*} - \zeta^{n+1} \quad (5a)$$

$$\lambda_{\ell 2}(\psi^{n+1,\ell+1} - \psi^{n+1,*}) = \psi_{xx}^{n+1,\ell+1} + \psi_{yy}^{n+1,*} - \zeta^{n+1} \quad (5b)$$

$$\frac{2}{\Delta t}(T^{*} - T^{n}) + u^{\alpha} T_x^{n} + v^{\alpha} T_y^{*} = a_3(T_{xx}^{n} + T_{yy}^{*}) \quad (6a)$$

$$\frac{2}{\Delta t}(T^{n+1} - T^{*}) + u^{\alpha} T_x^{n+1} + v^{\alpha} T_y^{*} = a_3(T_{xx}^{n+1} + T_{yy}^{*}) \quad (6b)$$

in which the accuracy of the approximation depends on the evaluations of the non linear terms and of the boundary conditions on ζ as indicated in Table 2 :

Variable	Formulation	Reference	Accuracy	
u^{α}	u^{n}	Sun and Hanratty (1969)	$O(\Delta t)$	(7a)
	$\frac{1}{2}(u^{n+1} + u^{n})$	Pearson (1965)	$O(\Delta t^2)$	(7b)
ζ_w^{n+1}	$(\psi_{\eta\eta})_w^{n}$	Wilkes and Churchill(1966)	$O(\Delta t)$	(7c)
	$(\psi_{\eta\eta})_w^{n+1}$	Pearson (1965)	$O(\Delta t^2)$	(7d)
ζ_w^{*}	$\frac{1}{2}(\zeta_w^{n+1} + \zeta_w^{n})$	Peaceman and Rachford (1955)	$O(\Delta t)$	(7e)
	$\frac{1}{2}(\zeta_w^{n+1}+\zeta_w^{n}) - \frac{a_1\Delta t}{4}(\zeta_{xx}^{n+1}-\zeta_{xx}^{n})$	Fairweather and Mitchell (1967)	$O(\Delta t^2)$	(7f)

Table 2

In our numerical experiments for Pearson's problem, the formulation (7e) do not destroy the second order accuracy of the scheme when used with (7b) and (7d). These formulations are kept for problem IV, while in the case of the external flows (V, VI and VII), the formulation (7a) has been used. For the steady flow problems (II and III), the formulations (7a) and (7c), which save computer memories, have been retained.

The coefficients $\lambda_{\ell 1}$ and $\lambda_{\ell 2}$ have been determined from the optimal values proposed by Wachpress (1966) for problems V to VII and have been optimized from numerical experiments for the other problems (Bontoux et al, 1978).

3.2. Spatial discretization

Equations (4-6) take the following standard form :

$$(8) \quad m_{1i}\, f_i + m_{2i}\, (f_x)_i + m_{3i}\, (f_{xx})_i + m_{4i} = 0 \quad , \qquad 1 < i < N$$

The spatial discretization will be considered for this standard one - dimensional equation.

3.2.1. Compact Hermitian methods

The method developped by Hirsh (1975), which consider the derivatives as supplementary unknowns, use the following hermitian relations :

$$(9) \quad (f_x)_{i+1} + 4\, (f_x)_i + (f_x)_{i-1} = \frac{3}{h}\,(f_{i+1} - f_{i-1}) + 0\,(h^4)$$

$$(10) \quad (f_{xx})_{i+1} + 10\,(f_{xx})_i + (f_{xx})_{i-1} = \frac{12}{h^2}\,(f_{i+1} - 2f_i + f_{i-1}) + 0\,(h^4)$$

$$1 < i < N$$

where h is the spatial step size. At the boundaries the closure of the system is ensured by using the "Padé approximant" (Hirsh, 1975) :

$$(11) \quad (f_{xx})_i - (f_{xx})_{i+1} + \frac{6}{h}\left[(f_x)_i + (f_x)_{i+1}\right] + \frac{12}{h^2}\left[f_i - f_{i+1}\right] = 0 + 0\,(h^3)$$

In the version given by Hirsh the equations (8-11) and the boundary conditions lead to a (3 x 3) block tridiagonal system. A second version, based on the elimination of the second derivatives has been proposed by Adam (1975) and leads to a (2 x 2) block tridiagonal system. When the coefficient of the first derivative in (8) is zero, as for the streamfunction equation, a third version can be used in which f is solution of a single tridiagonal system. However, for these two last versions, the term m_{4i} involves the two first derivatives of f as we deal with 2D problem, and f_x and f_{xx} must be calculated as solution of single tridiagonal systems.

3.2.2. Other Hermitian methods

There are other techniques using derivatives as additional unknowns. Among these we mention the Mehrstellen method developped by Krause et al (1976) in which f_x and f_{xx} are substituted in terms of f. In this way f becomes solution of a single tridiagonal system. But the calculation of the coefficients of this system requires additional computing time which has not been evaluated. We also mention the Spline methods proposed by Rubin and Khosla (1977) based on polynomial interpolation instead of Taylor series expansions.

3.2.3. Combined methods

One of the main difficulty of hermitian methods compared to the classical second order accurate schemes lies in the imposition of boundary conditions for the variables and for their derivatives. This is not a problem for the streamfunction equation when the field is bounded by rigid walls. For that réasons, a "combined" method has been proposed by Ta Phuoc Loc and Daube (1977) in which the hermitian method is only used for the streamfunction equation. In addition, the use of a classical scheme for the vorticity equation eases the extension of the combined method to the divergence form of the NS equations and to the inclusion of an upwind difference procedure for the transport terms, with a first or second order accuracy. The discretization of these transport terms is then :

$$(12) \qquad |(v\zeta)_y|^*_{ij} = \frac{1}{2h}\left(e_1\, v^n_{ij+1}\, \zeta^*_{ij+1} + e_2\, v^n_{ij}\, \zeta^*_{ij} + e_3\, v^n_{ij-1}\, \zeta^*_{ij-1}\right)$$

$$(13) \qquad |(u\zeta)_x|^{n+1}_{ij} = \frac{1}{2h}\left(e_1\, u^n_{i+1j}\, \zeta^{n+1}_{i+1j} + e_2\, u^n_{ij}\, \zeta^{n+1}_{ij} + e_3\, u^n_{i-1j}\, \zeta^{n+1}_{i-1j}\right)$$

where e_1, e_2, e_3 are coefficients which depend on the scheme and on the values of the velocity components (Bozeman and Dalton, 1973).

3.2.4. Vorticity at the rigid walls

The boundary conditions for the vorticity are derived from equation (2) and couple the vorticity and the streamfunction in the field, by one of the following relations given by Hirsh (1975) and Roache (1972)

$$(14a) \qquad \zeta_w = \frac{12}{h^2}(\psi_{w-1} - \psi_w) + \frac{6}{h}\left((\psi_\eta)_{w-1} + (\psi_\eta)_w\right) + (\psi_{\eta\eta})_{w-1} + O(h^3)$$

$$(14b) \qquad \zeta_w = \frac{3}{h^2}(\psi_{w-1} - \psi_w) + \frac{3}{h}(\psi_\eta)_w - \frac{1}{2}\,\zeta_{w-1} + O(h^2)$$

$$(14c) \qquad \zeta_w = \frac{1}{2h^2}(-\psi_{w-2} + 8\,\psi_{w-1} - 7\,\psi_w) + \frac{3}{h}(\psi_\eta)_w + O(h^2)$$

where w-1 and w-2 characterize the first two points near the wall. This coupling is the source of limitations for the stability and for the accuracy of the numerical solution (Bontoux, Gilly and Roux, 1978).

4. ACCURACY AND STABILITY OF THE SOLUTION

4.1. Accuracy of the solution of Pearson's problem

The model problem proposed by Pearson (1965) corresponds to the unsteady NS equations for Re = 1 ($a_1 = 1$, $a_2 = 0$). An exact solution of this problem for compatible boundary conditions, is :

$$\psi = \exp(-2\pi^2 t)\cos\pi x\cos\pi y \quad ; \quad \zeta = 2\pi^2\psi$$

$$u = -\pi\exp(-2\pi^2 t)\cos\pi x\sin\pi y \quad ;$$

$$v = \pi \exp(-2\pi^2 t) \sin \pi x \cos \pi y$$

The losses of accuracy introduced by the evaluations of the terms u, v and ζ_w in equations (4-7) are controlled by evaluating the residuals :

$$\varepsilon_f = \text{Max} \ |f^n - \bar{f}^n| \ / \ \text{Max} \ |\bar{f}^n|$$

(where $\bar{f}^n$ is the exact solution) for different number of mesh points in x and y directions, $N = 1/\Delta x = 1/\Delta y = 1/h$.

4.1.1. Influence of the coupling terms u and v

The residuals ε_ζ are presented versus N on figure 1, for the case where the variables u and v are fixed (ψ fixed) or where they are effectively calculated. The role of these non linear terms appears to be very important when the second order scheme is used, specially if N is increased, but it remains weak when the compact hermitian method is used, whether using (14a) or (14c) for ζ_w.

4.1.2. Influence of ζ_w

The influence of ζ_w is also shown on Fig. 1, by comparing the results obtained for ψ fixed in the different cases where ζ_w is fixed or calculated by (14a) or (14c). The role of ζ_w is very important even for the compact hermitian scheme. Fig. 1 shows the significant improvement obtained with (14a) instead of (14c). For the second order accurate scheme, the third order accurate expression (14a) gives meaningless results.

4.1.3. Comparisons of the hermitian, combined and second order methods

The calculations are made for ψ_w and $(\psi_\eta)_w$ given, which is a practical case. The residuals determined for the variables ζ, u and ψ are shown on Fig. 2. The importance of the use of the $O(h^3)$ evaluation (14a) of ζ_w is again clearly shown either for the hermitian or the combined methods. When the required level of accuracy is characterized by $\varepsilon_f < 10^{-4}$ the saving of the number of mesh points, N, is obvious when the hermitian scheme is used. However, when this level corresponds to $\varepsilon_f > 10^{-4}$, as for many numerical simulations, the combined method leads to the same accuracy as the hermitian method for the same value of N, and gives a good improvement compared to the second order method.

4.2. Divergence form of the equations

When applied to the convective form (4-6) of the NS equations, as in the case of the driven cavity problem for Re sufficiently high, the compact hermitian method does not escape to the classical cell Reynolds number limitation for the stability (Roache, 1972). As an example, results obtained for Re = 200, exhibit wiggles for h = 1/10 (Fig. 3a). These wiggles are confined near the driving wall for h = 1/20 (Fig. 3b) and disappear for h = 1/35 (Fig. 3c). Quite similar results are obtained either with the combined method or with the second order accurate method. It was shown by Bozeman and Dalton (1973), for the second order schemes, that the divergence form of the NS equations improves the accuracy. This points out an other advantage of the combined method which can be directly generalized to solve such a divergence form. The results obtained in this case, even for h = 1/10 (Fig. 3d), compare favorably with those given by the convective form with h = 1/35.

5. OPERATIONAL COUNTS

The computer time needed by the different methods considered herein (compact hermitian, combined or second order) depends mainly on the type of the p-Thomas algorithm (denoted T_p in the table below) which is used to solve the (pxp) block tridiagonal system described in paragraph 3. However, for the steady problems, it depends also on the rate of convergence, which is connected with the number of mesh points, N, needed to achieve a given accurary.

5.1. Operational counts (OC) for the p-Thomas algorithm

According to Isaacson and Keller (1966) the operational count (including only multiplication and division) for a p-Thomas algorithm is given as follows :

- for a single tridiagonal system (p = 1)

$$OC = 5N - 4$$

- for m systems

$$OC = (3N - 2)\, m + 2N - 2$$

- for a single (p x p) block tridiagonal system

$$OC = (3N - 2)\, p^2\, (p + 1)$$

The operational counts involved by the three versions of the compact hermitian method and by the second order method to solve the standard equation (8), are summarized in Table 3, taking into account the two following points :

- Calculation of the two first derivatives, f_x and f_{xx}, is necessary even in the second order accurate method as $m_{4i} = m_{4i}\,(f_y, f_{yy})$ for the 2D-NS equations. For the second order accurate methods such a calculation is explicit.

- Some computation time can be saved when the coefficients of the system are constant (starred symbols in Table 3).

Applying these results, when N >> 1, to one half step of the ADI method for the solution of the NS equations, we can evaluate the minimum total OC involved for the different methods (in their optimal version). The combined method, with the version 3 for ψ, appears to require an increase of only 15 % of OC, compared to the second order accurate method. An other result is that by using brutally the version 1, the hermitian method for equations ψ and ζ would require 216 N^2 operations instead of 46 N^2 (Table 4).

5.2. Rate of convergence

As mentioned before the rate of convergence which is the inverse of the number of iterations, N_i, necessary to bring the solution up to a given accuracy, depends on the number of mesh points, N. This is shown on the Fig. 4 in the case of Pearson's problem. The step size Δt was taken small enough as not to introduce significant temporal truncation errors. In this case the rate of convergence to reach $\varepsilon_\zeta \simeq 10^{-8}$ behaves approximatively like N^{-2} (or h^2). This shows the interest of a higher order accurate scheme which involves smaller values of N.

Equation (8)		Variable coefficients			m_{1i}, m_{2i}, m_{3i} constant (ψ)		
Variables		f	f_x	f_{xx}	f	f_x	f_{xx}
Compact Hermitian Schemes	Version 1	T_3			T_3		
	Version 2	T_2		T_1^*	T_2		Explicit
	Version 3	no			T_1^*	T_1^*	Explicit
Second order Scheme		T_1	Explicit	Explicit	T_1^*	Explicit	Explicit

Table 3

Method	Total OC
Hermitian (version 1 for ζ and ψ)	216 N^2
Hermitian (version 2 for ζ and version 3 for ψ)	46 N^2
Combined (version 3 for ψ)	14 N^2
Second order	12 N^2

Table 4

6. INTERNAL FLOWS

6.1. Driven cavity

Calculations have been carried out for the driven cavity problem. The comparisons have been made for Re = 100 by using either the convective or the divergence form of the NS equations. The results obtained for the maximum of ψ are given in Fig. 5. The hermitian method gives good results even for $h = 1/10$, in spite of using the convective form of the equations. The combined method gives also good results, even for $h = 1/10$, but only if the divergence form of the equations is used. Of course for the second order method also, the use of the divergence form improves the solution. Comparisons have been made with the results given by Burggraf (1966), Gosman et al (1969) and Rubin and Khosla (1977). They show a good coherence between all these results, specially when $h \to 0$.

Furthermore we found that the divergence form is necessary for $N < 35$ as soon as $Re > 400$.

For higher values of Re, we have to mention an important influence of the mesh size on the structure of the flow. In fact, using the combined method for the divergence form of the equations, the flow in the right corner of the bottom of the cavity (Fig. 6) is attached for $h = 1/20$, but it presents a vortex when the mesh size is refined ($h = 1/35$). This last result is in agreement with computed data published by De Vahl Davis and Mallinson (1976) and by Benazeth (1978).

6.2. Natural convection in rectangular box

Many calculations have been made for a differentially heated inclined layer, by using the hermitian method up to values of Rayleigh number, Ra, of the order of 10^5 (Roux et al, 1978). But for higher values of Ra, the divergence form has to be used. This has been made by using the combined method (Bontoux, 1978). Sample results for the streamlines and the isotherms are given on the Fig. 7 for $Ra = 300000$ and for $Ra = 900000$ at $\Omega = 90°$, that correspond to the conditions previously used by Seki et al (1978) in their calculations. A comparison with the experimental results of Schinkel and Hoogendoorn (1978) is given on Fig. 8, which shows that the form of the isotherms is quite well predicted for $Ra = 630000$ $Pr = 0.7$ and $\Omega = 70°$.

6.3. Oscillatory flow in a constricted channel

The qualities of the hermitian (or combined) method have been also shown in a study of the time dependence of the flow structure in a constricted channel (Forestier et al, 1977 and Grassi, 1978). The structure of such a flow which is generated by an oscillatory pressure gradient, ($P_x = A \sin \omega t$), is given for half a period on Fig. 9 and compares well with those obtained by Cheng et al (1974) at $Re = 12.5$, notwithstanding the use of a coarser mesh size in our case.

7. EXTERNAL FLOWS

In this part, we present some results obtained in the domain of external unsteady flows, by using the combined methods. In order to check the validity of the proposed method, we first study two cases for which experimental data are available :

- flow around an elliptic cylinder at incidence
- flow around an impulsively started circular cylinder.

Then, results concerning the flow around a heaving elliptic cylinder in an uniform wind are given.

There are some particular features of the numerical scheme,for external flows, concerning the field of calculation , the boundary conditions and the computational grid.

- The exterior of the body is mapped into a semi-infinite strip by the use of a conformal mapping $z = x + iy = F(\xi + i\eta) = F(Z)$, so that the Z-domain is $\xi > 0$, $0 < \eta < 2$. The line $\xi = 0$ represents the boundary of the body.
- Far away enough from the body we impose an irrotationnal flow of inviscid fluid. We have then immediatly the conditions on ψ and its derivatives which are necessary to solve the Poisson equation of the stream function by the hermitian scheme. On the body, the usual no-slip condition is applied.
- We use a grid in which the first and last nodes on a line j = constant overlap. At these nodes we impose a "periodicity" condition on ζ , ψ and its derivatives. When integrating on a line ξ = constant by the ADI scheme, the closure of the linear systems is then ensured.

7.1. Flow around an elliptic cylinder

We consider here the flow around an elliptic cylinder of 50 % thickness, with an angle of incidence of 20°. The Reynolds number is taken equal to 452. These characteristics were chosen in order to compare the results to flow structure visualizations by Taneda (1972). A good agreement is observed on Fig. 10.

7.2. Flow around an impulsively started circular cylinder

The main goal of this study is to analyze the appearance of the secondary vortices in the wake of a circular cylinder. Experimental visualizations carried out by Coutanceau and Bouard (1979) show that from Re = 300 one secondary vortex appears from Re = 300 (Fig. 11), while a pair of secondary vortices is exhibited from Re = 550 (Fig. 12). The appearance of this pair of secondary vortices has never been evidenced by a numerical simulation up to date.

The divergence and convective forms of the NS equations have been used. But our numerical experiments show that for "high" Reynolds number, the divergence form is preferable. Indeed, the appearance of the pair of secondary vortices was obtained only with this divergence form. When using the convective form, an unique secondary vortex is obtained (Fig. 13). The flow structure shown on Fig. 11, 12, 13 corresponds to a dimensionless time of 5 and to a grid of 41 x 41 nodes.

7.3. Flow around a heaving elliptic cylinder

The object of this work is to study the temporal evolution of the drag and lift coefficients for a heaving elliptic cylinder in an uniform wind.

For all the cases studied herein the thickness of the elliptic cylinder is 10 % and the Reynolds number are taken equal to 10^3 and 10^4. The heaving motion is sinusoïdal with different amplitudes and periods. Fig. 14 shows some properties of the flow which agree with experimental data :

- The drag coefficient is periodic with a period equal to T/2, while the lift coefficient is periodic with a period equal to T.

- All the characteristics of the flow become periodic after a time of about T.

- The drag is much less important when the cylinder is heaving than when it is only in translationnal motion. For instance, for a constant ratio of 10 between the period and the amplitude, the mean value of the drag is about 0.04 when the period is 2 and nearly 0 when the period is 5. We have indicated on Fig. 14 the well known value 0.2 of the drag coefficient for the cylinder in translation.

A grid of 41 x 41 is used for Re = 10^3. For Re = 10^4, stability condition and accuracy requirement lead to the use of a grid of 61 x 41 nodes. For Re = 1000 some exemples of flow pattern in the relative frame are presented in Fig. 15 (amplitude = 0.4, wave length = 2) and in Fig. 16 (amplitude = 1, wave length = 5).

8. CONCLUSION

In this paper we discuss the efficiency of a "combined" method proposed by Ta Phuoc Loc and Daube (1977) to solve the 2D Navier-Stokes equations in the streamfunction and vorticy formulation, for the numerical simulation of viscous incompressible flow.

We have shown, for a model problem proposed by Pearson, that :

- the number of mesh points, N, needed to reach a given accuracy by the combined method is comparable to those by the compact hermitian method as long as the level of accuracy does not exceed 10^{-4}. Of course, for a higher accuracy, hermitian method would be preferred.

- for N = 10, if an accurate formula (14a) is used for ζ_w, the accuracy of the combined method is 50 to 100 times higher than those of the classical second order method.

An evaluation of the operational counts shows that the combined method requires an increase of only 15 % of the computational time compared to a classical second order accurate method. This time is proportionnel to N^2.

Furthermore we have shown, for the Pearson's problem that the rate of convergence is roughly proportional to N^{-2}. This gives an additional advantage to the higher order accurate methods which involve smaller values of N, for a given accuracy.

Finally the use of the combined method is strongly recommended for solving the NS equations in the streamfunction and vorticity formulation.

In addition, this method is very flexible and has been adapted for a large variety of internal or external flows, in the steady or unsteady cases. In all cases, this method applied to the divergence form of the NS equations for high Reynolds or high Rayleigh numbers is shown to lead, even by using a reasonably coarse mesh size, results in agreement with the experiments or with the calculations previously reported.

The work reported here was partly supported by the Centre National de la Recherche Scientifique, particularly by a special grant given under the programme A.T.P. Mathématiques pour l'Ingénieur. The authors wish to warmly thanks R. Peyret for his many helpful comments during this work.

REFERENCES

ADAM, Y. 1975, J. Comp. Phys. 24, 10-22.

BENAZETH, J.C. 1978, Thèse de 3ème Cycle, Université de Paris VI.

BONTOUX, P. 1978, D. Thesis, Université d'Aix-Marseille.

BONTOUX P., FORESTIER, B. and,ROUX, B. 1978, J. Mec. Appl. 2,1-26.

BONTOUX, P., GILLY, B. and,ROUX, B. 1978, J. Comp. Phys., in press.

BOZEMAN, J.D. and DALTON, C. 1973, J. Comp. Phys. 12, 348-363.

BURGGRAF, O.R. 1966, J. Fluid Mech. 24, 113-151.

CHENG, L.C., ROBERTSON, J.M. and,CLARK, M.E. 1974, Comp. and Fluids 2, 363-380.

COUTANCEAU, M. and BOUARD, R. 1977, J. Fluid Mech. 79, 257-272.

COUTANCEAU, M. and BOUARD, R. 1979, C.R. Acad. Sciences, Serie B, 288, 46

DAUBE, O., and TA PHUOC LOC 1978, J. Mécanique 17, 651-678.

DAUBE, O., and TA PHUOC LOC 1979, GAMM Conf. on Num. Math. in Fluid Mech., Cologne (to appear).

DE VAHL DAVIS, G., and MALLINSON G.D. 1976, Comp. and Fluids, 4, 29-43.

FAIRWEATHER, G., and MITCHELL, A.R. 1967, SIAM J. Num. Anal. 4, 2.

FORESTIER, B., GRASSI, J.P., IMBERT, P., PELISSIER, R., and ROUX, B. 1977, INSERM - EUROMECH 92 - Cardiovascular and Pulmonary Dynamics, 71, 89-102.

GOSMAN, A.D., PUN, W.M., RUNCHAL, A.K., SPALDING, D.G. and WOLFSHTEIN, M. 1969, Heat and Mass Transfer in Recirc. Flows, Academic Press.

HIRSH, S.R. 1975, J. Comp. Phys. 19, 90-109.

ISAACSON, E., and KELLER, H.B. 1966, Analysis of Numerical Methods, John Wiley.

KRAUSE, E., HIRSCHEL, E.H., and KORDULLA, W. 1976, Comp. and Fluids, 4, 2, 77-92.

ORSZAG, S.A., and ISRAELI, M. 1974, Ann. Rev. of Fluid Mech. 6, 281-318.

PEACEMAN, D.W., and RACHFORD, H.H. 1955, J. Soc. Indust. Appl. Math. 3, 1.

PEARSON, C.E. 1965, J. Fluid Mech. 21, 611-622.

ROACHE, P.K. 1972, Comp. Fluid Dyn., Hermosa Ed., Albuquerque.

ROUX, B., GRONDIN, J.C., BONTOUX, P., and GILLY, B. 1978, Num. Heat Transfer 1, 331-349.

RUBIN, S.G., and KHOSLA, E.K. 1977, J. Comp. Phys. 24, 217-244.

SCHINKEL, W.M.M. and, HOOGENDOORN, G.J. 1978, 6th Int. Cong. on Heat Transfer, Toronto, MGZ, 287-292.

SEKI, N., FUKUSAKO, S., and INABA, H. 1978, J. Fluid Mech. 84, 695-704.

SUN, J.S., and HANRATTY, T.J. 1969, J. Fluid Mech. 35, 369-386.

TANEDA, S. 1972, J. Phys. Soc. Japan 33, 1706-1711.

TA PHUOC LOC, and DAUBE, O. 1977, C.R. Acad. Sciences, 284, A, 1241-1243.

TA PHUOC LOC, 1979, (to appear).

WASCHSPRESS, E.L. 1966, Iterative Solution of Elliptic Systems, Prentice Hall.

WILKES, J.O., and CHURCHILL, S.W. 1966, A.I. Ch. E. J. 12, 161-166.

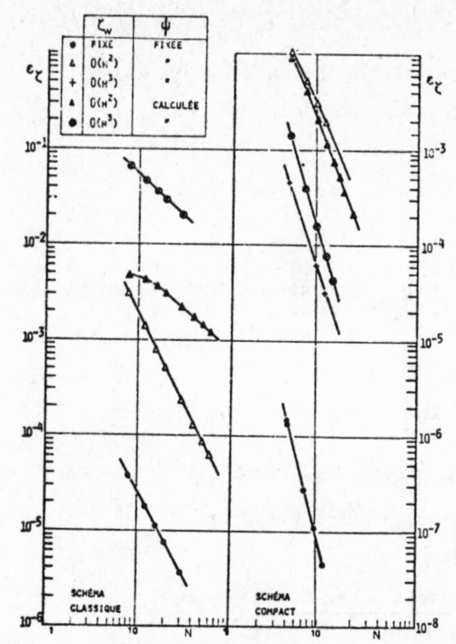

Fig.1 Influence of u,v, ζ_w on the accuracy : classical second order and compact hermitian schemes

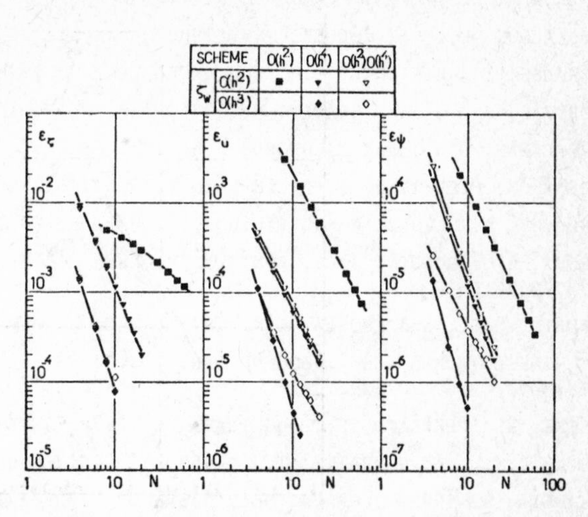

Fig. 2 : Accuracy for the classical second order, hermitian and combined schemes

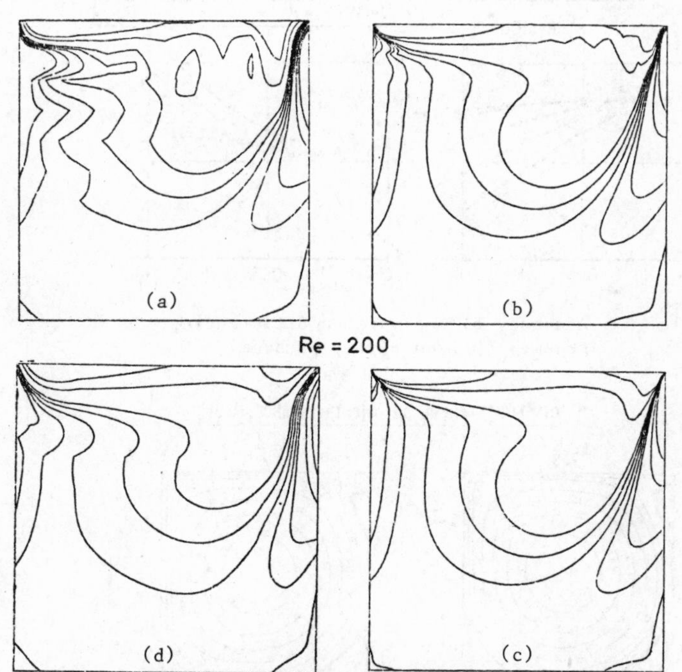

Fig. 3 : Vorticity contours for Re = 200 ; compact hermitian scheme (convective form), h = 1/10 (a), 1/20 (b), 1/35 (c) ; combined scheme (divergence form),h=1/10(d).

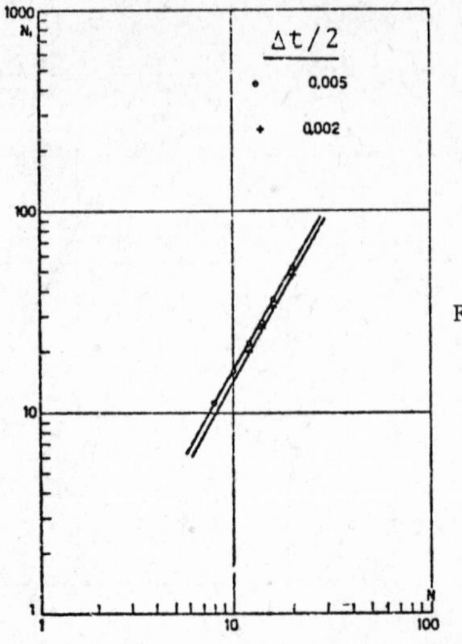

Fig. 4 : Rate of convergence versus N (Pearson's problem)

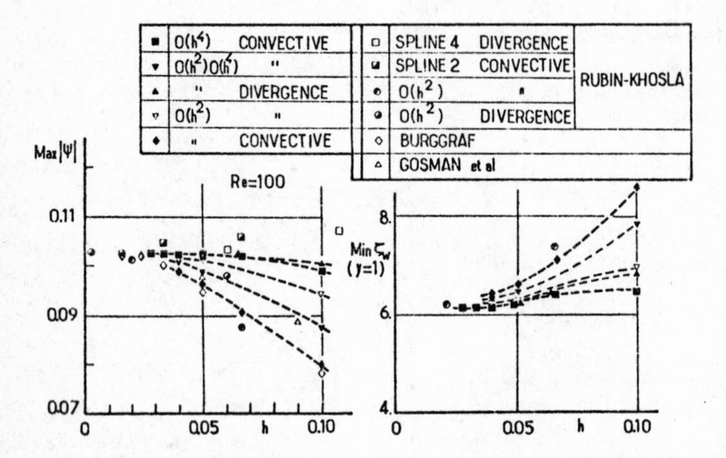

Fig. 5 : Accuracy of the various differencing schemes (Driven cavity problem)

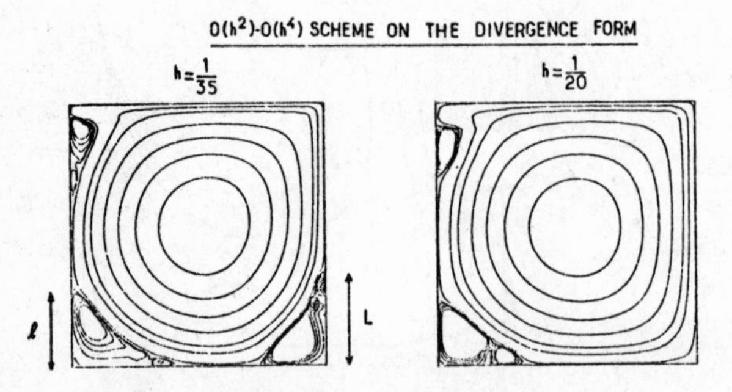

Fig. 6 : Influence of the spatial resolution on the flow structure (Driven cavity, Re = 4000).

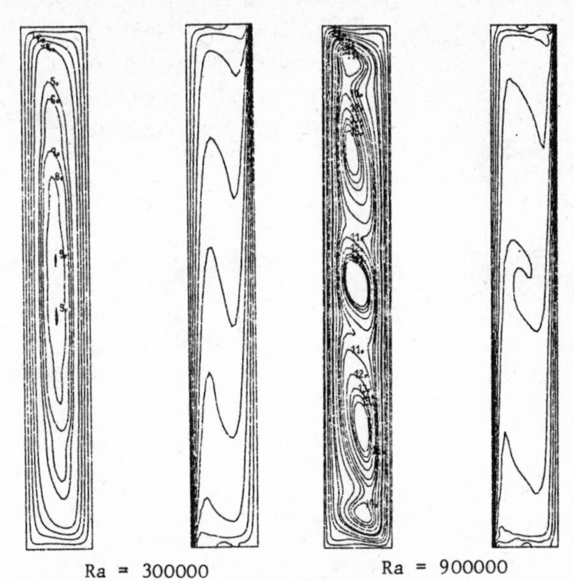

Ra = 300000 Ra = 900000

Fig. 7 : Streamlines and isotherms structures for a vertical differential heated layer (Pr = 480, Ω = 90°)

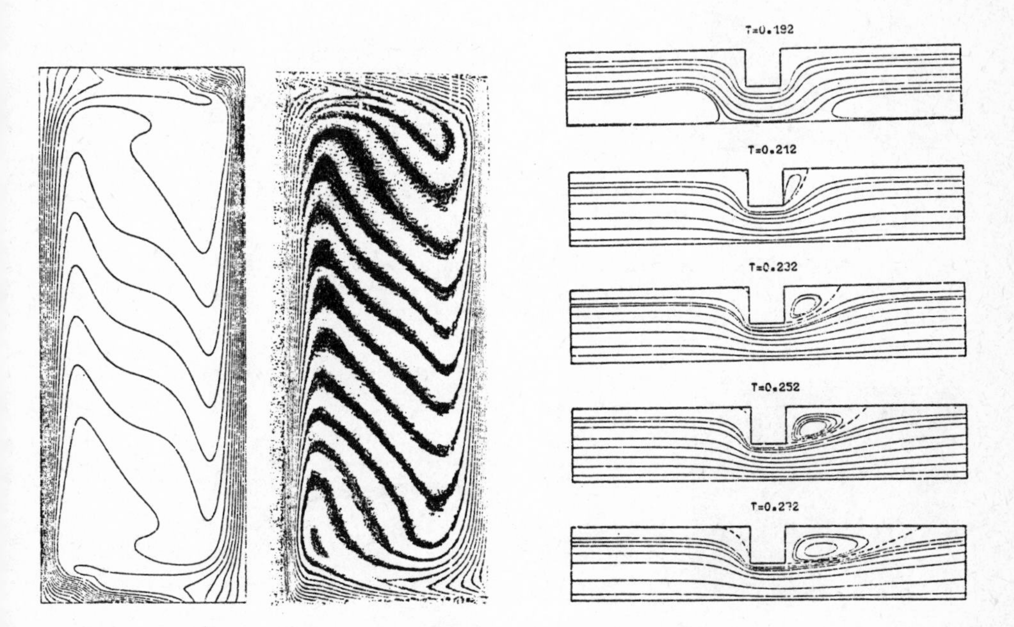

Fig. 8 : Isotherms contours from numerical and experimental results

(Ra = 630000, Pr = 0,7, Ω = 70°)

Fig. 9 : Evolution of the flow structure with time

(A = 1000, ω = 10π Re $\simeq$ 12,5)

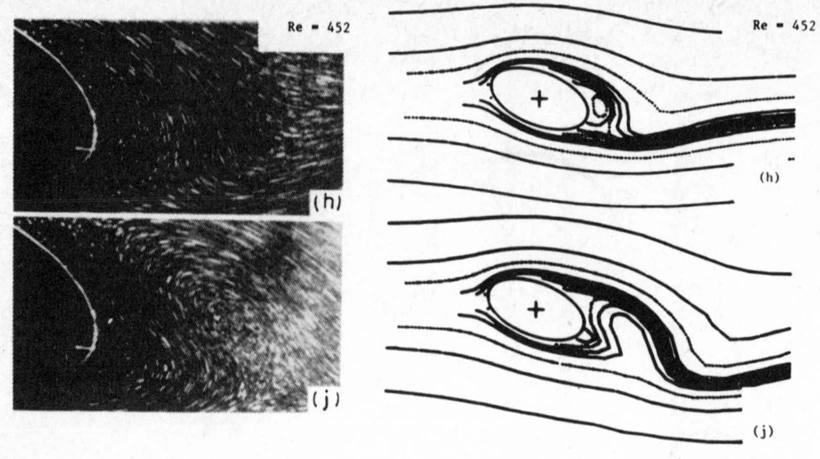

(From Taneda (1972))

Fig. 10 : Streamlines - Comparison between numerical results and experimental visualization of Von Karman vortices

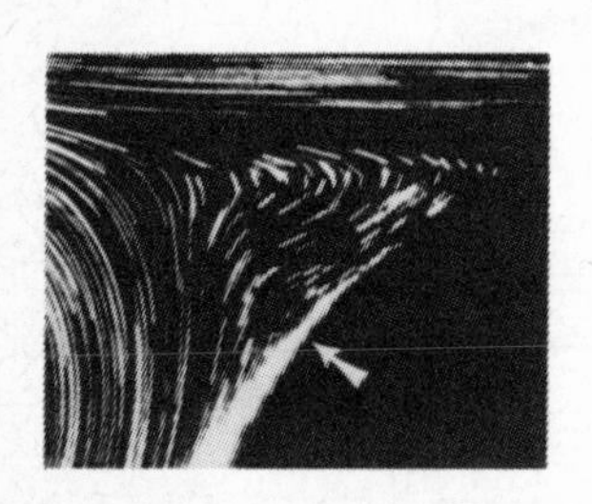

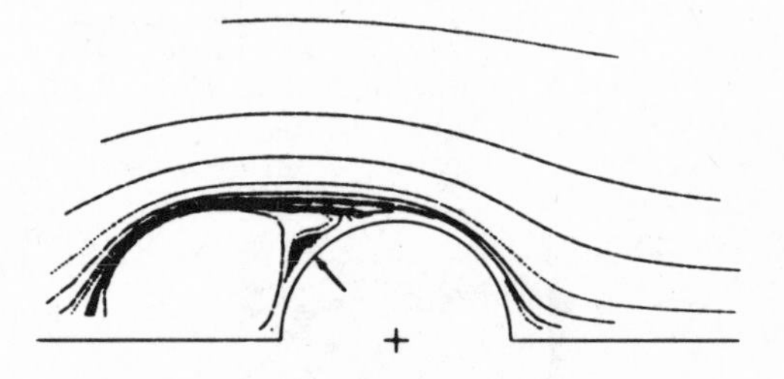

(From Coutanceau and Bouard (1979))

+Fig. 11 : Streamlines - Comparison between numerical results and experimental visualization (Re = 300, t = 5)

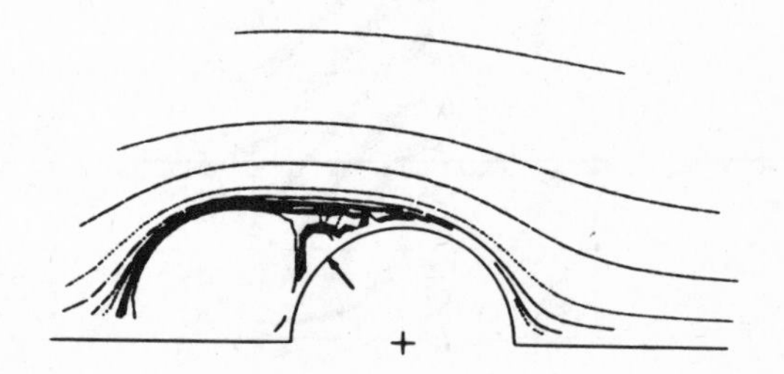

(From Coutanceau and Bouard (1979))

+Fig. 12 : Streamlines - Comparison between numerical results and experimental visualization (Re = 550, t=5)

+ Reproduced from: M. Coutanceau, R. Bouard C.R. Acad. Sciences, Serie B, 288 (1979) by permission of Gauthier Villars.

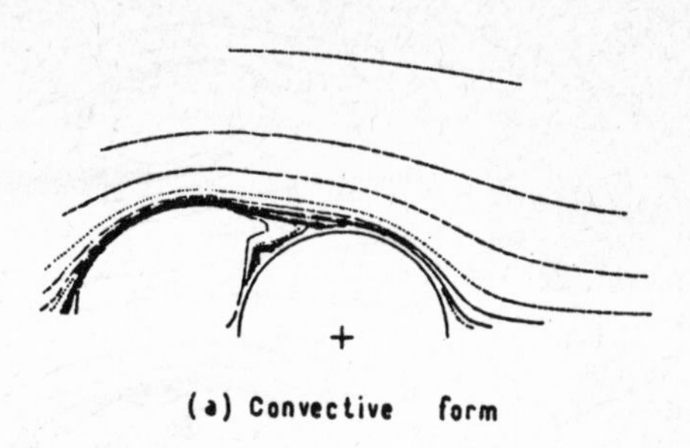

(a) Convective form

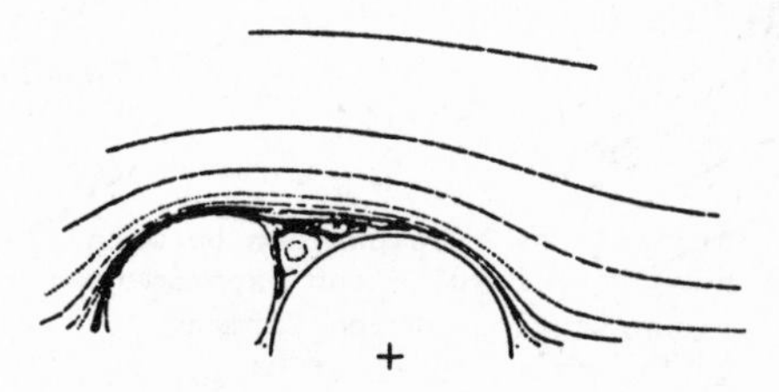

(b) Divergence form

Fig. 13 : Comparison between numerical results given by convective and divergence forms of transport terms in vorticity equation (Re = 1000, t = 5)

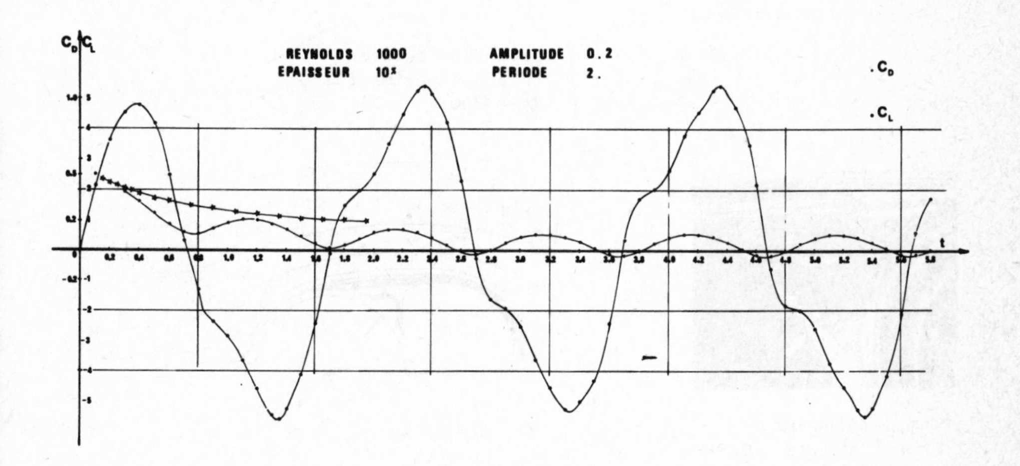

Fig. 14

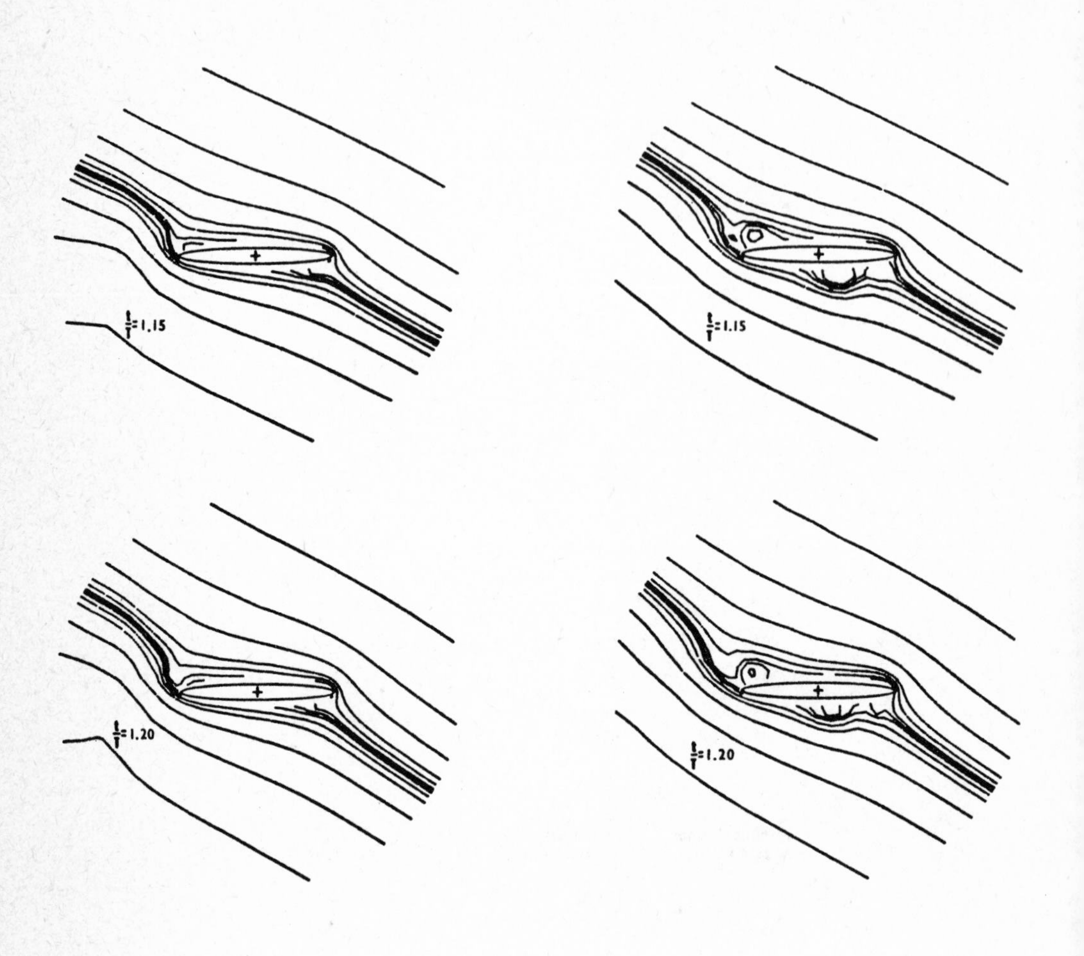

Fig. 15 : Flow pattern, Re = 1000, A = 0,4, λ = 2

Fig. 16 : Flow pattern, Re = 1000, A = 1, λ = 5

NAVIER-STOKES CALCULATIONS WITH A COUPLED STRONGLY IMPLICIT METHOD

PART II: SPLINE DEFERRED-CORRECTOR SOLUTIONS

S.G. Rubin* and P.K. Khosla*
Polytechnic Institute of New York
Farmingdale, New York 17735

ABSTRACT

The coupled strongly implicit (CSIP) method described in Part I of this study is combined with a deferred-corrector spline solver for the vorticity-stream function form of the Navier-Stokes equations. Solutions for cavity, channel and cylinder flows are obtained with the fourth-order spline 4 procedure. The strongly coupled spline corrector method converges as rapidly as the finite difference calculations of Part I and also allows for arbitrary large time increments for the Reynolds numbers considered, ($R_e \leq 1000$). In some cases fourth-order smoothing or filtering is required in order to suppress high frequency oscillations.

1. INTRODUCTION

In recent papers[1-3] the present authors have formulated various higher-order collocation techniques based on polynomial interpolation or Hermitian discretization procedures. Three-point formulations leading to fourth-order and sixth-order methods were considered. In order to obtain equal accuracy, these higher-order methods require fewer mesh points when compared with second-order accurate finite-difference techniques. This can mean a reduction in computer storage and time. In particular, the fourth-order accurate spline 4 method[1] generally requires one-quarter the number of points, in each coordinate direction, compared to finite-difference solutions of comparable accuracy.

In addition to improvements in accuracy, these methods have certain desirable properties. The discrete equations remain block-tridiagonal in character; in view of the Hermitian formulation, where the derivatives are treated as unknowns, the application of derivative boundary conditions is somewhat more straightforward, and with non-uniform grids the deterioration in accuracy is less severe than with conventional finite-difference methods. However, the resulting system of equations

---

* Current Address: University of Cincinnati, Cincinnati, Ohio 45221.

is more complex than those of the usual second-order accurate finite-difference procedures. For a single differential equation, finite-differences lead to a scalar tridiagonal system, whereas the higher-order techniques are block tridiagonal; specifically, 2x2 blocks are required for the fourth-order spline methods and 3x3 blocks for the sixth-order Hermite systems. Consequently, for a given mesh the storage requirements become larger. The solution procedure therefore requires more operations and is more time consuming than the scalar algorithm usually applied for the tridiagonal finite-difference equations. However, for equal accuracy the spline methods require fewer points and therefore are still more efficient.

In a recent study[4] some of the complexities associated with computer storage and the general spline solution procedure have been eliminated. This was accomplished by "uncoupling" the higher-order correction from the coupled lower-order ("finite-difference") system of equations. This deferred-corrector method can therefore result in a significant reduction in computer storage. A 2x2 block tridiagonal system is reduced to two scalar tridiagonal systems; 3x3 blocks reduce to a tridiagonal and a 2x2 block tridiagonal system. In particular, the method can be programmed into a subroutine package to be used with existing finite-difference codes in order to render higher-order accuracy to the final solutions. This simplicity is achieved by treating the higher-order accurate terms as explicit correctors to a modified finite-difference equation. This correction is evaluated in such a manner that the resulting procedure is consistent and unconditionally stable.[4] Related non-iterative methods of deferred-correction have been discussed by several authors for ordinary differential equations.[5] If these procedures are extended so as to achieve full convergence, for ODE's or for PDE's in a temporal or spatial marching procedure, an instability can result. The analysis of reference (4) corrects this deficiency. Finally, it is noteworthy that the Khosla-Rubin difference method,[6] which provides an explicit central-difference corrector to an implicit upwind-difference scheme, is in fact a lower-order unconditionally stable adaptation of the deferred-corrector method.

It should be noted that by uncoupling the spline or Hermite corrector and thereby reducing the size of the matrix blocks, it is now possible to obtain higher-order solutions and still maintain a considerable degree of coupling between the dependent variables in the governing systems of equations. In effect, the equations are coupled solely through the implicit or finite-difference-like terms. This is particularly important for the Navier-Stokes calculations, as the rate of convergence is increased notably when the equations are coupled. The rate of convergence does not appear to be as sensitive to the uncoupling of the higher-order spline terms. In this way we have been able to develop a coupled solver for the vorticity (ω)-stream function (ψ) description of the Navier-Stokes equations. Only (2x2) blocks must be inverted in order to obtain fourth-order accuracy. This procedure has been used with ADI, predictor-corrector, direct solver and strongly implicit two-dimensional solution techniques.[4] In the present study the coupled strongly implicit method is used exclusively to treat a variety of problems.

In Part I[7] of the present study the coupled strongly implicit (CSIP) method has been detailed. The importance of implicitly coupling the boundary conditions has been demonstrated. Second-order accurate central finite-difference solutions have been obtained for (i) the flow in a driven cavity, (ii) the temperature and heat transfer in the cavity, (iii) a channel with a rearward facing step, and (iv) a circular cylinder with a splitter plate. In the second part of the analysis presented here, these geometries are reconsidered with the deferred-corrector spline adaption of the coupled strongly implicit method.

In section 2 the spline collocation equations and analysis are reviewed. In section 3 the deferred-corrector technique of reference (4) is described for both the K-R and spline formulations. In section 4, the coupled strongly implicit spline corrector method is reviewed for the ψ-ω Navier-Stokes equations and a preliminary assessment of a conjugate gradient accelerator[9] is presented. Finally, the spline solutions for the problems of Part I are described.

2. SPLINE 4 FORMULATION

A variety of fourth and sixth-order spline and hermite procedures have been derived in references (1) and (2). The major advantages of these techniques over conventional three-point finite-difference methods are also discussed in these papers. In the present study we will be concerned only with the spline 4 procedure. The following two- and three-point spline relationships between the variable u_j and its first (ℓ_j) and second derivatives (L_j) in the y-coordinate direction (i fixed) are required for the execution of the iterative procedure.[1,2]

$$L_j = K_j + \frac{1+\sigma^3}{6\sigma(1+\sigma)^2}\left[K_{j+1} - (1+\sigma)\,K_j + \sigma K_{j-1}\right] \quad , \tag{1a}$$

$$\ell_j = \frac{h_j}{3}\,(K_j + .5K_{j-1}) + \frac{u_j - u_{j-1}}{h_j} = \frac{h_{j+1}}{3}\,(K_j + .5K_{j+1}) + \frac{u_{j+1} - u_j}{h_{j+1}} \tag{1b}$$

$$\sigma \ell_{j-1} + 2(1+\sigma)\ell_j + \ell_{j+1} = \frac{3}{h_j}\left(\frac{u_{j+1}}{\sigma} + \frac{\sigma^2-1}{\sigma} u_j - \sigma u_{j-1}\right) \tag{1c}$$

$$\sigma K_{j+1} + 2(1+\sigma)K_j + K_{j-1} = \frac{6}{\sigma h_j^2}\left[u_{j+1} - (1+\sigma)u_j + \sigma u_{j-1}\right] , \tag{1d}$$

$$K_j = \frac{2}{h_j}(\ell_{j-1} + 2\ell_j) - 6\,\frac{u_j - u_{j-1}}{h_j^2} = \frac{-2}{h_{j+1}}(\ell_{j+1} + 2\ell_j) + 6\,\frac{u_{j+1} - u_j}{h_{j+1}^2} , \tag{1e}$$

where $\sigma = h_{j+1}/h_j$ and h_j is the mesh width.

The truncation errors of the derivatives are given as:

$$\ell_j = (u_y)_j + \frac{\sigma(\sigma-1)}{2u} h_j^3 (u^{iv})_j + \frac{1-\sigma+\sigma^2}{18} h_j^4 (u^v)_j , \tag{2a}$$

$$L_j = (u_{yy})_j + \frac{7}{180}(1+\sigma^2)(\sigma-1) h_j^3 (u^v)_j$$
$$+ \left[\frac{\sigma^2}{360} + \frac{(\sigma-1)^2}{1080}(7\sigma^2 - 2\sigma + 7)\right] h_j^4 (u^{vl})_j \tag{2b}$$

Similar expressions are obtained for m_i, M_i representing (u_x), (u_{xx}), respectively.

It should be noted that with $\sigma = 1 + O(h_j)$, even the variable grid representation leads to fourth-order accuracy. For given values of u_j, a single inversion of a scalar tridiagonal matrix [equation (1d)] is required for the evaluation of ℓ_j and L_j. If u_j is to be evaluated from the solution of a differential equation, the spline 4 procedure leads to a 3x3 block-tridiagonal system for $(u,\ell,K)_j$; this can always be reduced to a 2x2 system with help of equations (1b). In references (1) and (2), this coupled procedure was used in the solution of a variety of flow problems. In reference (4), the spline 4 method was further simplified such that the 2x2 system for $(u,K)_j$ is reduced to the inversion of two scalar tridiagonal systems for u_j, K_j. This reduces the computer storage considerably and allows the higher-order improvement to be programmed as a simple corrector to a finite-difference code.

It is this deferred-corrector approach that is applied here for the solution of the (ψ-ω) Navier-Stokes equations. The procedure is completely general and can be extended to any higher-order method. In reference (3) solutions are obtained with the Hermite 6 procedure for turbulent boundary layers. The stability analysis is reported in reference (4).

3. ITERATIVE SOLUTION PROCEDURE: ψ-ω EQUATIONS

A general deferred-corrector procedure for several higher-order numerical methods has been described in reference (4), where the necessary stability conditions have also been presented. In this study, we focus specifically on the spline 4 procedure,[1] applied as a deferred-corrector, for the (ψ-ω) form of the Navier-Stokes equations. This approach is unconditionally stable, differs from earlier methods of deferred-correction,[5] and represents a natural extension of the second order K-R scheme[6] to higher-order differencing procedures. We are primarily concerned here with the steady-state solution procedure; however, the method is described and is applicable for the general unsteady equations. For the applications described herein, large time steps ($\Delta t \approx 10^6$) are used in all of the calculations.

For the (ψ-ω) Navier-Stokes equations, the general iterative procedure can be written, in the transformed variables (x,y) as follows:

Vorticity Transport Equation:

The difference approximation for the vorticity equation (3)

$$\frac{1}{J}\,\omega_t + (u\omega)_x + (v\omega)_y = \frac{1}{R_e}\,\nabla^2\omega \tag{3}$$

is

$$\begin{aligned}
&\frac{\omega^{n+1}_{ij} - \omega^{n}_{ij}}{J\Delta t} + A\Big[\mu_x\,\frac{(u\omega)_{i+1,j}-(u\omega)_{ij}}{\sigma_1 k_i} + (1-\mu_x)\,\frac{(u\omega)_{i,j}-(u\omega)_{i-1,j}}{k_i}\\
&\quad + \mu_y\,\frac{(v\omega)_{i,j+1}-(v\omega)_{ij}}{\sigma_2 h_j} + (1-\mu_y)\,\frac{(v\omega)_{i,j}-(v\omega)_{i,j-1}}{h_j}\Big]^{n+1}\\
&\quad + \frac{B}{R_e}\Big[\frac{2}{\sigma_1(1+\sigma_1)k_i^2}\{\omega_{i+1,j} - (1+\sigma_1)\omega_{i,j} + \sigma_1\omega_{i-1,j}\}\\
&\quad + \frac{2}{\sigma_2(1+\sigma_2)h_j^2}\{\omega_{i,j+1} - (1+\sigma_2)\omega_{i,j} + \sigma_2\omega_{i,j-1}\}\Big]^{n+1}\\
&\quad = \varepsilon\Big[CRx + CRy + \frac{1}{R_e}(VSx + VSy)\Big]^{n} \quad ,
\end{aligned} \tag{4a}$$

where

$$CRx = A\,\Big[\mu_x\,\frac{(u\omega)_{i+1,j}-(u\omega)_{i,j}}{\sigma_1 k_i} + (1-\mu_x)\,\frac{(u\omega)_{i,j}-(u\omega)_{i-1,j}}{R_i}\Big] - \tilde{m}_{i,j} \;, \tag{4b}$$

$$CRy = A\,[\mu_y \frac{(v\omega)_{i,j+1} - (v\omega)_{i,j}}{\sigma_2 h_j} + (1-\mu_y)\frac{(v\omega)_{i,j} - (v\omega)_{i,j-1}}{h_j}] - \tilde{\ell}_{i,j} \,, \tag{4c}$$

$$VSx = \frac{2B}{\sigma_1(1+\sigma_1)k_i^2}\{\omega_{i+1,j} - (1+\sigma_1)\omega_{i,j} + \sigma_1\omega_{i-1,j}\} - M_{i,j} \,, \tag{4d}$$

$$VSy = \frac{2B}{\sigma_2(1+\sigma_2)h_j^2}\{\omega_{i,j+1} - (1+\sigma_2)\omega_{i,j} + \sigma_2\omega_{i,j-1}\} - L_{i,j} \,, \tag{4e}$$

and

$$\mu_x = \mathrm{sgn}\,(\frac{u_{i,j}}{|u_{i,j}|}) \,, \qquad \mu_y = \mathrm{sgn}\,(\frac{v_{i,j}}{|v_{i,j}|}) \,. \tag{4f,g}$$

ℓ, m, $\tilde{\ell}$, $\tilde{m}$, L, M are the spline approximations to ω_y, ω_x, $(v\omega)_y$, $(u\omega)_x$, ω_{yy} and ω_{xx}, respectively, see section 2. J is the Jacobian of the mapping function to the (x,y) plane. The transformations will be reviewed for the various problems as they are presented, see Part I.[7] The constant ε is prescribed as zero or unity; A and B are constants, which are chosen in order that the iterative procedure is unconditionally stable. When complete convergence is achieved, the terms involving A and B cancel, so that with $\varepsilon = 1$, a higher-order numerical solution of the Navier-Stokes equations is recovered.

The convective terms on the left-hand side of equation (4a) are of the form obtained with an upwind finite-difference discretization. For $A = B = 1$ the left-hand side corresponds exactly with upwind differencing for convection and central differences for diffusion. The higher-order terms appear as explicit correctors on the right-hand side. If the spline correctors were replaced with only second-order correctors the K-R method[6] is recovered.

This implicit-explicit splitting of the convective terms is not unique, and is prescribed here to provide the appropriate five-point formulas required for the coupled 2x2 algorithm presented in Part I of this study.[7] The novel feature of the present procedure is the introduction of the A, B splitting. This is not found in the usual deferred-corrector methods.[5] The final A, B values are specified from stability considerations. The stability analysis[4] provides the minimum relative weights of implicit convection (A) and diffusion (B) required to achieve unconditional stability. For second-order schemes, equal weighting with $A = B = 1$ is unconditionally stable for $n \to \infty$; i.e., multiple iteration is convergent with the K-R method. On the contrary, in order to achieve higher-order accuracy, A and B must be non-equal and different from unity.[4] For the spline 4 method,

$A = 2$ and $B = 3$ are the mimimum integer values that lead to an unconditionally stable corrector. Therefore, higher-order accuracy is achieved iteratively with an implicit predictor finite-difference approximation, that in effect describes a flow at a different Reynolds number than that considered with the explicit corrector.

In certain cases it is desirable to use the deferred-corrector in a step-like fashion. An initial first-order accurate solution is obtained with $A = B = 1$, $\varepsilon = 0$, i.e., upwind-differencing. This can be corrected to second-order accuracy with the K-R corrector. Finally, with $\varepsilon = 1$, by selecting appropriate values of A and B, the order of the solution can be further upgraded. For spline 4, $A = 2$, $B = 3$, and the spline formulas of section 2 are applied. For some other higher-order methods the A, B values are given in reference (4). In this manner, a first-order accurate solution with considerable artificial viscosity is corrected to second- and then fourth-order. Alternatively, one could proceed directly from first to fourth-order or directly from second to higher-order, see reference (4). The higher-order solutions have little numerical diffusion.

Stream Function Equation

The stream function equation (5)

$$\nabla^2\psi = \omega/J \tag{5}$$

is approximated by

$$\frac{2}{\sigma_1(1+\sigma_1)k_1^2}[\psi_{i+1,j} - (1+\sigma_1)\psi_{i,j} + \sigma_1\psi_{i-1,j}]^{n+1}$$

$$+ \frac{2}{\sigma_2(1+\sigma_2)h_j^2}[\psi_{i,j+1} - (1+\sigma_2)\psi_{i,j} + \sigma_2\psi_{i,j-1}]^{n+1} \tag{6}$$

$$= (1-\varepsilon)\frac{\omega}{J} + \frac{2\varepsilon}{\sigma_1(1+\sigma_1)k_i^2}[\psi_{i+1,j} - (1+\sigma_1)\psi_{i,j} + \sigma_1\psi_{i-1,j}]^n$$

$$+ \frac{2\varepsilon}{\sigma_2(1+\sigma_2)h_j^2}[\psi_{i,j+1} - (1+\sigma_2)\psi_{i,j} + \sigma_2\psi_{i,j-1}]^n + \varepsilon C[\frac{\omega}{J} - \tilde{L}_{i,j} - \tilde{M}_{i,j}]^n$$

$\tilde{L}_{i,j}$ and $\tilde{M}_{i,j}$ are the spline approximations to ψ_{yy} and ψ_{xx}, respectively. The velocities u and v are related to ψ, $\tilde{L}_{i,j}$ and $\tilde{M}_{i,j}$ by the spline relationships given in section 2; the constant C plays a role similar to that of B in (4a) and ε is zero or one as before. For finite-differences $C = 1$ and for spline 4, $C \leq 1$. For sixth-order methods smaller C values are required.[4]

Computationally this iterative procedure has many advantages. The implicit inversion matrices are always strongly diagonally dominant. The initial spline correctors are obtained from reasonably smooth approximate solutions. In this way, it is possible to suppress spurious and bothersome oscillations in $\tilde{m}_{i,j}$, $\tilde{M}_j$, etc. arising from the use of arbitrary initial conditions, or inaccurate coarse grid central-difference solutions. The final spline result on the same course grid may be quite reasonable.

One of the major advantages of this approach is improved computational efficiency. With the higher-order deferred-corrector, it is possible to minimize the number of arithmetic operations so that only a marginal increase over the lower-order finite-difference requirement results. The evaluation of the correctors are incorporated in an efficient and independent subroutine. Moreover, the corrector curve fits need not be evaluated every iteration. A systematic analysis to determine the optimum corrector update procedure has not been rigorously considered. It has been found that the number of overall iterations for convergence is not significantly increased if the corrector and therefore the curve fits are evaluated every two to five iterations. It is through this optimization mechanism for achieving computational efficiency that the present application of higher-order deferred-correction techniques becomes even more attractive. It is interesting that the rate of convergence of this method is comparable to that found in the earlier coupled spline calculations.[1-3]

The higher-order approximations of the convective terms may sometimes lead to an increase in aliasing error and thus nonlinear instability.[8] The present iterative procedure will not alleviate this form of instability. In such cases, additional considerations of smoothing and or filtering are required.[8] These will be introduced, as necessary, in the present study in order to insure that the aliasing error remains bounded. The spline 4 solution for the flow in a channel with a backward step is one example in which this mode of instability is found. The introduction of smoothing in the spline curve fits, maintains the overall accuracy of the numerical scheme but eliminates the nonlinear instability.[8] This procedure will be discussed for the channel problem.

4.1 COUPLED STRONGLY IMPLICIT METHOD

The finite difference equations (4) and (6) are of the type:

$$A_{i,j}W_{i,j-1} + B_{i,j}W_{i,j} + C_{i,j}W_{i,j+1} + D_{i,j}W_{i-1,j} + E_{i,j}W_{i+1,j} = G_{i,j}$$

or

$$\tilde{L}\, W = G \ ,$$

The solution of these equations can be obtained by a variety of techniques, notably direct solution by elimination or iterative methods such as SOR, ADI, etc. A strongly implicit method,[7] which is more rapidly convergent and is based on an approximate LU decomposition of the matrix $\tilde{L}$, is considered here. Ordinarily the matrix $\tilde{L}$ is not separable into sparse upper and lower triangular forms; however, by an appropriate preconditioning, or in this case, addition of a matrix P, the following iterative procedure results:

$$(\tilde{L} + P)\, W^{n+1} = G + PW^{n}$$

where the superscript n refers to the iteration number. $(\tilde{L} + P)$ can now be written as replaced so that

$$LUW^{n+1} = G + PW^{n} \quad ,$$

or

$$L\tilde{V} = G + PW^{n} \quad ,$$

$$UW^{n+1} = \tilde{V} \quad ,$$

where L and U are as sparse as $\tilde{L}$.

This procedure leads to a simple two pass algorithm which converges rapidly when compared to some of the other iterative methods. This has been applied to the ψ-ω system. The algorithm is of the form:

$$\omega_{i,j} = GM_{1i,j} + T_{1i,j}\omega_{i,j+1} + T_{2i,j}\omega_{i+1,j} + T_{3i,j}\psi_{i,j+1} + T_{4i,j}\psi_{i+1,j}$$

$$\psi_{i,j} = GM_{2i,j} + T_{5i,j}\omega_{i,j+1} + T_{6i,j}\omega_{i+1,j} + T_{7i,j}\psi_{i,j+1} + T_{8i,j}\psi_{i+1,j}$$

This procedure converges rather rapidly and for large time steps, further acceleration is possible by a procedure described in section 4.3.

4.2 BOUNDARY CONDITIONS

For the spline calculations considered here, higher-order boundary conditions are required for ψ and ω; in addition, appropriate values for the corrector spline curve fits $m_{i,j}$, $M_{i,j}$, etc. must be specified. The various boundary conditions are as follows:

i) At a solid surface the stream function ψ is prescribed; the vorticity is evaluated from the no-slip condition, so that

$$\omega_{i,1} = \frac{3J}{h_2^2}(\psi_{i,2}-\psi_{i,1}) + J[-.5K_{i,2} + \frac{1+\sigma^3}{6\sigma(1+\sigma)^2}\{K_{i,3} - (1+\sigma)K_{i,2} + \sigma K_{i,1}\}] \quad ; \qquad (7a)$$

the portions of (7a) containing ψ are coupled directly into the 2x2 algorithm, see section 4.1. The higher-order spline correction in (7a) is treated explicitly. This is consistent with the deferred-correction treatment of the equations.

ii) At a far-field boundary, the conditions $\omega_y \to 0$ and $\psi_y \to 1$ are specified. With equations (1) these conditions are of the form:

$$\psi_{iN} = h_N + \psi_{i,N-1} - \frac{h_N^2}{3}(K_{iN} + .5\, K_{i,N-1}) \qquad (7b)$$

A similar expression obtains for ω_{iN}. As before, the K correctors are treated explicitly and the ψ terms implicitly.

iii) At the inflow boundary the values of ψ and ω are specified.

iv) At the outflow boundary the conditions $\psi_{xx} = \omega_{xx} = 0$ are imposed, either directly through extrapolation or indirectly by applying the boundary layer form of the governing equations. If extrapolation is used, the boundary values for ψ and ω are coupled implicitly into the solution algorithm. If the boundary-layer equations are assumed at the outflow, the spline 4 deferred-corrector procedure of section 3 is used to couple the solution of these equations to the interior flow. This is carried out by appropriate modification of the algorithm coefficients. There is no upstream influence from the outflow boundary with the boundary layer conditions.

Finally, the spline boundary conditions required for the curve fits of the M's, L's, etc., are obtained by satisfying the governing equations (3,5) at the surface; at the inflow/outflow the derivative conditions $K_{1j} = K_{2j}$ and $K_{Nj} = K_{N-1,j}$, etc., are specified. These conditions on the third derivative are less rigid than the direct specification of K. In many cases this leads to the suppression of high-frequency oscillations in the curve fits.

4.3 CONJUGATE GRADIENT ACCELERATOR

In the present paper a coupled strongly implicit method of solution has been considered because of the rapid influence of all the boundary values, the lack of sensititivy of grid aspect ratio and the stability properties. The rate of convergence depends upon the size of the coefficient matrix. Like most of the commonly used relaxation methods, SIP also suffers from one drawback, viz. the convergence rate decreases significantly after the high frequency Fourier components have converged (SIP is faster in eliminating the high frequency components). For large systems arising from two or three dimensional flow problems, this reduction in the rate of convergence represents an increase in the computational effort. When a dominant eigenvalue exists, over-relaxation or Chebyshev acceleration have been used to improve the rate of convergence of the iterative methods. However, both of these methods require apriori knowledge of the extreme eigenvalues of the coefficient matrix. In recent years the method of conjugate gradients has been realized as a viable procedure for accelerating the rate of convergence. This procedure leads to a finite termination property, it does not require an estimation of parameters, and takes into consideration the distribution of the eigenvalues of the coefficient matrix. Meijerink and Van der Vorst,[11] Kernshaw[12] and Golub et al.[13] have utilized the conjugate gradient method in conjunction with a variety of factorization techniques. The resulting procedures have performed quite favorably for model problems. The present authors[9] have utilized the conjugate gradient procedure to accelerate the rate of convergence of the strongly implicit method. For the system, AX = b, where A is any positive definite, the algorithm is summarized as follows:

$$X_{i+1} = X_i + \alpha p_i$$

$$\ell_{i+1} = \ell_i - \alpha B_i$$

$$p_{i+1} = \ell_{i+1} + \beta p_i$$

$$B_{i+1} = C\, \ell_{i+1} + \beta B_i$$

where $C = M^{-1}A$; $\alpha = \dfrac{p_i^T \ell_i}{p_i^T B_i}$ and $\beta = -\dfrac{\ell_{i+1}^T B_i}{p_i^T B_i}$.

The matrix M results from A through the strongly implicit procedure outlined previously. Although, this procedure has not been utilized for the 2x2 system given in the present paper, its application to the solution of potential subcritical flows have resulted in a dramatic increase in the rate of convergence. The results for the L_2 error for a substonic flow over a 60.0 thick airfoil are shown in Fig. 1. The preconditioning of the matrix A by M^{-1} has a highly desirable feature. It tries to redistribute the eigenvalues favorably for the application of the conjugate gradient method. Since the iteration matrix $C = M^{-1}A$ is close to the identity matrix, its eigenvalues are clustered together. A single application of conjugate gradients eliminates one such cluster. Thus, if

$$M = A - E$$

where the error matrix E is small, e.g. order h^n, this preconditioning will lead to a desirable solution procedure by the conjugate gradient method.

5.1 EXAMPLES

The various flows examined in Part I have been reinvestigated with more accurate higher-order spline 4 deferred-corrector strongly implicit procedure. In all the cases, solutions are obtained directly with $\Delta t = 10^6$. For the flow in a channel with a backward facing step, the spline deferred correction solution exhibits oscillations that lead to a non-linear instability for $Re \geq 1000$. This is an aliasing effect and can be eliminated by the introduction of filtering or smoothing of the spline curve fits in the axial directions. This effect will be detailed in a following section.

5.2 FLOW IN A DRIVEN CAVITY

This problem has been considered by many investigators using a variety of numerical techniques. It serves a particularly useful purpose here as the spline 4 solutions have already been obtained with an uncoupled time-dependent method[3] and therefore the effectiveness, stability and accuracy of the present coupled implicit corrector method can be assessed. Buneman's direct solver was applied in previous calculations[3] for the stream function and a predictor-corrector procedure for the vorticity equation was applied. Significantly, the present coupled strongly implicit method converges considerably faster than the uncoupled direct solver/predictor-corrector combination. An exact time comparison of the present procedure with the earlier spline 4 calculations is not possible as different computers were used. The uncoupled solutions typically required about 4 minutes on a CDC 6600 computer[3] for $R_e = 400$ and a 17x17 grid. The coupled strongly implicit deferred-corrector method converges in about 75 seconds on an IBM 360/65; this is about 3 to 8 times slower than the 6600. Furthermore, the present analysis allows for a solution of the steady state form of the equations. The final results, for a uniform 17x17 grid, are identical with those presented in reference (3). The number of iterations for the higher-order corrector method is approximately the same as was required for the finite-difference solution of Part I. Since the spline 4 calculations require only one-sixteenth the number of mesh points (in 2-dimensions) as the equally accurate finite-difference solutions of Part I, the higher-order spline computations require much less computer time. Furthermore, the deferred-corrector method is relatively simple to implement and the spline corrector can be added as a subroutine to an existing finite-difference.

5.3 HEAT TRANSFER IN A DRIVEN CAVITY

Once converged velocity profiles are obtained, the solution of the energy equation

$$\frac{1}{J} T_t + uT_x + vT_y = \frac{1}{P_e} \nabla^2 T \quad , \tag{7}$$

can be determined independently. The spline 4 discrete form of equation (7), utilizing the iterative procedure outlined in section 3, is the same as that for vorticity ω given by equation (4a); however, the non-conservation form of the energy equation is used here.

The thermal boundary conditions and the origin of the prescribed velocity distribution are explained in Part I. All of the thermal boundary conditions are implicitly coupled into Stone's solution algorithm. The spline boundary conditions are obtained by satisfying equation (7) at the walls of the cavity. The solution for a Peclet number $P_e = 50$ has been obtained on a 15x15 uniform grid. The heat transfer at the side walls is shown in Fig. 2 where the finite difference solution of Part 1 is also reproduced for comparison purposes. The temperature distribution at the upper and lower walls, as well as the mid-plane, is shown in Fig. 3. It can be seen that the spline 4 temperature distribution on the upper moving wall is considerably different from that obtained with the finite-difference procedure. Notably, there are no oscillations in the spline solutions.

5.4 FLOW IN A CHANNEL WITH A BACKWARD FACING STEP

The conformal transformation that maps the step geometry into a straight channel, and the prescribed grid distribution are described in Fig. 4. The spline 4 iterative procedure is initiated after 10 to 20 iterations of the central finite-difference calculation. At the outflow boundary, the spline 4 solution to the ψ-ω form of the boundary-layer equations was coupled into the 2x2 strongly implicit algorithm for the Navier-Stokes equations. In order to test the higher-order accuracy of the spline 4 method the fully developed flow in a straight channel was evaluated. The change from second to fourth-order accuracy was evident.

When the spline corrector was switched on, it was soon observed that the higher-order solution near the outflow boundary acquired a small oscillation. This subsequently developed into an instability. A number of reasons for this instability are possible. The most likely was the sudden transition from the fourth-order accurate representation of $(u\omega)_x$ in the interior to the first-order accurate boundary layer boundary condition at the outflow. In order to make this transition consistent a variety of boundary conditions for the spline curve fits were examined; however, the instability persisted. Subsequently, systematic numerical experimentation indicated that the source of this instability was closely related to the ψ_{xx} curve fit for the evaluation of v and the curve fit of $(u\omega)_x$ required in the deferred corrector in equation (4a). Based on some previous studies[8] of non-linear instability it was apparent that the root cause was the aliasing error due to the higher-order representation of the convective terms along the channel. The instability was eliminated by introducing smoothing only in the curve fits for the convective terms. For example, after K_1 is evaluated, a new $\tilde{K}_i$ is obtained from the three point smoothing formula

$$\tilde{K}_i = \frac{K_{i+1} + (1+\sigma_1)K_i + \sigma_1 K_{i-1}}{2(1+\sigma_1)} \qquad (8)$$

This is applied in the spline approximations for $(u\omega)_x$, ω_x and ψ_x only. It can be shown that for a variable grid, formal third-order accuracy of the convective as well as diffusive terms results. For uniform grids the smoothing is fourth-order accurate for ψ, ω. Converged spline 4 solutions for $R_e = 100$ and 1000 were now obtained. The vorticity along the lower and upper walls is shown in Figs. 5a,b and 5c where the finite-difference solution on the same grid is also depicted. Vector plots are presented in Fig. 6.

With the present refined grid, the finite-difference solution is quite accurate. Outside of the recirculation region, the changes in the flow variables are not large and therefore the finite-difference approximation is also reasonable in the fully developed flow. In the recirculation region where the grid is very fine the spline 4 solution shows only a marginal difference. The peak wall vorticity is slightly larger. Also, the point of separation is slightly different from that of the finite-difference calculations. Finally, for the fully developed region the spline 4 results agree with the known exact solution to three decimal places. For $R_e = 1$, the flow separates below the step.

The separation point moves towards the corner as R_e is increased.

5.5 FLOW OVER A CIRCULAR CYLINDER WITH A SPLITTER PLATE

The numerical evaluation of the flow over a circular cylinder has been the subject of many investigations. The preponderance of studies have used time-dependent methods to evaluate the transient and asymptotic steady state behavior. For large R_e flows, the solution for the wake region is apparently oscillatory and the question of the stability of a steady state solution remains open. In Part I, the ψ-ω equations were solved iteratively using the CSIP algorithm described herein. Oscillatory wake flow was observed for $R_e \geq 70$.

In the present study the spline deferred-corrector coupled 2x2 algorithm is used to reexamine the existence of steady flow solutions. The cylinder with a splitter plate is mapped into a semi-infinite flat plate. This is not optimal, but with a suitable non-uniform grid adequate resolution over the cylinder and in the wake region is possible, see Fig. 7.

As with the finite-difference procedure of Part I, spline corrector solutions have been obtained for a Reynolds number of fifty based on the cylinder diameter. These results for the vorticity along the cylinder and splitter plate are presented in Fig. 8. Significantly, the recirculation region, with the identical 33x17 variable grid prescribed in Part I, has become much smaller as separation occurs farther back on the cylinder. The finite-difference results agree quite well with the separation location predicted by others for a cylinder without a splitter plate. One should anticipate an influence on the separation location due to the appearance of the splitter plate; this has been seen in some experimental data. The higher-order accurate spline corrector results appear to show this effect. In order to test the validity of these solutions, additional spline and finite-difference calculations with finer meshes near the separation region are necessary. In addition, the effect of filtering on the spline derivatives must be further assessed. This evaluation has not been completed and therefore the present spline results for the cylinder are considered preliminary. The spline solution for $Re = 50$ have been obtained with $\Delta t \approx 10^6$; therefore the CSIP spline corrector algorithm can be applied as a steady state solver. Convergence rates for the strongly implicit method with the spline corrector are comparable to those found for the finite-difference calculations.

For a Reynolds number of 100, stable steady state solutions could not be obtained. Applying smaller time steps ($\Delta t \leq 10$), an oscillatory behavior similar to that found in the finite-difference solutions of Part I was observed. These results are typified by the vorticity patterns shown in Fig. 9 and lead us to believe that improved accuracy will not eliminate this stability problem. Either the numerical method fails in this case or a stable steady state solution does not exist, for $R_e = 100$. "Steady-state" solutions for $R_e = 100$ have been reported in the literature and therefore further evaluation of the present method is required. Significantly, the present results can be considered quasi-steady outside the separation bubble.

6. SUMMARY

A spline 4 deferred corrector strongly implicit method for the ψ-ω form of the Navier-Stokes equation has been investigated. In this iterative approach, both convective and diffusive terms in the governing equations are reordered such that at each level of the iteration a modified finite-difference form of the governing equations is solved, i.e., the implicit and explicit "Reynolds numbers" are different. The higher-order spline correction is treated explicitly. This approach has many distinct advantages. For example, (i) the computer program does not involve any significant additional complexity other than would occur for a second-order finite-difference program. The implicit block size remains unchanged. ii) The additional computational effort is associated solely with the evaluation of the explicit spline curve fits. Therefore, the existing finite-difference codes can easily be upgraded to higher-order accuracy. iii) The overall storage for the fourth-order method is considerably reduced. iv) Unlike the conditional stability of the method of deferred correction[5], the present technique is unconditionally stable.

Finally, with this coupled strongly implicit procedure, spline 4 "steady state" solutions (with $\Delta t = 10^6$) have been obtained for a variety of flow problems. It has been found that the number of iterations for the spline 4 deferred corrector procedure is comparable to that required for the finite-difference solutions presented in Part I. The method of solution converges rapidly when compared with some of the earlier uncoupled methods applied for the solution of the Navier-Stokes equations.

This coupled procedure is currently being applied to the primitive variable parabolized Navier-Stokes and boundary region equations. Several generalizations of the strongly implicit procedure have been explored[10] and preliminary results are quite encouraging. With the conjugate gradient accelerator,[9] which is particularly suitable to the preferred eigenvalue structure associated with the CSIP matrix preconditioning, even further improvements are predicted for both finite-difference and spline corrector methods.

ACKNOWLEDGEMENT

This research was sponsored in part by the NASA Langley Research Center Under Grant NSG-1244 and in part by AFOSR Contract F-49620-78-C-0020, Project No. 2307-A1.

REFERENCES

1. Rubin, S.G. and Khosla, P.K., AIAA Journal, 14, 7, pp. 851-588, (1976). Also NASA CR2653.
2. Rubin, S.G. and Khosla, P.K., J. Comp. Physics, 24, pp. 217-244, (1977). Also NASA CR2735.
3. Rubin, S.G. and Khosla, P.K., Computers and Fluids, 5, pp. 241-259, (1977).
4. Rubin, S.G. and Khosla, P.K., Proc. of 6th International Conference on Computational Fluid Mechanics, Tbilisi, USSR, June 20-25, Springer-Verlag, (1978).
5. Fox, L., "Numerical Solution of 2-Point Boundary Value Problems in Ordinary Differential Equations," Oxford Univ. Press, Fairlawn, New Jersey, (1957).
6. Khosla, P.K. and Rubin, S.G., Computers and Fluids, 2, pp. 207-209, (1974).
7. Rubin, S.G. and Khosla, P.K., AIAA Paper 79-0011, presented at the AIAA 17th Aerospace Sciences Meeting, New Orleans, La., January 15-17, (1979). (To appear in J. Computers and Fluids).
8. Khosla, P.K. and Rubin, S.G., J. of Engineering Mathematics, 13, 2, pp. 127-141, (1979).
9. Khosla, P.K. and Rubin, S.G., presented at Symposium on Computers in Aerodynamics, Polytechnic Institute of New York, June 4-5, (1979).
10. Lin, A., Private Communication.
11. Meijerink, J.A. and Van der Vorst, H.A., Math. Comp., 3, pp. 148-162, (1977).
12. Kershaw, D.S., J. Comp. Physics, 26, pp. 43-65, (1978).
13. Concus, P., Golub, G.H. and O'Leary, D.P., Computing, 19, pp. 321-339, (1978).

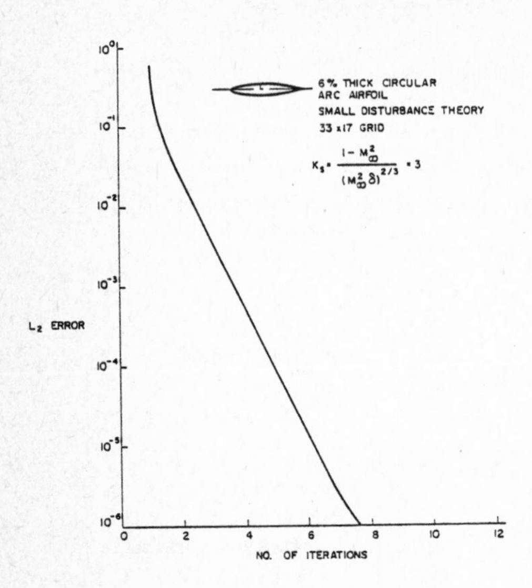

FIG.1 CONJUGATE GRADIENT-STRONGLY IMPLICIT SOLUTION

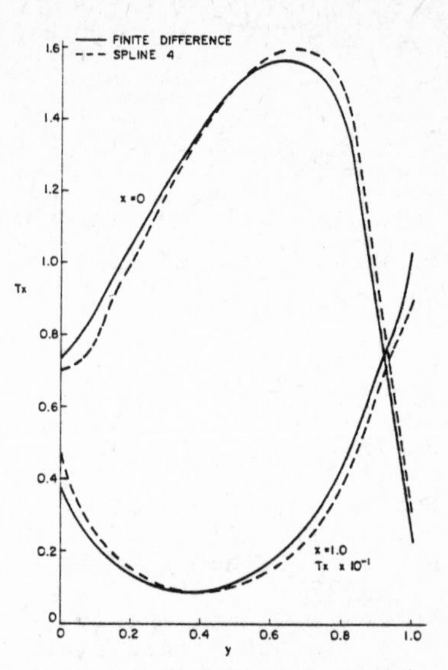

FIG. 2 HEAT TRANSFER ON SIDE WALLS OF CAVITY

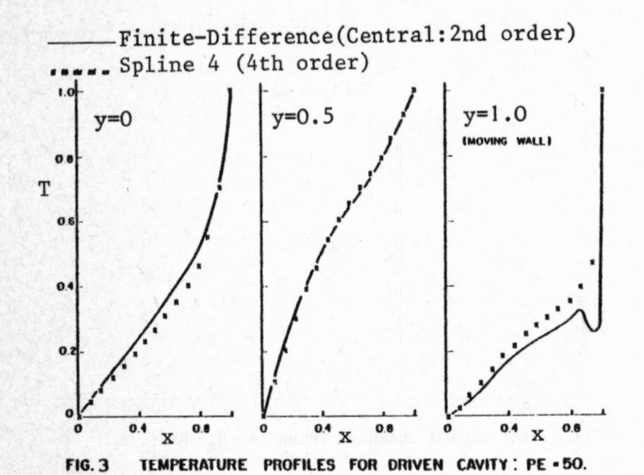

FIG. 3 TEMPERATURE PROFILES FOR DRIVEN CAVITY: PE = 50.

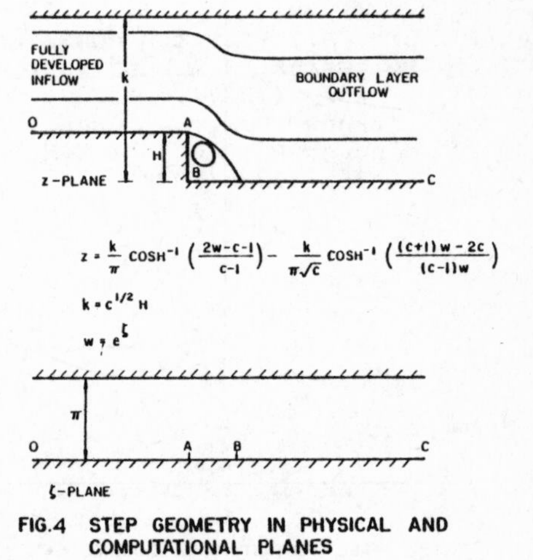

FIG.4 STEP GEOMETRY IN PHYSICAL AND COMPUTATIONAL PLANES

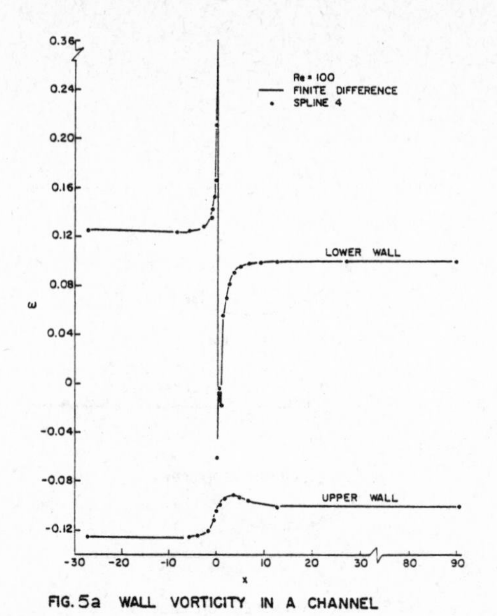

FIG. 5a WALL VORTICITY IN A CHANNEL

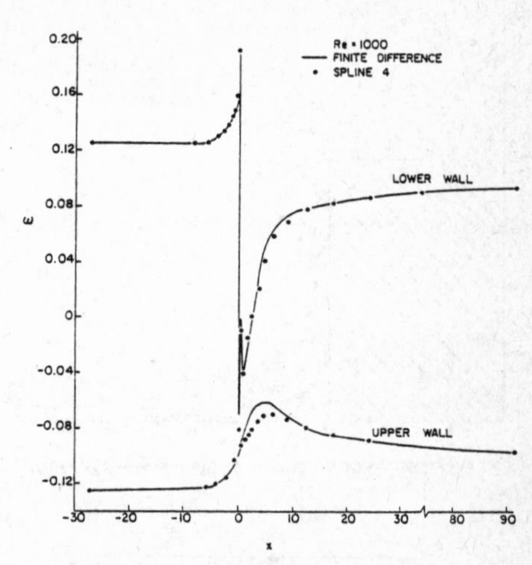

FIG. 5b WALL VORTICITY IN A CHANNEL

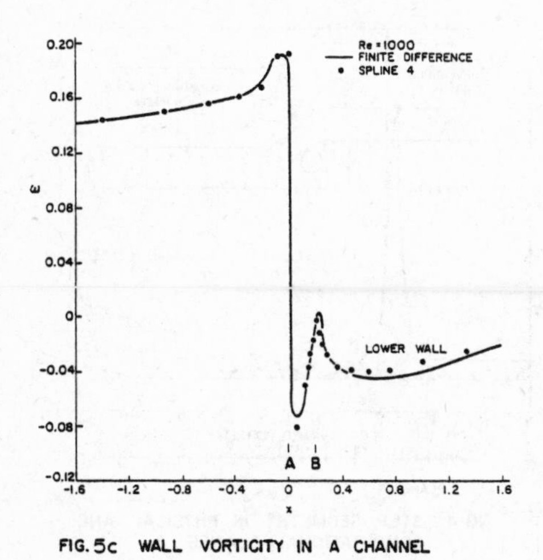

FIG. 5c WALL VORTICITY IN A CHANNEL

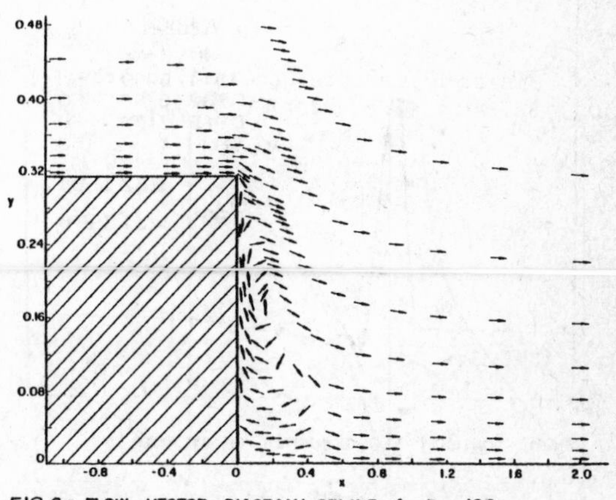

FIG. 6a FLOW VECTOR DIAGRAM: SPLINE 4 – R_e = 100

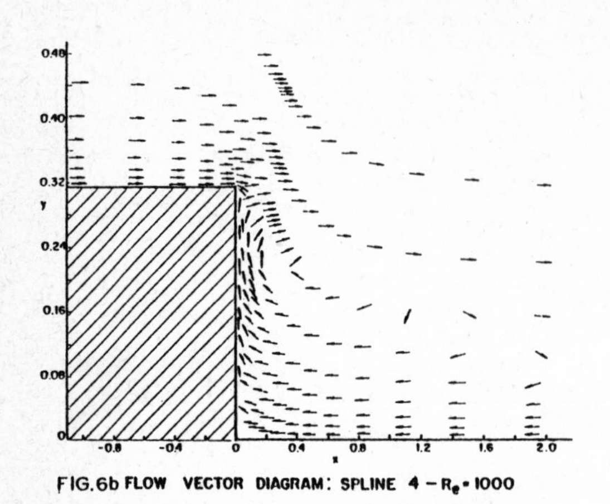

FIG.6b FLOW VECTOR DIAGRAM: SPLINE 4 – R_e = 1000

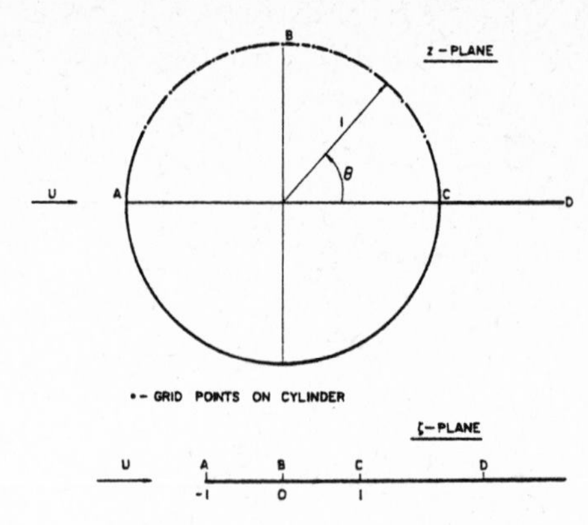

FIG. 7 CYLINDER GEOMETRY

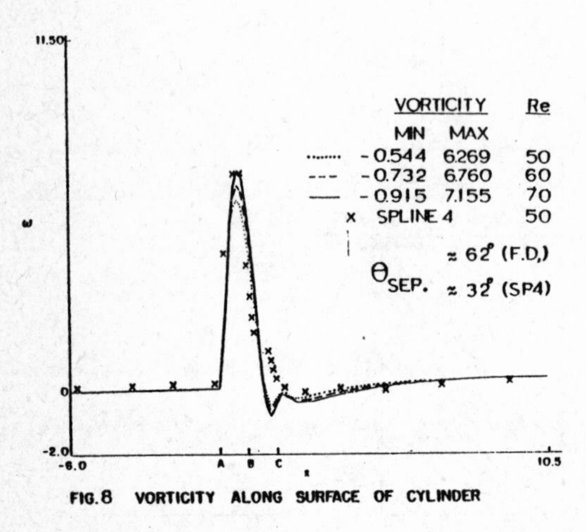

FIG.8 VORTICITY ALONG SURFACE OF CYLINDER

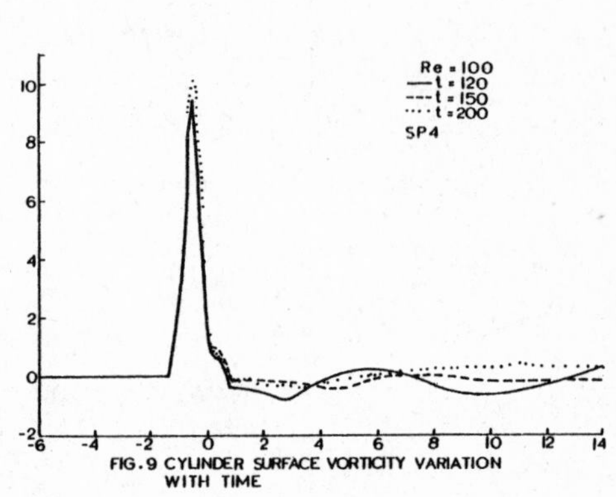

FIG.9 CYLINDER SURFACE VORTICITY VARIATION WITH TIME

STRANGE ATTRACTORS AND CHARACTERISTIC EXPONENTS OF TURBULENT FLOWS

David RUELLE

IHES. 91440 Bures-sur-Yvette. France

Consider a time evolution defined by a system of ordinary differential equations

$$\frac{dx_i}{dt} = F_i(x_1,\ldots,x_m) \qquad i = 1,\ldots,m$$

Solutions of such a system may be asymptotically constant or periodic. Another possibility is that, when $t \to \infty$, their behavior be neither periodic nor even quasi-periodic. This happens when the solutions are asymptotic to a <u>strange attractor</u> and have sensitive dependence with respect to initial condition. Examples of this are known theoretically, and from computer work following the discovery of the Lorenz attractor [4].

The sensitive dependence with respect to initial condition is seen by studying the growth in time of a small error $\delta x(t) = (\delta x_1,\ldots,\delta x_m)$. The multiplicative ergodic theorem of Oseledec [5], [6] asserts that, given a probability measure ρ on $\mathbb{R}^m$ invariant under time evolution, then $|\delta x(t)|$ grows exponentially for ρ-almost all initial conditions. The rate of exponential growth is called a <u>characteristic exponent</u>. If a characteristic exponent is larger than zero, there is sensitive dependence on initial condition.

For theoretical and numerical studies, it is often convenient to consider systems with a discrete time n rather than the continuous time t. The evolution equations are then

$$x_i(n+1) = f_i(x_1(n),\ldots,x_m(n)) \qquad i = 1,\ldots,m$$

Henceforth I shall mostly discuss the discrete time case, but everything carries over to continuous time.

Consider the map $f : \mathbb{R}^m \mapsto \mathbb{R}^m$ defined by the f_i, and let ρ be an f-invariant probability measure on $\mathbb{R}^m$ (with compact support, say). The multiplicative ergodic theorem says that for ρ-almost all $x \in \mathbb{R}^m$,

$$\lim_{n\to\infty} \frac{1}{n} \log \| (\frac{\partial f^n}{\partial x}) u \| = \chi(x,u)$$

exists, where $u = \delta x \in \mathbb{R}^m$ and $\frac{\partial f}{\partial x}$ denotes the matrix of partial derivatives. The values taken by $\chi(x,u)$ for $u \neq 0$ are the characteristic exponents, and we have sensitive dependence on initial condition if the upper exponent is > 0.

If ρ is ergodic (= indecomposable), the ordinary ergodic theorem implies that for all continuous functions φ

$$\lim_{n\to\infty} \frac{1}{n} \sum_{k=o}^{n-1} \varphi(f^k x) = \int \rho(dy)\varphi(y)$$

for ρ-almost all x. In other words

$$\frac{1}{n} \sum_{k=o}^{n-1} \delta_{f^k x} \tag{*}$$

where δ_x is the unit mass at x (Dirac delta), tends (weakly) towards μ.

At this point we have to face the fact that our definition of characteristic exponents, and of sensitive dependence on initial condition, depend on the choice of an invariant measure ρ. In many cases there are continuously many different ergodic measures, and all invariant measures are singular with respect to the Lebesgue measure (they do not have a density with respect to $dx_1 \ldots dx_m$). Can the notion of sensitive dependence of initial condition then have a natural meaning ? Experimentally, the answer is often yes. Experiment means here experiment with a physical system or with a computer.

Let us consider the time evolution in $\mathbb{R}^2$ defined by

$$x_1(n+1) = x_2(n)+1 - ax_1(n)^2$$
$$x_2(n+1) = b\, x_1(n)$$

This was introduced and first studied by computer by Hénon (for $a = 1.4$ and $b = .3$) . One finds experimentally that for x in a certain region of $\mathbb{R}^2$ (basin of attraction of the Hénon attractor) the measures (*) tend always to the same limit ρ . In a picture showing $f^k x$ for $k = 0,\ldots,10000$, the density of ρ is roughly the density of points $f^k x$. Such a picture also shows the "strange attractor" on which the "asymptotic measure" ρ sits (see [2]). In particular, it is clear that ρ is a singular measure. The upper characteristic exponent is here found to be $\approx .4$.

How is the measure ρ determined ? No general answer to this question is known, but a theorem valid for a large class of maps (Axiom A diffeomorphisms) gives an idea of what happens. The theorem says that near an Axiom A attractor, for almost every initial condition x with respect to <u>Lebesgue measure</u> $dx_1 \ldots dx_m$, the measure (*) tends to a unique asymptotic measure ρ on the attractor (see [7], [1]). The measure ρ is characterized by the fact that, <u>on unstable manifolds</u> it is absolutely continuous with respect to Lebesque measure. The unstable manifolds are the sets which contract under iterates f^{-n} of f^{-1} , for instance the very visible lines on the Hénon picture [2] are unstable manifolds [9] . They have dimension in general $\mu < m$, and ρ restricted to these manifolds should have a density with respect to μ-dimensional Lebesgue measure. As far as one can judge by inspection the measure on the Hénon attractor conforms to the above description, even though the Hénon map is not Axiom A, and the theorem does not apply to it.

Another theorem, again valid for Axiom A attractors, concerns stability under small stochastic perturbations. If ρ_ε is a stationary measure for a stochastic process obtained by adding to f a little noise term (described by a continuous kernel) then, as the noise tends to zero, ρ_ε tends to the same asymptotic measure discussed above [3] .

Stability under small stochastic perturbation is relevant to computer studies (where roundoff errors are important) and to experiments. For instance in hydrodynamic turbulence one can argue that the sensitive dependence on initial conditions amplifies even the small thermal fluctuations to macroscopic level in a relatively short time [10] .

In view of the above discussion it is not unreasonable to think that there are measures describing hydrodynamic turbulence, and characterized by absolute continuity on unstable manifolds [8] . To establish such a fact mathematically will require extension to general differentiable dynamical systems of results known for Axiom A systems, and also extension from finite dimension to Hilbert or Banach spaces. (This second extension may be easier than the first).

Even when a mathematical characterization of the measures describing turbulence is established, it will be a long and hard way to the study of physical turbulence at high Reynolds numbers. It may nevertheless be easier to proceed when an acceptable conceptual formulation of the foundations has been obtained.

References.

[1] R. Bowen and D. Ruelle. The ergodic theory of Axiom A flows. Inventiones Math. 29, 181-202 (1975).

[2] M. Hénon. A two-dimensional mapping with a strange attractor. Commun. math. Phys. 50, 69-77 (1976).

[3] Ju. I. Kifer. On small random perturbations of some smooth dynamical systems. Izv. Akad. Nauk SSSR. Ser. Mat. 38 (5), 1091-1115 (1974). English translation Math. USSR Izv. 8, 1083-1107 (1974).

[4] E.N. Lorenz. Deterministic nonperiodic flow. J. atmos. Sci. 20, 130-141 (1963).

[5] V.I. Oseledec. A multiplicative ergodic theorem. Ljapunov characteristic numbers for dynamical systems. Trudy Moskov. Mat. Obšč. 19, 179-210 (1968). English translation Trans. Moscow Math. Soc. 19, 197-231 (1968).

[6] M.S. Raghunathan. A proof of Oseledec' multiplicative ergodic theorem. Israel J. Math. To appear.

[7] D. Ruelle. A measure associated with Axiom A attractors. Amer. J. Math. 98, 619-654 (1976).

[8] D. Ruelle. Sensitive dependence on initial condition and turbulence behavior of dynamical systems. Ann. N.Y. Acad. Sci. 316, 408-416 (1978).

[9] D. Ruelle. Ergodic theory of differentiable dynamical systems. IHES. Publ. Math. To appear.

[10] D. Ruelle. Microscopic fluctuations and turbulence. Phys. Lett. 72A, 81-82 (1979).

Selection Mechanisms
in Symmetry Breaking Phenomena

by

D.H. Sattinger
Minneapolis, Minnesota 55455

1. Physical Examples of Symmetry Breaking

As we all know, when a layer of fluid is heated from below, convective instabilities set in when the temperature drop across the layer exceeds a certain critical value, and the convective motions which evolve sometimes display a striking cellular structure. This is not an isolated example in nature, but rather typifies a broad range of phenomena which are called "symmetry breaking instabilities". We can describe the appearance of cellular structures in the following mathematical terms. Consider the infinite plane layer model for the Bénard problem. Prior to the onset of instability the solutions are invariant under the entire group of rigid motions in the plane. The convective solutions, however, are invariant only under a crystallographic subgroup. Thus we speak of "broken symmetry", the bifurcating solutions are invariant under a smaller symmetry group.

Such situations, as I say, accompany the transition to instability in a wide variety of natural phenomena. Other physical problems in which symmetry breaking plays a role are buckling problems in elasticity, pattern formation in reaction-diffusion processes (see Auchmuty and Nicolis, Fife, and Turing); neurobiological problems, and physical chemistry (see Larter and Ortoleva).

Cowan and Ermentrout, in their theoretical investigations into the nature of hallucinatory phenomena, based their analysis on the plausible contention that "simple hallucinations arise from an instability of the resting state leading to concomitant spatial patterns of activity in the cortex." Furthermore, an experimental investigation into mescaline-induced hallucinations by Klüver showed that most simple hallucinations could be classified into one of four structural types: "grating, cobweb, or filigree; cobweb; funnel or cone; spiral." Cowan and Ermentrout observe that these patterns are obtained from the familiar crystallographic patterns common in the Bénard instabilities if one took into account the conformal nature of the transformation from the visual (reticular) field to the cortical field. Thus the analysis of pattern formation in the Bénard problem seems to have a good deal of relevance in the analysis of hallucinatory phenomena as well. In fact, what is at issue mathematically in both cases is the breaking of Euclidean invariance in the plane.

In the case of the Bénard problem much effort has been directed at the question of selection mechanisms. At the onset of instability, in the infinite plane layer model, there is a very high degree of "nonuniqueness" of the bifurcating solutions. Even under the restriction of "doubly periodic" solutions, there is an infinite degeneracy. Actually, the restriction even to doubly periodic solutions is entirely ad hoc; there is no mathematical justification for such an assumption. These mathematical difficulties arise primarily because the Euclidean group is not compact, and there does not seem to be any easy way around the difficulty. Nevertheless, one can restrict oneself to the doubly periodic solutions and ask whether there are any conditions under which a unique bifurcating pattern is selected. The answer to this question is a partial "yes" (see Busse [5] and Sattinger [13]); and we shall discuss some results of this type in the section IV.

Another example of symmetry breaking is that of intercellular localization processes which lead to asymmetric differentiation of the Fucus egg. Larter and Ortoleva [12] treat this process as a "transformation to a self-electrophoretic state [by] a symmetry breaking instability in the state of symmetrically distributed membrane potential." They go on to state "these self-sustaining gradients can exist in systems without inherent asymmetry, and are proposed as an example of Turing's physico-chemical morphogenesis hypothesis." Larter and Ortoleva take as a model for the electrophoretic effects a system of reactive-transport equations with isotropic rate and transport laws:

$$\Delta V = \frac{-4\pi\mathfrak{F}}{\epsilon} \sum_{i=1}^{n} z_i C_i$$

$$\frac{\partial C_i}{\partial t} = \operatorname{div} \sum_{j=1}^{n} \{\mathcal{D}_{ij} \operatorname{grad} C_j + M_{ij} C_j \operatorname{grad} V\} + \mathcal{R}_i(C)$$

where $\mathcal{D}_{ij}$ is the matrix of diffusion coefficients, M_{ij} is a matrix of mobilities, z_i are valences for the ith species, and $\mathcal{R}_i(C)$ is a rate reactions term (nonlinear). The homogeneity of the rate and transport terms (i.e. their invariance under rotations) means that the above system of partial differential equations is covariant with respect to the group of rigid motions. When differentiation occurs the symmetry is broken "spontaneously" - there is nothing implicit in the equations themselves to cause the symmetry to break.

In the Fucus problem, the relevant symmetry group is the group of rotations $O(3)$. Prior to differentiation, in the Larter-Ortoleva model, the egg is spherical, and so are the equilibrium solutions of the rate / transport equations. Experimental observations have shown that in the course of its development, the egg makes a transition from a spherically symmetric state of the membrane potential to a state of overall polarization with a non-zero cellular current running from pole to pole. This steady self-sustaining current is termed "self-electrophoresis". The asymmetric potential gradient is "believed to be essential in the development of the asymmetry which leads to dramatically different rhizoid and thallus cells after the first division of the egg." [12]

The breaking of rotational symmetry is also important in the context of geophysical applications - for example the pattern of convective motion in the earth's core is of crucial importance in the Dynamo problem (the maintenance of the earth's magnetic field). See Busse [7].

2. Group theory, bifurcation theory, and selection mechanisms.

Having discussed some of the diverse physical problems in which spontaneous symmetry breaking plays a key role, let us discuss the manner in which these bifurcation problems may be attacked via methods of group representation theory. To state the situation mathematically, we suppose that the equilibrium states of a physical system are determined as solutions of a functional equation

$$G(\lambda,u) = 0 \tag{1}$$

where λ is a parameter, u is an element of a Banach space $\mathcal{E} \times \Lambda$ to $\mathfrak{F}$. Let us say that $u \equiv 0$ is always a solution of (1); then a bifurcation point $(\lambda_c, 0)$ is a critical value of λ at which several non-trivial solutions of (1) confluesce. Possible bifurcation points are determined by the condition that the linear operator $G_u(\lambda_c, 0) = L_0$ be a Fredholm operator with a non-trivial kernel. That is, L_0 has a closed range of finite codimension equal to the dimension of the kernel $\mathcal{N}$. In that case, the problem of finding all solutions of (1) in the neighborhood of the branch point $(\lambda_c, 0)$ is reduced, via the Lyapounov-Schmidt procedure, to a finite system of equations

$$F(\lambda,v) = 0 \quad v \in \mathcal{N}, \; F : \Lambda \times \mathcal{N} \to \mathcal{N} \tag{2}$$

or

$$F_i(\lambda,z_1,\ldots,z_n) = 0 \quad i = 1,\ldots,n \tag{2b}$$

if one introduces a basis for the vector space $\mathcal{N}$.

The equations (1) are covariant with respect to a group $\mathcal{G}$ if there is a representation T_g of $\mathcal{G}$ on $\mathcal{E}$ and $\mathfrak{F}$ such that

$$T_g G(\lambda,u) = G(\lambda,T_g u) \tag{3}$$

Condition (3) is a natural one in physical theories; it is a mathematical expression of the axiom that the equations of mathematical physics be independent of the observer. For example, the Navier-Stokes equations

$$\Delta u_k - \frac{\partial p}{\partial x_k} = u_j \frac{\partial u_k}{\partial x_j}$$

$$\frac{\partial u_j}{\partial x_j} = 0$$

are covariant with respect to the group of rigid motions if one takes for the representation

$$T_g \begin{pmatrix} u_1 \\ u_2 \\ u_3 \\ p \end{pmatrix} (x) = \begin{pmatrix} & O & & 0 \\ & & & 0 \\ & & & 0 \\ 0 & 0 & 0 & 1 \end{pmatrix} \begin{pmatrix} u_1 \\ u_2 \\ u_3 \\ p \end{pmatrix} (g^{-1} \underline{x})$$

where $g = \{O,a\}$, O being a 3×3 orthogonal matrix and a being a vector in R^3 . (Thus $g\underline{x} = O\underline{x} + a$) The Boussinesq equations, which govern convective processes are similarly covariant with respect to the group of rigid motions. The nonlinear wave equation

$$u_{tt} - \Delta u + f(u) = 0 ,$$

on the other hand, is covariant with respect to the Lorentz group.

Theorem 1. *Let equations (1) be covariant with respect to a group representation* T_g . *Then* $T_g L_0 = L_0 T_g$, *the kernel* $\mathcal{N}$ *is invariant under* T_g , *and the bifurcation equations (2) are covariant with respect to the finite-dimensional representation obtained by restricting* T_g *to* $\mathcal{N}$: $T_g F(\lambda,v) = F(\lambda,T_g v)$.

As a consequence of theorem 1, the structure of the bifurcation equations can be determined, up to some scalar constants, by purely group-theoretic methods once the finite dimensional representation $T_{g|\mathcal{N}}$ is known.

Let us expand F in a power series in v :

$$F(\lambda,v) = A(\lambda)v + B_2(\lambda,v,v) + B_3(\lambda,v,v) + \dots$$

Then we must have,

$$T_g A(\lambda) = A(\lambda) T_g \tag{4}$$

$$T_g B_2(\lambda,v,w) = B_2(\lambda,T_g v,T_g W) \tag{5}$$

If we assume that $\mathcal{N}$ is irreducible under T_g , then from Equation 4 it follows by Schur's lemma that $A(\lambda) = \sigma(\lambda)I$, where I is the identity. Suppose for convenience $\lambda_0 = 0$ and $\sigma(\lambda) = C_1\lambda + C_2\lambda^2 + \dots$. Then by various scaling arguments, the bifurcation problem can be reduced to an analysis of the equations,

$$\lambda w = B_k(w) \tag{6}$$

where B_k is the first nonvanishing term in F , homogeneous of degree k . Call (6) reduced bifurcation equations. It can be shown that the stability of the bifurcating solutions can be determined to lowest-order from an analysis of the Jacobian of (6) at a solution ([15], Chapter IV).

The group theoretic approach, then, is to compute the lowest nonvanishing terms B_k , find all solutions of (6), and determine their stability in the neighborhood of the branch point. This attack not only allows us to bypass the numerical difficulties inherent in the Liapunov-Schmidt procedure; but it also provides us with a systematic approach to bifurcation at multiple eigenvalues and with a way of classifying multiple eigenvalue bifurcation points.

It turns out that, depending on the representation $T_g|\eta$, there may be more than one covariant term of lowest degree k. In that case we arrive at a system of reduced equations of the form,

$$\lambda w = \mathcal{a}_1 B_k^{(1)}(w) + \mathcal{a}_2 B_k^{(2)}(w) + \ldots + \mathcal{a}_\ell B_k^{(\ell)}(w) \tag{6a}$$

where the coefficients $\mathcal{a}_1 \ldots \mathcal{a}_\ell$ are parameters that depend on the original parameters of the systems. The multiplicity ℓ of covariant terms of degree k can be computed directly from a knowledge of the representation $T_g|\eta$, and does not depend on the particular structure of the equations at hand. When there are multiple covariant tensors, as in Equation 6a the possibility of selection mechanisms arises. The stability of the various bifurcating solutions depends on the relative sizes of the parameters $\mathcal{a}_1, \ldots, \mathcal{a}_\ell$. In the Bénard problem, for example, $\ell = 2$ when $k = 3$, and there occurs a selection mechanism for the stability of rolls and hexagons. (See §4)

Detailed methods for carrying out this program are given in [15], although there are still many unsolved algebraic problems.

3. An Example; bifurcating waves.

Herschkowitz-Kaufman and Erneux [11], in an analysis of instability phenomena in reaction diffusion equations in a circular geometry, discovered that when time dependent instabilities set in, two distinct wave-like phenomena were possible. One was standing waves, which in polar coordinates took the form

$$\cos(\theta - ct) + \cos(\theta + ct) = \operatorname{Re} e^{i(\theta + ct)}$$

and the other mode was a rotating wave:

$$\cos(\theta - ct) \ .$$

Their calculations are somewhat complicated, but one can readily see the possibility of these two distinct modes if one attacks the bifurcation problem from a group theoretic point of view.

Suppose that, instead of the circular geometry, we consider the bifurcation of wave-trains periodic in space and time for a system of equations defined on the real line. We assume the rest state loses stability by virtue of a pair of complex-conjugate eigenvalues crossing the imaginary axis and that the corresponding eigenfunctions are $e^{(\pm ikx \pm \omega t)}$. Thus, the null space is four dimensional and is spanned by

$$\Psi_1 = e^{i(kx + \omega t)}, \quad \Psi_2 = e^{-i(kx + \omega t)}, \quad \Psi_3 = e^{i(kx - \omega t)},$$

$$\Psi_4 = e^{-i(kx - \omega t)}$$

(Actually by vector multiples of these, but it is sufficient to consider these wave functions alone.) The group operations are

$$T_\gamma : x \to x + \gamma \ , \quad R : x \to -x \ , \quad S_\delta : t \to T + \delta \ .$$

Relative to the basis $\Psi_1, \dots, \Psi_4$ these operations have the matrix representations

$$T_\gamma = \begin{pmatrix} e^{ik\gamma} & & & \\ & e^{-ik\gamma} & & 0 \\ & & e^{ik\gamma} & \\ 0 & & & e^{-ik\gamma} \end{pmatrix} \qquad R = \begin{pmatrix} & & & 1 \\ & 0 & 1 & \\ & 1 & & 0 \\ 1 & & & \end{pmatrix}$$

$$S_\delta = \begin{pmatrix} e^{i\omega\delta} & & & \\ & e^{-i\omega\delta} & & 0 \\ & & e^{-i\omega\delta} & \\ 0 & & & e^{i\omega\delta} \end{pmatrix}$$

The bifurcation equations are

$$F_i(\lambda, \omega, z_1, z_2, z_3, z_4) = 0, \; i = 1, 2, 3, 4 .$$

Since we are looking for real solutions we want

$$z_2 = \overline{z}_1, \; z_4 = \overline{z}_3$$

and

$$F_2 = \overline{F}_1, \; F_4 = \overline{F}_3 \quad .$$

The symmetry $RF = FR$ leads to

$$F_4(\lambda, \omega, z_1, z_2, z_3, z_4) = F_1(\lambda, \omega, z_4, z_3, z_2, z_1)$$

$$F_3(\lambda, \omega, z_1, \dots) = F_2(\lambda, \omega, z_4, \dots) .$$

Therefore it is enough to find F_1. Suppose F_1 contains a term

$$z_1^a z_2^b z_3^c z_4^d \quad .$$

The symmetry $T_\gamma F = FT_\gamma$ implies

$$e^{ik\gamma}(z_1^a z_2^b z_3^c z_4^d)$$

$$= (e^{ik\gamma} z_1)^a \dots$$

$$= e^{ik\gamma(a-b+c-d)}(z_1^a z_2^b z_3^c z_4^d)$$

hence

$$a - b + c - d = 1 .$$

Similarly, $S_\delta F = FS_\delta$ leads to

$$a - b - c + d = 1 .$$

Combining these two equations we get

$$a = b + 1$$

$$c = d$$

so F_1 has the general term

$$z_1(z_1z_2)^b(z_3z_4)^d$$

or, putting $z = z_1$ and $w = z_3$,

$$z(|z|^{2b})(|w|^{2d}) .$$

The most general such F_1 is therefore

$$F_1(\lambda, \omega, z, \overline{z}, w, \overline{w}) = zg(\lambda, \omega, |z|^2, |w|^2) ,$$

and the bifurcation equations take the form

$$\begin{aligned} z\,g(\lambda, \omega, |z|^2, |w|^2) &= 0 \\ w\,g(\lambda, \omega, |w|^2, |z|^2) &= 0 . \end{aligned} \tag{7}$$

Expanding g in a power series in $|z|^2$ and $|w|^2$, we get

$$g = a_0 + a_1|z|^2 + a_2|w|^2 + \ldots$$

where $a_0, a_1, a_2, \ldots$ depend on λ and ω. The two parameters a_1 and a_2 signify the existence of two independent covariant terms of third order (the actual mapping was generated by zg).

Now there are two distinct solution types to equations (7):

(i) $|z| = |w|$. Then we must have

$$g(\lambda, \omega, |z|^2, |z|^2) = 0$$

(ii) $w = 0$; then we get

$$g(\lambda, \omega, |z|^2, 0) = 0 .$$

The case $z = 0$ is equivalent to $w = 0$ under the reflection R.

If $|z| = |w|$ we take $z > 0$ by a suitable translation in x and $w = ze^{i\gamma}$ (take $z = 1$ and $w = e^{i\gamma}$). Then the solution to first order is

$$\Psi_1 + \Psi_2 + e^{i\gamma}\Psi_3 + e^{-i\gamma}\Psi_4$$

$$= 2\cos(kx + \omega_0 t) + 2\cos(kx - \omega_0 t + \gamma) .$$

By suitable translation in x and t this is equivalent to

$$4\cos kx \cos \omega_0 t : \text{standing waves} .$$

In case (ii) with $c = 0$ we get

$$\Psi_1 + \Psi_2 = 2\cos(kx + \omega t) \quad \text{traveling waves.}$$

These two solution types exist simultaneously regardless of the values of the parameters a_1 and a_2. We conjecture, however, that these two modes cannot be simultaneously stable, and that in fact the stability depends on the relative sizes of the parameters a_1 and a_2. Herschkowitz-Kaufmann and Erneux have carried out the stability analysis in their case, and have found such a selection mechanism.

In the case of time independent bifurcation theory the stability of the bifurcating solutions can be determined by an analysis of the Jacobian of the reduced bifurcation equations (see [15]), Chapter IV):, but this theory has so far not been extended to the time dependent case.

The bifurcation of such periodic wave-trains in translation invariant problems arises in mathematical models of neuron networks analyzed by Cowan and Ermentrout[8]. They model space-time behavior of neuronal networks by a system of non-linear integral equations. The uniform rest state loses stability to wave-like disturbances which are periodic both in space and time. They demonstrate secondary bifurcation when two pair of complex conjugate eigenvalues cross the imaginary axis simultaneously; the analysis is considerably more intricate than the example discussed above. They relate their analysis of the bifurcating wave trains to events occuring in epileptic seizures:

> "There is considerable evidence that epileptic oscillations are in fact traveling waves. Furthermore it has been proposed that this transition occurs because of an increase in extracellular potassium as the seizure progresses."

Again, in their case there is a variety of wavelike structures, including a transition of modal wave types, which models the transition in epilepsy from tonic to clonic seizures.

Bifurcating waves have also been discussed by Auchmuty[3], though not from a group-theoretic viewpoint.

4. Pattern Selection in the Bénard Problem.

The bifurcation equations in the case of the Bénard problem can be obtained by group theoretic methods [13],[15]. For example, for the hexagonal lattice the equations are generated by

$$F_1(\lambda, z_1, \dots, z_6) = \lambda z_1 + c z_2 z_6 + a z_1^2 z_4 + b z_1 (z_2 z_3 + z_3 z_6) \qquad (8)$$

as follows: Let $\alpha = (1\,2\,3\,4\,5\,6)$ and let $F = (F_1, \dots, F_6)$. Then the full equations are generated by F_1 by using the symmetry $\alpha F = F\alpha$. Thus

$$F_2(\lambda, z_1, \dots, z_6) = F_1(\lambda, z_2, z_3, \dots, z_6, z_1),$$

etc. The bifurcation equations for the square and rhombic lattices are obtained similarly. Solutions over the hexagonal lattice are solutions u which are doubly periodic in the plane:

$$u(\underline{x} + \underline{\omega}) = u(\underline{x})$$

where $\underline{\omega}$ is any vector in the hexagonal lattice $\{\underline{\omega} \mid \underline{\omega} = n\underline{\omega}_1 + m\underline{\omega}_2\}$ where $\underline{\omega}_1$ and $\underline{\omega}_2$ are two vectors of critical length which make an angle of 60° with one another. One can prove the following:

Theorem 2. Let the bifurcation equations for solutions over the hexagonal lattice be generated by the scalar function (8). The necessary conditions for stability are

$$\text{stable rolls:} \quad b < a < 0$$
$$\text{stable hexagons:} \quad a < b\,, \quad a + 2b < 0$$

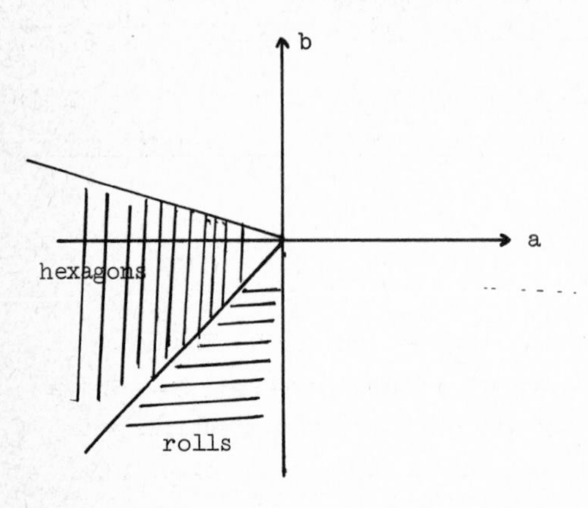

Theorem 2 does not give us a complete picture of the bifurcation phenomena associated with the Bénard problem, for we have arbitrarily restricted ourselves to bifurcating solutions (and disturbances) which are invariant only under translations in a hexagonal lattice in the plane. Furthermore the stability analysis applies only to disturbances which lie in this same hexagonal lattice. The result does illustrate clearly, however, the mathematical ideas involved. The two parameters a and b signify the existence of two covariant third order terms in the bifurcation equations. Depending on the relative sizes of a and b either hexagons or rolls are stable. Of course, if a and b lie outside the indicated regions, then there are no stable bifurcating solutions at all.

The results depicted in Theorem 2 are not really restricted to the Bénard problem, but apply whenever Euclidean invariance is broken. Similar results can be obtained for the rhombic and square lattices. Then one can go on and try to compare the stability of solutions simultaneously in two distinct lattices. This is not so easy to do, but partial resluts are described in [13].

Many of the results in the Bénard problem are unsatisfactory from the mathematical point of view. This is because the symmetry group involved is not compact. On the other hand, the plane layer model for the Bénard problem is rather a singular approximation to the actual physical situation anyway. When one considers bifurcation in a spherical geometry, the symmetry group involved is compact, so the difficulties which arise in the Bénard problem are not present. On the other hand, the algebraic problems of giving a complete solution of the bifurcation equations are much more difficult, and the bifurcation problem is not fully resolved in this case either. See Busse[6], Sattinger[14].

V. Conclusions.

By selection mechanisms for pattern formation I mean specifically that there exist multiple covariant mappings at the lowest order (i.e. in the reduced bifurcation equations). The actual multiplicity can be computed from the methods of group representation theory. If $\Gamma = T_g|_{\eta}$, where η is the kernel of the linearized equations, then the multiplicity is given by the number of times the representation Γ is continued in the tensor product representation $(\Gamma^{\otimes k})_s$, where the subscript s means that $\Gamma^{\otimes k}$ is restricted to the completely symmetric tensors. A generating function for these multiplicities is given in [14].

Theorem 3. *Let* Γ *be a representation of a compact group* $\mathcal{G}$ *and let* $c_n = c_n(\Gamma, \mathcal{G})$ *denote the number of completely symmetric n-linear operators* B *on* $\mathcal{H}^{\otimes n}$ *to* $\mathcal{H}$ *which are covariant with respect to* Γ. *Then a generating function for the coefficients* c_n *is*

$$\sum_{n=0}^{\infty} c_n(\Gamma, \mathcal{G}) z^n = \int_{\mathcal{G}} \det(I - z\Gamma(g))^{-1} \bar{\chi} \, d\mu(g)$$

where $d\mu(g)$ is the normalized invariant measure on $\mathcal{G}$ and $\chi(g) = \mathrm{Tr}\,\Gamma(g)$.

Thus, if the reduced bifurcation equations are quadratic and $c_2 > 1$, we have the possibility of selection mechanisms - that is, of bifurcating solutions with distinct symmetry groups and whose stability depends on the relative size of the c_2 parameters mulitplying the independent covariant mappings.

Physically we would discover bifurcating states with distinct symmetries whose stability depended on the various parameters of the problem.

References

1. Auchmuty, J.F.G. and G. Nicolis, "Bifurcation analysis of nonlinear reaction diffusion equations", Bull. of Math. Biol. 37, 323-363.

2. ______________________________, "Dissipative structures, catastrophes, and pattern formation: a bifurcation analysis", Proc. Nat. Acad. Sci. U.S.A. 71, 2748-2751.

3. Auchmuty, J.F.G., "Bifurcating waves", in Bifurcation Theory and Applications in Scientific Discipline, Annals, New York Academy of Sciences, 316 (1979), 263-278.

4. Birman, J.L., "Symmetry changes, phase transitions, and ferroelectricity", Ferroelectricity, Elsevier, 1967.

5. Busse, F., "The stability of finite amplitude cellular convection and its relation to an extremum principle", Jour. Fluid. Mech. 30 (1967), 625-650.

6. ________, "Patterns of convection in spherical shells", Jour. Fluid. Mech. 72 (1975), 67-85.

7. ________, "Mathematical problems of dynamo theory", in Applications of Bifurcation Theory, Academic Press, New York, 1977.

8. Cowan, J.D. and Ermentrout, G.B., "Secondary bifurcation in neuronal nets", to appear SIAM Jour. of Applied Math.

9. ________________________________, "A mathematical theory of visual hallucination patterns", to appear, Biological Cybernetics.

10. Fife, P., "Pattern formation in reacting and diffusing systems", Jour. Chem. Phys. 64 (1976), 554-564.

11. Hercshkowitz-Kaufman, M. and T. Erneux, "The bifurcation diagram of model chemical reactions", Annals of the New York Academy of Science, op. cit. 316 (1979), 296-313.

12. Larter, R. and P. Ortoleva, "Self electrophoretic phenomena", preprint, Indiana University, 1979.

13. Sattinger, D.H., "Selection mechanisms for pattern formation", Arch. Rat. Mech. Anal. 66 (1977), 31-42.

14. ______________, "Bifurcation from rotationally invariant states", Jour. Math. Phys. 19 (1978), 1720-1732.

15. ______________, "Group theoretic methods in bifurcation theory", to appear, Springer lecture notes in mathematics.

16. Turing, A.M. "The chemical basis of morphogenesis", Phil. Trans. Roy. Soc. London B 237 (1952), 37-72.

HIGH REYNOLDS-NUMBER FLOWS

K. Stewartson
Department of Mathematics
University College London

Until quite recent times the structural theory of high Reynolds-number flows was in a relatively undeveloped state largely because of its inability to deal with finite bodies. If they were smooth then the standard procedure, which is conveniently called hierarchical, was first to neglect viscosity, solve the resulting simpler Euler equations, and attempt to remove the non-uniformity at the body, occasioned by a failure to satisfy a no-slip condition there, by the introduction of a boundary layer. For the solution of the Euler equations there were two main possibilities. One could assume attached flow in which case the boundary layer developed an unacceptable singularity at the point where the skin friction vanished, usually called separation, and which has to occur before the rear stagnation point of the body is reached. This phenomenon terminated the theory, and indeed raised serious doubts about its relevance. Or one could assume that the main flow detached from the body at some point P as suggested first by Kirchhoff. The role of the boundary layer here was limited to the fixing of P as the point of maximum velocity on the body according to the Euler equations. For otherwise either the free streamline from P intersects the body or the boundary layer separates upstream of P leading to the unacceptable singularity mentioned earlier. However, with P so chosen the description of the nature of detachment was mysterious; for example, the boundary layer evolves as far as P in a favourable pressure gradient which does not provoke the separation needed at P to set up the detached streamline. Again, for sharp bodies such as a flat plate there were uncertainties even when separation does not occur. Thus the flow properties near the trailing edge were not at all understood and, as soon as the plate is set at incidence to the mainstream, the traditional boundary-layer theory failed completely on the upper, suction side.

Since these flows are of great importance in many practical problems there had nevertheless been developed a semi-empirical theory which had a considerable measure of success for real flows, usually turbulent, provided the separated region was very limited. This theory was based on some of the ideas of the more formal theory but did include a significant amount of curve fitting which opened it up to some trenchant criticisms. Incidentally the handling of extensive regions of turbulent separated flows still presents major problems to the theoretical engineer.

The seminal paper for overcoming the impasse of the separation singularity was due to Lighthill (1953) by his introduction of what might be called the three-fold way by analogy with the eight-fold way of strange particles in theoretical physics but which I prefer to refer to as the triple-deck (Fig.1) as being more evocative of

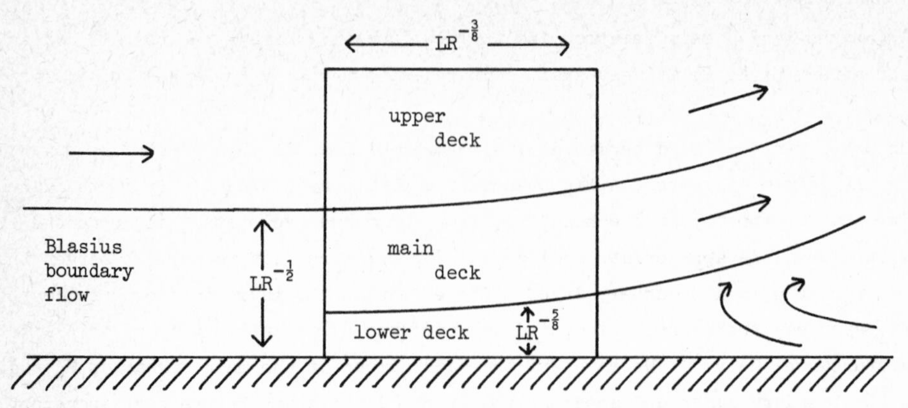

Fig.1. Sketch of the triple-deck region near separation on a flat plate.

the interlocking structure which plays a crucial role in the phenomenon. A second pair of papers of equal significance is due to Lees & Reeves (1964) and Mangler & Catherall (1966) which showed that the boundary layer need not be singular at separation provided one does not insist on the peculiar, but hitherto well entrenched, requirement that the pressure gradient be prescribed. Those papers have in fact been fatal to the hierarchical theory of high Reynolds number flows with separation.

The Goldstein singularity of classical boundary-layer theory (1948) contains a signal of great importance to the numerical analyst and theoretician, namely that the whole theory on which it is based, including the notion that separation occurs after a finite interval of adverse pressure gradient, is probably irrelevant to the behaviour of fluids moving past bodies at high Reynolds number and that they would do well to turn elsewhere for understanding. The alternative theory (reviewed by Stewartson 1974, Messiter 1978) starts from the premise that the separation phenomena must be interactive, i.e. the boundary layer must exert a significant restraining influence on the inviscid flow field just outside. This is only possible if the relevant longitudinal distance over which it occurs is small and in turn this means that viscous forces are confined to a narrow region near the wall (the lower deck) the remainder of the boundary layer behaving in an inviscid and largely passive manner. The conventional boundary-layer equations, namely

$$uu_x + vu_y = - p_x/\rho_w + \nu_w u_{yy} \quad , \quad u_x + v_y = 0 \quad , \tag{1}$$

in the usual notation, hold in the lower deck at the bottom of the boundary layer during the interaction region and the no-slip conditions hold at the wall, but elsewhere the boundary conditions are novel. Above and upstream of the triple deck the lower deck must be compatible with the conventional boundary layer from which it develops so that we must have

$$u \to \lambda y \quad \text{as} \quad x \to -\infty \quad , \quad u \to \lambda(y + A(x)) \quad \text{as} \quad y \to \infty \tag{2}$$

where λ is the Blasius skin friction and the new function A plays the role of a displacement effect being relatively small compared with y when y is large - on the scale of the lower deck. The effect of A on the main deck, comprising the remainder of the boundary layer, can be determined by a theorem of Prandtl which describes how a shift in the origin of y, here A, is transmitted to the mainstream. It is equivalent to saying that at the top of the boundary layer the normal velocity v is augmented by $A'(x)$ which, when the appropriate scaling is adopted, turns out to be much larger than that of the conventional boundary layer. Its effect on the inviscid flow outside the boundary layer may therefore be isolated and computed by an application of Ackeret's small disturbance theory. We have, if the Mach number of the main stream is M_∞ just outside the boundary layer and ahead of the triple deck, that the pressure increment is p where

$$p \propto -(M_\infty^2 - 1)^{-\frac{1}{2}} A'(x) \quad \text{if} \quad M_\infty > 1,$$
$$p \propto \pi^{-1}(1 - M_\infty^2)^{-\frac{1}{2}} \int_{-\infty}^{\infty} (x - x_1)^{-1} A'(x_1)\, dx_1 \quad \text{if} \quad M_\infty < 1 \tag{3}$$

and this provides the driving pressure in (1). The appropriate scaling laws may now be worked out in a straightforward way. Thus the longitudinal scaling of the triple-deck region is $O(R^{-\frac{3}{8}}L)$ where L is a representative length of the boundary layer and R is the Reynolds number based on L and local inviscid conditions. Further the latitudinal length scales of the lower, main and upper decks are $O(R^{-\frac{5}{8}}L)$, $O(R^{-\frac{1}{2}}L)$, $O(R^{-\frac{3}{8}}L)$ respectively and the numerical and physical constants in (1), (2), (3) may be replaced by unity if desired. The Lighthill theory is appropriate to small values of $(\lambda y - u)$ and $M_\infty > 1$.

Let us look at supersonic flow first. It turns out that a self-generated solution of the triple-deck can be found i.e. different from $u = \lambda y$ and not forced by any local external feature. The underlying reason was given by Oswatitsch & Wieghardt in 1948 - a growing displacement thickness provokes an increasing pressure in the supersonic stream which tends to thicken the boundary layer further. Thus a supersonic boundary layer can spontaneously evolve through a triple-deck and can eventually separate. Further if the separation point is fixed the solution upstream is unique - a feature well known to experimenters and discussed in physical terms by Chapman (1958).

Beyond separation, which is quite regular, the main boundary layer moves away from the wall at an angle $\propto R^{-\frac{1}{4}}$ and, according to triple-deck theory, a slow inflow develops below. The computation of this part of the flow field is not easy because the standard methods become unstable when $u < 0$ since they integrate in the direction of x increasing. The best method for open reversed flows is due to Flügge-Lotz & Reyhner (1968) (FLARE) with modifications due to Williams (1974) (ASPRO and DUIT). In FLARE the integration for $x > x_s$ is performed with the approximation $uu_x = 0$ if $u < 0$. This leads to a stable incorrect solution. ASPRO is the asymptotic profile for large x deduced by Neiland (1971) & Stewartson & Williams (1973) independently and is used to integrate upstream in the reversed flow region keeping its boundary and the pressure gradient the same as that given by FLARE. The value of uu_x when $u < 0$ may now be regarded as known in a reapplication of FLARE. The whole process is repeated (DUIT) until convergence is achieved.

Further downstream a plateau is reached in the pressure while the boundary layer takes on the characteristics of a mixing layer between a moving stream and stagnant fluid until the feature is reached that provoked the initial separation, say a shock or a ramp. Without that feature a contradiction would be achieved, with the downstream pressure preventing the onset of separation anywhere.

While on the subject of the calculations of the solution in regions of reversed flow, it is worth mentioning some other interesting methods. Global iteration using upwind and downwind derivatives has been applied effectively by Dennis in another related problem and by Carter (1972) for the inverse problem of computing the pressure, given the displacement thickness. Unsteady methods have been used by Rizzetta, Burggraf and Jenson (1978) and in my view have probably the best potential for the future.

Applications of the theory have been made to flows past corners and to problems of strong injection which have been reviewed in detail elsewhere. A major message of the theory is that, in laminar and probably also turbulent interactions, boundary layer theory may be used even when extensive regions of separating flow occur and Werle & Vatsa (1974) have obtained very good results in laminar shock-wave boundary-layer problems by honouring the triple-deck length scales. For turbulent flows, the main handicap to progress is the construction of suitable models for eddy viscosity, dissipation and length scales rather than a deficiency in the boundary-layer concept.

Let us now turn to subsonic flow where much exciting work has been done recently. The great feature of supersonic triple-decks which enables the separated flow to get started is the possibility of a self-generated solution, i.e. one in which

$$p \to 0, \quad u \to \lambda y \quad \text{as} \quad x \to -\infty.$$

In subsonic flow this is not possible and in any case we must expect that the detaching streamline will influence the flow upstream of separation. We need a pressure distribution when $|x| >> 1$ which is constant if $x > 0$ and gives a decaying pressure gradient if $x < 0$ while the corresponding form for A is increasing if $x > 0$ and tends to zero as $x \to -\infty$. Since the flow outside the boundary layer is essentially irrotational, an appropriate form may be derived from the $z^{\frac{1}{2}}$ solution of potential theory and is

$$p \to 0, \quad A \simeq \frac{2}{3}\alpha x^{3/2} \quad \text{as} \quad x \to \infty \quad ,$$

$$p \simeq \alpha(-x)^{\frac{1}{2}}, \quad A \to 0 \text{ as } \quad x \to -\infty$$

where α is a constant. If such a solution is feasible then we see that upstream of separation there is an adverse pressure gradient and attached flow, while downstream there is a detached streamline and below it a virtually constant pressure region.

In physical terms the adverse pressure gradient is $p^*_x \propto \alpha^*(x^*_s - x^*)^{-\frac{1}{2}}$ where $\alpha^* = O(\alpha R^{-1/16})$ and the free streamline is $y^* \propto \alpha^*(x^*-x^*_s)^{3/2}$ i.e. barely clearing the surface. So we could neglect the pressure gradient away from the triple deck and think of the free streamline as being tangential to the body at separation. The question is: does such a solution exist? This was answered by Smith (1977) in the affirmative in a quite convincing manner. He estimated that

$$\alpha = 0.44 .$$

He used FLARE which is of course somewhat inaccurate but in view of other experiences it is unlikely even that α is much in error, let alone the existence of the eigen-solution.

His numerical scheme is worth describing as representative of a departure from classical iteration routines in boundary-layer computations. Traditionally the pressure gradient is taken as fixed and the boundary layer computed even though this leads to a terminating singularity at separation. Smith uses inverse methods, i.e. he prescribes $A(x)$ and determines $p(x)$ which enables him to pass through separation easily. The iterative cycle is completed by using the Hilbert integral to compute $A(x)$ given $p(x)$. Large underrelaxation is necessary to achieve convergence. Of course, it is also necessary to get α right. If α was too large (up to 0.6) the separation moved upstream with each iteration and if α was too small ($\sim$ 0.25) downstream. The final result appears to be unique, just as in the supersonic case, certainly for $0 < \alpha < 0.6$. In view of the examples of non-uniqueness that have been recently found for incomplete physical systems, it would be rash to claim uniqueness for Smith's solution.

We can now make a confident claim about the nature of laminar flow past smooth bodies at high Reynolds number. The correct limit is the Kirchhoff-Sychev flow in which a streamline springs tangentially from the body, with zero curvature, and beyond it the fluid is at rest. The contribution of Kirchhoff to this picture is well-known. That of Sychev (1967,1972) was to propose an asymptotic structure, including the effects of viscosity, that made the picture unique. Sychev went on to criticise the classical boundary-layer concept of separation being preceded by a finite interval of adverse pressure as irrelevant to the asymptotic structure and I am strongly of the opinion that he is correct.

There seems little doubt, that for the circular cylinder at least, this theory is right and even that at values of $R \sim 100$ it predicts flow properties in good agreement with numerical and experimental results. In Fig.2 we display the pressure variation near separation both from Dennis and Chang's (1970) numerical results and the asymptotic theory. This theory has been extended (Sychev 1967, Smith 1979b) to cover many of the principal aspects of the whole flow field and shows that, in spite of the very small power of R^{-1}, namely 1/16, appearing in the asymptotic expansion, a good agreement can be obtained between this theory on the one hand and numerical and experimental work on the other. The discussion is involved and extended and I can only

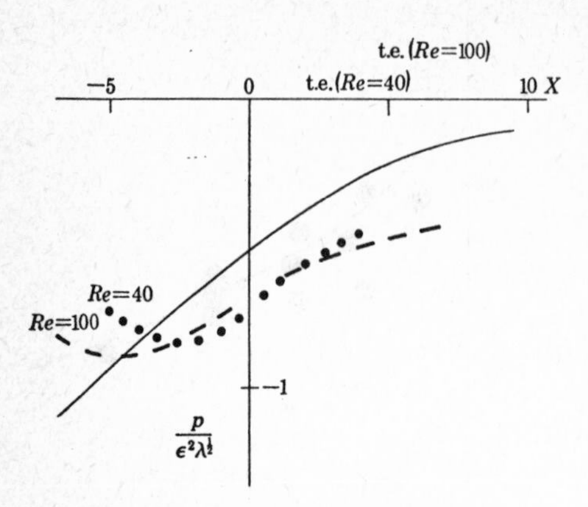

Fig.2. Comparison of the triple-deck pressure curve with solutions of the Navier-Stokes equations taken from Dennis & Chang (1970) (...Re = 40, - -Re = 100).t.e. signifies the positions of the rear stagnation point on this scale.

summarise the main ideas. On the large scale O(R) in x there is a bubble of slowly moving fluid behind the cylinder and bounded by the streamlines springing from the separation points and meeting again at a point C. The boundary layers on the cylinder continue along the neighbourhood of these streamlines and beyond C form the final wake. The shape of these free streamlines is known at separation and may be expressed in terms of the drag on the cylinder. In order to achieve this slow flow inside the bubble the pressure there must virtually be constant and so the shape of the bubble must be elliptical; hence the pressure inside is given by

$$p_B = -(4C_D/L)^{\frac{1}{2}}R^{-\frac{1}{2}} + \dots$$

where LR is the length of the bubble. Further, at reattachement C, the interaction between the two shear layers must be largely inviscid as discussed by Burggraf (1970, 1975), Messiter (1973) and others. Hence the entrainment of the free shear layer from S to C, which can be calculated using the Chapman mixing layer theory (1950), must be restored and the remainder carries on downstream in an evolving wake. The final form of this wake can be related to the drag on the body, as has been known for many years, and this gives a condition which determines the length of the bubble SC. Smith obtains $1.36\ C_D^2R = 0.34R$ for the length of the bubble and hence $p_B \simeq -1.37R^{-\frac{1}{2}}$. As we can see from Fig.3 taken from Smith (1979b), the agreement with experiment and numerical computation for $R \sim 50$ is remarkable, not only in the circumstances.

Let us now examine the flow past a finite flat plate in some detail. Up to a point it is now thoroughly understood but then the theory comes to a full stop, as I shall show. As usual with traditional views of high Reynolds number asymptotic theory, the expansion proceeds by the hierarchical method beginning with a uniform flow if the plate is symmetrically disposed to the oncoming flow and going on to the boundary layer on the plate together with its continuation, the wake. This procedure fails at the leading and trailing edges. The region near the leading edge may be considered separately by rescaling distance with R^{-1} but then we are reduced to solving the full Navier-Stokes equations for a semi-infinite flat plate. Fortunately, this has been done, by Van de Vooren and Dijkstra (1970), and we can now see how a smooth transition is effected between the virtually undisturbed flow upstream of the leading edge to the Blasius boundary layer on the plate. At the trailing edge the boundary layer develops a singularity on the downstream side with, for example, the velocity

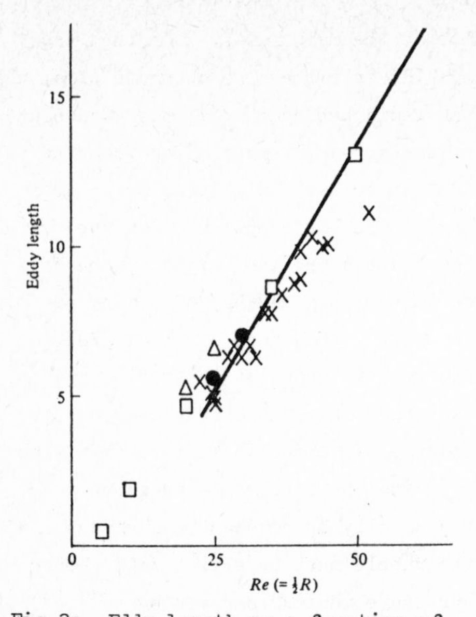

Fig.3a. Eddy length as a function of Reynolds number. The solid line is the asymptotic result apart from an origin shift. The crosses are experimental and the others are numerical values.

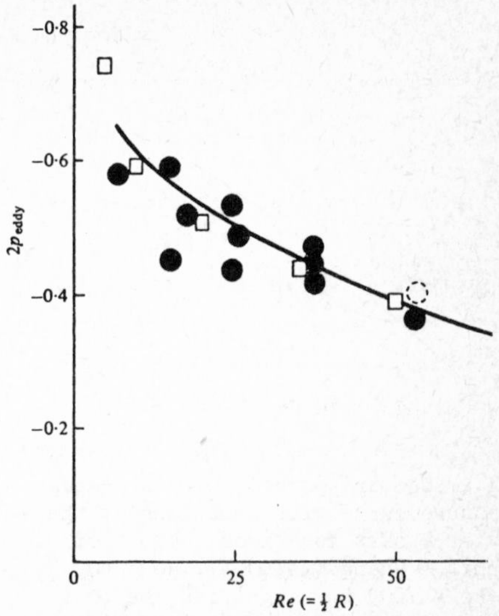

Fig.3b. Variation of the eddy pressure with Reynolds number. The solid line is the asymptotic result, the squares are numerical values and the remainder are experimental.

on the wake centre line being $\propto x^{1/3}$, where x measures distance from the trailing edge. This speeding up of the fluid causes a dramatic thinning out of the boundary layer and so a negative normal velocity, just outside, proportional to $R^{-\frac{1}{2}}x^{-2/3}$. The pressure gradient induced thereby is highly singular and the consequent increase of skin friction non-integrable, i.e. $\propto R^{-1}(-x)^{-4/3}$. It is not surprising that a non-hierarchical but interactive approach is needed to handle the flow near the trailing edge and that a triple-deck is appropriate. The scalings are the same as for separation and again the critical deck is the lower deck, right near the plate, also with the same equation as before. The only change is in the boundary conditions on the plate now

$$u = v = 0 \quad x < 0,\ y = 0 : v = u_y = 0 \quad x > 0,\ y = 0.$$

The numerical solution of the equation was first worked out for incompressible flow by Jobe & Burggraf (1974) and later by others. The most accurate solution is probably due to Melnik & Chow (1976) and we have

$$1.328R^{-\frac{1}{2}} + 2.668R^{-\frac{7}{8}} + O(R^{-1}) \qquad *$$

for the drag coefficient. The agreement with experiment and numerical studies by Dennis (see Stewartson 1975) is remarkable even when $R \sim 1$.

A comparable result to * has been obtained by Daniels (1974) for supersonic flow. The problem is easier because upstream of the trailing edge the solution is an expansion free interaction and, apart from one parameter, is fixed independently of what happens downstream. He gets

$$1.328R^{-\frac{1}{2}} + 2.026\,(M_\infty - 1)^{-\frac{3}{8}}(T_w/T_\infty)^{3/2}R^{-\frac{7}{8}}\,.$$

What happens when the plate is set at incidence? At the leading edge it may be shown (Cebeci, Khattab & Stewartson 1979) that irrecoverable separation occurs if

$$\alpha^* > 2.24(2r_c/L)^{\frac{1}{2}}$$

where α^* is the angle of incidence and r_c is the radius of curvature of the nose. This is because the fluid velocity overshoots in going round the nose and so there is an adverse pressure gradient just downstream which easily provokes separation. We shall see that the trailing-edge solution exists if $\alpha^* \sim R^{-1/16}$ and so we may maintain attached flow if $r_c \lesssim R^{-\frac{1}{8}}$ i.e. if the flat plate is replaced by a thin airfoil which approaches it as $R \to \infty$.

At the trailing edge for incompressible flow, even if we apply the Kutta condition, the inviscid pressure gradient on the suction side is proportional to $\alpha^*(-x)^{-\frac{1}{2}}$ and provokes separation unless the induced and favourable pressure gradient from the triple deck compensates, which can happen only if $\alpha^* \lesssim R^{-1/16}$. In order to elucidate the structure of this triple deck, we need to consider the lower-deck boundary layers on the two sides and in the wake where they join, insisting that the pressures are equal in the wake. The problem is very similar to the Sychev separation problem discussed earlier but more complicated because of the two boundary layers in $x < 0$ and the change of conditions at $x = 0$. The numerical work is difficult but has been completed by Melnik & Chow (1976).

They found that on the top side the pressure gradient is adverse far from the trailing edge and favourable near the trailing edge when α is small - on triple-deck scale of course - but that as α increases it becomes completely adverse, (Fig.4), a

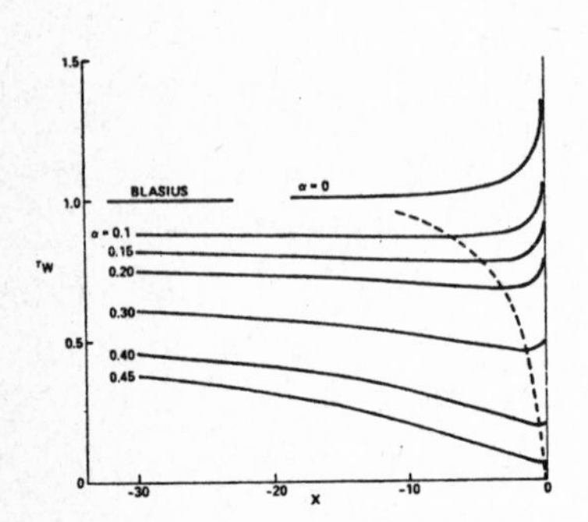

Fig.4a. Skin friction distribution on the plate (top).

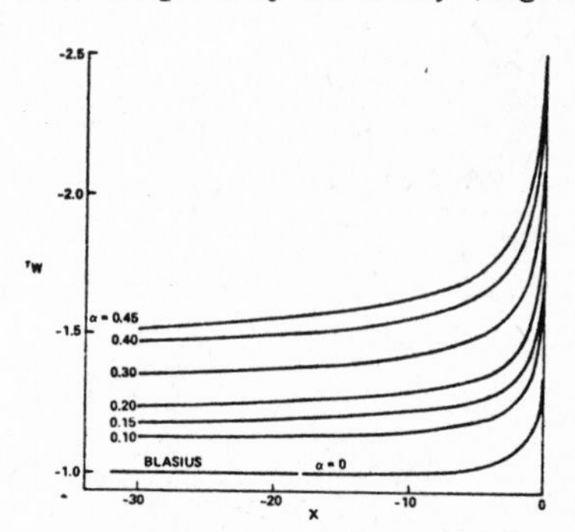

Fig.4b. Skin friction distribution on the plate (bottom).

skin friction minimum τ_M develops at x_M and τ_M, $x_M \to 0$ together at which point $\alpha \simeq 0.47$ and the solutions terminate. The reason is that the solution cannot proceed beyond $x = 0$ if the skin frictions on the two sides of the plate have opposite signs. For the streamfunction near this point takes on the form

$$\psi = x^{2/3} g(\eta) \, , \; \eta = y/x^{1/3}$$

where

$$g'''' + gg''' = 0, \; g' \to \eta \text{ as } \eta \to \infty \, , \; g' \to \tfrac{1}{2}\bar{\lambda}\eta \text{ as } \eta \to -\infty$$

and if $\bar{\lambda} < 0$ there is no solution for g.

When a solution exists the lift coefficient

$$C_L = 2\pi\alpha \left[1 - \frac{8a(\alpha)}{R^{\frac{3}{8}}} - \frac{0.285\log R}{R^{\frac{1}{2}}} + \ldots \right] , \; 0.52 < a < 1 + .$$

One interpretation of the failure of the solution for $\alpha > 0.47$ is that there is a switch to a Kirchhoff free-streamline solution with centre far from the trailing edge on triple-deck scale. But now the characteristic value of α is 0.44 according to Smith and this suggests that the move must be dramatic - to where the curvature is different from that at the trailing edge - possibly to the nose region. This would then simulate a stall situation. Note also a hysteresis phenomenon and possible non-uniqueness of solution as α varies around 0.45.

It might be argued that the numbers 0.44 and 0.47 are not significantly different. After all Smith used FLARE to get 0.44 and Melnik did not compute any solution beyond 0.45. However, Daniels (1974) has carried out a like investigation for supersonic flow where a much higher accuracy can be achieved. The corresponding numbers are 2.052 for the last attached flow and 1.80 for the free-interaction separation so the hysteresis is clear in that instance and by inference also in subsonic flow.

In relation to these flat-plate flows, the effect of injection is interesting - suction on the other hand is straightforward. The elementary results for injection velocities $O(R^{-\frac{1}{2}})$ i.e. comparable with the normal velocities in the conventional boundary layer are:

1) Separation does not occur in a favourable pressure gradient (Pretsch 1944). Instead the boundary-layer thickness increases as $LCR^{-\frac{1}{2}}$ as $C \to \infty$, where the injection velocity $V_w = CU_\infty R^{-\frac{1}{2}}$ (Fig.5a).

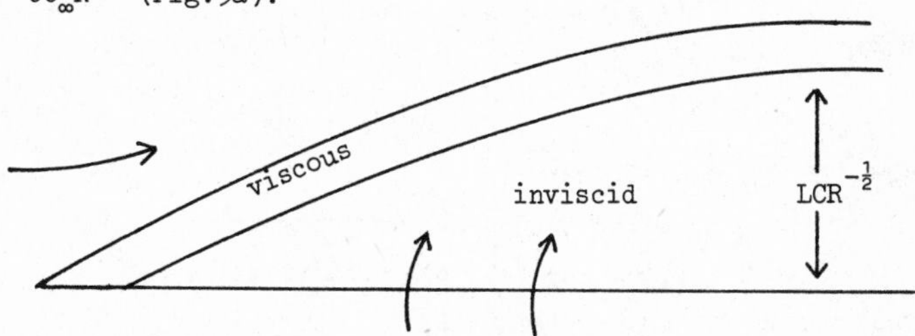

Fig.5a. Injection with velocity $CU_\infty R^{-\frac{1}{2}}$ into a favourable pressure gradient.

2) With a uniform mainstream and uniform injection separation occurs at (Catherall, Stewartson and Williams 1965)

$$0.75U_\infty^2 LR^{-1} V_w^{-2}$$

3) With a uniform mainstream and injection velocity $V_w = CU_\infty (L/2xR)^{\frac{1}{2}}$ solution of the boundary layer can be found if $C < 0.876 = C_o$ (Emmons and Leigh 1953). Sepa-

ration occurs if $C > C_0$.

Once separation occurs the flow field develops some new features of interest. Suppose to begin with the mainstream is supersonic. Then the situation in 3) may be converted into 1) by interacting the boundary layer with the mainstream, for which it must thicken enough so that the induced pressure gradient can drive some of the fluid downstream. In equilibrium there is an inviscid region near the plate of thickness $\delta \propto R^{-1/3}(C - C_0)^{2/3}$ in which the streamwise velocity $\propto R^{-1/6}(C - C_0)^{1/3} U_\infty$, being driven by a pressure gradient $\propto R^{-1/3}(C - C_0)^{2/3} U_\infty^2/L$. Between this layer and the mainstream there is a thinner viscous layer of thickness $\propto LR^{-\frac{1}{2}}$ entraining fluid at a rate $C_0 U_\infty (L/2xR)^{\frac{1}{2}}$ (see Fig.5b).

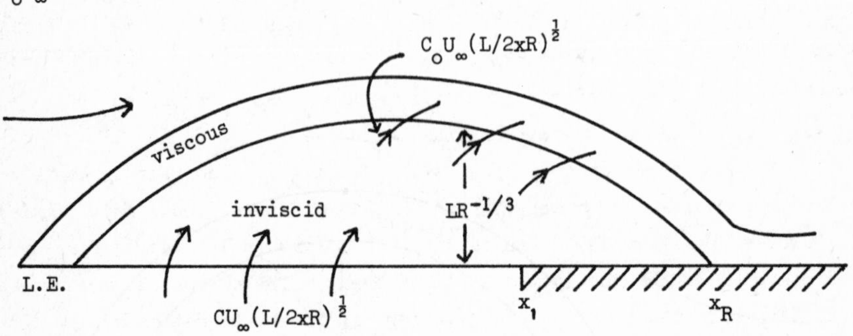

Fig.5b. Injection into a uniform stream with $C > C_0$ i.e. too much to be immediately absorbed into the viscous region. Notice the thickening of the interaction regime.

If the injection is cut off, at $x = x_1$ say, the free-shear layer must come back to the plate which it reaches just as the last of the injected fluid is absorbed into it, the boundary layer further downstream being of thickness $O(LR^{-\frac{1}{2}})$. Even though the mainstream is supersonic it is possible for the flow field to adjust to this requirement in $x < x_1$ since the solution there contains an arbitrary parameter (Diver and Stewartson 1978). Thus for $x \ll L$

$$R^{1/3}\delta \sim x^{1/3} + Ax^m \quad ..... \quad \text{where } 1.333 < m < 1.403$$

and A is arbitrary, until it is fixed by applying the closure condition at reattachment. There is very little recirculation at reattachment but a significant pressure rise is required to turn the shear layer to be parallel to the plate. As usual in these theories the recirculation problem is still largely open.

An obvious generalisation of this work is to uniform injection 2). Here separation occurs after some distance x_s the leading edge and for $x < x_s$ classical theory applies but it terminates with a singularity of great subtlety. For $x > x_s$ the boundary layer is detached and its main properties can be worked out using similar arguments to those of 3) (Diver 1979). Again the solution in $x_s < x < x_1$ contains an arbitrary constant, namely the slope of the free shear layer as $x \to x_s +$, and its value is finally determined at reattachment where the same conditions hold as before.

The details of the transition region near $x = x_s$ are however still open.

Another is to subsonic flow (Diver 1979) the main change being to replace the Ackeret condition connecting p and δ by the Hilbert integral. The numerical work is greatly increased and, for finite regions of injection, a further arbitrary constant appears in the solution which has to be removed by appeal to a minimum singularity principle but the broad features of the flow are the same as for supersonic mainstreams.

The injection velocity V_w may also be increased until it is strong, usually defined by $\Lambda = O(1)$ where

$$V_w = \Lambda U_\infty R^{-\frac{3}{8}}$$

Suppose we start uniform blowing at $x = x_s$ with $V_w = \beta R^{-\frac{1}{2}} U_\infty$. Then the previous discussion refers to $\beta = O(1)$ (Fig.5c).

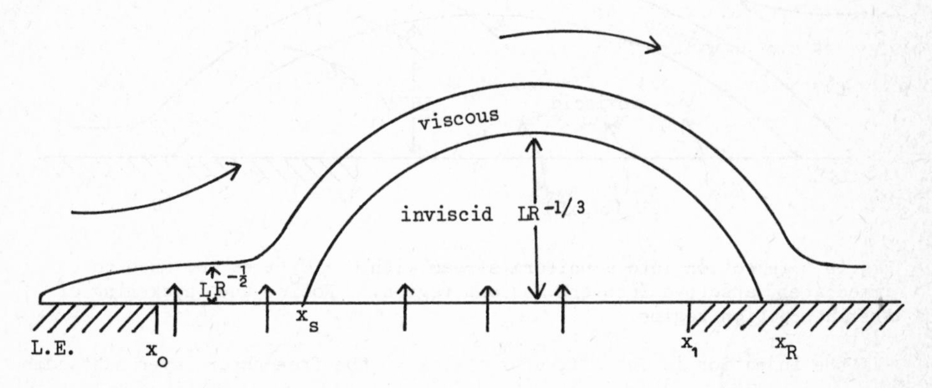

Fig.5c. Flow structure for a boundary layer separating and reattaching due to a finite region of injection.

As β increases $x_s \to x_o$ and x_R increases. As β reaches $O(R^{+\frac{1}{8}})$, x_s moves upstream of x_o and separation becomes a free interaction. With further increase of β, x_s moves up to the neighbourhood of the leading edge and a solution of the full Navier-Stokes equations is required to describe the phenomenon. Even at smaller values of β there are still a number of open questions left but nothing to make me doubt the essential correctness of the theory.

In addition to the rather detailed studies described so far there are a large number of other applications of the basic ideas of interactive boundary layers which through shortage of time and fear of indigestion on the part of the listener, I shall not go into in detail, but would like to mention some of them. First, a thorough calculation of the upstream effect of a ramp in supersonic flow has been carried out by Rizzetta <u>et al</u> 1978). The flow situation is shown in Fig.6.

Although again the asymptotic theory of the recirculating region is still not clear, they have been able to derive good estimates of the length of the bubble. Further work needs to be done on the post-reattachment pressure rise but mathematically this seems to be feasible (Daniels 1979).

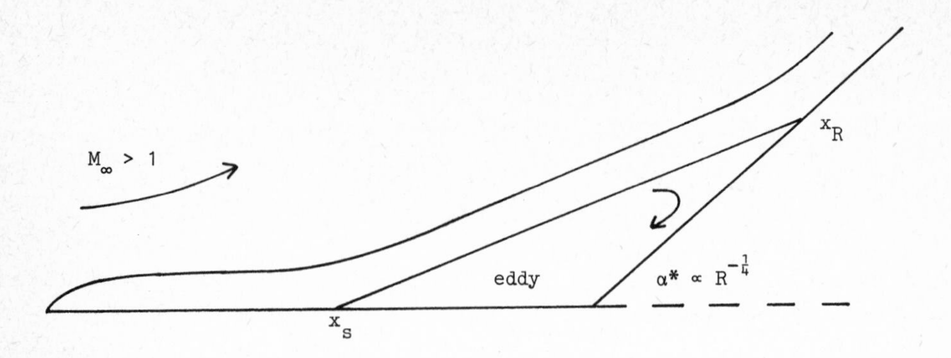

Fig. 6. Separation and reattachment of a supersonic boundary layer on a ramp.

Convex corners, hypersonic flow and unsteadiness have all been treated successfully. I would, however, like especially to mention Frank Smith's work on flow in channels with constrictions (1979a and references therein). A distinction has to be drawn between symmetric and anti-symmetric constrictions. The latter lead to more dramatic effects and I would like to concentrate on these here. A typical configuration is shown in Fig. 7. The basic ideas are similar to those for the triple deck.

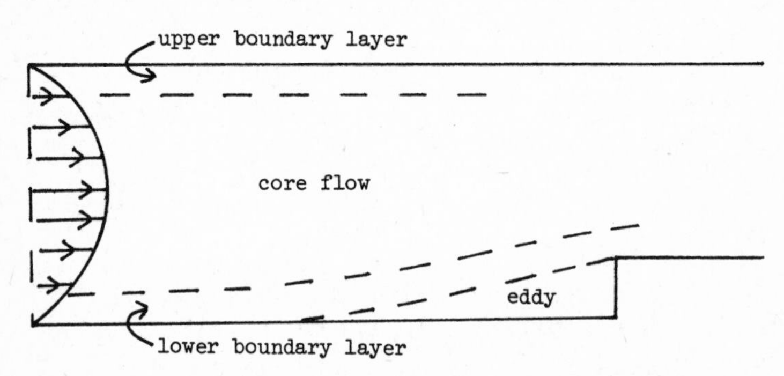

Fig. 7. The structure of the boundary layers which anticipate a constriction in a two-dimensional channel.

The chief difference is that the pressure variation across the core flow can no longer be neglected as it is in the main deck of the boundary layer. If R is a representative Reynolds number then we find that there is a free interaction region i.e. a spontaneous evolution a distance $O(R^{1/7}a)$ upstream of the constriction. After defining

$$X = R^{-1/7}x,$$

we may set

$$u = \tfrac{1}{2}(y - y^2) + R^{-2/7}A(X)(\tfrac{1}{2} - y) + \ldots \qquad ,$$

$$p = R^{-3/7}\left[P(X) + A''(X)\int_0^y \left\{\tfrac{1}{2}(y - y)\right\}^2 dy\right] + \ldots$$

in the core flow. Hence in the boundary layers top and bottom we get the standard lower-deck equations except that the two pressures are linked by

$$P_+(X) = P_-(X) + A''(X)/120$$

and this gives us a possibility of a self-induced evolution of the two boundary layers, one of which leads to separation and the other is increasingly compressed. The initial behaviour, i.e. as $X \to -\infty$, is very similar to that for supersonic free interactions. The free interaction is terminated by a singularity in P - just right in principle for matching with a plausible flow at distance O(a) upstream of the constriction.

Let me now summarise the present position. Apart from the question of recirculation flow in any bubble, we now have the main ingredients of a viable mathematical asymptotic theory of high Reynolds number flow in two dimensions. Thus we are in a position to make mathematical conjectures about the solution which, were the correct techniques available, could be settled. This is a very satisfying position to have reached even though 70 years have elapsed since Prandtl made the first significant step towards setting up a uniformly valid approximation to the solution of the Navier-Stokes equations. There are unresolved questions as yet; chiefly the completion of the description of the recirculation region set out by Burggraf several years ago. Also some of the transition regions are incompletely understood. Nevertheless the base for future development is now very firm and so different from the exasperating state of the subject twenty years ago when the separation singularity seemed to present an impassable barrier.

REFERENCES

O.R. Burggraf, 1970 U.S. Air Force Aerospace Lab. Rep. ARL 70-0275.
O.R. Burggraf, 1975 AGARD Paper 168.
J.E. Carter, 1972 NASA Tech. Rep. R-385.
D. Catherall, K. Stewartson & P.G. Williams, 1965 Proc. Roy. Soc. A 284, 270.
D. Catherall & K.W. Mangler, 1966 J. Fluid Mech. 26, 183.
T. Cebeci, A.K. Khattab & K. Stewartson, 1979 AIAA Paper 79-0138.
D.R. Chapman, 1950 NACA Rep. 958.
D.R. Chapman, D.M. Kuehn & H.K. Larsen, 1958 NACA Rep. 1356.
R. Chow & R.E. Melnik, 1976 Lecture Notes in Physics 59, 135
J.D. Cole & J. Aroesty, 1968 Int. J. Heat & Mass Transfer 11, 1167.
P.G. Daniels, 1974 Quart. J. Mech. Appl. Math. 27, 175.
P.G. Daniels, 1979 J. Fluid Mech. 90, 289.
S.C.R. Dennis & G.Z. Chang, 1970 J. Fluid Mech 42, 471.
S.C.R. Dennis, 1972 J. Inst. Maths. Appl. 10, 305.
C. Diver & K. Stewartson, 1978 J. Fluid Mech. 88, 115.
C. Diver, 1979 Ph.D. Thesis, London (in preparation).
H.W. Emmons & D.C. Leigh, 1953 A.R.C. Current Paper 157.
S. Goldstein, 1948 Quart. J. Mech. Appl. Maths. 1, 43.
C.E. Jobe & O.R. Burggraf, 1974 Proc. Roy. Soc. A 340, 91.
L. Lees & B.L. Reeves, 1964 AIAA J. 2, 1907.
M.J. Lighthill, 1953 Proc. Roy. Soc. A 217, 478.

A.F. Messiter, G.R. Hough & A. Feo, 1973 J. Fluid Mech. 60,605.
A.F. Messiter, 1978 Proc. 8th Nat. Cong. Appl. Mech. U.S.A.
V.Y. Neiland, 1969 Izv. Akad.Nauk SSSR Mekh. Zh. Gaza No. 4, 53.
V.Y. Neiland, 1971 Izv. Akad.Nauk SSSR Mekh. Zh. Gaza No. 3, 19.
K. Oswatitsch & K. Wieghardt, 1946 A.R.C. Rep. 10378 London.
J. Pretsch, 1944 Unter. Mitt. Dent. Luftfahrt Rep. 3091.
T.A. Reyhner & I. Flügge-Lotz, 1968 Int. J. Nonlinear Mech. 3, 173.
D.P. Rizzetta, O.R. Burggraf & R. Jenson, 1978 J. Fluid Mech. 89, 535.
F.T. Smith, 1977 Proc. Roy. Soc. A 356, 443.
F.T. Smith, 1979a J. Fluid Mech. 90, 725.
F.T. Smith, 1979b J. Fluid Mech. 92, 171.
K. Stewartson, 1974 Advances in Applied Mech. 14, 145.
K. Stewartson, 1975 SIAM J. Appl. Math. 28, 501.
K. Stewartson & P.G. Williams, 1973 Mathematika 20, 98.
V.Y. Sychev, 1967 Rep. to 8th Symp. Recent Problems in Mech. Liquids & Gases, Tarda, Poland.
V.Y. Sychev, 1972 Izv. Akad. Nauk SSSR., Mekh. Zh. Gaza 3, 47.
M.J. Werle & V.N. Vatsa, 1974 AIAA J. 12, 1491.
P.G. Williams, 1974 Lecture Notes in Physics 35, 445.

APPLICATION OF SPECTRAL METHODS TO THE SOLUTION OF NAVIER-STOKES EQUATIONS

Thomas D. Taylor and John W. Murdock
The Aerospace Corporation
Los Angeles, California 90009

In this paper a number of topics that arise in application of Chebyshev expansion methods to the solution of the Navier-Stokes equations are addressed. These include the equivalence of finite differences and finite elements approaches, a new pressure velocity formulation that permits easy extension to three dimensions, evaluation of the importance of pressure boundary conditions and the virtues of collocation over the tau method for satisfying boundary conditions. Example results from the pressure velocity formulation which eliminate a non-linear momentum equation in favor of the linear continuity equation are presented. The results are for a 2-D unsteady flow on a flat plate at large Reynolds numbers. The behavior of an unsteady disturbance in such a flow is examined and compared with previous stream-function vorticity results of Murdock.

Introduction

The solution of the Navier-Stokes equations by spectral methods is of growing interest. The technique has been extensively investigated by Orszag and co-workers. The principal references are available in [1]. Publications by other groups are also beginning to appear in [2-6]. The application of spectral methods offers promise in that they appear to yield advantages in cost and accuracy over finite difference approaches for high Reynolds number flow calculations. In addition, they could possibly be combined with finite element theory to yield a powerful tool for irregular geometry problems. Current applications have been limited to incompressible flows, but similar advantages may be possible in the compressible flow area.

In the application of the spectral methods to solution of the flow equations, three issues are of key importance. These are 1) use of series expansions that permit application of fast Fourier transforms for inversion, 2) proper satisfaction of non-periodic boundary conditions, and 3) methods for integration of the time-dependent equations for the spectral coefficients. Also,the form of the Navier-Stokes equations that one employs in the solution of a problem is of interest.

In this paper, the authors discuss their experience in applying the spectral method to incompressible flows and indicate the areas of difficulty as well. The discussion includes comments on derivation of equivalent finite difference methods by a spectral scheme. Also, the concept of solving an incompressible pressure velocity system of equations with a momentum equation eliminated is presented. The concept is of interest since it reduces the difficulties associated with integration of a nonlinear equation.

The Spectral Method

The application of spectral methods to solve flow equations is not a fully developed technology. Solution of periodic boundary condition problems is rather straightforward, but the application to non-periodic boundary condition problems is a subject under development. The spectral method can be applied in two ways. The first is by performing calculations directly in spectral space and the second is performing calculations in real space by employing difference equivalents derived by spectral expansions. In order to apply the method efficiently in spectral space, it is necessary that one be able to take advantage of the rapid computation speeds of fast Fourier transforms. This requires that the spectral expansion used in a problem be a representation of a finite Fourier transform. This can be accomplished by use of sines, cosines or Chebyshev polynomial expansions. For problems with nonperiodic boundary conditions, the Chebyshev expansions appear to work best. These polynomials hold only for finite regions so a problem must, therefore, be mapped or scaled into a set of finite regions for their application. This set of finite regions is effectively finite elements and therefore the use of Chebyshev polynomials can be viewed as a finite element method of one or more elements. This is important for a user to note since it permits one to reduce the number of expansion terms per element by adding elements or vice versa. This flexibility can be used to minimize computation costs for a given problem or geometry.

For real space application, one can derive finite difference equivalents to the original equations by spectral expansions. For problems which do not demand extreme accuracy or resolution, the finite difference approach may be desirable. Also, since Chebyshev expansions are known to give rapidly convergent expansions, finite difference schemes derived from these expansions should be preferable to schemes derived from Taylor's series. Peyret [7] has demonstrated this to be the case for difference methods derived by Hermitian expansions. In the next section, we indicate how finite difference equivalents can be derived from spectral expansions. We will not, however, cover the details of application of the method in spectral space since the details are available in [1].

Finite Difference Schemes Derived from Chebyshev Expansions

Chebyshev expansions can be utilized to derive finite difference approximations to differential equations in the following manner. If one assumes that a function $u(x,t)$ can be expanded in the form

$$u(x,t) = \sum_{n=o}^{N}{}'' a_n(t) T_n(x) \tag{1}$$

where a_n is an unknown function of t and $T_n(x)$ is the nth order Chebyshev polynomial then it can be shown [9] that

$$a_n(t) = \frac{2}{N} \sum_{k=o}^{N}{}'' u(x_k,t) T_n(x_k) \qquad x_k = \cos(\frac{k\pi}{N}) \tag{2a}$$

or

$$a_n(t) = \frac{2}{N+1} \sum_{k=o}^{N}{}'' u(x_k,b) T_n(x_k) \qquad x_k = \cos(\frac{2k+1}{N+1} \cdot \frac{\pi}{2}) \tag{2b}$$

where double prime means both a_o and a_n terms must be divided by 2. This formula gives a convenient way to derive the finite difference approximations for a differential equation. Consider, for example, the equation

$$\frac{\partial u}{\partial t} + \frac{\partial u}{\partial x} = \alpha \frac{\partial^2 u}{\partial x^2} \tag{3}$$

where $u = \sum_{n=o}^{N}{}'' \; a_n(t)\, T_n(x)$

One can then show that[1]

$$c_n \frac{da_n}{dt} = -2 \sum_{\substack{k=n+1 \\ k+n \text{ odd}}}^{N}{}'' \; k a_k + \sum_{\substack{k=n+2 \\ k+n \text{ even}}}^{N}{}'' \; k(k^2-n^2)a_k \qquad c_o = 2, \; c_n = 1 \; n > 0 \tag{4}$$

Inserting Eq. (2a) or (2b) into Eq. (4) yields a set of finite difference equations for $u(x_k,t)$.

These equations can be solved subject to the boundary conditions at $x = -1$ and $x = +1$. When Eq. (2a) is used with Eq. (4), then only the values $u(t,-1) = U_{k=o}$ and $u(t,1) = U_{k=N}$ are required. When Eq. (2b) is employed, the values of x_k do not extend to the boundaries so one must use the relations

$$U_{k=o} = \sum_{n=o}^{N}{}'' \; a_n(t) \; (-1)^n \tag{5}$$

$$U_{k=N} = \sum_{n=o}^{N}{}'' \; a_n(t) \; (+1)^n \tag{6}$$

For the simple case of $N = 2$, the difference equations for Eq. (4) are

$$\frac{d}{dt}(u_{ave}) + \frac{\delta u}{2} = \alpha\, \delta^2 u \tag{7}$$

$$\frac{d}{dt}(\delta u) + \delta^2 u = 0 \tag{8}$$

$$\frac{d}{dt}(\delta^2 u) = 0 \tag{9}$$

where the terms are given in Table 1.

	Eq. 2a	Eq. 2b
u_{Ave}	$\frac{u(-1)+2u(0)+u(1)}{4}$	$\frac{u(-\frac{\sqrt{3}}{2})+u(0)+u(\frac{\sqrt{3}}{2})}{3}$
δu	$[u(1)-u(-1)]$	$\frac{2}{3}\left[u(\frac{\sqrt{3}}{2})-u(-\frac{\sqrt{3}}{2})\right]$
$\delta^2 u$	$[u(1)-2u(0)+u(-1)]$	$\frac{1}{3}\left[u(\frac{\sqrt{3}}{2})-2u(0)+u(-\frac{\sqrt{3}}{2})\right]$

Table 1

Examining these equations, it is clear that Eqs. (8) and (9) do not satisfy conservation laws and this is because of truncation of the original expansion. As a result, they should be discarded and the boundary conditions used to replace them. After this is accomplished, it is interesting to compare Eq. (7) with a standard Taylor's series finite difference equivalent to Eq. (1). If we do this (noting that Δx is normalized), the δu is the standard second order convective difference and $\delta^2 u$ is the standard second order diffusion term difference. u_{AVE}, however, is a three point average which makes the scheme implicit in x even if an explicit time integration scheme is used. Since only a three point inversion is required, solution of this set of equations is straightforward. The interesting result, however, is that the second order Chebyshev expansion yields a difference scheme which varies principally from a standard Taylor's series scheme in the time-dependent term. The authors have not studied the full importance of this variation, but it is an area for further research.

In the finite difference application of spectral expansions, the application of the boundary conditions is clear and straightforward, but when one computes in spectral space these applications become more subtle. The authors addressed this question and the results are presented in the next section.

Application of Boundary Conditions in a Chebyshev Spectral Method

In the discussion of finite differences, we notice that the last two equations [(8) and (9)] of the example finite difference method were incomplete. This happens because the second derivative of a Chebyshev expansion reduces the order by two, thus reducing the accuracy of the last two equations due to series truncation. For inviscid flows with first derivatives only, the last equation would be inconsistent. The tau method was developed by Lanczos [8] to remove this difficulty and also to satisfy the boundary conditions. In effect, what one does is modify the equations for the unknown a_{N-1} and a_N in a way that will satisfy the boundary conditions (see [9] for details). If, however, you attempt this approach on the following problem

$$\frac{\partial u}{\partial t} + \frac{\partial u}{\partial x} = 0 \tag{10}$$

with

$$u=1 \quad x<0 \quad u=0 \quad x>0 \quad \text{at} \quad t=0$$
$$u=1 \quad x=0 \quad \text{for} \quad t>0$$

The tau scheme gives a very poor early time representation of the solution, as shown in Figure 1. An alternate approach which improves on this is the collocation method [1]. This method can best be understood, if one recognizes that in Fourier series solutions to p.d.e's nonhomogeneous boundary conditions are usually satisfied by adding a solution which can be made to satisfy the nonhomogeneous boundary conditions. This same procedure can be extended to Chebyshev expansions, i.e., assume for a one-dimensional problem that

$$u(x,t) = \sum_{n=o}^{N} a_n(t)T_n(x) + u_o(x,t) \tag{11}$$

where $u_o(x,t)$ is an arbitrary function which is chosen to satisfy the nonhomogeneous boundary conditions of the problem. The a_n portion of the problem then satisfies homogeneous boundary conditions. u_o, however, can be expanded in a series so that one can write

$$u_o(x,t) = \sum_{n=o}^{N} \left[a_n(t) + b_n(t)\right] T_n(x) \tag{12}$$

In this form it is easily observed that the boundary conditions influence all the a_n's and not just a_N and a_{N-1}. This follows because when one substitues this expansion into equation (10) to obtain equations for the a_n's you obtain nonhomogeneous terms from the b_n's. This method of satisfying boundary conditions appears to be more satisfactory than the tau method since it distributes the boundary condition influence over all the a_n's. The authors employed this technique to solve Eq. (10) in place of the tau method by assuming

$$u_o(x,t) = b(t)\delta(x) \tag{13}$$

where δ is the delta function.

The results are also shown on Figure 1. Note the improvement over the tau approach. As a result of these studies, the authors selected the collocation approach for problem solution with explicit time differencing.

The discussion thus far has attempted to answer some questions related to spectral method application. This knowledge permits one to attack solution of the Navier-Stokes equations.

Navier-Stokes Solutions by Chebyshev Spectral Expansions

The authors examined solution of the 2-D unsteady incompressible Navier-Stokes equations by Chebyshev expansions for two separate formulations. The first formulation examined was a stream-function vorticity (reported in [2],[3]) and the second was a velocity-pressure system. The velocity-pressure formulation was studied to permit easy advancement of calculations from two to three dimensions and also to reduce the nonlinear complexity. The results of the velocity-pressure study are reported in this section. The formulations employed in the velocity-pressure study is

$$v = - \int_o^y u_x dy \tag{14}$$

$$u_t + F = P_x + \frac{1}{R} \nabla^2 u, \quad F = (u^2)_x + (uv)_y \tag{15}$$

$$\nabla^2 P = g = - \left[u_x^2 + 2 u_y v_x + v_y^2 \right] \tag{16}$$

Note that in this formulation the v momentum equation is not employed and the v velocity is computed from the continuity equation which is linear.

The equations were time differenced in the following form

$$\frac{u(t+\Delta t)-u(t)}{\Delta t} + \frac{3}{2} F(t) - \frac{1}{2} F(t-\Delta t) = - P_x(t+\Delta t) + \frac{1}{R}\left[\frac{1}{2}\nabla^2 u(t+\Delta t) + \frac{1}{2}\nabla^2 u(t)\right] \tag{17}$$

$$\nabla^2 P(t+\Delta t) = - \frac{3}{2} g(t) + \frac{1}{2} g(t-\Delta t) \tag{18}$$

v is computed by numerical integration of equation (14). And the Poisson equation for P was solved by the tensor product method presented in [3]. For the solution, each function was expanded in the form

$$f = \sum_{n=0}^{N} \sum_{m=0}^{M} a_{n,m}(t)\, T_n^* \left(\frac{x-x_1}{x_2-x_1}\right) T_m^* (\Psi) \tag{19}$$

where

$$\Psi = \frac{\exp(-y/y_e) - \exp(y_{max}/y_e)}{1 - \exp(y_{max}/y_e)}$$

with y_{max} as the outer edge of the grid and y_e as a scaling parameter selected to give reasonable resolution of the calculation. Typically, y_e is of order 1. $T_n^*(x)$ are the Chebyshev polynomials with the range $0 \leq x \leq 1$.

These expansions were then used in conjunction with the time difference equations to find the values of the $a_{n,m}$'s for u, v and P. This equation set has been solved for flow over a flat plate with the boundaries shown in Figure 2. The boundary conditions were specified as follows. For the upstream boundary the flow was pescribed to be Blasius plus a time periodic solution of the Orr-Sommerfeld equation [2]i.e.

$$u = u_{Blasius} + A \operatorname{Re}[\phi_y(x,y) \exp(-i\omega\tau)] \tag{20}$$

$$P = P_{Blasius} + P_{Orr-Sommerfeld}$$

where ϕ is the Orr-Sommerfeld solution for a wave number α and a frequency ω. Tau is a dimensionless time defined by $\tau = \frac{tV_\infty}{\ell}$ where V_∞ is the characteristic velocity and ℓ is the length from the leading edge of the flat plate to the boundary of the computation. The other boundary conditions are

$$u=v=0 \quad \text{and} \quad \frac{\partial P}{\partial y} = -\frac{1}{R}\frac{\partial^2 u}{\partial x \partial y} \quad \text{at } y = 0 \tag{21}$$

$$u=1 \qquad P=P_{Blasius} \quad \text{at} \quad y = \infty \tag{22}$$

For this study, the authors parabolized the equations by neglecting $\frac{\partial^2 u}{\partial x^2}$ in the u momentum equations so a downstream boundary condition is not required on u. For the pressure downstream, the authors utilized the condition

$$\frac{\partial^2 P}{\partial x^2} = 0 \quad \text{at} \quad x = x_2 \tag{23}$$

The solution of the specified problem is straightforward except for coupling between the pressure and velocity which occurs because of the surface boundary conditions. The authors investigated the significance of this coupling by employing the correct surface pressure boundary condition as well as the condition

$$\frac{\partial P}{\partial y} = 0$$

The results are discussed in the next section.

Computational Results

The procedure which has been outlined was used by the authors to compute 2-D flow over a flat plate for Reynolds numbers in the range $1.2\times10^5 \leq R \leq 3.8\times10^5$. This problem was chosen to evaluate the method since results were available from an independent stream function vorticity calculation of Murdock [3]. In addition, if the method proved useful it could be used to extend the calculation to three dimensions and study the growth of small disturbances in a nonlinear high Reynolds number flow.

In the study the authors examined small and large amplitude disturbances as upstream inputs to the flow. The disturbances were perturbation about the Blasius profile shown in Figure 3 . The total velocity is given by

$$u = u_{Blasius} + u'$$

The small disturbance was chosen to be 0.06% of the freestream and the large disturbance 5% of the freestream. For the small disturbance case we compared the Fourier Amplitude of u' from the p-v model with the stream-function vorticity results of Murdock (3). The results are displayed on Figures 4 and 5. In the figures the boundary layer variable $\eta = y/2(R x)^{1/2}$ is employed. Note that results are the Tollmien-Schlichting wave amplitude for the flow. The frequency of the initial wave for the calculation was $\frac{\omega \nu}{V_\infty^2} = 56\times10^{-6}$.

The figures show reasonable agreement between the results of the two formulations considering the approximations and magnitudes of u'. These results proved useful from two points of view. First, they served to substantiate the v-p formulation and, secondly the confirm Murdock's calculations [3,6] for his boundary layer stability studies which are difficult to compare with experiments due to the limited data.

For the large amplitude results, the authors examined both the increased effects of nonlinearity and of pressure boundary conditions. Figure 6 displays a result obtained for the total velocity u as a function of x. Also shown is the Blasius velocity. The amplitude of the wave above the Blasius profile appears to decay as the wave propagates downstream, but no significant distortions are observed due to the nonlinear effects. The results computed from the v-p formulation for the Fourier Amplitude of the first mode of u' are shown in figures 7 and 8. The results do not show any significant change from those obtained for small amplitudes except for the magnitudes.

Also shown on figures 7 and 8 are results for the $\frac{\partial P}{\partial y} = 0$ boundary condition at the surface. There is some change in the results, but not anything significant. The next higer order mode was obtained for the nonlinear case, but it was not significantly changed by the pressure boundary condition and, therefore, is not shown. It is important to note, however, that the computer cost was reduced by approximately a factor of two when this boundary conditions was applied.

Conclusions

From the studies which the authors conducted it appears that a v-p formulation of the Navier-Stokes equations which eliminates a momentum equation in favor of continuity can be effectively applied to solve high Reynolds number flows. In addition, it appears that collocation offers improved accuracy over the tau method for satisfying boundary conditions in problems with sharp boundary gradients.

References

1. Gottlieb, D. and Orszag, S.A., "Numerical Analysis of Spectral Methods", Soc. Ind. and App. Math., Phil., Pa., 1977.

2. Murdock, J.W. and Taylor, T.D., "Numerical Investigation of Nonlinear Wave Interaction in a Two-Dimensional Boundary Layer," AGARD Symposium on Laminar Turbulent Transition, May 1977, Copenhagen, Denmark.

3. Murdock, J.W., "A Numerical Study of Nonlinear Effects on Boundary Layer Stability," AIAA Journal, No. 8, August 1977, pp. 1167.

4. Rogallo, R.S. "An ILLIAC Program for the Numerical Simulation of Homogeneous Incompressible Turbulence", NASA TM-73, 203, Nov. 77.

5. Moin, P. and Kim, J., "On the Numerical Solution of Time Dependent Viscous Incrompressible Flows Involving Solid Boundaries", to appear in Journal Comp. Physics.

6. Murdock, J.W., "Generation of a Tollmien-Schlichting Wave by a Sound Wave", to appear in IUTAM Proceedings of Symposium on Laminar-Turbulent Transition, Stuttgart, Germany, Sept. 1979.

7. Peyret, R.,"A Hermitian Finite Difference Method for the Solution of the Navier-Stokes Equations", Proc. International Conf. on Numerical Methods for Laminar and Turbulent Flow, Swansea (G.B.), July, 1978, also ONERA T.P. No. 1978-29.

8. Lanczos, C., Applied Analysis, Prentice Hall, Englewood Cliffs, N.J., 1956.

9. Fox, L. and Parker, I.B., Chebyshev Polynomials in Numerical Analysis, Oxford University Press, London, (1968).

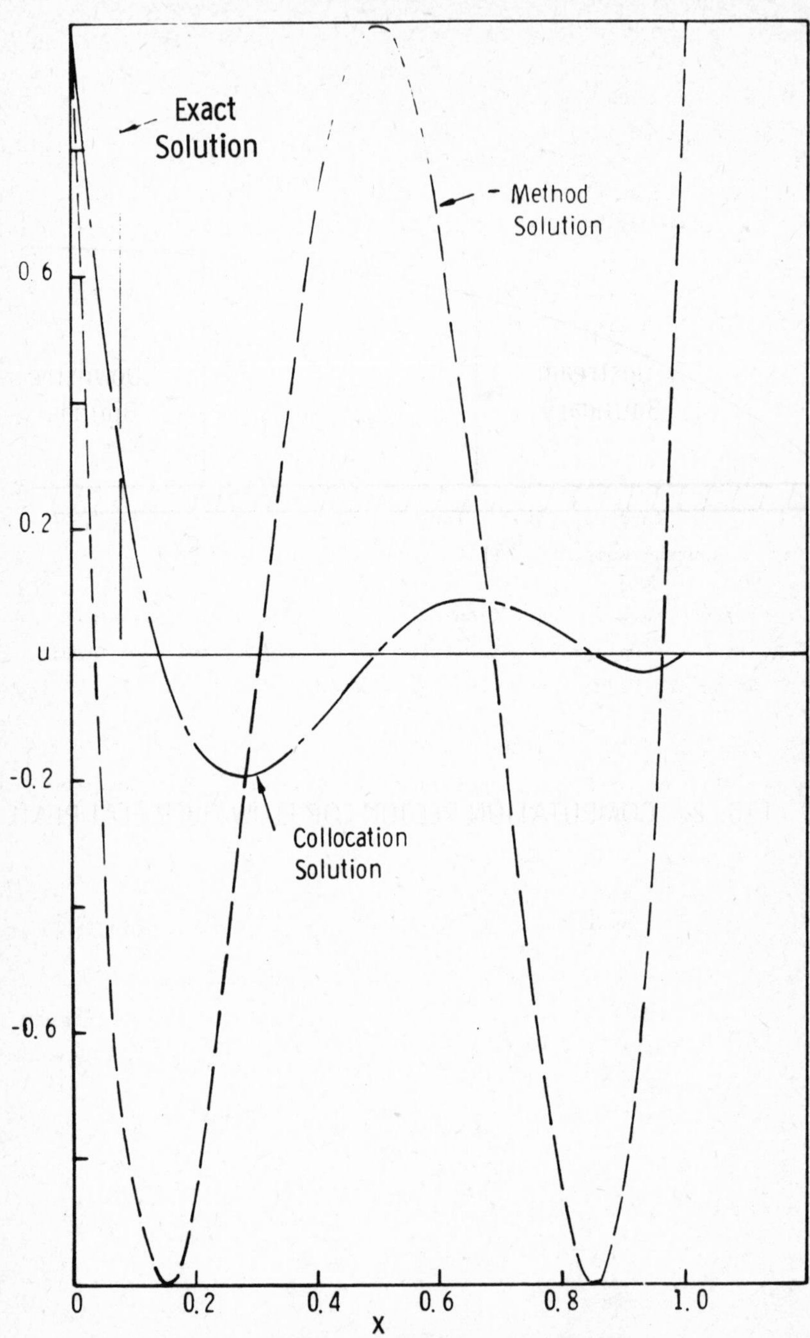

FIG. 1: COMPARISON OF SOLUTIONS PRODUCED BY TAU AND COLLOCATION METHODS $\Delta t = 1/2$

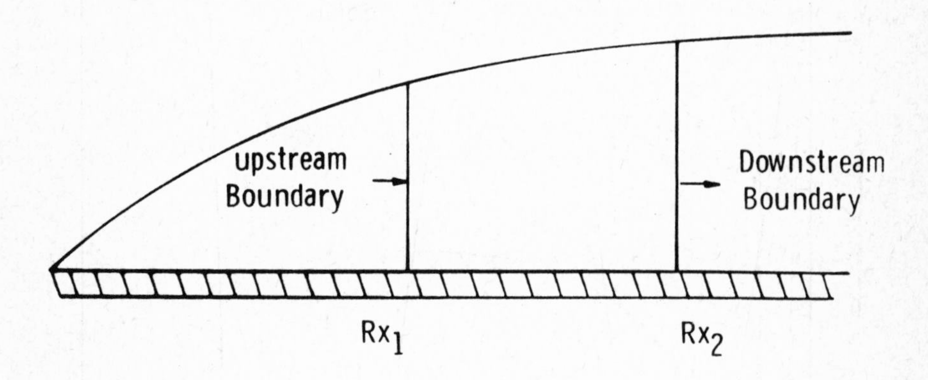

FIG. 2: COMPUTATION REGION FOR FLOW OVER FLAT PLATE

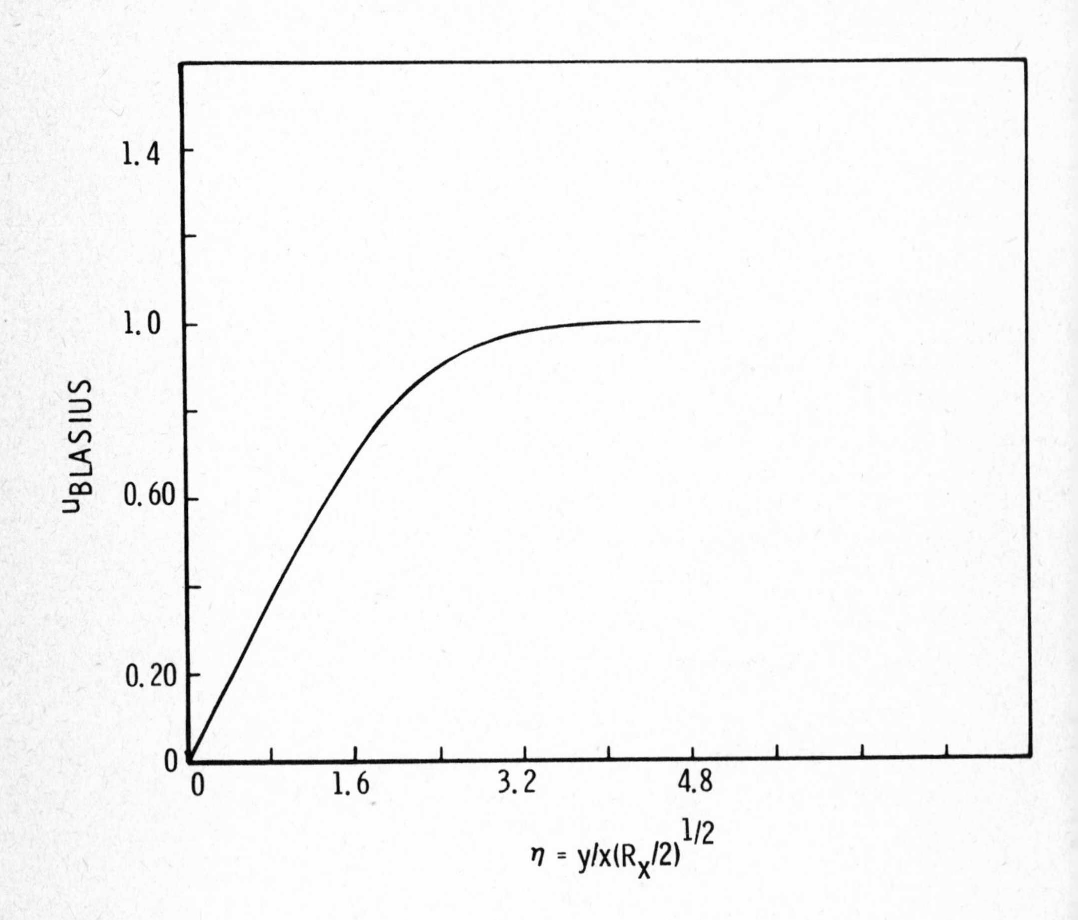

FIG. 3: BLASIUS VELOCITY PROFILE

FIG. 4: COMPARISON OF WAVE AMPLITUDE COMPUTED WITH SPECTRAL CODE

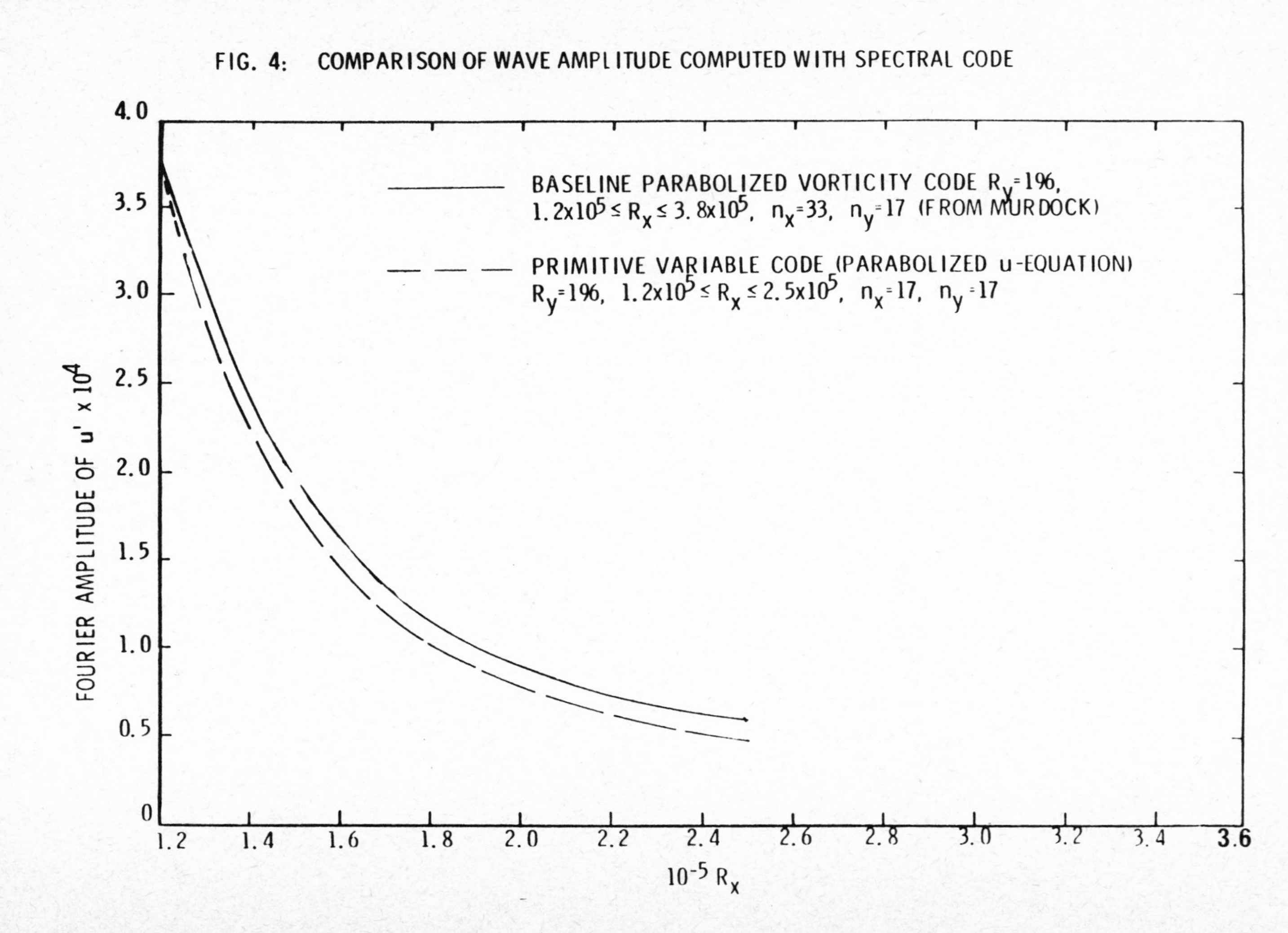

FIG. 5 : COMPARISON OF WAVE AMPLITUDES AT $R_x = 2 \times 10^5$

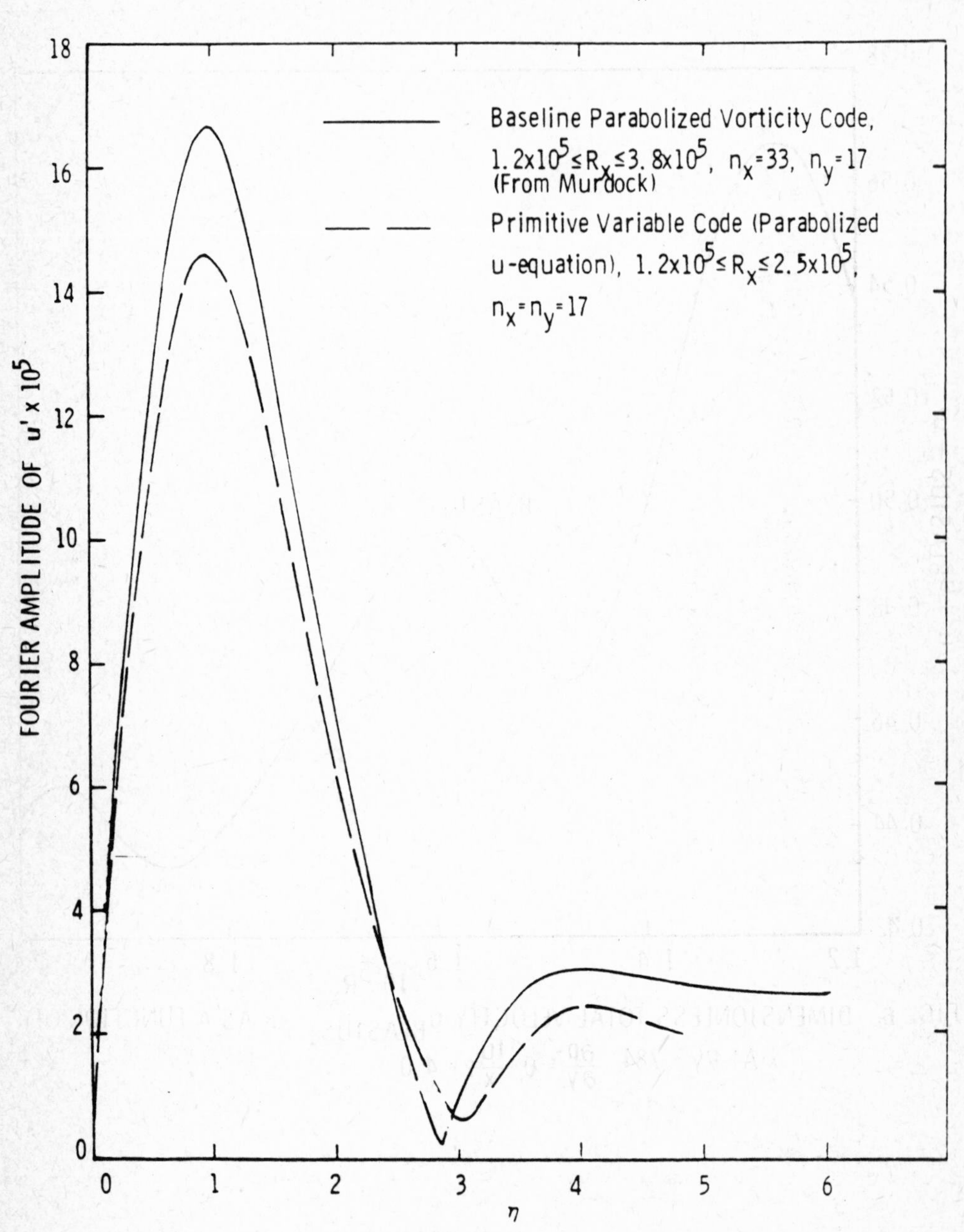

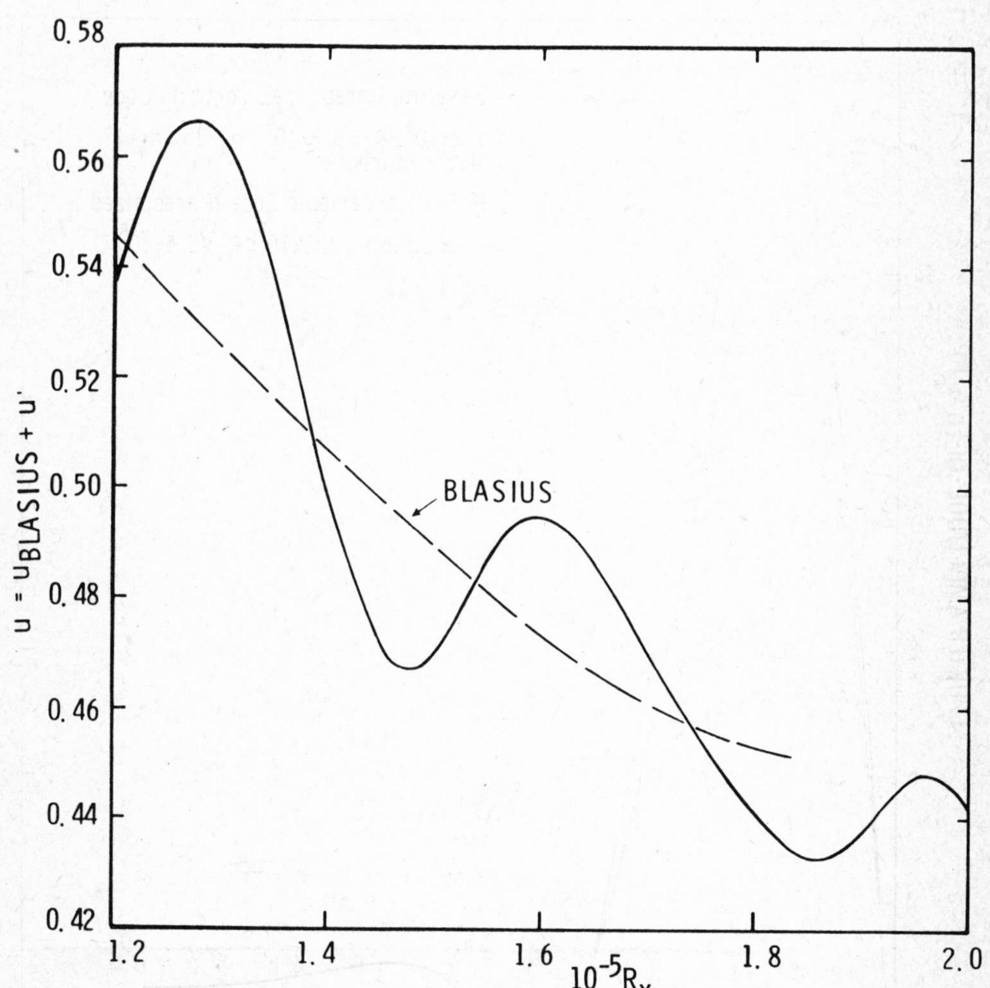

FIG. 6: DIMENSIONLESS TOTAL VELOCITY $u_{BLASIUS} + u'$ AS A FUNCTION OF x AT Ry = 784, $\frac{\partial p}{\partial y} = 0$, $\frac{tu_\infty}{x} = 4.0$

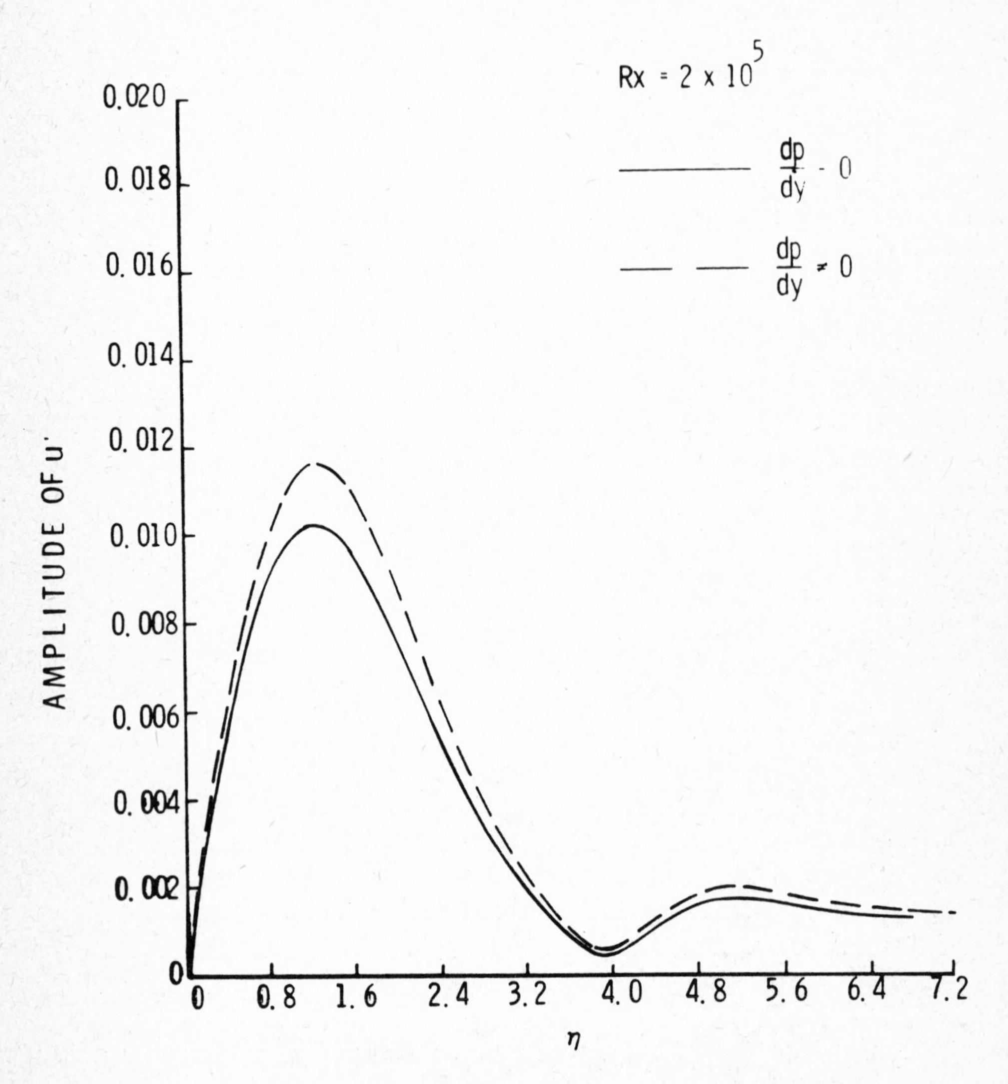

FIG. 7: FOURIER AMPLITUDE OF u' VS η FOR DIFFERENT PRESSURE BOUNDARY CONDITIONS (FIRST MODE)

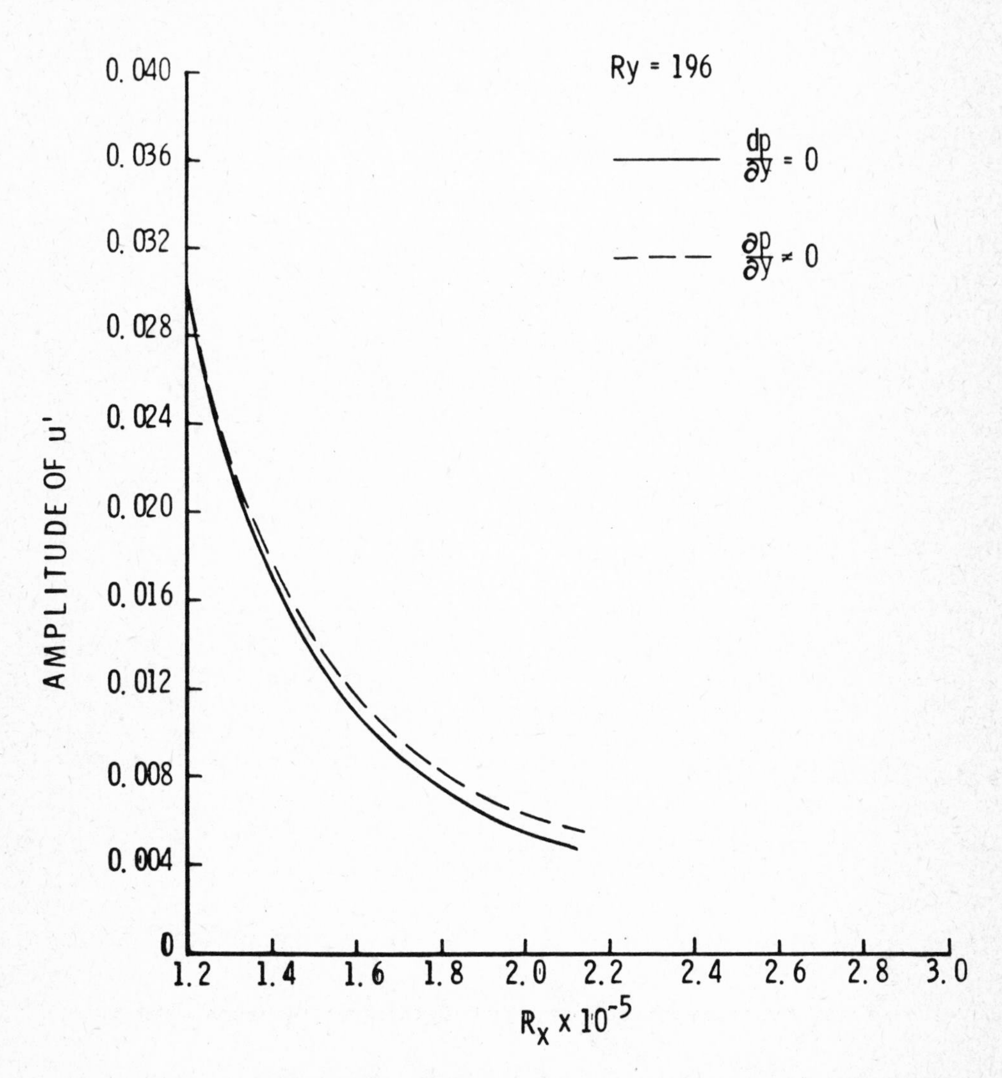

FIG. 8: FOURIER AMPLITUDE OF u' (FIRST MODE) AS A FUNCTION X FOR TWO PRESSURE BOUNDARY CONDITIONS

Regularity Questions for the Navier-Stokes Equations

Wolf von Wahl

Fachbereich Mathematik und Physik
der Universität Bayreuth
Postfach 3008
D-8580 Bayreuth
Federal Republic of Germany

We study the Navier-Stokes (shortly NS)-equations

$$\begin{aligned} u' - \nu\Delta u + u\circ\nabla u + \nabla p &= f, \\ \nabla\circ u &= 0, \\ u|\partial\Omega &= 0, \\ u(0) &= \varphi \end{aligned}$$

over a cylindrical domain $(0,T)\times\Omega$. Ω is a sufficiently nice (say C^3-boundary) bounded domain of $\mathbb{R}^n$, $n \geq 2$.

In this note we want to use the fact that the nonlinearity of the NS-equations is in a sense to be precised later on an analytic mapping between suitable Banach-spaces.

Let $\Omega \subset \mathbb{R}^n$ be a domain as before. Then $L_p(\Omega)$ can be decomposed into the direct sum

$$L_p(\Omega) = H_p(\Omega) + \{\nabla g \,|\, g \in L_{p,loc}(\Omega),\ \nabla g \in L_p(\Omega)\},$$

where $H_p(\Omega)$ is the closure of the $C_0^\infty(\Omega)$-vector fields ψ with $\nabla\circ\psi = 0$ with respect to the $L_p(\Omega)$-norm. p is an arbitrary number > 1. Let $f \equiv 0$. Then, if P is the projection of $L_p(\Omega)$ onto $H_p(\Omega)$, the NS-equations can be transformed into the ordinary differential equation

$$\begin{aligned} u' + Au + P(u\circ\nabla u) &= 0, \\ u(0) &= \varphi, \end{aligned}$$

in the Banach Space $B = H_p(\Omega)$, where $\varphi \in H_p$, $A = -\nu P\Delta$. A generates an analytic semi-group e^{-tA} and has as domain of definition $H^{2,p}(\Omega) \cap$ $\cap \overset{o}{H}{}^{1,p}(\Omega) \cap \{\psi | \psi \in H^{1,p}(\Omega),\ \nabla\cdot\psi = 0\}$.

First we assume that $p > n$, say $p = n+\varepsilon$. Using Sobolev's inequalities one can show then that the nonlinearity $M(u) = P(u\cdot\nabla u)$ is a mapping from $D(A^{1-\rho_1})$ into $B = H_p(\Omega)$ satisfying the estimate

$$\text{(I)}\quad \|M(u)\| \leq c\|A^{1-\rho_1}u\|\,\|u\|,$$

ρ_1 is fixed number with $0 < \rho_1 < 1$. The nonlinearity of the NS-equations being an entire function of its variables it is not difficult to show that it is an analytic mapping from $D(A^{1-\rho_1})$ (equipped with the graph-norm) into $H_p(\Omega)$ in the sense of E. Hille. Furthermore M is a mapping from $D(A)$ into $H^{1,p}(\Omega) \cap B$ since P preserves differentiability <u>but not boundary values</u>. M is also analytic and fulfils the estimate

$$\text{(II)}\quad \|M(u)\|_{H^{1,p}(\Omega)} \leq c\|Au\|^{1+\rho_2}(\|u\|+1)$$

where $\rho_2 > 0$. For s, $0 < s \leq 1$, we have that $\overset{o}{H}{}^{s,p}(\Omega) \cap B$ is equal to the closure of the $C_o^\infty(\Omega)$-vector fields ψ with $\nabla\circ\psi = 0$ in the $H^{s,p}(\Omega)$-norm. The latter closure being contained in $D(A^\rho)$, $\rho < \frac{s}{2}$, we finally get that

$$M(u) \in D(A^\rho),\ \rho < \frac{1}{2p},$$

$$\|A^\rho M(u)\| \leq c\|Au\|^{1+\rho_2}(\|u\|+1),$$

since $H^{s,p}(\Omega) = \overset{o}{H}{}^{s,p}(\Omega)$, $s \leq \frac{1}{p}$.

Using a general theorem on analytic mappings ([2], Kapitel II) we arrive at the following nonlinear interpolation property: Let $\gamma = (1-\sigma)\cdot 1+\sigma(1-\rho_1)$, $\delta = (1-\sigma)\circ\rho$, $0 \leq \sigma \leq 1$. Let $\eta > 0$ be arbitrarily small but fixed. Then $M(u)$ is in $D(A^{\delta-\eta})$ for $u \in D(A^{\gamma+\eta})$ and satisfies the following estimate:

$$\|A^{\delta-\eta}M(u)\| \leq c(\eta)\|A^{\gamma+\eta}u\|^{1+(1-\sigma)\rho_2}\circ(\|A^{\gamma+\eta-(1-\rho_1)}u\|^\sigma+1)\circ$$

$$\circ(\|u\|^{1-\sigma}+1).$$

What is important now is the following: Choosing q sufficiently small we have as a consequence that for every ρ, $0 < \rho < \frac{1}{2p}$, there exist $\rho_1', \rho_2' > 0$ with the following property:

$$\text{(III)} \quad \|A^{\rho}M(u)\| \leq c\|A^{1-\rho_1'}u\|^{1+\rho_2'}(\|u\|+1).$$

For $\varphi \in B$ we have by (I) a unique local solution of the integral equation

$$\text{(IV)} \quad u(t) = e^{-tA}\varphi + \int_0^t e^{-(t-s)A}M(u(s))\,ds,$$

i.e. on a maximal interval $(0,T_{max})$ we have an $u \in C^0([0,T_{max}),B)$, $A^{1-\rho_1}u(.) \in C^0((0,T_{max}),B)$ fulfilling the integral equation (IV). If $\|\varphi\|$ is small then $T_{max} = \infty$. It is standard to show that

$$u \in \bigcap_{0<\varepsilon'<1} C^0((0,T_{max}),D(A^{1-\varepsilon'})).$$

We try if $u(t) \in D(A^{1+\rho})$, $\rho < \frac{1}{2p}$. Let $\varepsilon'' > 0$ so small that $\rho+\varepsilon'' < \frac{1}{2p}$. We have

$$A^{1+\rho}u(t) = A^{1+\rho}e^{-tA}u(\tilde{\eta}) + \int_{\tilde{\eta}}^t A^{1-\varepsilon''}e^{-(t-s)A}A^{\rho+\varepsilon''}M(u(s))\,ds,$$

$0 < \tilde{\eta} \leq t < T_{max}$. Because of (III) we have $A^{\rho+\varepsilon}M(u(.)) \in C^0([\tilde{\eta},T_{max}))$ if $u \in C^0([\tilde{\eta},T_{max}), D(A^{1-\varepsilon'}))$, ε' sufficiently small. The latter being in fact fulfilled we get: $u(.) \in C^0((0,T_{max}),D(A^{1+\rho}))$ and therefore $u \in C^1((0,T_{max}),B)$, T_{max} being the same as before. Using the fact $p > n$ it is not difficult to show that

$$u_t, D^2u \in C^0((0,T_{max}),C^{\frac{1}{n+1}}(\bar{\Omega})),$$

i.e. we have a classical solution. Observe that the Hölder exponent does not depend on the ε in $p = n+\varepsilon$.

This result was announced in the paper [1] by Kato and Fujita, but under an additional boundary condition on φ. If one wants to consider an external force f it has to be from $L^{\frac{1}{\rho_1}+\varepsilon}((0,T),B) \cap C^{\frac{1}{2p}}((0,T],B)$ for the $L_p(\Omega)$-theory. If one wants a classical solution for $t > 0$, one must have also $f \in L^{1/\rho_1+\varepsilon}((0,T),B)$, but for $t > 0$ the force f must fulfil some additional Hölder conditions with respect to t and x, the details of which will be carried through in a forthcoming paper on analytic mappings in Banach spaces and their applications to nonlinear differential equations.

Let us now turn to Serrin's regularity result and the paper of Kato-Fujita cited above.

If one has a weak solution $\tilde{u}$ of the NS-equations with $\tilde{u} \in L^{\infty}((0,T), L^{n+\varepsilon}(\Omega))$, $u(0) = \varphi \in PL^{n+\varepsilon}(\Omega)$ then this solution coincides on $(0,\min(T,T_{max}))$ with the solution u of the integral equation (IV). We assume: $T_{max} \leq T$. $\sup_{0<t<T_{max}} \|u(t)\|_{L^{n+\varepsilon}(\Omega)}$ being bounded, one can continue u, thus getting $T_{max} > T$. The solution of the integral equation being classical in $(0,T_{max})\times\overline{\Omega}$ we have shown that $\tilde{u}$ is classical in $(0,T)\times\overline{\Omega}$. Serrin proved interior regularity in x only.

Let A be the generator of an analytic semigroup e^{-tA} in a Banach space B. We now study the abstract integral equation

$$u(t) = e^{-tA}\varphi + \int_0^t e^{-(t-s)A}M(u(s))\, ds$$

for an initial value $\varphi \in D(A^{\delta})$, where δ is a number > 0. The assumptions on M are as follows:

$$M: D(A^{1-\rho_1}) \to B \text{ analytic with}$$

$$\|M(u)\| \leq c\|A^{1-\rho_1}u\|^{1+\rho_0}(\|A^{\delta}u\|+1)$$

for a ρ_1 with $1-\rho_1 > \delta$.

$$M: D(A) \to D(A^{\rho}) \text{ analytic with}$$

a ρ, $1 > \rho > 0$, and with

$$\|A^{\rho}M(u)\| \leq c\|Au\|^{1+\rho_2}(\|u\|+1).$$

If $1 - \rho_1 - \frac{\rho_1}{\rho_0} \leq \delta$ we get a solution of the integral equation on $(0,T_{max})$, and in the same interval $A^{1+\rho}u(t) \in C^0((0,T_{max}),B)$. The method is similar as before. Specialising on p = 2, $B = H_2(\Omega)$, n = 3 and the NS-equations, we get Kato-Fujitas result: $\delta = \frac{1}{4}$. In this case $\rho_1 = \frac{1}{4}$,

$\rho_o = \frac{1}{2}$. This means: the initial value has to be from $A^{\frac{1}{4}}$.

Literature

[1] Fujita, H., Kato, T.: On the Navier-Stokes Initial Value Problem. I. Archive Rat. Mech. Anal. 16, 269-315(1964).

[2] Wahl, Wolf von: Analytische Abbildungen und semilineare Differentialgleichungen in Banachräumen, to appear in Nachr. Ak. Wiss. Göttingen.

NUMERICAL EXPERIMENTS WITH A MULTIPLE GRID AND A PRECONDITIONED LANCZOS TYPE METHOD

P. Wesseling and P. Sonneveld
Dept. of Mathematics, Delft University of Technology,
Julianalaan 132, 2600 AJ Delft, The Netherlands.

1. Introduction

Numerical experiments will be described with two methods for the numerical solution of difference schemes for general non-self-adjoint elliptic boundary value problems, namely a multiple grid method (MG method) and a preconditioned Lanczos type method (PIDR method). The interest of these relatively novel methods lies in the fact, that they seem potentially to be significantly more efficient than the methods that are currently popular. Our purpose is to investigate the efficiency of these methods by carrying out numerical experiments on two problems that are representative of the sort of problem that is encountered in engineering applications, namely the convection-diffusion equation, and the driven cavity problem for the Navier-Stokes equations.

2. Two testproblems

Our first testproblem is the convection-diffusion equation:

$$-(a_{ij}\phi_{,i})_{,j} + (b_i\phi)_{,i} + cu = f, \quad x_i \in \Omega, \qquad \phi\big|_{\partial\Omega} = g, \tag{2.1}$$

where Cartesian tensor notation has been used. The equation is assumed to be uniformly elliptic, i.e. there exist constants C_1, $C_2 > 0$ such that $C_1\xi_i\xi_i \leq a_{ij}\xi_i\xi_j \leq C_2\xi_i\xi_i$, $\forall\ \xi_i \in \mathbb{R}$. Our numerical experiments have been performed for the following special case:

$$-\varepsilon\phi_{,ii} + x_1\phi_{,1} = f = 2\varepsilon(x_1+x_2-x_ix_i) + x_1x_2(1-2x_1)(1-x_2),\ g = 0 \tag{2.2}$$

in two dimensions, with $\Omega = (0,1) \times (0,1)$. The exact solution is: $\phi = (x_1-x_1^2)(y_1-y_1^2)$. Keeping engineering applications in mind, special attention has been paid to the situation $\varepsilon << 1$. For simplicity we have restricted ourselves to Dirichlet boundary conditions, but this restriction is not necessary for the applicability of the MR and PIDR methods.

Our second testproblem is the vorticity-streamfunction formulation of the Navier-Stokes equations in two dimensions:

$$\psi_{,ii} = \omega, \qquad x_i \in \Omega,$$

$$u_i\omega_{,i} = Re^{-1}\omega_{,ii}\ , \quad u_1 = \psi_{,2}\ , \quad u_2 = -\psi_{,1}\ , \qquad x_i \in \Omega, \tag{2.3}$$

$$\psi\big|_{\partial\Omega} = 0\ , \qquad \psi_{,n}\big|_{\partial\Omega} = g,$$

with ψ the streamfunction, ω the vorticity, Re the Reynoldsnumber, and $\psi_{,n}$ the normal derivative of ψ.
Furthermore, $\Omega = (0,1) \times (0,1)$, and $g = 1$ for $x_2 = 1$, $g = 0$ elsewhere. This is the so-called driven-cavity problem.

Finite difference discretizations are chosen as follows. An equidistant computational grid Ω^ℓ with stepsize h is defined by:

$$\Omega^\ell \equiv \{(x_1,x_2)|\ x_i = m_i h,\ m_i = 1(1)2^\ell+1,\ h = (2+2^\ell)^{-1} \tag{2.4}$$

Note that Ω^ℓ lies a distance h within $\partial\Omega$. This is because the boundary conditions are substituted in the difference scheme. A more general stepsize may be chosen just as well for the PIDR method, but the programming of the MG method would become much more complicated. Forward and backward difference operators Δ_i and ∇_i are defined by:

$$(\Delta_1\phi)_{ij} \equiv (\phi_{i+1,j}-\phi_{ij})/h,\ (\nabla_1\phi)_{ij} \equiv (\phi_{ij}-\phi_{i-1,j})/h, \tag{2.5}$$

and similarly for Δ_2 and ∇_2; the subscripts i,j here indicate a point of Ω^ℓ in the usual way.

Equation (2.2) is discretized by means of the Il'in scheme [16] (twice rediscovered [4,7]):

$$-\gamma\Delta_1\nabla_1\phi - \varepsilon\nabla_2\Delta_2\phi + \tfrac{1}{2}x_1(\Delta_1+\nabla_1)\phi = f\ , \tag{2.6}$$

with $\gamma \equiv x_1 h \quad \coth(x_1 h/\varepsilon)$. The Il'in scheme has certain advantages when $\varepsilon << 1$ over a straightforward central or upwind difference scheme. These advantages do not interest us here; for us (2.6) is merely an example of a system on which to test the efficiency of the MR and PIDR methods. Near the boundary (2.6) is modified by substitution of the boundary conditions in the equation.

Equation (2.3) is discretized by means of central differences:

$$\Delta_i\nabla_i\psi = h^{-2}\bar{\omega}\ ,$$
$$\tfrac{1}{2}u_i(\Delta_i+\nabla_i)\bar{\omega} = Re^{-1}\Delta_i\nabla_i\bar{\omega}\ , \tag{2.7}$$

with

$$u_1 = \tfrac{1}{2}(\Delta_2+\nabla_2)\psi,\quad u_2 = -\tfrac{1}{2}(\Delta_1+\nabla_1)\psi.$$

The boundary contitions are:

$$\bar{\omega}_w + \tfrac{1}{2}\bar{\omega}_{w+1} - 3\psi_{w+1} = 3hg,\quad \psi_w = 0. \tag{2.8}$$

Here $\bar{\omega} = h^2\omega$. This rescaling of ω is introduced in order to get coefficients of O(1) in (2.8). For the derivation of this well-known approximation, see e.g. [21]. The subscript w indicates a point of $\partial\Omega$, the subscript w+1 indicates its nearest neighbour in Ω^{ℓ} in the direction of the normal.

The nonlinear system (2.7) has to be solved by some iterative method. We have chosen the Newton method:

$$\Delta_i\nabla_i\psi = h^{-2}\bar{\omega} ,$$

$$\tfrac{1}{2}\hat{u}_i(\Delta_i+\nabla_i)\bar{\omega} + \tfrac{1}{2}u_i(\Delta_i+\nabla_i)\hat{\bar{\omega}} = Re^{-1}\Delta_i\nabla_i\hat{\bar{\omega}} + \tfrac{1}{2}\hat{u}_i(\Delta_i+\nabla_i)\hat{\bar{\omega}}, \qquad (2.9)$$

$$\bar{\omega}_w + \tfrac{1}{2}\bar{\omega}_{w+1} - 3\psi_{w+1} = 3hg, \quad \psi_w = 0,$$

with ^ indicating values obtained from the preceding iteration. We do not claim that the use of the Newton method is always advisable; for us (2.9) is merely an interesting system on which to test the MG and PIDR methods.

Naturally a comparison with other methods would be of interest. A discussion of other methods is deferred to a later chapter. One method will be described here, namely the method called LAD method in [22]. This method is defined as follows:

$$Re^{-1}\Delta_i\nabla_i\bar{\omega} = \tfrac{1}{2}\hat{u}_i(\Delta_i+\nabla_i)\hat{\bar{\omega}} ,$$

$$\Delta_i\nabla_i\psi = h^{-2}\bar{\omega} , \qquad (2.10)$$

$$\bar{\omega}_w = r(-\tfrac{1}{2}\bar{\omega}_{w+1}+3\psi_{w+1}+3hg) + (1-r)\hat{\bar{\omega}}_w ,$$

with r a relaxation factor for which a good value has to be found by trial and error. Hence, the problem is reduced to two Poisson equations on a rectangle, which can be solved very efficiently with a fast Poisson solver.

Because of its great speed one is tempted to also use a fast Poisson solver for (2.6). This can be done iteratively as follows:

$$\varepsilon\Delta_i\nabla_i\phi = (\varepsilon-\gamma)\Delta_1\nabla_1\hat{\phi} + \tfrac{1}{2}x_1(\Delta_1+\nabla_1)\hat{\phi} - f . \qquad (2.11)$$

This method will be referred to as the FP method.

The iteration processes defined by (2.9), (2.10) and (2.11) will sometimes be referred to as outer iterations; the MG and PIDR methods are also of an iterative nature, and will sometimes be referred to as inner iterations.

3. A multiple grid method

For the intuitive background of multiple grid methods we refer to [6]. We start with a formal description of a specific multiple grid method which we shall call MG method. Then we discuss the relation between the MG method and other multiple grid methods, and make a few historical remarks.

In addition to the computational grid Ω^{ℓ} a hierarchy of computational grids Ω^k,

$k < \ell$ is defined by replacing ℓ by k in (2.4). The simultaneous use of such computational grids Ω^k is the reason for the appellation "multi-level methods" (cf. [6]), "multigrid methods" (cf. [12]) or "multiple grid methods" (cf. [19]). The d-dimensional vector of unknowns in each point of Ω^k is denoted by u^k, with d the number of unknowns in the partial differential equations that are to be solved. For example, for equations (2.6) or (2.9) $d = 1$ or $d = 2$ respectively. The set of functions U^k is defined by:

$$U^k \equiv \{u^k: \Omega^k \to \mathbb{R}^d\}$$

Furthermore, a prolongation operator $p^k : U^{k-1} \to U^k$ is defined by linear interpolation:

$$\begin{aligned}
(p^k u^{k-1})_{2i,2j} &= u^{k-1}_{ij} , \\
(p^k u^{k-1})_{2i+1,2j} &= \tfrac{1}{2}(u^{k-1}_{ij} + u^{k-1}_{i+1,j}) , \\
(p^k u^{k-1})_{2i,2j+1} &= \tfrac{1}{2}(u^{k-1}_{ij} + u^{k-1}_{i,j+1}) , \\
(p^k u^{k-1})_{2i+1,2j+1} &= \tfrac{1}{4}(u^{k-1}_{ij} + u^{k-1}_{i+1,j} + u^{k-1}_{i,j+1} + u^{k-1}_{i+1,j+1}).
\end{aligned} \tag{3.2}$$

The restriction operator $r^k : U^k \to U^{k-1}$ is defined to be the adjoint of p^k:

$$\begin{aligned}
(r^k u^k)_{ij} = \tfrac{1}{4}u^k_{2i,2j} &+ \tfrac{1}{8}(u^k_{2i+1,2j} + u^k_{2i-1,2j} + u^k_{2i,2j+1} + u^k_{2i,2j-1}) + \\
&\tfrac{1}{16}(u^k_{2i+1,2j+1} + u^k_{2i-1,2j+1} + u^k_{2i+1,2j-1} + u^k_{2i-1,2j-1}).
\end{aligned} \tag{3.3}$$

The linear algebraic system to be solved (e.g. (2.6) or (2.9)) is denoted by:

$$A^\ell u^\ell = f^\ell . \tag{3.4}$$

The matrix A^ℓ is approximated on coarser grids by $A^{\ell-1}, A^{\ell-2},\ldots,$ defined as follows:

$$A^{k-1} \equiv r^k A^k p^k , \quad k = \ell, \ell-1, \ldots \tag{3.5}$$

With each matrix A^k an approximate LU-decomposition $L^k U^k$ is associated; L^k and U^k will be defined in section 5.

The MG method is an iterative method. One iteration is defined as follows, in quasi-Algol:

<u>Algorithm 3.1</u>

$e^\ell := f^\ell - A^\ell u^\ell$;

<u>for</u> $k := \ell(-1)2$ <u>do</u> $e^{k-1} := r^k e^k$;

$v^1 := (L^1 U^1)^{-1} e^1$;

<u>for</u> $k := 2(1)\ell-1$ <u>do</u>

$\underline{\text{begin}}$ $v^k := p^k v^{k-1}$;

$e^k := e^k - A^k v^k$;

$v^k := v^k + (L^k U^k)^{-1} e^k$

$\underline{\text{end}}$;

$v^\ell := p^\ell v^{\ell-1}$;

$u^\ell := u^\ell + v^\ell$;

$e^\ell := f^\ell - A^\ell u^\ell$;

$u^\ell := u^\ell + (L^\ell U^\ell)^{-1} e^\ell$;

We proceed to estimate the computational complexity (number of operations) needed for this method. Every operation from the set {+,-,*,/,sqrt} is counted is one operation. Computation of pointers etc. is neglected. The computational complexity thus defined is easy to determine and is independent of the machine and the skill of the programmer, but gives only a rough indication of the actual computer time. However, for the comparison of methods this is a very useful quantity, and we feel that this kind of information should be provided more often.

It turns out that if A^ℓ results from a difference scheme based upon a 5-point or a 9-point difference molecule, then the structure of A^k, $k = \ell-1, \ell-2, \ldots$ corresponds to a 9-point molecule. One arrives at the following computational complexity:

	d = 1	d = 2
$A^k := r^{k+1} A^{k+1} p^{k+1}$	254,278	1016,1112
L^k and U^k	25,41	246,406
$v^k := (L^k U^k)^{-1} e^k$	16,20	58,72
$e^k := r^{k+1} e^{k+1}$	18,18	36,36
$v^k := p^k v^{k-1}$	1.5,1.5	3,3
$e^k := e^k - A^k v^k$	15,27	60,108
$u^k := u^k + v^k$	1,1	2,2

Table 3.1 Computational complexity for portions of MG-algorithm, divided by $(2^k+1)^2$.

The figure before the comma refers to the case that A^k (in rows 2,3 and 6) or A^{k+1} (in row 1) has 5-point structure, the figure after the comma corresponds to a 9-point structure (in our examples A^ℓ has 5-point structure, A^k, $k < \ell$ have 9-point structures). We recall that d is the number of unknowns in the given differential equation. In the computation of A^k only terms proportional to $(2^k+1)^2$ have been taken into account. In addition there is a (negligible) constant amount of work.

The computation of A^k, L^k and U^k has to be performed only once for a given problem

(3.4); the associated computational work will be called "preliminary work", whereas the work to be performed for each iteration will be called "iteration work". For the complete MG method we then find:

	ℓ	2	3	4	5
d = 1	p.w.	$3.3.10^3$	$1.2.10^4$	$4.2.10^4$	$1.5.10^5$
	i.w.	$1.6.10^3$	$6.0.10^3$	$2.2.10^4$	$8.1.10^4$
d = 2	p.w.	$1.9.10^4$	$6.9.10^4$	$2.4.10^5$	$8.5.10^5$
	i.w.	$5.6.10^3$	$2.1.10^4$	$7.8.10^4$	$3.1.10^5$

Table 3.2 Computational complexity for MG-method. p.w. = preliminary work, i.w. = iteration work.

With the approximation $(2^k+1)^2 \cong 4^k$ one may deduce:

$$\begin{aligned} d=1: &\quad \text{preliminary work} \cong 126.4^{\ell}, \\ &\quad \text{iteration work} \cong 72.4^{\ell}. \\ d=2: &\quad \text{preliminary work} \cong 728.4^{\ell}. \\ &\quad \text{iteration work} \cong 259.4^{\ell}. \end{aligned} \tag{3.6}$$

Hence, 1 MG-iteration has a computational complexity of $O(4^{\ell})$. As we will see, the distinguishing feature of multiple grid methods is that their rate of convergence is independent of the mesh-size of the computational grid, i.e. independent of ℓ. Requiring a precision of $O(4^{-\ell})$ (of the same order as the discretization error) then results in an overall computational complexity of $O(\ell 4^{\ell})$.

The MG method described here is but one example of a multiple grid method; many different multiple grid methods are possible and have been proposed. The above algorithm is a variant of a multiple grid method used by Frederickson [10] for the Poisson equation. Frederickson did not use approximate LU decomposition, but an approximate inverse. We have tried this also, but in our applications approximate LU was significantly faster (roughly by a factor 2). The other multiple grid methods that we know of do not use approximate LU decomposition or an approximate inverse, but some form of relaxation iteration, such as Jacobi- or line-iteration. Except [12] the other methods define A^k, $k > \ell$ not by (3.5) but by a discretization of the differential equations. Also, p^k and r^k are often different: for p^k higher order interpolation may be used, and r^k may simply be given by canonical injection: $(r^k u^k)_{ij} = u^k_{2i,2j}$. The freedom one has in these matters is both a blessing and a curse: by making careful decisions ([6] contains much useful advice) one may obtain a very efficient method for a given problem, but the rate of convergence may also be disappointing if wrong decisions are made. We feel that in this respect algorithm (3.1) has the advantage, that it is completely defined and that it has worked without fail for the cases that we have tried. Furthermore, for a closely related algorithm it has been proven [26] that for (2.1) in its full generality with Ω a square the computational complexity is $O(\ell 4^{\ell})$.

The intuitive ideas underlying multiple grid methods can already be found in the work of Southwell [24]. A multiple grid method was first proposed by Fedorenko [8]. A^k is defined by discretization, p^k by cubic interpolation and r^k by canonical injection. For the Poisson equation on a rectangle Fedorenko has shown in [9] that the asymptotic computational complexity is $O(\ell 4^{\ell})$. This was generalized to (2.1) on a rectangle by Bakhvalov [3]. In the same period a two-grid method has been proposed by Wachspress [25]. Apparently these methods did not find widespread use at the time although encouraging results were described in [25]. The present period of widespread interest in multiple grid methods was inaugurated by Brandt's pioneering paper [5]. His multiple grid methods are of the same type as the method of Fedorenko. For the Poisson equation a multiple grid method with A^k defined by (3.5) has been developed by Frederickson [10], who proves $O(\ell 4^{\ell})$ asymptotic computational complexity. The generalization to (2.1) on a rectangle is given in [26].

The multiple grid methods have been discussed here as linear equation solvers. They can also be used as special nonlinear equation solvers (see e.g. [6,13]), eliminating the need for the Newton-iterations (2.9). Furthermore, they can be used to construct adaptive discretizations [6]. These aspects fall outside the scope of the present paper.

4. A preconditioned Lanczos type method

A Lanczos type method for solving a linear system of equations is a method which is based essentially on the bi-orthogonalisation method of Lanczos finding the characteristic polynomial of a matrix. Lanczos type methods are of interest for the solution of equations with large sparse matrices, because they only involve the matrix in computing matrix-vector products.

A Lanczos type method is in principle finite, however, due to roundoff, this is not at all true in practice. If the matrix satisfies certain requirements, then Lanczos type methods can be constructed which behave very much like iterative methods and have quite competitive rates of convergence. An example is the conjugate gradient method (c.g. method) for symmetric positive definite systems. For this method it has been proven that the number of iterations necessary for gaining one decimal digit is:

$$\nu_{dig} \leq \tfrac{1}{2} \sqrt{\mathrm{cond}(A)}\, \ln(10) \qquad (4.1)$$

with cond(A) the spectral condition number of the matrix, i.e. $\mathrm{cond}(A) = \lambda\max/\lambda\min$. (see for example [1]). For the matrix of the 5-point discretization of Poisson's equation, this is the same as for the successive overrelaxation method with optimal relaxation factor. The spectral condition number is related to the order of the differential equation and to the stepsize:

$$\text{cond}(A) = \alpha . h^{-2k} \tag{4.2}$$

for a 2k-th order operator.
In a D-dimensional domain, h is related to the total number of equations by $N = \beta h^{-D}$, resulting in:

$$\nu_{dig} = O(N^{k/D}) \tag{4.3}$$

This means for a second order elliptic equation on a domain in $\mathbb{R}^2$, that the computational complexity for a fixed accuracy is:

$$w_{f\ dig} = O(N^{1.5}), \tag{4.4}$$

since each c.g step has computational complexity O(N).

One can try to speed up the convergence of a Lanczos type method by replacing A by a suitably chosen matrix $\tilde{A}$, which has a smaller spectral condition number. This is called preconditioning. This can be done by premultiplication by some suitable matrix C:

$$Ax = b \rightarrow CAx = Cb \leftrightarrow \tilde{A}x = \tilde{b}.$$

"Suitable" means here, that the computational work for the construction of C and calculating CAx must be of the same order as the work for calculating Ax.

In section 5 we shall describe a simple but very powerful example of preconditioning, the so called incomplete LU decomposition. For the Poisson equation and a special type of preconditioning it has been proven [11], that this preconditioning method reduces the spectral condition number to its square root. Other variants also transform the spectrum of A in such way that the convergence is substantially faster then is guarantueed by (4.1).

We will now describe a Lanczos type method that is suitable for non symmetric systems. The following theorem, proven in [23], demonstrates the principle on which the algorithm is based:

<u>Theorem 4.1</u> 1) Let A be an $N * N$ matrix;

2) Let P be a proper subspace of $\mathbb{R}^N$;

3) Let f be any vector in $\mathbb{R}^N$;

4) Let G_o be the Krylov space associated with f and A, i.e.:

$$G_o = \bigoplus_{n \geq 0} A^n f;$$

5) Let the sequence of spaces G_n be defined by

$$G_n = (1-\alpha A)\{G_{n-1} \cap P\} , \quad n \geq 1.$$

where $\{\alpha_n\}$ is a sequence of real numbers, $\alpha_n \neq 0$.
Then $M \leq N$ exists such that

$$G_n \equiv G_M \subset P , \quad \forall\, n \geq M.$$

In fact, G_n is a proper subspace of G_{n-1} for all n, until G_n becomes itself a subspace of P. In general G_M will be the null-space, unless P contains an eigenvector of A, which is not very likely.

An algorithm can be based on theorem 4.1 as follows. Let the system to be solved be given by $Ax = b$. If preconditioning is used it is assumed that A has already been preconditioned. A sequence $\{x_n\}$ is constructed iteratively such that $f_n = Ax_n - b \subset G_k$; we try to make k increase as fast as possible with n. Because the dimension of G_k decreases as k increases this method has been called induced dimension reduction (IDR) method by the second author [23]. When preconditioning is used the method is called PIDR method. The algorithm is defined as follows. Let x_0, $p \in \mathbb{R}^N$ be given.

<u>Algorithm 4.1</u> (IDR method)

```
n:=ω_n:=0; f_n:=Ax_n-b, dg_n:=dy_n:=0;
for n:=n+1 do
begin s_n:=f_{n-1}+ω_{n-1}dg_{n-1};t_n:=As_n;
    if n=1 V n is even then α~_n:=(t_n,s_n)/(t_n,t_n)
                       else α~_n:=α~_{n-1};
    dx_n:=ω_{n-1}dy_{n-1}-α~_n s_n;
    df_n:=ω_{n-1}dg_{n-1}-α~_n t_n;
    x_n:=x_{n-1}+dx_n;f_n:=f_{n-1}+df_n;
    if n is even then
       begin dg_n:=dg_{n-1};dy_n:=dy_{n-1}
       end else
       begin dg_n:=df_n;dy_n:=dx_n
       end;
    ω_n:=-(p,f_n)/(p,dg_n)
end of IDR method;
```

The relation between $\tilde{\alpha}_n$ in algorithm 4.1 and α_n in theorem 4.1 is: $\tilde{\alpha}_{2n} = \tilde{\alpha}_{2n+1} = \tilde{\alpha}_n$.

It will be shown that algorithm 4.1 is finite. Define $P = \{x \in \mathbb{R}^N | (p,x) = 0\}$. It is easy to show that $f_n = (1-\tilde{\alpha}_n A)s_{n-1}$, so that theorem 4.1 is applicable:

$$s_n \in G_r \cap P \rightarrow f_n \in G_{r+1}. \tag{4.5}$$

By induction we will show:

$$f_{2n}, f_{2n+1} \in G_n. \tag{4.6}$$

Obviously (4.6) holds for $n = 0$. Suppose (4.6) holds for some $n > 0$.

Then $s_{2n+2} \in G_n \cap P$, and with (4.5): $f_{2n+2} \in G_{n+1}$. Similarly, $f_{2n+3} \in G_{n+1}$, so that (4.6) is established for general n. From theorem 4.1 it follows that

$$f_{2n},\ f_{2n+1} \in G_M\ ,\quad n \geq M.$$

and in general $G_M = \{0\}$, so that the algorithm terminates when $n = M$.

Usually $p = f_0$ is a good choice (the best if A is symmetric positive definite). Because of rounding errors the algorithm is not finite in practice. The strategy for choosing $\{\alpha_n\}$ is such as to make $||(1-\alpha_n A)s_n||$ as small as possible.

The algorithm breaks down when P contains eigenvectors of A, but this is highly unlikely. More serious is the possibility of (almost) zero division when computing ω_n. For A symmetric positive definite is has been shown [23] that this does not occur, and that the following relation holds between the IDR method and the classical c.g. method:

$$f_{2n} = (1-\tilde{\alpha}_2 A)(1-\tilde{\alpha}_4 A)(1-\tilde{\alpha}_6 A)\ldots(1-\tilde{\alpha}_{2n} A)r_n \tag{4.7}$$

with r_n the nth residual of the c.g. method. In the numerical experiments reported here breakdown did not occur.

The computational complexity of one iteration with the PIDR method is easily determined as follows. Matrix * vector, scalar * vector, vector ± vector, (vector, vector) each take respectively W(A), 1, 1, 2 operations per unknown, with W(A) the number of operations per unknown for the preconditioned matrix-vector multiplication; W(A) will be further discussed in section 5. The number of unknowns is $(2^\ell+1)^2$ or $2(2^\ell+1)^2$ respectively for the convection-diffusion and the Navier-Stokes problem. One arrives at the following computational complexity per unknown for the PIDR method:

$$\begin{aligned} \text{p.w.} &= W(LU) + W(A) + 1, \\ \text{i.w.} &= W(A) + 14. \end{aligned} \tag{4.8}$$

with p.w. the preliminary work and i.w. the work per iteration. W(LU) is the computational complexity per unknown of the incomplete LU-decomposition; this will be discussed in section 5.

An additional test problem was solved with the PIDR method, namely the Navier-Stokes problem (2.9) with $\bar{\omega}$ eliminated, i.e. $\bar{\omega}$ everywhere replaced by $h^2\Delta_i\nabla_i\psi$. This will be called the fourth order formulation. The three test cases for the PIDR method will be denoted by PIDR(1), PIDR(2) and PIDR(3), corresponding to the convection-diffusion problem, the ω-ψ formulation and the fourth order formulation of the Navier-Stokes problem respectively.

5. Approximate LU decomposition

An algorithm to decompose an N * N matrix A into the product of a lower triangular matrix L and an upper triangular matrix U is the following (in quasi-Algol):

Algorithm 5.1

$A^o := A$;

for $r := 1(1)N$ do

begin $a^r_{rr} := \mathrm{sqrt}(a^{r-1}_{rr})$;

  for $j \geq r$ do $a^r := a^{r-1}_{rj} / a^r_{rr}$;

  for $i > r$ do $a^r_{ir} := a^{r-1}_{ir} / a^r_{rr}$;

  for $i > r \wedge j > r$ do

    $a^r_{ij} := a^{r-1}_{ij} - a^r_{ir} a^r_{rj}$

end LU decomposition;

We have $\ell_{ij} = a^N_{ij}$, $j \leq i$ and $u_{ij} = a^N_{ij}$, $j \geq i$. This special version of LU decomposition has $\ell_{ii} = u_{ii}$. It only works as long as $a^{r-1}_{rr} > 0$.

If A is a bandmatrix, i.e. $a_{ij} = 0$ for $|i-j| > b$ (the bandwidth), then L and U share this property, which results in the following computational complexity c.c. of algorithm 5.1:

$$\text{c.c.} \cong N(2b^2+b) \qquad (5.1)$$

For the difference scheme (2.6) $b = O(N^{\frac{1}{2}})$, which results in c.c. $= O(N^2)$. A disadvantage of this algorithm is that sparsity of A inside the bandwidth is not used. Approximate LU decomposition does not have this weakness. A nonzero pattern P is defined, usually consisting of the nonzero pattern of A plus perhaps only a few more entries. Elements of L and U outside P are per definition zero and need not be computed. Of course we no longer have A = LU. It turns out that P can be chosen such that Lanczos-type methods for $\tilde{A} \equiv L^{-1}AU^{-1}$ have a much faster convergence than for A; see for example [11,18] for the case that A is symmetric positive definite.

An imcomplete LU decomposition algorithm is the following. Note the resemblance to algorithm 5.1.

Algorithm 5.2

$A^o := A$;

for $r := 1(1)N$ do

begin $a^r_{rr} := \mathrm{sqrt}(a^{r-1}_{rr})$; (5.2)

  for $j > r \wedge (r,j) \in P$ do $a^r_{rj} := a^{r-1}_{rj} / a^r_{rr}$; (5.3)

  for $i > r \wedge (i,r) \in P$ do $a^r_{ir} := a^{r-1}_{ir} / a^r_{rr}$; (5.4)

  for $(i,j) \in P \wedge i > r \wedge j > r \wedge (i,r) \in P \wedge (r,j) \in P$ do

    $a^r_{ij} := a^{r-1}_{ij} - a^r_{ir} a^r_{rj}$

end incomplete LU decomposition;

In another version the neglected quantities $-a^r_{ir} a^r_{rj}$, $(i,j) \notin P$ are added to the diagonal:

Algorithm 5.3 As algorithm 5.2, with the last for-statement replaced by:

$$\begin{array}{l} \underline{\text{for}}\ i>r \wedge j>r \wedge (i,r)\in P \wedge (r,j)\in P\ \underline{\text{do}} \\ \underline{\text{begin}}\ \text{quant}:=-a^r_{ir}a^r_{rj}; \\ \quad \underline{\text{if}}\ (i,j)\in P\ \underline{\text{then}}\ a^r_{ij}:=a^{r-1}_{ij}+\text{quant} \\ \quad \underline{\text{else}}\ a^r_{ii}:=a^r_{ii}+\text{quant} \\ \underline{\text{end}}; \end{array} \tag{5.5}$$

For algorithm 5.3 it has been shown [11]that for the Poisson equation cond $(\tilde{A})$ = $O(\sqrt{\text{cond (A)}})$ with cond the spectral condition number.

For the MG method algorithm 2 was used. The nonzero pattern P was chosen as follows. If A corresponds to a 5-point difference molecule (as on the finest grid) then P corresponds to the following 7-point difference molecule: ** *** **
If A corresponds to a 9-point molecule (as on the coarser grids) then P corresponds to the same difference molecule. For the Navier-Stokes problem there are two unknowns, and the matrix is four times as large. The matrix may be thought to consist of four equal parts, the first two related to the first equation of (2.9) and the second two to the second equation of (2.9), and the first part of each pair corresponding to ψ, the second part corresponding to $\bar{\omega}$. For each of these four parts P is the same as for the convection-diffusion equation.

For PIDR(2) algorithm 5.3 was used as preconditioning. For PIDR(3) algorithm 5.2 was used, while algorithm 5.3 did not always work, because of the occurrence of negative diagonal elements. The computational complexity of algorithm 5.3 is easily determined. Define

$$\sigma_i^+ \equiv \sum_{j>i,(i,j)\in P'} 1 \tag{5.6}$$

i.e. σ_i^+ is the number of non-zero elements to the right of the diagonal in the ith row. Let P be symmetric. The following table gives a breakdown of the computational complexity (c.c.) for the rth elimination step.

statement	c.c.
(5.2)	1
(5.3)	σ_r^+
(5.4)	σ_r^+
(5.5)	$2(\sigma_r^+)^2$

Table 5.1 Computational complexity of rth elimination step of algorithm 5.3.

Hence the total computational complexity of algorithm 5.3 is:

$$\text{c.c. algorithm 5.3} = \sum_{r=1}^{N} [1+2\sigma_r^+(1+\sigma_r^+)]. \tag{5.7}$$

A general formula for the c.c. of algorithm 5.2 is not easy to find. But for a given P the c.c. is easily determined; results are given in table 3.1.

The following table gives W(A) and W(LU) (which occur in (4.8)).

	P(A)	P(LU)	alg.	W(LU)	W(A)	p.w.	i.w.
1	* *** *	*** *** ***	5.3	41	28	70	42
2	* * *** *** * * * * *** *** * *	*** * *** *** *** * * *** *** *** * ***	5.3	106	48	155	62
3	* *** ***** *** *	**** **** ******* **** ****	5.2	201	72	274	86

Table 5.2 W(LU), W(A) p.w. and i.w. for PIDR(α).

Table 5.2 gives the c.c. per unknown of the construction of L and U, a preconditioned matrix-vector multiplication, the preliminary work p.w. and the work per iteration i.w. For α = 1,2,3 the number of unknowns is, respectively: $(2^\ell+1)^2$, $2(2^\ell+1)^2$, $(2^\ell+1)^2$.

6. Numerical experiments

For every computation, as initial iterand the gridfunction zero was chosen. The initial iterand for the inner iterations was the result of the preceding outer iteration. The inner iterations were terminated when, in the notation of eq. (3.4), $||A^\ell u^{\ell,\nu}-f^\ell|| \le 10^{-6}||f^\ell||$, $u^{\ell,\nu}$ being the result of the νth and last inner iteration and $||.||$ the maximum norm. The rate of convergence was monitored by the quantity

$$\mu_\nu \equiv \{||A^\ell u^{\ell,\nu}-f^\ell||/||A^\ell u^{\ell,0}-f^\ell||\}^{1/\nu}. \tag{6.1}$$

It was found that μ_ν is a weak function of ν for the MG method, but varies appreciably with ν for the PIDR method (this is typical for this type of method: during a few iterations one may notice hardly any decrease in $||A^\ell u^{\ell,\nu}-f^\ell||$, and then an iteration comes along in which the residue is reduced appreciably; cf. c.g. methods). We have listed

$$\mu \equiv \{\mu_\nu |\ \nu \text{ the number of the last inner iteration}\}. \tag{6.2}$$

A reasonable estimate of the number of inner iterations needed for one extra decimal digit is: $\ln 0.1/\ln \mu$.

The rate of convergence of the outer iterations was monitored by means of

$||u^{\ell,\nu+1}-u^{\ell,\nu}||$, $u^{\ell,\nu}$ now being the result of the νth outer iteration. The Newton method (2.9) was always found to be quadratically convergent. The outer iterations (2.10) and (2.11) were not always convergent. When they converged the convergence seemed linear, with a somewhat variable value of $||u^{\ell,\nu+1}-u^{\ell,\nu}||/||u^{\ell,\nu}-u^{\ell,\nu-1}||$, with ν counting outer iterations. The quantity

$$\sigma_\nu \equiv \{||u^{\ell,\nu+1}-u^{\ell,\nu}||/||u^{\ell,1}-u^{\ell,0}||\}^{1/\nu} \tag{6.3}$$

was found to be almost independent of ν for $\nu \geq 10$ when there was convergence; therefore we have listed $\sigma \equiv \sigma_{10}$ as an indicator of the rate of convergence. This is not true for the LAD method, where we worked with $\sigma \equiv \sigma_{25}$; even $\nu = 25$ is sometimes not quite high enough for this method.

Obviously the computational complexity can be reduced by making the termination criterion dependent on the accuracy of the outer iterand, but we have not done this: our sole aim is to investigate the efficiency of the MG and PIDR methods as linear equation solvers.

For the convection-diffusion equation the following results were obtained:

ε	ℓ	2	3	4	5
1	MG	4, .030, 9.7'3	5, .052, 4.2'4	5, .061, 1.5'5	5, .063, 5.5'5
	PIDR(1)	4, .027, 6.0'3	6, .097, 2.6'4	11, .29, 1.5'5	18, .45, 9.0'5
	FP	5, .067, 2.0'3	5, .061, 1.5'4	5, .058, 6.3'4	5, .055, 2.6'5
0.1	MG	5, .032, 1.1'4	5, .055, 4.2'4	5, .054, 1.5'5	5, .054, 5.5'5
	PIDR(1)	4, .031, 6.0'3	7, .11, 2.9'4	11, .29, 1.5'5	20, .49, 9.9'5
	FP	66, .81, 2.6'4	46, .74, 1.4'5	36, .68, 4.6'5	31, .64, 1.7'6
0.01	MG	5, .043, 1.1'4	5, .040, 4.2'4	5, .062, 1.5'5	5, .052, 5.5'5
	PIDR(1)	4, .025, 6.0'3	4, .027, 1.9'4	6, .071, 9.3'4	8, .16, 4.4'5
	FP	divergent			

Table 6.1 Results for the convection-diffusion equation.

The first number of each entry is the number of iterations, the second number is μ (for MG and PIDR(1) or σ (for FP), and the third number is the computational complexity. For the fast Poisson solver the Fourier analysis cyclic reduction method (FACR) of Hockney [14,15] was chosen with 0 cyclic reductions. The computational complexity for one application of this fast Poisson solver is:

$$(2^\ell+1)^2(10+14.42 \ln(2^\ell+1)) \tag{6.4}$$

(in the computations this fast Poisson solver was not actually used, but the MG or

PIDR(1) method, but (6.4) was used for the determination of the computational complexity). For the FP method only 10 iterations were carried out in order to determine σ; the listed number of iterations (say, n) is determined from $\sigma^n < 10^{-6}$.

As $\varepsilon \downarrow 0$ the system (2.6) tends to a triangular matrix, for which the approximate LU-decomposition becomes exact. This explains why the rate of convergence of the MG and especially the PIDR(1) methods becomes larger as $\varepsilon \downarrow 0$. This would not happen for (2.1) in general.

Table 6.1 confirms that the rate of convergence of the MG method is independent of the mesh-size (independent of ℓ), so that the computational complexity is $O(\ell 4^{\ell})$ for a precision of $O(4^{-\ell})$. This is independent of ε, so that for $\varepsilon \downarrow 0$, $\ell \to \infty$ this method is a clear winner. However, for $\ell = 5$ the advantage of MG is not yet very great, and because of the much easier programming we feel, that for moderately large problems of this type the PIDR(1) method is preferable over the MG method. For $\varepsilon = 1$ the FP method is clearly attractive, but as one would expect FP gets worse as $\varepsilon \downarrow 0$; already for $\varepsilon = 0.1$ the other methods are faster. FP may be improved by convergence acceleration techniques, but for the very small values of ε that one encounters in engineering practice it seems highly unlikely that FP can be made competitive.

The data of table 6.1 indicate an asymptotic computational complexity of PIDR(1) of $O(N^{\alpha})$ with $\alpha \in [1.25,1.4]$ and N the number of unknowns; cf. the theoretical $O(N^{1.25})$ result for Poisson's equation, see section 7.

For the Navier-Stokes equations the following results were obtained for the MG-method:

Re	n\ℓ	2	3	4	5
10	1	5, .039, 4.7'4	5, .047, 1.7'5	5, .044, 6.3'5	5, .043, 2.4'6
	2	4, .032, 4.1'4	4, .038, 1.5'5	4, .044, 5.5'5	4, .045, 2.1'6
	3	2, .036, 3.0'4	2, .051, 1.1'5	1, .041, 3.2'5	1, .11, 1.2'6
	4	1, .051, 2.5'4	-	-	-
50	1	5, .039, 4.7'4	5, .047, 1.7'5	5, .044, 6.3'5	5, .043, 2.4'6
	2	6, .062, 5.3'4	5, .054, 1.7'5	5, .053, 6.3'5	4, .044, 2.1'6
	3	5, .065, 4.7'4	4, .075, 1.5'5	3, .051, 4.7'5	3, .067, 1.8'6
	4	2, .11, 3.0'4	1, .079, 0.9'5	1, .12, 3.2'5	1, .084, 1.2'6
100	1	5, .039, 4.7'4	5, .047, 1.7'5	5, .044, 6.3'5	5, .043, 2.4'6
	2	7, .12, 5.8'4	7, .13, 2.2'5	5, .079, 6.3'5	5, .072, 2.4'6
	3	7, .16, 5.8'4	5, .088, 1.7'5	3, .059, 4.7'5	4, .068, 2.1'6
	4	5, .15, 4.7'4	3, .099, 1.3'5	1, .078, 3.2'5	1, .085, 1.2'6
	5	2, .23, 3.0'4	1, .11, 0.9'5	-	-

Table 6.2 Results for the MG method applied to the Navier-Stokes problem.

The symbol n is the Newton-iteration number. The first number in each entry is the number of multiple grid iterations, the second number is μ (defined by 6.2). It is clear that the rate of convergence of the MG method is independent of ℓ, so that the computational complexity is $O(\ell 4^{\ell})$ for this problem also. The comparatively large values of μ for Re = 100, ℓ = 2 are probably due to the fact that for this case the matrix is not diagonally dominant, except for the first Newton iteration. In all cases we had $||u^{\ell,\nu}-u^{\ell,\nu-1}|| < 10^{-5}$, with ν the number of the last Newton iteration.

The following table gives the results for PIDR(2): the ω-ψ formulation of the Navier-Stokes problem solved with the PIDR method.

Re	n	2	3	4	
10	1	8, .16, 3.2'4	11, .25, 1.4'5	19, .48, 7.7'5	
	2	7, .16, 2.9'4	10, .28, 1.3'5	15, .47, 6.3'5	
	3	3, .15, 1.7'4	4, .26, 6.5'4	5, .45, 2.7'5	
	4	1, .20, 1.1'4	1, .42, 3.5'4	1, .49, 1.3'5	
50	1	8, .16, 3.2'4	11, .25, 1.4'5	19, .48, 7.7'5	
	2	10, .25, 3.9'4	13, .35, 1.6'5	21, .54, 8.4'5	
	3	9, .30, 3.6'4	11, .42, 1.4'5	14, .54, 5.9'5	
	4	5, .48, 2.3'4	2, .22, 4.5'4	1, .39, 1.3'5	
100	1	8, .16, 3.2'4	11, .25, 1.4'5	19, .48, 7.7'5	42, .72, 6.0'6
	2	13, .31, 4.8'4	18, .47, 2.1'5	26, .60, 1.0'6	51, .78, 7.2'6
	3	13, .38, 4.8'4	14, .44, 1.7'5	21, .61, 8.4'5	48, .83, 6.8'6
	4	9, .42, 3.6'4	11, .51, 1.4'5	-	27, .87, 4.0'6

Table 6.3 Results for PIDR(2).

In order to save computer time the outer iterations were sometimes terminated earlier than for table 6.2. For the same reason, for ℓ = 5 only the case Re = 100 was computed. The same holds for the following table, which gives the results for PIDR(3).

Apparently PIDR(3) is roughly a factor 2 fasterthan PIDR(2). For large problems MG (in this case $\ell > 4$) is faster than PIDR, as is to be expected. Just as in the case of the convection-diffusion problem we conclude that for medium-sized problems the PIDR methods seem preferable.

Re	n\ℓ	2	3	4	5
10	1	4, .01, 1.5'4	6, .083, 6.4'4	11, .28, 3.53'5	
	2	3, .008, 1.3'4	5, .077, 5.7'4	10, .30, 3.28'5	
	3	2, .007, 1.1'4	2, .035, 3.6'4	3, .17, 1.54'5	
50	1	4, .01 1.5'4	6, .083, 6.4'4	11, .28, 3.53'5	
	2	4, .025, 1.5'4	6, .104, 6.4'4	11, .30, 3.53'5	
	3	3, .022, 1.3'4	4, .059, '4	8, .33, 2.78'5	
	4	1, .029, '3	1, .012, 2.9'4	1, .05, 1.04'5	
100	1	4, .01, 1.5'4	6, .083, 6.4'4	11, .28, 3.53'5	33, .66, 3.4'6
	2	5, .037, 1.8'4	8, .154, 7.8'4	12, .31, 3.77'5	29, .64, 3.0'6
	3	4, .023, 1.5'4	6, .117, 6.4'4	10, .35, 3.28'5	27, .70, 2.8'6
	4	2, .019, 1.1'4	3, .125, 4.3'4	6, .45, 2.28'5	8, .59, 1.0'6

Table 6.4 Results for PIDR(3)

Some results with the LAD method (defined by eq. (2.10)) are given in the following table.

Re\ℓ	2	3	4
10	.276, 21, 3.5'4	.175, 26, 1.8'5	.092, 36, 1.0'6
20	-	.170, 27, 1.8'5	-
40	-	.089, 54, 3.7'5	-

Table 6.5 Results for LAD method.

The entries are, respectively, r (see eq. (2.10)), number of iterations, computational complexity. The values of r were the best we could find.
Each iteration takes two applications of the fast Poisson solver. We have measured $\sigma \equiv \sigma_{25}$ as defined in (6.3), and the number of iterations ν in table 6.5 follows from:

$$\sigma^{\nu}||u^{\ell,1}-u^{\ell,0}|| \leq 10^{-5}/(1-\sigma)$$

which more or less guarantees $||u^{\ell,\nu}-u^{\ell}|| \leq 10^{-5}$. This makes comparison with the MG results possible, because here the Newton iterations were terminated when $||u^{\ell,\nu}-u^{\ell,\nu-1}|| \leq 10^{-5}$, which for this quadratic process also guarantees that $||u^{\ell,\nu}-u^{\ell}|| \leq 10^{-5}$. With LAD, we were not able to get convergence for Re = 50 and Re = 100.

Clearly the LAD method is only competitive for Re $\lesssim$ 10. Roache [21] gives reasons to believe that LAD is more efficient than many methods currently in practical use.

The convergence difficulties for Re >> 1 are due to the ω-ψ splitting. In [28] Roache and Ellis therefore solve the fourth-order formulation by iterating with biharmonic solvers with asymptotic complexity $O(N^{3/2})$. Because no actual computational complexity is presented a direct comparison is not possible. The convergence of this method gets worse as Re $\to \infty$, which is not true for the MG and PIDR methods. We conclude that for Re >> 1 Newton-iteration combined with fast solvers for general elliptic non-self-adjoint systems, such as MG and PIDR, is preferable.

7. Discussion

There are a great many methods for solving partial differential equations by means of finite difference methods, and we have just highlighted two relatively novel ones. There is no general consensus on the relative merits of these methods, but a few global remarks can be made. In the following (non-exhaustive) table several important (classes of) methods are listed, together

Gauss elimination		$3N^2$
SOR	c.s.a.	$12.8\ N^{1.5}\ln N$
ADI/PR	c.s.a.	$75N(\ln N)^2$
FACR	c.s.a.	$(10+7.21\ \ln N)N$
MG		$O(N \ln N)$
PCG	c.s.a.	$O(N^{1.25}\ln N)$
PCG	s.a.	?
PIDR		?

Table 7.1 Applicability and operations count of several solution methods

with some information about their applicability to (2.1) and their computational complexity. The region Ω is rectangular; the computational grid is assumed to have N points (in the preceding sections, $N = (2^{\ell}+1)^2$). The third column gives the computational complexity. For the iterative methods in table 7.1 it has been assumed that the iterations are terminated when the residue has been reduced by a factor 1/N, which is proportional to the discretization error. The second column gives restrictions (if any) on the method. No entry in column 2 means no restrictions;c means constant coefficients; s.a. means self-adjoint, i.e. $b_i = 0$. SOR is the successive overrelaxation method [27] with optimal ω, which is only known for special cases including the c.s.a. case. ADI/PR is the ADI method with optimal Peaceman-Rachford parameters [20], known in the c.s.a. case. FACR is the method mentioned in section 6. Strictly speaking c in eq. (2.1) should be zero, but the method can be generalized to non-zero constant c and even to the case where c and a_{ij} depend on x_1 or x_2; the computational complexity remains $O(N \ln N)$. For the MG method a proof is given in [26] that the computational complexity is $O(N \ln N)$ for (2.1) with Ω a rectangle. Our present results indicate that also for the Navier-Stokes equations we have computational complexity = $O(N \ln N)$. This makes the MG-method the most general and for large N the most efficient method of the methods listed in table 7.1. PDG stands for the class of preconditioned conjugate gradient methods. These methods have recently seen much development, and are often very efficient, see for example the application of the ICCG-method [18] in [17]. In [11] a proof is given that the asymptotic computational complexity is $O(N^{1.25})$ for a method of PCG type applied to the Poisson equation on a square. This is not much more than $O(N \ln N)$, and practical experience will have to show whether multiple grid or PCG methods are to be preferred. PCG methods are easy to program, but for large N multiple grid methods might be significantly faster. It seems likely that in the future SOR and ADI will be superseded by PCG. Efforts are under way to develop methods similar to PCG for the non-self-adjoint case, and PIDR may be regarded as one such method; another example is to be found in [2]. At the moment a rigorous theoretical estimate of the operations count of the PIDR method is not available. Our numerical experiments indicate that the operations count is $O(N^{\alpha})$ with $\alpha \in [1.25, 1.40]$.

From our numerical experiments one may conclude that for large N (in our examples $N \sim 289$) MG is rather more efficient than PIDR. But PIDR is much easier to program, and the programming does not become more complicated when Ω has an arbitrary shape. There is every reason to believe that the multiple grid approach works for arbitrary regions, see e.g. [12], but the programming gets very complicated.

References

1. O. Axelsson: Solution of linear systems of equations: iterative methods. In: V.A. Barker (ed.): Sparse matrix techniques. Lecture Notes in Mathematics 572, 1977, Berlin etc., Springer-Verlag.

2. O. Axelsson, I. Gustafsson: A modified upwind scheme for convective transport equations and the use of a conjugate gradient method for the solution of non-symmetric systems of equations. J. Inst. Math. Appl. 23, 321-338, 1979.

3. N.S. Bakhvalov: On the convergence of a relaxation method with natural constraints on the elliptic operator. USSR Comp. Math. Math. Phys. 6 no. 5, 101-135, 1966.

4. K.E. Barrett: The numerical solution of singular-perturbation boundary-value problems. J. Mech. Appl. Math. 27, 57-68, 1974.

5. A. Brandt: Multi-level adaptive technique (MLAT) for fast numerical solution to boundary-value problems. Proc. 3rd Internat. Conf. on Numerical Methods in Fluid Mechanics, Paris, 1972. Lecture Notes in Physics 180, 82-89, Springer-Verlag 1973.

6. A. Brandt: Multi-level adaptive solutions to boundary value problems. Math. Comp. 31, 333-390, 1977.

7. J.C. Chien: A general finite difference formulation with application to the Navier-Stokes equations. Comp. Fl. 5, 15-31, 1977.

8. R.P. Fedorenko: A relaxation method for solving elliptic difference equations. USSR Comp. Math. Math. Phys. 1, 1092-1096, 1962.

9. R.P. Fedorenko: The speed of convergence of one iterative process USSR Comp. Math. Math. Phys. 4, no. 3, 227-235, 1964.

10. P.O. Frederickson: Fast approximate inversion of large sparse linear systems. Mathematics Report 7-75, 1975, Lakehead University.

11. I. Gustafsson: A class of first order factorization methods. BIT 18, 142-156, 1978.

12. W. Hackbusch: On the multi-grid method applied to difference equations. Computing 20, 291-306, 1978.

13. W. Hackbusch: On the fast solutions of nonlinear elliptic equations. Num. Math. 32, 83-95, 1979.

14. R.W. Hockney: A fast direct solution of Poisson's equation using Fourier analysis. J. Ass. Comput. Mach. 12, 95-113, 1965.

15. R.W. Hockney: The potential calculation and some applications. Methods in Comp. Phys. 9, 135, 1970.

16. A.M. Il'in: Differencing scheme for a differential equation with a small parameter affecting the highest derivative. Math. Notes Acad. Sc. USSR 6, 596-602, 1969.

17. D.S. Kershaw: The incomplete Cholesky-conjugate gradient method for the iterative solution of systems of linear equations. J. Comp. Phys. 26, 43-65, 1978.

18. J.A. Meijerink and H.A. van der Vorst: An iterative solution method for linear systems of which the coefficient matrix is a symmetric M-matrix. Math. Comp. 31, 148-162, 1977.

19. R.A. Nicolaides: On multiple grid and related techniques for solving discrete elliptic systems. J. Comp. Phys. 19, 418-431, 1975.

20. D.W. Peaceman and H.H. Rachford Jr.: The numerical solution of parabolic and elliptic differential equations. JSIAM 3, 28-41, 1955.

21. P.J. Roache: Computational fluid dynamics. Hermosa Publishers, Albuquerque, 1972.

22. P.J. Roache: The LAD, NOS and Split NOS methods for the steady-state Navier-Stokes equations. Computers and Fluids 3, 179-196, 1975.

23. P. Sonneveld: The method of induced dimension reduction, an iterative solver for non-symmetric linear systems. Publication in preparation.

24. R.V.Southwell: Stress calculation in frameworks by the method of systematic relaxation of constraints. Proc. Roy. Soc. London A 151, 56-95, 1935.

25. E.L. Wachspress: Iterative solution of elliptic systems. Prentice-Hall, Inc., Englewood Cliffs, N.J., 1966.

26. P. Wesseling: The rate of convergence of a multiple grid method. To appear in the Proceedings of the Biennial Conference on Numerical Analysis, Dundee, 1979, Lecture Notes in Mathematics, Springer-Verlag.

27. D.M. Young: Iterative methods for solving partial difference methods of elliptic type. Trans. Amer. Math. Soc. 76, 92-111, 1954.

28. P.J. Roache and M.A. Ellis: The BID method for the steady-state Navier-Stokes equations. Computers and Fluids 3, 305-321, 1975.

New solutions of the Karman problem for rotating flows

by

P.J. Zandbergen

Twente University of Technology, The Netherlands

1. *Introduction.*

During the last decade there has been a wide interest in the study of solutions of the Navier-Stokes equations for the case of a rotating fluid above an infinite disk or between two disks which are itself rotating.

For the one disk problem the essential parameter is the ratio s of the angular velocity of the fluid far from the disk and the angular velocity of the disk.

In 1921 Von Karman showed that the problem could be reduced to the solution of a system of two ordinary differential equations.

As it appeared later there is a range of s values where no solutions can be found: $-0.16054 \geq s \geq -1.4355$.

As has been shown by Bodonyi [1] the solution becomes singular for $s = -1.4355$. For $s = -0.16054$ branching of the solution occurs as has been shown by Zandbergen and Dijkstra [2]. From a further study in [3] it is quite clear that there are an infinite number of branches wrapping themselves around $s = 0$. Also for the two disk problem multiple solutions have been reported as for instance by Holodniok, Kubicek and Hlavacek [4].

The question arises whether there is a connection between the multiple solutions for the one disk problem and those for the two disk problem. Also the behaviour of the solutions on the multiple branches poses some questions since they are characterized by the occurrence of ever increasing large inviscid "humps" whereas also the number of humps increases.

These questions become rather intriguing when one realizes, as will be shown, that all solutions for $s \neq 0$ asymptotically have to coincide with a solution arbitrarily near to $s = 1$ with a certain amount of blowing or suction through the disk applied. It is precisely this last fact which is the starting point of the investigations to be described in this paper. It will be shown that it is possible to obtain a full characterization of any solution in this way. But it also provides a very efficient means for the actual computation of solutions.

A key role is played by solutions of the Karman equations which have a tangential velocity only for one part of space and axial and radial velocities only for the other part, the two regions connected by a layer which might be called a shock.

As will be shown two different families of solutions can be obtained which are in a sense closely connected, one of the families being identical with the multiple solutions already obtained, the other giving rise to a completely new set of solutions wrapping itself around two singular points namely $s = -1.43553$ and $s = +1.43553$. Some of the characteristics of the solutions will be given.

So far the investigations can be termed as exact calculations of solutions of the Navier-Stokes equations. Problems of using approximate methods arise when one is interested in the study of the stability of the solutions by using time dependent methods. The conditions at infinity will now be applied as point conditions in a point considered far enough away from the disk. This however can lead to very curious results as will be demonstrated, necessitating the search for better methods. In the sequel there will first be given a brief description of the earlier results. Thereafter the items described above will be considered in more detail.

2. *A survey of earlier results.*

In the first place the mathematical formulation of the problem will be given. In a cylindrical coordinate system (r,ϕ,z) the disk is the plane $z = 0$ and the velocities are given by

$$u = r\Omega f'(x), \quad v = r\Omega g(x), \quad w = -2(\nu\Omega)^{\frac{1}{2}} f(x) \qquad (2.1)$$

where Ω is the angular velocity of the disk and $x = z(\Omega/\nu)^{\frac{1}{2}}$ (see fig. 1)

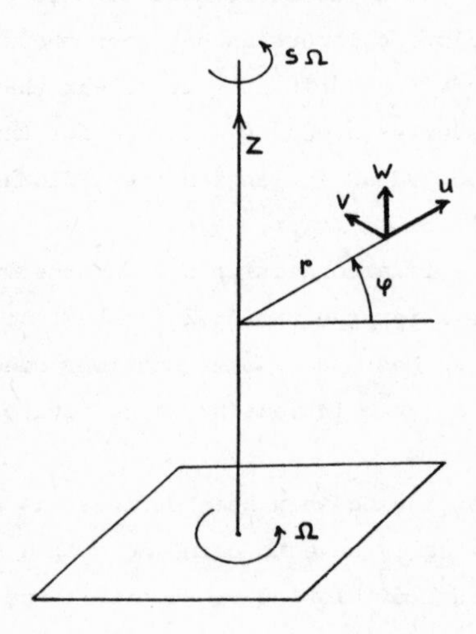

fig. 1. The velocities and coordinates.

The Navier-Stokes equations reduce in this case to

$$f''' + 2ff'' = f'^2 + s^2 - g^2 \qquad (2.2)$$

$$g'' + 2fg' = 2f'g \qquad (2.3)$$

The boundary conditions are

$$f = f_o, f' = 0 \quad g = 1 \qquad \text{for } x = 0 \qquad (2.4)$$

$$f' \to 0 \quad g' \to s \qquad \text{for } x \to \infty$$

The asymptotic behaviour of the solution (see [2]) can be represented as

$$f = f_\infty + e^{px} \{\frac{bp + cq}{p^2 + q^2} \sin qx + \frac{cp - bq}{p^2 + q^2} \cos qx\}$$

$$f' = e^{px} \{b \sin qx + c \cos qx\} \tag{2.5}$$

$$g = s + e^{px} \{c \sin qx - b \cos qx\}$$

In refs. [2] an [3] solutions to the above problem have been obtained by using a so called inverse shooting method. There exist two versions of this method. Either the values of f_∞, b and c are chosen and with the help of eq. (2.5), starting conditions are provided for $x = x_m$ and the integration is performed in negative x direction towards x = 0 where the conditions of eq. (2.4) should be fulfilled. The value of s is then fixed and we have a three parameter iteration. Or the value of f_∞ is fixed and b and c are iterated such that f = 0 and f' = 0 for x = 0. The actual value of s follows then from the value of g(0).
In this way the solution has been obtained for $f_o = 0$, that is no suction through the plate. It can be characterized by giving the value of f_∞ as a function of s as is done in fig. 2.

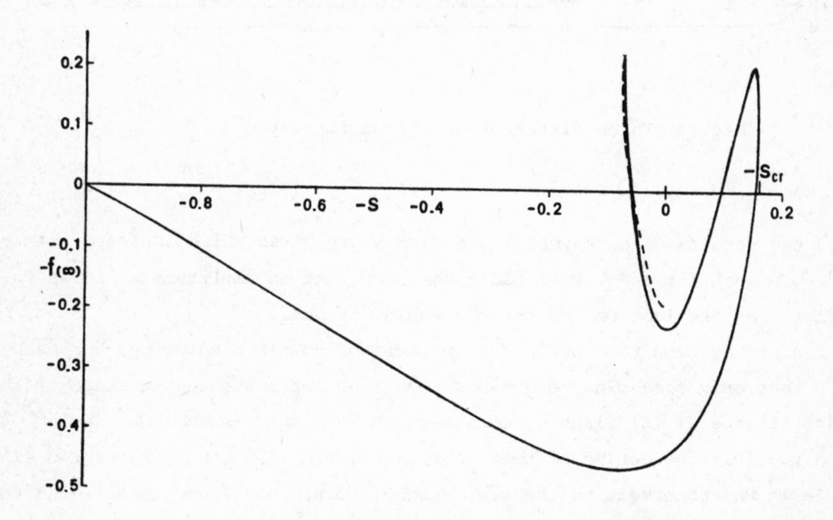

fig. 2. The Karman solutions around s = 0

It is clear from this figure that the branches not only wrap around s = 0 but also cut the line $f_\infty = 0$ for every branch twice. As has been indicated in the introduction also the behaviour of the solution itself on the various branches deserves attention due to the occurence of large humps. To illustrate this, in fig. 3, the function f(x) has been given for the three solutionsindicated in fig. 2.

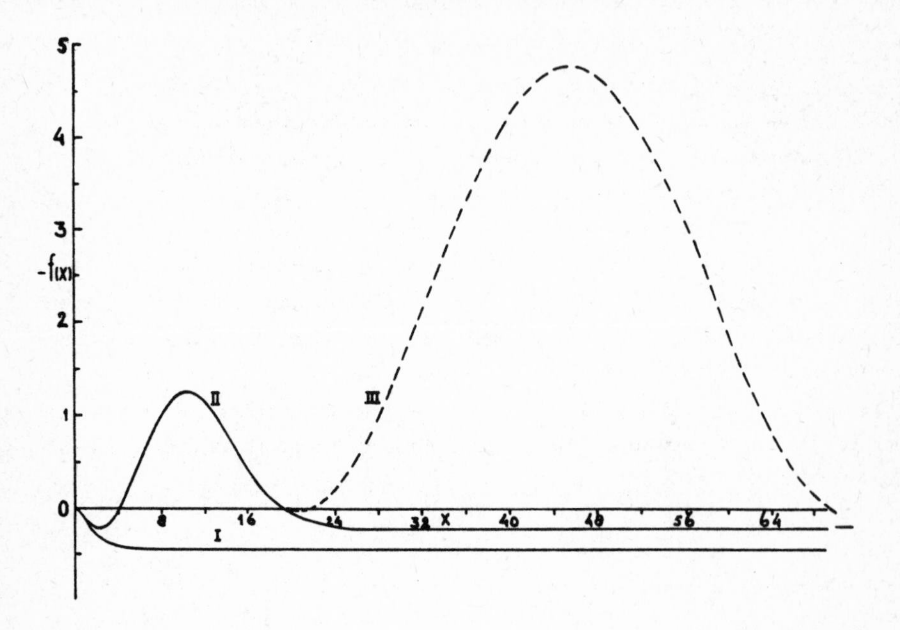

Fig. 3. Three different solutions for s = 0

In ref [3] the results of an asymptotic analysis are given which indicate that each new solution branch for s = 0 will add a new hump with an amplitude which is four times at large as the last one of the preceeding branch.
Being so far it may seem that now a full picture of possible solutions has been obtained and that only some minor questions remain as for instance the shape of the limit branch in the (f_∞,s) plane or more insight into the precise behaviour of the humps or a possible connection of these solutions with solutions for the two disk problem. As it was observed, on the new branches points could be identified - near to $f_\infty = 0$ - where the solution consists of two cells. This mean that for a certain value of $x \neq 0$ there also holds $f = f' = 0$. As will be clear these are solutions for the one disk problem which consist in fact of a certain solution for the two disk problem together with another solution for the one disk problem attached to it. This last solution could be identified with a point on the solution branch near to s = 1. But then naturally the idea arises to study the solution around s = 1 in more detail and trying to obtain general solutions by continuation of these solutions in negative

x-direction. In the next section therefore it will be shown that any solution for $s \neq 0$ can be considered as the continuation of a solution arbitrarily close to $s = 1$.

3. Asymptotic properties of solutions.

As has been shown by McLeod in ref [5] the asymptotic formulae (2.5) are rigorously valid for any value of s. Now it is observed that there always can be chosen a point $x = x_n$ where $f' = 0$ for $s \neq 0$ and where according to eq (2.5) for f' may be written

$$f' = s\,\varepsilon\, e^{p(x-x_n)} \sin q(x-x_n) \tag{3.1}$$

with ε arbitrarily small.

It follows then immediately that also is valid

$$g = s - s\,\varepsilon\, e^{p(x-x_n)} \cos q(x-x_n) \tag{3.2}$$

Now since $g(x_n) = s(1-\varepsilon)$ and $g(\infty) = s$, in fact the solution for values of $x > x_n$ can be viewed upon as one for which the parameter s takes the value

$$s^* = \frac{g(\infty)}{g(x_n)} = \frac{1}{1-\varepsilon} \approx 1+\varepsilon \tag{3.3}$$

to first order in ε.

But this is precisely a case which is as close to $s = 1$ as one pleases.

Now it is know from McLeod [6] that there is a surrounding of $s = 1$ where the problem (2.2)-(2.4) has a solution for any value of the suction parameter $f_o = a$. A direct proof by means of a series expansion has been given by Dijkstra [7].

For future reference here the first two terms are given of this series for arbitrary values of a.

If $h = f' + i\,g$, then there holds

$$h = i + \varepsilon i\,\{1-e^{(\alpha+\beta i)x}\} + \varepsilon^2\{-\delta e^{(\alpha+\beta i)x} + dx\, e^{(\alpha+\beta i)x} + \delta\, e^{2\alpha x}\} \tag{3.4}$$

where

$$\alpha = -a - \frac{1}{2}\sqrt{2}\,\{\sqrt{a^4+4} + a^2\}^{\frac{1}{2}} \tag{3.5a}$$

$$\beta = \frac{1}{\alpha+a} \tag{3.5b}$$

$$\delta_1 = \frac{\beta^2}{2}\,\frac{4a+\alpha}{2a+5\alpha} \tag{3.5c}$$

$$\delta_2 = \frac{\beta^3}{2\alpha}\,\frac{a-2\alpha}{2a+5\alpha} \tag{3.5d}$$

$$d_1 = \frac{\alpha\beta(3a+2\alpha)}{2(a^2+2\beta^2)} \tag{3.5e}$$

$$d_2 = \frac{\alpha\beta^2}{(2(a^2+\beta^2)} \tag{3.5f}$$

For small values of ε - say of order 10^{-6} - with eq.(3.4) very accurate solutions can be obtained. From the equation follows that for $x = 0$ there is valid

$$f = a \quad f' = 0 \quad g = 1 \quad f'' = \varepsilon\beta + \varepsilon^2 \{\alpha\delta_1 + \beta\delta_2 + d_1\}$$
$$g' = -\varepsilon\alpha + \varepsilon^2\{\alpha\delta_2 - \beta\delta_1 + d_2\} \tag{3.6}$$

he value of f_∞ is to first order in ε

$$f_\infty = a + \frac{\beta^2}{2\alpha}\varepsilon \tag{3.7}$$

The values as given by eq.(3.6) can be used as starting conditions for the integration in negative x direction. In the case that $a = 0$ much more terms of the series in ε have been calculated as f.i. by Van Hulzen [8].

$$f(0)'' = -\varepsilon - 0.4\,\varepsilon^2 + 0.0625\,\varepsilon^3 - 0.01784049773756\,\varepsilon^4$$
$$+ 0.00608335920385\,\varepsilon^5 - 0.00224692498047\,\varepsilon^6 \qquad (3.8)\text{a}$$
$$g'(0) + \varepsilon + 0.3\,\varepsilon^2 - 0.0475\,\varepsilon^3 + 0.01651583710407\,\varepsilon^4$$
$$- 0.00778855668137\,\varepsilon^5 + 0.00431163181563\,\varepsilon^6 \qquad (3.8)\text{b}$$
$$f(\infty) = -0.5\,\varepsilon + 0.1\,\varepsilon^2 - 0.00625\,\varepsilon^3 - 0.02238291855204\,\varepsilon^4$$
$$+ 0.02801441613476\,\varepsilon^5 - 0.02557639460144\,\varepsilon^6 \qquad (3.8)\text{c}$$

It should be observed, that in total there have been calculated up to thirty terms of this series and that the whole first branch, nearly up to the branching point, can be calculated by this series. But if the attention is only for values near to $s = 1$, the six terms of eqs (3.8) will suffice.

As will be clear by now, it is quite natural to use eqs (3.4) or (3.8) as the basic formulae to obtain solutions. For in that case there is no need to worry about the conditions at infinity, because it is known beforehand, that they are met.

The only thing to be done, is to continue the solution in negative x direction, to locate points where $f' = 0$ (if any) and there is obtained a new solution. It will be obvious that in general f will have a certain value as well as g. The new solution obtained then can be characterized by a value of s, which of course is related to the value of g in the point where $f' = 0$, say at $x = -x^*$.

If

$$s = \frac{1+\varepsilon}{g(-x^*)}, \quad \sigma = \sqrt{|g(-x^*)|} \tag{3.9}$$

$$f = \sigma\bar{f}, \quad g = g(-x^*)\,\bar{g}$$

then a solution $\bar{f}$, $\bar{g}$ is obtained which fulfills eqs (2.2) - (2.4) with the given value of s. and with $f_o = f(-x^*)/\sigma$

A few remarks can be made. The first is, that starting with (3.4) or (3.8) means that there is only one parameter, ε. So if one is interested in a certain value of $f = f_o$ at $x = -x^*$ then there is an iteration only with respect to ε; This of course means much less work than the iteration with respect to two or three parameters as required when using eqs (2.5).

Another very obvious remark is that every solution to the Karman problem for $s \neq 0$, contains an infinite set of solutions to the same problem, for it is known that f' goes to zero oscillating, which means that there is an infinite sequence of points where $f' = 0$, and the above explained procedure will give a new solution. But if this argument is reserved, it is seen that for ε small, only a finite range of ε values will have to be considered, because thereafter the same solution is obtained. This is immediately clear when considering first order terms of eqs (3.4).

$$\begin{aligned} f' &= + \quad e^{\alpha x} \sin \beta x \\ f &= a + \varepsilon\, e^{\alpha x} \{\frac{\beta}{2} \sin \beta x - \frac{\beta^2}{2\alpha} \cos \beta x\} + \frac{\beta^2}{2\alpha} \\ g &= 1 + \varepsilon\, \{1 - e^{\alpha x} \cos \beta x\} \end{aligned} \tag{3.10}$$

As is evident for $x = -k\,\pi/\beta$, $f' = 0$. The value of g is then $g = 1 + \varepsilon\{1 - e^{-k\pi\frac{\alpha}{\beta}}\}$. But this is a solution with a value of s which is equal to

$$1 + \varepsilon_K = \frac{1+\varepsilon}{1+\varepsilon\{1-e^{-k\pi\frac{\alpha}{\beta}}\}} \approx 1 + \varepsilon\, e^{-k\pi\frac{\alpha}{\beta}} \tag{3.11}$$

The value of the axial velocity is then

$$a + \varepsilon \frac{\beta^2}{2\alpha} (1 - e^{-k\pi\frac{\alpha}{\beta}}) \tag{3.12}$$

So far small values of ε compared to a, it may be said that a case where $\varepsilon_k = \varepsilon\, e^{-k\pi\frac{\alpha}{\beta}}$ is virtually the same as the ε case, which proves the argument. For a is zero a somewhat different situation is present, which will be considered the next section.

4. *An abundance of cell solutions.*

As has been observed earlier in section 2, on each of the solution branches found in [2] points could be determined where the solution consists of two cells. This is of course precisely the situation where $a = 0$ and $f_o = 0$. Hence use can be made of eqs (3.8) and for an appropriate value of ε, there will be most certainly a point where $f'(-x^*) = f(-x^*) = 0$. The corresponding value of f_∞ will be given by eq (3.8)c. These two-cell solutions are of course also characterized by the occurrence of large inviscid humps for the higher solution branches. As is clear from the analysis given in section 3 for ε values which are given by

$$\varepsilon_k = \varepsilon\, e^{-k\pi\frac{\alpha}{\beta}} \tag{4.1}$$

precisely the same solution is obtained if now instead of $a = 0$, a is taken equal to

$$\frac{\beta^2}{2\alpha}\,\varepsilon\,\{1-(-1)^k\,e^{-k\pi\frac{\alpha}{\beta}}\}$$

As it turns out, if ε_k is taken according to eq (4.1) and a equal to zero, then a very slight shift in ε_k will again produce a two-cell solution, which is very near to the original one, but where k oscillations have been added. This is illustrated in fig. 4 where two two-cell solutions have been given on the fourth solution branch.

The first one is for $\varepsilon_1 = 9.12159.10^{-4}$

The second one for $\varepsilon_2 = 1.696025.10^{-6}$

The corresponding values of s and f_∞ are

$s_1 = 4.09788 \,.\, 10^{-2}$ $\qquad f_\infty = -\,9.24278 \times 10^{-5}$

$s_2 = 4.09722 \,.\, 10^{-2}$ $\qquad f_\infty = -\,1.71648 \times 10^{-7}$

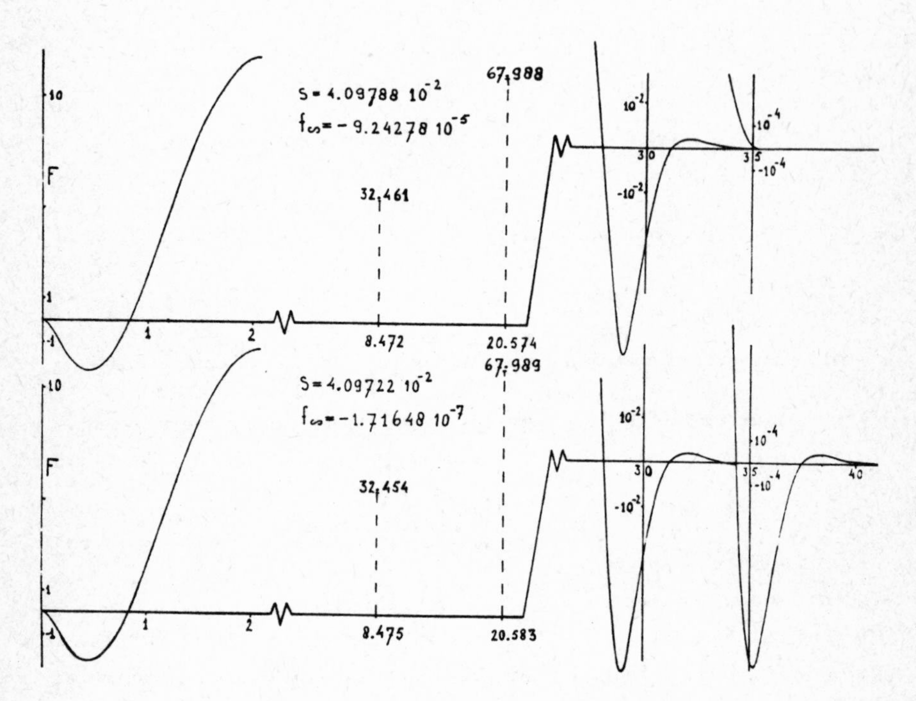

fig. 4. Twe nearly equal two-cell solutions.

The ratio of ε_1 and ε_2 is only 5 $^o/oo$ different from $e^{-2\pi\frac{\alpha}{\beta}}$

Since ε is negative for odd values of k and hence f_∞ positive, and positive for even values with f_∞ negative, it can be stated as a result of these investigations that there is an infinity of two-cell solutions grouped around the point where on a branch $f_\infty = 0$.

But when performing the calculations necessary for obtaining these two cell solutions, it was found that by a slight shift in ε another family of two cell solutions could be obtained, which must belong to a hitherto unknown family.

Of course immediately the question arises of how these solutions are precisely located in the (s,f_∞) plane. Before this question will be answered, something more has to be said about the characteristics of possible solutions in the (a,ε) plane. As will be shown, there is a very special kind of solutions which play a key roll in the understanding of what is really happening.

5. *A short characterization of possible solutions.*

As has been explained, it is possible to obtain any solution of the problem (2.2)-(2.4) by considering solutions near to s = 1 and to continue the solution in negative x-direction. The question which arises now is what has to be expected for the behaviour of these solutions. This has been thouroughly investigated, on which will be reported elsewhere. Here only a short survey is given.

According to McLeod [6] the quantity

$$\phi = f''^2 + g'^2 \tag{5.1}$$

is a strictly decreasing function towards $x \to \infty$. But this means that this function is strictly increasing in negative x direction. There are of course two possibilities. Either ϕ becomes unbounded for a finite or infinite value of x or ϕ remains bounded for $x \to \infty$.

When the first possibility occurs, the solution of the equations becomes singular. As can be shown near to the singularity the behaviour is

$$f \approx \frac{2}{x-x_o} - \frac{C^2}{16} + \frac{C^2}{352}\cos 2\beta - \frac{5\sqrt{7}}{352}C^2 \sin 2\beta + D\sqrt{x-x_o}\cos\rho \tag{5.2)a}$$

$$g \approx \frac{C}{(x-x_o)^{3/2}}\sin\beta + O\left(\frac{C^3}{(x-x_o)^{1/2}}\right) \tag{5.2)b}$$

with

$$\beta = \frac{1}{2}\sqrt{7}\ \ln\ (x-x_o) + \psi_1(\text{constant}) \tag{5.2)c}$$

$$\rho = \frac{1}{2}\sqrt{15}\ \ln\ (x-x_o) + \psi_2(\text{constant}) \tag{5.2)d}$$

This singular solution is characterized by the five constants x_o, C, D, ψ_1 and ψ_2. The behaviour as sketched here will normally occur, but the second possibility is in fact the more interesting. It is readily found that if ϕ has to be finite for $x \to \infty$ then there must hold

$$f \sim \bar{A}(x + \frac{B}{2\bar{A}})^2 + \frac{s^2}{4\bar{A}} \tag{5.3)a$$

$$g \sim 0 \tag{5.3)b$$

More insight into the behaviour of g can be obtained by inserting eq (5.3)a into the linear differential equation for g, eq (2.3).

There is obtained to leading order for $x \to -\infty$

$$g_{xx} + 2\bar{A}x^2 g_x - 4\bar{A}xg = 0 \tag{.5.4}$$

Now equation (5.4) which is an equation for the confluent hypergeometric functions will have two fundamental solutions, one decaying exponentially for $x \to -\infty$ and one behaving as x^2.

To meet eq (5.3)b the coefficient of x has to be zero, and since this coefficient is dependent on the pair (a,ε), we have to look for the zeros of a function $C(a,\varepsilon)$. It proves possible indeed to construct a curve in the (a,ε) plane where these solutions occur. In fig. 5 part of this curve has been given

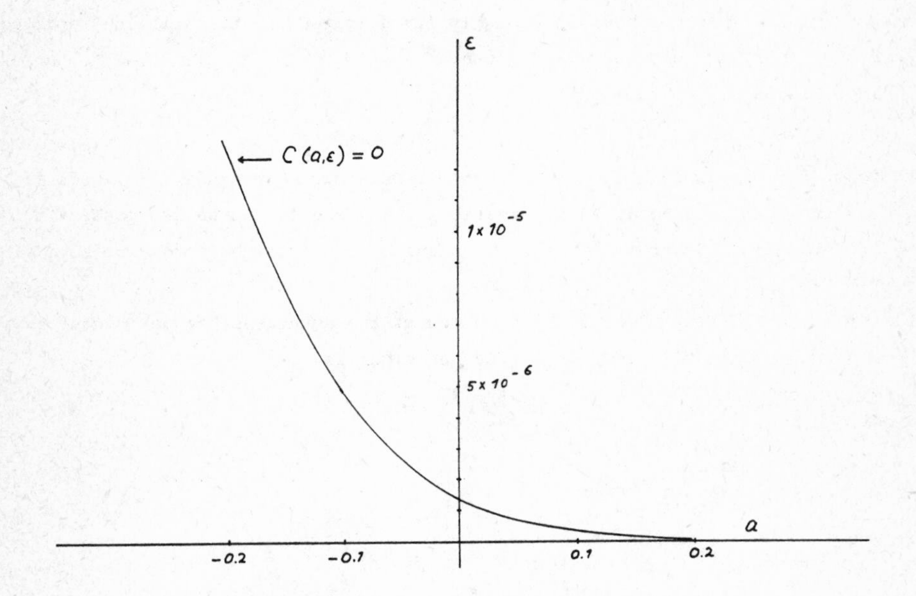

Fig. 5. The curve for $C(a,\varepsilon) = 0$.

An example of the solution is given in fig. 6 for the case where a = 0 and $\varepsilon = -1.585843 \times 10^{-2}$. The typical exponential decay of g is clearly visible. Of course also here the periodicity in ε as explained in section 3 occurs.

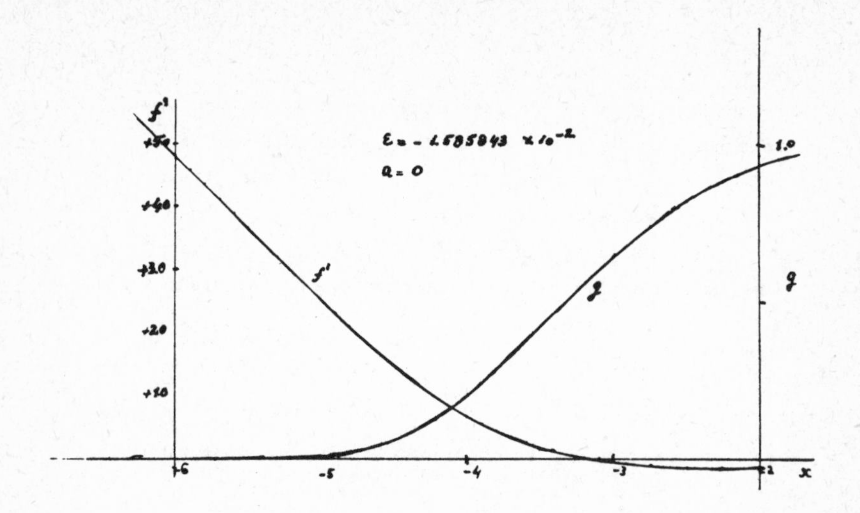

Fig. 6. The tangential and radial velocity in a circulation shock.

An important remark can be made here on the characteristics of these solutions. It is an easy matter to deduce from section 2, that in the physical plane

$$u \approx 2\,\Omega^{3/2}\,\nu^{-1/2}\,Azr \text{ and } w \approx -\,2\,\Omega^{3/2}\,\nu^{-1/2}\,Az^2 \tag{5.5}$$

It follows directly from eqs (5.5) that the streamlines are given by hyperbolas

$$z = \frac{\lambda}{r} \tag{5.6}$$

Along the streamlines the velocity is directed in positive x direction for small r and is given as

$$\sqrt{u^2+w^2} = 2\,\Omega^{3/2}\,\nu^{-1/2}\,|A|\,\sqrt{\lambda^2 + \frac{\lambda 4}{r4}} \tag{5.7}$$

and is thus clearly decelarating. This means that this is a solution for the Navier-Stokes equations which may be called a circulation shock, since in half of space the fluid is moving practically as a rigid continuum with an angular velocity Ω. Around the origin a swift change takes place and at the other side of the region the fluid moves as a kind of decelerated jet-flow without any angular velocity.
The phenomenon seems related to the breakdown of vorties as observed in leading edge vortices along delta wings and as investigated by Brook Benjamin (see f.i. ref [9]). Having established the existance of these solutions, the next important question is what happens if the parameters are slightly changed. Now if a is taken constant and ε varied by a very small amount, the coefficient $C(a,\varepsilon)$ of the fundamental solution of eq (5.4) with x^2 as leading term is varying away from zero. In the surrounding of the origin the behaviour will not be influenced very much. So f will climb rapidly

and g is at first becoming small. But then the term with x^2 in g will make itself felt in eq (2.2) by means of the g^2 term. The result of this is that the "infinite hump" is now brought back to a cosine like hump with an amplitude inversely proportional to the square of the coefficient $C(a,\varepsilon)$. The question of course is what is to be expected to occur after this first large hump. The analysis as given in [3] for s = 0 suggests that large inviscid humps are connected to each other by a viscid region in between. Dijkstra in ref. [10] recently investigated this problem for general values of s and gives elegant asymptotic results. His results can be summarized as follows. For a large hump asymptotically is valid

$$f \sim -\frac{\lambda}{2}\cos^2 \mu x \tag{5.8)a}$$

$$g \sim -\lambda\mu \cos^2 \mu x \tag{5.8)b}$$

with $\lambda \to \infty$, $\mu \to 0$ such that $\lambda\mu^2 = A$ is constant. For two adjacent humps it is valid then that

$$\frac{\lambda_2}{\lambda_1} = \frac{1}{\rho^2} \qquad \frac{\mu_2}{\mu_1} = |\rho| \tag{5.9}$$

where ρ is dependent on the ratio $s^2/A^{4/3}$ and behaves as is given in fig. 7.

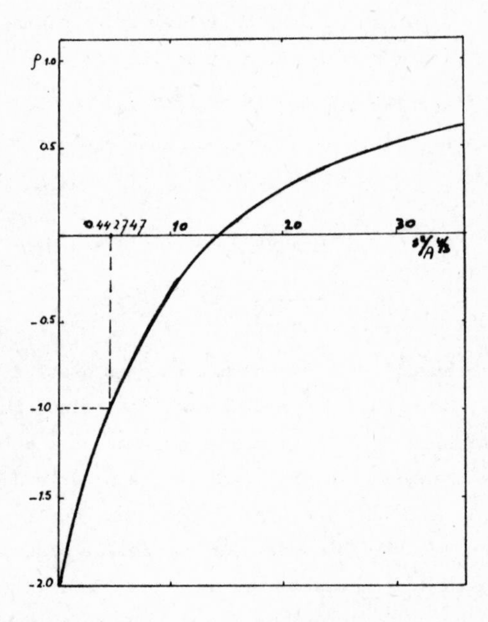

fig. 7. The function ρ for the interconnection of two large "humps".

The maximum value of f between the two humps is given by

$$f_{max} = -\frac{s^2}{2A} \tag{5.10}$$

As it turns out near to s = 1 and for values of a not too large positive the function ρ is between -1 and -2, that is to say, after the first large hump a smaller one will appear and so on till finally eq.(5.8)a is no longer valid and the solution will end in the singularity described in eq.(5.2). This means that the smaller $C(a,\varepsilon)$ is the more humps will occur before the singularity is reached,reversely, the more $C(a,\varepsilon)$ is away from zero, the lesser humps will occur, till finally the solution proceeds to the singularity without any hump at all. This behaviour is precisely what can be found also by numerical methods.

As long as eq.(5.8)a is valid it follows from eq.(5.10) that f is everywhere negative. So only at the end of the chain of humps f_{max} may become positive due to second order effects. The general behaviour is as depicted in figure 8.

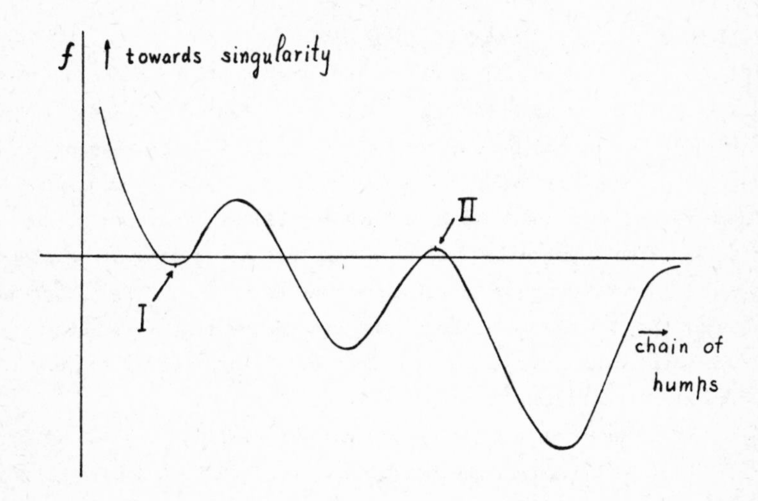

Fig. 8. The behaviour of the solution at the end of a chain of large humps.

From this figure it follows at once that there are two possibilities where both f and f' may be equal to zero, and hence solutions to the problem (2.2)-(2.4) with $f_o = 0$. In the place indicated with I the quantity ε may be determined such, that f = f' = 0. As it turns out this possibility gives the solutions as presented earlier, that is to say those who wrap around s = 0. The number of the branch in this case is regulated by the number of humps in the chain of humps. So it follows that the nearer ε is chosen to the ε of the "circulation shock", the higher the number of the solution curve will be.

But it is also clear from fig. 8 that both f and f' = 0 may be obtained at the place indicated with II. This proves indeed to be true. Thereby in the (f_∞,s) plane a totally different and hitherto unknown family of solution curves is ontained, which also consists of an infinite number of solution branches.

The difference in the (f_∞,s) plane is caused by the behaviour of the function g which changes rapidly in the surrounding of place I, wheras g changes only little in the surrounding of place II. With the aid of eq.(3.9) then the real value of s follows. So it has been established by now that it is possible to construct the whole set of solutions to the Karman problem for a = 0 at the disk by considering solution in the vicinity of a particular (a,ε) value such that a circulation shock is obtained. Two different families may be obtained, which are however closely related in the (a,ε) plane. In the next section the results will be discussed.

6. The solution of the Karman problem for a = 0.

Using the principles as explained so far calculations have been performed to obtain solutions. It should be emphasized that special care had to be taken, for in the large humps the equations can become very stiff indeed. Therefore a procedure was developed which automatically would detect the occurrence of a large hump. The characteristic values of λ and μ as used in eqs (5.8) will be established and the difference between the asymptotic values and the actual value will be claculated. This makes use of a procedure for stiff equations as given in [11] programmed for double precision. At the end of a large hump the computation will automatically return to the Diffsys integration method as used in [2]. In this way it is very easy to calculate also higher solution branches.

In fig. 9 five branches have been given for the surrounding of s = 0. Also small parts of the twelfth branch in the surrounding of $f_\infty = 0$ have been given. It suggests that this is already rather close to the limit curve.

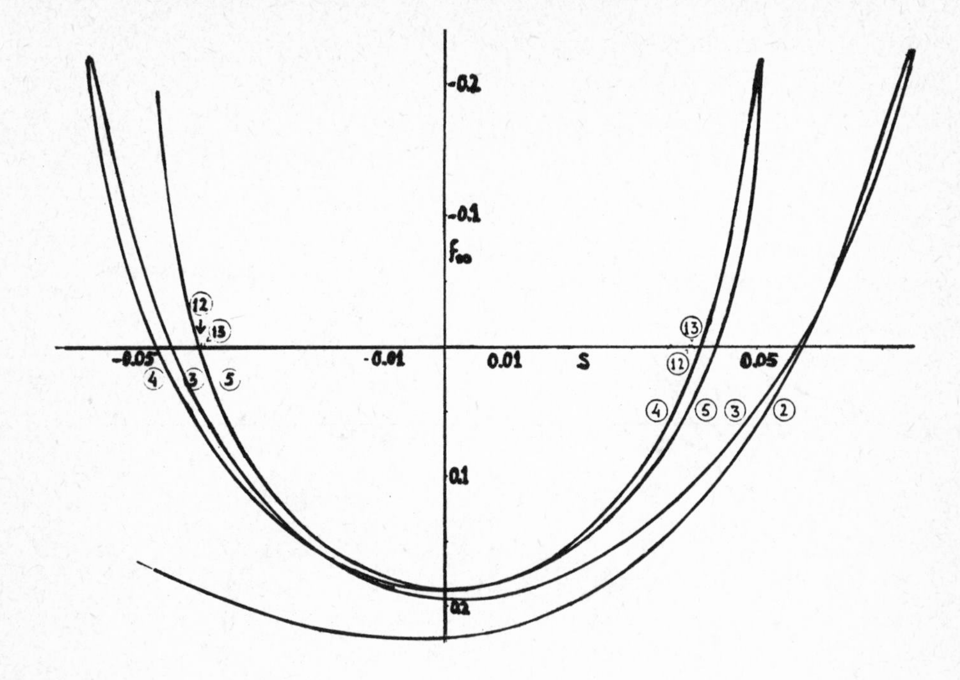

fig. 9. Detailed information on the solution branches around s = 0.

The points for s = 0 have been calculated by the original procedure as explained in ref.[2]. For the fifth branch the ratio of f_∞ and the asymptotic result as given in ref.[3] is

$$f(\infty) \; : \; 0.6048 \; (\lambda\mu^2)^{1/3} = 0.9973 \qquad *)$$

In fig. 10 and 11 the new solution branches have been given. They wrap around the lines of singularity s = - 1.4355 and s = 1.4355 and are confined to a small region in the (s, f_∞) plane. As it turns out the analysis of Bodonyi [1] is of course also valid for s = 1.4355, since it gives only the value of s^2. The new family cuts the original constructed solution of the Karman problem, so that in the vicinity of s = 1.4355 an infinite number of bifurcation points occurs as is indicated in fig. 11.

*) Note.
For the case s = 0 second order asymptotic values are available which make it possible to assess the value of $f(\infty)$ for the limit branch. This quantity can be indicated by $f_\infty(\infty)$. From the results obtained for the fifth branch, this value can be calculated with an accuracy of five digits

$$f_\infty(\infty) = 0.18498$$

For more details the reader is referred to ref.[10].

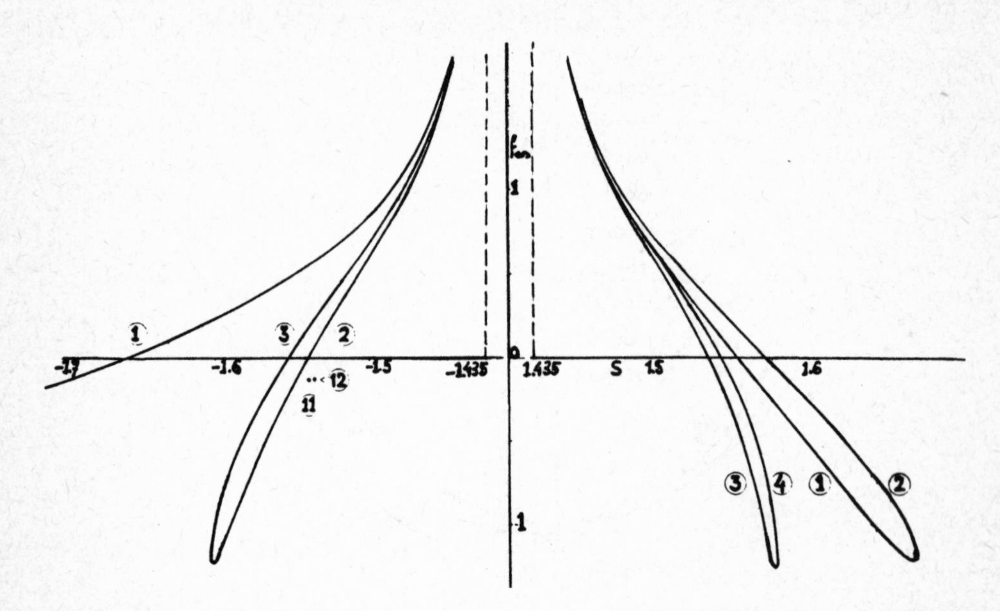

Fig.10. The new solutions in the (s,f_∞) plane for negative and positive s values,

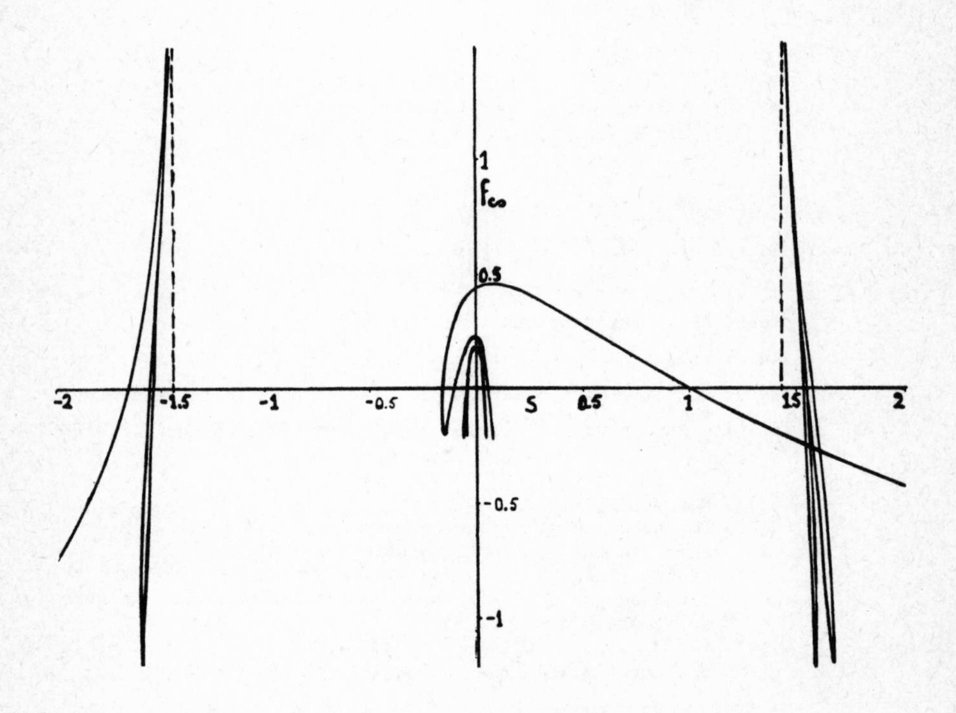

Fig. 11. The (s,f_∞) plane for $f_o = 0$.

What is interesting is the occurrence of the lobes as well in fig. 9 as in fig. 10. Can they be explained? To do this it is interesting to look at the behaviour of the solution when moving along a lobe. As appears the behaviour for positive values of $f(\infty)$ is clearly as given qualitatively in section 5; that is: the occurrence of a chain of large humps, apparently with a factor $|\rho|$ according to eq.(5.9) of about 1,4 and larger. Now for negative values of f_∞ the factor $|\rho|$ is becoming smaller and near to the minimum value of f_∞ it may become even slightly less than one. This means that the humps are now of growing amplitude. Apparently second order effects can be so strong as to reduce the amplitude of the humps when the number of humps is not too large. Where the tangent is vertical, $|\rho|$ is practically one. As will be seen a small hump has been starting to grow and will rapidly become the largest hump when again moving in the direction of positive f_∞. The behaviour has been given in fig.12.

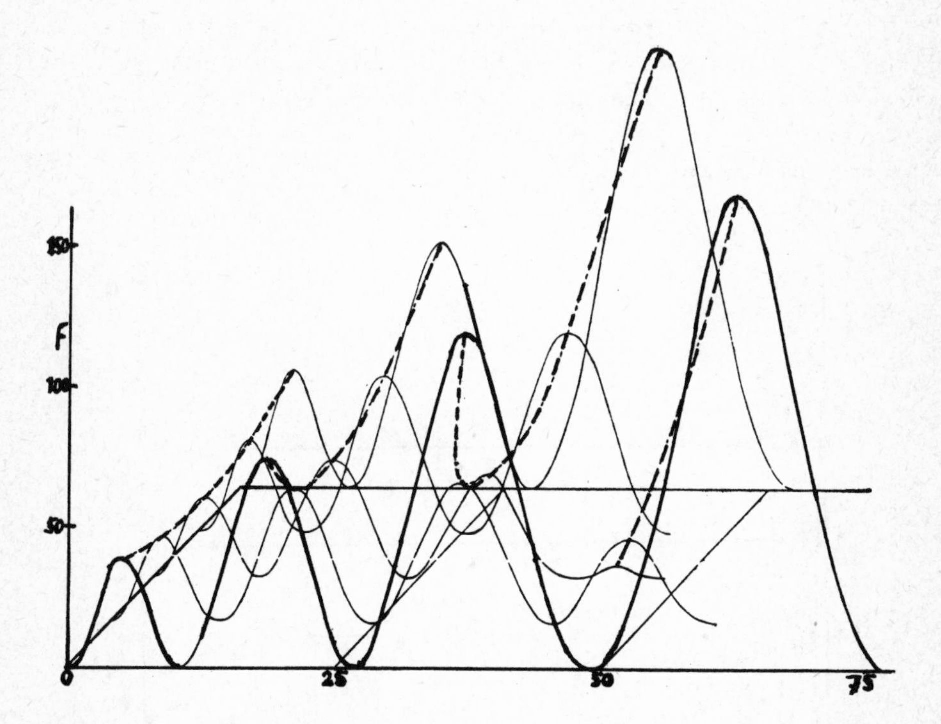

Fig. 12. The behaviour of the solution along the 3-4 lobe for the new solutions.

7. Some remarks on stability.

So far no investigations have been done on the stability of these new solutions.

But as studies on the stability in ref [12] have revealed it may be expected that all these solutions are instable. When this is true, then this should be valid also for the original solution curve for the Bödewadt case, indicated with 1 in fig. 10. But then the question remains, if the whole of curve 1 is instable, and if not what happens precisely. This question has considered already by Bodonyi in ref.[13]. However the investigations of these problems is rather awkward, as may be explained by the following discussion. In ref.[12] use has been made of a difference method to integrate the time dependent equations for the rotating flow, with essentially the same boundary conditions as given in eq.(2.4), However these boundary conditions will be required in the computation procedure at $z = z_m$ instead at infinity.
So as an indication for what may happens the following example was calculated. As a starting condition the f, g values were taken for s = -0,15 on the first branch. A range z was used with $x_m = 3.6$ and a stepwidth $\Delta x = 0.2$. The time step was chosen $\Delta t = 0.2$. The value of s from the second time step on at $x = x_m$ was taken as s= -1.55. The time dependent solution is convergent to a stable solution,but as it appears from fig. 12 to a totally different solution than would be expected. This can be inferred from the behaviour of f and g. In fact the solution is the solution for s = 1.55 with a free boundary layer at $x = x_m$ where the flow changes rapidly the sense of rotation. So this solution is totally inacceptable.

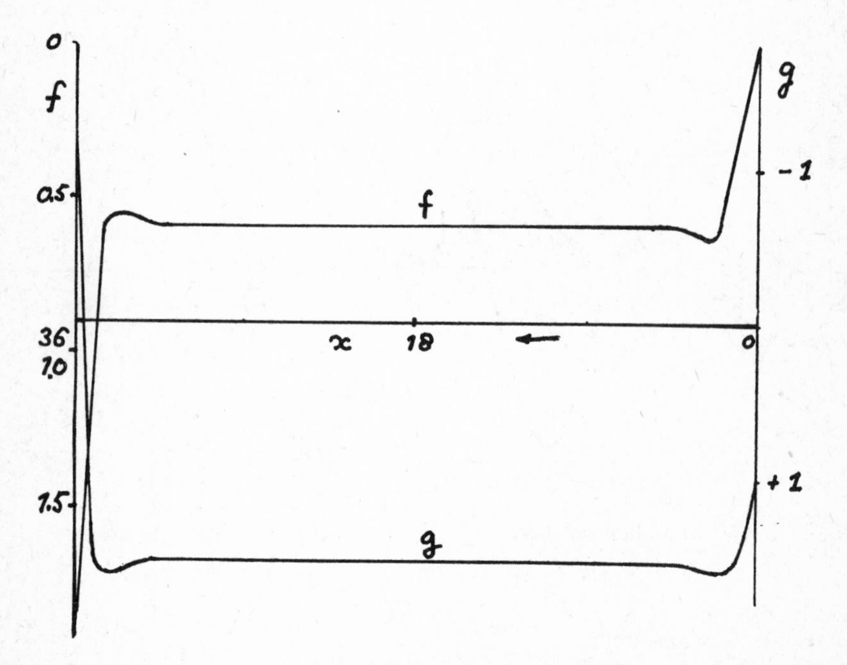

Fig. 13. The velocities f and g for the case s = -1.55 with boundary conditions at $x_n = 36$.

This is due to the fact that the boundary conditions at infinity, which really should induce a behaviour have been replaced by point conditions. Therefore it is better to look for a form of boundary conditions which more faithfully deal with the conditions at infinity.

Acknowledgement.

The author wishes to acknowledge Mr. E.H. Derks for programming and performing all the necessary calculations. He also is very grateful to Dr. Dijkstra for many stimulating discussions, much to the benefit of this paper.

References.

[1] Bodonyi, R.J. On rotationally symmetric flow above an infinite rotating disk. J.F.M., 67, 657 (1975).

[2] Zandbergen, P.J. and Dijkstra, D. Non unique solutions of the Navier-Stokes equations for the Karman swirling flow. J.Eng.Math., 11. 167 (1977).

[3] Dijkstra, D. and Zandbergen P.J. Some further investigations on non unique solutions of the Navier-Stokes equations for the Karman swirling flow. Archives of Mechanics (Archiv. Mechanoki Slosowany), 30, 411 (1978).

[4] Holodniok, M.; Kubicek, M. and Hlavacek, V. Computation of the flow between two rotating coaxial disks. J.F.M., 81 (1977).

[5] McLeod, J.B. The asymptotic form of solutions of Von Karman's swirling flow problem. Quarterly Journal of Mathematics (Oxford), 20 (1969).

[6] McLeod, J.B. The existence of axially symmetric flow above a rotating disk. Proceedings of the Royal Society, A324 (1971).

[7] Dijkstra, D. Some contributions to the solution of the rotating disk problem. Memorandum nr. 205, THT, TW (1978).

[8] Van Hulzen, J.A. Production of exact Taylor coefficients for the rotating disk problem using an algebra system. Memorandum nr. 183, THT, TW (1977).

[9] Benjamin, J.B. Theory of the vortex breakdown phenomenon. J.F.M. 14, 593 (1962).

[10] Dijkstra, D. On the relation between adjacent inviscid cell type solutions to the rotating disk equations. Paper to be published shortly.

[11] Craigie, J.A.I. A variable order multistep method for stiff systems of ordinary differential equations University of Manchester, N.A. Report 11, (1975).

[12] Dijkstra, D. Schippers, H. and Zandbergen, J.P. On certain solutions of the non-stationary equations for rotating flow. Proceedings of the Sixth Intern. Conf. on Num. Methods in Fluid Dynamics. Tbilisi, (1978).

[13] Bodonyi, R.J. On the unsteady similarity equations for the flow above a rotating disk in a rotating fluid. Q.Jl.Mech.appl.Math. Vol. XXXI, 461, (1978).

Vol. 609: General Topology and Its Relations to Modern Analysis and Algebra IV. Proceedings 1976. Edited by J. Novák. XVIII, 225 pages. 1977.

Vol. 610: G. Jensen, Higher Order Contact of Submanifolds of Homogeneous Spaces. XII, 154 pages. 1977.

Vol. 611: M. Makkai and G. E. Reyes, First Order Categorical Logic. VIII, 301 pages. 1977.

Vol. 612: E. M. Kleinberg, Infinitary Combinatorics and the Axiom of Determinateness. VIII, 150 pages. 1977.

Vol. 613: E. Behrends et al., L^p-Structure in Real Banach Spaces. X, 108 pages. 1977.

Vol. 614: H. Yanagihara, Theory of Hopf Algebras Attached to Group Schemes. VIII, 308 pages. 1977.

Vol. 615: Turbulence Seminar, Proceedings 1976/77. Edited by P. Bernard and T. Ratiu. VI, 155 pages. 1977.

Vol. 616: Abelian Group Theory, 2nd New Mexico State University Conference, 1976. Proceedings. Edited by D. Arnold, R. Hunter and E. Walker. X, 423 pages. 1977.

Vol. 617: K. J. Devlin, The Axiom of Constructibility: A Guide for the Mathematician. VIII, 96 pages. 1977.

Vol. 618: I. I. Hirschman, Jr. and D. E. Hughes, Extreme Eigen Values of Toeplitz Operators. VI, 145 pages. 1977.

Vol. 619: Set Theory and Hierarchy Theory V, Bierutowice 1976. Edited by A. Lachlan, M. Srebrny, and A. Zarach. VIII, 358 pages. 1977.

Vol. 620: H. Popp, Moduli Theory and Classification Theory of Algebraic Varieties. VIII, 189 pages. 1977.

Vol. 621: Kauffman et al., The Deficiency Index Problem. VI, 112 pages. 1977.

Vol. 622: Combinatorial Mathematics V, Melbourne 1976. Proceedings. Edited by C. Little. VIII, 213 pages. 1977.

Vol. 623: I. Erdelyi and R. Lange, Spectral Decompositions on Banach Spaces. VIII, 122 pages. 1977.

Vol. 624: Y. Guivarc'h et al., Marches Aléatoires sur les Groupes de Lie. VIII, 292 pages. 1977.

Vol. 625: J. P. Alexander et al., Odd Order Group Actions and Witt Classification of Innerproducts. IV, 202 pages. 1977.

Vol. 626: Number Theory Day, New York 1976. Proceedings. Edited by M. B. Nathanson. VI, 241 pages. 1977.

Vol. 627: Modular Functions of One Variable VI, Bonn 1976. Proceedings. Edited by J.-P. Serre and D. B. Zagier. VI, 339 pages. 1977.

Vol. 628: H. J. Baues, Obstruction Theory on the Homotopy Classification of Maps. XII, 387 pages. 1977.

Vol. 629: W.A. Coppel, Dichotomies in Stability Theory. VI, 98 pages. 1978.

Vol. 630: Numerical Analysis, Proceedings, Biennial Conference, Dundee 1977. Edited by G. A. Watson. XII, 199 pages. 1978.

Vol. 631: Numerical Treatment of Differential Equations. Proceedings 1976. Edited by R. Bulirsch, R. D. Grigorieff, and J. Schröder. X, 219 pages. 1978.

Vol. 632: J.-F. Boutot, Schéma de Picard Local. X, 165 pages. 1978.

Vol. 633: N. R. Coleff and M. E. Herrera, Les Courants Résiduels Associés à une Forme Méromorphe. X, 211 pages. 1978.

Vol. 634: H. Kurke et al., Die Approximationseigenschaft lokaler Ringe. IV, 204 Seiten. 1978.

Vol. 635: T. Y. Lam, Serre's Conjecture. XVI, 227 pages. 1978.

Vol. 636: Journées de Statistique des Processus Stochastiques, Grenoble 1977, Proceedings. Edité par Didier Dacunha-Castelle et Bernard Van Cutsem. VII, 202 pages. 1978.

Vol. 637: W. B. Jurkat, Meromorphe Differentialgleichungen. VII, 194 Seiten. 1978.

Vol. 638: P. Shanahan, The Atiyah-Singer Index Theorem, An Introduction. V, 224 pages. 1978.

Vol. 639: N. Adasch et al., Topological Vector Spaces. V, 125 pages. 1978.

Vol. 640: J. L. Dupont, Curvature and Characteristic Classes. X, 175 pages. 1978.

Vol. 641: Séminaire d'Algèbre Paul Dubreil, Proceedings Paris 1976–1977. Edité par M. P. Malliavin. IV, 367 pages. 1978.

Vol. 642: Theory and Applications of Graphs, Proceedings, Michigan 1976. Edited by Y. Alavi and D. R. Lick. XIV, 635 pages. 1978.

Vol. 643: M. Davis, Multiaxial Actions on Manifolds. VI, 141 pages. 1978.

Vol. 644: Vector Space Measures and Applications I, Proceedings 1977. Edited by R. M. Aron and S. Dineen. VIII, 451 pages. 1978.

Vol. 645: Vector Space Measures and Applications II, Proceedings 1977. Edited by R. M. Aron and S. Dineen. VIII, 218 pages. 1978.

Vol. 646: O. Tammi, Extremum Problems for Bounded Univalent Functions. VIII, 313 pages. 1978.

Vol. 647: L. J. Ratliff, Jr., Chain Conjectures in Ring Theory. VIII, 133 pages. 1978.

Vol. 648: Nonlinear Partial Differential Equations and Applications, Proceedings, Indiana 1976–1977. Edited by J. M. Chadam. VI, 206 pages. 1978.

Vol. 649: Séminaire de Probabilités XII, Proceedings, Strasbourg, 1976–1977. Edité par C. Dellacherie, P. A. Meyer et M. Weil. VIII, 805 pages. 1978.

Vol. 650: C*-Algebras and Applications to Physics. Proceedings 1977. Edited by H. Araki and R. V. Kadison. V, 192 pages. 1978.

Vol. 651: P. W. Michor, Functors and Categories of Banach Spaces. VI, 99 pages. 1978.

Vol. 652: Differential Topology, Foliations and Gelfand-Fuks-Cohomology, Proceedings 1976. Edited by P. A. Schweitzer. XIV, 252 pages. 1978.

Vol. 653: Locally Interacting Systems and Their Application in Biology. Proceedings, 1976. Edited by R. L. Dobrushin, V. I. Kryukov and A. L. Toom. XI, 202 pages. 1978.

Vol. 654: J. P. Buhler, Icosahedral Golois Representations. III, 143 pages. 1978.

Vol. 655: R. Baeza, Quadratic Forms Over Semilocal Rings. VI, 199 pages. 1978.

Vol. 656: Probability Theory on Vector Spaces. Proceedings, 1977. Edited by A. Weron. VIII, 274 pages. 1978.

Vol. 657: Geometric Applications of Homotopy Theory I, Proceedings 1977. Edited by M. G. Barratt and M. E. Mahowald. VIII, 459 pages. 1978.

Vol. 658: Geometric Applications of Homotopy Theory II, Proceedings 1977. Edited by M. G. Barratt and M. E. Mahowald. VIII, 487 pages. 1978.

Vol. 659: Bruckner, Differentiation of Real Functions. X, 247 pages. 1978.

Vol. 660: Equations aux Dérivée Partielles. Proceedings, 1977. Edité par Pham The Lai. VI, 216 pages. 1978.

Vol. 661: P. T. Johnstone, R. Paré, R. D. Rosebrugh, D. Schumacher, R. J. Wood, and G. C. Wraith, Indexed Categories and Their Applications. VII, 260 pages. 1978.

Vol. 662: Akin, The Metric Theory of Banach Manifolds. XIX, 306 pages. 1978.

Vol. 663: J. F. Berglund, H. D. Junghenn, P. Milnes, Compact Right Topological Semigroups and Generalizations of Almost Periodicity. X, 243 pages. 1978.

Vol. 664: Algebraic and Geometric Topology, Proceedings, 1977. Edited by K. C. Millett. XI, 240 pages. 1978.

Vol. 665: Journées d'Analyse Non Linéaire. Proceedings, 1977. Edité par P. Bénilan et J. Robert. VIII, 256 pages. 1978.

Vol. 666: B. Beauzamy, Espaces d'Interpolation Réels: Topologie et Géometrie. X, 104 pages. 1978.

Vol. 667: J. Gilewicz, Approximants de Padé. XIV, 511 pages. 1978.

Vol. 668: The Structure of Attractors in Dynamical Systems. Proceedings, 1977. Edited by J. C. Martin, N. G. Markley and W. Perrizo. VI, 264 pages. 1978.

Vol. 669: Higher Set Theory. Proceedings, 1977. Edited by G. H. Müller and D. S. Scott. XII, 476 pages. 1978.

Vol. 670: Fonctions de Plusieurs Variables Complexes III, Proceedings, 1977. Edité par F. Norguet. XII, 394 pages. 1978.

Vol. 671: R. T. Smythe and J. C. Wierman, First-Passage Perculation on the Square Lattice. VIII, 196 pages. 1978.

Vol. 672: R. L. Taylor, Stochastic Convergence of Weighted Sums of Random Elements in Linear Spaces. VII, 216 pages. 1978.

Vol. 673: Algebraic Topology, Proceedings 1977. Edited by P. Hoffman, R. Piccinini and D. Sjerve. VI, 278 pages. 1978.

Vol. 674: Z. Fiedorowicz and S. Priddy, Homology of Classical Groups Over Finite Fields and Their Associated Infinite Loop Spaces. VI, 434 pages. 1978.

Vol. 675: J. Galambos and S. Kotz, Characterizations of Probability Distributions. VIII, 169 pages. 1978.

Vol. 676: Differential Geometrical Methods in Mathematical Physics II, Proceedings, 1977. Edited by K. Bleuler, H. R. Petry and A. Reetz. VI, 626 pages. 1978.

Vol. 677: Séminaire Bourbaki, vol. 1976/77, Exposés 489-506. IV, 264 pages. 1978.

Vol. 678: D. Dacunha-Castelle, H. Heyer et B. Roynette. Ecole d'Eté de Probabilités de Saint-Flour. VII-1977. Edité par P. L. Hennequin. IX, 379 pages. 1978.

Vol. 679: Numerical Treatment of Differential Equations in Applications, Proceedings, 1977. Edited by R. Ansorge and W. Törnig. IX, 163 pages. 1978.

Vol. 680: Mathematical Control Theory, Proceedings, 1977. Edited by W. A. Coppel. IX, 257 pages. 1978.

Vol. 681: Séminaire de Théorie du Potentiel Paris, No. 3, Directeurs: M. Brelot, G. Choquet et J. Deny. Rédacteurs: F. Hirsch et G. Mokobodzki. VII, 294 pages. 1978.

Vol. 682: G. D. James, The Representation Theory of the Symmetric Groups. V, 156 pages. 1978.

Vol. 683: Variétés Analytiques Compactes, Proceedings, 1977. Edité par Y. Hervier et A. Hirschowitz. V, 248 pages. 1978.

Vol. 684: E. E. Rosinger, Distributions and Nonlinear Partial Differential Equations. XI, 146 pages. 1978.

Vol. 685: Knot Theory, Proceedings, 1977. Edited by J. C. Hausmann. VII, 311 pages. 1978.

Vol. 686: Combinatorial Mathematics, Proceedings, 1977. Edited by D. A. Holton and J. Seberry. IX, 353 pages. 1978.

Vol. 687: Algebraic Geometry, Proceedings, 1977. Edited by L. D. Olson. V, 244 pages. 1978.

Vol. 688: J. Dydak and J. Segal, Shape Theory. VI, 150 pages. 1978.

Vol. 689: Cabal Seminar 76-77, Proceedings, 1976-77. Edited by A.S. Kechris and Y. N. Moschovakis. V, 282 pages. 1978.

Vol. 690: W. J. J. Rey, Robust Statistical Methods. VI, 128 pages. 1978.

Vol. 691: G. Viennot, Algèbres de Lie Libres et Monoïdes Libres. III, 124 pages. 1978.

Vol. 692: T. Husain and S. M. Khaleelulla, Barrelledness in Topological and Ordered Vector Spaces. IX, 258 pages. 1978.

Vol. 693: Hilbert Space Operators, Proceedings, 1977. Edited by J. M. Bachar Jr. and D. W. Hadwin. VIII, 184 pages. 1978.

Vol. 694: Séminaire Pierre Lelong - Henri Skoda (Analyse) Année 1976/77. VII, 334 pages. 1978.

Vol. 695: Measure Theory Applications to Stochastic Analysis, Proceedings, 1977. Edited by G. Kallianpur and D. Kölzow. XII, 261 pages. 1978.

Vol. 696: P. J. Feinsilver, Special Functions, Probability Semigroups, and Hamiltonian Flows. VI, 112 pages. 1978.

Vol. 697: Topics in Algebra, Proceedings, 1978. Edited by M. F. Newman. XI, 229 pages. 1978.

Vol. 698: E. Grosswald, Bessel Polynomials. XIV, 182 pages. 1978.

Vol. 699: R. E. Greene and H.-H. Wu, Function Theory on Manifolds Which Possess a Pole. III, 215 pages. 1979.

Vol. 700: Module Theory, Proceedings, 1977. Edited by C. Faith and S. Wiegand. X, 239 pages. 1979.

Vol. 701: Functional Analysis Methods in Numerical Analysis, Proceedings, 1977. Edited by M. Zuhair Nashed. VII, 333 pages. 1979.

Vol. 702: Yuri N. Bibikov, Local Theory of Nonlinear Analytic Ordinary Differential Equations. IX, 147 pages. 1979.

Vol. 703: Equadiff IV, Proceedings, 1977. Edited by J. Fábera. XIX, 441 pages. 1979.

Vol. 704: Computing Methods in Applied Sciences and Engineering, 1977, I. Proceedings, 1977. Edited by R. Glowinski and J. L. Lions. VI, 391 pages. 1979.

Vol. 705: O. Forster und K. Knorr, Konstruktion verseller Familien kompakter komplexer Räume. VII, 141 Seiten. 1979.

Vol. 706: Probability Measures on Groups, Proceedings, 1978. Edited by H. Heyer. XIII, 348 pages. 1979.

Vol. 707: R. Zielke, Discontinuous Čebyšev Systems. VI, 111 pages. 1979.

Vol. 708: J. P. Jouanolou, Equations de Pfaff algébriques. V, 255 pages. 1979.

Vol. 709: Probability in Banach Spaces II. Proceedings, 1978. Edited by A. Beck. V, 205 pages. 1979.

Vol. 710: Séminaire Bourbaki vol. 1977/78, Exposés 507-524. IV, 328 pages. 1979.

Vol. 711: Asymptotic Analysis. Edited by F. Verhulst. V, 240 pages. 1979.

Vol. 712: Equations Différentielles et Systèmes de Pfaff dans le Champ Complexe. Edité par R. Gérard et J.-P. Ramis. V, 364 pages. 1979.

Vol. 713: Séminaire de Théorie du Potentiel, Paris No. 4. Edité par F. Hirsch et G. Mokobodzki. VII, 281 pages. 1979.

Vol. 714: J. Jacod, Calcul Stochastique et Problèmes de Martingales. X, 539 pages. 1979.

Vol. 715: Inder Bir S. Passi, Group Rings and Their Augmentation Ideals. VI, 137 pages. 1979.

Vol. 716: M. A. Scheunert, The Theory of Lie Superalgebras. X, 271 pages. 1979.

Vol. 717: Grosser, Bidualräume und Vervollständigungen von Banachmoduln. III, 209 pages. 1979.

Vol. 718: J. Ferrante and C. W. Rackoff, The Computational Complexity of Logical Theories. X, 243 pages. 1979.

Vol. 719: Categorial Topology, Proceedings, 1978. Edited by H. Herrlich and G. Preuß. XII, 420 pages. 1979.

Vol. 720: E. Dubinsky, The Structure of Nuclear Fréchet Spaces. V, 187 pages. 1979.

Vol. 721: Séminaire de Probabilités XIII. Proceedings, Strasbourg, 1977/78. Edité par C. Dellacherie, P. A. Meyer et M. Weil. VII, 647 pages. 1979.

Vol. 722: Topology of Low-Dimensional Manifolds. Proceedings, 1977. Edited by R. Fenn. VI, 154 pages. 1979.

Vol. 723: W. Brandal, Commutative Rings whose Finitely Generated Modules Decompose. II, 116 pages. 1979.

Vol. 724: D. Griffeath, Additive and Cancellative Interacting Particle Systems. V, 108 pages. 1979.

Vol. 725: Algèbres d'Opérateurs. Proceedings, 1978. Edité par P. de la Harpe. VII, 309 pages. 1979.

Vol. 726: Y.-C. Wong, Schwartz Spaces, Nuclear Spaces and Tensor Products. VI, 418 pages. 1979.

Vol. 727: Y. Saito, Spectral Representations for Schrödinger Operators With Long-Range Potentials. V, 149 pages. 1979.

Vol. 728: Non-Commutative Harmonic Analysis. Proceedings, 1978. Edited by J. Carmona and M. Vergne. V, 244 pages. 1979.

Vol. 729: Ergodic Theory. Proceedings, 1978. Edited by M. Denker and K. Jacobs. XII, 209 pages. 1979.

Vol. 730: Functional Differential Equations and Approximation of Fixed Points. Proceedings, 1978. Edited by H.-O. Peitgen and H.-O. Walther. XV, 503 pages. 1979.

Vol. 731: Y. Nakagami and M. Takesaki, Duality for Crossed Products of von Neumann Algebras. IX, 139 pages. 1979.

Vol. 732: Algebraic Geometry. Proceedings, 1978. Edited by K. Lønsted. IV, 658 pages. 1979.

Vol. 733: F. Bloom, Modern Differential Geometric Techniques in the Theory of Continuous Distributions of Dislocations. XII, 206 pages. 1979.

Vol. 734: Ring Theory, Waterloo, 1978. Proceedings, 1978. Edited by D. Handelman and J. Lawrence. XI, 352 pages. 1979.

Vol. 735: B. Aupetit, Propriétés Spectrales des Algèbres de Banach. XII, 192 pages. 1979.

Vol. 736: E. Behrends, M-Structure and the Banach-Stone Theorem. X, 217 pages. 1979.

Vol. 737: Volterra Equations. Proceedings 1978. Edited by S.-O. Londen and O. J. Staffans. VIII, 314 pages. 1979.

Vol. 738: P. E. Conner, Differentiable Periodic Maps. 2nd edition, IV, 181 pages. 1979.

Vol. 739: Analyse Harmonique sur les Groupes de Lie II. Proceedings, 1976–78. Edited by P. Eymard et al. VI, 646 pages. 1979.

Vol. 740: Séminaire d'Algèbre Paul Dubreil. Proceedings, 1977–78. Edited by M.-P. Malliavin. V, 456 pages. 1979.

Vol. 741: Algebraic Topology, Waterloo 1978. Proceedings. Edited by P. Hoffman and V. Snaith. XI, 655 pages. 1979.

Vol. 742: K. Clancey, Seminormal Operators. VII, 125 pages. 1979.

Vol. 743: Romanian-Finnish Seminar on Complex Analysis. Proceedings, 1976. Edited by C. Andreian Cazacu et al. XVI, 713 pages. 1979.

Vol. 744: I. Reiner and K. W. Roggenkamp, Integral Representations. VIII, 275 pages. 1979.

Vol. 745: D. K. Haley, Equational Compactness in Rings. III, 167 pages. 1979.

Vol. 746: P. Hoffman, τ-Rings and Wreath Product Representations. V, 148 pages. 1979.

Vol. 747: Complex Analysis, Joensuu 1978. Proceedings, 1978. Edited by I. Laine, O. Lehto and T. Sorvali. XV, 450 pages. 1979.

Vol. 748: Combinatorial Mathematics VI. Proceedings, 1978. Edited by A. F. Horadam and W. D. Wallis. IX, 206 pages. 1979.

Vol. 749: V. Girault and P.-A. Raviart, Finite Element Approximation of the Navier-Stokes Equations. VII, 200 pages. 1979.

Vol. 750: J. C. Jantzen, Moduln mit einem höchsten Gewicht. III, 195 Seiten. 1979.

Vol. 751: Number Theory, Carbondale 1979. Proceedings. Edited by M. B. Nathanson. V, 342 pages. 1979.

Vol. 752: M. Barr, *-Autonomous Categories. VI, 140 pages. 1979.

Vol. 753: Applications of Sheaves. Proceedings, 1977. Edited by M. Fourman, C. Mulvey and D. Scott. XIV, 779 pages. 1979.

Vol. 754: O. A. Laudal, Formal Moduli of Algebraic Structures. III, 161 pages. 1979.

Vol. 755: Global Analysis. Proceedings, 1978. Edited by M. Grmela and J. E. Marsden. VII, 377 pages. 1979.

Vol. 756: H. O. Cordes, Elliptic Pseudo-Differential Operators – An Abstract Theory. IX, 331 pages. 1979.

Vol. 757: Smoothing Techniques for Curve Estimation. Proceedings, 1979. Edited by Th. Gasser and M. Rosenblatt. V, 245 pages. 1979.

Vol. 758: C. Năstăsescu and F. Van Oystaeyen; Graded and Filtered Rings and Modules. X, 148 pages. 1979.

Vol. 759: R. L. Epstein, Degrees of Unsolvability: Structure and Theory. XIV, 216 pages. 1979.

Vol. 760: H.-O. Georgii, Canonical Gibbs Measures. VIII, 190 pages. 1979.

Vol. 761: K. Johannson, Homotopy Equivalences of 3-Manifolds with Boundaries. 2, 303 pages. 1979.

Vol. 762: D. H. Sattinger, Group Theoretic Methods in Bifurcation Theory. V, 241 pages. 1979.

Vol. 763: Algebraic Topology, Aarhus 1978. Proceedings, 1978. Edited by J. L. Dupont and H. Madsen. VI, 695 pages. 1979.

Vol. 764: B. Srinivasan, Representations of Finite Chevalley Groups. XI, 177 pages. 1979.

Vol. 765: Padé Approximation and its Applications. Proceedings, 1979. Edited by L. Wuytack. VI, 392 pages. 1979.

Vol. 766: T. tom Dieck, Transformation Groups and Representation Theory. VIII, 309 pages. 1979.

Vol. 767: M. Namba, Families of Meromorphic Functions on Compact Riemann Surfaces. XII, 284 pages. 1979.

Vol. 768: R. S. Doran and J. Wichmann, Approximate Identities and Factorization in Banach Modules. X, 305 pages. 1979.

Vol. 769: J. Flum, M. Ziegler, Topological Model Theory. X, 151 pages. 1980.

Vol. 770: Séminaire Bourbaki vol. 1978/79 Exposés 525–542. IV, 341 pages. 1980.

Vol. 771: Approximation Methods for Navier-Stokes Problems. Proceedings, 1979. Edited by R. Rautmann. XVI, 581 pages. 1980.

This series reports new developments in mathematical research and teaching – quickly, informally and at a high level. The type of material considered for publication includes:

1. Preliminary drafts of original papers and monographs
2. Lectures on a new field or presentations of a new angle in a classical field
3. Seminar work-outs
4. Reports of meetings, provided they are
   a) of exceptional interest and
   b) devoted to a single topic.

Texts which are out of print but still in demand may also be considered if they fall within these categories.

The timeliness of a manuscript is more important than its form, which may be unfinished or tentative. Thus, in some instances, proofs may be merely outlined and results presented which have been or will later be published elsewhere. If possible, a subject index should be included. Publication of Lecture Notes is intended as a service to the international mathematical community, in that a commercial publisher, Springer-Verlag, can offer a wide distribution of documents which would otherwise have a restricted readership. Once published and copyrighted, they can be documented in the scientific literature.

Manuscripts

Manuscripts should be no less than 100 and preferably no more than 500 pages in length.
They are reproduced by a photographic process and therefore must be typed with extreme care. Symbols not on the typewriter should be inserted by hand in indelible black ink. Corrections to the typescript should be made by pasting in the new text or painting out errors with white correction fluid. Authors receive 75 free copies and are free to use the material in other publications. The typescript is reduced slightly in size during reproduction; best results will not be obtained unless the text on any one page is kept within the overall limit of 18 x 26.5 cm (7 x 10½ inches). On request, the publisher will supply special paper with the typing area outlined.

Manuscripts should be sent to Prof. A. Dold, Mathematisches Institut der Universität Heidelberg, Im Neuenheimer Feld 288, 6900 Heidelberg/Germany, Prof. B. Eckmann, Eidgenössische Technische Hochschule, CH-8092 Zürich/Switzerland, or directly to Springer-Verlag Heidelberg.

Springer-Verlag, Heidelberger Platz 3, D-1000 Berlin 33
Springer-Verlag, Neuenheimer Landstraße 28–30, D-6900 Heidelberg 1
Springer-Verlag, 175 Fifth Avenue, New York, NY 10010/USA

ISBN 3-540-**09734**-1
ISBN 0-387-**09734**-1